RECIPROCALS OF BASIC FUNCTIONS

18. $\displaystyle\int \frac{1}{1 \pm \sin u}\, du = \tan u \mp \sec u + C$

19. $\displaystyle\int \frac{1}{1 \pm \cos u}\, du = -\cot u \pm \csc u + C$

20. $\displaystyle\int \frac{1}{1 \pm \tan u}\, du = \tfrac{1}{2}(u \pm \ln|\cos u \pm \sin u|) + C$

21. $\displaystyle\int \frac{1}{\sin u \cos u}\, du = \ln|\tan u| + C$

22. $\displaystyle\int \frac{1}{1 \pm \cot u}\, du = \tfrac{1}{2}(u \mp \ln|\mathrm{si}$

23. $\displaystyle\int \frac{1}{1 \pm \sec u}\, du = u + \cot u \mp$

24. $\displaystyle\int \frac{1}{1 \pm \csc u}\, du = u - \tan u \pm$

25. $\displaystyle\int \frac{1}{1 \pm e^u}\, du = u - \ln(1 \pm e^u) + C$

POWERS OF TRIGONOMETRIC FUNCTIONS

26. $\displaystyle\int \sin^2 u\, du = \tfrac{1}{2}u - \tfrac{1}{4}\sin 2u + C$

27. $\displaystyle\int \cos^2 u\, du = \tfrac{1}{2}u + \tfrac{1}{4}\sin 2u + C$

28. $\displaystyle\int \tan^2 u\, du = \tan u - u + C$

29. $\displaystyle\int \sin^n u\, du = -\frac{1}{n}\sin^{n-1} u \cos u + \frac{n-1}{n}\int \sin^{n-2} u\, du$

30. $\displaystyle\int \cos^n u\, du = \frac{1}{n}\cos^{n-1} u \sin u + \frac{n-1}{n}\int \cos^{n-2} u\, du$

31. $\displaystyle\int \tan^n u\, du = \frac{1}{n-1}\tan^{n-1} u - \int \tan^{n-2} u\, du$

32. $\displaystyle\int \cot^2 u\, du = -\cot u - u + C$

33. $\displaystyle\int \sec^2 u\, du = \tan u + C$

34. $\displaystyle\int \csc^2 u\, du = -\cot u + C$

35. $\displaystyle\int \cot^n u\, du = -\frac{1}{n-1}\cot^{n-1} u - \int \cot^{n-2} u\, du$

36. $\displaystyle\int \sec^n u\, du = \frac{1}{n-1}\sec^{n-2} u \tan u + \frac{n-2}{n-1}\int \sec^{n-2} u\, du$

37. $\displaystyle\int \csc^n u\, du = -\frac{1}{n-1}\csc^{n-2} u \cot u + \frac{n-2}{n-1}\int \csc^{n-2} u\, du$

PRODUCTS OF TRIGONOMETRIC FUNCTIONS

38. $\displaystyle\int \sin mu \sin nu\, du = -\frac{\sin(m+n)u}{2(m+n)} + \frac{\sin(m-n)u}{2(m-n)} + C$

39. $\displaystyle\int \cos mu \cos nu\, du = \frac{\sin(m+n)u}{2(m+n)} + \frac{\sin(m-n)u}{2(m-n)} + C$

40. $\displaystyle\int \sin mu \cos nu\, du = -\frac{\cos(m+n)u}{2(m+n)} - \frac{\cos(m-n)u}{2(m-n)} + C$

41. $\displaystyle\int \sin^m u \cos^n u\, du = -\frac{\sin^{m-1} u \cos^{n+1} u}{m+n} + \frac{m-1}{m+n}\int \sin^{m-2} u \cos^n u\, du$

$\displaystyle\qquad = \frac{\sin^{m+1} u \cos^{n-1} u}{m+n} + \frac{n-1}{m+n}\int \sin^m u \cos^{n-2} u\, du$

PRODUCTS OF TRIGONOMETRIC AND EXPONENTIAL FUNCTIONS

42. $\displaystyle\int e^{au}\sin bu\, du = \frac{e^{au}}{a^2+b^2}(a \sin bu - b \cos bu) + C$

43. $\displaystyle\int e^{au}\cos bu\, du = \frac{e^{au}}{a^2+b^2}(a \cos bu + b \sin bu) + C$

POWERS OF u MULTIPLYING OR DIVIDING BASIC FUNCTIONS

44. $\displaystyle\int u \sin u\, du = \sin u - u \cos u + C$

45. $\displaystyle\int u \cos u\, du = \cos u + u \sin u + C$

46. $\displaystyle\int u^2 \sin u\, du = 2u \sin u + (2 - u^2)\cos u + C$

47. $\displaystyle\int u^2 \cos u\, du = 2u \cos u + (u^2 - 2)\sin u + C$

48. $\displaystyle\int u^n \sin u\, du = -u^n \cos u + n\int u^{n-1}\cos u\, du$

49. $\displaystyle\int u^n \cos u\, du = u^n \sin u - n\int u^{n-1}\sin u\, du$

50. $\displaystyle\int u^n \ln u\, du = \frac{u^{n+1}}{(n+1)^2}[(n+1)\ln u - 1] + C$

51. $\displaystyle\int u e^u\, du = e^u(u-1) + C$

52. $\displaystyle\int u^n e^u\, du = u^n e^u - n\int u^{n-1}e^u\, du$

53. $\displaystyle\int u^n a^u\, du = \frac{u^n a^u}{\ln a} - \frac{n}{\ln a}\int u^{n-1}a^u\, du + C$

54. $\displaystyle\int \frac{e^u\, du}{u^n} = -\frac{e^u}{(n-1)u^{n-1}} + \frac{1}{n-1}\int \frac{e^u\, du}{u^{n-1}}$

55. $\displaystyle\int \frac{a^u\, du}{u^n} = -\frac{a^u}{(n-1)u^{n-1}} + \frac{\ln a}{n-1}\int \frac{a^u\, du}{u^{n-1}}$

56. $\displaystyle\int \frac{du}{u \ln u} = \ln|\ln u| + C$

POLYNOMIALS MULTIPLYING BASIC FUNCTIONS

57. $\displaystyle\int p(u)e^{au}\, du = \frac{1}{a}p(u)e^{au} - \frac{1}{a^2}p'(u)e^{au} + \frac{1}{a^3}p''(u)e^{au} - \cdots$ [signs alternate: $+ - + - \cdots$]

58. $\displaystyle\int p(u)\sin au\, du = -\frac{1}{a}p(u)\cos au + \frac{1}{a^2}p'(u)\sin au + \frac{1}{a^3}p''(u)\cos au - \cdots$ [signs alternate in pairs after first term: $+ + - - + + - - \cdots$]

59. $\displaystyle\int p(u)\cos au\, du = \frac{1}{a}p(u)\sin au + \frac{1}{a^2}p'(u)\cos au - \frac{1}{a^3}p''(u)\sin au - \cdots$ [signs alternate in pairs: $+ + - - + + - - \cdots$]

Calculus

A New Horizon

Volume II

Calculus

A New Horizon

Volume II

Sixth Edition

Howard Anton
Drexel University

JOHN WILEY & SONS, INC.
New York Chichester Brisbane Toronto Singapore

Mathematics Editor: Barbara Holland
Associate Editor: Sharon Smith
Senior Developmental Editor: Madalyn Stone
Senior Photo Editor: Hilary Newman
Senior Marketing Manager: Leslie Hines
Copy Editor: Lilian Brady
Production and Text Design: HRS Electronic Text Management
Electronic Illustration: Techsetters, Inc.
Typesetting: Techsetters, Inc.
Cover Design: Madelyn Lesure
Cover Photo: © Dann Coffey/The Image Bank

This book was set in Times Roman by Techsetters, Inc., and printed and bound by Von Hoffmann Press, Inc. The cover was printed by The Phoenix Color Corp.

Recognizing the importance of preserving what has been written, it is a policy of John Wiley & Sons, Inc. to have books of enduring value published in the United States printed on acid-free paper, and we exert our best efforts to that end.

The paper on this book was manufactured by a mill whose forest management programs include sustained yield harvesting of its timberlands. Sustained yield harvesting principles ensure that the numbers of trees cut each year does not exceed the amount of new growth.

ISBN 0-471-24348-5

Printed in the United States of America

10 9 8 7 6 5 4 3 2

About Howard Anton

Howard Anton obtained his B.A. from Lehigh University, his M.A. from the University of Illinois, and his Ph.D. from the Polytechnic University of Brooklyn, all in mathematics. In the early 1960s he worked for Burroughs Corporation and Avco Corporation at Cape Canaveral, Florida, where he was involved with missile tracking problems for the manned space program. In 1968 he joined the Mathematics Department at Drexel University, where he taught full time until 1983. Since that time he has been an adjunct professor at Drexel and has devoted the majority of his time to textbook writing and activities for mathematical associations. Dr. Anton was President of the EPADEL Section of the Mathematical Association of America (MAA), served on the board of Governors of that organization, and guided the creation of the Student Chapters of the MAA. He has published numerous research papers in Functional Analysis, Approximation Theory, and Topology, as well as pedagogical papers on applications of mathematics. He is best known for his textbooks in mathematics, which are among the most widely used in the world. There are currently more than ninety versions of his books, including translations into Spanish, Arabic, Portuguese, Italian, Indonesian, French, Japanese, Chinese, Hebrew, and German. Dr. Anton has an avid interest in computer technology as it relates to mathematical education and publishing. He has developed pedagogical software for teaching calculus and linear algebra as well as various software programs for the publishing industry that automate the production of four-color mathematical text and art. For relaxation he enjoys traveling and photography.

To

My Wife Pat

My Children Brian, David, and Lauren

In Memory of

My Mother Shirley

Stephen Girard (1750–1831)—Benefactor

Albert Herr—Esteemed Colleague and Contributor

A Note from the Author

When I began writing the first edition of this calculus text almost 25 years ago, the task, though daunting, was straightforward in that the content and organization of a standard calculus course was nearly universal—the challenge for me at that time was to present the material in a livelier style and with greater clarity than my predecessors. Since this calculus text is still among the most widely used in the world, I take comfort that the goals I set for myself as a young writer and mathematician have been achieved.

However, times are changing, and the era of a standard and universal calculus course seems destined for the repository of slide rules and three-cent stamps. We are witnessing a lot of experimentation with the content, organization, and goals of calculus—some of which has been successful and some of which has not. Thus, my challenge in writing the sixth edition has been to create a text that has all of the strengths of the earlier editions, yet incorporates those new ideas that are clearly important and have withstood the objective scrutiny of skilled and thoughtful teachers.

In preparing for this edition, I sought advice from outstanding teachers at a wide variety of institutions. Needless to say, I received a diversity of opinions—some reviewers advised against any major changes, arguing that the book was already clearly written and working well in the classroom, while others felt that major changes were required to incorporate technology and make the book more contemporary. I listened carefully, and the lively discussions that followed helped me formulate my philosophy for the new edition. Many of the specific changes are itemized in the preface, but here are some of the general goals:

- Add graphing calculator and CAS materials to the text in a way that will allow students who have those tools to use them but that will not prevent the text from being used by those students who do not have access to that technology.
- Place more emphasis on mathematical modeling and applications.
- Incorporate new examples and exercises that will be meaningful to today's students and will more accurately convey the role of calculus in the real world.
- Widen the variety of exercises to focus more on *conceptual understanding* through conjecture, multi-step analysis, expository writing, and what-if analysis.

In addition, I wanted to provide some optional innovative materials that would capture the student's interest and provide the kind of problem-solving experience that he or she might find in a research or industrial setting. This gave birth to an exciting set of modules that we have called *Expanding the Calculus Horizon*. These modules appear at the ends of selected chapters and each has an optional Internet component that we hope will grow dynamically over time with input from teachers and students.

In developing this edition I have stood firm on two principles that were adhered to in earlier editions:

- The text material is presented at a mathematical level that is suitable for students who will embark on careers in engineering and science.
- It remains a primary goal of the text to teach the student clear, logical, mathematical thinking. Informal discussions play an important motivational role in the exposition and are used extensively, but eventually I want the student to be able to read and understand the language of mathematics.

Although this edition has many changes and new features, they have been implemented in a *flexible* way that will accommodate a wide variety of teaching philosophies. Thus, I am confident that professors who have had positive experiences with earlier editions will be comfortable with this revision, and I am hopeful that those professors who are looking for a contemporary text with an established history of success in the classroom will be pleased with the innovations in this new edition.

Sincerely,

Howard Anton

Howard Anton

From Student to Student

At times the words of a complete stranger are difficult to accept. That is why I am about to take this first opportunity to introduce myself. Hopefully by revealing a bit about myself and how I relate to this textbook may help you find these words more compelling.

Hello, my name is Ajay Arora and I am an Electrical Engineering student at McMaster University in Hamilton, Canada. I too was in your place when I began my entry into the much dreaded field of *CALCULUS*. The vast amounts of rate of change and antiderivative problems were overwhelming. With a little struggle and hard work, I successfully completed that course only to be faced with three more advanced level calculus courses. What I am about to write is the unbiased truth on how you can be successful in calculus and how this textbook will assist you on your journey.

I have been a member of the Student Advisory Board for this textbook for over a year now. The committee came together as a venture from the authors and publisher to get more student input in the development stages instead of simply focusing on feedback when the book was published. After a chapter was completed by the author, each student committee member evaluated, commented, and in some cases, recommended alternative approaches. These tasks involved lots of special deliveries, E-mails, faxes, telephone calls, conference calls, and of course, a whole lot of calculus! But in the end it was a total rewarding experience.

How many times have you asked yourself, "Is math really useful?" Or how about, "Will I ever use calculus in the real world?" I know I have! This textbook will definitely help you answer some of these questions with true applications of the theories you learn. The modules entitled Expanding the Calculus Horizon have been included for precisely that purpose. Every module has been critiqued extensively by the Student Advisory Board, and I encourage you to try them. Not only will these applications of calculus surprise you, but they may actually help give you direction in a field that you might want to pursue after college.

I wish you success in this course, as well as the many others you will face during your college career. Good Luck!

Sincerely,

Ajay Arora
McMaster University
A.Arora@ieee.org

Best Wishes for Success from the Student Advisory Board

Dan Arndt, *University of Texas at Dallas*
Ajay Arora, *McMaster University*
Scott E. Barnett, *Wayne State University*
Fatenah Issa, *Loyola University of Chicago*
Laurie Haskell Messina, *University of Oklahoma*
Steven E. Pav, *Alfred University*

PREFACE

ABOUT THIS EDITION

This is a major revision. In keeping with current trends in calculus, the goal for this edition is to focus more on ***conceptual understanding*** and ***applicability*** of the subject matter. In designing this edition, we worked closely with a talented team of reviewers to ensure that the book is sufficiently ***flexible*** that it will continue to meet the needs of those using the last edition and at the same time provide a fresh approach for those instructors who are taking their calculus course in a new direction. Some of the more significant changes are as follows:

Technology This edition provides extensive materials for instructors who want to use graphing calculators or computer algebra systems. However, these materials are implemented in a way that allows the text to be used in courses where technology is used heavily, moderately, or not at all. To provide a sound foundation for the technology material, I have added a new section entitled Graphing Functions on Calculators and Computers; Computer Algebra Systems (Section 1.3).

Horizon Modules Selected chapters end with modules called Expanding the Calculus Horizon. As the name implies, these modules are intended to take the student a step beyond the traditional calculus text. The modules, all of which are optional, can be assigned either as individual or group projects and can be used by instructors to tailor the calculus course to meet their specific needs and teaching philosophies. For example, there are modules that touch on iteration and dynamical systems, modeling from experimental data by curve fitting, applications, expository report writing, and so forth.

Mathematical Modeling Mathematical modeling plays a more prominent role in this edition. The concept of a mathematical model is introduced in Section 1.5, and the terminology of modeling is used extensively thereafter. The optional Horizon module for Chapter 5 discusses how to obtain various kinds of mathematical models from experimental data, and Chapter 10 discusses mathematical modeling using differential equations.

Applicability of Calculus One of the goals in this edition is to link calculus more closely to the real world and to the student's own experience. This theme starts with the Introduction and is carried through in the modules, examples, and exercises. Applications appearing in exercises and examples are carefully chosen to be sufficiently simple that they do not divert time from learning important mathematical fundamentals. More extensive applications appear in various Horizon modules.

Earlier Differential Equations Basic ideas about differential equations, initial-value problems, direction fields, and integral curves are introduced concurrently with integration and then revisited in more detail in Chapter 10.

Quicker Entry to Functions Chapter 1 begins immediately with functions, and the precalculus material that formed the first chapter in earlier editions has been moved to the appendix.

For the Reader This element is new. At various points in the exposition the student is assigned a brief task. Some tasks are appropriate for all readers, while others are appropriate only for readers who have a graphing calculator or a CAS. The tasks for all readers are designed to immerse the student more deeply into the text by asking them to think about an idea and reach some conclusion; the tasks for students using technology are designed to familiarize them with the procedures for using that technology by asking them to read their documentation and perform some text-related computation. Some instructors may want to make these tasks part of their assignments.

Earlier Logarithms and Exponentials Logarithmic and exponential functions are introduced in Chapter 4 from the exponent point of view and then revisited in Section 7.9 from the integral point of view. This provides a richer variety of functions to work with earlier in the text, fits in better with the discussions of modeling, and makes for a less fragmented presentation of the analysis of graphs. However, for instructors who prefer a later presentation of logarithmic and exponential functions, there is an instructor's guide that explains how to move this material to Chapter 7 and provides a reference list of those exercises in Chapters 4, 5, and 6 that involve logarithmic and exponential functions.

Early Parametric Option There is a new option for introducing parametric curves in Section 1.7 of Chapter 1 and revisiting the material in Chapter 12, where calculus-related issues are discussed. Instructors who prefer the traditional late discussion of parametric equations will have no problem teaching Section 1.7 as part of Chapter 12, as in earlier editions.

More Variety in Exercises The exercise sets have been revised extensively to create a richer variety—there are many more exercises that include conjecture, exploration, multistep analysis, and expository writing. The goal has been to put more focus on *conceptual understanding*. There are also many new exercises that are intended to be solved using a graphing calculator or a CAS. These are marked with icons for easy identification.

Analysis of Functions The old "curve-sketching" material has been replaced by Sections 5.1–5.3 on the Analysis of Functions. The name change reflects a more contemporary approach to the material—there is more emphasis on the interplay between technology and calculus and more focus on the problem of finding a *complete graph*, that is, a graph that contains all of the significant features of concern.

Principles of Integral Evaluation The old "Techniques of Integration" has been renamed Principles of Integral Evaluation to reflect its more contemporary approach to the material. The chapter has been condensed and there is now more emphasis on general methods and less on tricks for evaluating complicated or obscure integrals. The section entitled Using Integral Tables and Computer Algebra Systems has been expanded and rewritten extensively.

Supplementary Exercises Supplementary exercises have been added at the ends of chapters.

New Appendix on Solving Polynomial Equations Appendix F, entitled Solving Polynomial Equations, is new. It reviews the Factor Theorem, the Remainder Theorem, and procedures for finding rational roots. Many students are weak on this material, yet it plays an important role in determining whether a polynomial graph generated on a calculator or computer is complete.

Rule of Four The "rule of four" refers to the presentation of material from the verbal, algebraic, visual, and numerical points of view. It is used more extensively in this edition, where appropriate.

Internet An internet site http://www.wiley.com/college/anton has been established to complement the text. This site contains additional Horizon modules and technology materials. The site is experimental, but we expect it to grow dynamically over time.

OTHER FEATURES

Flexibility This edition has a built-in flexibility that is designed to serve a broad spectrum of calculus philosophies, ranging from traditional to reform. Graphing technology can be used heavily, moderately, or not at all; and the order of presentation of many sections can be permuted to accommodate specific course needs.

Trigonometry Review Deficiencies in trigonometry plague many students, so I have included a substantial trigonometry review in Appendix E.

Historical Notes The biographies and historical notes have been a hallmark of this text from its first edition and have been maintained in this edition. All of the biographical materials have been distilled from standard sources with the goal of capturing the personalities of the great mathematicians and bringing them to life for the student.

Graded Exercise Sets Section Exercise Sets are graded to begin with routine problems and progress gradually toward problems of greater difficulty. However, in the Supplementary Exercises I have opted not to grade the exercises by level of difficulty to avoid giving the student a predisposition about the level of effort required.

Rigor The challenge of writing a good calculus book is to strike the right balance between rigor and clarity. My goal is to present precise mathematics to the fullest extent possible for the freshman audience, but where clarity and rigor conflict I choose clarity. However, I believe it to be essential that the student understand the difference between a careful proof and an informal argument, so I try to make it clear to the reader when arguments are informal. Theory involving δ-ϵ arguments appear in separate sections, so they can be bypassed if desired.

Mathematical Level This book is written at a mathematical level that is suitable for students planning on careers in engineering or science.

Student Review A Student Advisory Board was actively involved in the development process of this edition to provide information on pedagogical clarity and to advise on the development of examples, exercises, and modules that students would find interesting and relevant.

Some Organization Changes from Fifth Edition

- Much of the precalculus material has been moved to appendices to allow for an earlier presentation of functions. However, where appropriate, we have included quick summaries of review material in the body of the text.
- The material on logarithmic and exponential functions has been reorganized, so it can be covered in the first semester (an early transcendental presentation). There is a guide on the next page for implementing a late transcendental presentation.
- The first 13 chapters of the fifth edition are covered in the first 12 chapters of the sixth edition.
- The first 7 chapters of the fifth edition correspond to the first 9 chapters of the sixth edition. However, ***the number of sections is about the same***, so there is ***no increase in the number of lectures required to cover the material***. The new subdivision is more natural in that the chapter titles now reflect the chapter content more accurately.
- In the sixth edition, as in the fifth edition, instructors teaching on the semester system should have no trouble covering material on integration in the first semester.
- Chapter 11 on Infinite Series has been condensed from 12 sections to 10, and the material has been reorganized so that Taylor polynomials and Taylor series appear earlier. This makes it possible to cover these topics without covering the entire chapter.
- The material on analytic geometry and polar coordinates, which occupied Chapters 12 and 13 in the fifth edition, is covered in Chapter 12 of the sixth edition.
- L'Hôpital's rule was moved to an earlier position, so it can be used to analyze the end-behavior of logarithmic and exponential functions.
- The two parts to the Fundamental Theorem of Calculus, which appeared in separate sections of the fifth edition, now appear together in the same section (Section 7.6).

LATE TRANSCENDENTAL OPTION

In keeping with current trends, Volumes I and II of this text are organized so that the basic material on logarithmic and exponential functions is covered in the first semester (commonly called an "early transcendental" presentation). This is achieved by introducing logarithms informally from the exponent point of view and deferring the integral representation of the natural logarithm (Section 7.9). However, we have included the following guide for instructors who prefer to cover logarithmic and exponential functions in the second semester (as in the fifth edition). Depending on your preference, you can place the deferred material after Chapter 7 or after Chapter 8. The guide shows how to place it after Chapter 8. To place it after Chapter 7, ignore the exercise modifications listed for Chapter 8.

	Section	Text Modifications (bulleted)	Exercise Modifications
1	1.1	Functions and Analysis of Graphical Information	
2	1.2	Properties of Functions	
3	1.3	Graphing Functions on Calculators and Computers	
4	1.4	New Functions from Old	
5	1.5	Mathematical Models; Linear Models	
6	1.6	Families of Functions	
7	1.7	Parametric Equations	
8	2.1	Limits (Intuitive)	
9	2.2	Limits (Computational)	
10	2.3	Limits (Rigorous)	
11	2.4	Continuity	
12	2.5	Limits / Continuity of Trigonometric Functions	
13	3.1	Tangent Lines and Rates of Change	
14	3.2	The Derivative	
15	3.3	Techniques of Differentiation	
16	3.4	Derivatives of Trigonometric Functions	
17	3.5	The Chain Rule	
18	3.6	Local Linear Approximation; Differentials	
19	4.3	Implicit Differentiation • Defer the concluding subsection on derivatives of inverse functions (pp. 252–253).	Defer Exercises 10, 53–56.
20	4.6	Related Rates • Defer the alternative solution to Example 3 at the bottom of p. 272.	Defer Exercise 37. Defer Supplementary Exercises 1–6, 8–14, 16–24.
21	5.1	Analysis I: Increase, Decrease, Concavity • Defer Examples 6(a) and 6(c) on p. 295.	Defer Exercises 21–24, 38, 41, 53.
22	5.2	Analysis II: Relative Extrema	Defer Exercises 15, 31, 32, 39–42, 50, 51.
23	5.3	Analysis III: Applying Technology • Defer Example 8 and the discussion of logistic curves that follows it (pp. 316–319). • Defer the Horizon Module for Chapter 5.	Defer Exercises 39–48, 53–55, 69, 70. Defer Supplementary Exercises 17–24, 33, 37–39.
24	6.1	Absolute Maxima and Minima	Defer Exercises 31, 32, 44.
25	6.2	Applied Maximum and Minimum Problems	Defer Exercise 15.
26	6.3	Rectilinear Motion	Defer Exercise 16.
27	6.4	Newton's Method	Defer Exercises 14, 16
28	6.5	Rolle's Theorem; Mean-Value Theorem	Defer Exercise 36 Defer Supplementary Exercises 7(d), 8(d), 22.

29	7.1	An Overview of the Area Problem	Defer Exercise 9.
30	7.2	Indefinite Integral; Integral Curves; Direction Fields • Defer integration formulas (6), (10), (11) in Table 7.2.1 on p. 384. • Defer the last (fifth) integral in Example 2 on p. 385.	Defer Exercises l(b), 19, 20, 25, 34, 39(b).
31	7.3	Integration by Substitution • Defer Example 5, p. 393. • Defer Example 7, p. 394.	Defer Exercises 2(e), 3(c), (d), (e), 5, 6, 19–22, 27, 28, 35, 36, 45–48, 54.
32	7.4	Sigma Notation	Defer Exercises 2(c), (e), 39(b).
33	7.5	The Definite Integral	Defer Exercises 6, 13, 14, 33(b), 38(a), 45(a).
34	7.6	The Fundamental Theorem of Calculus • Defer the first three of the four integrals in Example 5, p. 419.	Defer Exercises 7, 8, 19, 20, 24, 28(b), 45(b), 46(b), 55(b), 59.
35	7.7	Rectilinear Motion Revisited	Defer Exercises 13(b), 14(a), 23, 24, 27, 28, 53, 54.
36	7.8	Evaluating Definite Integrals by Substitution • Defer Example 2(a), p. 422. • Defer Example 3, p. 443.	Defer Exercises 2(a, b), 11, 12, 16, 21, 37, 38, 45–48. Defer Supplementary Exercises 12, 13, 14(c), 37, 39–41, 49.
37	8.1	Area Between Two Curves	Defer Exercise 13.
38	8.2	Volumes by Slicing; Disks and Washers	Defer Exercises 11, 12, 38.
39	8.3	Volumes by Cylindrical Shells	Defer Exercise 11.
40	8.4	Length of a Plane Curve	Defer Exercises 7, 13, 14, 15, 16.
41	8.5	Surface Area	Defer Exercises 13, 15, 22.
42	8.6	Work	
43	8.7	Fluid Pressure and Force	
44	4.1	Inverse Functions • Pick up material deferred from Section 4.3.	**Pick up Exercises:** Section 4.3 (53–56) Chapter 4 Supplementary (1, 6).
45	4.2	Logarithmic and Exponential Functions	**Pick up Exercises:** Section 4.3 (10) Chapter 4 Supplementary (5, 9, 17, 20–22).
46	4.4	Derivatives of Log and Exponential Functions • Pick up material deferred from Section 7.2. • Pick up material deferred from Section 5.3.	**Pick up Exercises:** Section 4.6 (37) Chapter 4 Supplementary (10, 12, 14, 16, 19, 23, 24) Section 5.1 (21–24, 38, 41, 53) Section 5.2 (15, 31, 32, 39–42, 50, 51) Section 5.3 (44, 53–55) Chapter 5 Supplementary (20, 33, 39) Section 6.1 (31, 32, 44) Section 6.2 (15) Section 6.3 (16) Section 6.4 (14, 16) Section 6.5 (36) Chapter 6 Supplementary (7(d), 8(d), 22). Chapter 7 (all deferred) Section 5.3 (69, 70) Chapter 8 (all deferred).
47	4.5	Derivatives of Inverse Trig Functions • Pick up material deferred from Section 4.6.	**Pick up Exercises:** Chapter 4 Supplementary (2, 8, 13, 18).
48	7.9	Logarithmic Functions; Integral Point of View	
49	8.8	Hyperbolic Functions and Hanging Cables	
50	4.7	L'Hôpital's Rules	**Pick up Exercises:** Chapter 4 Supplementary (3, 4, 11) Section 5.3 (39–43, 45–48) Chapter 5 Supplementary (17–19, 21–24, 37, 38).
		• Pick up Horizon module from Chapter 5.	

Supplements

Resources for the Student

Student Resource and Survival CD

0-471-24608-5

This CD for IBM compatibles or Macintosh platforms provides students with an electronic form of detailed solutions to odd-numbered exercises, multiple choice and true–false sample tests for each section and chapter of the text, precalculus review material, and a brief introduction to those aspects of linear algebra that are of immediate concern to the calculus student. Two demonstration modules from the Windows-based multimedia calculus program *Calculus Connections, A Multimedia Adventure* are also available on this CD.

Student Resource Manual

0-471-24616-6

This manual provides students with detailed solutions to odd-numbered exercises and multiple choice and true–false sample tests for each section and chapter of the text.

Resources for the Instructor

Hard copy and electronic resources are available for the instructor. These can be obtained by sending a request on your institutional letterhead to Leslie Hines, Senior Marketing Manager, John Wiley & Sons, Inc., 605 Third Avenue, New York, NY 10158-0012, or by requesting them from your local Wiley representative.

Other Resources

We are proud to offer special pricing of the student educational versions of *MAPLE*™ or *MATHEMATICA*™ packaged with the sixth edition of Howard Anton's *Calculus* textbook. For pricing information, you can contact your local Wiley representative, email us at math@wiley.com, or call us at (800) 225-5945.

Acknowledgments

It has been my good fortune to have the advice and guidance of many talented people whose knowledge and skills have enhanced this book in many ways. For their valuable help I thank:

Reviewers and Contributors to Earlier Editions

Edith Ainsworth, *University of Alabama*
Loren Argabright, *Drexel University*
David Armacost, *Amherst College*
John Bailey, *Clark State Community College*
Robert C. Banash, *St. Ambrose University*
George R. Barnes, *University of Louisville*
Larry Bates, *University of Calgary*
John P. Beckwith, *Michigan Technological University*
Joan E. Bell, *Northeastern Oklahoma State University*
Irl C. Bivens, *Davidson College*
Harry N. Bixler, *Bernard M. Baruch College, CUNY*
Marilyn Blockus, *San Jose State University*
Ray Boersma, *Front Range Community College*
Barbara Bohannon, *Hofstra University*
David Bolen, *Virginia Military Institute*
Daniel Bonar, *Denison University*
George W. Booth, *Brooklyn College*
Phyllis Boutilier, *Michigan Technological University*
Mark Bridger, *Northeastern University*
John Brothers, *Indiana University*
Stephen L. Brown, *Olivet Nazarene University*
Virginia Buchanan, *Hiram College*
Robert C. Bueker, *Western Kentucky University*
Robert Bumcrot, *Hofstra University*
Christopher Butler, *Case Western Reserve University*
Carlos E. Caballero, *Winthrop University*
James Caristi, *Valparaiso University*
Stan R. Chadick, *Northwestern State University*
Hongwei Chen, *Christopher Newport University*
Chris Christensen, *Northern Kentucky University*
Robert D. Cismowski, *San Bernardino Valley College*
Patricia Clark, *Rochester Institute of Technology*
Hannah Clavner, *Drexel University*
David Clydesdale, *Sauk Valley Community College*
David Cohen, *University of California, Los Angeles*
Michael Cohen, *Hofstra University*
Pasquale Condo, *University of Lowell*
Robert Conley, *Precision Visuals*
Cecil J. Coone, *State Technical Institute at Memphis*
Norman Cornish, *University of Detroit*
Terrance Cremeans, *Oakland Community College*
Lawrence Cusick, *California State University–Fresno*
Michael Dagg, *Numerical Solutions, Inc.*
Stephen L. Davis, *Davidson College*
A. L. Deal, *Virginia Military Institute*
Charles Denlinger, *Millersville University*
William H. Dent, *Maryville College*
Blaise DeSesa, *Drexel University*
Dennis DeTurck, *University of Pennsylvania*
Jacqueline Dewar, *Loyola Marymount University*
Preston Dinkins, *Southern University*
Irving Drooyan, *Los Angeles Pierce College*
Tom Drouet, *East Los Angeles College*
Clyde Dubbs, *New Mexico Institute of Mining and Technology*
Della Duncan, *California State University–Fresno*
Ken Dunn, *Dalhousie University*
Sheldon Dyck, *Waterloo Maple Software*
Hugh B. Easler, *College of William and Mary*
Scott Eckert, *Cuyamaca College*
Joseph M. Egar, *Cleveland State University*
Judith Elkins, *Sweet Briar College*
Brett Elliott, *Southeastern Oklahoma State University*
Garret J. Etgen, *University of Houston*
Benny Evans, *Oklahoma State University*
James H. Fife, *Educational Testing Service*
Dorothy M. Fitzgerald, *Golden West College*
Barbara Flajnik, *Virginia Military Institute*
Daniel Flath, *University of South Alabama*
Ernesto Franco, *California State University–Fresno*
Nicholas E. Frangos, *Hofstra University*
Katherine Franklin, *Los Angeles Pierce College*
Marc Frantz, *Indiana University–Purdue University at Indianapolis*

Michael Frantz, *University of La Verne*
Susan L. Friedman, *Bernard M. Baruch College, CUNY*
William R. Fuller, *Purdue University*
Daniel B. Gallup, *Pasadena City College*
Mahmood Ghamsary, *Long Beach City College*
G. S. Gill, *Brigham Young University*
Michael Gilpin, *Michigan Technological University*
Kaplana Godbole, *Michigan Technological Institute*
S. B. Gokhale, *Western Illinois University*
Morton Goldberg, *Broome Community College*
Mardechai Goodman, *Rosary College*
Sid Graham, *Michigan Technological University*
Raymond Greenwell, *Hofstra University*
Gary Grimes, *Mt. Hood Community College*
Jane Grossman, *University of Lowell*
Michael Grossman, *University of Lowell*
Diane Hagglund, *Waterloo Maple Software*
Douglas W. Hall, *Michigan State University*
Nancy A. Harrington, *University of Lowell*
Kent Harris, *Western Illinois University*
Jim Hefferson, *St. Michael College*
Albert Herr, *Drexel University*
Peter Herron, *Suffolk County Community College*
Warland R. Hersey, *North Shore Community College*
Konrad J. Heuvers, *Michigan Technological University*
Robert Higgins, *Quantics Corporation*
Rebecca Hill, *Rochester Institute of Technology*
Edwin Hoefer, *Rochester Institute of Technology*
Louis F. Hoelzle, *Bucks County Community College*
Robert Homolka, *Kansas State University–Salina*
Jerry Johnson, *University of Nevada–Reno*
John M. Johnson, *George Fox College*
Wells R. Johnson, *Bowdoin College*
Herbert Kasube, *Bradley University*
Phil Kavanaugh, *Mesa State College*
Maureen Kelley, *Northern Essex Community College*
Harvey B. Keynes, *University of Minnesota*
Richard Krikorian, *Westchester Community College*
Paul Kumpel, *SUNY, Stony Brook*
Fat C. Lam, *Gallaudet University*
Leo Lampone, *Quantics Corporation*
James F. Lanahan, *University of Detroit–Mercy*
Bruce Landman, *University of North Carolina at Greensboro*
Kuen Hung Lee, *Los Angeles Trade–Technology College*
Marshall J. Leitman, *Case Western Reserve University*
Benjamin Levy, *Lexington H.S., Lexington, Mass.*
Darryl A. Linde, *Northeastern Oklahoma State University*
Phil Locke, *University of Maine, Orono*
Leland E. Long, *Muscatine Community College*
John Lucas, *University of Wisconsin–Oshkosh*
Stanley M. Lukawecki, *Clemson University*
Nicholas Macri, *Temple University*
Melvin J. Maron, *University of Louisville*
Mauricio Marroquin, *Los Angeles Valley College*
Majid Masso, *Brookdale Community College*
Larry Matthews, *Concordia College*
Thomas McElligott, *University of Lowell*
Phillip McGill, *Illinois Central College*
Judith McKinney, *California State Polytechnic University, Pomona*
Joseph Meier, *Millersville University*
Aileen Michaels, *Hofstra University*
Janet S. Milton, *Radford University*
Robert Mitchell, *Rowan College of New Jersey*
Marilyn Molloy, *Our Lady of the Lake University*
Ron Moore, *Ryerson Polytechnical Institute*
Barbara Moses, *Bowling Green State University*
David Nash, *VP Research, Autofacts, Inc.*
Kylene Norman, *Clark State Community College*
Roxie Novak, *Radford University*
Richard Nowakowski, *Dalhousie University*
Stanley Ocken, *City College–CUNY*
Donald Passman, *University of Wisconsin*
David Patterson, *West Texas A & M*
Walter M. Patterson, *Lander University*
Edward Peifer, *Ulster County Community College*
Robert Phillips, *University of South Carolina at Aiken*
Mark A. Pinsky, *Northeastern University*
Catherine H. Pirri, *Northern Essex Community College*
Father Bernard Portz, *Creighton University*
David Randall, *Oakland Community College*
Richard Remzowski, *Broome Community College*
Guanshen Ren, *College of Saint Scholastica*
William H. Richardson, *Wichita State University*
David Rollins, *University of Central Florida*
Naomi Rose, *Mercer County Community College*
Sharon Ross, *DeKalb College*
David Ryeburn, *Simon Fraser University*

David Sandell, *U.S. Coast Guard Academy*
Ned W. Schillow, *Lehigh County Community College*
Dennis Schneider, *Knox College*
Dan Seth, *Morehead State University*
George Shapiro, *Brooklyn College*
Parashu R. Sharma, *Grambling State University*
Donald R. Sherbert, *University of Illinois*
Howard Sherwood, *University of Central Florida*
Bhagat Singh, *University of Wisconsin Centers*
Martha Sklar, *Los Angeles City College*
John L. Smith, *Rancho Santiago Community College*
Wolfe Snow, *Brooklyn College*
Ian Spatz, *Brooklyn College*
Jean Springer, *Mount Royal College*
Norton Starr, *Amherst College*
Richard B. Thompson, *The University of Arizona*
William F. Trench, *Trinity University*
Walter W. Turner, *Western Michigan University*
Richard C. Vile, *Eastern Michigan University*
David Voss, *Western Illinois University*
Shirley Wakin, *University of New Haven*
James Warner, *Precision Visuals*
Peter Waterman, *Northern Illinois University*
Evelyn Weinstock, *Glassboro State College*
Candice A. Weston, *University of Lowell*
Bruce F. White, *Lander University*
Gary L. Wood, *Azusa Pacific University*
Yihren Wu, *Hofstra University*
Richard Yuskaitis, *Precision Visuals*
Michael Zeidler, *Milwaukee Area Technical College*
Michael L. Zwilling, *Mount Union College*

Development Team for the Sixth Edition

The following people critiqued and reviewed various parts of the manuscript and suggested many of the ideas that found their way into this new edition:

Judith Broadwin, *Jericho High School*
Christopher D. Butler, *Case Western Reserve University*
Larry Cusick, *California State University–Fresno*
Philip Farmer, *Diablo Valley College*
Sally E. Fischbeck, *Rochester Institute of Technology*
J. Derrick Head, *University of Minnesota–Morris*
Tommie Ann Hill-Natter, *Prairie View A&M University*
Holly Hirst, *Appalachian State University*
Dan Kemp, *South Dakota State University*
Holly A. Kresch, *Diablo Valley College*
Marshall Leitman, *Case Western Reserve University*
Thomas W. Mason, *Florida A&M University*
Gary L. Peterson, *James Madison University*
Douglas Quinney, *University of Keele*
William H. Richardson, *Wichita State University*
Lila F. Roberts, *Georgia Southern University*
Avinash Sathaye, *University of Kentucky*
Mary Margaret Shoaf-Grubbs, *College of New Rochelle*
Mark Stevenson, *Oakland Community College*
John A. Suvak, *Memorial University of Newfoundland*
Skip Thompson, *Radford University*
Bruce R. Wenner, *University of Missouri–Kansas City*

The following people participated in phone surveys that helped to answer important questions about organization, philosophy, technology, and content:

Linda Bridge, *Long Beach City College*
Ted Clinkenbeard, *Des Moines Area Community College*
Victor Feser, *University of Maryland*
David Gross, *University of Connecticut*
Dennis Hadah, *Saddleback Community College*
Henry Horton, *University of West Florida*
Emmett Johnson, *Grambling State University*
Bill Kavanagh, *Mesa College*
Phil Locke, *University of Maine*
Thomas W. Mason, *Florida A&M University*
Ralph Okojie, *Elizabeth City State University*
David Robbins, *Trinity College*
Skip Thompson, *Radford University*
Paul Vesce, *University of Missouri–Kansas City*
Ronald Wagoner, *California State University–Fresno*

The following people read the sixth edition at various stages for mathematical and pedagogical accuracy and/or assisted with the critically important job of preparing answers to exercises:

Larry Cusick, *California State University–Fresno*
Stephen Davis, *Davidson College*
Blaise DeSesa, *Allentown College of St. Francis de Sales*
Thomas Vanden Eynden, *Thomas More College*
Beverly Fusfield, *Techsetters, Inc.*
Susan Gerstein
Konrad Heuvers, *Michigan Technological University*
Dean Hickerson
Majid Masso, *University of Delaware*
Kylene Norman, *Clark State Community College*
Irwin Pressman, *Carleton University*
David Ryeburn, *Simon Fraser University*
Shirley Wakin, *University of New Haven*
Neil Wigley, *University of Windsor*

The following students are members of the Student Advisory Board that critiqued the manuscript for clarity and provided valuable advice on making material interesting and relevant to today's students:

Dan Arndt, *University of Texas at Dallas*
Ajay Arora, *McMaster University*
Scott E. Barnett, *Wayne State University*
Fatenah Issa, *Loyola University of Chicago*
Laurie Haskell Messina, *University of Oklahoma*
Steven E. Pav, *Alfred University*

The following people created materials for tests and other supplements:

William H. Barker, *Bowdoin College*
Henry Smith, *Southeastern Louisiana University*
James E. Ward, *Bowdoin College*
Neil Wigley, *University of Windsor*

I would also like to thank Gary S. Stoudt of the *University of Indiana of Pennsylvania* for his assistance in locating various obscure historical materials.

Content Contributions

The following people assisted in the creation of modules and exercises or contributed valuable ideas that helped in the development of those materials:

Mary Ann Connors, *U.S. Military Academy at West Point*
Art Davis, *San Jose State University*
Gloria S. Dion, *Educational Testing Service*
Iris Brann Feta, *Clemson University*
Dixie Griffin, Jr., *Louisiana Tech University*
Dean Hickerson
Hugh E. Huntley, *University of Michigan*
Lynn Kiaer, *Rose-Hulman Institute of Technology*
Cecilia Knoll, *Florida Institute of Technology*
Michael Magill, *Purdue University*
Robert Meitz, *Arizona State University*
John Rickert, *Rose-Hulman Institute of Technology*
David Ryeburn, *Simon Fraser University*
Michael D. Shaw, *Florida Institute of Technology*
P. Narayana Swamy, *Southern Illinois University*
Josef Torok, *Rochester Institute of Technology*

I gratefully acknowledge the permission to adapt various exercises, examples, and text materials from the following publications:

Physics, 3rd ed., Cutnell and Johnson, John Wiley & Sons, Inc., 1995.
Fundamentals of Physics, 4th ed., Halliday, Resnick, and Walker, John Wiley & Sons, Inc., 1993.
Applications of Calculus, Philip Straffin, Ed., MAA Notes Number 29, The Mathematical Association of America, 1993.
Calculus Problems for a New Century, Robert Fraga, Ed., MAA Notes Number 28, Vol. 2, The Mathematical Association of America, 1993.
Engineering Mechanics, Meriam and Kraige, 3rd ed., Vol. 2, John Wiley & Sons, Inc., 1992.

Special Contributions

A special debt of gratitude to:

Barbara Holland, my editor, for sharing a vision and having unwavering faith in my work.

Madalyn Stone, for her infectious enthusiasm.

Ann Berlin, Pam Kennedy, and Charlotte Hyland of the Wiley Production Department for a scheduling miracle.

Sharon Smith, for again successfully coordinating a complex set of supplements.

Maddy Lesure, for capturing the "horizon" theme so beautifully in the cover.

Hilary Newman, for unearthing my obscure photographic requests.

Lilian Brady, for her unerring eye for aesthetics and typography.

The group at HRS, for a beautiful design and their patience with an (occasionally) cranky author.

The group at Techsetters, for superb composition and illustration and their devotion to excellence.

Lynn Kiaer, John Rickert, and Mary Ann Connors for their creative contributions to the modules.

Neil Wigley, for enjoyable E-mail humor and an outstanding job in writing the solutions manuals.

David Ryeburn, for his attention to detail and valued advice on applications of calculus.

My assistant, Dolores Morgan, whose superb organizational skills played a major role in keeping me on schedule.

CONTENTS

CHAPTER 7. INTEGRATION 377

CHAPTER 8. APPLICATIONS OF THE DEFINITE INTEGRAL IN GEOMETRY, SCIENCE, AND ENGINEERING 461

CHAPTER 9. PRINCIPLES OF INTEGRAL EVALUATION 513

Chapter 10. Mathematical Modeling with Differential Equations 579

Chapter 11. Infinite Series 615

Chapter 12. Analytic Geometry in Calculus 699

VOLUME 1

VOLUME 3

Brief Edition

Complete Edition

For the Student

Calculus is a compilation of ideas that provides a way of viewing and analyzing the physical world. As with all mathematics courses, calculus involves equations and formulas. However, if you successfully learn to use all of the formulas and solve all of the problems in this text but don't master the underlying ideas, you will have missed the most important part of calculus. Keep in mind that every single problem in this text has already been solved by somebody, so your ability to solve those problems gives you nothing unique. However, if you master the ideas of calculus, then you will have the tools to go beyond what other people have done, limited only by your own talents and creativity.

Before starting your studies, you may find it helpful to leaf through this text to get a general feeling for its different parts.

- At the beginning of each chapter you will find a page that gives an overview of the chapter, and at the beginning of each section you will find an introduction that gives an overview of that section. To help you locate specific information, sections are divided into topics described by headings in the margin.
- Each section ends with a set of exercises. The answers to most odd-numbered exercises appear in the back of the book. Worked-out solutions to the odd-numbered exercises are given in the *Student Resource Manual* and on a CD, which are available as supplements to the text.
- Some of the exercises are tagged with icons to indicate that some kind of technology is required for their solution. If your calculus course does not incorporate the use of technology, then your instructor will probably not assign these. Those exercises tagged with the icon [graphing icon] require graphing technology, which might be either a graphing calculator or a computer program that produces graphs from equations. Those exercises tagged with the icon C require a computer algebra system (called a CAS), which is a program that can perform symbolic as well as numerical calculations. The most common CAS programs are *Mathematica*, *Maple*, and *Derive*. Some of the newer calculators incorporate CAS capabilities.
- Each chapter ends with a set of supplementary exercises, many of which involve a combination of ideas from various sections within the chapter.
- Near the end of the text you will find seven appendices. Appendices A–F review some precalculus material, including trigonometry, and Appendix G contains some proofs that may or may not be part of your course.
- There is also reference material on the endpapers that are inside the front and back covers of the text.
- Illustrations in the exposition are referenced using a triple-number system. For example, Figure 1.6.3 is the third figure in Section 1.6, and Figure 7.2.5 is the fifth figure in Section 7.2. The same numbering system is used for theorems and definitions. Illustrations in the exercises are identified by the exercise number with which they are associated. For example, in a particular exercise set, Figure Ex-7 would be associated with Exercise 7.

- The ideas in this text were created by real people with interesting personalities and backgrounds. Pictures and signatures of many of these people appear on the opening pages of the chapters, and biographical sketches of various mathematicians appear throughout the text as footnotes.
- At various places in the text you will see elements labeled "For the Reader," which are designed to reinforce ideas in the text. Some of these ask you to think about an idea, some ask you to perform a computation, and (for students using technology) some ask you to read your reference manual and then use the technology to perform a computation or to generate a graph.

As you read through this book, you will find some ideas that you understand immediately, others that you don't understand until you have read them several times, and others that you do not understand, even after numerous readings. Don't become discouraged—some calculus ideas take time to "percolate," and you may well find that the idea suddenly becomes clear later when you least expect it.

If you find that your answer to an exercise does not match that in the back of the book, do not presume immediately that your answer is incorrect—there may be more than one way to express the answer. For example, if your answer is $\sqrt{3}/3$ and the text answer is $1/\sqrt{3}$, then both are correct, since your answer can be obtained by rationalizing the text answer. In general, if your answer does not match that in the text, then your best first step is to look for an algebraic manipulation or a trigonometric identity that relates the two answers. In cases where the answer is a decimal approximation, your answer may differ from that in the text because of different choices in the number of decimal places used in the computations.

Some exercises require a verbal answer. Express those answers in complete, correctly punctuated, logical sentences—not fragmented phrases and formulas.

It is *not* essential to have graphing technology to read and use this text. Exercises requiring technology have been tagged with icons precisely so they can be omitted if necessary. Text elements requiring technology are relegated to the "For the Reader," so they can be omitted as well. If you have graphing technology, then you may want to use it as you read the text or to check your work in exercises that are not tagged with icons. However, it is not essential.

7

INTEGRATION

Gottfried Leibniz

Traditionally, that portion of calculus concerned with finding tangent lines and rates of change is called ***differential calculus*** and that portion concerned with finding areas is called ***integral calculus***. However, we will see in this chapter that the two problems are so closely related that the distinction between differential and integral calculus is often hard to discern.

In this chapter we will begin with an overview of the problem of finding areas—we will discuss what the term "area" means, and we will outline two approaches to defining and calculating areas. Following this overview, we will discuss the "Fundamental Theorem of Calculus", which is the theorem that relates the problems of finding tangent lines and areas, and we will discuss techniques for calculating areas. Finally, we will use the ideas in this chapter to continue our study of rectilinear motion and to reexamine the concept of a natural logarithm.

7.1 AN OVERVIEW OF THE AREA PROBLEM

In this introductory section we will give an overview of the problem of defining and calculating areas of plane regions with curvilinear boundaries. All of the results in this section will be reexamined in more detail later in this chapter, so our purpose here is to introduce the fundamental concepts.

DEFINING AREA

The main goal of this chapter is to study the following major problem of calculus:

> **7.1.1** THE AREA PROBLEM. Given a function f that is continuous and nonnegative on an interval $[a, b]$, find the area between the graph of f and the interval $[a, b]$ on the x-axis (Figure 7.1.1).

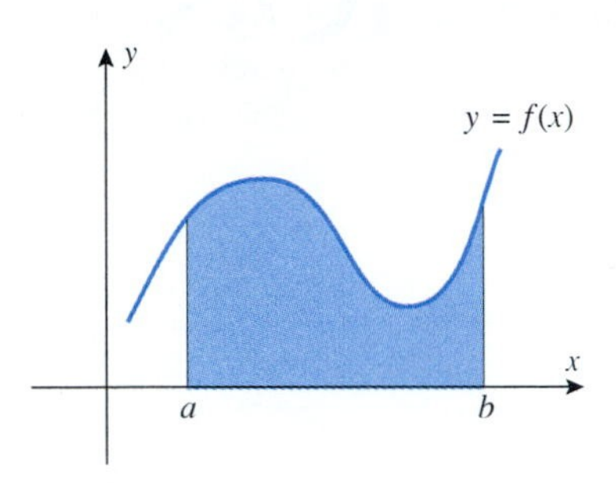

Figure 7.1.1

Area formulas for basic geometric figures, such as rectangles, polygons, and circles, date back to the earliest written records of mathematics. The first real advance beyond the elementary level of area computation was made by the Greek mathematician, Archimedes,* who devised an ingenious but cumbersome technique, called the *method of exhaustion*, for finding areas of regions bounded by parabolas, spirals, and various other curves.

---

* ARCHIMEDES (287 B.C.–212 B.C.). Greek mathematician and scientist. Born in Syracuse, Sicily, Archimedes was the son of the astronomer Pheidias and possibly related to Heiron II, king of Syracuse. Most of the facts about his life come from the Roman biographer, Plutarch, who inserted a few tantalizing pages about him in the massive biography of the Roman soldier, Marcellus. In the words of one writer, "the account of Archimedes is slipped like a tissue-thin shaving of ham in a bull-choking sandwich."

Archimedes ranks with Newton and Gauss as one of the three greatest mathematicians who ever lived, and he is certainly the greatest mathematician of antiquity. His mathematical work is so modern in spirit and technique that it is barely distinguishable from that of a seventeenth-century mathematician, yet it was all done without benefit of algebra or a convenient number system. Among his mathematical achievements, Archimedes developed a general method (exhaustion) for finding areas and volumes, and he used the method to find areas bounded by parabolas and spirals and to find volumes of cylinders, paraboloids, and segments of spheres. He gave a procedure for approximating π and bounded its value between $3\frac{10}{71}$ and $3\frac{1}{7}$. In spite of the limitations of the Greek numbering system, he devised methods for finding square roots and invented a method based on the Greek myriad (10,000) for representing numbers as large as 1 followed by 80 million billion zeros.

Of all his mathematical work, Archimedes was most proud of his discovery of the method for finding the volume of a sphere—he showed that the volume of a sphere is two-thirds the volume of the smallest cylinder that can contain it. At his request, the figure of a sphere and cylinder was engraved on his tombstone.

In addition to mathematics, Archimedes worked extensively in mechanics and hydrostatics. Nearly every schoolchild knows Archimedes as the absent-minded scientist who, on realizing that a floating object displaces its weight of liquid, leaped from his bath and ran naked through the streets of Syracuse shouting, "Eureka, Eureka!"—(meaning, "I have found it!"). Archimedes actually created the discipline of hydrostatics and used it to find equilibrium positions for various floating bodies. He laid down the fundamental postulates of mechanics, discovered the laws of levers, and calculated centers of gravity for various flat surfaces and solids. In the excitement of discovering the mathematical laws of the lever, he is said to have declared, "Give me a place to stand and I will move the earth."

Although Archimedes was apparently more interested in pure mathematics than its applications, he was an engineering genius. During the second Punic war, when Syracuse was attacked by the Roman fleet under the command of Marcellus, it was reported by Plutarch that Archimedes' military inventions held the fleet at bay for three years. He invented super catapults that showered the Romans with rocks weighing a quarter ton or more, and fearsome mechanical devices with iron "beaks and claws" that reached over the city walls, grasped the ships, and spun them against the rocks. After the first repulse, Marcellus called Archimedes a "geometrical Briareus (a hundred-armed mythological monster) who uses our ships like cups to ladle water from the sea."

Eventually the Roman army was victorious and contrary to Marcellus' specific orders the 75-year-old Archimedes was killed by a Roman soldier. According to one report of the incident, the soldier cast a shadow across the sand in which Archimedes was working on a mathematical problem. When the annoyed Archimedes yelled, "Don't disturb my circles," the soldier flew into a rage and cut the old man down.

With his death the Greek gift of mathematics passed into oblivion, not to be fully resurrected again until the sixteenth century. Unfortunately, there is no known accurate likeness or statue of this great man.

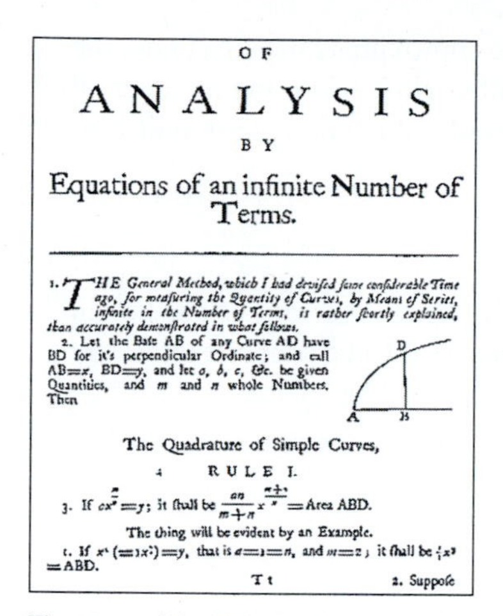

OF

ANALYSIS

BY

Equations of an infinite Number of Terms.

1. *THE General Method, which I had devised some considerable Time ago, for measuring the Quantity of Curves, by Means of Series, infinite in the Number of Terms, is rather shortly explained, than accurately demonstrated in what follows.*

2. Let the Base AB of any Curve AD have BD for it's perpendicular Ordinate; and call AB=x, BD=y, and let a, b, c, &c. be given Quantities, and m and n whole Numbers. Then

The Quadrature of Simple Curves,

RULE I.

3. If $ax^{\frac{m}{n}} = y$; it shall be $\frac{an}{m+n} x^{\frac{m+n}{n}}$ = Area ABD.

The thing will be evident by an Example.

1. If $x^2 (= 1x^{\frac{2}{1}}) = y$, that is $a = 1 = n$, and $m = 2$; it shall be $\frac{1}{3}x^3$ = ABD.

T t 2. Suppose

First page of the 1745 English translation of Newton's *De Analysi*

By the seventeenth century, several mathematicians had discovered how to obtain such areas more simply by calculating limits. However, the method of exhaustion and its successors lacked generality—for each different problem one had to devise special procedures. The major breakthrough in obtaining a general method for calculating areas was made independently by Newton and Leibniz, both of whom discovered that areas could be obtained by reversing the process of differentiation. This discovery, which is regarded as the beginning of calculus, was circulated by Newton in 1669 and published in 1711 in a paper entitled, *De Analysi per Aequationes Numero Terminorum Infinitas* (*On the Analysis by Means of Equations with Infinitely Many Terms*); and it was discovered by Leibniz around 1673 and stated in an unpublished manuscript dated November 11, 1675.

Before one can talk logically about methods for calculating areas, it is necessary to have a precise definition of what the term *area* means. To avoid a lot of mathematical formality, let us assume that the areas of geometric figures with straight boundaries, such as rectangles, triangles, and polygons, are defined and computed using the standard formulas for such figures. However, the problem of defining and computing areas of figures with *curvilinear* boundaries is more complicated and will require various limiting processes. For example, in the introductory section of this text we showed that the area of a circle could be viewed as a limit of areas of inscribed polygons (Figure 7 in the Introduction). Thus, once a definition is established for the area of a polygon, the area of a circle can be *defined* as a limit of areas of polygons.

THE RECTANGLE METHOD FOR FINDING AREAS

There are two basic methods for finding the area of the region having the form shown in Figure 7.1.1—the *rectangle method* and the *antiderivative method*. The idea behind the rectangle method is as follows:

- Divide the interval $[a, b]$ into n equal subintervals, and over each subinterval construct a rectangle that extends from the x-axis to any point on the curve $y = f(x)$ that is above the subinterval; the particular point does not matter—it can be above the center, above an endpoint, or above any other point in the subinterval. In Figure 7.1.2 it is above the center.
- For each n, the total area of the rectangles can be viewed as an *approximation* to the exact area under the curve over the interval $[a, b]$. Moreover, it is evident intuitively that as n increases these approximations will get better and better and will approach the exact area as a limit (Figure 7.1.3).

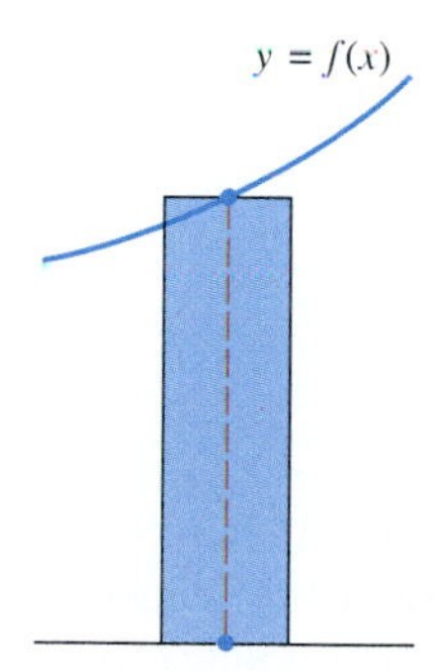

Figure 7.1.2

This procedure serves both as a mathematical definition and a method of computation—we can *define* the area under $y = f(x)$ over the interval $[a, b]$ as the limit of the areas of the approximating rectangles, and we can use the method itself to approximate this area.

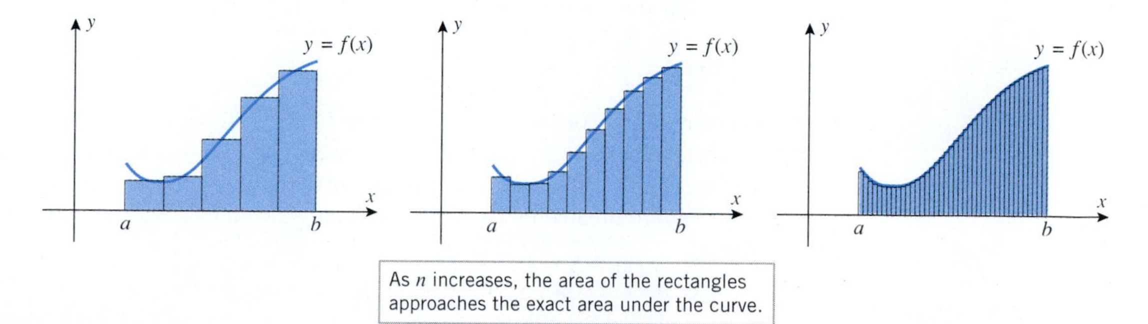

Figure 7.1.3

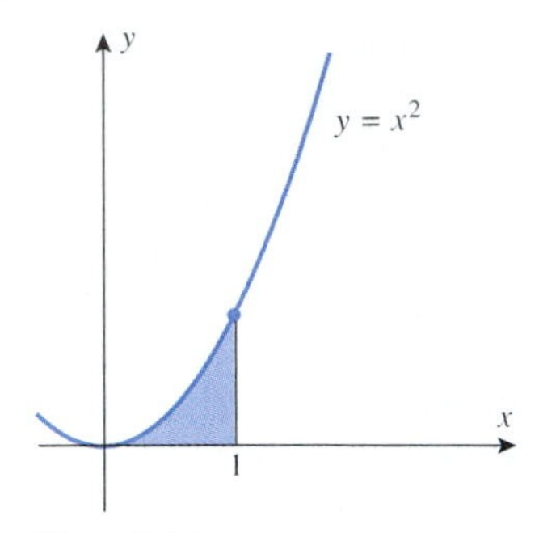

Figure 7.1.4

To illustrate this idea, we will use the rectangle method to approximate the area under the curve $y = x^2$ over the interval [0, 1] (Figure 7.1.4). We will begin by dividing the interval [0, 1] into n equal subintervals, from which it follows that each subinterval has length $1/n$; the endpoints of the subintervals occur at

$$0, \quad \frac{1}{n}, \quad \frac{2}{n}, \quad \frac{3}{n}, \ldots, \quad \frac{n-1}{n}, \quad 1$$

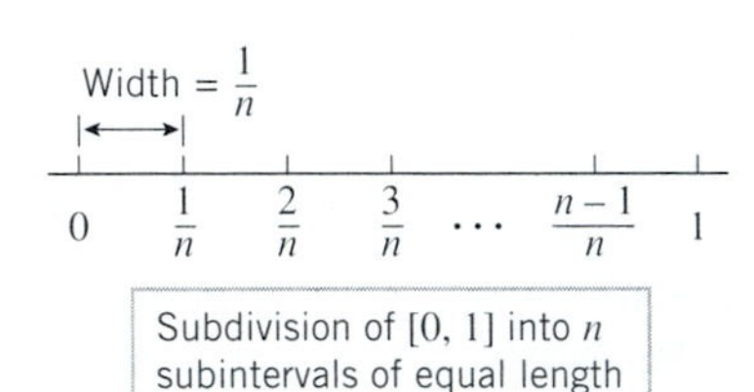

Figure 7.1.5

(Figure 7.1.5). We want to construct a rectangle over each of these intervals whose height is the value of the function $f(x) = x^2$ at any point in the interval. To be specific, let us use the right endpoints, in which case the heights of our rectangles will be

$$\left(\frac{1}{n}\right)^2, \quad \left(\frac{2}{n}\right)^2, \quad \left(\frac{3}{n}\right)^2, \ldots, \quad 1$$

and since each rectangle has a base of width $1/n$, the total area A_n of the n rectangles will be

$$A_n = \left[\left(\frac{1}{n}\right)^2 + \left(\frac{2}{n}\right)^2 + \left(\frac{3}{n}\right)^2 + \cdots + 1^2\right]\left(\frac{1}{n}\right) \tag{1}$$

For example, if $n = 4$, then the total area of the four approximating rectangles would be

$$A_4 = \left[\left(\tfrac{1}{4}\right)^2 + \left(\tfrac{2}{4}\right)^2 + \left(\tfrac{3}{4}\right)^2 + 1^2\right]\left(\tfrac{1}{4}\right) = \tfrac{15}{32} = 0.46875$$

Table 7.1.1 shows the result of evaluating (1) on a computer for some increasingly large values of n. These computations suggest that the exact area is close to $\frac{1}{3}$.

Table 7.1.1

n	4	10	100	1000	10,000	100,000
A_n	0.468750	0.385000	0.338350	0.333834	0.333383	0.333338

FOR THE READER. Use your calculating utility to confirm the value of A_{10} given in Table 7.1.1.

THE ANTIDERIVATIVE METHOD FOR FINDING AREAS

The antiderivative method for finding areas reflects the genius of Newton and Leibniz—they suggested that to find the area under the curve in Figure 7.1.1, one should first consider the more general problem of finding the area $A(x)$ under the curve from the point a to an arbitrary point x in the interval $[a, b]$ (Figure 7.1.6). Newton and Leibniz discovered independently that the *derivative* of the function $A(x)$ is easy to find, so that if one can figure out how to find $A(x)$ from $A'(x)$, then the area under the curve from a to b can be obtained by substituting $x = b$ in the area formula $A(x)$.

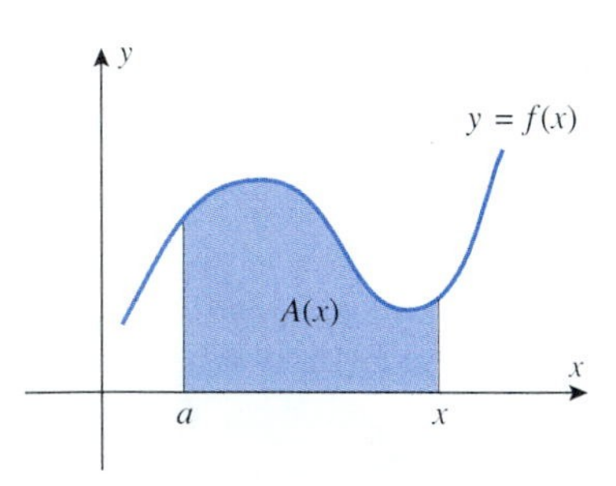

Figure 7.1.6

To illustrate how all of this works, let us begin with the problem of finding

$$A'(x) = \lim_{h \to 0} \frac{A(x+h) - A(x)}{h} \tag{2}$$

For simplicity, consider the case where $h > 0$. The numerator on the right side of (2) is the difference of two areas: the area between a and $x + h$ minus the area between a and x (Figure 7.1.7a). If we let c be the midpoint between x and $x + h$, then this difference of areas can be approximated by the area of a rectangle with base h and height $f(c)$ (Figure 7.1.7b). Thus,

$$\frac{A(x+h) - A(x)}{h} \approx \frac{f(c) \cdot h}{h} = f(c) \tag{3}$$

It seems plausible from Figure 7.1.7b that the error in approximation (3) will approach

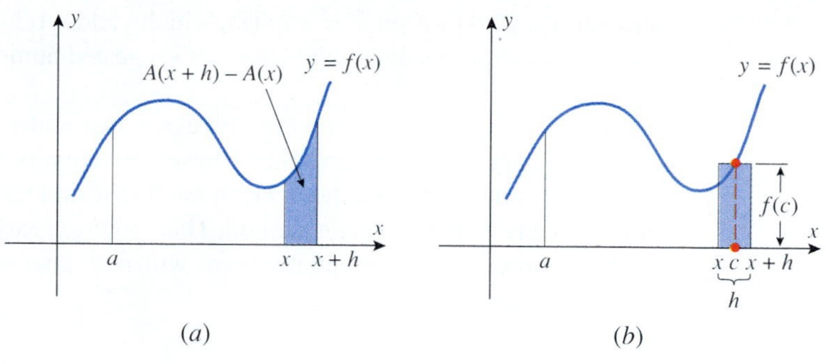

Figure 7.1.7

approach zero as $h \to 0$. If we accept this to be so, then it follows from (2) and (3) that

$$A'(x) = \lim_{h \to 0} \frac{A(x+h) - A(x)}{h} = \lim_{h \to 0} f(c) \tag{4}$$

Since c is the midpoint between x and $x + h$, it follows that $c \to x$ as $h \to 0$. But we have assumed f to be a continuous function, so $f(c) \to f(x)$ as $c \to x$. Therefore,

$$\lim_{h \to 0} f(c) = f(x)$$

Thus, it follows from (4) that

$$A'(x) = f(x) \tag{5}$$

This is the result we were looking for; it tells us that *the derivative of the area function* $A(x)$ *is the function whose graph forms the upper boundary of the region.*

To illustrate how the antiderivative method works, let us apply it to the same problem we investigated with the rectangle method—finding the area under $y = x^2$ over the interval $[0, 1]$. The upper boundary of the region is the graph of $f(x) = x^2$, so it follows from (5) that the derivative of the area function is

$$A'(x) = x^2 \tag{6}$$

Thus, to find $A(x)$ we must look for a function whose derivative is x^2. This is called an ***antidifferentiation*** problem because we are trying to find $A(x)$ by "undoing" a differentiation. By simply guessing we see that

$$A(x) = \tfrac{1}{3}x^3$$

is one solution to (6). But this is not the only solution, since it follows from Theorem 6.5.3 that

$$A(x) = \tfrac{1}{3}x^3 + C \tag{7}$$

also satisfies (6) for any real value of C. We still have some work to do since this formula involves an unknown constant C that must be determined. This is where the decision to solve the area problem for a general right-hand endpoint helps. If we consider the case where $x = 0$, then the interval $[0, x]$ reduces to a single point. If we agree that the area above a single point should be taken as zero, then it follows on substituting $x = 0$ in (7) that

$$A(0) = 0 + C = 0 \quad \text{or} \quad C = 0$$

so (7) simplifies to

$$A(x) = \tfrac{1}{3}x^3 \tag{8}$$

which is the formula for the area under $y = x^2$ over the interval $[0, x]$. For the area over the interval [0, 1] we set $x = 1$ in (8), which yields $A(1) = \frac{1}{3}$ for the exact area under the curve. This confirms definitely what was suggested numerically in Table 7.1.1.

REMARK. Our success in finding the exact area under the curve $y = x^2$ hinged on our ability to guess at a function $A(x)$ whose derivative is x^2. Had we not been able to find such a function, then the antiderivative method would have failed and we would have been forced to rely on the rectangle method. Thus, whereas earlier in this text we were concerned with the process of differentiation, we will now also be concerned with the process of antidifferentiation.

EXERCISE SET 7.1

In Exercises 1–4, use an appropriate formula from plane geometry to find the exact area between the graph of f and the given interval; and then use the rectangle method to make a table of approximations $A_1, A_2, \ldots, A_{10}$ to the exact area, where A_n is the approximation that results by dividing the interval into n subintervals and constructing a rectangle over each subinterval whose height is the y-coordinate of the curve $y = f(x)$ at the right endpoint.

1. $f(x) = x$; $[0, 1]$

2. $f(x) = 4 - 2x$; $[0, 2]$

3. $f(x) = 6x + 2$; $[0, 2]$

4. $f(x) = \sqrt{1 - x^2}$; $[0, 1]$

5. Let $A(x) = x^2/2$. Confirm that $A'(x) = x$, and use the antiderivative method to find the exact area in Exercise 1.

6. Let $A(x) = 4x - x^2$. Confirm that $A'(x) = 4 - 2x$, and use the antiderivative method to find the exact area in Exercise 2.

7. Let $A(x) = 3x^2 + 2x$. Confirm that $A'(x) = 6x + 2$, and use the antiderivative method to find the exact area in Exercise 3.

8. Let $A(x) = \frac{1}{2}x\sqrt{1 - x^2} + \frac{1}{2}\sin^{-1} x$. Then confirm that $A'(x) = \sqrt{1 - x^2}$, and use the antiderivative method to find the exact area in Exercise 4.

9. Use the antiderivative method to find the exact area between the curve $y = e^x$ and the interval $[0, 1]$.

10. Use the antiderivative method to find the exact area between the curve $y = \sin x$ and the interval $[0, \pi]$.

7.2 THE INDEFINITE INTEGRAL; INTEGRAL CURVES AND DIRECTION FIELDS

In the last section we saw that antidifferentiation plays an important role in finding exact areas. In this section we will develop some fundamental results about antidifferentiation that will ultimately lead us to systematic procedures for finding a function from its derivative.

7.2.1 DEFINITION. A function F is called an ***antiderivative*** of a function f on a given interval I if $F'(x) = f(x)$ for all x in the interval.

For example, the function $F(x) = \frac{1}{3}x^3$ is an antiderivative of $f(x) = x^2$ on the interval $(-\infty, +\infty)$ because for each x in this interval

$$F'(x) = \frac{d}{dx}\left[\tfrac{1}{3}x^3\right] = x^2 = f(x)$$

However, this is not the only antiderivative of F on this interval. If we add any constant C to $\frac{1}{3}x^3$, then the function $F(x) = \frac{1}{3}x^3 + C$ is also an antiderivative of f on $(-\infty, +\infty)$, since

$$F'(x) = \frac{d}{dx}\left[\tfrac{1}{3}x^3 + C\right] = x^2 + 0 = f(x)$$

In general, once any single antiderivative of a function is known, other antiderivatives can be obtained by adding constants to the known antiderivative. Thus,

$$\tfrac{1}{3}x^3, \quad \tfrac{1}{3}x^3+2, \quad \tfrac{1}{3}x^3-5, \quad \tfrac{1}{3}x^3+\sqrt{2}$$

are all antiderivatives of $f(x)=x^2$.

WARNING. Do not confuse derivatives and antiderivatives—the *derivative* of the function $f(x)=x^2$ is $f'(x)=2x$, but the functions $F(x)=\frac{1}{3}x^3+C$ are *antiderivatives* of f.

It is reasonable to ask if there are antiderivatives of a function f that cannot be obtained by adding some constant to a known antiderivative F. The answer is *no*—once a single antiderivative of f on an interval I is known, all other antiderivatives on that interval are obtainable by adding constants to that antiderivative. This is so because Theorem 6.5.3 tells us that if two functions have the same derivative on an interval, then they differ by a constant on that interval. The following theorem summarizes these observations.

> **7.2.2 THEOREM.** *If $F(x)$ is any antiderivative of $f(x)$ on an interval I, then for any constant C the function $F(x)+C$ is also an antiderivative of $f(x)$ on that interval. Moreover, each antiderivative of $f(x)$ on the interval I can be expressed in the form $F(x)+C$ by choosing the constant C appropriately.*

THE INDEFINITE INTEGRAL

Extract from the manuscript of Leibniz dated October 29, 1675 in which the integral sign first appeared.

The process of finding antiderivatives is called ***antidifferentiation*** or ***integration***. Thus, if

$$\frac{d}{dx}[F(x)]=f(x)$$

then integrating (or antidifferentiating) $f(x)$ produces the antiderivatives $F(x)+C$. We denote this by writing

$$\int f(x)\,dx = F(x)+C \tag{1}$$

For example, the antiderivatives of $f(x)=x^2$ are the functions $F(x)=\frac{1}{3}x^3+C$, so

$$\int x^2\,dx = \tfrac{1}{3}x^3+C$$

The "elongated s" that appears on the left side of (1) is called an ***integral sign**** or an ***indefinite integral***, the function $f(x)$ is called the ***integrand***, and the constant C is called the ***constant of integration***. You should read Equation (1) as "the integral of $f(x)$ with respect to x is equal to $F(x)+C$." The adjective "indefinite" emphasizes that the integration process does not produce a *definite* function, but rather a whole set of functions.

The dx symbols in the differentiation and antidifferentiation operations

$$\frac{d}{dx}[\ \ ] \quad \text{and} \quad \int [\ \ ]\,dx$$

serve to identify the independent variable. If an independent variable other than x is used, say t, then the notation must be adjusted appropriately. Thus,

$$\frac{d}{dt}[F(t)]=f(t) \quad \text{and} \quad \int f(t)\,dt = F(t)+C$$

are equivalent statements.

---

*This notation was devised by Leibniz. In his early papers Leibniz used the notation "omn." (an abbreviation for the Latin word "omnes") to denote integration. Then on October 29, 1675 he wrote, "It will be useful to write $\int$ for omn., thus $\int \ell$ for omn. ℓ . . . ." Two or three weeks later he refined the notation further and wrote $\int[\ \ ]\,dx$ rather than $\int$ alone. This notation is so useful and so powerful that its development by Leibniz must be regarded as a major milestone in the history of mathematics and science.

Example 1

DERIVATIVE FORMULA	EQUIVALENT INTEGRATION FORMULA
$\frac{d}{dx}[x^3] = 3x^2$	$\int 3x^2\,dx = x^3 + C$
$\frac{d}{dx}[\sqrt{x}] = \frac{1}{2\sqrt{x}}$	$\int \frac{1}{2\sqrt{x}}\,dx = \sqrt{x} + C$
$\frac{d}{dt}[\tan t] = \sec^2 t$	$\int \sec^2 t\,dt = \tan t + C$
$\frac{d}{du}[u^{3/2}] = \frac{3}{2}u^{1/2}$	$\int \frac{3}{2}u^{1/2}\,du = u^{3/2} + C$

◀

For simplicity, the dx is sometimes absorbed into the integrand. For example,

$$\int 1\,dx \quad \text{can be written as} \quad \int dx$$

$$\int \frac{1}{x^2}\,dx \quad \text{can be written as} \quad \int \frac{dx}{x^2}$$

INTEGRATION FORMULAS

Integration is essentially educated guesswork—given the derivative of a function f, one tries to guess what the function f is. However, many basic integration formulas can be obtained directly from their companion differentiation formulas. Some of the most important ones are given in Table 7.2.1.

Table 7.2.1

DIFFERENTIATION FORMULA	INTEGRATION FORMULA
1. $\frac{d}{dx}[x] = 1$	$\int dx = x + C$
2. $\frac{d}{dx}\left[\frac{x^{r+1}}{r+1}\right] = x^r \quad (r \neq -1)$	$\int x^r\,dx = \left[\frac{x^{r+1}}{r+1}\right] + C \quad (r \neq -1)$
3. $\frac{d}{dx}[\sin x] = \cos x$	$\int \cos x\,dx = \sin x + C$
4. $\frac{d}{dx}[-\cos x] = \sin x$	$\int \sin x\,dx = -\cos x + C$
5. $\frac{d}{dx}[\tan x] = \sec^2 x$	$\int \sec^2 x\,dx = \tan x + C$
6. $\frac{d}{dx}[-\cot x] = \csc^2 x$	$\int \csc^2 x\,dx = -\cot x + C$
7. $\frac{d}{dx}[\sec x] = \sec x \tan x$	$\int \sec x \tan x\,dx = \sec x + C$
8. $\frac{d}{dx}[-\csc x] = \csc x \cot x$	$\int \csc x \cot x\,dx = -\csc x + C$
9. $\frac{d}{dx}[e^x] = e^x$	$\int e^x\,dx = e^x + C$
10. $\frac{d}{dx}\left[\frac{b^x}{\ln b}\right] = b^x$	$\int b^x\,dx = \frac{b^x}{\ln b} + C$
11. $\frac{d}{dx}[\ln \lvert x\rvert] = \frac{1}{x}$	$\int \frac{dx}{x} = \ln \lvert x\rvert + C$

Example 2

The second integration formula in this table will be easy to remember if you express it in words: *to integrate a power of* x *(other than* -1*), add* 1 *to the power and divide by the new power*. Here are some examples:

$$\int x^2\,dx = \frac{x^3}{3} + C \qquad \boxed{r=2}$$

$$\int x^3\,dx = \frac{x^4}{4} + C \qquad \boxed{r=3}$$

$$\int \frac{1}{x^5}\,dx = \int x^{-5}\,dx = \frac{x^{-5+1}}{-5+1} + C = -\frac{1}{4x^4} + C \qquad \boxed{r=-5}$$

$$\int \sqrt{x}\,dx = \int x^{\frac{1}{2}}\,dx = \frac{x^{\frac{1}{2}+1}}{\frac{1}{2}+1} + C = \tfrac{2}{3}x^{\frac{3}{2}} + C = \tfrac{2}{3}(\sqrt{x})^3 + C \qquad \boxed{r=\tfrac{1}{2}}$$

$$\int x^{-1}\,dx = \int \frac{dx}{x} = \ln|x| + C$$

◀

PROPERTIES OF THE INDEFINITE INTEGRAL

If we differentiate an antiderivative of $f(x)$, we obtain $f(x)$ back again. Thus,

$$\frac{d}{dx}\left[\int f(x)\,dx\right] = f(x) \tag{2}$$

This result is helpful for proving the following basic properties of antiderivatives.

7.2.3 THEOREM.

(*a*) *A constant factor can be moved through an integral sign; that is,*

$$\int cf(x)\,dx = c\int f(x)\,dx$$

(*b*) *An antiderivative of a sum is the sum of the antiderivatives; that is,*

$$\int [f(x) + g(x)]\,dx = \int f(x)\,dx + \int g(x)\,dx$$

(*c*) *An antiderivative of a difference is the difference of the antiderivatives; that is,*

$$\int [f(x) - g(x)]\,dx = \int f(x)\,dx - \int g(x)\,dx$$

Proof. In each part we must show that the expression on the right side of the equation is an antiderivative of the integrand on the left side of the equation. This can be done using (2) as follows:

$$\frac{d}{dx}\left[c\int f(x)\,dx\right] = c\frac{d}{dx}\left[\int f(x)\,dx\right] = cf(x)$$

$$\begin{aligned}\frac{d}{dx}\left[\int f(x)\,dx + \int g(x)\,dx\right] &= \frac{d}{dx}\left[\int f(x)\,dx\right] + \frac{d}{dx}\left[\int g(x)\,dx\right]\\ &= f(x) + g(x)\end{aligned}$$

$$\begin{aligned}\frac{d}{dx}\left[\int f(x)\,dx - \int g(x)\,dx\right] &= \frac{d}{dx}\left[\int f(x)\,dx\right] - \frac{d}{dx}\left[\int g(x)\,dx\right]\\ &= f(x) - g(x)\end{aligned}$$

■

When applying Theorem 7.2.3, it is best to put in the constant of integration at the *very end* of the computations to obtain the simplest form of the answer. This is illustrated in the following example.

Example 3

Evaluate

(a) $\int 4\cos x\,dx$ (b) $\int (x+x^2)\,dx$

Solution (a).

$$\int 4\cos x\,dx = 4\int \cos x\,dx = 4(\sin x + C) = 4\sin x + 4C$$

Theorem 7.2.3(*a*) Table 7.2.1

Since C is an arbitrary constant, so is $4C$. However, this latter form is unnecessarily complicated and can be avoided by deferring the insertion of the constant until the end of the computations; this procedure yields

$$\int 4\cos x\,dx = 4\int \cos x\,dx = 4\sin x + C$$

Solution (b).

$$\int (x+x^2)\,dx = \int x\,dx + \int x^2\,dx = \frac{x^2}{2} + \frac{x^3}{3} + C$$

Theorem 7.2.3(*b*) Table 7.2.1

◀

Parts (*b*) and (*c*) of Theorem 7.2.3 can be extended to more than two functions, which in combination with part (*a*) results in the following general formula:

$$\int [c_1 f_1(x) + c_2 f_2(x) + \cdots + c_n f_n(x)]\,dx = c_1\int f_1(x)\,dx + c_2\int f_2(x)\,dx + \cdots + c_n\int f_n(x)\,dx \tag{3}$$

Example 4

$$\int (3x^6 - 2x^2 + 7x + 1)\,dx = 3\int x^6\,dx - 2\int x^2\,dx + 7\int x\,dx + \int 1\,dx$$
$$= \frac{3x^7}{7} - \frac{2x^3}{3} + \frac{7x^2}{2} + x + C$$

◀

Sometimes it is useful to rewrite an integrand in a different form before performing the integration.

Example 5

Evaluate

(a) $\int \frac{\cos x}{\sin^2 x}\,dx$ (b) $\int \frac{t^2 - 2t^4}{t^4}\,dt$

Solution (a).

$$\int \frac{\cos x}{\sin^2 x}\,dx = \int \frac{1}{\sin x}\frac{\cos x}{\sin x}\,dx = \int \csc x \cot x\,dx = -\csc x + C$$

Formula 8 in Table 7.2.1

Solution (b).

$$\int \frac{t^2 - 2t^4}{t^4}\,dt = \int \left(\frac{1}{t^2} - 2\right) dt = \int (t^{-2} - 2)\,dt$$

$$= \frac{t^{-1}}{-1} - 2t + C = -\frac{1}{t} - 2t + C$$

◀

INTEGRAL CURVES

Graphs of antiderivatives of a function f are called ***integral curves*** of f. We know from Theorem 7.2.2 that if $y = F(x)$ is any integral curve of $f(x)$, then all other integral curves are vertical translations of this curve, since they have equations of the form $y = F(x) + C$. For example, $y = \frac{1}{3}x^3$ is one integral curve for $f(x) = x^2$, so all the other integral curves have equations of the form $y = \frac{1}{3}x^3 + C$; conversely, the graph of any equation of this form is an integral curve (Figure 7.2.1).

In many problems one is interested in finding a function whose derivative satisfies specified conditions. The following example illustrates a geometric problem of this type.

Example 6

Suppose that a point moves along some unknown curve $y = f(x)$ in the xy-plane in such a way that at each point (x, y) on the curve, the tangent line has slope x^2. Find an equation for the curve given that it passes through the point (2, 1).

Solution. We know that $dy/dx = x^2$, so

$$y = \int x^2\,dx = \tfrac{1}{3}x^3 + C$$

Since the curve passes through (2, 1), a specific value for C can be found by using the fact that $y = 1$ if $x = 2$. Substituting these values in the above equation yields

$$1 = \tfrac{1}{3}(2^3) + C \quad \text{or} \quad C = -\tfrac{5}{3}$$

so the curve is $y = \frac{1}{3}x^3 - \frac{5}{3}$. ◀

Observe that in this example the requirement that the unknown curve pass through the point (2, 1) enabled us to determine a specific value for the constant of integration, thereby isolating the single integral curve $y = \frac{1}{3}x^3 - \frac{5}{3}$ from the family $y = \frac{1}{3}x^3 + C$ (Figure 7.2.2).

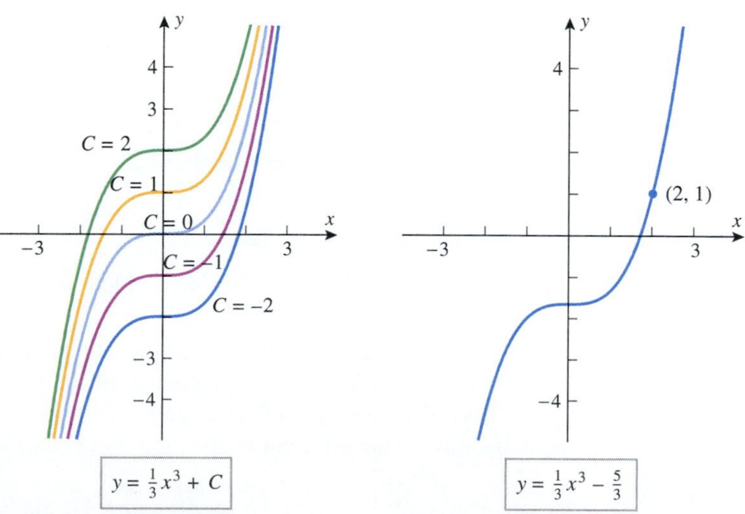

Figure 7.2.1

Figure 7.2.2

INTEGRATION FROM THE VIEWPOINT OF DIFFERENTIAL EQUATIONS

We will now consider another way of looking at integration that will be useful in our later work. Suppose that $f(x)$ is a known function and we are interested in finding a function $F(x)$ such that $y = F(x)$ satisfies the equation

$$\frac{dy}{dx} = f(x) \tag{4}$$

The solutions of this equation are the antiderivatives of $f(x)$, and we know that these can be obtained by integrating $f(x)$. For example, the solutions of the equation

$$\frac{dy}{dx} = x^2 \tag{5}$$

are

$$y = \int x^2\,dx = \frac{x^3}{3} + C$$

Equation (4) is called a ***differential equation*** because it involves a derivative of an unknown function. Differential equations are different from the kinds of equations we have encountered so far in that the unknown is a *function* and not a *number* as in an equation such as $x^2 + 5x - 6 = 0$.

Sometimes we will not be interested in finding all of the solutions of (4), but rather we will want only the solution whose integral curve passes through a specified point (x_0, y_0). For example, in Example 6 we solved (5) for the integral curve that passed through the point (2, 1).

For simplicity, it is common in the study of differential equations to denote a solution of $dy/dx = f(x)$ as $y(x)$ rather than $F(x)$, as earlier. With this notation, the problem of finding a function $y(x)$ whose derivative is $f(x)$ and whose integral curve passes through the point (x_0, y_0) is expressed as

$$\frac{dy}{dx} = f(x), \quad y(x_0) = y_0 \tag{6}$$

For reasons that will be explained later, this is called an ***initial-value problem***, and the requirement that $y(x_0) = y_0$ is called the ***initial condition*** for the problem.

Example 7

Solve the initial-value problem

$$\frac{dy}{dx} = \cos x, \quad y(0) = 1$$

Solution. The solution of the differential equation is

$$y = \int \cos x\,dx = \sin x + C \tag{7}$$

The initial condition $y(0) = 1$ implies that $y = 1$ if $x = 0$; substituting these values in (7) yields

$$1 = \sin(0) + C \quad \text{or} \quad C = 1$$

Thus, the solution of the initial-value problem is $y = \sin x + 1$. ◀

DIRECTION FIELDS

If we interpret dy/dx as the slope of a tangent line, then at a point (x, y) on an integral curve of the equation $dy/dx = f(x)$, the slope of the tangent line is $f(x)$. What is interesting about this is that the slopes of the tangent lines to the integral curves can be obtained without actually solving the differential equation. For example, if

$$\frac{dy}{dx} = \sqrt{x^2 + 1}$$

then we know without solving the equation that at the point where $x = 1$ the tangent line to an integral curve has slope $\sqrt{1^2+1} = \sqrt{2}$; and more generally, at a point where $x = a$, the tangent line to an integral curve has slope $\sqrt{a^2+1}$.

A geometric description of the integral curves of a differential equation $dy/dx = f(x)$ can be obtained by choosing a rectangular grid of points in the xy-plane, calculating the slopes of the tangent lines to the integral curves at the gridpoints, and drawing small portions of the tangent lines at those points. The resulting picture, which is called a ***direction field*** or ***slope field*** for the equation, shows the "direction" of the integral curves at the gridpoints. With sufficiently many gridpoints it is often possible to visualize the integral curves themselves; for example, Figure 7.2.3*a* shows a direction field for the differential equation $dy/dx = x^2$, and Figure 7.2.3*b* shows that same field with the integral curves imposed on it—the more gridpoints that are used, the more completely the direction field reveals the shape of the integral curves. However, the amount of computation can be considerable, so computers are usually used when direction fields with many gridpoints are needed.

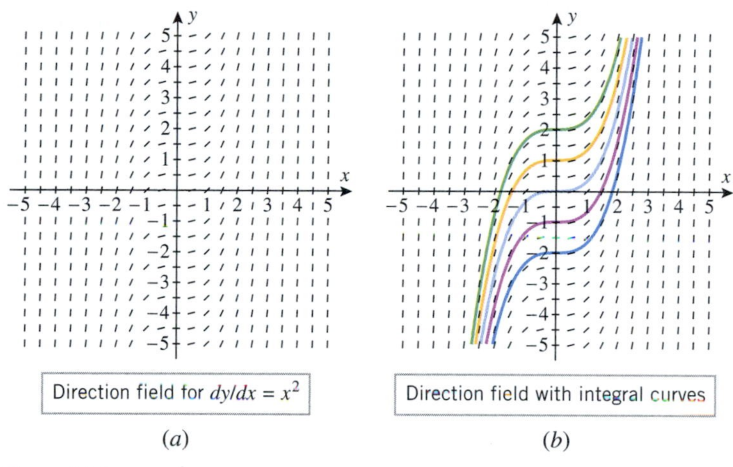

Figure 7.2.3

EXERCISE SET 7.2 Graphing Calculator C CAS

1. In each part, confirm that the formula is correct, and state a corresponding integration formula.
   (a) $\frac{d}{dx}[\sqrt{1+x^2}\,] = \frac{x}{\sqrt{1+x^2}}$
   (b) $\frac{d}{dx}[xe^x] = (x+1)e^x$

2. In each part, confirm that the stated formula is correct by differentiating.
   (a) $\int x \sin x\, dx = \sin x - x\cos x + C$
   (b) $\int \frac{dx}{(1-x^2)^{3/2}} = \frac{x}{\sqrt{1-x^2}} + C$

In Exercises 3–6, find the derivative and state a corresponding integration formula.

3. $\frac{d}{dx}[\sqrt{x^3+5}\,]$

4. $\frac{d}{dx}\left[\frac{x}{x^2+3}\right]$

5. $\frac{d}{dx}[\sin(2\sqrt{x}\,)]$

6. $\frac{d}{dx}[\sin x - x\cos x]$

In Exercises 7 and 8, evaluate the integral by rewriting the integrand appropriately, if required, and then applying Formula 2 in Table 7.2.1.

7. (a) $\int x^8\, dx$ (b) $\int x^{5/7}\, dx$ (c) $\int x^3\sqrt{x}\, dx$

8. (a) $\int \sqrt[3]{x^2}\, dx$ (b) $\int \frac{1}{x^6}\, dx$ (c) $\int x^{-7/8}\, dx$

In Exercises 9–12, evaluate the integral by applying Theorem 7.2.3 and Formula 2 in Table 7.2.1 appropriately.

9. (a) $\int \frac{1}{2x^3}\, dx$ (b) $\int (u^3 - 2u + 7)\, du$

10. $\int (x^{2/3} - 4x^{-1/5} + 4)\, dx$

11. $\int (x^{-3} + \sqrt{x} - 3x^{1/4} + x^2)\, dx$

12. $\int \left(\frac{7}{y^{3/4}} - \sqrt[3]{y} + 4\sqrt{y} \right) dy$

In Exercises 13–30, evaluate the integral, and check your answer by differentiating.

13. $\int x(1 + x^3)\, dx$

14. $\int (2 + y^2)^2\, dy$

15. $\int x^{1/3}(2 - x)^2\, dx$

16. $\int (1 + x^2)(2 - x)\, dx$

17. $\int \frac{x^5 + 2x^2 - 1}{x^4}\, dx$

18. $\int \frac{1 - 2t^3}{t^3}\, dt$

19. $\int \left[\frac{2}{x} + 3e^x \right] dx$

20. $\int \left[\frac{1}{2t} - \sqrt{2}e^t \right] dt$

21. $\int [4 \sin x + 2 \cos x]\, dx$

22. $\int [4 \sec^2 x + \csc x \cot x]\, dx$

23. $\int \sec x(\sec x + \tan x)\, dx$

24. $\int \sec x(\tan x + \cos x)\, dx$

25. $\int \left[\frac{1}{\theta} - 2e^\theta - \csc^2 \theta \right] d\theta$

26. $\int \frac{dy}{\csc y}$

27. $\int \frac{\sin x}{\cos^2 x}\, dx$

28. $\int \left[\phi + \frac{2}{\sin^2 \phi} \right] d\phi$

29. $\int [1 + \sin^2 \theta \csc \theta]\, d\theta$

30. $\int \frac{\sin 2x}{\cos x}\, dx$

31. Evaluate the integral

$$\int \frac{1}{1 + \sin x}\, dx$$

by multiplying the numerator and denominator by an appropriate expression.

C **32.** For each of the integrals you evaluated in Exercises 13–31, use a CAS to check your answer. If the answer produced by the CAS does not match yours, show that the two answers are equivalent.

33. (a) Graph some representative integral curves of $f(x) = x$.
(b) Find an equation for the integral curve that passes through the point $(4, 7)$.

34. (a) Graph some representative integral curves of the function $f(x) = e^x/2$.
(b) Find an equation for the integral curve that passes through the point $(0, 1)$.

35. Use a graphing utility to generate some representative integral curves of the function $f(x) = 5x^4 - \sec^2 x$ over the interval $(-\pi/2, \pi/2)$.

36. Use a graphing utility to generate some representative integral curves of $f(x) = (x - 1)/x$ over the interval $(0, 5)$.

37. Suppose that a point moves along a curve $y = f(x)$ in the xy-plane in such a way that at each point (x, y) on the curve the tangent line has slope $-\sin x$. Find an equation for the curve, given that it passes through the point $(0, 2)$.

38. Suppose that a point moves along a curve $y = f(x)$ in the xy-plane in such a way that at each point (x, y) on the curve the tangent line has slope $(x + 1)^2$. Find an equation for the curve, given that it passes through the point $(-2, 8)$.

In Exercises 39 and 40, solve the initial-value problems.

39. (a) $\frac{dy}{dx} = \sqrt[3]{x},\ y(1) = 2$ (b) $\frac{dy}{dt} = \frac{1}{t},\ y(-1) = 5$
(c) $\frac{dy}{dx} = \frac{x + 1}{\sqrt{x}},\ y(1) = 0$

40. (a) $\frac{dy}{dx} = \frac{1}{(2x)^3},\ y(1) = 0$
(b) $\frac{dy}{dt} = \sec^2 t - \sin t,\ y\left(\frac{\pi}{4}\right) = 1$
(c) $\frac{dy}{dx} = x^2\sqrt{x^3},\ y(0) = 0$

41. Find the general form of a function whose second derivative is $\sqrt{x}$. [*Hint:* Solve the equation $f''(x) = \sqrt{x}$ for $f(x)$ by integrating both sides twice.]

42. Find a function f such that $f''(x) = x + \cos x$ and such that $f(0) = 1$ and $f'(0) = 2$. [*Hint:* Integrate both sides of the equation twice.]

In Exercises 43–45, find an equation of the curve that satisfies the given conditions.

43. At each point (x, y) on the curve the slope is $2x + 1$; the curve passes through the point $(-3, 0)$.

44. At each point (x, y) on the curve the slope equals the square of the distance between the point and the y-axis; the point $(-1, 2)$ is on the curve.

45. At each point (x, y) on the curve, y satisfies the condition $d^2y/dx^2 = 6x$; the line $y = 5 - 3x$ is tangent to the curve at the point where $x = 1$.

46. Suppose that a uniform metal rod 50 cm long is insulated laterally, and the temperatures at the exposed ends are maintained at 25°C and 85°C, respectively. Assume that an x-axis is chosen as in the accompanying figure and that the temperature $T(x)$ at each point x satisfies the equation

$$\frac{d^2T}{dx^2} = 0$$

Find $T(x)$ for $0 \leq x \leq 50$.

Figure Ex-46

47. (a) Show that

$$F(x) = \tfrac{1}{6}(3x+4)^2 \quad \text{and} \quad G(x) = \tfrac{3}{2}x^2 + 4x$$

differ by a constant by showing that they are antiderivatives of the same function.

(b) Find the constant C such that $F(x) - G(x) = C$ by evaluating $F(x)$ and $G(x)$ at some point x_0.

(c) Check your answer in part (b) by simplifying the expression $F(x) - G(x)$ algebraically.

48. Follow the directions of Exercise 47 with

$$F(x) = \frac{x^2}{x^2+5} \quad \text{and} \quad G(x) = -\frac{5}{x^2+5}$$

In Exercises 49 and 50, use a trigonometric identity to help evaluate the integral.

49. $\int \tan^2 x\,dx$ **50.** $\int \cot^2 x\,dx$

51. Use the identities $\cos 2\theta = 1 - 2\sin^2\theta = 2\cos^2\theta - 1$ to help evaluate the integrals

(a) $\int \sin^2(x/2)\,dx$ (b) $\int \cos^2(x/2)\,dx$

52. Let F and G be the functions defined piecewise by

$$F(x) = \begin{cases} x, & x > 0 \\ -x, & x < 0 \end{cases} \quad \text{and} \quad G(x) = \begin{cases} x+2, & x > 0 \\ -x+3, & x < 0 \end{cases}$$

(a) Show that F and G have the same derivative.

(b) Show that $G(x) \neq F(x) + C$ for any constant C.

(c) Do parts (a) and (b) violate Theorem 7.2.2? Explain.

53. The speed of sound in air at 0°C (or 273 K on the Kelvin scale) is 1087 ft/s, but the speed v increases as the temperature T rises. Experimentation has shown that the rate of change of v with respect to T is

$$\frac{dv}{dT} = \frac{1087}{2\sqrt{273}}T^{-1/2}$$

where v is in feet per second and T is in kelvins (K). Find a formula that expresses v as a function of T.

7.3 INTEGRATION BY SUBSTITUTION

In this section we will study a technique, called ***substitution****, that can often be used to transform complicated integration problems into simpler ones.*

u-SUBSTITUTION

The method of substitution can be motivated by examining the chain rule from the viewpoint of antidifferentiation. For this purpose, suppose that F is an antiderivative of f and that g is a differentiable function. The chain rule implies that the derivative of $F(g(x))$ can be expressed as

$$\frac{d}{dx}[F(g(x))] = F'(g(x))g'(x)$$

which we can write in integral form as

$$\int F'(g(x))g'(x)\,dx = F(g(x)) + C \tag{1}$$

or since F is an antiderivative of f,

$$\int f(g(x))g'(x)\,dx = F(g(x)) + C \tag{2}$$

For our purposes it will be useful to let $u = g(x)$ and to write $du/dx = g'(x)$ in the differential form $du = g'(x)\,dx$. With this notation (1) can be expressed as

$$\int f(u)\,du = F(u) + C \tag{3}$$

The process of evaluating an integral of form (2) by converting it into form (3) with the substitution

$$u = g(x) \quad \text{and} \quad du = g'(x)\,dx$$

is called the ***method of u-substitution***. The following example illustrates how the method works.

Example 1

Evaluate $\int (x^2+1)^{50} \cdot 2x\,dx$.

Solution. If we let $u = x^2+1$, then $du/dx = 2x$, which implies that $du = 2x\,dx$. Thus, the given integral can be written as

$$\int (x^2+1)^{50} \cdot 2x\,dx = \int u^{50}\,du = \frac{u^{51}}{51} + C = \frac{(x^2+1)^{51}}{51} + C$$ ◀

It is important to realize that in the method of u-substitution you have control over the choice of u, but once you make that choice you have no control over the resulting expression for du. Thus, in the last example we *chose* $u = x^2+1$ but $du = 2x\,dx$ was *computed.* Fortunately, our choice of u, combined with the computed du, worked out perfectly to produce an integral involving u that was easy to evaluate. However, in general, the method of u-substitution will fail if the chosen u and the computed du do not produce an integrand in which no expressions involving x remain, or if you cannot evaluate the resulting integral. Thus, for example, the substitution $u = x^2+1$, $du = 2x\,dx$ will not work for the integral

$$\int (x^2+1)^{50} \cdot 2x \cos x\,dx$$

because this substitution results in the integral

$$\int u^{50} \cos x\,du$$

which still contains an expression involving x.

In general, there are no hard and fast rules for choosing u, and in some problems no choice of u will work. In such cases other methods need to be used, some of which will be discussed later. Making appropriate choices for u will come with experience, but you may find the following guidelines, combined with a mastery of the basic integrals in Table 7.2.1, helpful.

Integration by Substitution

Step 1. Make a choice for u, say $u = g(x)$.

Step 2. Compute $du/dx = g'(x)$.

Step 3. Make the substitution $u = g(x)$, $du = g'(x)\,dx$.

At this stage, the *entire* integral must be in terms of u; no x's should remain. If this is not the case, try a different choice of u.

Step 4. Evaluate the resulting integral, if possible.

Step 5. Replace u by $g(x)$, so that the final answer is in terms of x.

Example 2

The easiest substitutions occur when the integrand is the derivative of a known function, except for a constant added to or subtracted from the independent variable. For example,

$$\int \sin(x+9)\,dx = \int \sin u\,du = -\cos u + C = -\cos(x+9) + C$$

$u = x+9$
$du = 1 \cdot dx = dx$

$$\int (x-8)^{23}\,dx = \int u^{23}\,du = \frac{u^{24}}{24} + C = \frac{(x-8)^{24}}{24} + C$$ ◀

$u = x-8$
$du = 1 \cdot dx = dx$

Another easy u-substitution occurs when the integrand is the derivative of a known function, except for a constant that multiplies or divides the independent variable. The following example illustrates two ways to evaluate such integrals.

Example 3

Evaluate $\int \cos 5x\, dx$.

Solution.

$$\int \cos 5x\, dx = \int (\cos u) \cdot \frac{1}{5} du = \frac{1}{5} \int \cos u\, du = \frac{1}{5} \sin u + C = \frac{1}{5} \sin 5x + C$$

$u = 5x$
$du = 5\, dx$ or $dx = \frac{1}{5}\, du$

Alternative Solution. There is a variation of the preceding method that some people prefer. The substitution $u = 5x$ requires $du = 5\, dx$. If there were a factor of 5 in the integrand, then we could group the 5 and dx together to form the du required by the substitution. Since there is no factor of 5, we will insert one and compensate by putting a factor of $\frac{1}{5}$ in front of the integral. The computations are as follows:

$$\int \cos 5x\, dx = \frac{1}{5} \int \cos 5x \cdot 5\, dx = \frac{1}{5} \int \cos u\, du = \frac{1}{5} \sin u + C = \frac{1}{5} \sin 5x + C \quad ◀$$

$u = 5x$
$du = 5\, dx$

Example 4

Evaluate $\int \sin^2 x \cos x\, dx$.

Solution. If we let $u = \sin x$, then

$$\frac{du}{dx} = \cos x, \quad \text{so} \quad du = \cos x\, dx$$

Thus,

$$\int \sin^2 x \cos x\, dx = \int u^2\, du = \frac{u^3}{3} + C = \frac{\sin^3 x}{3} + C \quad ◀$$

Example 5

Evaluate $\int \frac{e^{\sqrt{x}}}{\sqrt{x}}\, dx$.

Solution. If we let $u = \sqrt{x}$, then

$$\frac{du}{dx} = \frac{1}{2\sqrt{x}}, \quad \text{so} \quad du = \frac{1}{2\sqrt{x}}\, dx \quad \text{or} \quad 2\, du = \frac{1}{\sqrt{x}}\, dx$$

Thus,

$$\int \frac{e^{\sqrt{x}}}{\sqrt{x}}\, dx = \int 2e^u\, du = 2 \int e^u\, du = 2e^u + C = 2e^{\sqrt{x}} + C \quad ◀$$

Example 6

$$\int \frac{dx}{\left(\frac{1}{3}x - 8\right)^5} = \int \frac{3\, du}{u^5} = 3 \int u^{-5}\, du = -\frac{3}{4} u^{-4} + C = -\frac{3}{4}\left(\frac{1}{3}x - 8\right)^{-4} + C \quad ◀$$

$u = \frac{1}{3}x - 8$
$du = \frac{1}{3}\, dx$ or $dx = 3\, du$

Example 7

With the help of Theorem 7.2.3, a complicated integral can sometimes be computed by expressing it as a sum of simpler integrals. For example,

$$\int \left(\frac{1}{x} + \sec^2 \pi x\right) dx = \int \frac{dx}{x} + \int \sec^2 \pi x \, dx = \ln|x| + \int \sec^2 \pi x \, dx$$

$$= \ln|x| + \frac{1}{\pi} \int \sec^2 u \, du$$

$u = \pi x$
$du = \pi \, dx$ or $dx = \frac{1}{\pi} du$

$$= \ln|x| + \frac{1}{\pi} \tan u + C = \ln|x| + \frac{1}{\pi} \tan \pi x + C$$ ◀

Example 8

Evaluate $\int t^4 \sqrt[3]{3 - 5t^5} \, dt$.

Solution. After some possible false starts most readers would eventually hit on the following substitution:

$$\int t^4 \sqrt[3]{3 - 5t^5} \, dt = -\frac{1}{25} \int \sqrt[3]{u} \, du = -\frac{1}{25} \int u^{1/3} \, du$$

$u = 3 - 5t^5$
$du = -25t^4 \, dt$ or $-\frac{1}{25} du = t^4 \, dt$

$$= -\frac{1}{25} \frac{u^{4/3}}{4/3} + C = -\frac{3}{100} \left(3 - 5t^5\right)^{4/3} + C$$ ◀

Example 9

Evaluate $\int x^2 \sqrt{x - 1} \, dx$.

Solution. Let

$$u = x - 1 \quad \text{so that} \quad du = dx \tag{4}$$

From the first equality in (4)

$$x^2 = (u + 1)^2 = u^2 + 2u + 1$$

so that

$$\int x^2 \sqrt{x - 1} \, dx = \int (u^2 + 2u + 1)\sqrt{u} \, du = \int (u^{5/2} + 2u^{3/2} + u^{1/2}) \, du$$

$$= \tfrac{2}{7} u^{7/2} + \tfrac{4}{5} u^{5/2} + \tfrac{2}{3} u^{3/2} + C$$

$$= \tfrac{2}{7}(x - 1)^{7/2} + \tfrac{4}{5}(x - 1)^{5/2} + \tfrac{2}{3}(x - 1)^{3/2} + C$$ ◀

REMARK. Not every function can be integrated in terms of familiar functions using u-substitutions. For example, you will not find any u-substitution that will integrate

$$\int \sin(x^2) \, dx$$

in terms of functions encountered thus far in this text (try).

INTEGRATION USING COMPUTER ALGEBRA SYSTEMS

The advent of computer algebra systems has made it possible to evaluate many kinds of integrals that would be laborious to evaluate by hand. For example, *Mathematica*, *Maple*, and *Derive* all produce the following result in a matter of seconds:

$$\int \sqrt{2x - x^2}\, dx = \tfrac{1}{2}(x-1)\sqrt{2x - x^2} - \tfrac{1}{2}\sin^{-1}(1-x) + C$$

However, just as one would not want to rely on a calculator to compute $2 + 2$, so one would not want to use a CAS to integrate a simple function such as $f(x) = x^2$. Thus, even if you have a CAS, you will want to develop a reasonable level of competence in evaluating basic integrals. Moreover, the mathematical techniques that we will introduce for evaluating basic integrals are precisely the techniques that computer algebra systems use to evaluate more complicated integrals.

FOR THE READER. If you have a CAS, use it to calculate the integrals in the examples of this section. If your CAS produces a form of the answer that is different from the one in the text, then confirm algebraically that the two answers agree. Your CAS has various commands for simplifying answers. Explore the effect of using the CAS to simplify the expressions it produces for the integrals.

EXERCISE SET 7.3  

In Exercises 1–4, evaluate the integrals by making the indicated substitutions.

1. (a) $\int 2x\left(x^2+1\right)^{23} dx;\ u = x^2+1$

(b) $\int \cos^3 x \sin x\, dx;\ u = \cos x$

(c) $\int \frac{1}{\sqrt{x}} \sin\sqrt{x}\, dx;\ u = \sqrt{x}$

(d) $\int \frac{3x\, dx}{\sqrt{4x^2+5}};\ u = 4x^2+5$

(e) $\int \frac{x^2}{x^3-4}\, dx;\ u = x^3-4$

2. (a) $\int \sec^2(4x+1)\, dx;\ u = 4x+1$

(b) $\int y\sqrt{1+2y^2}\, dy;\ u = 1+2y^2$

(c) $\int \sqrt{\sin \pi\theta} \cos \pi\theta\, d\theta;\ u = \sin \pi\theta$

(d) $\int (2x+7)(x^2+7x+3)^{4/5}\, dx;\ u = x^2+7x+3$

(e) $\int \frac{e^x}{1+e^x}\, dx;\ u = 1+e^x$

3. (a) $\int \cot x \csc^2 x\, dx;\ u = \cot x$

(b) $\int (1+\sin t)^9 \cos t\, dt;\ u = 1+\sin t$

(c) $\int \frac{dx}{x \ln x};\ u = \ln x$

(d) $\int e^{-5x}\, dx;\ u = -5x$

(e) $\int \frac{\sin 3\theta}{1+\cos 3\theta}\, d\theta;\ u = 1+\cos 3\theta$

4. (a) $\int x^2\sqrt{1+x}\, dx;\ u = 1+x$

(b) $\int [\csc(\sin x)]^2 \cos x\, dx;\ u = \sin x$

(c) $\int e^{\tan x} \sec^2 x\, dx;\ u = \tan x$

(d) $\int e^{2t}\sqrt{1+e^{2t}}\, dt;\ u = 1+e^{2t}$

(e) $\int \frac{5x^4}{x^5+1}\, dx;\ u = x^5+1$

In Exercises 5–36, evaluate the integrals by making appropriate substitutions.

5. $\int e^{2x}\, dx$

6. $\int \frac{dx}{2x}$

7. $\int x\left(2-x^2\right)^3 dx$

8. $\int (3x-1)^5\, dx$

9. $\int \cos 8x\, dx$

10. $\int \sin 3x\, dx$

11. $\int \sec 4x \tan 4x\, dx$

12. $\int \sec^2 5x\, dx$

13. $\int t\sqrt{7t^2+12}\, dt$

14. $\int \frac{x}{\sqrt{4-5x^2}}\, dx$

15. $\int \frac{x^2}{\sqrt{x^3+1}}\, dx$

16. $\int \frac{1}{(1-3x)^2}\, dx$

17. $\int \frac{x}{(4x^2+1)^3}\, dx$

18. $\int x\cos(3x^2)\, dx$

19. $\int e^{\sin x}\cos x\, dx$

20. $\int x^3 e^{x^4}\, dx$

21. $\int x^2 e^{-2x^3}\, dx$

22. $\int \frac{e^x+e^{-x}}{e^x-e^{-x}}\, dx$

23. $\int \frac{\sin(5/x)}{x^2}\, dx$

24. $\int \frac{\sec^2(\sqrt{x})}{\sqrt{x}}\, dx$

25. $\int x^2\sec^2(x^3)\, dx$

26. $\int \cos^3 2t \sin 2t\, dt$

27. $\int \frac{dx}{e^x}$

28. $\int \sqrt{e^x}\, dx$

29. $\int \sin^5 3t \cos 3t\, dt$

30. $\int \frac{\sin 2\theta}{(5+\cos 2\theta)^3}\, d\theta$

31. $\int \cos 4\theta\sqrt{2-\sin 4\theta}\, d\theta$

32. $\int \tan^3 5x \sec^2 5x\, dx$

33. $\int \sec^3 2x \tan 2x\, dx$

34. $\int [\sin(\sin\theta)]\cos\theta\, d\theta$

35. $\int \frac{e^{\sqrt{y}}}{\sqrt{y}}\, dy$

36. $\int \frac{dy}{\sqrt{y}e^{\sqrt{y}}}$

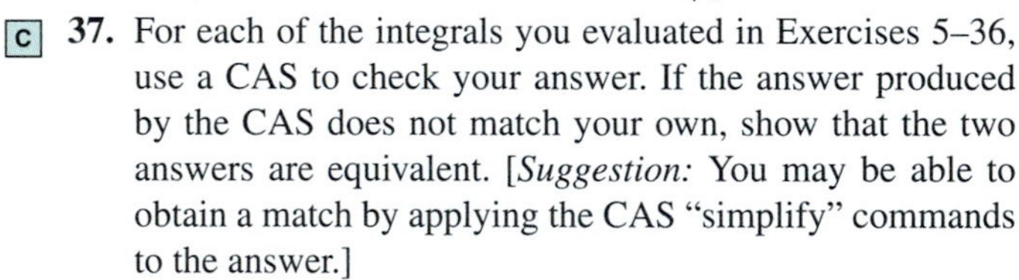

C **37.** For each of the integrals you evaluated in Exercises 5–36, use a CAS to check your answer. If the answer produced by the CAS does not match your own, show that the two answers are equivalent. [*Suggestion:* You may be able to obtain a match by applying the CAS "simplify" commands to the answer.]

In Exercises 38 and 39, evaluate the integrals assuming that n is a positive integer and $b \neq 0$.

38. $\int \sqrt[n]{a+bx}\, dx \quad (b \neq 0)$

39. $\int \sin^n(a+bx)\cos(a+bx)\, dx$

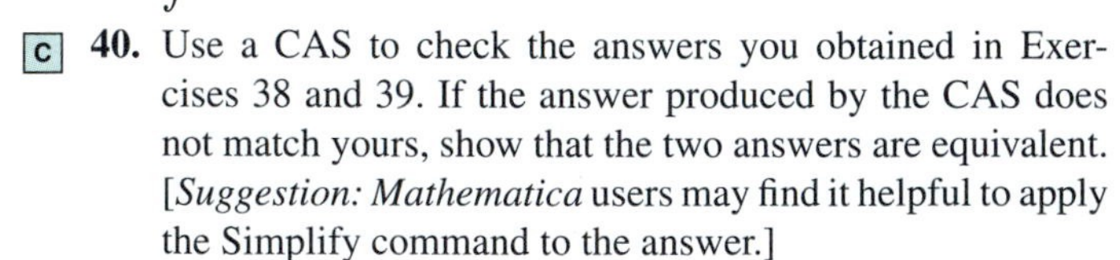

C **40.** Use a CAS to check the answers you obtained in Exercises 38 and 39. If the answer produced by the CAS does not match yours, show that the two answers are equivalent. [*Suggestion: Mathematica* users may find it helpful to apply the Simplify command to the answer.]

In Exercises 41 and 42, evaluate the integrals by making the indicated substitutions.

41. $\int x\sqrt{x-3}\, dx;\ u = x-3$

42. $\int \frac{y\, dy}{\sqrt{y+1}};\ u = y+1$

The integrals in Exercises 43–48 are a little trickier than those you have encountered thus far. To evaluate these integrals you will have to apply a trigonometric identity or modify the form of the integrand algebraically before making a substitution.

43. $\int \tan^2 3\theta\, d\theta$

44. $\int \sin^3 2\theta\, d\theta$

45. $\int \frac{t+1}{t}\, dt$

46. $\int e^{2\ln x}\, dx$

47. $\int [\ln(e^x)+\ln(e^{-x})]\, dx$

48. $\int \cot x\, dx$

49. (a) Evaluate the integral $\int \sin x\cos x\, dx$ by two methods: first by letting $u = \sin x$, then by letting $u = \cos x$.
(b) Explain why the two apparently different answers obtained in part (a) are really equivalent.

50. (a) Evaluate $\int (5x-1)^2\, dx$ by two methods: first square and integrate, then let $u = 5x-1$.
(b) Explain why the two apparently different answers obtained in part (a) are really equivalent.

In Exercises 51 and 52, solve the initial-value problems.

51. $\frac{dy}{dx} = \sqrt{3x+1};\ y(1) = 5$

52. $\frac{dy}{dx} = 6 - 5\sin 2x;\ y(0) = 3$

53. Find a function f such that the slope of the tangent line at a point (x, y) on the curve $y = f(x)$ is $\sqrt{3x+1}$, and the curve passes through the point $(0, 1)$.

54. Use a graphing utility to generate some typical integral curves of $f(x) = x/(x^2+1)$ over the interval $(-5, 5)$.

55. Suppose that a population p of frogs is estimated at the start of 1995 to be 100,000, and the growth model for the population assumes that the rate of growth (in thousands) after t years will be $p'(t) = (4+0.15t)^{3/2}$. Estimate the projected population at the start of the year 2000.

7.4 SIGMA NOTATION

In this section we will digress briefly from the main theme of this chapter to introduce a notation that can be used to write lengthy sums in a compact form. This material will be needed in many of the later chapters.

SIGMA NOTATION

The notation we will discuss in this section is called ***sigma notation*** or ***summation notation*** because it uses the uppercase Greek letter Σ (sigma) to denote various kinds of sums. To illustrate how this notation works, consider the sum

$$1^2 + 2^2 + 3^2 + 4^2 + 5^2$$

in which each term is of the form k^2, where k is one of the integers from 1 to 5. In sigma notation this sum can be written as

$$\sum_{k=1}^{5} k^2$$

which is read "the summation of k^2, where k runs from 1 to 5." The notation tells us to form the sum of the terms that result when we substitute successive integers for k in the expression k^2, starting with $k = 1$ and ending with $k = 5$.

More generally, if $f(k)$ is a function of k, and if m and n are integers such that $m \leq n$, then

$$\sum_{k=m}^{n} f(k) \tag{1}$$

denotes the sum of the terms that result when we substitute successive integers for k, starting with $k = m$ and ending with $k = n$ (Figure 7.4.1).

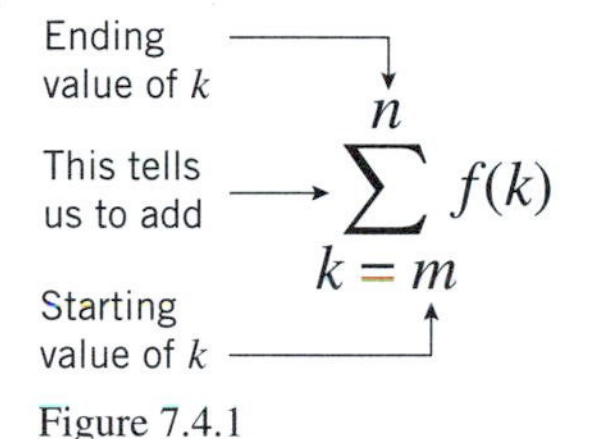

Figure 7.4.1

Example 1

$$\sum_{k=4}^{8} k^3 = 4^3 + 5^3 + 6^3 + 7^3 + 8^3$$

$$\sum_{k=1}^{5} 2k = 2 \cdot 1 + 2 \cdot 2 + 2 \cdot 3 + 2 \cdot 4 + 2 \cdot 5 = 2 + 4 + 6 + 8 + 10$$

$$\sum_{k=0}^{5} (2k + 1) = 1 + 3 + 5 + 7 + 9 + 11$$

$$\sum_{k=0}^{5} (-1)^k (2k + 1) = 1 - 3 + 5 - 7 + 9 - 11$$

$$\sum_{k=-3}^{1} k^3 = (-3)^3 + (-2)^3 + (-1)^3 + 0^3 + 1^3 = -27 - 8 - 1 + 0 + 1$$

$$\sum_{k=1}^{3} k \sin\left(\frac{k\pi}{5}\right) = \sin\frac{\pi}{5} + 2\sin\frac{2\pi}{5} + 3\sin\frac{3\pi}{5}$$ ◀

The numbers m and n in (1) are called, respectively, the ***lower*** and ***upper limits of summation***; and the letter k is called the ***index of summation***. It is not essential to use k as the index of summation; any letter not reserved for another purpose will do. For example,

$$\sum_{i=1}^{6} \frac{1}{i}, \quad \sum_{j=1}^{6} \frac{1}{j}, \quad \text{and} \quad \sum_{n=1}^{6} \frac{1}{n}$$

all denote the sum

$$1+\frac{1}{2}+\frac{1}{3}+\frac{1}{4}+\frac{1}{5}+\frac{1}{6}$$

If the upper and lower limits of summation are the same, then the "sum" in (1) reduces to one term. For example,

$$\sum_{k=2}^{2} k^3 = 2^3 \quad \text{and} \quad \sum_{i=1}^{1} \frac{1}{i+2} = \frac{1}{1+2} = \frac{1}{3}$$

In the sums

$$\sum_{i=1}^{5} 2, \quad \sum_{k=3}^{6} 7, \quad \text{and} \quad \sum_{j=0}^{2} x^3$$

the expression to the right of the Σ sign does not involve the index of summation. In such cases, we take all the terms in the sum to be the same, with one term for each allowable value of the summation index. Thus,

$$\sum_{i=1}^{5} 2 = 2+2+2+2+2$$

$$\sum_{k=3}^{6} 7 = 7+7+7+7$$

$$\sum_{j=0}^{2} x^3 = x^3+x^3+x^3$$

A sum can be written in more than one way with sigma notation by changing the limits of summation. For example, the sum of the first five positive even integers can be written in the following ways:

$$\sum_{k=1}^{5} 2k = 2+4+6+8+10$$

$$\sum_{k=0}^{4} (2k+2) = 2+4+6+8+10$$

$$\sum_{k=2}^{6} (2k-2) = 2+4+6+8+10$$

CHANGING THE INDEX OF SUMMATION

On occasion we will want to change the sigma notation for a given sum to a sigma notation with different limits of summation. The following example illustrates a method for doing this.

Example 2

Express

$$\sum_{k=3}^{7} 5^{k-2}$$

in sigma notation so that the lower limit of summation is 0 rather than 3.

Solution. If we define a new summation index j by means of the formula

$$j = k - 3 \tag{2}$$

then j runs from 0 up to 4 as k runs from 3 up to 7. From (2), $k = j + 3$, so

$$\sum_{k=3}^{7} 5^{k-2} = \sum_{j=0}^{4} 5^{(j+3)-2} = \sum_{j=0}^{4} 5^{j+1}$$

As a check, the reader can verify that

$$\sum_{j=0}^{4} 5^{j+1} \quad \text{and} \quad \sum_{k=3}^{7} 5^{k-2}$$

both denote the sum $5 + 5^2 + 5^3 + 5^4 + 5^5$. ◀

REMARK. In the solution of Example 2 the summation index was changed from k to j. If it is desirable to keep the same symbol for the summation index, we can change the j back to k *at the very end* and express the final result as

$$\sum_{k=0}^{4} 5^{k+1} \quad \text{instead of} \quad \sum_{j=0}^{4} 5^{j+1}$$

When we want to represent a general sum we will use letters with subscripts. For example, a general sum with five terms might be written as

$$a_1 + a_2 + a_3 + a_4 + a_5$$

or in sigma notation as

$$\sum_{k=1}^{5} a_k, \quad \sum_{j=1}^{5} a_j, \quad \text{or} \quad \sum_{m=1}^{5} a_m$$

A general sum with n terms might be written as

$$b_1 + b_2 + \cdots + b_n$$

or in sigma notation as

$$\sum_{k=1}^{n} b_k, \quad \sum_{j=1}^{n} b_j, \quad \text{or} \quad \sum_{m=1}^{n} b_m$$

PROPERTIES OF SIGMA NOTATION

The following properties of sigma notation will help to manipulate sums:

7.4.1 THEOREM.

(*a*) $$\sum_{k=1}^{n} ca_k = c\sum_{k=1}^{n} a_k$$

(*b*) $$\sum_{k=1}^{n} (a_k + b_k) = \sum_{k=1}^{n} a_k + \sum_{k=1}^{n} b_k$$

(*c*) $$\sum_{k=1}^{n} (a_k - b_k) = \sum_{k=1}^{n} a_k - \sum_{k=1}^{n} b_k$$

We will prove parts (*a*) and (*b*) and leave part (*c*) as an exercise.

Proof (a).

$$\sum_{k=1}^{n} ca_k = ca_1 + ca_2 + \cdots + ca_n = c(a_1 + a_2 + \cdots + a_n) = c\sum_{k=1}^{n} a_k$$

Proof (b).

$$\sum_{k=1}^{n}(a_k + b_k) = (a_1 + b_1) + (a_2 + b_2) + \cdots + (a_n + b_n)$$
$$= (a_1 + a_2 + \cdots + a_n) + (b_1 + b_2 + \cdots + b_n)$$
$$= \sum_{k=1}^{n} a_k + \sum_{k=1}^{n} b_k$$

REMARK. Loosely phrased, this theorem states: *A constant factor can be moved through a sigma sign; sigma of a sum equals the sum of the sigmas; and sigma of a difference equals the difference of the sigmas.*

SUMMATION FORMULAS

The following formulas will be used in our later work.

7.4.2 THEOREM.

(*a*) $$\sum_{k=1}^{n} k = 1 + 2 + 3 + \cdots + n = \frac{n(n+1)}{2}$$

(*b*) $$\sum_{k=1}^{n} k^2 = 1^2 + 2^2 + 3^2 + \cdots + n^2 = \frac{n(n+1)(2n+1)}{6}$$

(*c*) $$\sum_{k=1}^{n} k^3 = 1^3 + 2^3 + 3^3 + \cdots + n^3 = \left[\frac{n(n+1)}{2}\right]^2$$

We will prove parts (*a*) and (*b*) and leave part (*c*) as an exercise.

Proof (a). If we write the terms of

$$\sum_{k=1}^{n} k = 1 + 2 + 3 + \cdots + (n-2) + (n-1) + n \tag{3}$$

in the opposite order, we obtain

$$\sum_{k=1}^{n} k = n + (n-1) + (n-2) + \cdots + 3 + 2 + 1 \tag{4}$$

Adding (3) and (4) term by term yields

$$2\sum_{k=1}^{n} k = \underbrace{(n+1) + (n+1) + (n+1) + \cdots + (n+1)}_{n \text{ terms}} = n(n+1)$$

Thus,

$$\sum_{k=1}^{n} k = \frac{n(n+1)}{2}$$

Proof (b). This proof begins with a trick. Since

$$(k+1)^3 - k^3 = k^3 + 3k^2 + 3k + 1 - k^3 = 3k^2 + 3k + 1$$

we obtain

$$\sum_{k=1}^{n} [(k+1)^3 - k^3] = \sum_{k=1}^{n} (3k^2 + 3k + 1) \tag{5}$$

Writing out the left side of (5) yields

$$[2^3 - 1^3] + [3^3 - 2^3] + [4^3 - 3^3] + \cdots + [(n+1)^3 - n^3] \tag{6}$$

Observe that in (6) the 2^3 in the first term cancels out the -2^3 in the second term, the 3^3 in the second term cancels out the -3^3 in the third term, and so forth, so that the entire sum collapses like a folding telescope (hence, is called a ***telescoping sum***), leaving only $-1^3 + (n+1)^3$. Thus, (5) can be rewritten as

$$-1 + (n+1)^3 = \sum_{k=1}^{n} (3k^2 + 3k + 1) \tag{7}$$

or, from Theorem 7.4.1,

$$-1 + (n+1)^3 = 3\sum_{k=1}^{n} k^2 + 3\sum_{k=1}^{n} k + \sum_{k=1}^{n} 1 \tag{8}$$

But

$$\sum_{k=1}^{n} 1 = \underbrace{1 + 1 + \cdots + 1}_{n \text{ terms}} = n$$

and by part (*a*) of this theorem

$$\sum_{k=1}^{n} k = \frac{n(n+1)}{2}$$

Thus, (8) can be written as

$$-1 + (n+1)^3 = 3\sum_{k=1}^{n} k^2 + 3\frac{n(n+1)}{2} + n$$

Therefore,

$$\begin{aligned} \sum_{k=1}^{n} k^2 &= \frac{1}{3}\left[(n+1)^3 - 3\frac{n(n+1)}{2} - (n+1)\right] \\ &= \frac{n+1}{6}[2(n+1)^2 - 3n - 2] \\ &= \frac{n+1}{6}(2n^2 + n) = \frac{n(n+1)(2n+1)}{6} \end{aligned}$$

■

Example 3

Evaluate $\sum_{k=1}^{30} k(k+1)$.

Solution.

$$\begin{aligned} \sum_{k=1}^{30} k(k+1) &= \sum_{k=1}^{30} (k^2 + k) = \sum_{k=1}^{30} k^2 + \sum_{k=1}^{30} k \\ &= \frac{30(31)(61)}{6} + \frac{30(31)}{2} = 9920 \end{aligned}$$

Theorem 7.4.2(*a*), (*b*)

◀

REMARK. In formulas such as

$$\sum_{k=1}^{n} k^2 = \frac{n(n+1)(2n+1)}{6}$$

or

$$1^2 + 2^2 + \cdots + n^2 = \frac{n(n+1)(2n+1)}{6}$$

the left side of the equality is said to express the sum in ***open form*** and the right side is said to express it in ***closed form***; the open form just indicates the terms to be added, while the closed form is an explicit formula for their sum.

Example 4

Express $\sum_{k=1}^{n}(3+k)^2$ in closed form.

Solution.

$$\begin{aligned}\sum_{k=1}^{n}(3+k)^2 &= \sum_{k=1}^{n}(9+6k+k^2) = \sum_{k=1}^{n}9 + 6\sum_{k=1}^{n}k + \sum_{k=1}^{n}k^2 \\ &= 9n + 6\frac{n(n+1)}{2} + \frac{n(n+1)(2n+1)}{6} \\ &= \frac{1}{3}n^3 + \frac{7}{2}n^2 + \frac{73}{6}n\end{aligned}$$

◀

FOR THE READER. Your numerical calculating utility probably provides some way of evaluating sums that can be expressed in sigma notation. Check your documentation to find out how to do this, and then use your utility to confirm that the numerical result obtained in Example 3 is correct. If you have access to a CAS, then it provides some method for finding closed forms for sums such as those in Theorem 7.4.2. Use your CAS to confirm the formulas in that theorem, and then find closed forms for

$$\sum_{k=1}^{n}k^4 \quad\text{and}\quad \sum_{k=1}^{n}k^5$$

EXERCISE SET 7.4 [C] CAS

1. Evaluate

(a) $\sum_{k=1}^{3}k^3$ (b) $\sum_{j=2}^{6}(3j-1)$ (c) $\sum_{i=-4}^{1}(i^2-i)$

(d) $\sum_{n=0}^{5}1$ (e) $\sum_{k=0}^{4}(-2)^k$ (f) $\sum_{n=1}^{6}\sin n\pi$.

2. Evaluate

(a) $\sum_{k=1}^{4}k\sin\frac{k\pi}{2}$ (b) $\sum_{j=0}^{5}(-1)^j$ (c) $\sum_{i=7}^{20}e^2$

(d) $\sum_{m=3}^{5}2^{m+1}$ (e) $\sum_{n=1}^{6}\ln n$ (f) $\sum_{k=0}^{10}\cos k\pi$.

In Exercises 3–12, write each expression in sigma notation, but do not evaluate.

3. $1+2+3+\cdots+10$

4. $3\cdot1+3\cdot2+3\cdot3+\cdots+3\cdot20$

5. $1\cdot2+2\cdot3+3\cdot4+\cdots+49\cdot50$

6. $1+2+2^2+2^3+2^4$

7. $2+4+6+8+\cdots+20$

8. $1+3+5+7+\cdots+15$

9. $1-3+5-7+9-11$

10. $1-\frac{1}{2}+\frac{1}{3}-\frac{1}{4}+\frac{1}{5}$

11. $-1+\frac{1}{2}-\frac{1}{3}+\frac{1}{4}-\frac{1}{5}$

12. $1+\cos\frac{\pi}{7}+\cos\frac{2\pi}{7}+\cos\frac{3\pi}{7}$

13. (a) Express the sum of the even integers from 2 to 100 in sigma notation.
(b) Express the sum of the odd integers from 1 to 99 in sigma notation.

14. Express in sigma notation.
(a) $a_1-a_2+a_3-a_4+a_5$
(b) $-b_0+b_1-b_2+b_3-b_4+b_5$
(c) $a_0+a_1x+a_2x^2+\cdots+a_nx^n$
(d) $a^5+a^4b+a^3b^2+a^2b^3+ab^4+b^5$

In Exercises 15–22, use Theorem 7.4.2 to evaluate the sums, and check your answers using the summation feature of a calculating utility.

15. $\sum_{k=1}^{100}k$ **16.** $\sum_{k=3}^{100}k$ **17.** $\sum_{k=1}^{20}k^2$

18. $\sum_{k=1}^{100}(7k+1)$ **19.** $\sum_{k=1}^{6}(4k^3-2k+1)$ **20.** $\sum_{k=4}^{20}k^2$

21. $\sum_{k=1}^{30} k(k-2)(k+2)$ **22.** $\sum_{k=1}^{6} (k-k^3)$

In Exercises 23–28, express the sums in closed form.

23. $\sum_{k=1}^{n} (4k-3)$ **24.** $\sum_{k=1}^{n-1} k^2$ **25.** $\sum_{k=1}^{n} \frac{3k}{n}$

26. $\sum_{k=1}^{n-1} \frac{k^2}{n}$ **27.** $\sum_{k=1}^{n-1} \frac{k^3}{n^2}$ **28.** $\sum_{k=1}^{n} \left(\frac{5}{n} - \frac{2k}{n}\right)$

C **29.** For each of the sums that you obtained in Exercises 23–28, use a CAS to check your answer. If the answer produced by the CAS does not match your own, show that the two answers are equivalent.

30. Let

$$S = \sum_{k=0}^{n} ar^k$$

Show that $S - rS = a - ar^{n+1}$ and hence that

$$\sum_{k=0}^{n} ar^k = \frac{a - ar^{n+1}}{1-r} \quad (r \neq 1)$$

(A sum of this form is called a ***geometric sum***.)

31. In each part, rewrite the sum, if necessary, so that the lower limit is 0, and then use the formula derived in Exercise 30 to evaluate the sum. Check your answers using the summation feature of a calculating utility.

(a) $\sum_{k=1}^{20} 3^k$ (b) $\sum_{k=5}^{30} 2^k$ (c) $\sum_{k=0}^{100} (-1)^{k+1}\frac{1}{2^k}$

C **32.** In each part, make a conjecture about the limit by using a CAS to evaluate the sum for $n = 10$, 20, and 50; and then check your conjecture by using the formula in Exercise 30 to express the sum in closed form, and then finding the limit exactly.

(a) $\lim_{n\to+\infty} \sum_{k=0}^{n} \frac{1}{2^k}$ (b) $\lim_{n\to+\infty} \sum_{k=1}^{n} \left(\frac{3}{4}\right)^k$

In Exercises 33–36, express the function of n in closed form, and then use L'Hôpital's rule to find the limit. [*Note:* L'Hôpital's rule was derived for functions of a real-valued variable x, whereas here the variable n assumes only integer values. Thus, strictly speaking, L'Hôpital's rule cannot be used without justifying that it applies to functions of integer-valued variables. We will do this later in the text.]

33. $\lim_{n\to+\infty} \frac{1+2+3+\cdots+n}{n^2}$

34. $\lim_{n\to+\infty} \frac{1^2+2^2+3^2+\cdots+n^2}{n^3}$

35. $\lim_{n\to+\infty} \sum_{k=1}^{n} \frac{5k}{n^2}$ **36.** $\lim_{n\to+\infty} \sum_{k=1}^{n-1} \frac{2k^2}{n^3}$

37. Express $1+2+2^2+2^3+2^4+2^5$ in sigma notation with
(a) $j = 0$ as the lower limit of summation
(b) $j = 1$ as the lower limit of summation
(c) $j = 2$ as the lower limit of summation.

38. Express

$$\sum_{k=5}^{9} k2^{k+4}$$

in sigma notation with
(a) $k = 1$ as the lower limit of summation
(b) $k = 13$ as the upper limit of summation.

39. Change the limits of summation appropriately to simplify

(a) $\sum_{k=11}^{28} \sin\left(\frac{\pi}{k-10}\right)$ (b) $\sum_{k=6}^{12} e^{k-6}$

40. Show that the sum of the first n consecutive positive odd integers is n^2.

41. The accompanying figure shows a square that is n units by n units that has been subdivided into a one-unit square and $n-1$ "L-shaped" regions. Use this figure to derive the result in Exercise 40.

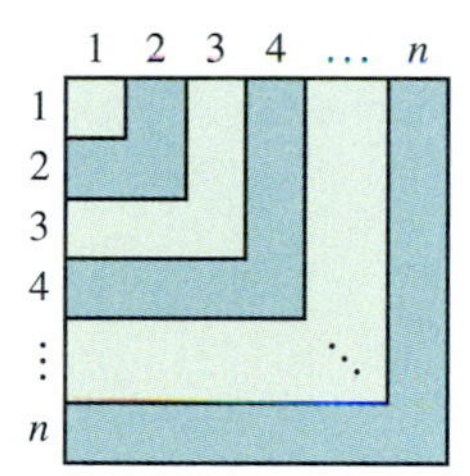

Figure Ex-41

42. Solve the equation $\sum_{k=1}^{n} k = 465$.

When part of each term of a sum cancels part of the next term, leaving only portions of the first and last terms at the end, the sum is said to ***telescope***. In Exercises 43–46, evaluate the telescoping sum.

43. $\sum_{k=5}^{17} (3^k - 3^{k-1})$ **44.** $\sum_{k=1}^{50} \left(\frac{1}{k} - \frac{1}{k+1}\right)$

45. $\sum_{k=2}^{20} \left(\frac{1}{k^2} - \frac{1}{(k-1)^2}\right)$ **46.** $\sum_{k=1}^{100} (2^{k+1} - 2^k)$

47. (a) Show that

$$\frac{1}{1\cdot 3} + \frac{1}{3\cdot 5} + \cdots + \frac{1}{(2n-1)(2n+1)} = \frac{n}{2n+1}$$

$$\left[\textit{Hint: } \frac{1}{(2n-1)(2n+1)} = \frac{1}{2}\left(\frac{1}{2n-1} - \frac{1}{2n+1}\right).\right]$$

(b) Use the result in part (a) to find

$$\lim_{n\to+\infty} \sum_{k=1}^{n} \frac{1}{(2k-1)(2k+1)}$$

48. (a) Show that

$$\frac{1}{1\cdot 2}+\frac{1}{2\cdot 3}+\frac{1}{3\cdot 4}+\cdots+\frac{1}{n(n+1)}=\frac{n}{n+1}$$

$$\left[\textit{Hint: } \frac{1}{n(n+1)}=\frac{1}{n}-\frac{1}{n+1}.\right]$$

(b) Use the result in part (a) to find

$$\lim_{n\to+\infty}\sum_{k=1}^{n}\frac{1}{k(k+1)}$$

49. By writing out the sums, determine whether the following are valid identities.

(a) $\displaystyle\int\left[\sum_{i=1}^{n} f_i(x)\right]dx=\sum_{i=1}^{n}\left[\int f_i(x)\,dx\right]$

(b) $\displaystyle\frac{d}{dx}\left[\sum_{i=1}^{n} f_i(x)\right]=\sum_{i=1}^{n}\left[\frac{d}{dx}[f_i(x)]\right]$

50. Which of the following are valid identities?

(a) $\displaystyle\sum_{i=1}^{n} a_i b_i=\sum_{i=1}^{n} a_i\sum_{i=1}^{n} b_i$

(b) $\displaystyle\sum_{i=1}^{n}\frac{a_i}{b_i}=\sum_{i=1}^{n} a_i\Big/\sum_{i=1}^{n} b_i$

(c) $\displaystyle\sum_{i=1}^{n} a_i^2=\left(\sum_{i=1}^{n} a_i\right)^2$

51. Let $\bar{x}$ denote the arithmetic average of the n numbers $x_1, x_2, \ldots, x_n$. Use Theorem 7.4.1 to prove that

$$\sum_{i=1}^{n}(x_i-\bar{x})=0$$

52. Prove part (*c*) of Theorem 7.4.1.

53. Prove part (*c*) of Theorem 7.4.2. [*Hint:* Begin with the difference $(k+1)^4-k^4$ and follow the steps used to prove part (*b*) of the theorem.]

54. An artist wants to create a rough triangular design using uniform square tiles glued edge to edge. She places n tiles in a row to form the base of the triangle and then makes each successive row two tiles shorter than the preceding row. Find a formula for the number of tiles used in the design. [*Hint:* Your answer will depend on whether n is even or odd.]

55. An artist wants to create a sculpture by gluing together uniform spheres. She creates a rough rectangular base that has 50 spheres along one edge and 30 spheres along the other. She then creates successive layers by gluing spheres in the grooves of the preceding layer. How many spheres will there be in the sculpture?

7.5 THE DEFINITE INTEGRAL

Recall from the informal discussion in Section 7.1 that if a function f is continuous and nonnegative on an interval $[a, b]$, then the area under the graph of f over the interval $[a, b]$ can be obtained by either the "rectangle method" or "the antiderivative method." In this section we will discuss the rectangle method in more detail, and we will introduce the concept of a "definite integral," which will link the concept of area to other important concepts such as length, volume, density, probability, and work.

A DEFINITION OF AREA

Our first goal in this section is to define formally what we mean by the area of a region R that is bounded below by the x-axis, bounded on the sides by the vertical lines $x = a$ and $x = b$, and bounded above by the curve $y = f(x)$, where f is continuous and nonnegative on the interval $[a, b]$ (Figure 7.5.1). We will start by defining the area of a rectangle to be the product of its length and width and defining the area of a region composed of finitely many rectangles to be the sum of the areas of those rectangles. To define the area of the region R, we will use these definitions and the rectangle method of Section 7.1. The basic idea is as follows (Figure 7.5.2):

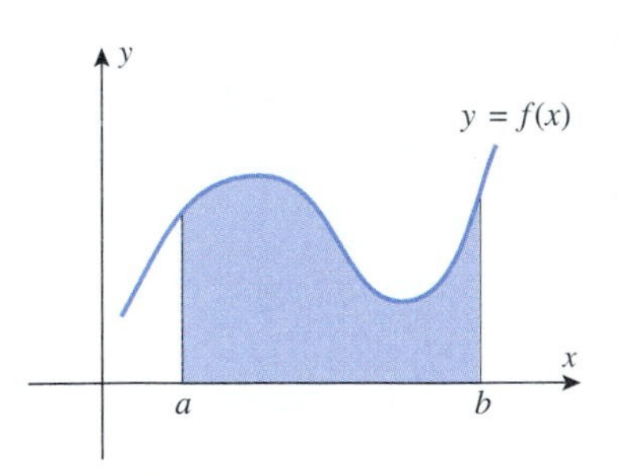

Figure 7.5.1

- Divide the interval $[a, b]$ into n equal subintervals.
- Over each subinterval construct a rectangle whose height is the value of f at any point in the subinterval.
- The union of these rectangles forms a region R_n whose area can be regarded as an approximation to the "area" A of the region R.
- Repeat the process using more and more subdivisions.

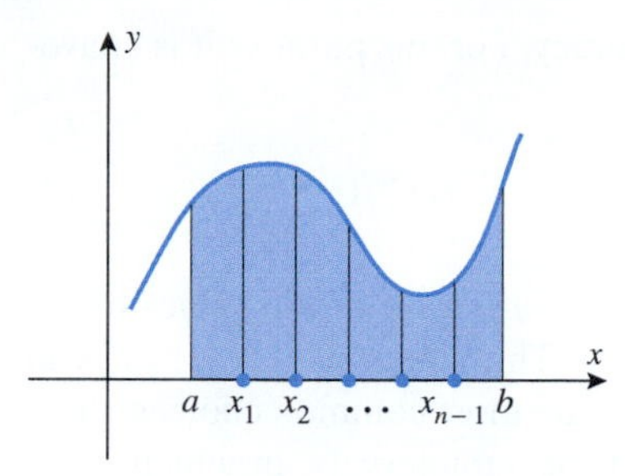

Figure 7.5.2

- Define the area of R to be the limit of the areas of the approximating regions, R_n; that is,

$$A = \text{area}(R) = \lim_{n \to +\infty} [\text{area}(R_n)] \tag{1}$$

To make all of this more precise, it will be helpful to capture this procedure in mathematical notation. For this purpose, suppose that we divide the interval $[a, b]$ into n subintervals by inserting $n - 1$ equally spaced points between a and b, say

$$x_1, x_2, \ldots, x_{n-1}$$

(Figure 7.5.2). Each of these intervals has width $(b - a)/n$, which it is customary to denote by

$$\Delta x = \frac{b - a}{n}$$

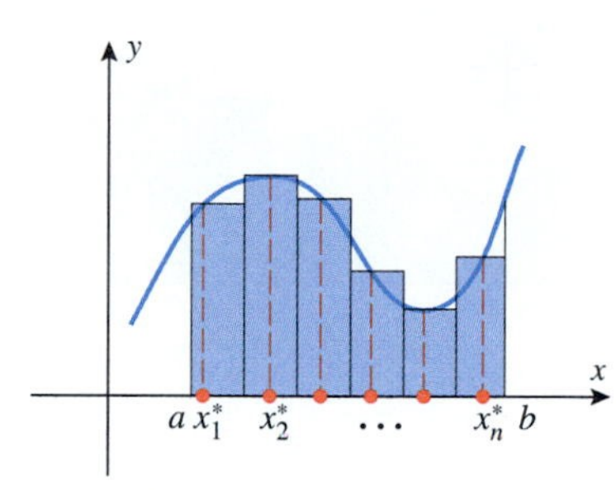

Figure 7.5.3

In each subinterval we need to choose a point at which to evaluate the function f to determine the height of a rectangle over that interval. If we denote those points by

$$x_1^*, x_2^*, \ldots, x_n^*$$

(Figure 7.5.3), then the areas of the rectangles constructed over these intervals will be

$$f(x_1^*)\Delta x, \quad f(x_2^*)\Delta x, \ldots, \quad f(x_n^*)\Delta x$$

(Figure 7.5.4), and the total area of the region R_n will be

$$\text{area}(R_n) = f(x_1^*)\Delta x + f(x_2^*)\Delta x + \cdots + f(x_n^*)\Delta x$$

or in sigma notation,

$$\text{area}(R_n) = \sum_{k=1}^{n} f(x_k^*)\Delta x$$

With this notation (1) can be expressed as

$$A = \lim_{n \to +\infty} \sum_{k=1}^{n} f(x_k^*)\Delta x$$

which suggests the following definition of the area of the region R.

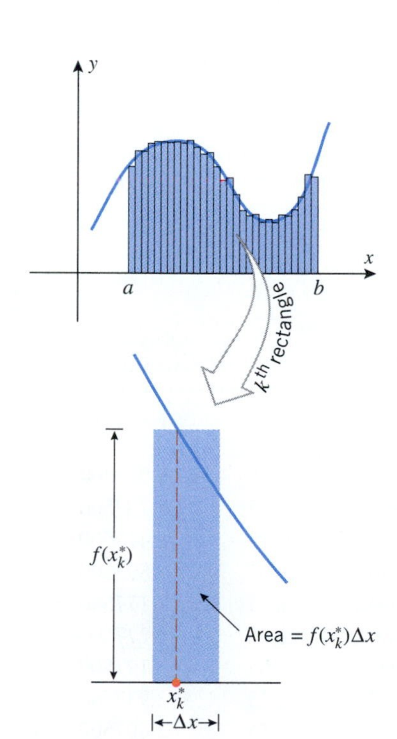

Figure 7.5.4

> **7.5.1** DEFINITION (***Area Under a Curve***). If the function f is continuous on $[a, b]$ and if $f(x) \geq 0$ for all x in $[a, b]$, then the ***area*** under the curve $y = f(x)$ over the interval $[a, b]$ is defined by
>
> $$A = \lim_{n \to +\infty} \sum_{k=1}^{n} f(x_k^*)\Delta x \tag{2}$$

REMARK. Although this definition is satisfactory for our present purposes, there are some issues that would have to be resolved before it could be regarded as a rigorous mathematical definition. For example, we would have to prove that the limit actually exists and that its value does not depend on how the points $x_1^*, x_2^*, \ldots, x_n^*$ are chosen. It can be proved that this is true if f is continuous on $[a, b]$, but the details are beyond the scope of this text.

The limit in Formula (2) is often difficult or impossible to find, so that when an *exact* area is needed the antiderivative method, which we will discuss in the next section, is the method of choice. However, if an *approximation* to the area will suffice, then instead of taking the limit we can approximate the area as

$$A \approx \sum_{k=1}^{n} f(x_k^*)\Delta x$$

where n is sufficiently large to produce the required accuracy. For this purpose it is convenient to rewrite this sum as

$$\sum_{k=1}^{n} f(x_k^*)\Delta x = \Delta x \sum_{k=1}^{n} f(x_k^*) = \Delta x[f(x_1^*) + f(x_2^*) + \cdots + f(x_n^*)] \tag{3}$$

where $\Delta x = (b-a)/n$. The calculation here involves only the sum of the values of the function at n points, followed by a multiplication by Δx. The points $x_1^*, x_2^*, \ldots, x_n^*$ can be chosen arbitrarily in successive subintervals; however, the most common choices are at the left endpoints, the right endpoints, or the centers of the subintervals, in which cases Formula (3) is called the ***left endpoint approximation***, the ***right endpoint approximation***, or the ***midpoint approximation*** of the exact area (Figure 7.5.5).

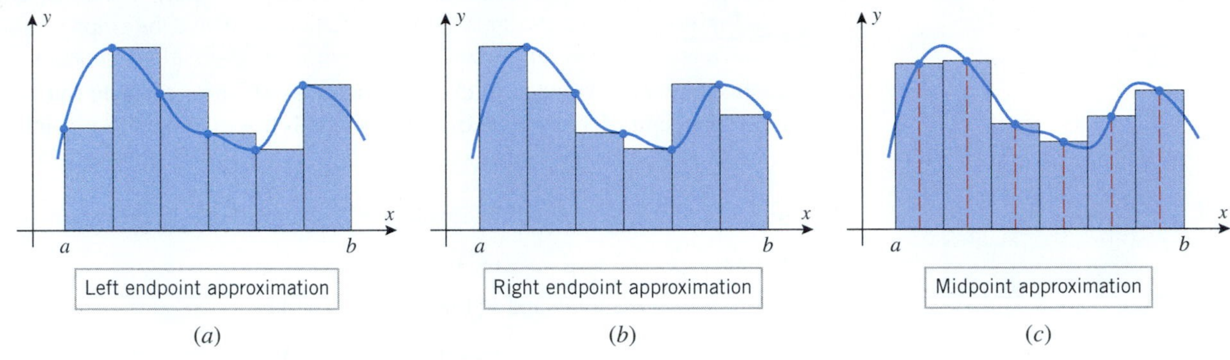

Figure 7.5.5

Example 1

Find the left endpoint, right endpoint, and midpoint approximations of the area under the curve $y = 9 - x^2$ over the interval $[0, 3]$ with $n = 10$, $n = 20$, and $n = 50$ (Figure 7.5.6).

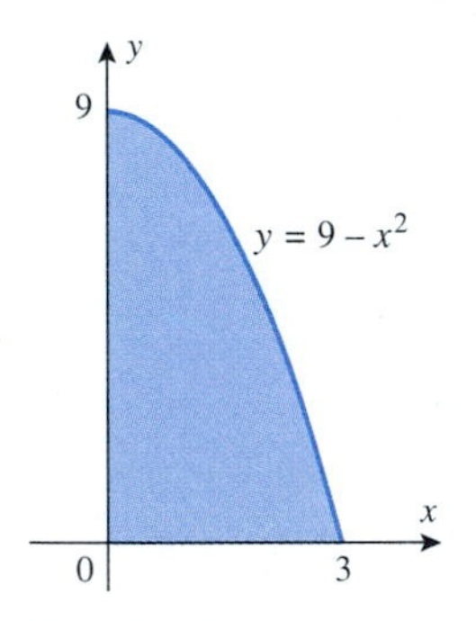

Figure 7.5.6

Solution. Details of the computations for the case $n = 10$ are shown to six decimal places in Table 7.5.1 and the results of all computations are given in Table 7.5.2. ◀

Table 7.5.1

$n = 10$, $\Delta x = (b-a)/n = (3-0)/10 = 0.3$

	LEFT ENDPOINT APPROXIMATION		RIGHT ENDPOINT APPROXIMATION		MIDPOINT APPROXIMATION	
k	x_k^*	$9-(x_k^*)^2$	x_k^*	$9-(x_k^*)^2$	x_k^*	$9-(x_k^*)^2$
1	0.0	9.000000	0.3	8.910000	0.15	8.977500
2	0.3	8.910000	0.6	8.640000	0.45	8.797500
3	0.6	8.640000	0.9	8.190000	0.75	8.437500
4	0.9	8.190000	1.2	7.560000	1.05	7.897500
5	1.2	7.560000	1.5	6.750000	1.35	7.177500
6	1.5	6.750000	1.8	5.760000	1.65	6.277500
7	1.8	5.760000	2.1	4.590000	1.95	5.197500
8	2.1	4.590000	2.4	3.240000	2.25	3.937500
9	2.4	3.240000	2.7	1.710000	2.55	2.497500
10	2.7	1.710000	3.0	0.000000	2.85	0.877500
		64.350000		55.350000		60.075000
$\Delta x \sum_{k=1}^{n} f(x_k^*)$		(.3)(64.350000) = 19.305000		(.3)(55.350000) = 16.605000		(.3)(60.075000) = 18.022500

Table 7.5.2

n	LEFT ENDPOINT APPROXIMATION	RIGHT ENDPOINT APPROXIMATION	MIDPOINT APPROXIMATION
10	19.305000	16.605000	18.022500
20	18.663750	17.313750	18.005625
50	18.268200	17.728200	18.000900

REMARK. We will show in the next section that the exact area under $y = 9 - x^2$ over the interval $[0, 3]$ is 18 (i.e., 18 square units), so that in the preceding example the midpoint approximation is more accurate than either of the endpoint approximations. This can also be seen geometrically from the approximating rectangles: Since the graph of $y = 9 - x^2$ is decreasing over the interval $[0, 3]$, each left endpoint approximation overestimates the area, each right endpoint approximation underestimates the area, and each midpoint approximation falls between the overestimate and the underestimate (Figure 7.5.7). This is consistent with the values in Table 7.5.2. Later in the text we will investigate the error that results when an area is approximated by the midpoint rule.

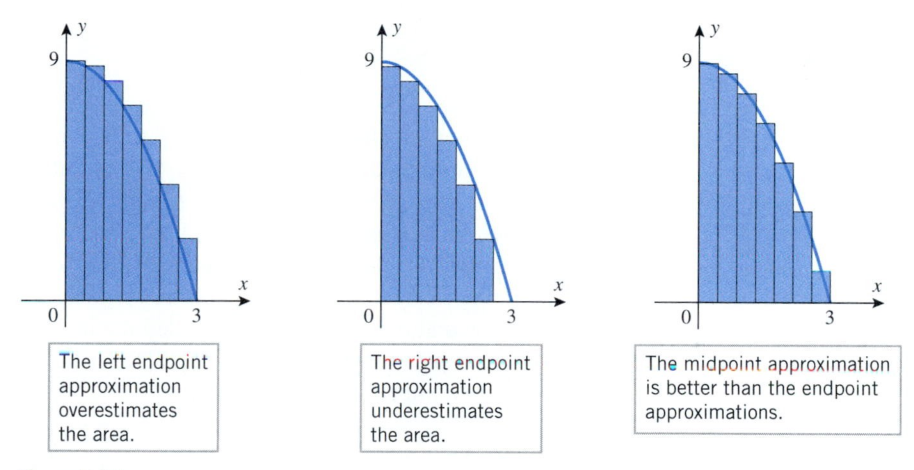

Figure 7.5.7

THE DEFINITE INTEGRAL OF A CONTINUOUS FUNCTION

In Definition 7.5.1 we assumed that f is continuous and nonnegative on the interval $[a, b]$. If f is continuous and assumes both positive and negative values on $[a, b]$, then the limit

$$\lim_{n\to+\infty} \sum_{k=1}^{n} f(x_k^*)\Delta x \tag{4}$$

no longer represents the area between the curve $y = f(x)$ and the interval $[a, b]$; rather it represents a difference of areas—the area of the region that is above the interval $[a, b]$ and below the curve $y = f(x)$ minus the area of the region that is below the interval $[a, b]$ and above the curve $y = f(x)$. We call this the ***net signed area*** between the graph of $y = f(x)$ and the interval $[a, b]$. For example, in Figure 7.5.8a, the net signed area between the curve $y = f(x)$ and the interval $[a, b]$ is

$$(A_I + A_{III}) - A_{II} = \big[\text{area above } [a, b]\big] - \big[\text{area below } [a, b]\big]$$

To explain why the limit in (4) represents this net signed area, let us subdivide the interval $[a, b]$ in Figure 7.5.8a into n equal subintervals and examine the terms in the sum

$$\sum_{k=1}^{n} f(x_k^*)\Delta x \tag{5}$$

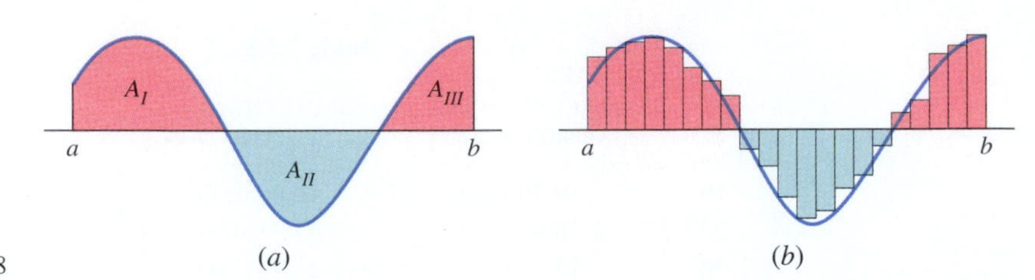

Figure 7.5.8 (a) (b)

If $f(x_k^*)$ is positive, then the product $f(x_k^*)\Delta x$ represents the area of the rectangle with height $f(x_k^*)$ and base Δx (the beige rectangles in Figure 7.5.8*b*). However, if $f(x_k^*)$ is negative, then the product $f(x_k^*)\Delta x$ is the *negative* of the area of the rectangle with height $|f(x_k^*)|$ and base Δx (the green rectangles in Figure 7.5.8*b*). Thus, (5) represents the total area of the beige rectangles minus the total area of the green rectangles. As n increases, the beige rectangles fill out the regions with areas A_I and A_{III} and the green rectangles fill out the region with area A_{II}, which explains why the limit in (4) represents the signed area between $y = f(x)$ and the interval $[a, b]$.

The limit in (4) is so important that there is some terminology and notation associated with it. We will denote this limit by the symbol

$$\int_a^b f(x)\,dx = \lim_{n\to+\infty} \sum_{k=1}^{n} f(x_k^*)\Delta x \tag{6}$$

which is called the ***definite integral*** of f from a to b. Geometrically, the definite integral represents the signed area between $y = f(x)$ and the interval $[a, b]$, and in the case where $f(x)$ is nonnegative on the interval $[a, b]$, the definite integral represents the area under the curve over the interval $[a, b]$. The numbers a and b are called the ***lower limit of integration*** and ***upper limit of integration***, respectively, and $f(x)$ is called the ***integrand***. The reason for the integral sign will become clear in the next section, where we will establish a link between the definite integral and the indefinite integral studied earlier.

In the simplest cases, definite integrals can be calculated using formulas from plane geometry to compute the signed areas.

Example 2

Sketch the region whose area is represented by the definite integral, and evaluate the integral using an appropriate formula from geometry.

(a) $\displaystyle\int_1^4 2\,dx$ (b) $\displaystyle\int_{-1}^2 (x+2)\,dx$ (c) $\displaystyle\int_0^1 \sqrt{1-x^2}\,dx$

Solution (a). The graph of the integrand is the horizontal line $y = 2$, so the region is a rectangle of height 2 extending over the interval from 1 to 4 (Figure 7.5.9*a*). Thus,

$$\int_1^4 2\,dx = (\text{area of rectangle}) = 2(3) = 6$$

Solution (b). The graph of the integrand is the line $y = x + 2$, so the region is a trapezoid whose base extends from $x = -1$ to $x = 2$ (Figure 7.5.9*b*). Thus,

$$\int_{-1}^2 (x+2)\,dx = (\text{area of trapezoid}) = \tfrac{1}{2}(3)(1+4) = \tfrac{15}{2}$$

Solution (c). The graph of $y = \sqrt{1-x^2}$ is the upper semicircle of radius 1, centered at the origin, so the region is the right quarter-circle extending from $x = 0$ to $x = 1$ (Figure 7.5.9*c*).

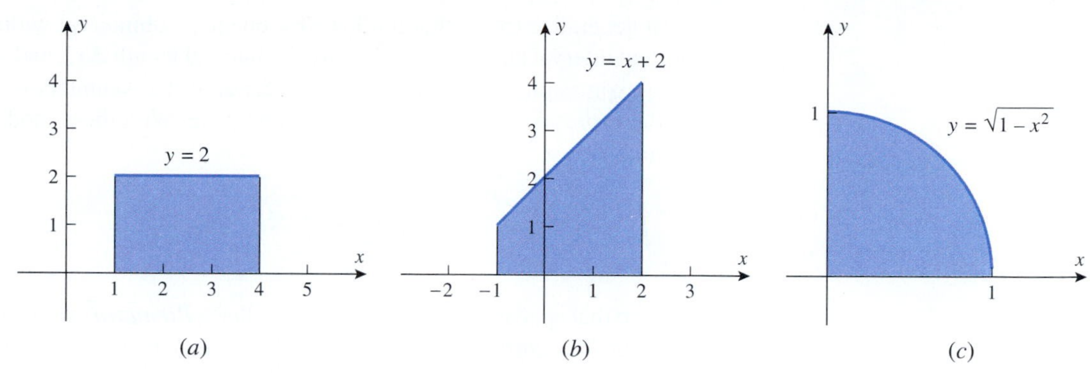

Figure 7.5.9

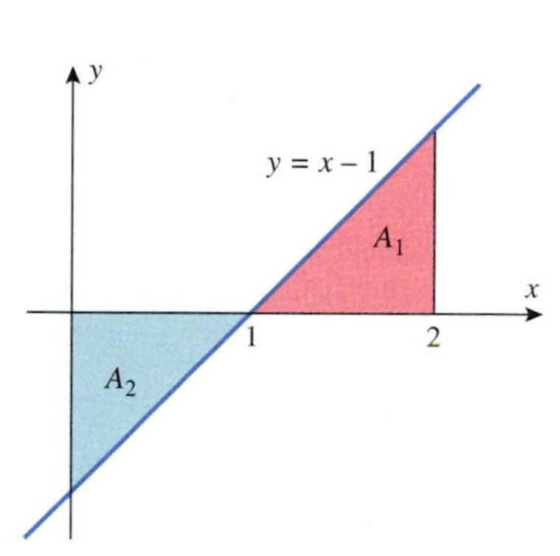

Figure 7.5.10

Thus,

$$\int_0^1 \sqrt{1-x^2}\,dx = (\text{area of quarter-circle}) = \tfrac{1}{4}\pi(1^2) = \frac{\pi}{4}$$ ◀

Example 3

Evaluate

(a) $\displaystyle\int_0^2 (x-1)\,dx$ (b) $\displaystyle\int_0^1 (x-1)\,dx$

Solution. The graph of $y = x - 1$ is shown in Figure 7.5.10, and we leave it for you to verify that the shaded triangular regions both have area $\frac{1}{2}$. Over the interval $[0, 2]$ the net signed area is $A_1 - A_2 = \frac{1}{2} - \frac{1}{2} = 0$, and over the interval $[0, 1]$ the net signed area is $-A_2 = -\frac{1}{2}$. Thus,

$$\int_0^2 (x-1)\,dx = 0 \quad \text{and} \quad \int_0^1 (x-1)\,dx = -\tfrac{1}{2}$$ ◀

THE RIEMANN INTEGRAL

It is assumed in (6) that the function f is continuous on the interval $[a, b]$ and that for each n this interval is subdivided into n subintervals of equal length to create bases for the approximating rectangles. Although equal lengths are useful for computations, this restriction is not essential. That is, the signed area between $y = f(x)$ and $[a, b]$ can be obtained using rectangles with different widths provided that successive subdivisions are constructed in such a way that the widths of the rectangles approach zero as n increases (Figure 7.5.11). Thus, we must preclude the kind of situation that occurs in Figure 7.5.12 in which the right half of the interval is never subdivided. If this kind of subdivision were allowed, the error in the approximation would not approach zero as n increased.

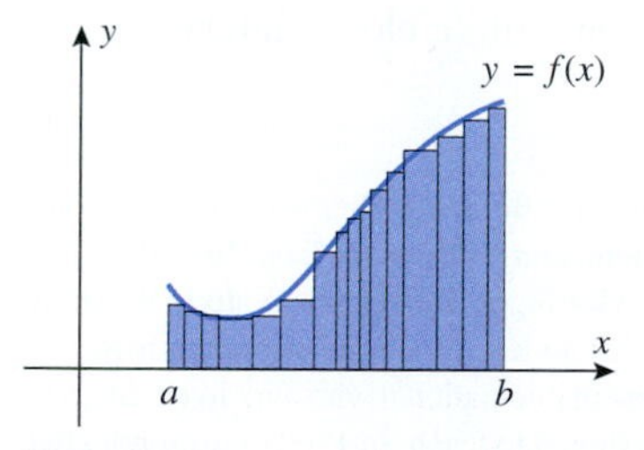

Figure 7.5.11

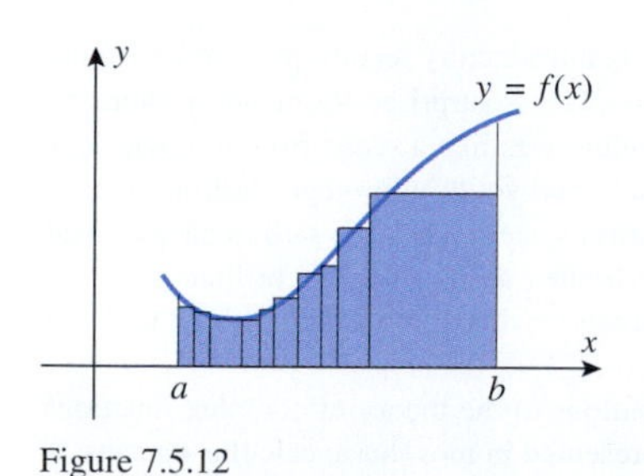

Figure 7.5.12

To provide for the added generality of unequal intervals, suppose that the interval $[a, b]$ is subdivided into n subintervals whose widths are

$$\Delta x_1, \Delta x_2, \ldots, \Delta x_n$$

and let $\max \Delta x_k$ denote the largest of the subinterval widths, which is read "the maximum of the Δx_k's." The subintervals are said to form a ***partition*** of the interval $[a, b]$, and $\max \Delta x_k$ is called the ***mesh size*** of the partition. For example, Figure 7.5.13 shows a partition of the interval $[0, 6]$ into four subintervals with a mesh size of 2.

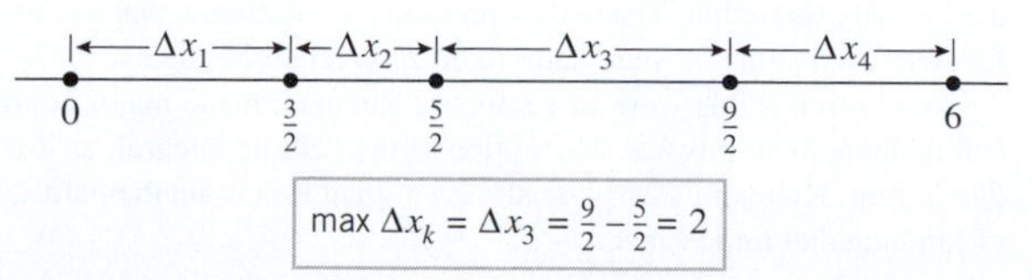

Figure 7.5.13

To generalize (6) so that it allows for unequal subinterval widths, we must replace the constant interval length Δx by the variable interval length Δx_k, and we must replace $n \to +\infty$ by an expression to specify that the lengths of all the subintervals approach zero. We will use the expression $\max \Delta x_k \to 0$ for this purpose. With these modifications in notation (6) becomes

$$\int_a^b f(x)\,dx = \lim_{\max \Delta x_k \to 0} \sum_{k=1}^{n} f(x_k^*)\Delta x_k \tag{7}$$

The sum that appears in this expression is called a ***Riemann****[*] ***sum***, and the limit is sometimes called the ***Riemann integral*** in honor of the German mathematician Bernhard Riemann who formulated many of the basic concepts of integration.

REMARK. Some writers use the symbol $\|\Delta\|$ rather than $\max \Delta x_k$ for the mesh size of the partition, in which case (7) would be written as

$$\int_a^b f(x)\,dx = \lim_{\|\Delta\| \to 0} \sum_{k=1}^{n} f(x_k^*)\Delta x_k$$

INTEGRABILITY

Because the definite integral is defined as a limit, it is possible that the limit may not exist, in which case the definite integral would not exist. Thus, we make the following definition:

7.5.2 DEFINITION. A function f is said to be ***Riemann integrable*** or more simply ***integrable*** on a finite closed interval $[a, b]$ if the limit

$$\int_a^b f(x)\,dx = \lim_{\max \Delta x_k \to 0} \sum_{k=1}^{n} f(x_k^*)\Delta x_k$$

exists and does not depend on choice of the partitions or on the points x_k^* in the subintervals.

At the end of this section we will discuss various conditions that ensure integrability, but for now suffice it to say that a function that is continuous on a finite closed interval $[a, b]$ is integrable on that interval.

---

[*] GEORG FRIEDRICH BERNHARD RIEMANN (1826–1866). German mathematician. Bernhard Riemann, as he is commonly known, was the son of a Protestant minister. He received his elementary education from his father and showed brilliance in arithmetic at an early age. In 1846 he enrolled at Göttingen University to study theology and philology, but he soon transferred to mathematics. He studied physics under W. E. Weber and mathematics under Karl Friedrich Gauss, whom some people consider to be the greatest mathematician who ever lived. In 1851 Riemann received his Ph.D. under Gauss, after which he remained at Göttingen to teach. In 1862, one month after his marriage, Riemann suffered an attack of pleuritis, and for the remainder of his life was an extremely sick man. He finally succumbed to tuberculosis in 1866 at age 39.

An interesting story surrounds Riemann's work in geometry. For his introductory lecture prior to becoming an associate professor, Riemann submitted three possible topics to Gauss. Gauss surprised Riemann by choosing the topic Riemann liked the least, the foundations of geometry. The lecture was like a scene from a movie. The old and failing Gauss, a giant in his day, watching intently as his brilliant and youthful protégé skillfully pieced together portions of the old man's own work into a complete and beautiful system. Gauss is said to have gasped with delight as the lecture neared its end, and on the way home he marveled at his student's brilliance. Gauss died shortly thereafter. The results presented by Riemann that day eventually evolved into a fundamental tool that Einstein used some 50 years later to develop relativity theory.

In addition to his work in geometry, Riemann made major contributions to the theory of complex functions and mathematical physics. The notion of the definite integral, as it is presented in most basic calculus courses, is due to him. Riemann's early death was a great loss to mathematics, for his mathematical work was brilliant and of fundamental importance.

PROPERTIES OF THE DEFINITE INTEGRAL

It is assumed in Definition 7.5.2 that $[a, b]$ is a finite closed interval with $a < b$, and hence the upper limit of integration in the definite integral is greater than the lower limit of integration. However, it will be convenient to extend this definition to allow for cases in which the upper and lower limits of integration are equal or the lower limit of integration is greater than the upper limit of integration. For this purpose we make the following special definitions.

7.5.3 DEFINITION.

(a) If a is in the domain of f, we define

$$\int_a^a f(x)\,dx = 0$$

(b) If f is integrable on $[a, b]$, then we define

$$\int_b^a f(x)\,dx = -\int_a^b f(x)\,dx$$

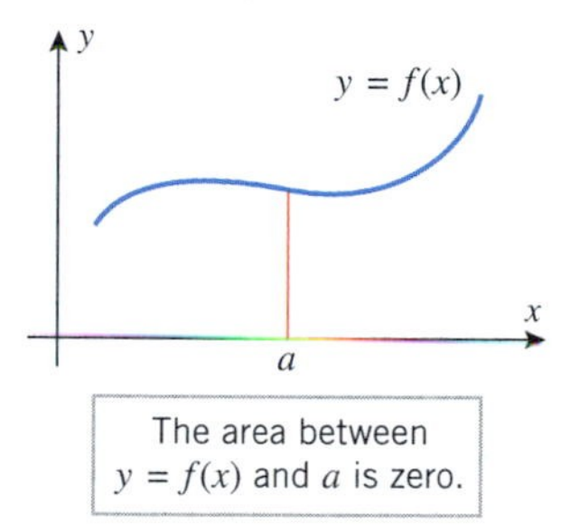

The area between $y = f(x)$ and a is zero.

Figure 7.5.14

REMARK. Part (a) of this definition is consistent with the intuitive idea that the area between a point on the x-axis and a curve $y = f(x)$ should be zero (Figure 7.5.14). Part (b) of the definition is simply a useful convention; it states that interchanging the limits of integration reverses the sign of the integral.

Example 4

(a) $\displaystyle\int_1^1 x^2\,dx = 0$

(b) $\displaystyle\int_1^0 \sqrt{1-x^2}\,dx = -\int_0^1 \sqrt{1-x^2}\,dx = -\frac{\pi}{4}$ ◀

Example 2(c)

Because definite integrals are defined as limits, they inherit many of the properties of limits. For example, we know that constants can be moved through limit signs and that the limit of a sum or difference is the sum or difference of the limits. Thus, you should not be surprised by the following theorem, which we state without formal proof.

7.5.4 THEOREM. *If f and g are integrable on $[a, b]$ and if c is a constant, then cf, $f + g$, and $f - g$ are integrable on $[a, b]$ and*

(a) $\displaystyle\int_a^b cf(x)\,dx = c\int_a^b f(x)\,dx$

(b) $\displaystyle\int_a^b [f(x) + g(x)]\,dx = \int_a^b f(x)\,dx + \int_a^b g(x)\,dx$

(c) $\displaystyle\int_a^b [f(x) - g(x)]\,dx = \int_a^b f(x)\,dx - \int_a^b g(x)\,dx$

Part (b) of this theorem can be extended to more than two functions. More precisely,

$$\int_a^b [f_1(x) + f_2(x) + \cdots + f_n(x)]\, dx = \int_a^b f_1(x)\, dx + \int_a^b f_2(x)\, dx + \cdots + \int_a^b f_n(x)\, dx \tag{8}$$

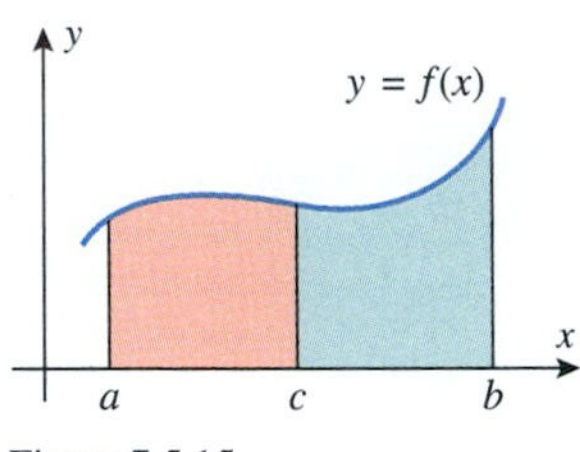

Figure 7.5.15

Some properties of definite integrals can be motivated by interpreting the integral as an area. For example, if f is continuous and nonnegative on the interval $[a, b]$, and if c is a point between a and b, then the area under $y = f(x)$ over the interval $[a, b]$ can be split into two parts and expressed as the area under the graph from a to c plus the area under the graph from c to b (Figure 7.5.15), that is,

$$\int_a^b f(x)\, dx = \int_a^c f(x)\, dx + \int_c^b f(x)\, dx$$

This is a special case of the following theorem about definite integrals, which we state without proof.

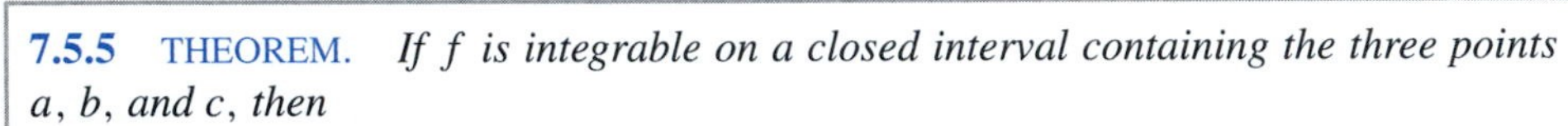

7.5.5 THEOREM. *If f is integrable on a closed interval containing the three points a, b, and c, then*

$$\int_a^b f(x)\, dx = \int_a^c f(x)\, dx + \int_c^b f(x)\, dx \tag{9}$$

no matter how the points are ordered.

The following theorem, which we state without formal proof, can also be motivated by interpreting definite integrals as areas.

7.5.6 THEOREM.

(a) *If f is integrable on $[a, b]$ and $f(x) \geq 0$ for all x in $[a, b]$, then*

$$\int_a^b f(x)\, dx \geq 0$$

(b) *If f and g are integrable on $[a, b]$ and $f(x) \geq g(x)$ for all x in $[a, b]$, then*

$$\int_a^b f(x)\, dx \geq \int_a^b g(x)\, dx$$

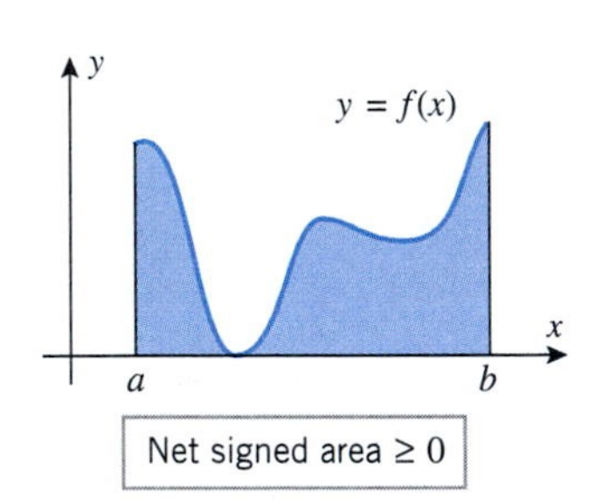

Figure 7.5.16

Geometrically, part (a) of this theorem states the obvious fact that if f is nonnegative on $[a, b]$, then the net signed area between the graph of f and the interval $[a, b]$ is also nonnegative (Figure 7.5.16). Part (b) has its simplest interpretation when f and g are nonnegative on $[a, b]$, in which case the theorem states that if the graph of f does not go below the graph of g, then the area under the graph of f is at least as large as the area under the graph of g (Figure 7.5.17).

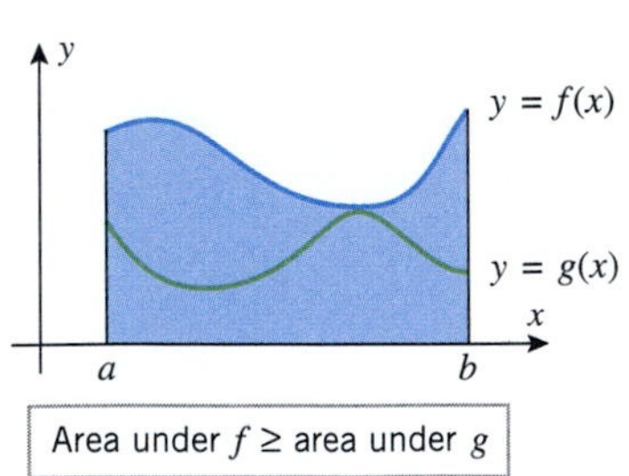

Figure 7.5.17

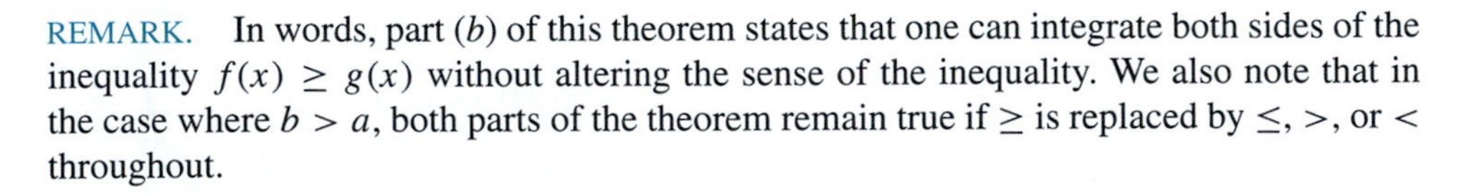

REMARK. In words, part (b) of this theorem states that one can integrate both sides of the inequality $f(x) \geq g(x)$ without altering the sense of the inequality. We also note that in the case where $b > a$, both parts of the theorem remain true if $\geq$ is replaced by $\leq$, $>$, or $<$ throughout.

Example 5

Evaluate

$$\int_0^1 (5 - 3\sqrt{1 - x^2})\, dx$$

Solution. From parts (*a*) and (*c*) of Theorem 7.5.4 we can write

$$\int_0^1 (5 - 3\sqrt{1 - x^2})\, dx = \int_0^1 5\, dx - \int_0^1 3\sqrt{1 - x^2}\, dx = \int_0^1 5\, dx - 3\int_0^1 \sqrt{1 - x^2}\, dx$$

The first integral can be interpreted as the area of a rectangle of height 5 and base 1, so its value is 5, and from Example 2 the value of the second integral is $\pi/4$. Thus,

$$\int_0^1 (5 - 3\sqrt{1 - x^2})\, dx = 5 - 3\left(\frac{\pi}{4}\right) = 5 - \frac{3\pi}{4}$$

◀

CONDITIONS FOR INTEGRABILITY

The problem of determining precisely which functions are integrable is quite complex and beyond the scope of this text. However, there are a few basic results about integrability that are important to know; we begin with a definition.

7.5.7 DEFINITION. A function f is said to be ***bounded*** on an interval I if there is a positive number M such that

$$-M \le f(x) \le M$$

for all x in the interval I. Geometrically, this means that the graph of f over the interval I lies between the lines $y = -M$ and $y = M$.

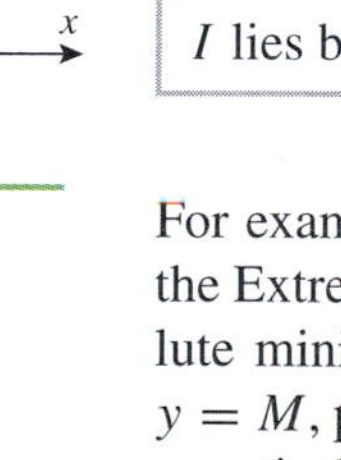

f is bounded on $[a, b]$.

Figure 7.5.18

For example, a continuous function f is bounded on *every* finite closed interval because the Extreme-Value Theorem (6.1.3) implies that f has an absolute maximum and an absolute minimum on the interval; hence, its graph will lie between the line $y = -M$ and $y = M$, provided we make M large enough (Figure 7.5.18). In contrast, a function that has a vertical asymptote inside of an interval is not bounded on that interval because its graph over the interval cannot be made to lie between the lines $y = -M$ and $y = M$, no matter how large we make the value of M (Figure 7.5.19).

The following theorem, which we state without proof, lists three of the most important facts about integrability.

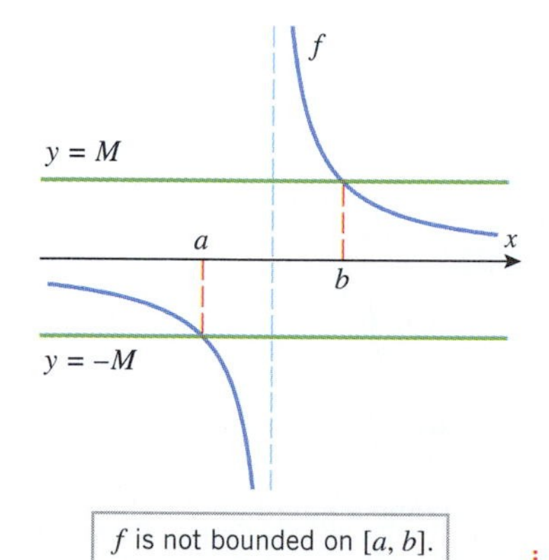

f is not bounded on $[a, b]$.

Figure 7.5.19

7.5.8 THEOREM. *Let f be a function that is defined at all points in the finite closed interval $[a, b]$.*

(*a*) *If f is continuous on $[a, b]$, then f is integrable on $[a, b]$.*

(*b*) *If f has finitely many points of discontinuity on $[a, b]$ but is bounded on $[a, b]$, then f is integrable on $[a, b]$.*

(*c*) *If f is not bounded on $[a, b]$, then f is not integrable on $[a, b]$.*

FOR THE READER. Sketch the graph of a function over the interval $[0, 1]$ that has the properties stated in part (*b*) of this theorem.

EXERCISE SET 7.5 C CAS

1. (a) Use an appropriate geometric formula to find the exact area A under the line $x + y = 4$ over the interval $[0, 4]$.
   (b) Sketch the rectangles for the left endpoint approximation to the area A using $n = 4$ subintervals. Is that approximation greater than, less than, or equal to A? Explain your reasoning, and check your conclusion by calculating the left endpoint approximation.
   (c) Sketch the rectangles for the right endpoint approximation to the area A using $n = 4$ subintervals. Is that approximation greater than, less than, or equal to A? Explain your reasoning, and check your conclusion by calculating the right endpoint approximation.
   (d) Sketch the rectangles for the midpoint approximation to the area A using $n = 4$ subintervals. Is that approximation greater than, less than, or equal to A? Explain your reasoning, and check your conclusion by calculating the midpoint approximation.

2. Follow the directions of Exercise 1 for the area A under the line $y = 3x$ over the interval $[2, 6]$.

3. Find the left endpoint, right endpoint, and midpoint approximations of the area under the curve $y = x^2 + 1$ over the interval $[0, 5]$ using $n = 5$ subintervals.

4. Find the left endpoint, right endpoint, and midpoint approximations of the area under the curve $y = x^3$ over the interval $[1, 6]$ using $n = 5$ subintervals.

5. Find the left endpoint, right endpoint, and midpoint approximations of the area under the curve $y = \cos x$ over the interval $[-\pi/2, \pi/2]$ using $n = 4$ subintervals.

6. Find the left endpoint, right endpoint, and midpoint approximations of the area under the curve $y = e^x$ over the interval $[0, 5]$ using $n = 5$ subintervals.

7. The accompanying figure shows five points on the graph of an unknown function f. Devise a strategy for using the known points to approximate the area A under the graph of $y = f(x)$ over the interval $[1, 5]$. Describe your strategy, and use it to approximate A.

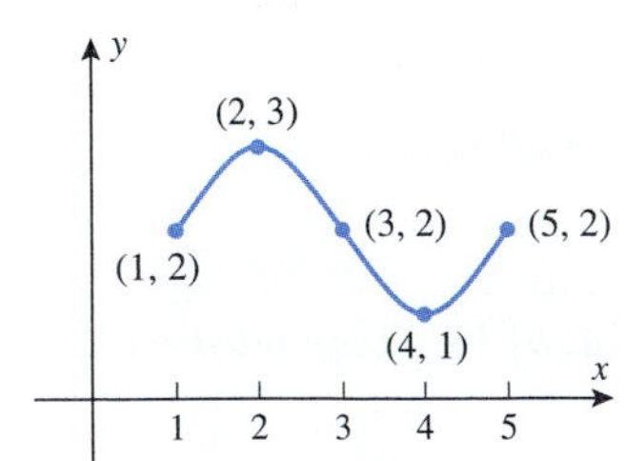

Figure Ex-7

8. (a) Use an appropriate geometric formula to find the exact area A under the line $y = 3x + 1$ over the interval $[1, 5]$.
   (b) Show that the exact area is equal to the average value of the left endpoint and right endpoint approximations of A obtained using $n = 4$ subintervals.
   (c) What is the explanation of the result in part (b)?

In Exercises 9–14, use a calculating utility to find the left endpoint, right endpoint, and midpoint approximations to the area under the curve $y = f(x)$ over the stated interval using $n = 10$ subintervals.

9. $y = 1/x$; $[1, 2]$
10. $y = 1/x^2$; $[1, 3]$
11. $y = \sin x$; $[0, \pi/2]$
12. $y = \sqrt{x}$; $[0, 4]$
13. $y = \ln x$; $[1, 2]$
14. $y = e^x$; $[0, 1]$

15. If you have a programmable calculator, create a program for calculating the midpoint approximation of the area under a curve $y = f(x)$ over an interval $[a, b]$ using n subintervals, and use the program to find midpoint approximations in Exercises 9–14 with
   (a) $n = 25$ (b) $n = 50$ (c) $n = 100$.

C 16. If you have a CAS, devise a procedure for using it to calculate the midpoint approximation of the area under a curve $y = f(x)$ over an interval $[a, b]$ using n subintervals, and use the procedure to find the midpoint approximations in Exercises 9–14 with
   (a) $n = 25$ (b) $n = 50$ (c) $n = 100$.

In Exercises 17–20, sketch the region whose signed area is represented by the definite integral, and evaluate the integral using an appropriate formula from geometry, where needed.

17. (a) $\int_0^3 x\,dx$ (b) $\int_{-2}^{-1} x\,dx$
   (c) $\int_{-1}^4 x\,dx$ (d) $\int_{-5}^5 x\,dx$

18. (a) $\int_0^2 \left(1 - \frac{1}{2}x\right) dx$ (b) $\int_{-1}^1 \left(1 - \frac{1}{2}x\right) dx$
   (c) $\int_2^3 \left(1 - \frac{1}{2}x\right) dx$ (d) $\int_0^3 \left(1 - \frac{1}{2}x\right) dx$

19. (a) $\int_0^5 2\,dx$ (b) $\int_0^\pi \cos x\,dx$
   (c) $\int_{-1}^2 |2x - 3|\,dx$ (d) $\int_{-1}^1 \sqrt{1 - x^2}\,dx$

20. (a) $\int_{-10}^{-5} 6\,dx$ (b) $\int_{-\pi/3}^{\pi/3} \sin x\,dx$
   (c) $\int_0^3 |x - 2|\,dx$ (d) $\int_0^2 \sqrt{4 - x^2}\,dx$

21. Use the areas shown in the accompanying figure to find
   (a) $\int_a^b f(x)\,dx$ (b) $\int_b^c f(x)\,dx$
   (c) $\int_a^c f(x)\,dx$ (d) $\int_a^d f(x)\,dx$.

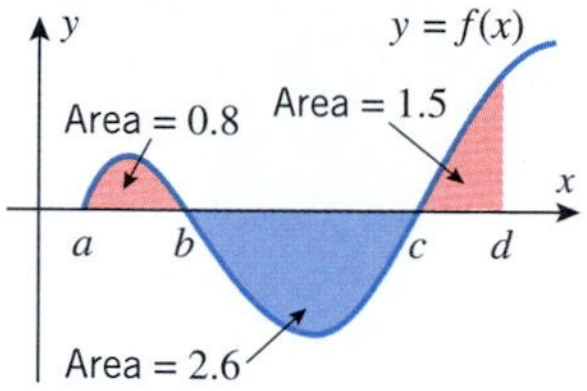

Figure Ex-21

22. In each part, evaluate the integral, given that

$$f(x) = \begin{cases} 2x, & x \leq 1 \\ 2, & x > 1 \end{cases}$$

(a) $\int_0^1 f(x)\,dx$ (b) $\int_{-1}^1 f(x)\,dx$

(c) $\int_1^{10} f(x)\,dx$ (d) $\int_{1/2}^5 f(x)\,dx$

23. Find $\int_{-1}^2 [f(x) + 2g(x)]\,dx$ if

$$\int_{-1}^2 f(x)\,dx = 5 \quad \text{and} \quad \int_{-1}^2 g(x)\,dx = -3$$

24. Find $\int_1^4 [3f(x) - g(x)]\,dx$ if

$$\int_1^4 f(x)\,dx = 2 \quad \text{and} \quad \int_1^4 g(x)\,dx = 10$$

25. Find $\int_1^5 f(x)\,dx$ if

$$\int_0^1 f(x)\,dx = -2 \quad \text{and} \quad \int_0^5 f(x)\,dx = 1$$

26. Find $\int_3^{-2} f(x)\,dx$ if

$$\int_{-2}^1 f(x)\,dx = 2 \quad \text{and} \quad \int_1^3 f(x)\,dx = -6$$

In Exercises 27 and 28, use Theorem 7.5.4 and appropriate formulas from geometry to evaluate the integrals.

27. (a) $\int_0^1 (x + 2\sqrt{1 - x^2})\,dx$ (b) $\int_{-1}^3 (4 - 5x)\,dx$

28. (a) $\int_{-3}^0 (2 + \sqrt{9 - x^2})\,dx$ (b) $\int_{-2}^2 (1 - 3|x|)\,dx$

In Exercises 29 and 30, use Theorem 7.5.6 to determine whether the value of the integral is positive or negative.

29. (a) $\int_2^3 \frac{\sqrt{x}}{1 - x}\,dx$ (b) $\int_0^4 \frac{x^2}{3 - \cos x}\,dx$

30. (a) $\int_{-3}^{-1} \frac{x^4}{\sqrt{3 - x}}\,dx$ (b) $\int_{-2}^2 \frac{x^3 - 9}{|x| + 1}\,dx$

In Exercises 31 and 32, evaluate the integrals by completing the square and applying appropriate formulas from geometry.

31. $\int_0^{10} \sqrt{10x - x^2}\,dx$ **32.** $\int_0^3 \sqrt{6x - x^2}\,dx$

In Exercises 33 and 34, express the limits as definite integrals over the interval $[a, b]$. Do not try to evaluate the integrals.

33. (a) $\lim_{\max \Delta x_k \to 0} \sum_{k=1}^n 4x_k^*(1 - 3x_k^*)\Delta x_k;\ a = -3, b = 3$

(b) $\lim_{\max \Delta x_k \to 0} \sum_{k=1}^n e^{x_k^*}\,\Delta x_k;\ a = 0, b = 1$

34. (a) $\lim_{\max \Delta x_k \to 0} \sum_{k=1}^n (x_k^*)^3 \Delta x_k;\ a = 1, b = 2$

(b) $\lim_{\max \Delta x_k \to 0} \sum_{k=1}^n (\sin^2 x_k^*)\Delta x_k;\ a = 0, b = \pi/2$

In Exercises 35 and 36, evaluate the limit over the interval $[a, b]$ by expressing it as a definite integral and applying an appropriate formula from geometry.

35. $\lim_{\max \Delta x_k \to 0} \sum_{k=1}^n (3x_k^* + 1)\Delta x_k;\ a = 0, b = 1$

36. $\lim_{\max \Delta x_k \to 0} \sum_{k=1}^n \sqrt{4 - (x_k^*)^2}\,\Delta x_k;\ a = -2, b = 2$

In Exercises 37 and 38, use Formula (7) to express the integrals as limits of Riemann sums. Do not try to evaluate the integrals.

37. (a) $\int_1^2 2x\,dx$ (b) $\int_0^1 \frac{x}{x + 1}\,dx$

38. (a) $\int_1^2 \ln x\,dx$ (b) $\int_{-\pi/2}^{\pi/2} (1 + \cos x)\,dx$

39. In this exercise you will find the area A under the graph of $y = x$ over the interval $[1, 2]$ by calculating the limit of right endpoint approximations. For this particular problem, the area can be found much more easily using a formula from geometry, so our purpose here is not to provide a practical method for calculating the area, but rather to illustrate the idea that underlies the concept of a definite integral.

(a) Suppose that the interval $[1, 2]$ is subdivided into n equal subintervals of length $\Delta x = 1/n$ and that the points $x_1^*, x_2^*, \ldots, x_n^*$ are the right endpoints of the subintervals. Show that the right endpoint of the kth subinterval is

$$x_k^* = 1 + \frac{k}{n}$$

[*Suggestion:* Find x_1^*, x_2^*, and x_3^*, and then look for the pattern.]

(b) Show that with n subintervals the right endpoint approximation of the area A is

$$\sum_{k=1}^{n} f(x_k^*)\Delta x = \sum_{k=1}^{n}\left[\left(1+\frac{k}{n}\right)\frac{1}{n}\right]$$

(c) Use Theorem 7.4.2 to show that the right endpoint approximation can be expressed as

$$\sum_{k=1}^{n} f(x_k^*)\Delta x = \frac{3}{2} + \frac{1}{2n}$$

(d) From (2), the area A is

$$A = \lim_{n\to+\infty} \sum_{k=1}^{n} f(x_k^*)\Delta x$$

Find this limit, and check your answer by using a formula from geometry to calculate A.

40. Find the area A in Exercise 39 as a limit of left endpoint approximations.

In Exercises 41–44, use the method of Exercise 39 to find the area under the curve $y = f(x)$ over the interval $[a, b]$ as a limit of right and left endpoint approximations.

41. $y = x^2$; $a = 0, b = 1$

42. $y = 4 - \frac{1}{4}x^2$; $a = 0, b = 3$

43. $y = x^3$; $a = 2, b = 6$

44. $y = 1 - x^3$; $a = -3, b = -1$

45. In each part, use Theorem 7.5.8 to determine whether the function f is integrable on the interval $[-1, 1]$.

(a) $f(x) = e^x \cos x$

(b) $f(x) = \begin{cases} x/|x|, & x \neq 0 \\ 0, & x = 0 \end{cases}$

(c) $f(x) = \begin{cases} 1/x^2, & x \neq 0 \\ 0, & x = 0 \end{cases}$

(d) $f(x) = \begin{cases} \sin 1/x, & x \neq 0 \\ 0, & x = 0 \end{cases}$

46. It can be shown that every interval contains both rational and irrational numbers. Accepting this to be so, do you believe that the function

$$f(x) = \begin{cases} 1 & \text{if} \quad x \text{ is rational} \\ 0 & \text{if} \quad x \text{ is irrational} \end{cases}$$

is integrable on a closed interval $[a, b]$? Explain your reasoning.

47. It can be shown that the limit in Formula (7) has all of the limit properties stated in Theorem 2.2.2. Accepting this to be so, show that

(a) $\displaystyle\int_a^b cf(x)\,dx = c\int_a^b f(x)\,dx$

(b) $\displaystyle\int_a^b [f(x) + g(x)]\,dx = \int_a^b f(x)\,dx + \int_a^b g(x)\,dx$

48. Find the smallest and largest values that the Riemann sum

$$\sum_{k=1}^{3} f(x_k^*)\Delta x_k$$

can have on the interval $[0, 4]$ if $f(x) = x^2 - 3x + 4$ and $\Delta x_1 = 1$, $\Delta x_2 = 2$, $\Delta x_3 = 1$.

7.6 THE FUNDAMENTAL THEOREM OF CALCULUS

In this section we will establish two basic relationships between definite and indefinite integrals that together constitute a result called the Fundamental Theorem of Calculus. One part of this theorem will relate the rectangle and antiderivative methods for calculating areas, and the second part will provide a powerful method for evaluating definite integrals using antiderivatives.

THE FUNDAMENTAL THEOREM OF CALCULUS

To motivate the results we are looking for, let us begin by assuming that f is nonnegative and continuous on the interval $[a, b]$, in which case the area A under the graph of f over the interval $[a, b]$ is represented by the definite integral

$$A = \int_a^b f(x)\,dx \tag{1}$$

(Figure 7.6.1).

Recall from our discussion of the antiderivative method in Section 7.1 that if $A(x)$ is the area under the graph of f from a to x (Figure 7.6.2), then:

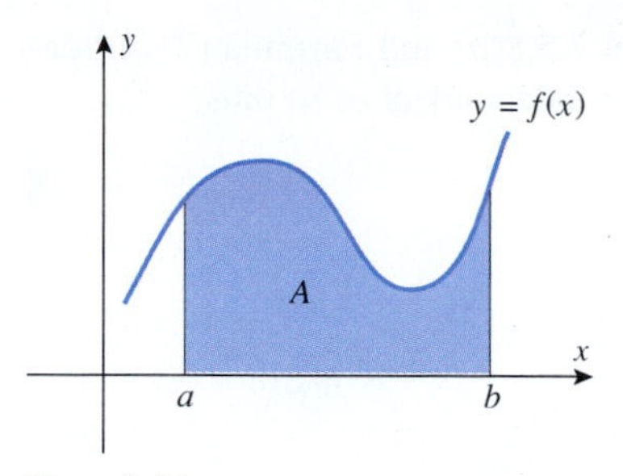

Figure 7.6.1

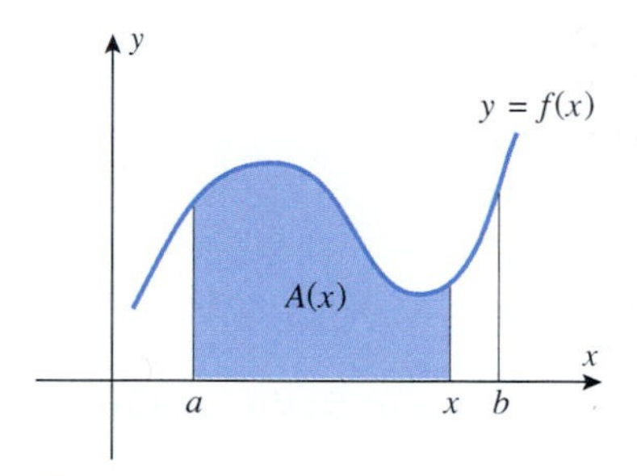

Figure 7.6.2

- $A'(x) = f(x)$
- $A(a) = 0$ The area under the curve from a to a is the area above the single point a, and hence is zero.
- $A(b) = A$ The area under the curve from a to b is A.

The formula $A'(x) = f(x)$ states that $A(x)$ is an antiderivative of $f(x)$, which implies that every other antiderivative of $f(x)$ can be obtained by adding a constant to $A(x)$. Accordingly, let

$$F(x) = A(x) + C$$

be any antiderivative of $f(x)$, and consider what happens when we subtract $F(a)$ from $F(b)$. We obtain

$$F(b) - F(a) = [A(b) + C] - [A(a) + C] = A(b) - A(a) = A - 0 = A$$

and hence (1) can be expressed as

$$\int_a^b f(x)\,dx = F(b) - F(a)$$

In words, this equation states that the definite integral can be evaluated by finding any antiderivative of the integrand and then subtracting the value of this antiderivative at the lower limit of integration from its value at the upper limit of integration. Although we derived this result subject to the assumption that f is nonnegative on $[a, b]$, this assumption is not essential, as we will prove in the following theorem, which is the main tool used to evaluate definite integrals.

> **7.6.1 THEOREM (*The Fundamental Theorem of Calculus, Part 1*).** *If f is continuous on $[a, b]$, and if F is any antiderivative of f on $[a, b]$, then*
>
> $$\int_a^b f(x)\,dx = F(b) - F(a) \tag{2}$$

Proof. Let $x_1, x_2, \ldots, x_{n-1}$ be any points in $[a, b]$ such that

$$a < x_1 < x_2 < \cdots < x_{n-1} < b$$

These points divide $[a, b]$ into n subintervals

$$[a, x_1], [x_1, x_2], \ldots, [x_{n-1}, b] \tag{3}$$

whose lengths, as usual, we denote by

$$\Delta x_1, \Delta x_2, \ldots, \Delta x_n$$

By hypothesis, $F'(x) = f(x)$ for all x in $[a, b]$, so F satisfies the hypotheses of the Mean-Value Theorem (6.5.2) on each subinterval in (3). Hence, we can find points $x_1^*, x_2^*, \ldots, x_n^*$ in the respective subintervals in (3) such that

$$\begin{aligned}
F(x_1) - F(a) &= F'(x_1^*)(x_1 - a) = f(x_1^*)\Delta x_1 \\
F(x_2) - F(x_1) &= F'(x_2^*)(x_2 - x_1) = f(x_2^*)\Delta x_2 \\
F(x_3) - F(x_2) &= F'(x_3^*)(x_3 - x_2) = f(x_3^*)\Delta x_3 \\
&\vdots \\
F(b) - F(x_{n-1}) &= F'(x_n^*)(b - x_{n-1}) = f(x_n^*)\Delta x_n
\end{aligned}$$

Adding the preceding equations yields

$$F(b) - F(a) = \sum_{k=1}^{n} f(x_k^*)\Delta x_k \tag{4}$$

Let us now increase n in such a way that $\max \Delta x_k \to 0$. Since f is assumed to be continuous, the right side of (4) approaches $\int_a^b f(x)\,dx$, by Theorem 7.5.8(*a*) and Formula (7) of Section 7.5. However, the left side of (4) is a constant that is independent of n; thus,

$$F(b) - F(a) = \lim_{\max \Delta x_k \to 0} \sum_{k=1}^{n} f(x_k^*)\Delta x_k = \int_a^b f(x)\,dx$$ ■

It is standard to denote the difference $F(b) - F(a)$ as

$$F(x)\Big]_a^b = F(b) - F(a) \quad \text{or} \quad \Big[F(x)\Big]_a^b = F(b) - F(a)$$

For example, using the first of these notations we can express (2) as

$$\int_a^b f(x)\,dx = F(x)\Big]_a^b \tag{5}$$

Example 1

Evaluate $\displaystyle\int_1^2 x\,dx$.

Solution. The function $F(x) = \frac{1}{2}x^2$ is an antiderivative of $f(x) = x$; thus, from (2)

$$\int_1^2 x\,dx = \frac{1}{2}x^2\Big]_1^2 = \frac{1}{2}(2)^2 - \frac{1}{2}(1)^2 = 2 - \frac{1}{2} = \frac{3}{2}$$ ◀

Example 2

In Example 1 of the last section we approximated the area under the graph of $y = 9 - x^2$ over the interval $[0, 3]$ using left endpoint, right endpoint, and midpoint approximations, all of which produced an approximation of roughly 18 (square units); and in the remark following that example we stated without proof that the exact area A is 18 (square units). We can now confirm this using the Fundamental Theorem of Calculus as follows:

$$A = \int_0^3 (9 - x^2)\,dx = 9x - \frac{x^3}{3}\Big]_0^3 = \left(27 - \frac{27}{3}\right) - 0 = 18$$ ◀

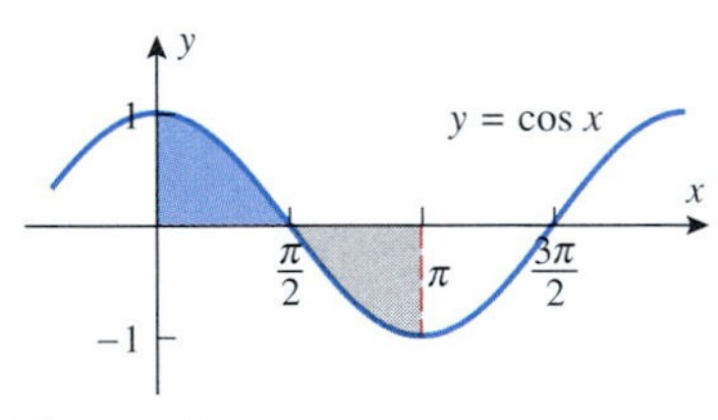

Figure 7.6.3

Example 3

(a) Find the area under the curve $y = \cos x$ over the interval $[0, \pi/2]$ (Figure 7.6.3).

(b) Make a conjecture about the value of the integral

$$\int_0^{\pi} \cos x\,dx$$

and confirm your conjecture using the Fundamental Theorem of Calculus.

Solution (a). Since $\cos x \geq 0$ over the interval $[0, \pi/2]$, the area A under the curve is

$$A = \int_0^{\pi/2} \cos x\,dx = \sin x\Big]_0^{\pi/2} = \sin\frac{\pi}{2} - \sin 0 = 1$$

Solution (b). The given integral can be interpreted as the signed area between the graph of $y = \cos x$ and the interval $[0, \pi]$. The graph in Figure 7.6.3 suggests that over the interval $[0, \pi]$ the portion of area above the x-axis is the same as the portion of area below the x-axis, so we conjecture that the signed area is zero; this implies that the value of the integral is zero. This is confirmed by the computations

$$\int_0^{\pi} \cos x\,dx = \sin x\Big]_0^{\pi} = \sin\pi - \sin 0 = 0$$ ◀

THE RELATIONSHIP BETWEEN DEFINITE AND INDEFINITE INTEGRALS

Observe that in the preceding examples we did not include a constant of integration in the antiderivatives. In general, when applying the Fundamental Theorem of Calculus there is no need to include a constant of integration because it will drop out anyhow. To see that this is so, let F be any antiderivative of the integrand on $[a, b]$, and let C be any constant; then

$$\int_a^b f(x)\,dx = F(x) + C\Big]_a^b = [F(b) + C] - [F(a) + C] = F(b) - F(a)$$

Thus, for purposes of evaluating a definite integral we can omit the constant of integration in

$$\int f(x)\,dx = F(x) + C$$

and express (5) as

$$\int_a^b f(x)\,dx = \left[\int f(x)\,dx\right]_a^b \tag{6}$$

which relates the definite and indefinite integrals.

Example 4

$$\int_1^9 \sqrt{x}\,dx = \int \sqrt{x}\,dx\Big]_1^9 = \int x^{1/2}\,dx\Big]_1^9 = \frac{2}{3}x^{3/2}\Big]_1^9 = \frac{2}{3}(27-1) = \frac{52}{3}$$ ◀

REMARK. Usually, we will dispense with the step of displaying the indefinite integral explicitly and write the antiderivative immediately, as in our first three examples.

Example 5

Table 7.2.1 will be helpful for the following computations.

$$\int_0^{\ln 3} 5e^x\,dx = 5\int_0^{\ln 3} e^x\,dx = 5e^x\Big]_0^{\ln 3} = 5(e^{\ln 3} - e^0) = 5(3-1) = 10$$

$$\int_1^2 \frac{1}{x}\,dx = \ln|x|\Big]_1^2 = \ln|2| - \ln|1| = \ln 2 - \ln 1 = \ln 2$$

$$\int_{-2}^{-1} \frac{1}{x}\,dx = \ln|x|\Big]_{-2}^{-1} = \ln|-1| - \ln|-2| = \ln 1 - \ln 2 = -\ln 2$$

$$\int_{-\pi/4}^{\pi/4} \sec x \tan x\,dx = \sec x\Big]_{-\pi/4}^{\pi/4} = \sec\left(\frac{\pi}{4}\right) - \sec\left(-\frac{\pi}{4}\right) = \frac{2}{\sqrt{2}} - \frac{2}{\sqrt{2}} = 0$$ ◀

WARNING. The requirement in the Fundamental Theorem of Calculus that f be continuous on $[a, b]$ is important to keep in mind, for if you attempt to apply this theorem in cases where the integrand is not continuous on the interval of integration, then you may obtain erroneous results. For example, the function $f(x) = 1/x^2$ has a discontinuity at $x = 0$, so the Fundamental Theorem of Calculus cannot be used to integrate f on any interval that contains $x = 0$. However, if we ignore this and blindly apply the theorem over the interval $[-1, 1]$, we obtain

$$\int_{-1}^1 \frac{1}{x^2}\,dx = -\frac{1}{x}\Big]_{-1}^1 = -[1 - (-1)] = -2$$

which is clearly erroneous because $f(x) = 1/x^2$ is a nonnegative function and hence cannot possibly produce a negative definite integral.

FOR THE READER. If you have a CAS, read the documentation on evaluating definite integrals, and then check the results in the preceding examples.

The Fundamental Theorem of Calculus can be applied without modification to definite integrals in which the lower limit of integration is greater than or equal to the upper limit of integration.

Example 6

$$\int_1^1 x^2\,dx = \frac{x^3}{3}\Big]_1^1 = \frac{1}{3} - \frac{1}{3} = 0$$

$$\int_4^0 x\,dx = \frac{x^2}{2}\Big]_4^0 = \left[\frac{0}{2} - \frac{16}{2}\right] = -8$$

The latter result is consistent with the result that would be obtained by first reversing the limits of integration in accordance with Definition 7.5.3(b):

$$\int_4^0 x\,dx = -\int_0^4 x\,dx = -\frac{x^2}{2}\Big]_0^4 = -\left[\frac{16}{2} - \frac{0}{2}\right] = -8 \quad ◀$$

To integrate a continuous function that is defined piecewise on an interval $[a, b]$, split this interval into subintervals at the breakpoints of the function, and integrate separately over each subinterval in accordance with Theorem 7.5.5.

Example 7

Evaluate $\int_0^6 f(x)\,dx$ if

$$f(x) = \begin{cases} x^2, & x < 2 \\ 3x - 2, & x \geq 2 \end{cases}$$

Solution. From Theorem 7.5.5

$$\int_0^6 f(x)\,dx = \int_0^2 f(x)\,dx + \int_2^6 f(x)\,dx = \int_0^2 x^2\,dx + \int_2^6 (3x-2)\,dx$$

$$= \frac{x^3}{3}\Big]_0^2 + \left[\frac{3x^2}{2} - 2x\right]_2^6 = \left(\frac{8}{3} - 0\right) + (42 - 2) = \frac{128}{3} \quad ◀$$

Example 8

Evaluate $\int_{-1}^2 |x|\,dx$.

Solution. Since $|x| = x$ when $x \geq 0$ and $|x| = -x$ when $x \leq 0$,

$$\int_{-1}^2 |x|\,dx = \int_{-1}^0 |x|\,dx + \int_0^2 |x|\,dx$$

$$= \int_{-1}^0 (-x)\,dx + \int_0^2 x\,dx$$

$$= -\frac{x^2}{2}\Big]_{-1}^0 + \frac{x^2}{2}\Big]_0^2 = \frac{1}{2} + 2 = \frac{5}{2} \quad ◀$$

DUMMY VARIABLES

To evaluate a definite integral using the Fundamental Theorem of Calculus, one needs to be able to find an antiderivative of the integrand; thus, it is important to know what kinds of functions have antiderivatives. It is our next objective to show that all continuous functions have antiderivatives, but to do this we will need some preliminary results.

Formula (6) shows that there is a close relationship between the integrals

$$\int_a^b f(x)\,dx \quad \text{and} \quad \int f(x)\,dx$$

However, the definite and indefinite integrals differ in some important ways. For one thing, the two integrals are different kinds of objects—the definite integral is a *number* (the signed area between the graph of $y = f(x)$ and the interval $[a, b]$), whereas the indefinite integral is a *function*, or more accurately a set of functions [the antiderivatives of $f(x)$]. However, the two types of integrals also differ in the role played by the variable of integration. In an indefinite integral, the variable of integration is "passed through" to the antiderivative in the sense that integrating a function of x produces a function of x, integrating a function of t produces a function of t, and so forth. For example,

$$\int x^2\,dx = \frac{x^3}{3} + C \quad \text{and} \quad \int t^2\,dt = \frac{t^3}{3} + C$$

In contrast, the variable of integration in a definite integral is not passed through to the end result, since the end result is a number. Thus, integrating a function of x over an interval and integrating the same function of t over the same interval of integration produces the same value for the integral. For example,

$$\int_1^3 x^2\,dx = \left.\frac{x^3}{3}\right]_{x=1}^{3} = \frac{27}{3} - \frac{1}{3} = \frac{26}{3} \quad \text{and} \quad \int_1^3 t^2\,dt = \left.\frac{t^3}{3}\right]_{t=1}^{3} = \frac{27}{3} - \frac{1}{3} = \frac{26}{3}$$

However, this latter result should not be surprising, since the area under the graph of the curve $y = f(x)$ over an interval $[a, b]$ on the x-axis is the same as the area under the graph of the curve $y = f(t)$ over the interval $[a, b]$ on the t-axis (Figure 7.6.4).

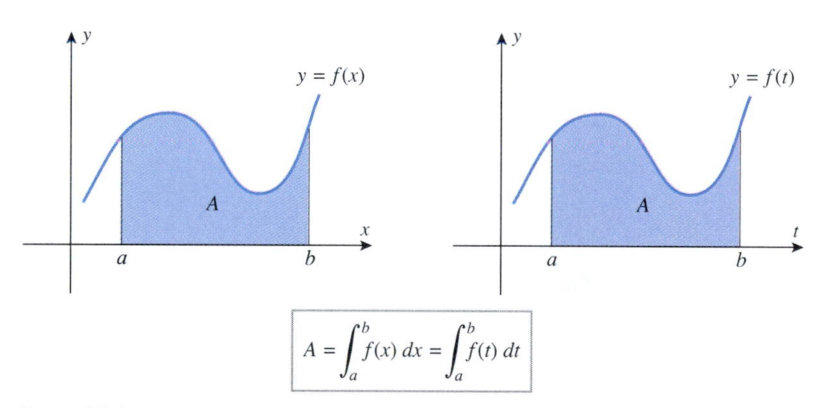

Figure 7.6.4

Because the variable of integration in a definite integral plays no role in the end result, it is often referred to as a ***dummy variable***. In summary:

> *Whenever you find it convenient to change the letter used for the variable of integration in a definite integral, you can do so without changing the value of the integral.*

THE MEAN-VALUE THEOREM FOR INTEGRALS

To reach our goal of showing that continuous functions have antiderivatives, we will need to develop a basic property of definite integrals, known as the *Mean-Value Theorem for Integrals*. In the next section we will use this theorem to extend the familiar idea of "average value" so that it applies to continuous functions, but here we will need it as a tool for developing other results.

Let f be a continuous nonnegative function on $[a, b]$, and let m and M be the minimum and maximum values of $f(x)$ on this interval. Consider the rectangle of heights m and M

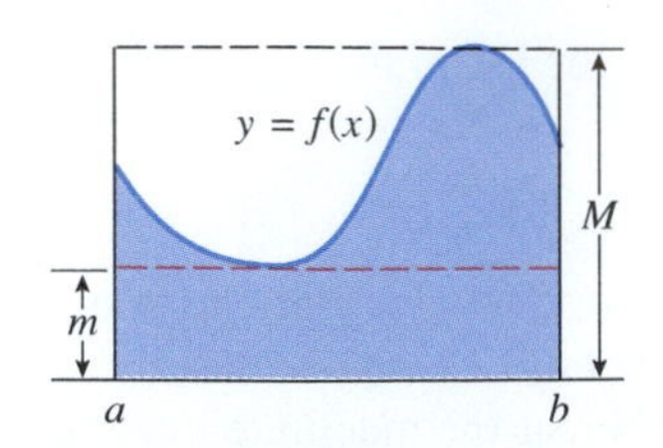

Figure 7.6.5

over the interval $[a, b]$ (Figure 7.6.5). It is clear geometrically from this figure that the area

$$A = \int_a^b f(x)\,dx$$

under $y = f(x)$ is at least as large as the area of the rectangle of height m and no larger than the area of the rectangle of height M. It seems reasonable, therefore, that there is a rectangle over the interval $[a, b]$ of some appropriate height $f(x^*)$ between m and M whose area is precisely A; that is,

$$\int_a^b f(x)\,dx = f(x^*)(b-a)$$

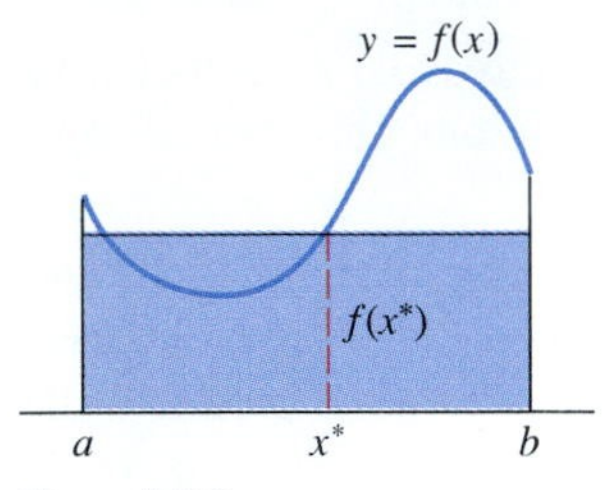

Figure 7.6.6

(Figure 7.6.6). This is a special case of the following result.

7.6.2 THEOREM (*The Mean-Value Theorem for Integrals*). *If f is continuous on a closed interval $[a, b]$, then there is at least one number x^* in $[a, b]$ such that*

$$\int_a^b f(x)\,dx = f(x^*)(b-a) \tag{7}$$

Proof. By the Extreme-Value Theorem (6.1.3), f assumes a maximum value M and a minimum value m on $[a, b]$. Thus, for all x in $[a, b]$,

$$m \le f(x) \le M$$

and from Theorem 7.5.6(*b*)

$$\int_a^b m\,dx \le \int_a^b f(x)\,dx \le \int_a^b M\,dx$$

or

$$m(b-a) \le \int_a^b f(x)\,dx \le M(b-a) \tag{8}$$

or

$$m \le \frac{1}{b-a}\int_a^b f(x)\,dx \le M$$

This implies that

$$\frac{1}{b-a}\int_a^b f(x)\,dx \tag{9}$$

is a number between m and M, and since $f(x)$ assumes the values m and M on $[a, b]$, it follows from the Intermediate-Value Theorem (2.4.8) that $f(x)$ must assume the value (9) at some point x^* in $[a, b]$; that is,

$$\frac{1}{b-a}\int_a^b f(x)\,dx = f(x^*) \quad \text{or} \quad \int_a^b f(x)\,dx = f(x^*)(b-a)$$

■

Example 9

Since $f(x) = x^2$ is continuous on the interval $[1, 4]$, the Mean-Value Theorem for Integrals guarantees that there is a number x^* in $[1, 4]$ such that

$$\int_1^4 x^2\,dx = f(x^*)(4-1) = (x^*)^2(4-1) = 3(x^*)^2$$

But

$$\int_1^4 x^2\,dx = \frac{x^3}{3}\bigg]_1^4 = 21$$

so that

$$3(x^*)^2 = 21 \quad \text{or} \quad (x^*)^2 = 7 \quad \text{or} \quad x^* = \pm\sqrt{7}$$

Thus, $x^* = \sqrt{7} \approx 2.65$ is the number in the interval $[1, 4]$ whose existence is guaranteed by the Mean-Value Theorem for Integrals. ◀

PART 2 OF THE FUNDAMENTAL THEOREM OF CALCULUS

In Section 7.1 we gave an informal argument to show that if f is continuous and nonnegative on $[a, b]$, and if $A(x)$ is the area under the graph of $y = f(x)$ over the interval $[a, x]$ (Figure 7.6.2), then $A'(x) = f(x)$. But $A(x)$ can be expressed as the definite integral

$$A(x) = \int_a^x f(t)\,dt$$

(where we have used t rather than x as the variable of integration to avoid a conflict with the x that appears as the upper limit of integration). Thus, the relationship $A'(x) = f(x)$ can be expressed as

$$\frac{d}{dx}\left[\int_a^x f(t)\,dt\right] = f(x)$$

This is a special case of the following more general result, which applies even if f has negative values.

7.6.3 THEOREM (***The Fundamental Theorem of Calculus, Part 2***). *If f is continuous on an interval I, then f has an antiderivative on I. In particular, if a is any point in I, then the function F defined by*

$$F(x) = \int_a^x f(t)\,dt$$

is an antiderivative of f on I; that is, $F'(x) = f(x)$ for each x in I, or in an alternative notation

$$\frac{d}{dx}\left[\int_a^x f(t)\,dt\right] = f(x) \tag{10}$$

Proof. We will show first that $F(x)$ is defined at each point x in the interval I. If $x > a$ and x is in the interval I, then Theorem 7.5.8(a) applied to the interval $[a, x]$ and the continuity of f on I ensures that $F(x)$ is defined; and if x is in the interval I and $x \le a$, then Definition 7.5.3(b) combined with Theorem 7.5.8(a) ensures that $F(x)$ is defined. Thus, $F(x)$ is defined for all x in I.

Next we will show that $F'(x) = f(x)$ for each x in the interval I. If x is not an endpoint of I, then it follows from the definition of a derivative that

$$\begin{aligned}
F'(x) &= \lim_{h \to 0} \frac{F(x+h) - F(x)}{h} \\
&= \lim_{h \to 0} \frac{1}{h}\left[\int_a^{x+h} f(t)\,dt - \int_a^x f(t)\,dt\right] \\
&= \lim_{h \to 0} \frac{1}{h}\left[\int_a^{x+h} f(t)\,dt + \int_x^a f(t)\,dt\right] \\
&= \lim_{h \to 0} \frac{1}{h}\int_x^{x+h} f(t)\,dt \qquad \boxed{\text{Theorem 7.5.5}}
\end{aligned}$$

Applying the Mean-Value Theorem for Integrals (7.6.2) to the last expression, we obtain

$$F'(x) = \lim_{h \to 0} \frac{1}{h}[f(t^*) \cdot h] = \lim_{h \to 0} f(t^*) \tag{11}$$

where t^* is some number between x and $x+h$. Because t^* is between x and $x+h$, it follows that $t^* \to x$ as $h \to 0$. Thus, $f(t^*) \to f(x)$ as $h \to 0$, since f is assumed continuous at x. Therefore, it follows from (11) that $F'(x) = f(x)$. If x is an endpoint of the interval I, then the two-sided limits in the proof must be replaced by the appropriate one-sided limits, but otherwise the arguments are identical. ∎

In words, Formula (10) states:

If a definite integral has a variable upper limit of integration and a continuous integrand, then the derivative of the integral with respect to its upper limit is equal to the integrand evaluated at the upper limit.

Example 10

Find

$$\frac{d}{dx}\left[\int_1^x t^3\,dt\right]$$

by applying Part 2 of the Fundamental Theorem of Calculus, and then confirm the result by performing the integration and then differentiating.

Solution. The integrand is a continuous function, so from (10)

$$\frac{d}{dx}\left[\int_1^x t^3\,dt\right] = x^3$$

Alternatively, evaluating the integral and then differentiating yields

$$\int_1^x t^3\,dt = \frac{t^4}{4}\bigg]_{t=1}^x = \frac{x^4}{4} - \frac{1}{4}, \quad \frac{d}{dx}\left[\frac{x^4}{4} - \frac{1}{4}\right] = x^3$$

so the two methods for differentiating the integral agree. ◀

Example 11

Since

$$f(x) = \frac{\sin x}{x}$$

is continuous on any interval that does not contain the origin, it follows from (10) that on the interval $(0, +\infty)$ we have

$$\frac{d}{dx}\left[\int_1^x \frac{\sin t}{t}\,dt\right] = \frac{\sin x}{x}$$

Unlike the preceding example, there is no way to evaluate the integral in terms of familiar functions, so Formula (10) provides the only simple method for finding the derivative. ◀

DIFFERENTIATION AND INTEGRATION ARE INVERSE PROCESSES

The two parts of the Fundamental Theorem of Calculus, when taken together, tell us that differentiation and integration are inverse processes in the sense that each undoes the effect of the other. To see why this is so, note that Part 1 of the Fundamental Theorem of Calculus (7.6.1) implies that

$$\int_a^x f'(t)\,dt = f(x) - f(a)$$

which tells us that if the value of $f(a)$ is known, then function f can be recovered from its derivative f' by integrating. Conversely, Part 2 of the Fundamental Theorem of Calculus

(7.6.3) states that

$$\frac{d}{dx}\left[\int_a^x f(t)\,dt\right] = f(x)$$

which tells us that the function f can be recovered from its integral by differentiating. Thus, differentiation and integration can be viewed as inverse processes.

It is common to treat parts 1 and 2 of the Fundamental Theorem of Calculus as a single theorem, and refer to it simply as the *Fundamental Theorem of Calculus*. This theorem ranks as one of the greatest discoveries in the history of science, and its formulation by Newton and Leibniz is generally regarded to be the "discovery of calculus."

EXERCISE SET 7.6 Graphing Calculator C CAS

1. In each part, use a definite integral to find the area of the region, and check your answer using an appropriate formula from geometry.

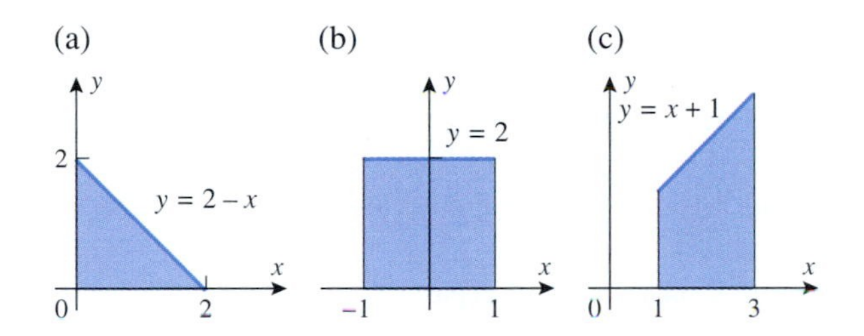

2. In each part, use a definite integral to find the area under the curve $y = f(x)$ over the stated interval, and check your answer using an appropriate formula from geometry.
   (a) $f(x) = x;\ [0, 5]$
   (b) $f(x) = 5;\ [3, 9]$
   (c) $f(x) = x + 3;\ [-1, 2]$

In Exercises 3–8, find the area under the curve $y = f(x)$ over the stated interval.

3. $f(x) = x^3;\ [2, 3]$
4. $f(x) = x^4;\ [-1, 1]$
5. $f(x) = \sqrt{x};\ [1, 9]$
6. $f(x) = x^{-3/5};\ [1, 4]$
7. $f(x) = e^x;\ [1, 3]$
8. $f(x) = \dfrac{1}{x};\ [1, 5]$

In Exercises 9–24, evaluate the integrals using Part 1 of the Fundamental Theorem of Calculus.

9. $\displaystyle\int_{-3}^{0} (x^2 - 4x + 7)\,dx$
10. $\displaystyle\int_{-1}^{2} x(1 + x^3)\,dx$
11. $\displaystyle\int_{1}^{3} \frac{1}{x^2}\,dx$
12. $\displaystyle\int_{1}^{2} \frac{1}{x^6}\,dx$
13. $\displaystyle\int_{4}^{9} 2x\sqrt{x}\,dx$
14. $\displaystyle\int_{1}^{8} (5x^{2/3} - 4x^{-2})\,dx$
15. $\displaystyle\int_{-\pi/2}^{\pi/2} \sin\theta\,d\theta$
16. $\displaystyle\int_{0}^{\pi/4} \sec^2\theta\,d\theta$
17. $\displaystyle\int_{-\pi/4}^{\pi/4} \cos x\,dx$
18. $\displaystyle\int_{0}^{1} (x - \sec x\tan x)\,dx$
19. $\displaystyle\int_{\ln 2}^{3} 5e^x\,dx$
20. $\displaystyle\int_{1/2}^{1} \frac{1}{2x}\,dx$
21. $\displaystyle\int_{1}^{4} \left(\frac{3}{\sqrt{t}} - 5\sqrt{t} - t^{-3/2}\right) dt$
22. $\displaystyle\int_{4}^{9} (4y^{-1/2} + 2y^{1/2} + y^{-5/2})\,dy$
23. $\displaystyle\int_{\pi/6}^{\pi/2} \left(x + \frac{2}{\sin^2 x}\right) dx$
24. $\displaystyle\int_{1}^{2} (x^{-1} + \sqrt{2}e^x - \csc x\cot x)\,dx$

C 25. For each of the integrals you evaluated in Exercises 9–24, use a CAS to check your answer. [*Note:* CAS programs have commands for evaluating definite integrals exactly or approximately. Use the exact evaluation here.]

C 26. Use a CAS to evaluate the integral

$$\int_a^{4a} (a^{1/2} - x^{1/2})\,dx$$

and check the answer by hand.

In Exercises 27–29, use Theorem 7.5.5 to evaluate the given integrals.

27. (a) $\displaystyle\int_0^2 |2x - 3|\,dx$ (b) $\displaystyle\int_0^{3\pi/4} |\cos x|\,dx$
28. (a) $\displaystyle\int_{-1}^2 \sqrt{2 + |x|}\,dx$ (b) $\displaystyle\int_{-1}^1 |e^x - 1|\,dx$

29. $\int_{-2}^{3} f(x)\,dx$, where $f(x) = \begin{cases} -x, & x \geq 0 \\ x^2, & x < 0 \end{cases}$

C **30.** CAS programs provide methods for entering functions that are defined piecewise. Check your documentation to see how this is done, and then use the CAS to evaluate

$$\int_0^4 f(x)\,dx, \quad \text{where} \quad f(x) = \begin{cases} \sqrt{x}, & 0 \leq x < 1 \\ 1/x^2, & x \geq 1 \end{cases}$$

Check the answer by hand.

In Exercises 31–33, use a calculating utility to find the midpoint approximation of the integral using $n = 20$ subintervals, and then find the exact value of the integral using Part 1 of the Fundamental Theorem of Calculus.

31. $\int_1^3 \frac{1}{x^2}\,dx$ **32.** $\int_0^{\pi/2} \sin x\,dx$ **33.** $\int_1^3 \frac{1}{x}\,dx$

C **34.** Compare the answers obtained by the midpoint rule in Exercises 31–33 to those obtained using the numerical (approximate) integration command of a CAS.

35. Find the area under the curve $y = x^2 + 1$ over the interval $[0, 3]$. Make a sketch of the region.

36. Find the area that is above the x-axis, but below the curve $y = (1 - x)(x - 2)$. Make a sketch of the region.

37. Find the area under the curve $y = 3\sin x$ over the interval $[0, 2\pi/3]$. Sketch the region.

38. Find the area below the interval $[-2, -1]$, but above the curve $y = x^3$. Make a sketch of the region.

39. Find the total area between the curve $y = x^2 - 3x - 10$ and the interval $[-3, 8]$. Make a sketch of the region. [*Hint:* Find the portion of area above the interval and the portion of area below the interval separately.]

40. (a) Use a graphing utility to generate the graph of

$$f(x) = \frac{1}{100}(x + 2)(x + 1)(x - 3)(x - 5)$$

and use the graph to make a conjecture about the sign of the integral

$$\int_{-2}^{5} f(x)\,dx$$

(b) Check your conjecture by evaluating the integral.

41. (a) Let f be an odd function; that is, $f(-x) = -f(x)$. Invent a theorem that makes a statement about the value of an integral of the form

$$\int_{-a}^{a} f(x)\,dx$$

(b) Confirm that your theorem works for the integrals

$$\int_{-1}^{1} x^3\,dx \quad \text{and} \quad \int_{-\pi/2}^{\pi/2} \sin x\,dx$$

(c) Let f be an even function; that is, $f(-x) = f(x)$. Invent a theorem that makes a statement about the relationship between the integrals

$$\int_{-a}^{a} f(x)\,dx \quad \text{and} \quad \int_0^{a} f(x)\,dx$$

(d) Confirm that your theorem works for the integrals

$$\int_{-1}^{1} x^2\,dx \quad \text{and} \quad \int_{-\pi/2}^{\pi/2} \cos x\,dx$$

C **42.** Use the theorem you invented in Exercise 41(a) to evaluate the integral

$$\int_{-5}^{5} \frac{x^7 - x^5 + x}{x^4 + x^2 + 7}\,dx$$

and check your answer with a CAS.

43. Define $F(x)$ by

$$F(x) = \int_1^x (t^3 + 1)\,dt$$

(a) Use Part 2 of the Fundamental Theorem of Calculus to find $F'(x)$.

(b) Check the result in part (a) by first integrating and then differentiating.

44. Define $F(x)$ by

$$F(x) = \int_{\pi/4}^{x} \cos 2t\,dt$$

(a) Use Part 2 of the Fundamental Theorem of Calculus to find $F'(x)$.

(b) Check the result in part (a) by first integrating and then differentiating.

In Exercises 45–48, use Part 2 of the Fundamental Theorem of Calculus to find the derivative.

45. (a) $\frac{d}{dx}\int_1^x \sin(\sqrt{t}\,)\,dt$ (b) $\frac{d}{dx}\int_0^x e^{t^2}\,dt$

46. (a) $\frac{d}{dx}\int_0^x \frac{dt}{1 + \sqrt{t}}$ (b) $\frac{d}{dx}\int_1^x \ln t\,dt$

47. $\frac{d}{dx}\int_x^0 \frac{t}{\cos t}\,dt$ [*Hint:* Use Definition 7.5.3(b).]

48. $\frac{d}{du}\int_0^u |x|\,dx$

49. Let $F(x) = \int_2^x \sqrt{3t^2 + 1}\,dt$. Find

(a) $F(2)$ (b) $F'(2)$ (c) $F''(2)$

50. Let $F(x) = \int_0^x \frac{\cos t}{t^2 + 3}\,dt$. Find

(a) $F(0)$ (b) $F'(0)$ (c) $F''(0)$

51. Let $F(x) = \int_0^x \frac{t - 3}{t^2 + 7}\,dt$ for $-\infty < x < +\infty$.

(a) Find the value of x where F attains its minimum value.

(b) Find intervals over which F is only increasing or only decreasing.
(c) Find open intervals over which F is only concave up or only concave down.

C **52.** Use the plotting and numerical integration commands of a CAS to generate the graph of the function F in Exercise 51 over the interval $-20 \le x \le 20$, and confirm that the graph is consistent with the results obtained in that exercise.

53. (a) Over what open interval does the formula

$$F(x) = \int_1^x \frac{dt}{t}$$

represent an antiderivative of $f(x) = 1/x$?
(b) Find a point where the graph of F crosses the x-axis.

54. (a) Over what open interval does the formula

$$F(x) = \int_1^x \frac{1}{t^2-9}\,dt$$

represent an antiderivative of

$$f(x) = \frac{1}{x^2-9}?$$

(b) Find a point where the graph of F crosses the x-axis.

In Exercises 55 and 56, find all values of x^* in the stated interval that satisfy Equation (7) in the Mean-Value Theorem for Integrals (7.6.2), and explain what these numbers represent.

55. (a) $f(x) = \sqrt{x}$; $[0, 9]$ (b) $f(x) = 1/x$; $[1, e]$

56. (a) $f(x) = \sin x$; $[-\pi, \pi]$ (b) $f(x) = 1/x^2$; $[1, 3]$

It was shown in the proof of the Mean-Value Theorem for Integrals that if f is continuous on $[a, b]$, and if $m \le f(x) \le M$ on $[a, b]$, then

$$m(b-a) \le \int_a^b f(x)\,dx \le M(b-a)$$

[see (8)]. These inequalities make it possible to obtain bounds on the size of a definite integral from bounds on the size of its integrand. This is illustrated in Exercises 57–59.

57. Find the maximum and minimum values of $\sqrt{x^3+2}$ for $0 \le x \le 3$, and use these values to find bounds on the value of the integral

$$\int_0^3 \sqrt{x^3+2}\,dx$$

58. Find values of m and M such that $m \le x\sin x \le M$ for $0 \le x \le \pi$, and use these values to find bounds on the value of the integral

$$\int_0^\pi x\sin x\,dx$$

59. Show that

$$0 \le \int_1^5 \ln x\,dx \le 4\ln 5$$

60. Prove:
(a) $[cF(x)]_a^b = c[F(x)]_a^b$
(b) $[F(x)+G(x)]_a^b = F(x)]_a^b + G(x)]_a^b$
(c) $[F(x)-G(x)]_a^b = F(x)]_a^b - G(x)]_a^b$.

7.7 RECTILINEAR MOTION REVISITED; AVERAGE VALUE

In Section 6.3 we used the derivative to define the notions of instantaneous velocity and acceleration for a particle moving along a line. In this section we will resume the study of such motion using the tools of integration. We will also investigate the general problem of integrating a rate of change, and we will show how the definite integral can be used to define the average value of a continuous function. More applications of integration will be given in Chapter 8.

FINDING POSITION AND VELOCITY BY INTEGRATION

Recall from Definitions 6.3.1 and 6.3.2 that if $s(t)$ is the position function of a particle moving on a coordinate line, then the instantaneous velocity and acceleration of the particle are given by the formulas

$$v(t) = s'(t) = \frac{ds}{dt} \quad \text{and} \quad a(t) = v'(t) = \frac{dv}{dt} = \frac{d^2s}{dt^2}$$

It follows from these formulas that $s(t)$ is an antiderivative of $v(t)$ and $v(t)$ is an antideriva-

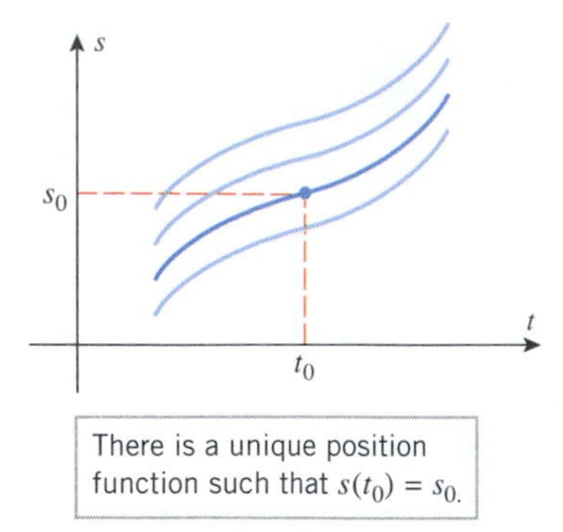

Figure 7.7.1

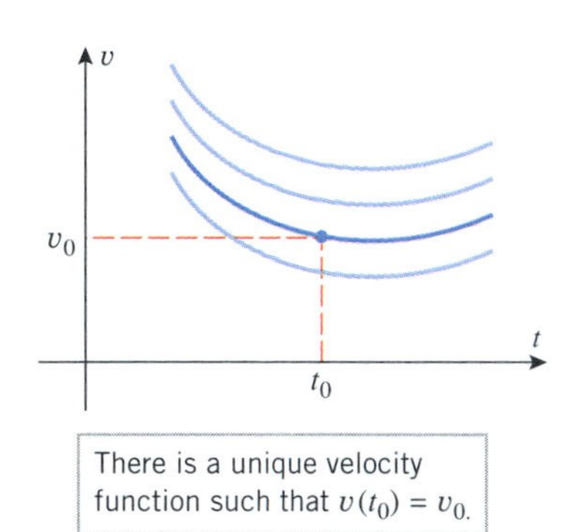

Figure 7.7.2

tive of $a(t)$; that is,

$$s(t) = \int v(t)\,dt \quad \text{and} \quad v(t) = \int a(t)\,dt \tag{1–2}$$

Thus, if the velocity of a particle is known, then its position function can be obtained from (1) by integration, provided there is sufficient additional information to determine the constant of integration. In particular, we can determine the constant of integration if we know the position s_0 of the particle at some time t_0, since this information determines a unique antiderivative $s(t)$ (Figure 7.7.1). Similarly, if the acceleration function of the particle is known, then its velocity function can be obtained from (2) by integration if we know the velocity v_0 of the particle at some time t_0 (Figure 7.7.2).

Example 1

Find the position function of a particle that is moving with velocity $v(t) = \cos \pi t$ along a coordinate line, assuming that the particle has coordinates $s = 4$ at time $t = 0$.

Solution. The position function is

$$s(t) = \int v(t)\,dt = \int \cos \pi t\,dt = \frac{1}{\pi}\sin \pi t + C$$

Since $s = 4$ when $t = 0$, it follows that

$$4 = s(0) = \frac{1}{\pi}\sin 0 + C = C$$

Thus,

$$s(t) = \frac{1}{\pi}\sin \pi t + 4$$ ◀

UNIFORMLY ACCELERATED MOTION

One of the most important cases of rectilinear motion occurs when a particle has constant acceleration. We call this ***uniformly accelerated motion***.

We will show that if a particle moves with constant acceleration along an s-axis, and if the position and velocity of the particle are known at some point in time, say when $t = 0$, then it is possible to derive formulas for the position $s(t)$ and the velocity $v(t)$ at any time t. To see how this can be done, suppose that the particle has constant acceleration

$$a(t) = a \tag{3}$$

and

$$s = s_0 \quad \text{when} \quad t = 0 \tag{4}$$

$$v = v_0 \quad \text{when} \quad t = 0 \tag{5}$$

where s_0 and v_0 are known. We call (4) and (5) the ***initial conditions*** for the motion.

With (3) as a starting point, we can integrate $a(t)$ to obtain $v(t)$, and we can integrate $v(t)$ to obtain $s(t)$, using an initial condition in each case to determine the constant of integration. The computations are as follows:

$$v(t) = \int a(t)\,dt = \int a\,dt = at + C_1 \tag{6}$$

To determine the constant of integration C_1 we apply initial condition (5) to this equation to obtain

$$v_0 = v(0) = a \cdot 0 + C_1 = C_1$$

Substituting this in (6) and putting the constant term first yields

$$v(t) = v_0 + at$$

Since v_0 is constant, it follows that

$$s(t) = \int v(t)\,dt = \int (v_0 + at)\,dt = v_0 t + \tfrac{1}{2}at^2 + C_2 \tag{7}$$

To determine the constant C_2 we apply initial condition (4) to this equation to obtain

$$s_0 = s(0) = v_0 \cdot 0 + \tfrac{1}{2}a \cdot 0 + C_2 = C_2$$

Substituting this in (7) and putting the constant term first yields

$$s(t) = s_0 + v_0 t + \tfrac{1}{2}at^2$$

In summary, we have the following result.

7.7.1 UNIFORMLY ACCELERATED MOTION. *If a particle moves with constant acceleration a along an s-axis, and if the position and velocity at time $t = 0$ are s_0 and v_0, respectively, then the position and velocity functions of the particle are*

$$s(t) = s_0 + v_0 t + \tfrac{1}{2}at^2 \tag{8}$$

$$v(t) = v_0 + at \tag{9}$$

FOR THE READER. How can you tell from the velocity versus time curve whether a particle moving along a line has uniformly accelerated motion?

Example 2

Suppose that an intergalactic spacecraft uses a sail and the "solar wind" to produce a constant acceleration of 0.032 m/s^2. Assuming that the spacecraft has a velocity of 10,000 m/s when the sail is first raised, how far will the spacecraft travel in 1 hour, and what will its velocity be at that time?

Solution. In this problem the choice of a coordinate axis is at our discretion, so we will choose it to make the computations as simple as possible. Accordingly, let us introduce an s-axis whose positive direction is in the direction of motion, and let us take the origin to coincide with the position of the spacecraft at the time $t = 0$ when the sail is raised. Thus, the Formulas (8) and (9) for uniformly accelerated motion apply with

$$s_0 = s(0) = 0, \quad v_0 = v(0) = 10{,}000, \quad \text{and} \quad a = 0.032$$

Since 1 hour corresponds to $t = 3600$ s, it follows from (8) that in 1 hour the spacecraft travels a distance of

$$s(3600) = 10{,}000(3600) + \tfrac{1}{2}(0.032)(3600)^2 \approx 36{,}207{,}400 \text{ m}$$

and it follows from (9) that after 1 hour its velocity is

$$v(3600) = 10{,}000 + (0.032)(3600) \approx 10{,}115 \text{ m/s}$$ ◀

Example 3

A bus has stopped to pick up riders, and a woman is running at a constant velocity of 5 m/s to catch it. When she is 11 m behind the front door the bus pulls away with a constant acceleration of 1 m/s^2. From that point in time, how long will it take for the woman to reach the front door of the bus if she keeps running with a velocity of 5 m/s?

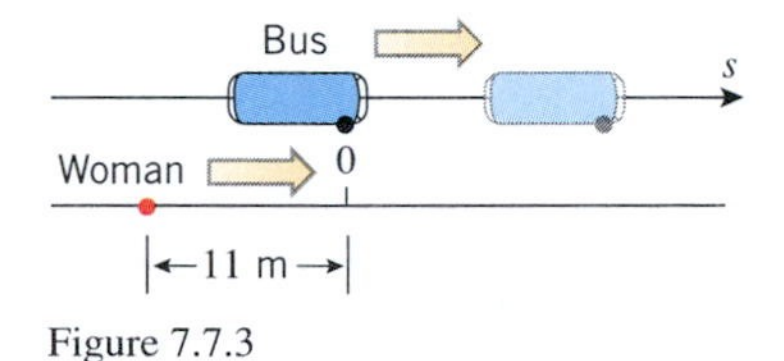

Figure 7.7.3

Solution. As shown in Figure 7.7.3, choose the s-axis so that the bus and the woman are moving in the positive direction, and the front door of the bus is at the origin at the time $t = 0$ when the bus begins to pull away. To catch the bus at some later time t, the woman will have to cover a distance $s_w(t)$ that is equal to 11 m plus the distance $s_b(t)$ traveled by the bus; that is, the woman will catch the bus when

$$s_w(t) = s_b(t) + 11 \tag{10}$$

Since the woman has a constant velocity of 5 m/s, the distance she travels in t seconds is $s_w(t) = 5t$. Thus, (10) can be written as

$$s_b(t) = 5t - 11 \tag{11}$$

Since the bus has a constant acceleration of $a = 1\ \text{m/s}^2$, and since $s_0 = v_0 = 0$ at time $t = 0$ (why?), it follows from (8) that

$$s_b(t) = \tfrac{1}{2}t^2$$

Substituting this equation into (11) and reorganizing the terms yields the quadratic equation

$$\tfrac{1}{2}t^2 - 5t + 11 = 0 \quad \text{or} \quad t^2 - 10t + 22 = 0$$

Solving this equation for t using the quadratic formula yields two solutions:

$$t = 5 - \sqrt{3} \approx 3.3 \quad \text{and} \quad t = 5 + \sqrt{3} \approx 6.7$$

(verify). Thus, the woman can reach the door at two different times, $t = 3.3$ s and $t = 6.7$ s. The reason that there are two solutions can be explained as follows: When the woman first reaches the door, she is running faster than the bus and can run past it if the driver does not see her. However, as the bus speeds up, it eventually catches up to her, and she has another chance to flag it down. ◀

THE FREE-FALL MODEL

In Section 6.3 we discussed the free-fall model of motion near the surface of the Earth with the promise that we would derive Formula (5) of that section later in the text; we will now show how to do this. As stated in 6.3.4 and illustrated in Figure 6.3.7, we will assume that the object moves on an s-axis whose origin is at the surface of the Earth and whose positive direction is up; and we will assume that the position and velocity of the object at time $t = 0$ are s_0 and v_0, respectively.

It is a fact of physics that a particle moving on a vertical line near the Earth's surface and subject only to the force of the Earth's gravity moves with constant acceleration. The magnitude of this constant, denoted by the letter g, is approximately $9.8\ \text{m/s}^2$ or $32\ \text{ft/s}^2$, depending on whether distance is measured in meters or feet.*

Recall that a particle is speeding up when its velocity and acceleration have the same sign and is slowing down when they have opposite signs. Thus, because we have chosen the positive direction to be up, it follows that the acceleration $a(t)$ of a particle in free fall is negative for all values of t. To see that this is so, observe that an upward-moving particle (positive velocity) is slowing down, so its acceleration must be negative; and a downward-moving particle (negative velocity) is speeding up, so its acceleration must also be negative. Thus, we conclude that

$$a(t) = -g$$

and hence it follows from (8) and (9) that the position and velocity functions of an object in free fall are

$$s(t) = s_0 + v_0 t - \tfrac{1}{2}gt^2 \tag{12}$$

$$v(t) = v_0 - gt \tag{13}$$

FOR THE READER. Had we chosen the positive direction of the s-axis to be down, then the acceleration would have been $a(t) = g$ (why?). How would this have affected Formulas (12) and (13)?

Example 4

A ball is thrown directly upward with an initial velocity of 49 m/s and is released from a point that is 8 m above the ground. Assuming that the free-fall model applies, how high will the ball travel?

* Strictly speaking, the constant g varies with the latitude and the distance from the Earth's center. However, for motion at a fixed latitude and near the surface of the Earth, the assumption of a constant g is satisfactory for many applications.

Solution. Since distance is in meters, we take $g = 9.8$ m/s^2. Initially, we have $s_0 = 8$ and $v_0 = 49$, so from (12) and (13)

$$v(t) = -9.8t + 49$$
$$s(t) = -4.9t^2 + 49t + 8$$

The ball will rise until $v(t) = 0$, that is, until $-9.8t + 49 = 0$ or $t = 5$. At this instant the height above the ground will be

$$s(5) = -4.9(5)^2 + 49(5) + 8 = 130.5 \text{ m}$$ ◀

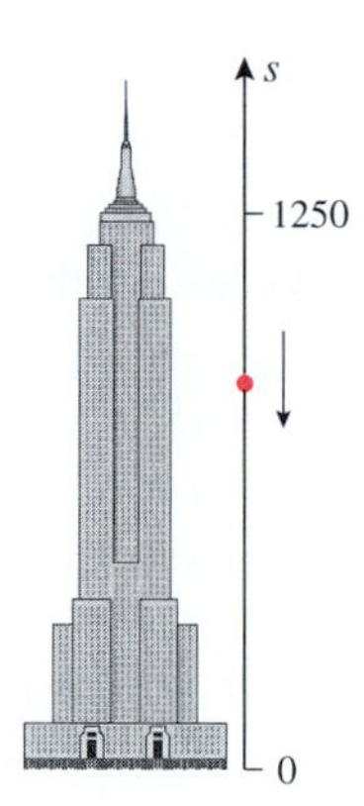

Figure 7.7.4

Example 5

A penny is released from rest near the top of the Empire State Building at a point that is 1250 ft above the ground (Figure 7.7.4). Assuming that the free-fall model applies, how long does it take for the penny to hit the ground, and what is its speed at the time of impact?

Solution. Since distance is in feet, we take $g = 32$ ft/s^2. Initially, we have $s_0 = 1250$ and $v_0 = 0$, so from (12)

$$s(t) = -16t^2 + 1250 \tag{14}$$

Impact occurs when $s(t) = 0$. Solving this equation for t, we obtain

$$-16t^2 + 1250 = 0$$
$$t^2 = \frac{1250}{16} = \frac{625}{8}$$
$$t = \pm\frac{25}{\sqrt{8}} \approx \pm 8.8 \text{ s}$$

Since $t \geq 0$, we can discard the negative solution and conclude that it takes $25/\sqrt{8} \approx 8.8$ s for the penny to hit the ground. To obtain the velocity at the time of impact, we substitute $t = 25/\sqrt{8}$, $v_0 = 0$, and $g = 32$ in (13) to obtain

$$v\left(\frac{25}{\sqrt{8}}\right) = 0 - 32\left(\frac{25}{\sqrt{8}}\right) = -200\sqrt{2} \approx -282.8 \text{ ft/s}$$

Thus, the speed at the time of impact is

$$\left|v\left(\frac{25}{\sqrt{8}}\right)\right| = 200\sqrt{2} \approx 282.8 \text{ ft/s}$$

which is more than 192 mi/h. ◀

INTEGRATING RATES OF CHANGE

The Fundamental Theorem of Calculus

$$\int_a^b f(x)\,dx = F(b) - F(a) \tag{15}$$

has a useful interpretation that can be seen by rewriting it in a slightly different form. Since F is an antiderivative of f on the interval $[a, b]$, we can use the relationship $F'(x) = f(x)$ to rewrite (15) as

$$\int_a^b F'(x)\,dx = F(b) - F(a) \tag{16}$$

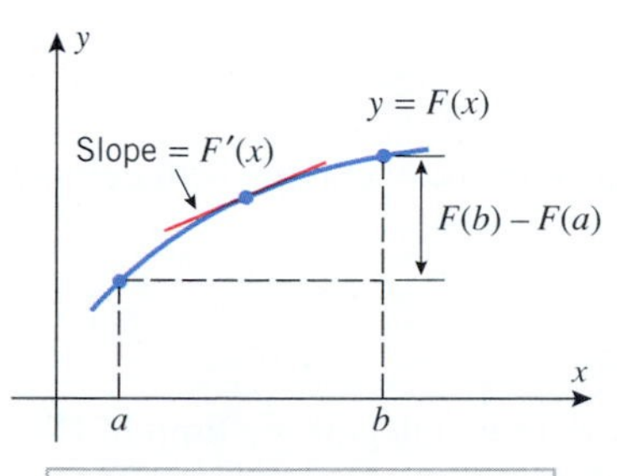

Integrating the slope of $y = F(x)$ over the interval $[a, b]$ produces the change $F(b) - F(a)$ in the value of $F(x)$.

Figure 7.7.5

In this formula we can view $F'(x)$ as the rate of change of $F(x)$ with respect to x, and we can view $F(b) - F(a)$ as the *change* in the value of $F(x)$ as x increases from a to b (Figure 7.7.5). Thus, we have the following useful principle.

7.7.2 INTEGRATING A RATE OF CHANGE. Integrating the rate of change of $F(x)$ with respect to x over an interval $[a, b]$ produces the change in the value of $F(x)$ that occurs as x increases from a to b.

Here are some examples of this idea:

- If $P(t)$ is a population (e.g., plants, animals, or people) at time t, then $P'(t)$ is the rate at which the population is changing at time t, and

$$\int_{t_1}^{t_2} P'(t)\,dt = P(t_2) - P(t_1)$$

is the change in the population between times t_1 and t_2.

- If $A(t)$ is the area of an oil spill at time t, then $A'(t)$ is the rate at which the area of the spill is changing at time t, and

$$\int_{t_1}^{t_2} A'(t)\,dt = A(t_2) - A(t_1)$$

is the change in the area of the spill between times t_1 and t_2.

- If $P'(x)$ is the marginal profit that results from producing and selling x units of a product (see Section 6.2), then

$$\int_{x_1}^{x_2} P'(x)\,dx = P(x_2) - P(x_1)$$

is the change in the profit that results when the production level increases from x_1 units to x_2 units.

DISPLACEMENT IN RECTILINEAR MOTION

As another application of (16), suppose that $s(t)$ and $v(t)$ are the position and velocity functions of a particle moving on a coordinate line. Since $v(t)$ is the rate of change of $s(t)$ with respect to t, it follows from the principle in 7.7.2 that integrating $v(t)$ over an interval $[t_0, t_1]$ will produce the change in the value of $s(t)$ as t increases from t_0 to t_1; that is,

$$\int_{t_0}^{t_1} v(t)\,dt = \int_{t_0}^{t_1} s'(t)\,dt = s(t_1) - s(t_0) \tag{17}$$

The expression $s(t_1) - s(t_0)$ in this formula is called the ***displacement*** or ***change in position*** of the particle over the time interval $[t_0, t_1]$. For a particle moving horizontally, the displacement is positive if the final position of the particle is to the right of its initial position, negative if it is to the left of its initial position, and zero if it coincides with the initial position (Figure 7.7.6).

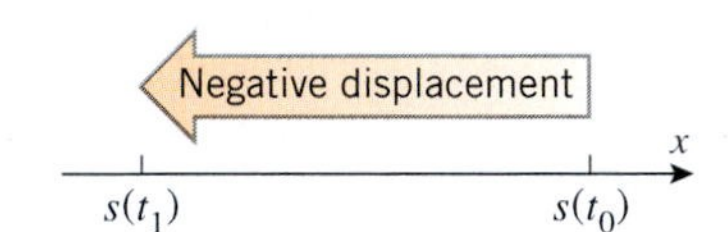

Figure 7.7.6

REMARK. In physical problems it is important to associate the correct units with definite integrals. In general, the units for the definite integral

$$\int_a^b f(x)\,dx$$

will be units of $f(x)$ times units of x. This is because the definite integral is a limit of Riemann sums each of whose terms is a product of the form $f(x) \cdot \Delta x$. For example, if time is measured in seconds (s) and velocity is measured in meters per second (m/s), then integrating velocity over a time interval will produce a result whose units are in meters, since $\text{m/s} \times \text{s} = \text{m}$. Note that this is consistent with Formula (17), since displacement has units of length.

DISTANCE TRAVELED IN RECTILINEAR MOTION

In general, the displacement of a particle is not the same as the distance traveled by the particle. For example, a particle that travels 100 units in the positive direction and then 100 units in the negative direction travels a distance of 200 units but has a displacement of zero, since it returns to its starting point. The only case in which the displacement and the distance traveled are the same occurs when the particle moves in the positive direction without reversing the direction of its motion.

FOR THE READER. What is the relationship between the displacement of a particle and the distance it travels if the particle moves in the negative direction without reversing the direction of motion?

From (17), integrating the velocity function of a particle over a time interval yields the displacement of a particle over that time interval. In contrast, to find the *total distance* traveled by the particle over the time interval (the distance traveled in the positive direction plus the distance traveled in the negative direction), we must integrate the *absolute value* of the velocity function; that is, we must integrate the speed:

$$\begin{bmatrix} \text{total distance} \\ \text{traveled during} \\ \text{time interval} \\ [t_0, t_1] \end{bmatrix} = \int_{t_0}^{t_1} |v(t)|\, dt \tag{18}$$

Example 6

A particle moves on a coordinate line so that its velocity at time t is $v(t) = t^2 - 2t$ m/s.

(a) Find the displacement of the particle during the time interval $0 \le t \le 3$.

(b) Find the distance traveled by the particle during the time interval $0 \le t \le 3$.

Solution (a). From (17) the displacement is

$$\int_0^3 v(t)\, dt = \int_0^3 (t^2 - 2t)\, dt = \left[\frac{t^3}{3} - t^2\right]_0^3 = 0$$

Thus, the particle is at the same position at time $t = 3$ as at $t = 0$.

Solution (b). The velocity can be written as $v(t) = t^2 - 2t = t(t-2)$, from which we see that $v(t) \le 0$ for $0 \le t \le 2$ and $v(t) \ge 0$ for $2 \le t \le 3$. Thus, it follows from (18) that the distance traveled is

$$\begin{aligned}
\int_0^3 |v(t)|\, dt &= \int_0^2 -v(t)\, dt + \int_2^3 v(t)\, dt \\
&= \int_0^2 -(t^2 - 2t)\, dt + \int_2^3 (t^2 - 2t)\, dt \\
&= -\left[\frac{t^3}{3} - t^2\right]_0^2 + \left[\frac{t^3}{3} - t^2\right]_2^3 = \frac{4}{3} + \frac{4}{3} = \frac{8}{3} \text{ m}
\end{aligned}$$

◀

ANALYZING THE VELOCITY VERSUS TIME CURVE

In Section 6.3 we showed how to use the position versus time curve to obtain information about the behavior of a particle moving on a coordinate line (Table 6.3.1). Similarly, there is valuable information that can be obtained from the *velocity versus time curve*. For example, the integral in (17) can be interpreted geometrically as the net signed area between the graph of $v(t)$ and the interval $[t_0, t_1]$, and it can be interpreted physically as the displacement of the particle over this interval. Thus, we have the following result.

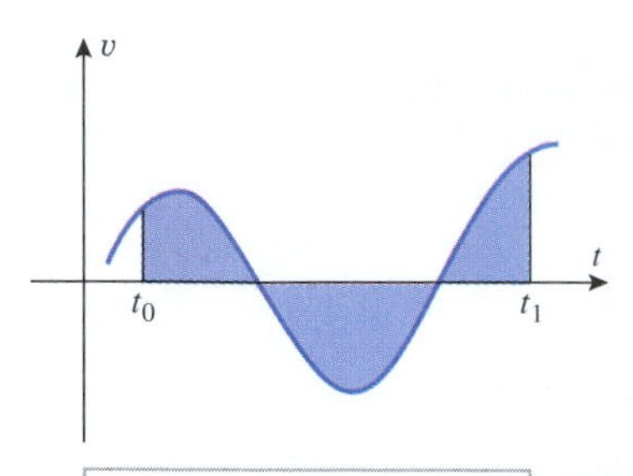

The net signed area is the displacement of the particle during the interval $[t_0, t_1]$.

Figure 7.7.7

7.7.3 FINDING DISPLACEMENT FROM THE VELOCITY VERSUS TIME CURVE. For a particle in rectilinear motion, the net signed area between the velocity versus time curve and an interval $[t_0, t_1]$ on the t-axis represents the displacement of the particle over that time interval (Figure 7.7.7).

Example 7

Figure 7.7.8 shows three velocity versus time curves for a particle in rectilinear motion along a horizontal line. In each case, find the displacement of the particle over the time interval $0 \leq t \leq 4$, and explain what it tells you about the motion of the particle.

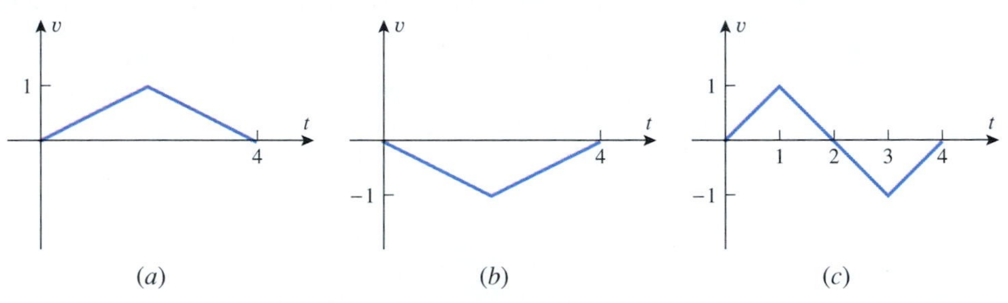

Figure 7.7.8

Solution. In part (a) of Figure 7.7.8 the net signed area under the curve is 2, so the particle is 2 units to the right of its starting point at the end of the time period. In part (b) the net signed area under the curve is -2, so the particle is 2 units to the left of its starting point at the end of the time period. In part (c) the net signed area under the curve is 0, so the particle is back at its starting point at the end of the time period. ◀

Sometimes we will not want the net signed area between a curve $y = f(x)$ and an interval $[a, b]$, but rather the total area between the curve and the interval. This can be found by integrating $|f(x)|$ rather than $f(x)$ over the interval $[a, b]$.

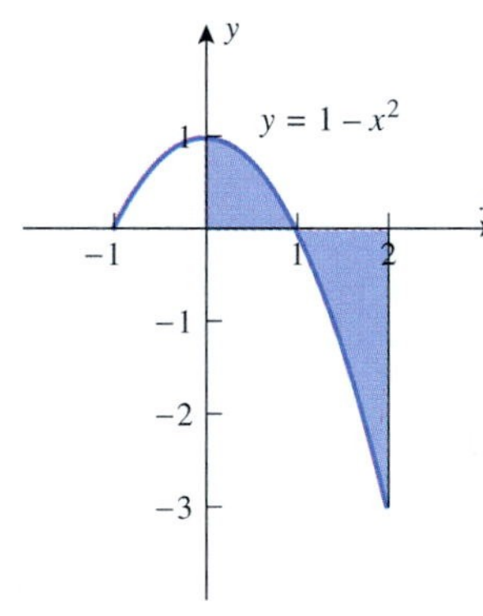

Figure 7.7.9

Example 8

Find the total area between the curve $y = 1 - x^2$ and the x-axis over the interval $[0, 2]$ (Figure 7.7.9).

Solution. The area A is given by

$$A = \int_0^2 |1 - x^2|\, dx = \int_0^1 (1 - x^2)\, dx + \int_1^2 -(1 - x^2)\, dx$$

$$= \left[x - \frac{x^3}{3}\right]_0^1 - \left[x - \frac{x^3}{3}\right]_1^2$$

$$= \frac{2}{3} - \left(-\frac{4}{3}\right) = 2$$ ◀

From (18), integrating the speed $|v(t)|$ over a time interval $[t_0, t_1]$ produces the distance traveled by the particle during the time interval. However, we can also interpret the integral in (18) as the total area between the velocity versus time curve and the interval $[t_0, t_1]$ on the t-axis. Thus, we have the following result.

7.7.4 FINDING DISTANCE TRAVELED FROM THE VELOCITY VERSUS TIME CURVE.. For a particle in rectilinear motion, the total area between the velocity versus time curve and an interval $[t_0, t_1]$ on the t-axis represents the distance traveled by the particle over that time interval.

Example 9

For each of the velocity versus time curves in Figure 7.7.8 find the total distance traveled by the particle over the time interval $0 \leq t \leq 4$.

Solution. In all three parts of Figure 7.7.8 the total area between the curve and the interval $[0, 4]$ is 2, so the particle travels a distance of 2 units during the time period in all three cases, even though the displacement is different in each case, as discussed in Example 7. ◀

AVERAGE VALUE OF A CONTINUOUS FUNCTION

In scientific work, numerical information is often summarized by computing some sort of *average* or *mean* value of the observed data. There are various kinds of averages, but the most common is the ***arithmetic mean*** or ***arithmetic average***, which is formed by adding the data and dividing by the number of data points. Thus, the arithmetic average $\overline{a}$ of n numbers $a_1, a_2, \ldots, a_n$ is

$$\overline{a} = \frac{1}{n}(a_1 + a_2 + \cdots + a_n) = \frac{1}{n}\sum_{k=1}^{n} a_k$$

In the case where the a_k's are values of a function f, say,

$$a_1 = f(x_1), a_2 = f(x_2), \ldots, a_n = f(x_n)$$

then the arithmetic average $\overline{a}$ of these function values is

$$\overline{a} = \frac{1}{n}\sum_{k=1}^{n} f(x_k)$$

We will now show how to extend this concept so that we can compute not only the arithmetic average of finitely many function values but an average of *all* values of $f(x)$ as x varies over a closed interval $[a, b]$. For this purpose recall the Mean-Value Theorem for Integrals (7.6.2), which states that if f is continuous on the interval $[a, b]$, then there is at least one point x^* in this interval such that

$$\int_a^b f(x)\,dx = f(x^*)(b - a)$$

The quantity

$$f(x^*) = \frac{1}{b-a}\int_a^b f(x)\,dx \tag{19}$$

will be our candidate for the average value of f over the interval $[a, b]$. To explain what motivates this, divide the interval $[a, b]$ into n subintervals of equal length

$$\Delta x = \frac{b-a}{n} \tag{20}$$

and choose arbitrary points $x_1^*, x_2^*, \ldots, x_n^*$ in successive subintervals. Then the arithmetic average of the numbers $f(x_1^*), f(x_2^*), \ldots, f(x_n^*)$ is

$$\text{ave} = \frac{1}{n}[f(x_1^*) + f(x_2^*) + \cdots + f(x_n^*)]$$

or from (20)

$$\text{ave} = \frac{1}{b-a}[f(x_1^*)\Delta x + f(x_2^*)\Delta x + \cdots + f(x_n^*)\Delta x] = \frac{1}{b-a}\sum_{k=1}^{n} f(x_k^*)\Delta x$$

Taking the limit as $n \to +\infty$ yields

$$\lim_{n\to+\infty} \frac{1}{b-a}\sum_{k=1}^{n} f(x_k^*)\Delta x = \frac{1}{b-a}\int_a^b f(x)\,dx$$

Since this equation describes what happens when we compute the average of "more and more" values of $f(x)$, we are led to the following definition.

7.7.5 DEFINITION. If f is continuous on $[a, b]$, then the ***average value*** (or ***mean value***) of f on $[a, b]$ is defined to be

$$f_{\text{ave}} = \frac{1}{b-a} \int_a^b f(x)\,dx \tag{21}$$

REMARK. When f is nonnegative on $[a, b]$, the quantity f_{ave} has a simple geometric interpretation, which can be seen by writing (21) as

$$f_{\text{ave}} \cdot (b-a) = \int_a^b f(x)\,dx$$

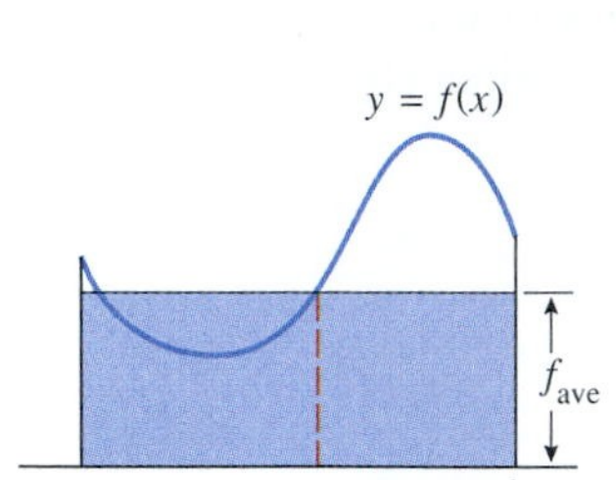

Figure 7.7.10

The left side of this equation is the area of a rectangle with a height of f_{ave} and base of length $b - a$, and the right side is the area under $y = f(x)$ over $[a, b]$. Thus, f_{ave} is the height of a rectangle constructed over the interval $[a, b]$, whose area is the same as the area under the graph of f over that interval (Figure 7.7.10). Note also that the Mean-Value Theorem, when expressed in form (21), ensures that there is always at least one point x^* in $[a, b]$ at which the value of f is equal to the average value of f over the interval.

Example 10

Find the average value of the function $f(x) = \sqrt{x}$ over the interval $[1, 4]$, and find all points in the interval at which the value of f is the same as the average.

Solution.

$$f_{\text{ave}} = \frac{1}{b-a} \int_a^b f(x)\,dx = \frac{1}{4-1} \int_1^4 \sqrt{x}\,dx = \frac{1}{3}\left[\frac{2x^{3/2}}{3}\right]_1^4$$

$$= \frac{1}{3}\left[\frac{16}{3} - \frac{2}{3}\right] = \frac{14}{9} \approx 1.6$$

The x-values at which $f(x) = \sqrt{x}$ is the same as the average satisfy $\sqrt{x} = 14/9$, from which we obtain $x = 196/81 \approx 2.4$ (Figure 7.7.11). ◀

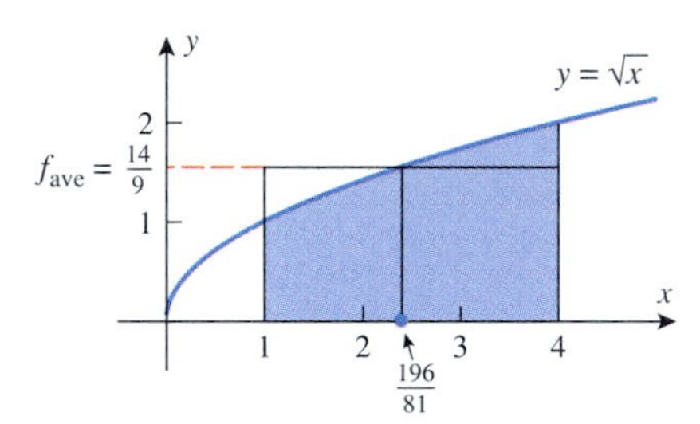

Figure 7.7.11

AVERAGE VELOCITY REVISITED

In Section 3.1 we considered the motion of a particle moving in the *positive direction* along a coordinate line, and we motivated the concept of instantaneous velocity in that special case by viewing it as the limit of average velocities over smaller and smaller time intervals. That discussion led us to conclude that the average velocity of the particle over a time interval could be interpreted as the slope of a secant line and the instantaneous velocity as the slope of a tangent line to the position versus time curve (Figure 3.1.5). We will now show that the same results are true in the more general case where the particle can move in either direction along the coordinate line.

For this purpose, suppose that $s(t)$ and $v(t)$ are the position and velocity functions of such a particle, and let us use Formula (21) to calculate the average velocity of the particle over a time interval $[t_0, t_1]$. This yields

$$v_{\text{ave}} = \frac{1}{t_1 - t_0}\int_{t_0}^{t_1} v(t)\,dt = \frac{1}{t_1 - t_0}\int_{t_0}^{t_1} s'(t)\,dt = \frac{s(t_1) - s(t_0)}{t_1 - t_0}$$

Thus, *the average velocity over a time interval is the displacement divided by the elapsed time*. Geometrically, this is the slope of the secant line shown in Figure 7.7.12. Moreover, if we allow t_1 to approach t_0, then the slopes of the secant lines approach the slope of the tangent line at t_0, which is the instantaneous velocity at that instant. Thus, the relationship between average and instantaneous velocity developed in Section 3.1 also applies to general rectilinear motion.

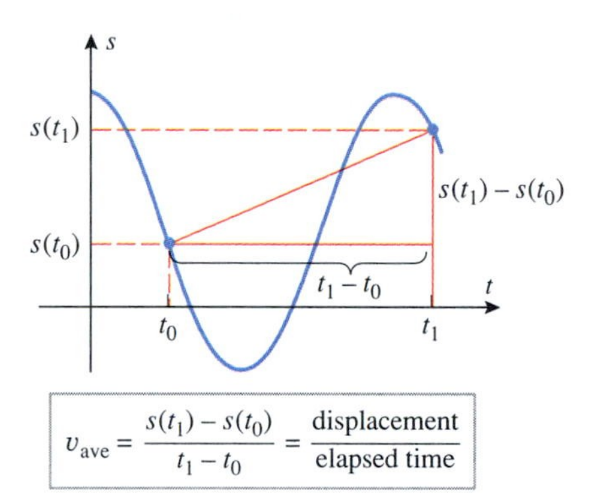

Figure 7.7.12

EXERCISE SET 7.7 Graphing Calculator C CAS

1. (a) If $h'(t)$ is the rate of change of a child's height measured in inches per year, what does the integral $\int_0^{10} h'(t)\,dt$ represent, and what are its units?

(b) If $r'(t)$ is the rate of change of the radius of a spherical balloon measured in centimeters per second, what does the integral $\int_1^2 r'(t)\,dt$ represent, and what are its units?

(c) If $H(t)$ is the rate of change of the speed of sound with respect to temperature measured in ft/s per °F, what does the integral $\int_{32}^{100} H(t)\,dt$ represent, and what are its units?

(d) If $v(t)$ is the velocity of a particle in rectilinear motion, measured in cm/h, what does the integral $\int_{t_1}^{t_2} v(t)\,dt$ represent, and what are its units?

2. (a) Suppose that sludge is emptied into a river at the rate of $V(t)$ gallons per minute, starting at time $t = 0$. Write an integral that represents the total volume of sludge that is emptied into the river during the first hour.

(b) Suppose that the tangent line to a curve $y = f(x)$ has slope $m(x)$ at the point x. What does the integral $\int_{x_1}^{x_2} m(x)\,dx$ represent?

3. In each part, the velocity versus time curve is given for a particle moving along a line. Use the curve to find the displacement and the distance traveled by the particle over the time interval $0 \le t \le 3$.

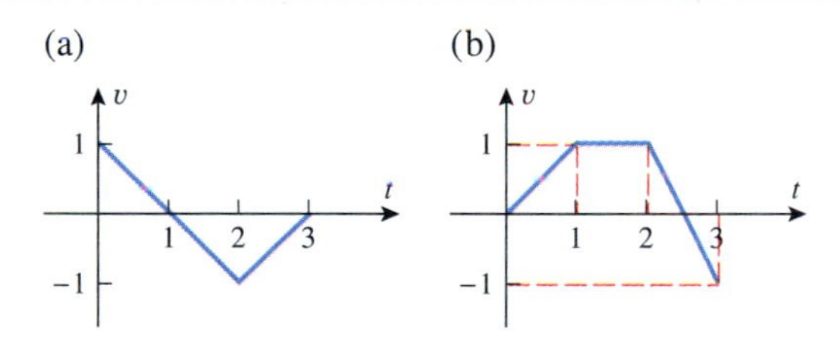

4. Sketch a velocity versus time curve for a particle that travels a distance of 5 units along a coordinate line during the time interval $0 \le t \le 10$ and has a displacement of 0 units.

5. The accompanying figure shows the acceleration versus time curve for a particle moving along a coordinate line. If the initial velocity of the particle is 20 m/s, estimate
(a) the velocity at time $t = 4$ s
(b) the velocity at time $t = 6$ s.

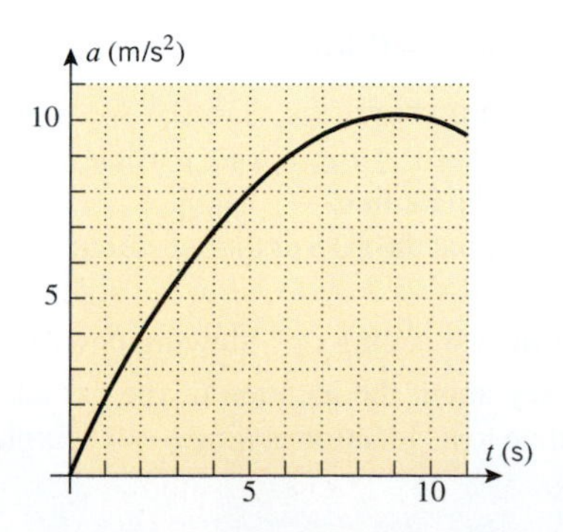

Figure Ex-5

6. Determine whether the particle in Exercise 5 is speeding up or slowing down at times $t = 4$ s and $t = 6$ s.

In Exercises 7–10, a particle moves along an s-axis. Use the given information to find the position function of the particle.

7. (a) $v(t) = t^3 - 2t^2 + 1;\ s(0) = 1$
   (b) $a(t) = 4\cos 2t;\ v(0) = -1;\ s(0) = -3$

8. (a) $v(t) = 1 + \sin t;\ s(0) = -3$
   (b) $a(t) = t^2 - 3t + 1;\ v(0) = 0;\ s(0) = 0$

9. (a) $v(t) = 2t - 3;\ s(1) = 5$
   (b) $a(t) = \cos t;\ v(\pi/2) = 2;\ s(\pi/2) = 0$

10. (a) $v(t) = t^{2/3};\ s(8) = 0$
    (b) $a(t) = \sqrt{t};\ v(4) = 1;\ s(4) = -5$

In Exercises 11–14, a particle moves with a velocity of $v(t)$ m/s along an s-axis. Find the displacement and the distance traveled by the particle during the given time interval.

11. (a) $v(t) = \sin t;\ 0 \le t \le \pi/2$
    (b) $v(t) = \cos t;\ \pi/2 \le t \le 2\pi$

12. (a) $v(t) = 2t - 4;\ 0 \le t \le 6$
    (b) $v(t) = |t - 3|;\ 0 \le t \le 5$

13. (a) $v(t) = t^3 - 3t^2 + 2t;\ 0 \le t \le 3$
    (b) $v(t) = e^t - 2;\ 0 \le t \le 3$

14. (a) $v(t) = \frac{1}{2} - 1/t;\ 1 \le t \le 3$
    (b) $v(t) = 3/\sqrt{t};\ 4 \le t \le 9$

In Exercises 15–18, a particle moves with acceleration $a(t)$ m/s^2 along an s-axis and has velocity v_0 m/s at time $t = 0$. Find the displacement and the distance traveled by the particle during the given time interval.

15. $a(t) = -2;\ v_0 = 3;\ 1 \le t \le 4$

16. $a(t) = t - 2;\ v_0 = 0;\ 1 \le t \le 5$

17. $a(t) = 1/\sqrt{5t+1};\ v_0 = 2;\ 0 \le t \le 3$

18. $a(t) = \sin t;\ v_0 = 1;\ \pi/4 \le t \le \pi/2$

19. In each part use the given information to find the position, velocity, speed, and acceleration at time $t = 1$.
    (a) $v = \sin \frac{1}{2}\pi t;\ s = 0$ when $t = 0$
    (b) $a = -3t;\ s = 1$ and $v = 0$ when $t = 0$

20. The accompanying figure shows the velocity versus time curve over the time interval $1 \le t \le 5$ for a particle moving along a horizontal coordinate line.
    (a) What can you say about the sign of the acceleration over the time interval?
    (b) When is the particle speeding up? Slowing down?
    (c) What can you say about the location of the particle at time $t = 5$ relative to its location at time $t = 1$? Explain your reasoning.

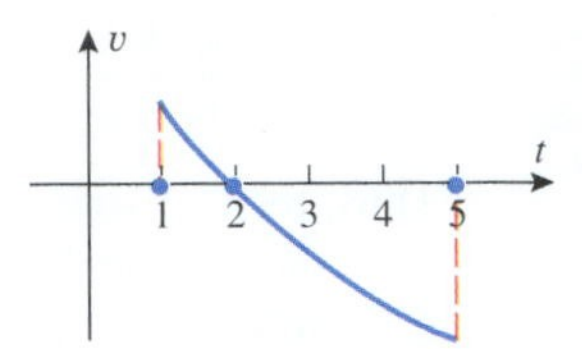

Figure Ex-20

In Exercises 21–24, sketch the curve and find the total area between the curve and the given interval on the x-axis.

21. $y = x^2 - 1;\ [0, 3]$

22. $y = \sin x;\ [0, 3\pi/2]$

23. $y = e^x - 1;\ [-1, 1]$

24. $y = \dfrac{x-1}{x};\ [\frac{1}{2}, 2]$

25. Suppose that the velocity function of a particle moving along an s-axis is $v(t) = 20t^2 - 100t + 50$ ft/s and that the particle is at the origin at time $t = 0$. Use a graphing utility to generate the graphs of $s(t)$, $v(t)$, and $a(t)$ for the first 6 s of motion.

26. Suppose that the acceleration function of a particle moving along an s-axis is $a(t) = 4t - 30$ m/s and that the position and velocity at time $t = 0$ are $s_0 = -5$ m and $v_0 = 3$ m/s. Use a graphing utility to generate the graphs of $s(t)$, $v(t)$, and $a(t)$ for the first 25 s of motion.

C 27. Let the velocity function for a particle that is at the origin initially and moves along an s-axis be $v(t) = 0.5 - te^{-t}$.
    (a) Generate the velocity versus time curve, and use it to make a conjecture about the sign of the displacement over the time interval $0 \le t \le 5$.
    (b) Use a CAS to find the displacement.

C 28. Let the velocity function for a particle that is at the origin initially and moves along an s-axis be $v(t) = t\ln(t + 0.1)$.
    (a) Generate the velocity versus time curve, and use it to make a conjecture about the sign of the displacement over the time interval $0 \le t \le 1$.
    (b) Use a CAS to find the displacement.

29. Suppose that at time $t = 0$ a particle is at the origin of an x-axis and has a velocity of $v_0 = 25$ cm/s. For the first 4 s thereafter it has no acceleration, and then it is acted on by a retarding force that produces a constant negative acceleration of $a = -10$ cm/s^2.
    (a) Sketch the acceleration versus time curve over the interval $0 \le t \le 12$.
    (b) Sketch the velocity versus time curve over the time interval $0 \le t \le 12$.
    (c) Find the x-coordinate of the particle at times $t = 8$ s and $t = 12$ s.
    (d) What is the maximum x-coordinate of the particle over the time interval $0 \le t \le 12$?

30. Formulas (8) and (9) for uniformly accelerated motion can be rearranged in various useful ways. For simplicity, let $s = s(t)$ and $v = v(t)$, and derive the following variations of those formulas.

(a) $a = \dfrac{v^2 - v_0^2}{2(s - s_0)}$ (b) $t = \dfrac{2(s - s_0)}{v_0 + v}$

(c) $s = s_0 + vt - \frac{1}{2}at^2$ [Note how this differs from (8).]

Exercises 31–38 involve uniformly accelerated motion. In these exercises assume that the object is moving in the positive direction of a coordinate line, and apply Formulas (8) and (9) or those from Exercise 30, as appropriate. In some of these problems you will need the fact that 88 ft/s = 60 mi/h.

31. (a) An automobile traveling on a straight road decelerates uniformly from 55 mi/h to 25 mi/h in 30 s. Find its acceleration in ft/s^2.

(b) A bicycle rider traveling on a straight path accelerates uniformly from rest to 30 km/h in 1 min. Find his acceleration in km/s^2.

32. A car traveling 60 mi/h along a straight road decelerates at a constant rate of 10 ft/s^2.

(a) How long will it take until the speed is 45 mi/h?

(b) How far will the car travel before coming to a stop?

33. Spotting a police car, you hit the brakes on your new Porsche to reduce your speed from 90 mi/h to 60 mi/h at a constant rate over a distance of 200 ft.

(a) Find the acceleration in ft/s^2.

(b) How long does it take for you to reduce your speed to 55 mi/h?

(c) At the acceleration obtained in part (a), how long would it take for you to bring your Porsche to a complete stop from 90 mi/h?

34. A particle moving along a straight line is accelerating at a constant rate of 3 m/s^2. Find the initial velocity if the particle moves 40 m in the first 4 s.

35. A motorcycle, starting from rest, speeds up with a constant acceleration of 2.6 m/s^2. After it has traveled 120 m, it slows down with a constant acceleration of -1.5 m/s^2 until it attains a speed of 12 m/s. What is the distance traveled by the motorcycle at that point?

36. A sprinter in a 100-m race explodes out of the starting block with an acceleration of 4.0 m/s^2, which she sustains for 2.0 s. Her acceleration then drops to zero for the rest of race.

(a) What is her time for the race?

(b) Make a graph of her distance from the starting block versus time.

37. A car that has stopped at a toll booth leaves the booth with a constant acceleration of 2 ft/s^2. At the time the car leaves the booth it is 5000 ft behind a truck traveling with a constant velocity of 50 ft/s. How long will it take for the car to catch the truck, and how far will the car be from the toll booth at that time?

38. In the final sprint of a rowing race the challenger is rowing at a constant speed of 12 m/s. At the point where the leader is 100 m from the finish line and the challenger is 15 m behind, the leader is rowing at 8 m/s but starts accelerating at a constant 0.5 m/s^2. Who wins?

In Exercises 39–48, assume that a free-fall model applies. Solve these exercises by applying Formulas (12) and (13) or, if appropriate, use those from Exercise 30 with $a = -g$. In these exercises take $g = 32\ \text{ft/s}^2$ or $g = 9.8\ \text{m/s}^2$, depending on the units.

39. A projectile is launched vertically upward from ground level with an initial velocity of 112 ft/s.

(a) Find the velocity at $t = 3$ s and $t = 5$ s.

(b) How high will the projectile rise?

(c) Find the speed of the projectile when it hits the ground.

40. A projectile fired downward from a height of 112 ft reaches the ground in 2 s. What is its initial velocity?

41. A projectile is fired vertically upward from ground level with an initial velocity of 16 ft/s.

(a) How long will it take for the projectile to hit the ground?

(b) How long will the projectile be moving upward?

42. A rock is dropped from the top of the Washington Monument, which is 555 ft high.

(a) How long will it take for the rock to hit the ground?

(b) What is the speed of the rock at impact?

43. A helicopter pilot drops a package when the helicopter is 200 ft above the ground and rising at a speed of 20 ft/s.

(a) How long will it take for the package to hit the ground?

(b) What will be its speed at impact?

44. A stone is thrown downward with an initial speed of 96 ft/s from a height of 112 ft.

(a) How long will it take for the stone to hit the ground?

(b) What will be its speed at impact?

45. A projectile is fired vertically upward with an initial velocity of 49 m/s from a tower 150 m high.

(a) How long will it take for the projectile to reach its maximum height?

(b) What is the maximum height?

(c) How long will it take for the projectile to pass its starting point on the way down?

(d) What is the velocity when it passes the starting point on the way down?

(e) How long will it take for the projectile to hit the ground?

(f) What will be its speed at impact?

46. A man drops a stone from a bridge. What is the height of the bridge if

(a) the stone hits the water 4 s later

(b) the sound of the splash reaches the man 4 s later? [Take 1080 ft/s as the speed of sound.]

47. In the final stages of a Moon landing, a lunar module fires its retrorockets and descends to a height of $h = 5$ m above the lunar surface (Figure Ex-47). At that point the retrorockets are cut off, and the module goes into free fall. Given that the Moon's gravity is 1/6 of the Earth's, find the speed of the module when it touches the lunar surface.

Figure Ex-47

48. Given that the Moon's gravity is 1/6 of the Earth's, how much faster would a projectile have to be launched upward from the surface of the Earth than from the surface of the Moon to reach a height of 1000 ft?

In Exercises 49–54, find the average value of the function over the given interval.

49. $f(x) = 3x;\ [1, 3]$

50. $f(x) = x^2;\ [-1, 2]$

51. $f(x) = \sin x;\ [0, \pi]$

52. $f(x) = \cos x;\ [0, \pi]$

53. $f(x) = 1/x;\ [1, e]$

54. $f(x) = e^x;\ [-1, \ln 5]$

55. (a) Find f_{ave} of $f(x) = x^2$ over $[0, 2]$.
(b) Find a point x^* in $[0, 2]$ such that $f(x^*) = f_{\text{ave}}$.
(c) Sketch the graph of $f(x) = x^2$ over $[0, 2]$ and construct a rectangle over the interval whose area is the same as the area under the graph of f over the interval.

56. (a) Find f_{ave} of $f(x) = 2x$ over $[0, 4]$.
(b) Find a point x^* in $[0, 4]$ such that $f(x^*) = f_{\text{ave}}$.
(c) Sketch the graph of $f(x) = 2x$ over $[0, 4]$ and construct a rectangle over the interval whose area is the same as the area under the graph of f over the interval.

57. (a) Suppose that the velocity function of a particle moving along a coordinate line is $v(t) = 3t^3 + 2$. Find the average velocity of the particle over the time interval $1 \le t \le 4$ by integrating.
(b) Suppose that the position function of a particle moving along a coordinate line is $s(t) = 6t^2 + t$. Find the average velocity of the particle over the time interval $1 \le t \le 4$ algebraically.

58. (a) Suppose that the acceleration function of a particle moving along a coordinate line is $a(t) = t + 1$. Find the average acceleration of the particle over the time interval $0 \le t \le 5$ by integrating.
(b) Suppose that the velocity function of a particle moving along a coordinate line is $v(t) = \cos t$. Find the average acceleration of the particle over the time interval $0 \le t \le \pi/4$ algebraically.

59. Water is run at a constant rate of 1 ft^3/min to fill a cylindrical tank of radius 3 ft and height 5 ft. Assuming that the tank is empty initially, make a conjecture about the average weight of the water in the tank over the time period required to fill it, and then check your conjecture by integrating. [Take the weight density of water to be 62.4 lb/ft^3.]

60. (a) The temperature of a 10-m-long metal bar is 15°C at one end and 30°C at the other end. Assuming that the temperature increases linearly from the cooler end to the hotter end, what is the average temperature of the bar?
(b) Explain why there must be a point on the bar where the temperature is the same as the average, and find it.

61. (a) Suppose that a reservoir supplies water to an industrial park at a constant rate of $r = 4$ gallons per minute (gal/min) between 8:30 A.M. and 9:00 A.M. How much water does the reservoir supply during that time period?
(b) Suppose that one of the industrial plants increases its water consumption between 9:00 A.M. and 10:00 A.M. and that the rate at which the reservoir supplies water increases linearly, as shown in the accompanying figure. How much water does the reservoir supply during that 1-hour time period?
(c) Suppose that from 10:00 A.M. to 12 noon the rate at which the reservoir supplies water is given by the formula $r(t) = 10 + \sqrt{t}$ gal/min, where $t = 0$ corresponds to 10:00 A.M. How much water does the reservoir supply during that 2-hour time period?

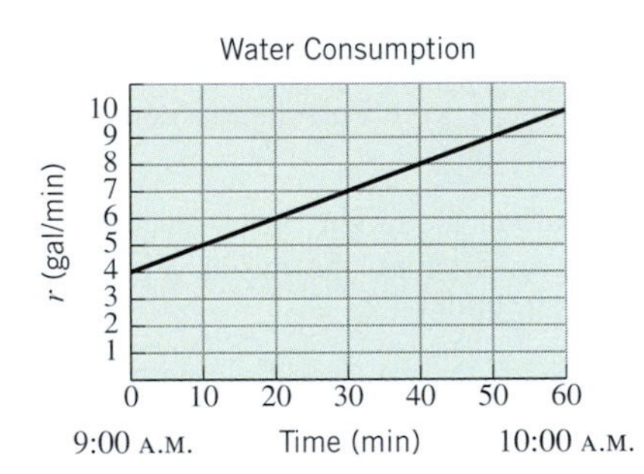

Figure Ex-61

62. A traffic engineer monitors the rate at which cars enter the main highway during the afternoon rush hour. From her data she estimates that between 4:30 P.M. and 5:30 P.M. the rate $R(t)$ at which cars enter the highway is given by the formula $R(t) = 100(1 - 0.0001t^2)$ cars per minute, where $t = 0$ corresponds to 4:30 P.M.
(a) When does the peak traffic flow into the highway occur?
(b) Find the number of cars that enter the highway during the rush hour.

63. (a) Prove: If f is continuous on $[a, b]$, then

$$\int_a^b [f(x) - f_{\text{ave}}]\,dx = 0$$

(b) Does there exist a constant $c \ne f_{\text{ave}}$ such that

$$\int_a^b [f(x) - c]\,dx = 0?$$

7.8 EVALUATING DEFINITE INTEGRALS BY SUBSTITUTION

In this section we will discuss two methods for evaluating definite integrals in which a substitution is required.

TWO METHODS FOR MAKING SUBSTITUTIONS IN DEFINITE INTEGRALS

Recall from Section 7.3 that indefinite integrals of the form

$$\int f(g(x))g'(x)\,dx$$

can sometimes be evaluated by making the u-substitution

$$u = g(x), \quad du = g'(x)\,dx \tag{1}$$

which converts the integral to the form

$$\int f(u)\,du$$

To apply this method to a definite integral of the form

$$\int_a^b f(g(x))g'(x)\,dx$$

we need to account for the effect that the substitution has on the x-limits of integration. There are two ways of doing this.

Method 1

First evaluate the indefinite integral

$$\int f(g(x))g'(x)\,dx$$

by substitution, and then use the relationship

$$\int_a^b f(g(x))g'(x)\,dx = \left[\int f(g(x))g'(x)\,dx\right]_a^b$$

to evaluate the definite integral. This procedure does not require any modification of the x-limits of integration.

Method 2

Make the substitution (1) directly in the definite integral, and then use the relationship $u = g(x)$ to replace the x-limits, $x = a$ and $x = b$, by corresponding u-limits, $u = g(a)$ and $u = g(b)$. This produces a new definite integral

$$\int_{g(a)}^{g(b)} f(u)\,du$$

that is expressed entirely in terms of u.

Example 1

Use the two methods above to evaluate $\int_0^2 x(x^2+1)^3\,dx$.

Solution by Method 1. If we let

$$u = x^2 + 1 \quad \text{so that} \quad du = 2x\,dx \tag{2}$$

then we obtain

$$\int x(x^2+1)^3\,dx = \frac{1}{2}\int u^3\,du = \frac{u^4}{8} + C = \frac{(x^2+1)^4}{8} + C$$

Thus,

$$\int_0^2 x(x^2+1)^3\,dx = \left[\int x(x^2+1)^3\,dx\right]_{x=0}^2 = \left.\frac{(x^2+1)^4}{8}\right]_{x=0}^2$$
$$= \frac{625}{8} - \frac{1}{8} = 78$$

Solution by Method 2. If we make the substitution $u = x^2 + 1$ in (2), then

$u = 1$ if $x = 0$
$u = 5$ if $x = 2$

Thus,

$$\int_0^2 x(x^2+1)^3\,dx = \frac{1}{2}\int_1^5 u^3\,du = \left.\frac{u^4}{8}\right]_{u=1}^5 = \frac{625}{8} - \frac{1}{8} = 78$$

which agrees with the result obtained by Method 1. ◀

The following theorem states precise conditions under which Method 2 can be used. The proof is a straightforward application of the chain rule and the Fundamental Theorem of Calculus, but we will omit the details.

7.8.1 THEOREM. *If g' is continuous on $[a, b]$ and f is continuous and has an antiderivative on an interval containing the values of $g(x)$ for $a \le x \le b$, then*

$$\int_a^b f(g(x))g'(x)\,dx = \int_{g(a)}^{g(b)} f(u)\,du$$

The choice of methods for evaluating definite integrals by substitution is generally a matter of taste, but in the following examples we will use the second method, since the idea is new.

Example 2

Evaluate

(a) $\displaystyle\int_0^{3/4} \frac{dx}{1-x}$ (b) $\displaystyle\int_0^{\pi/8} \sin^5 2x \cos 2x\,dx$

Solution (a). Let

$u = 1 - x$ so that $du = -dx$

With this substitution we have

$u = 1$ if $x = 0$
$u = \frac{1}{4}$ if $x = \frac{3}{4}$

Thus,

$$\int_0^{3/4} \frac{dx}{1-x} = -\int_1^{1/4} \frac{du}{u} = -\ln|u|\Big]_{u=1}^{1/4}$$
$$= -\left[\ln\left(\frac{1}{4}\right) - \ln(1)\right] = \ln 4$$

Solution (b). Let

$$u = \sin 2x \quad \text{so that} \quad du = 2\cos 2x\,dx \quad (\text{or } \tfrac{1}{2}\,du = \cos 2x\,dx)$$

With this substitution we have

$$u = \sin(0) = 0 \quad \text{if} \quad x = 0$$
$$u = \sin(\pi/4) = 1/\sqrt{2} \quad \text{if} \quad x = \pi/8$$

so

$$\int_0^{\pi/8} \sin^5 2x \cos 2x\,dx = \frac{1}{2}\int_0^{1/\sqrt{2}} u^5\,du = \frac{1}{2}\cdot\frac{u^6}{6}\Bigg]_0^{1/\sqrt{2}}$$
$$= \frac{1}{2}\left[\frac{1}{6(\sqrt{2})^6} - 0\right] = \frac{1}{96}$$ ◀

Example 3

In Example 8 of Section 4.4 we stated the following model for the temperature T in degrees Fahrenheit (°F) of a glass of lemonade t hours after being placed in a room with a constant temperature of 70°F, given that the initial temperature of the lemonade was 40°F:

$$T = 70 - 30e^{-0.5t}$$

Find the average temperature T_{ave} of the lemonade over the first 5 hours.

Solution. From Definition 7.7.5 the average value of T over the time interval [0, 5] is

$$T_{\text{ave}} = \frac{1}{5}\int_0^5 (70 - 30e^{-0.5t})\,dt \tag{3}$$

To evaluate this integral, we make the substitution

$$u = -0.5t \quad \text{so that} \quad du = -0.5\,dt \quad [\text{or } dt = -(1/0.5)\,du]$$

With this substitution we have

$$u = 0 \quad \text{if} \quad t = 0$$
$$u = -(0.5)5 = -2.5 \quad \text{if} \quad t = 5$$

Thus, (3) can be expressed as

$$T_{\text{ave}} = \frac{1}{5}\int_0^{-2.5} (70 - 30e^u)\left(-\frac{1}{0.5}\right)du = -\frac{1}{2.5}\int_0^{-2.5}(70 - 30e^u)\,du$$
$$= -\frac{1}{2.5}\Big[70u - 30e^u\Big]_{u=0}^{-2.5} = -\frac{1}{2.5}\Big[(-175 - 30e^{-2.5}) - (-30)\Big]$$
$$= 58 + 12e^{-2.5} \approx 58.99^\circ\text{F}$$ ◀

REMARK. Observe that the u-substitution in this example produced an integral in which the upper u-limit of integration was smaller than the lower u-limit of integration. In our computations we left the limits of integration in that order, but had we wanted to we could have reversed the order to put the larger limit on top and compensated by reversing the sign of the integral in accordance with Definition 7.5.3(b). The choice of procedures is a matter of taste; both produce the same result (verify).

FOR THE READER. If you have a CAS, use it to evaluate the integral in the last example. See whether it makes any difference in the form of the answer if you express the exponent as $-t/2$ rather than $-0.5t$.

EXERCISE SET 7.8 C CAS

In Exercises 1 and 2, express the integral in terms of the variable u, but do not evaluate it.

1. (a) $\int_0^2 (x+1)^7\,dx;\ u = x+1$

(b) $\int_{-1}^2 x\sqrt{8-x^2}\,dx;\ u = 8-x^2$

(c) $\int_{-1}^1 \sin(\pi\theta)\,d\theta;\ u = \pi\theta$

(d) $\int_0^3 (x+2)(x-3)^{20}\,dx;\ u = x-3$

2. (a) $\int_0^1 e^{2x-1}\,dx;\ u = 2x-1$

(b) $\int_e^{e^2} \frac{\ln x}{x}\,dx;\ u = \ln x$

(c) $\int_0^{\pi/4} \tan^2 x \sec^2 x\,dx;\ u = \tan x$

(d) $\int_0^1 x^3\sqrt{x^2+3}\,dx;\ u = x^2+3$

In Exercises 3–12, evaluate the definite integral two ways: first by a u-substitution in the definite integral and then by a u-substitution in the corresponding indefinite integral.

3. $\int_0^1 (2x+1)^4\,dx$

4. $\int_1^2 (4x-2)^3\,dx$

5. $\int_{-1}^0 (1-2x)^3\,dx$

6. $\int_1^2 (4-3x)^8\,dx$

7. $\int_0^8 x\sqrt{1+x}\,dx$

8. $\int_{-5}^0 x\sqrt{4-x}\,dx$

9. $\int_0^{\pi/2} 4\sin(x/2)\,dx$

10. $\int_0^{\pi/6} 2\cos 3x\,dx$

11. $\int_{-\ln 3}^{\ln 3} \frac{e^x}{e^x+4}\,dx$

12. $\int_0^{\ln 5} e^x(3-4e^x)\,dx$

In Exercises 13–16, evaluate the definite integral by expressing it in terms of u and evaluating the resulting integral using a formula from geometry.

13. $\int_0^{5/3} \sqrt{25-9x^2}\,dx;\ u = 3x$

14. $\int_0^2 x\sqrt{16-x^4}\,dx;\ u = x^2$

15. $\int_{\pi/3}^{\pi/2} \sin\theta\sqrt{1-4\cos^2\theta}\,d\theta;\ u = 2\cos\theta$

16. $\int_{e^{-6}}^{e^6} \frac{\sqrt{36-(\ln x)^2}}{x}\,dx;\ u = \ln x$

17. Find the area under the curve $y = \sin \pi x$ over the interval $[0, 1]$.

18. Find the area under the curve $y = 3\cos 2x$ over the interval $[0, \pi/8]$.

19. Find the area under the curve $y = 1/(x+5)^2$ over the interval $[3, 7]$.

20. Find the area under the curve $y = 1/(3x+1)^2$ over the interval $[0, 1]$.

21. Find the average value of $f(x) = e^{-2x}$ over the interval $[0, 4]$.

22. Find the average value of $f(x) = \sec^2 \pi x$ over the interval $[-\frac{1}{4}, \frac{1}{4}]$.

In Exercises 23–38, evaluate the integrals by any method.

23. $\int_0^1 \frac{dx}{\sqrt{3x+1}}$

24. $\int_1^2 \sqrt{5x-1}\,dx$

25. $\int_{-1}^1 \frac{x^2\,dx}{\sqrt{x^3+9}}$

26. $\int_{-1}^0 6t^2(t^3+1)^{19}\,dt$

27. $\int_1^3 \frac{x+2}{\sqrt{x^2+4x+7}}\,dx$

28. $\int_1^2 \frac{dx}{x^2-6x+9}$

29. $\int_{-3\pi/4}^{\pi/4} \sin x\cos x\,dx$

30. $\int_0^{\pi/4} \sqrt{\tan x}\sec^2 x\,dx$

31. $\int_0^{\sqrt{\pi}} 5x\cos(x^2)\,dx$

32. $\int_{\pi^2}^{4\pi^2} \frac{1}{\sqrt{x}}\sin\sqrt{x}\,dx$

33. $\int_{\pi/12}^{\pi/9} \sec^2 3\theta\,d\theta$

34. $\int_0^{\pi/2} \sin^2 3\theta\cos 3\theta\,d\theta$

35. $\int_0^1 \frac{y^2\,dy}{\sqrt{4-3y}}$

36. $\int_{-1}^4 \frac{x\,dx}{\sqrt{5+x}}$

37. $\int_0^e \frac{dx}{x+e}$

38. $\int_1^{\sqrt{2}} xe^{-x^2}\,dx$

C **39.** For each of the integrals you evaluated in Exercises 23–38, check your answer using a CAS.

C **40.** Use a CAS to find the exact value of the integral

$$\int_{-3}^1 \sqrt{3-2x-x^2}\,dx$$

and then confirm the result by hand calculation. [*Hint:* Complete the square.]

41. (a) Find $\int_0^1 f(3x+1)\,dx$ if $\int_1^4 f(x)\,dx = 5$.

(b) Find $\int_0^3 f(3x)\,dx$ if $\int_0^9 f(x)\,dx = 5$.

(c) Find $\int_{-2}^0 xf(x^2)\,dx$ if $\int_0^4 f(x)\,dx = 1$.

42. Given that m and n are positive integers, show that

$$\int_0^1 x^m(1-x)^n\,dx = \int_0^1 x^n(1-x)^m\,dx$$

by making a substitution. Do not attempt to evaluate the integrals.

43. Given that n is a positive integer, show that

$$\int_0^{\pi/2} \sin^n x\,dx = \int_0^{\pi/2} \cos^n x\,dx$$

by using a trigonometric identity and making a substitution. Do not attempt to evaluate the integrals.

44. Given that n is a positive integer, evaluate the integral

$$\int_0^1 x(1-x)^n\,dx$$

45. Suppose that at time $t=0$ there are 750 bacteria in a growth medium and the bacteria population $y(t)$ grows at the rate $y'(t) = 802.137e^{1.528t}$ bacteria per hour. How many bacteria will there be in 12 hours?

46. Suppose that the value of a yacht in dollars after t years of use is $V(t) = 275{,}000e^{-0.17t}$. What is the average value of the yacht over its first 10 years of use?

47. Suppose that a particle moving along a coordinate line has velocity $v(t) = 25 + 10e^{-0.05t}$ ft/s.

(a) What is the distance traveled by the particle from time $t=0$ to time $t=10$?

(b) Does the term $10e^{-0.05t}$ have much effect on the distance traveled by the particle over that time interval? Explain your reasoning.

48. Find a positive value of k such that the area under the graph of $y=e^{2x}$ over the interval $[0,k]$ is 3 square units.

49. Electricity is supplied to homes in the form of ***alternating current***, which means that the voltage has a sinusoidal waveform described by an equation of the form

$$V = V_p \sin(2\pi f t)$$

(see the accompanying figure). In this equation, V_p is called the ***peak voltage*** or ***amplitude*** of the current, f is called its ***frequency***, and $1/f$ is called its ***period***. The voltages V and V_p are measured in volts (V), the time t is measured in seconds (s), and the frequency is measured in hertz (Hz) or sometimes in cycles per second. (A ***cycle*** is the electrical term for one period of the waveform.) Alternating current voltmeters read what is called the ***rms*** or ***root-mean-square*** value of V. By definition, this is the square root of the average value of V^2 over one period.

(a) Show that

$$V_{\text{rms}} = \frac{V_p}{\sqrt{2}}$$

[*Hint:* Compute the average over the cycle from $t=0$ to $t=1/f$, and use the identity $\sin^2\theta = \frac{1}{2}(1-\cos 2\theta)$ to help evaluate the integral.]

(b) In the United States, electrical outlets supply alternating current with an rms voltage of 120 V at a frequency of 60 Hz. What is the peak voltage at such an outlet?

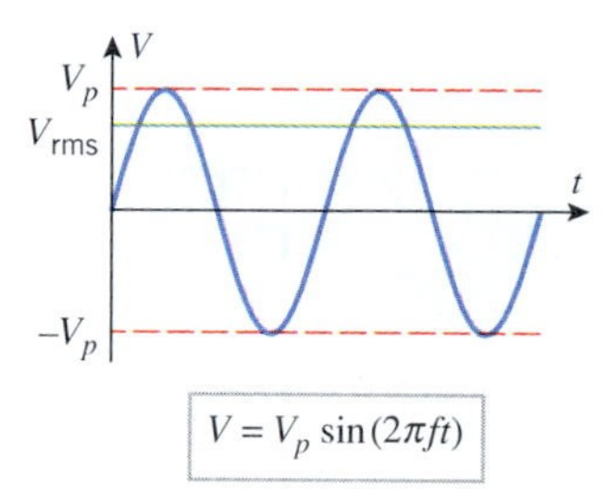

Figure Ex-49

50. Show that if f and g are continuous functions, then

$$\int_0^t f(t-x)g(x)\,dx = \int_0^t f(x)g(t-x)\,dx$$

51. (a) Let $I = \displaystyle\int_0^a \frac{f(x)}{f(x)+f(a-x)}\,dx$. Show that $I = a/2$.

[*Hint:* Let $u = a-x$, and then express the integrand as the sum of two fractions.]

(b) Use the result of part (a) to find

$$\int_0^3 \frac{\sqrt{x}}{\sqrt{x}+\sqrt{3-x}}\,dx$$

(c) Use the result of part (a) to find

$$\int_0^{\pi/2} \frac{\sin x}{\sin x + \cos x}\,dx$$

52. Let $I = \displaystyle\int_{-1}^1 \frac{1}{1+x^2}\,dx$. Show that the substitution $x = 1/u$ results in

$$I = -\int_{-1}^1 \frac{1}{1+u^2}\,du = -I$$

so $2I = 0$, which implies that $I = 0$. However, this is impossible since the integrand of the given integral is positive over the interval of integration. Where is the error?

53. Find the limit

$$\lim_{n\to+\infty} \sum_{k=1}^{n} \frac{\sin(k\pi/n)}{n}$$

by evaluating an appropriate definite integral over the interval $[0,1]$.

C **54.** Check your answer to Exercise 53 by evaluating the limit directly with a CAS.

55. (a) Prove that if f is an odd function, then

$$\int_{-a}^{a} f(x)\,dx = 0$$

and give a geometric explanation of this result. [*Hint:* One way to prove that a quantity q is zero is to show that $q = -q$.]

(b) Prove that if f is an even function, then

$$\int_{-a}^{a} f(x)\,dx = 2\int_{0}^{a} f(x)\,dx$$

and give a geometric explanation of this result. [*Hint:* Split the interval of integration from $-a$ to a into two parts at 0.]

7.9 LOGARITHMIC FUNCTIONS FROM THE INTEGRAL POINT OF VIEW

In Section 4.2 we discussed natural logarithms from the viewpoint of exponents; that is, we regarded $y = \ln x$ to mean that $e^y = x$. In this section we will show that $\ln x$ can also be expressed as an integral with a variable upper limit. This integral representation of $\ln x$ is important mathematically because it provides a convenient way of establishing properties such as differentiability and continuity. However, it is also important in applications because it provides a way of recognizing when integral solutions of problems can be expressed as natural logarithms.

THE LINK BETWEEN NATURAL LOGARITHMS AND INTEGRALS

The connection between natural logarithms and integrals was made in the middle of the seventeenth century in the course of investigating areas under the curve $y = 1/t$. The problem being considered was to find values of $t_1, t_2, t_3, \ldots, t_n, \ldots$ for which the areas $A_1, A_2, A_3, \ldots, A_n, \ldots$ in Figure 7.9.1*a* would be equal. Through the combined work of Isaac Newton, the Belgian Jesuit priest, Gregory of St. Vincent (1584–1667), and Gregory's student, Alfons A. de Sarasa (1618–1667), it was shown that by taking the points to be

$$t_1 = e,\quad t_2 = e^2,\quad t_3 = e^3, \ldots,\quad t_n = e^n, \ldots$$

each of the areas would be 1 (Figure 7.9.1*b*). Thus, in modern integral notation

$$\int_{1}^{e^n} \frac{1}{t}\,dt = n$$

which can be expressed as

$$\int_{1}^{e^n} \frac{1}{t}\,dt = \ln(e^n)$$

By comparing the upper limit of the integral and the expression inside the logarithm, it is a natural leap to the more general result

$$\int_{1}^{x} \frac{1}{t}\,dt = \ln x$$

which today we take as the formal definition of the natural logarithm.

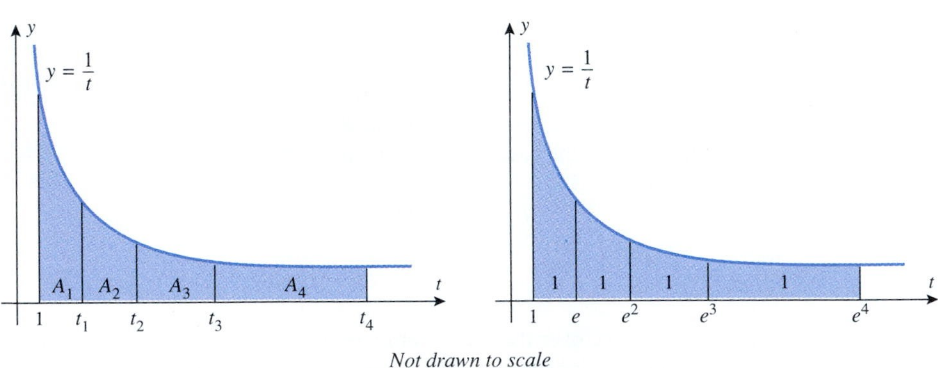

Figure 7.9.1

7.9.1 DEFINITION. The ***natural logarithm*** of x is denoted by $\ln x$ and is defined by the integral

$$\ln x = \int_1^x \frac{1}{t}\,dt, \quad x > 0 \tag{1}$$

Geometrically, $\ln x$ is the area under the curve $y = 1/t$ from $t = 1$ to $t = x$ when $x > 1$, and $\ln x$ is the negative of the area under the curve $y = 1/t$ from $t = x$ to $t = 1$ when $0 < x < 1$ (Figure 7.9.2). If $x = 1$, then $\ln x = 0$, since the upper and lower limits in (1) are the same. All of this is consistent with the computer-generated graph of $y = \ln x$ in Figure 4.2.4.

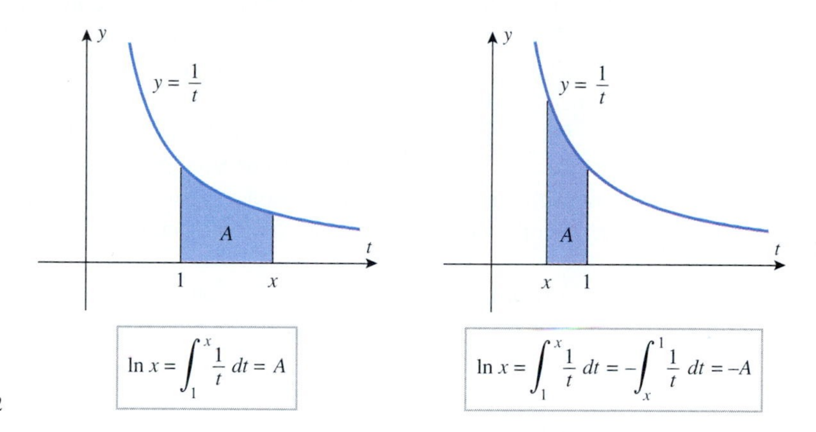

Figure 7.9.2

FOR THE READER. Review Theorem 7.5.8, and then explain why x is required to be positive in Definition 7.9.1.

APPROXIMATING ln x NUMERICALLY

For specific values of x, the value of $\ln x$ can be approximated numerically by approximating the definite integral in (1), say by using the midpoint approximation that was discussed in Section 7.5.

Table 7.9.1

$n = 10$
$\Delta t = (b - a)/n = (2 - 1)/10 = 0.1$

k	t_k^*	$1/t_k^*$
1	1.05	0.952381
2	1.15	0.869565
3	1.25	0.800000
4	1.35	0.740741
5	1.45	0.689655
6	1.55	0.645161
7	1.65	0.606061
8	1.75	0.571429
9	1.85	0.540541
10	1.95	0.512821
		6.928355

$$\Delta t \sum_{k=1}^{n} f(t_k^*) = (0.1)(6.928355) = 0.692836$$

Example 1

Approximate $\ln 2$ using the midpoint approximation with $n = 10$.

Solution. From (1), the exact value of $\ln 2$ is represented by the integral

$$\ln 2 = \int_1^2 \frac{1}{t}\,dt$$

The midpoint rule is given in Formula (3) of Section 7.5. Expressed in terms of t, that formula is

$$\int_a^b f(t)\,dt \approx \Delta t \sum_{k=1}^{n} f(t_k^*)$$

where Δt is the width of each subinterval and $t_1^*, t_2^*, \ldots, t_n^*$ are the midpoints. In this case we have 10 subintervals, so $\Delta t = (2-1)/10 = 0.1$. The computations to six decimal places are shown in Table 7.9.1. By comparison, a calculator set to display six decimal places gives $\ln 2 \approx 0.693147$, so the magnitude of the error in the midpoint approximation is about 0.000311. Greater accuracy in the midpoint approximation can be obtained by increasing n. For example, the midpoint approximation with $n = 100$ yields $\ln 2 \approx 0.693144$, which is correct to five decimal places. ◀

DIFFERENTIABILITY AND CONTINUITY OF ln x AND e^x

Definition 7.9.1 is not only useful for approximating values of $\ln x$, but it is the key to establishing many of the fundamental properties of the natural logarithm. For example, in Section 4.4 we obtained the derivative

$$\frac{d}{dx}[\ln x] = \frac{1}{x} \quad (x > 0) \tag{2}$$

by *assuming* that $f(x) = \ln x$ is differentiable for $x > 0$. However, now that we have Definition 7.9.1 to work with, both the differentiability of $\ln x$ and Formula (2) follow immediately from Part 2 of the Fundamental Theorem of Calculus (7.6.3). Moreover, since differentiable functions are continuous, this also shows that $\ln x$ is continuous for $x > 0$.

Although it is not our objective to prove all of the properties of the functions we encounter, it is worthwhile to understand in principle how the differentiability and continuity of $\ln x$ can be used to establish differentiability and continuity of other important functions. For example, since the exponential function e^x is the inverse of $\ln x$, it follows from Theorem 4.1.7, with $f(x) = \ln x$ and $f^{-1}(x) = e^x$, that e^x is differentiable at any point x where $f'(f^{-1}(x)) = 1/e^x \neq 0$. Since this holds for all x, it follows that e^x is differentiable and hence continuous everywhere.

The differentiability $\ln x$ for $x > 0$ can be used to prove the differentiability of $\log_b x$ for $x > 0$ by using Formula (9) of Section 4.2 to express $\log_b x$ in terms of $\ln x$, and the differentiability of e^x can be used to prove the differentiability of b^x by expressing b^x in terms of e^x as $b^x = e^{x \ln b}$. We omit the details.

THE DEFINITION OF e REVISITED

In Formulas (3), (4), and (5) of Section 4.2 we gave three limits for e, but at that time we did not have the mathematical tools to prove the existence of those limits; the following theorem does this.

> **7.9.2 THEOREM.**
>
> (*a*) $\lim_{x\to 0}(1+x)^{1/x} = e$ (*b*) $\lim_{x\to +\infty}\left(1+\frac{1}{x}\right)^x = e$ (*c*) $\lim_{x\to -\infty}\left(1+\frac{1}{x}\right)^x = e$

Proof. We will prove part (*a*), and leave the proofs of the other parts for the exercises. Our proof will build on the differentiability of $\ln x$, and more specifically on the derivative of $\ln x$ at the point $x = 1$, namely

$$\frac{d}{dx}[\ln x]\bigg|_{x=1} = \frac{1}{x}\bigg|_{x=1} = 1$$

If we express this relationship using the definition of a derivative, we obtain

$$1 = \lim_{h\to 0}\frac{\ln(1+h)-\ln 1}{h} = \lim_{h\to 0}\frac{\ln(1+h)}{h} = \lim_{h\to 0}\ln(1+h)^{1/h}$$

Thus, it follows that

$$e = e^{\lim_{h\to 0}\ln(1+h)^{1/h}}$$

which from the continuity of e^x can be written as

$$e = \lim_{h\to 0} e^{\ln(1+h)^{1/h}} = \lim_{h\to 0}(1+h)^{1/h}$$

Except for a difference in notation, this is what we wanted to prove. ∎

FUNCTIONS DEFINED BY INTEGRALS

The functions that we have dealt with thus far in this text are called ***elementary functions***; they include polynomials, rational functions, power functions, exponential functions, logarithmic functions, trigonometric functions, and all other functions that can be obtained from these by addition, subtraction, multiplication, division, root extraction, composition, and by taking inverses.

However, there are many important functions that do not fall into this category. Such functions occur in many ways, but they commonly arise in the course of solving initial-value problems of the form

$$\frac{dy}{dx} = f(x), \quad y(x_0) = y_0 \tag{3}$$

Recall from Example 7 of Section 7.2 and the discussion preceding it that the basic method for solving (3) is to integrate $f(x)$, and then use the initial condition to determine the constant of integration. It can be proved that if f is continuous, then (3) has a unique solution and that this procedure produces it. However, there is another approach: Instead of solving each initial-value problem individually, we can find a general formula for the solution of (3), and then apply that formula to solve specific problems. We will now show that

$$y(x) = y_0 + \int_{x_0}^{x} f(t)\,dt \tag{4}$$

is a formula for the solution of (3). To confirm that this is so we must show that $dy/dx = f(x)$ and that $y(x_0) = y_0$. The computations are as follows:

$$\frac{dy}{dx} = \frac{d}{dx}\left[y_0 + \int_{x_0}^{x} f(t)\,dt\right] = 0 + f(x) = f(x)$$

$$y(x_0) = y_0 + \int_{x_0}^{x_0} f(t)\,dt = y_0 + 0 = y_0$$

Example 2

In Example 7 of Section 7.2 we showed that the solution of the initial-value problem

$$\frac{dy}{dx} = \cos x, \quad y(0) = 1$$

is $y(x) = 1 + \sin x$. This initial-value problem can also be solved by applying Formula (4) with $f(x) = \cos x$, $x_0 = 0$, and $y_0 = 1$. This yields

$$y(x) = 1 + \int_0^x \cos t\,dt = 1 + \big[\sin t\big]_{t=0}^{x} = 1 + \sin x$$ ◀

In the last example we were able to perform the integration in Formula (4) and express the solution of the initial-value problem as an elementary function. However, sometimes this will not be possible, in which case the solution of the initial-value problem must be left in terms of an "unevaluated" integral. For example, from (4), the solution of the initial-value problem

$$\frac{dy}{dx} = e^{-x^2}, \quad y(0) = 1$$

is

$$y(x) = 1 + \int_0^x e^{-t^2}\,dt$$

However, it can be shown that there is no way to express the integral in this solution as an elementary function. Thus, we have encountered a *new* function, which we regard to be *defined* by the integral. A close relative of this function, known as the ***error function***, plays an important role in probability and statistics; it is denoted by erf(x) and is defined as

$$\operatorname{erf}(x) = \frac{2}{\sqrt{\pi}}\int_0^x e^{-t^2}\,dt \tag{5}$$

Indeed, many of the most important functions in science and engineering are defined as integrals that have special names and notations associated with them. For example, the functions defined by

$$S(x) = \int_0^x \sin\left(\frac{\pi t^2}{2}\right) dt \quad \text{and} \quad C(x) = \int_0^x \cos\left(\frac{\pi t^2}{2}\right) dt \tag{6–7}$$

are called the ***Fresnel sine and cosine functions***, respectively, in honor of the French physicist Augustin Fresnel (1788–1827), who first encountered them in his study of diffraction of light waves.

EVALUATING AND GRAPHING FUNCTIONS DEFINED BY INTEGRALS

The following values of $S(1)$ and $C(1)$ were produced by a CAS that has a built-in algorithm for approximating definite integrals:

$$S(1) = \int_0^1 \sin\left(\frac{\pi t^2}{2}\right) dt \approx 0.438259, \qquad C(1) = \int_0^1 \cos\left(\frac{\pi t^2}{2}\right) dt \approx 0.779893$$

To generate graphs of functions defined by integrals, computer programs choose a set of x-values in the domain, approximate the integral for each of those values, and then plot the resulting points. Thus, there is a lot of computation involved in generating such graphs, since each plotted point requires the approximation of an integral. The graphs of the Fresnel functions in Figure 7.9.3 were generated in this way using a CAS.

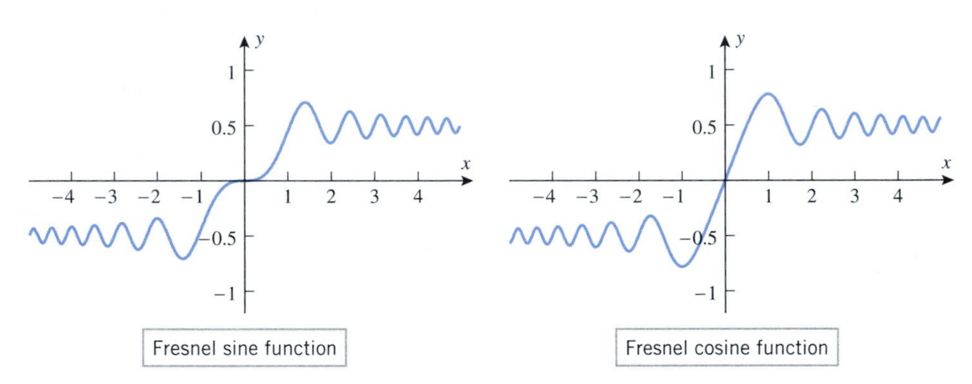

Figure 7.9.3

REMARK. Although it required a considerable amount of computation to generate the graphs of the Fresnel functions, the derivatives of $S(x)$ and $C(x)$ are easy to obtain using Part 2 of the Fundamental Theorem of Calculus (7.6.3); they are

$$S'(x) = \sin\left(\frac{\pi x^2}{2}\right) \quad \text{and} \quad C'(x) = \cos\left(\frac{\pi x^2}{2}\right) \tag{8–9}$$

These derivatives can be used to determine the locations of the relative extrema and inflection points and to investigate other properties of $S(x)$ and $C(x)$.

INTEGRALS WITH FUNCTIONS AS LIMITS OF INTEGRATION

Various applications can lead to integrals in which one or both of the limits of integration is a function of x. Some examples are

$$\int_x^1 \sqrt{\sin t}\, dt, \quad \int_{x^2}^{\sin x} \sqrt{t^3+1}\, dt, \quad \int_{\ln x}^{\pi} \frac{dt}{t^7-8}$$

We will complete this section by showing how to differentiate integrals of the form

$$\int_a^{g(x)} f(t)\, dt \tag{10}$$

where a is constant. Derivatives of other kinds of integrals with functions as limits of integration will be discussed in the exercises.

To differentiate (10) we can view the integral as a composition $F(g(x))$, where

$$F(x) = \int_a^x f(t)\,dt$$

If we now apply the chain rule, we obtain

$$\frac{d}{dx}\left[\int_a^{g(x)} f(t)\,dt\right] = \frac{d}{dx}[F(g(x))] = F'(g(x))g'(x) = f(g(x))g'(x)$$

Theorem 7.6.3

Thus,

$$\frac{d}{dx}\left[\int_a^{g(x)} f(t)\,dt\right] = f(g(x))g'(x) \tag{11}$$

REMARK. In words, *to differentiate an integral with a constant lower limit and a function as the upper limit, substitute the upper limit into the integrand, and multiply by the derivative of the upper limit.*

Example 3

$$\frac{d}{dx}\left[\int_1^{\sin x} (1-t^2)\,dt\right] = (1-\sin^2 x)\cos x = \cos^3 x$$

◀

EXERCISE SET 7.9 Graphing Calculator C CAS

1. Sketch the curve $y = 1/t$, and shade a region under the curve whose area is
(a) $\ln 2$ (b) $-\ln 0.5$ (c) 2.

2. Sketch the curve $y = 1/t$, and shade two different regions under the curve whose area is $\ln 1.5$.

3. Given that $\ln a = 2$ and $\ln c = 5$, find
(a) $\int_1^{ac} \frac{1}{t}\,dt$ (b) $\int_1^{1/c} \frac{1}{t}\,dt$
(c) $\int_1^{a/c} \frac{1}{t}\,dt$ (d) $\int_1^{a^3} \frac{1}{t}\,dt$.

4. Given that $\ln a = 4$, find
(a) $\int_1^{\sqrt{a}} \frac{1}{t}\,dt$ (b) $\int_1^{2a} \frac{1}{t}\,dt$
(c) $\int_1^{2/a} \frac{1}{t}\,dt$ (d) $\int_2^{a} \frac{1}{t}\,dt$.

5. Approximate $\ln 5$ using the midpoint rule with $n = 10$, and estimate the magnitude of the error by comparing your answer to that produced directly by a calculating utility.

6. Approximate $\ln 3$ using the midpoint rule with $n = 20$, and estimate the magnitude of the error by comparing your answer to that produced directly by a calculating utility.

7. Simplify the expression and state the values of x for which your simplification is valid.
(a) $e^{-\ln x}$ (b) $e^{\ln x^2}$
(c) $\ln\left(e^{-x^2}\right)$ (d) $\ln(1/e^x)$
(e) $\exp(3\ln x)$ (f) $\ln(xe^x)$
(g) $\ln\left(e^{x-\sqrt[3]{x}}\right)$ (h) $e^{x-\ln x}$

8. (a) Let $f(x) = e^{-2x}$. Find the simplest exact value of the function $f(\ln 3)$.
(b) Let $f(x) = e^x + 3e^{-x}$. Find the simplest exact value of the function $f(\ln 2)$.

In Exercises 9 and 10, express the given quantity as a power of e.

9. (a) 3^π (b) $2^{\sqrt{2}}$

10. (a) π^{-x} (b) x^{2x}, $x > 0$

In Exercises 11 and 12, find the limits by making appropriate substitutions in the limits given in Theorem 7.9.2.

11. (a) $\lim_{x\to+\infty}\left(1+\frac{1}{x}\right)^{2x}$ (b) $\lim_{x\to 0}(1+2x)^{1/x}$

12. (a) $\lim_{x\to+\infty}\left(1+\frac{1}{3x}\right)^x$ (b) $\lim_{x\to 0}(1+x)^{1/3x}$

In Exercises 13 and 14, find $g'(x)$ using Part 2 of the Fundamental Theorem of Calculus, and check your answer by evaluating the integral and then differentiating.

13. $g(x)=\int_1^x (t^2-t)\,dt$ **14.** $g(x)=\int_\pi^x (1-\cos t)\,dt$

In Exercises 15 and 16, find the derivative using Formula (11), and check your answer by evaluating the integral and then differentiating.

15. (a) $\frac{d}{dx}\int_1^{x^3}\frac{1}{t}\,dt$ (b) $\frac{d}{dx}\int_1^{\ln x} e^t\,dt$

16. (a) $\frac{d}{dx}\int_{-1}^{x^2}\sqrt{t+1}\,dt$ (b) $\frac{d}{dx}\int_\pi^{1/x}\sin t\,dt$

17. Let $F(x)=\int_0^x \frac{\cos t}{t^2+3}\,dt$. Find

(a) $F(0)$ (b) $F'(0)$ (c) $F''(0)$.

18. Let $F(x)=\int_2^x \sqrt{3t^2+1}\,dt$. Find

(a) $F(2)$ (b) $F'(2)$ (c) $F''(2)$.

C **19.** (a) Use Formula (11) to find

$$\frac{d}{dx}\int_1^{x^2} t\sqrt{1+t}\,dt$$

(b) Use a CAS to evaluate the integral and differentiate the resulting function.

(c) Use the simplification command of the CAS, if necessary, to confirm that answers in parts (a) and (b) are the same.

20. Show that

(a) $\frac{d}{dx}\left[\int_x^a f(t)\,dt\right]=-f(x)$

(b) $\frac{d}{dx}\left[\int_{g(x)}^a f(t)\,dt\right]=-f(g(x))g'(x)$.

In Exercises 21 and 22, use the results in Exercise 20 to find the derivative.

21. (a) $\frac{d}{dx}\int_x^1 \sin(t^2)\,dt$ (b) $\frac{d}{dx}\int_{\tan x}^3 \frac{t^2}{1+t^2}\,dt$

22. (a) $\frac{d}{dx}\int_x^0 (t^2+1)^{40}\,dt$ (b) $\frac{d}{dx}\int_{1/x}^\pi \cos^3 t\,dt$

23. Find

$$\frac{d}{dx}\left[\int_{3x}^{x^2}\frac{t-1}{t^2+1}\,dt\right]$$

by writing

$$\int_{3x}^{x^2}\frac{t-1}{t^2+1}\,dt=\int_{3x}^{0}\frac{t-1}{t^2+1}\,dt+\int_0^{x^2}\frac{t-1}{t^2+1}\,dt$$

24. Use Exercise 20(b) and the idea in Exercise 23 to show that

$$\frac{d}{dx}\int_{h(x)}^{g(x)} f(t)\,dt=f(g(x))g'(x)-f(h(x))h'(x)$$

25. Use the result obtained in Exercise 24 to perform the following differentiations:

(a) $\frac{d}{dx}\int_{x^2}^{x^3}\sin^2 t\,dt$ (b) $\frac{d}{dx}\int_{-x}^{x}\frac{1}{1+t}\,dt$.

26. Prove that the function

$$F(x)=\int_x^{3x}\frac{1}{t}\,dt$$

is constant on the interval $(0,+\infty)$ by using Exercise 24 to find $F'(x)$. What is that constant?

27. Let $F(x)=\int_0^x f(t)\,dt$, where f is the function whose graph is shown in the accompanying figure.

(a) Find $F(0)$, $F(3)$, $F(5)$, $F(7)$, and $F(10)$.

(b) On what subintervals of the interval $[0, 10]$ is F increasing? Decreasing?

(c) Where does F have its maximum value? Its minimum value?

(d) Sketch the graph of F.

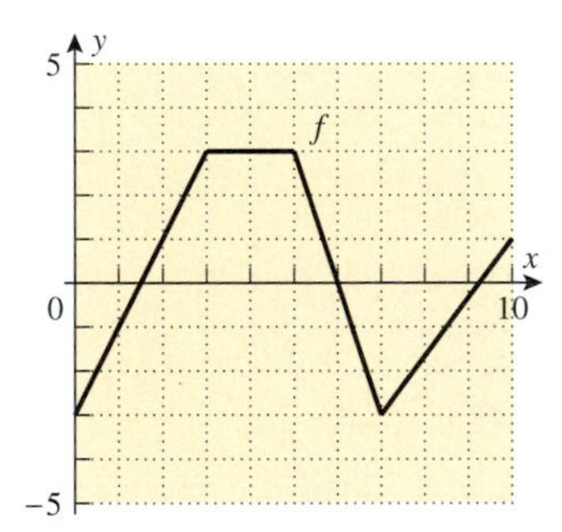

Figure Ex-27

28. Use the appropriate values found in part (a) of Exercise 27 to find the average value of f over the interval $[0, 10]$.

In Exercises 29 and 30, express $F(x)$ in a piecewise form that does not involve an integral.

29. $F(x)=\int_{-1}^x |t|\,dt$

30. $F(x)=\int_0^x f(t)\,dt$, where $f(x)=\begin{cases}x, & 0\le x\le 2\\ 2, & x>2\end{cases}$

In Exercises 31–34, use Formula (4) to solve the initial-value problem.

31. $\dfrac{dy}{dx} = \sqrt[3]{x};\ y(1) = 2$

32. $\dfrac{dy}{dx} = \dfrac{x+1}{\sqrt{x}};\ y(1) = 0$

33. $\dfrac{dy}{dx} = \sec^2 x - \sin x;\ y(\pi/4) = 1$

34. $\dfrac{dy}{dx} = xe^{x^2};\ y(0) = 0$

35. Suppose that at time $t = 0$ there are P_0 individuals who have disease X, and suppose that a certain model for the spread of the disease predicts that the disease will spread at the rate of $r(t)$ individuals per day. Write a formula for the number of individuals who will have disease X after x days.

36. Suppose that $v(t)$ is the velocity function of a particle moving along an s-axis. Write a formula for the coordinate of the particle at time T if the particle is at the point s_1 at time $t = 1$.

37. The accompanying figure shows the graphs of $y = f(x)$ and $y = \int_0^x f(t)\,dt$. Determine which graph is which, and explain your reasoning.

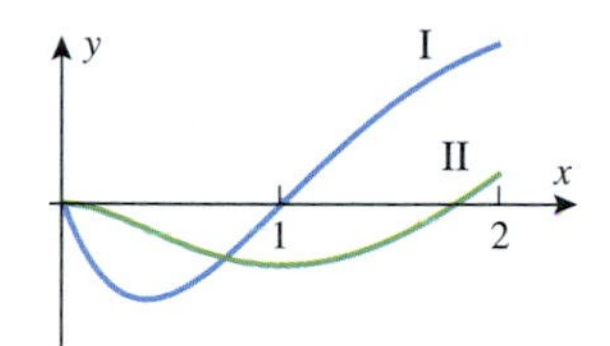

Figure Ex-37

38. (a) Make a conjecture about the value of the limit

$$\lim_{k \to 0} \int_1^x t^{k-1}\,dt \quad (x > 0)$$

(b) Check your conjecture by evaluating the integral, and then using L'Hôpital's rule to find the limit.

39. Let $F(x) = \int_0^x f(t)\,dt$, where f is the function graphed in the accompanying figure.

(a) Where do the relative minima of F occur?

(b) Where do the relative maxima of F occur?

(c) Where does the absolute maximum of F on the interval $[0, 5]$ occur?

(d) Where does the absolute minimum of F on the interval $[0, 5]$ occur?

(e) Where is F concave up? Concave down?

(f) Sketch the graph of F.

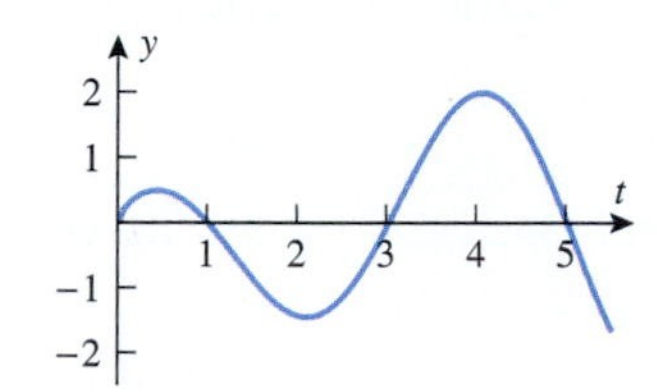

Figure Ex-39

C **40.** CAS programs have commands for working with most of the important nonelementary functions. Check your CAS documentation for information about the error function erf(x) [see Formula (5)], and then complete the following.

(a) Generate the graph of erf(x).

(b) Use the graph to make a conjecture about the existence and location of any relative maxima and minima of erf(x).

(c) Check your conjecture in part (b) using the derivative of erf(x).

(d) Use the graph to make a conjecture about the existence and location of any inflection points of erf(x).

(e) Check your conjecture in part (d) using the second derivative of erf(x).

(f) Use the graph to make a conjecture about the existence of horizontal asymptotes of erf(x).

(g) Check your conjecture in part (f) by using the CAS to find the limits of erf(x) as $x \to \pm\infty$.

41. The Fresnel sine and cosine functions $S(x)$ and $C(x)$ were defined in Formulas (6) and (7) and graphed in Figure 7.9.3. Their derivatives were given in Formulas (8) and (9).

(a) At what points does $C(x)$ have relative minima? Relative maxima?

(b) Where do the inflection points of $C(x)$ occur?

(c) Confirm that your answers in parts (a) and (b) are consistent with the graph of $C(x)$.

42. Find the limit

$$\lim_{h \to 0} \frac{1}{h} \int_x^{x+h} \ln t\,dt$$

43. Find a function f and a number a such that

$$2 + \int_a^x f(t)\,dt = e^{3x}$$

44. (a) Give a geometric argument to show that

$$\frac{1}{x+1} < \int_x^{x+1} \frac{1}{t}\,dt < \frac{1}{x}, \quad x > 0$$

(b) Use the result in part (a) to prove that

$$\frac{1}{x+1} < \ln\left(1 + \frac{1}{x}\right) < \frac{1}{x}, \quad x > 0$$

(c) Use the result in part (b) to prove that

$$e^{\frac{x}{x+1}} < \left(1 + \frac{1}{x}\right)^x < e, \quad x > 0$$

and hence that

$$\lim_{x \to +\infty} \left(1 + \frac{1}{x}\right)^x = e$$

(d) Use the inequality in part (c) to prove that

$$\left(1 + \frac{1}{x}\right)^x < e < \left(1 + \frac{1}{x}\right)^{x+1}, \quad x > 0$$

45. Use a graphing utility to generate the graph of

$$y = \left(1 + \frac{1}{x}\right)^{x+1} - \left(1 + \frac{1}{x}\right)^x$$

in the window $[0, 100] \times [0, 0.2]$, and use that graph and part (d) of Exercise 44 to make a rough estimate of the error in the approximation

$$e \approx \left(1 + \frac{1}{50}\right)^{50}$$

46. Prove: If f is continuous on an open interval I and a is any point in I, then

$$F(x) = \int_a^x f(t)\,dt$$

is continuous on I.

SUPPLEMENTARY EXERCISES

1. Write a paragraph that describes the *rectangle method* for defining the area under a curve $y = f(x)$ over an interval $[a, b]$.

2. What is an *integral curve* of a function f? How are two integral curves of a function f related?

3. The *definite integral* of f over the interval $[a, b]$ is defined as the limit

$$\int_a^b f(x)\,dx = \lim_{\max \Delta x_k \to 0} \sum_{k=1}^{n} f(x_k^*)\Delta x_k$$

Explain what the various symbols on the right side of this equation mean.

4. State the two parts of the Fundamental Theorem of Calculus, and explain what is meant by the phrase "differentiation and integration are inverse processes."

5. Derive the formulas for the position and velocity functions of a particle that moves with uniformly accelerated motion along a coordinate line.

6. (a) Devise a procedure for finding upper and lower estimates of the area of the region in the accompanying figure (in cm^2).
(b) Use your procedure to find upper and lower estimates of the area.
(c) Improve on the estimates you obtained in part (b).

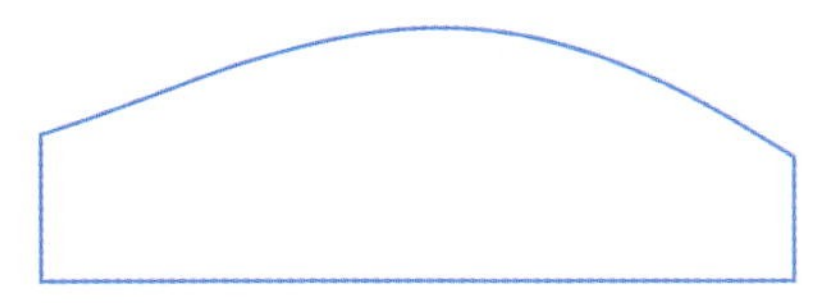

Figure Ex-6

7. Suppose that

$$\int_0^1 f(x)\,dx = \frac{1}{2}, \quad \int_1^2 f(x)\,dx = \frac{1}{4},$$

$$\int_0^3 f(x)\,dx = -1, \quad \int_0^1 g(x)\,dx = 2$$

In each part, use this information to evaluate the given integral, if possible. If there is not enough information to evaluate the integral, then say so.

(a) $\int_0^2 f(x)\,dx$ (b) $\int_1^3 f(x)\,dx$ (c) $\int_2^3 5f(x)\,dx$

(d) $\int_1^0 g(x)\,dx$ (e) $\int_0^1 g(2x)\,dx$ (f) $\int_0^1 [g(x)]^2\,dx$

8. In each part, use the information in Exercise 7 to evaluate the given integral. If there is not enough information to evaluate the integral, then say so.

(a) $\int_0^1 [f(x) + g(x)]\,dx$ (b) $\int_0^1 f(x)g(x)\,dx$

(c) $\int_0^1 \frac{f(x)}{g(x)}\,dx$ (d) $\int_0^1 [4g(x) - 3f(x)]\,dx$

9. In each part, evaluate the integral. Where appropriate, you may use a geometric formula.

(a) $\int_{-1}^1 1 + \sqrt{1 - x^2}\,dx$

(b) $\int_0^3 (x\sqrt{x^2 + 1} - \sqrt{9 - x^2})\,dx$

(c) $\int_0^1 x\sqrt{1 - x^4}\,dx$

10. Evaluate the integral $\int_0^1 |2x - 1|\,dx$, and sketch the region whose area it represents.

11. One of the numbers π, $\pi/2$, $35\pi/128$, $1 - \pi$ is the correct value of the integral

$$\int_0^\pi \sin^8 x\,dx$$

Use the accompanying graph of $y = \sin^8 x$ and a logical process of elimination to find the correct value. [Do not attempt to evaluate the integral.]

Figure Ex-11

12. Evaluate

$$\int \frac{e^{2x}}{e^x + 3}\, dx$$

[*Hint:* Divide $e^x + 3$ into e^{2x}.]

13. Give a convincing geometric argument to show that

$$\int_1^e \ln x \, dx + \int_0^1 e^x \, dx = e$$

14. In each part, find the limit by interpreting it as a limit of Riemann sums in which the interval $[0, 1]$ is divided into n subintervals of equal length.

(a) $\displaystyle\lim_{n\to+\infty} \frac{\sqrt{1} + \sqrt{2} + \sqrt{3} + \cdots + \sqrt{n}}{n^{3/2}}$

(b) $\displaystyle\lim_{n\to+\infty} \frac{1^4 + 2^4 + 3^4 + \cdots + n^4}{n^5}$

(c) $\displaystyle\lim_{n\to+\infty} \frac{e^{1/n} + e^{2/n} + e^{3/n} + \cdots + e^{n/n}}{n}$

15. (a) Divide the interval $[1, 2]$ into 5 subintervals of equal length, and use appropriate Riemann sums to show that

$$0.2\left[\tfrac{1}{1.2} + \tfrac{1}{1.4} + \tfrac{1}{1.6} + \tfrac{1}{1.8} + \tfrac{1}{2.0}\right] < \ln 2 < 0.2\left[\tfrac{1}{1.0} + \tfrac{1}{1.2} + \tfrac{1}{1.4} + \tfrac{1}{1.6} + \tfrac{1}{1.8}\right]$$

(b) Show that if the interval $[1, 2]$ is divided into n subintervals of equal length, then

$$\sum_{k=1}^{n} \frac{1}{n+k} < \ln 2 < \sum_{k=0}^{n-1} \frac{1}{n+k}$$

(c) Show that the difference between the two sums in part (b) is $1/2n$, and use this result to show that the sums in part (a) approximate $\ln 2$ with an error of at most 0.1.

(d) How large must n be to ensure that the sums in part (b) approximate $\ln 2$ to three decimal places?

16. The accompanying figure shows the direction field for a differential equation $dy/dx = f(x)$. Which of the following functions is most likely to be $f(x)$?

$\sqrt{x}$, $\sin x$, x^4, x

Explain your reasoning.

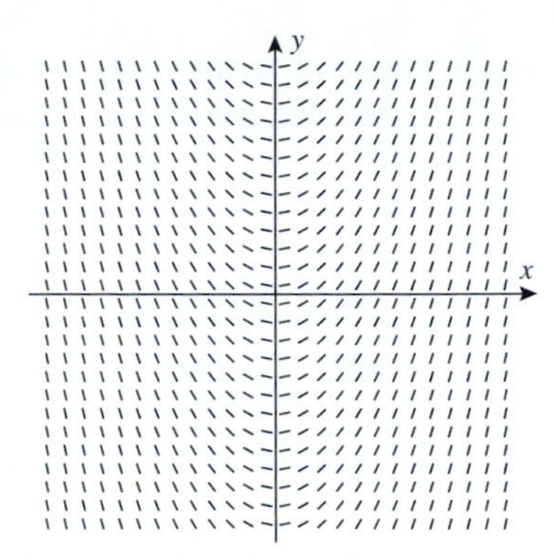

Figure Ex-16

17. In each part, confirm the stated equality.

(a) $1 \cdot 2 + 2 \cdot 3 + \cdots + n(n+1) = \frac{1}{3}n(n+1)(n+2)$

(b) $\displaystyle\lim_{n\to+\infty} \sum_{k=1}^{n-1} \left(\frac{9}{n} - \frac{k}{n^2}\right) = \frac{17}{2}$

(c) $\displaystyle\sum_{i=1}^{3} \left(\sum_{j=1}^{2} (i + j)\right) = 21$

18. Express

$$\sum_{k=4}^{18} k(k-3)$$

in sigma notation with

(a) $k = 0$ as the lower limit of summation

(b) $k = 5$ as the lower limit of summation.

19. (a) Show that the substitutions $u = \sec x$ and $u = \tan x$ produce different values for the integral

$$\int \sec^2 x \tan x \, dx$$

(b) Explain why both are correct.

20. Use the two substitutions in Exercise 19 to evaluate the definite integral

$$\int_0^{\pi/4} \sec^2 x \tan x \, dx$$

and confirm that they produce the same result.

21. Evaluate the integral

$$\int \sqrt{1 + x^{-2/3}}\, dx$$

by making the substitution $u = 1 + x^{2/3}$.

22. (a) Express Formula 8 of Section 7.5 in sigma notation.

(b) If $c_1, c_2, \ldots, c_n$ are constants and $f_1, f_2, \ldots, f_n$ are integrable functions on $[a, b]$, do you think it is always true that

$$\int_a^b \left(\sum_{k=1}^{n} c_k f_k(x)\right) dx = \sum_{k=1}^{n} \left[c_k \int_a^b f_k(x)\, dx\right]?$$

Explain your reasoning.

23. Find an integral formula for the antiderivative of $1/(1+x^2)$ on the interval $(-\infty, +\infty)$ whose value at $x = 1$ is (a) 0 and (b) 2.

C **24.** Let 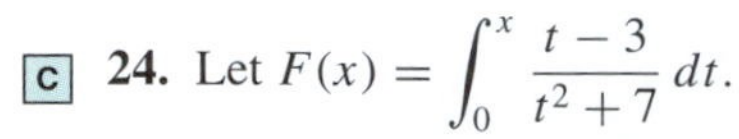 $F(x) = \int_0^x \frac{t-3}{t^2+7}\,dt$.

(a) Find the intervals on which F is increasing. Decreasing.

(b) Find the open intervals on which F is concave up. Concave down.

(c) Find the x-values, if any, at which the function F has absolute extrema.

(d) Use a CAS to graph F, and confirm that the results in parts (a), (b), and (c) are consistent with the graph.

25. Prove that the function

$$F(x) = \int_0^x \frac{1}{1+t^2}\,dt + \int_0^{1/x} \frac{1}{1+t^2}\,dt$$

is constant on the interval $(0, +\infty)$.

26. What is the natural domain of the function

$$F(x) = \int_1^x \frac{1}{t^2-9}\,dt?$$

Explain your reasoning.

27. In each part, determine the values of x for which $F(x)$ is positive, negative, or zero without performing the integration; explain your reasoning.

(a) $F(x) = \int_1^x \frac{t^4}{t^2+3}\,dt$ (b) $F(x) = \int_{-1}^x \sqrt{4-t^2}\,dt$

28. Find a formula (defined piecewise) for the upper boundary of the trapezoid shown in the accompanying figure, and then integrate that function to derive the formula for the area of the trapezoid given on the inside front cover of this text.

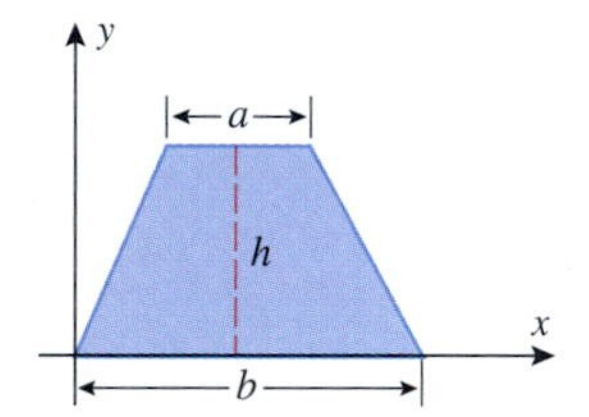

Figure Ex-28

29. An engineer studying the power consumption of a manufacturing plant determines that the plant's daily rate of electricity usage in kilowatts per hour (kW/h) can be reasonably modeled by the formula

$$R(t) = 2000e^{-t/48} + 500\sin\left(\tfrac{\pi}{12}t\right) \quad (0 \le t \le 24)$$

(a) How many kilowatts of electricity does the plant use in a 24-hour period?

(b) Find the average rate of electricity usage over the first 8 hours of operation.

(c) Generate the graph of $R(t)$ over the first 8-hour period, and use it to make a rough estimate of the maximum rate of electricity usage during that period and when it occurs.

(d) Determine the maximum rate of electricity usage during the first 8-hour period to two decimal places.

30. Suppose that a tumor grows at the rate of $r(t) = t/7$ grams (g) per week. When, during the second 26 weeks of growth, is the weight of the tumor the same as its average weight during that period?

31. The velocity of a particle moving along an s-axis is measured at 5-s intervals for 40 s, and the velocity function is modeled by a smooth curve drawn through the data points, as shown in the accompanying figure.

(a) Does the particle have constant acceleration? Explain your reasoning.

(b) Is there any 15-s time interval during which the acceleration is constant? Explain your reasoning.

(c) Estimate the average velocity of the particle over the 40-s time period.

(d) Estimate the distance traveled by the particle from time $t = 0$ to time $t = 40$.

(e) Is the particle ever slowing down during the 40-s time period? Explain your reasoning.

(f) Is there sufficient information for you to determine the s-coordinate of the particle at time $t = 10$? If so, find it. If not, explain what additional information you need.

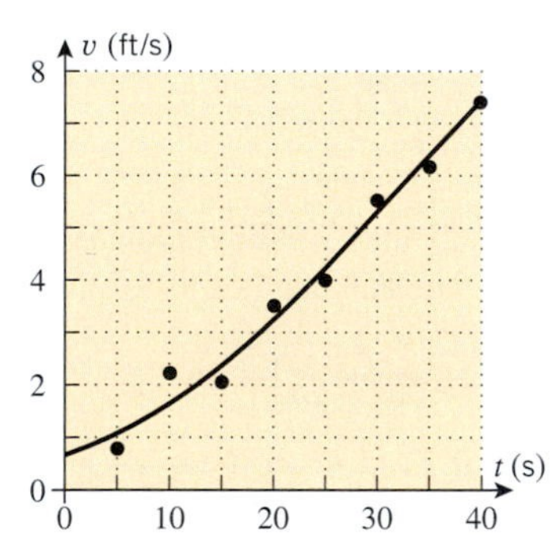

Figure Ex-31

32. Suppose that a particle moves along the x-axis so that its x-coordinate at time t is given by $x = ae^{kt} + be^{-kt}$.

(a) Show that the acceleration is proportional to x.

(b) Assuming that the velocity of the particle at time $t = 0$ is v_0, find a formula for the acceleration function in terms of a, b, x, and v_0.

In Exercises 33–42, evaluate the integrals by hand, and check your answers with a CAS if you have one.

33. $\int \frac{\cos 3x}{\sqrt{5+2\sin 3x}}\,dx$ **34.** $\int \frac{\sqrt{3+\sqrt{x}}}{\sqrt{x}}\,dx$

35. $\int \frac{x^2}{(ax^3+b)^2}\,dx$ **36.** $\int x\sec^2(ax^2)\,dx$

37. $\int [\ln(e^x) + \ln(e^{-x})]\,dx$

38. $\int_{-2}^{-1} \left(u^{-4} + 3u^{-2} - \frac{1}{u^5}\right) du$

39. $\int_{e}^{e^2} \frac{dx}{x \ln x}$

40. $\int_{0}^{1} \frac{dx}{\sqrt{e^x}}$

41. $\int_{0}^{\ln\sqrt{2}} \frac{1 + \cos(e^{-2x})}{e^{2x}}\,dx$

42. $\int_{0}^{1} \sin^2(\pi x)\cos(\pi x)\,dx$

C **43.** Use a CAS to approximate the area of the region in the first quadrant that lies below the curve $y = x + x^2 - x^3$ and above the x-axis.

C **44.** In each part, use a CAS to solve the initial-value problem.

(a) $\frac{dy}{dx} = x^2 \cos 3x;\ y(\pi/2) = -1$

(b) $\frac{dy}{dx} = \frac{x^3}{(4 + x^2)^{3/2}};\ y(0) = -2$

C **45.** In each part, use a CAS, where needed, to solve for k.

(a) $\int_{1}^{k} (x^3 - 2x - 1)\,dx = 0, \quad k > 1$

(b) $\int_{0}^{k} (x^2 + \sin 2x)\,dx = 3, \quad k \geq 0$

C **46.** Use a CAS to approximate the largest and smallest values of the integral

$$\int_{-1}^{x} \frac{t}{\sqrt{2 + t^3}}\,dt$$

for $1 \leq x \leq 3$.

C **47.** The function J_0 defined by

$$J_0(x) = \frac{1}{\pi}\int_{0}^{\pi} \cos(x \sin t)\,dt$$

is called the ***Bessel function of order zero***.

(a) Use a CAS to graph the equation $y = J_0(x)$ over the interval $0 \leq x \leq 8$.

(b) Find $J_0(1)$.

(c) Find the smallest positive zero of $J_0(x)$.

C **48.** Let A be the area under the curve $y = x^2$ over the interval $[0, 1]$.

(a) Find A by using Part 1 of the Fundamental Theorem of Calculus.

(b) Find A by computing the limit of the left endpoint approximations by hand, and then find the limit using a CAS.

(c) Find A by computing the limit of the right endpoint approximations by hand, and then find the limit using a CAS.

C **49.** In number theory, $\pi(n)$ denotes the number of prime numbers that are less than or equal to the positive integer n. For example, it can be shown with the help of a computer that $\pi(100{,}000) = 9592$; that is, there are 9592 prime numbers that are less than or equal to 100,000. There are two useful approximations to $\pi(n)$ that are appropriate for large values of n:

$$\pi(n) \approx \frac{n}{\ln n} \quad \text{and} \quad \pi(n) \approx \int_{2}^{n} \frac{1}{\ln t}\,dt$$

Use a CAS to determine which of these approximations produces the better estimate of $\pi(100{,}000)$.

EXPANDING THE CALCULUS HORIZON

Blammo the Human Cannonball

Blammo the Human Cannonball will be fired from a cannon and hopes to land in a small net at the opposite end of the circus arena. Your job as Blammo's manager is to do the mathematical calculations that will allow Blammo to perform his death-defying act safely. The methods that you will use are from the field of ***ballistics*** *(the study of projectile motion).*

The Problem

Blammo's cannon has a ***muzzle velocity*** of 35 m/s, which means that Blammo will leave the muzzle with that velocity. The muzzle opening will be 5 m above the ground, and Blammo's objective is to land in a net that is also 5 m above the ground and that extends a distance of 10 m between 90 m and 100 m from the cannon opening (Figure 1). Your mathematical problem is to determine the ***elevation angle*** α of the cannon (the angle from the horizontal to the cannon barrel) that will make Blammo land in the net.

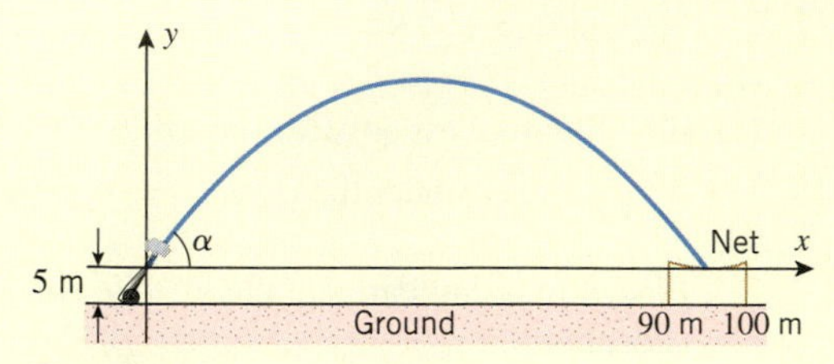

Figure 1

Modeling Assumptions

Blammo's trajectory will be determined by his initial velocity, the elevation angle of the cannon, and the forces that act on him after he leaves the muzzle. We will assume that the only force acting on Blammo after he leaves the muzzle is the downward force of the Earth's gravity. In particular, we will ignore the effect of air resistance. It will be convenient to introduce the xy-coordinate system shown in Figure 1 and to assume that Blammo is at the origin at time $t = 0$. We will also assume that Blammo's motion can be decomposed into two independent components, a horizontal component parallel to the x-axis and a vertical component parallel to the y-axis. We will analyze the horizontal and vertical components of Blammo's motion separately, and then we will combine the information to obtain a complete picture of his trajectory.

Blammo's Equations of Motion

We will denote the position and velocity functions for Blammo's horizontal component of motion by $x(t)$ and $v_x(t)$, and we will denote the position and velocity functions for his vertical component of motion by $y(t)$ and $v_y(t)$.

Since the only force acting on Blammo after he leaves the muzzle is the downward force of the Earth's gravity, there are no horizontal forces to alter his initial horizontal velocity $v_x(0)$. Thus, Blammo will have a constant velocity of $v_x(0)$ in the x-direction; this implies that

$$x(t) = v_x(0)t \tag{1}$$

In the y-direction Blammo is acted on only by the downward force of the Earth's gravity. Thus, his motion in this direction is governed by the free-fall model; hence, from (12) in Section 7.7 his vertical position function is

$$y(t) = y(0) + v_y(0)t - \tfrac{1}{2}gt^2$$

Taking $g = 9.8\ \text{m/s}^2$, and using the fact that $y(0) = 0$, this equation can be written as

$$y(t) = v_y(0)t - 4.9t^2 \tag{2}$$

Exercise 1 At time $t = 0$ Blammo's velocity is 35 m/s, and this velocity is directed at an angle α with the horizontal. It is a fact of physics that the initial velocity components $v_x(0)$ and $v_y(0)$ can be obtained geometrically from the muzzle velocity and the angle of elevation using the triangle shown in Figure 2. We will justify this later in the text, but for now use this fact to show that Equations (1) and (2) can be expressed as

$$x(t) = (35\cos\alpha)t$$

$$y(t) = (35\sin\alpha)t - 4.9t^2$$

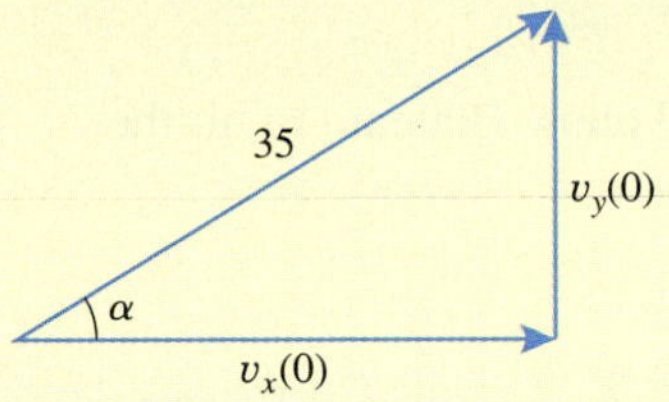

35

$v_y(0)$

α

$v_x(0)$

Figure 2

Exercise 2

(a) Use the result in Exercise 1 to find the velocity functions $v_x(t)$ and $v_y(t)$ in terms of the elevation angle α.

(b) Find the time t at which Blammo is at his maximum height above the x-axis, and show that this maximum height (in meters) is

$$y_{\max} = 62.5 \sin^2 \alpha$$

Exercise 3 The equations obtained in Exercise 1 can be viewed as parametric equations for Blammo's trajectory. Show, by eliminating the parameter t, that if $0 < \alpha < \pi/2$, then Blammo's trajectory is given by the equation

$$y = (\tan \alpha)x - \frac{0.004}{\cos^2 \alpha} x^2$$

Explain why Blammo's trajectory is a parabola.

Finding the Elevation Angle

Define Blammo's ***horizontal range*** R to be the horizontal distance he travels until he returns to the height of the muzzle opening ($y = 0$). Your objective is to find elevation angles that will make the horizontal range fall between 90 m and 100 m, thereby ensuring that Blammo lands in the net (Figure 3).

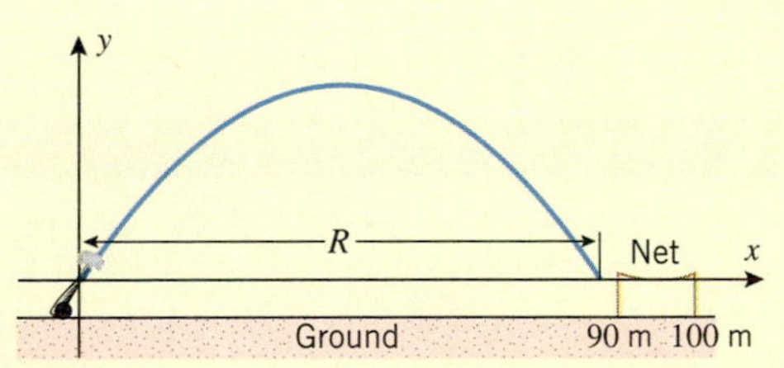

Figure 3

Exercise 4 Use a graphing utility and either the parametric equations obtained in Exercise 1 or the single equation obtained in Exercise 3 to generate Blammo's trajectories, taking elevation angles at increments of 10° from 15° to 85°. In each case, determine visually whether Blammo lands in the net.

Exercise 5 Find the time required for Blammo to return to his starting height ($y = 0$), and use that result to show that Blammo's range R is given by the formula

$$R = 125 \sin 2\alpha$$

Exercise 6

(a) Use the result in Exercise 5 to find two elevation angles that will allow Blammo to hit the midpoint of the net 95 m away.

(b) The tent is 55 m high. Explain why the larger elevation angle cannot be used.

Exercise 7 How much can the smaller elevation angle in Exercise 6 vary and still have Blammo hit the net between 90 m and 100 m?

Blammo's Shark Trick

Blammo is to be fired from 5 m above ground level with a muzzle velocity of 35 m/s over a flaming wall that is 20 m high and past a 5-m-high shark pool (Figure 4). To make the feat impressive, the pool will be made as long as possible. Your job as Blammo's manager is to determine the length of the pool, how far to place the cannon from the wall, and what elevation angle to use to ensure that Blammo clears the pool.

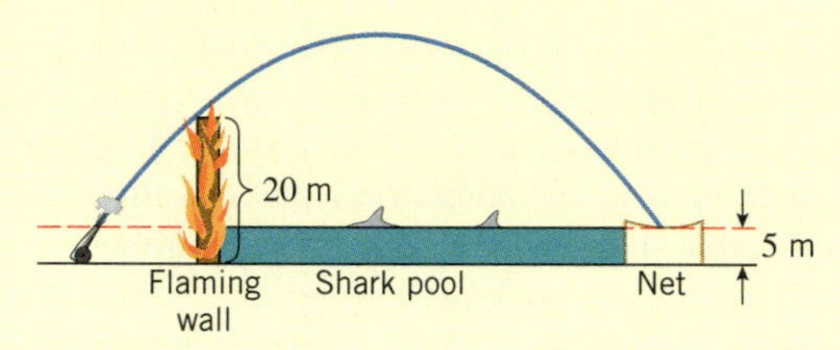

Figure 4

Exercise 8 Prepare a written presentation of the problem and your solution of it that is at an appropriate level for an engineer, physicist, or mathematician to read. Your presentation should contain the following elements: an explanation of all notation, a list and description of all formulas that will be used, a diagram that shows the orientation of any coordinate systems that will be used, a description of any assumptions you make to solve the problem, graphs that you think will enhance the presentation, and a clear step-by-step explanation of your solution.

Module by: John Rickert, Rose-Hulman Institute of Technology
Howard Anton, Drexel University

Additional material for this module can be found on the World Wide Web at http://www.wiley.com/college/anton

8

G.F.B. Riemann

APPLICATIONS OF THE DEFINITE INTEGRAL IN GEOMETRY, SCIENCE, AND ENGINEERING

In the last chapter we introduced the definite integral as the limit of Riemann sums in the context of finding areas. However, Riemann sums and definite integrals have applications that extend far beyond the area problem. In this chapter we will show how Riemann sums and definite integrals arise in such problems as finding the volume and surface area of a solid, finding the length of a plane curve, calculating the work done by a force, finding the pressure and force exerted by a fluid on a submerged object, and finding properties of suspended cables.

Although these problems are diverse, the required calculations can all be approached by the same procedure that we used to find areas—breaking the required calculation into "small parts," making an approximation that is good because the part is small, adding the approximations from the parts to produce a Riemann sum that approximates the entire quantity to be calculated, and then taking the limit of the Riemann sums to produce an exact result.

8.1 AREA BETWEEN TWO CURVES

In the last chapter we showed how to find the area between a curve $y = f(x)$ and an interval on the x-axis. Here we will show how to find the area between two curves.

A REVIEW OF RIEMANN SUMS

Before we consider the problem of finding the area between two curves it will be helpful to review the basic principle that underlies the calculation of area as a definite integral. Recall that if f is continuous and nonnegative on $[a, b]$, then the definite integral for the area A under $y = f(x)$ over the interval $[a, b]$ is obtained in four steps (Figure 8.1.1):

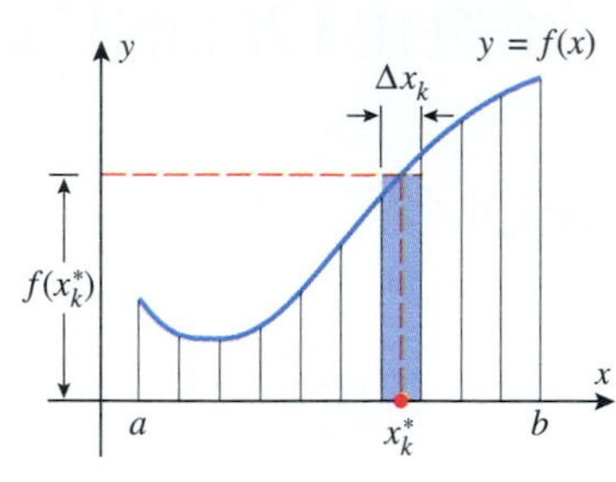

Figure 8.1.1

- Divide the interval $[a, b]$ into n subintervals, and use those subintervals to divide the area under the curve $y = f(x)$ into n strips.
- Assuming that the width of the kth strip is Δx_k, approximate the area of that strip by the area of a rectangle of width Δx_k and height $f(x_k^*)$, where x_k^* is any point in the kth subinterval.
- Add the approximate areas of the strips to approximate the entire area A by the Riemann sum:

$$A \approx \sum_{k=1}^{n} f(x_k^*)\Delta x_k$$

- Take the limit of the Riemann sums as the number of subintervals increases and their widths approach zero. This causes the error in the approximations to approach zero and produces the following definite integral for the exact area A:

$$A = \lim_{\max \Delta x_k \to 0} \sum_{k=1}^{n} f(x_k^*)\Delta x_k = \int_a^b f(x)\,dx$$

Observe the effect that the limit process has on the various parts of the Riemann sum:

- The quantity x_k^* in the Riemann sum becomes the variable x in the definite integral.
- The interval width Δx_k in the Riemann sum becomes the dx in the definite integral.
- The endpoints of the interval $[a, b]$ do not appear in the Riemann sum, but they become the limits of integration in the definite integral.

AREA BETWEEN $y = f(x)$ AND $y = g(x)$

We will now consider the following extension of the area problem.

8.1.1 FIRST AREA PROBLEM. Suppose that f and g are continuous functions on an interval $[a, b]$ and

$$f(x) \geq g(x) \quad \text{for} \quad a \leq x \leq b$$

[This means that the curve $y = f(x)$ lies above the curve $y = g(x)$ and that the two can touch but not cross.] Find the area A of the region bounded above by $y = f(x)$, below by $y = g(x)$, and on the sides by the lines $x = a$ and $x = b$ (Figure 8.1.2*a*).

To solve this problem we divide the interval $[a, b]$ into n subintervals, which has the effect of subdividing the region into n strips (Figure 8.1.2*b*). If we assume that the width of the kth strip is Δx_k, then the area of the strip can be approximated by the area of a rectangle of width Δx_k and height $f(x_k^*) - g(x_k^*)$, where x_k^* is any point in the kth subinterval. Adding these approximations yields the following Riemann sum that approximates the area A:

$$A \approx \sum_{k=1}^{n} [f(x_k^*) - g(x_k^*)]\Delta x_k$$

Taking the limit as n increases and the widths of the subintervals approach zero yields the

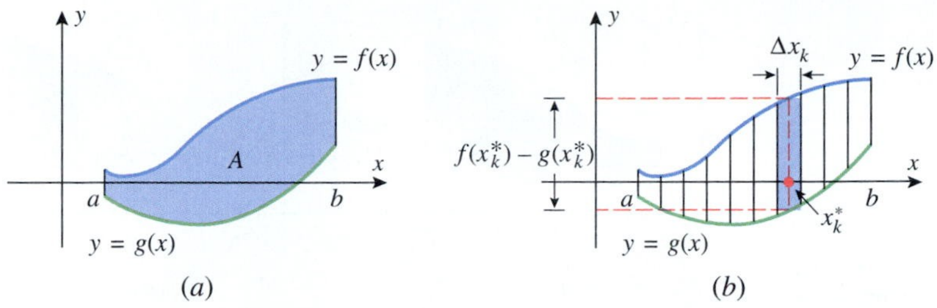

Figure 8.1.2

following definite integral for the area A between the curves:

$$A = \lim_{\max \Delta x_k \to 0} \sum_{k=1}^{n} [f(x_k^*) - g(x_k^*)]\Delta x_k = \int_a^b [f(x) - g(x)]\,dx$$

In summary, we have the following result:

8.1.2 AREA FORMULA. If f and g are continuous functions on the interval $[a, b]$, and if $f(x) \geq g(x)$ for all x in $[a, b]$, then the area of the region bounded above by $y = f(x)$, below by $y = g(x)$, on the left by the line $x = a$, and on the right by the line $x = b$ is

$$A = \int_a^b [f(x) - g(x)]\,dx \tag{1}$$

In the case where f and g are *nonnegative* on the interval $[a, b]$, the formula

$$A = \int_a^b [f(x) - g(x)]\,dx = \int_a^b f(x)\,dx - \int_a^b g(x)\,dx$$

states that the area A between the curves can be obtained by subtracting the area under $y = g(x)$ from the area under $y = f(x)$ (Figure 8.1.3).

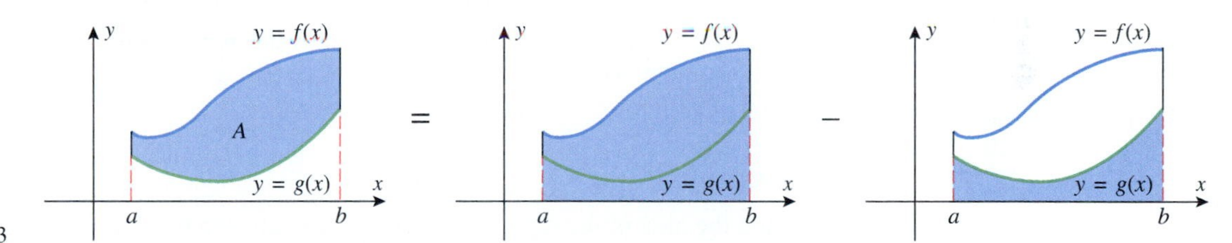

Figure 8.1.3

When the region is complicated, it may require some careful thought to determine the integrand and limits of integration in (1). Here is a systematic procedure that you can follow to set up this formula.

Step 1. Sketch the region and then draw a vertical line segment through the region at an arbitrary point x, connecting the top and bottom boundaries (Figure 8.1.4*a*).

Step 2. The top endpoint of the line segment sketched in Step 1 will be $f(x)$, the bottom one $g(x)$, and the length of the line segment will be $f(x) - g(x)$. This is the integrand in (1).

Step 3. To determine the limits of integration, imagine moving the line segment left and then right. The leftmost position at which the line segment intersects the region is $x = a$ and the rightmost is $x = b$ (Figures 8.1.4*b* and 8.1.4*c*).

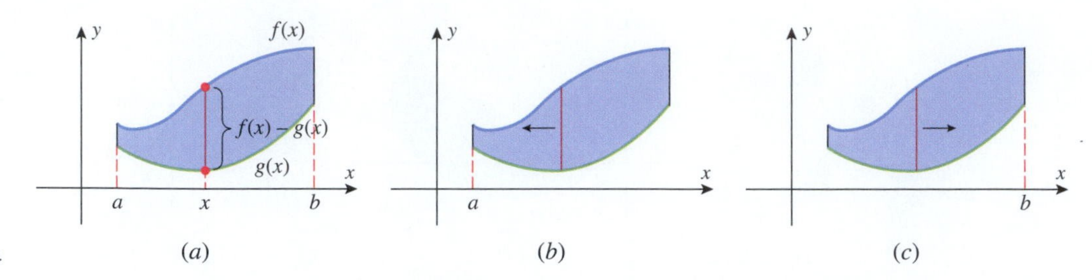

Figure 8.1.4

REMARK. It is not necessary to make an extremely accurate sketch in Step 1; the only purpose of the sketch is to determine which curve is the upper boundary and which is the lower boundary.

REMARK. There is a useful way of thinking about this procedure: If you view the vertical line segment as the "cross section" of the region at the point x, then Formula (1) states that the area between the curves is obtained by integrating the length of the cross section over the interval from a to b.

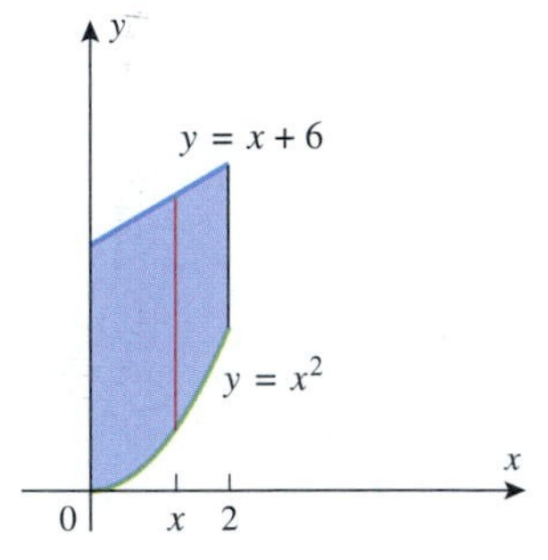

Figure 8.1.5

Example 1

Find the area of the region bounded above by $y = x + 6$, bounded below by $y = x^2$, and bounded on the sides by the lines $x = 0$ and $x = 2$.

Solution. The region and a cross section are shown in Figure 8.1.5. The cross section extends from $g(x) = x^2$ on the bottom to $f(x) = x + 6$ on the top. If the cross section is moved through the region, then its leftmost position will be $x = 0$ and its rightmost position will be $x = 2$. Thus, from (1)

$$A = \int_0^2 [(x+6) - x^2]\,dx = \left[\frac{x^2}{2} + 6x - \frac{x^3}{3}\right]_0^2 = \frac{34}{3} - 0 = \frac{34}{3} \qquad \blacktriangleleft$$

It is possible that the upper and lower boundaries of a region may intersect at one or both endpoints, in which case the sides of the region will be points, rather than vertical line segments (Figure 8.1.6). When that occurs you will have to determine the points of intersection to obtain the limits of integration.

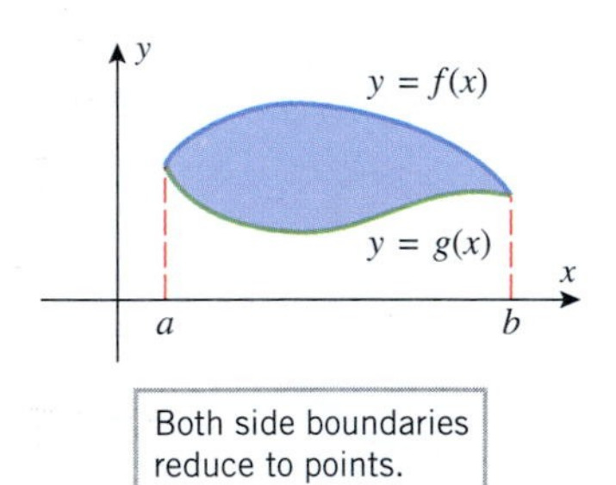

Both side boundaries reduce to points.

The left-hand boundary reduces to a point.

Figure 8.1.6

Example 2

Find the area of the region that is enclosed between the curves $y = x^2$ and $y = x + 6$.

Solution. A sketch of the region (Figure 8.1.7) shows that the lower boundary is $y = x^2$ and the upper boundary is $y = x + 6$. At the endpoints of the region, the upper and lower boundaries have the same y-coordinates; thus, to find the endpoints we equate

$$y = x^2 \quad \text{and} \quad y = x + 6 \tag{2}$$

This yields

$$x^2 = x + 6 \quad \text{or} \quad x^2 - x - 6 = 0 \quad \text{or} \quad (x+2)(x-3) = 0$$

from which we obtain

$$x = -2 \quad \text{and} \quad x = 3$$

Although the y-coordinates of the endpoints are not essential to our solution, they may be obtained from (2) by substituting $x = -2$ and $x = 3$ in either equation. This yields $y = 4$ and $y = 9$, so the upper and lower boundaries intersect at $(-2, 4)$ and $(3, 9)$.

From (1) with $f(x) = x + 6$, $g(x) = x^2$, $a = -2$, and $b = 3$, we obtain the area

$$A = \int_{-2}^3 [(x+6) - x^2]\,dx = \left[\frac{x^2}{2} + 6x - \frac{x^3}{3}\right]_{-2}^3 = \frac{27}{2} - \left(-\frac{22}{3}\right) = \frac{125}{6} \qquad \blacktriangleleft$$

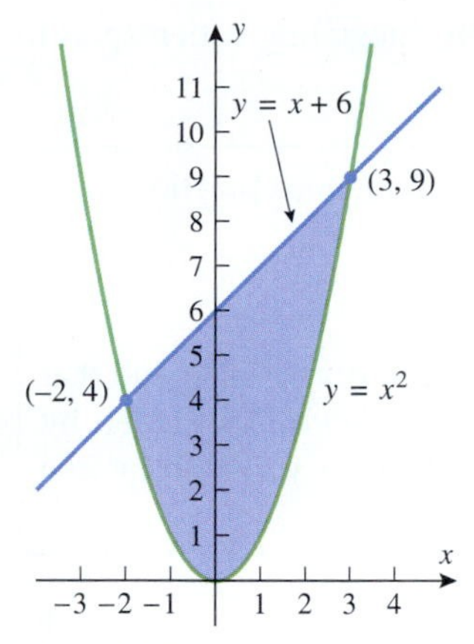

Figure 8.1.7

It is possible for the upper or lower boundary of a region to consist of two or more different curves, in which case it will be necessary to subdivide the region into smaller pieces in order to apply Formula (1). This is illustrated in the next example.

Example 3

Find the area of the region enclosed by $x = y^2$ and $y = x - 2$.

Solution. To make an accurate sketch of the region, we need to know where the curves $x = y^2$ and $y = x - 2$ intersect. In Example 2 we found intersections by equating the expressions for y. Here it is easier to rewrite the latter equation as $x = y + 2$ and equate the expressions for x, namely

$$x = y^2 \quad \text{and} \quad x = y + 2 \tag{3}$$

This yields

$$y^2 = y + 2 \quad \text{or} \quad y^2 - y - 2 = 0 \quad \text{or} \quad (y + 1)(y - 2) = 0$$

from which we obtain $y = -1$, $y = 2$. Substituting these values in either equation in (3) we see that the corresponding x-values are $x = 1$ and $x = 4$, respectively, so the points of intersection are $(1, -1)$ and $(4, 2)$ (Figure 8.1.8*a*).

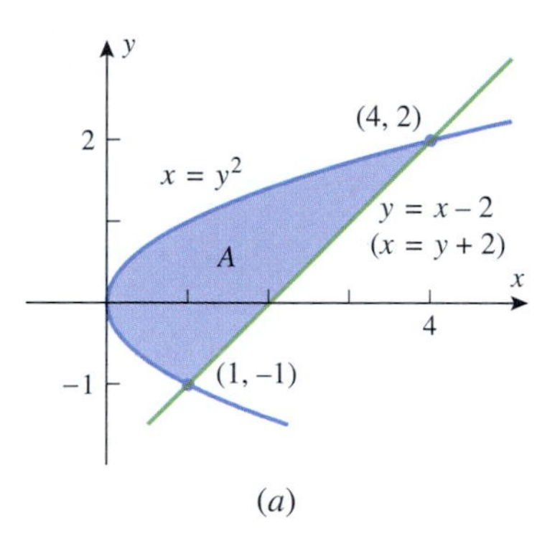

(*a*)

To apply Formula (1), the equations of the boundaries must be written so that y is expressed explicitly as a function of x. The upper boundary can be written as $y = \sqrt{x}$ (rewrite $x = y^2$ as $y = \pm\sqrt{x}$ and choose the $+$ for the upper portion of the curve). The lower portion of the boundary consists of two parts: $y = -\sqrt{x}$ for $0 \le x \le 1$ and $y = x - 2$ for $1 \le x \le 4$ (Figure 8.1.8*b*). Because of this change in the formula for the lower boundary, it is necessary to divide the region into two parts and find the area of each part separately.

From (1) with $f(x) = \sqrt{x}$, $g(x) = -\sqrt{x}$, $a = 0$, and $b = 1$, we obtain

$$A_1 = \int_0^1 [\sqrt{x} - (-\sqrt{x})]\,dx = 2\int_0^1 \sqrt{x}\,dx = 2\left[\frac{2}{3}x^{3/2}\right]_0^1 = \frac{4}{3} - 0 = \frac{4}{3}$$

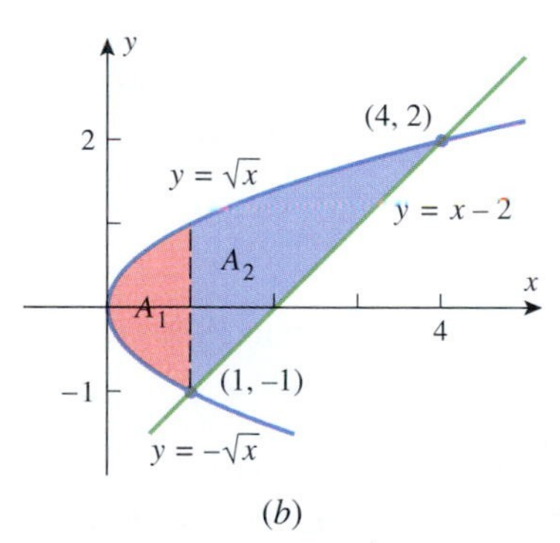

(*b*)

Figure 8.1.8

From (1) with $f(x) = \sqrt{x}$, $g(x) = x - 2$, $a = 1$, and $b = 4$, we obtain

$$A_2 = \int_1^4 [\sqrt{x} - (x - 2)]\,dx = \int_1^4 (\sqrt{x} - x + 2)\,dx$$

$$= \left[\frac{2}{3}x^{3/2} - \frac{1}{2}x^2 + 2x\right]_1^4 = \left(\frac{16}{3} - 8 + 8\right) - \left(\frac{2}{3} - \frac{1}{2} + 2\right) = \frac{19}{6}$$

Thus, the area of the entire region is

$$A = A_1 + A_2 = \frac{4}{3} + \frac{19}{6} = \frac{9}{2}$$

◀

FOR THE READER. It is assumed in Formula (1) that $f(x) \ge g(x)$ for all x in the interval $[a, b]$. What do you think that the integral represents if this condition is not satisfied, that is, the graphs of f and g cross one another over the interval? Explain your reasoning, and give an example to support your conclusion.

Example 4

Figure 8.1.9 shows velocity versus time curves for two race cars that move along a straight track, starting from rest at the same line. What does the area A between the curves over the interval $0 \le t \le T$ represent?

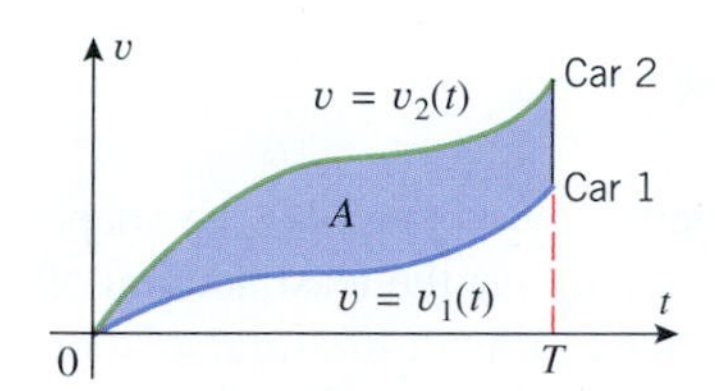

Figure 8.1.9

Solution. From (1)

$$A = \int_0^T [v_2(t) - v_1(t)]\,dt = \int_0^T v_2(t)\,dt - \int_0^T v_1(t)\,dt$$

But from 7.7.4, the first integral is the distance traveled by car 2 during the time interval, and the second integral is the distance traveled by car 1. Thus, A is the distance by which car 2 is ahead of car 1 at time T. ◀

REVERSING THE ROLES OF x AND y

Sometimes it is possible to avoid splitting a region into parts by integrating with respect to y rather than x. We will now show how this can be done.

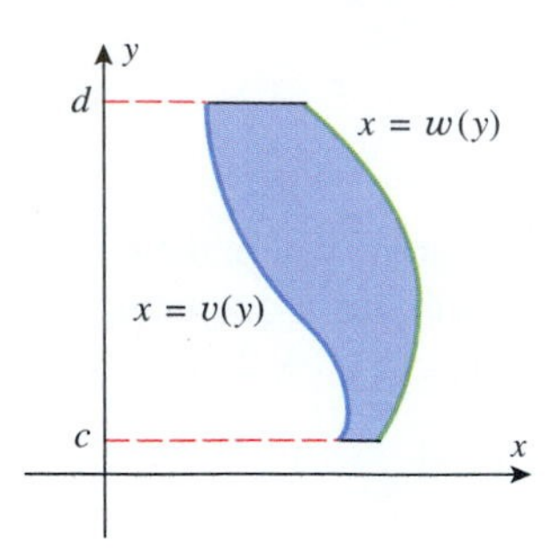

Figure 8.1.10

8.1.3 SECOND AREA PROBLEM. Suppose that w and v are continuous functions of y on an interval $[c, d]$ and that

$$w(y) \geq v(y) \quad \text{for} \quad c \leq y \leq d$$

[This means that the curve $x = w(y)$ lies to the right of the curve $x = v(y)$ and that the two can touch but not cross.] Find the area A of the region bounded on the left by $x = v(y)$, on the right by $x = w(y)$, and above and below by the lines $y = d$ and $y = c$ (Figure 8.1.10).

Proceeding as in the derivation of (1), but with the roles of x and y reversed, leads to the following analog of 8.1.2.

8.1.4 AREA FORMULA. If w and v are continuous functions and if $w(y) \geq v(y)$ for all y in $[c, d]$, then the area of the region bounded on the left by $x = v(y)$, on the right by $x = w(y)$, below by $y = c$, and above by $y = d$ is

$$A = \int_c^d [w(y) - v(y)]\, dy \tag{4}$$

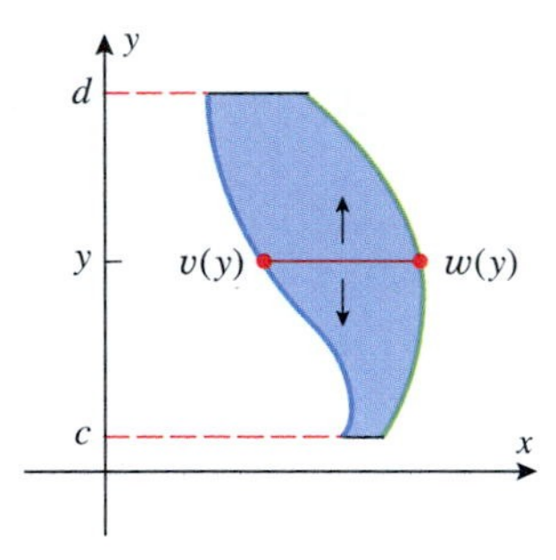

Figure 8.1.11

The guiding principle in applying this formula is the same as with (1): The integrand in (4) can be viewed as the length of the horizontal cross section at the point y, in which case Formula (4) states that the area can be obtained by integrating the length of the horizontal cross section over the interval $[c, d]$ on the y-axis (Figure 8.1.11).

In Example 3, where we integrated with respect to x to find the area of the region enclosed by $x = y^2$ and $y = x - 2$, we had to split the region into parts and evaluate two integrals. In the next example we will see that by integrating with respect to y no splitting of the region is necessary.

Example 5

Find the area of the region enclosed by $x = y^2$ and $y = x - 2$, integrating with respect to y.

Solution. From Figure 8.1.8 the left boundary is $x = y^2$, the right boundary is $y = x - 2$, and the region extends over the interval $-1 \leq y \leq 2$. However, to apply (4) the equations for the boundaries must be written so that x is expressed explicitly as a function of y. Thus, we rewrite $y = x - 2$ as $x = y + 2$. It now follows from (4) that

$$A = \int_{-1}^{2} [(y+2) - y^2]\, dy = \left[\frac{y^2}{2} + 2y - \frac{y^3}{3}\right]_{-1}^{2} = \frac{9}{2}$$

which agrees with the result obtained in Example 3. ◀

REMARK. The choice between Formulas (1) and (4) is generally dictated by the shape of the region, and one would usually choose the formula that requires the least amount of splitting. However, if the integral(s) resulting by one method are difficult to evaluate, then the other method might be preferable, even if it requires more splitting.

EXERCISE SET 8.1 Graphing Calculator C CAS

In Exercises 1–4, find the area of the shaded region.

1.

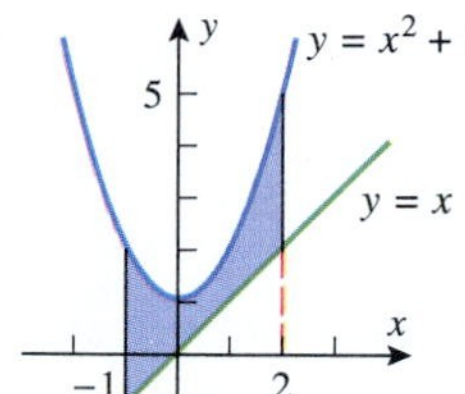

2.

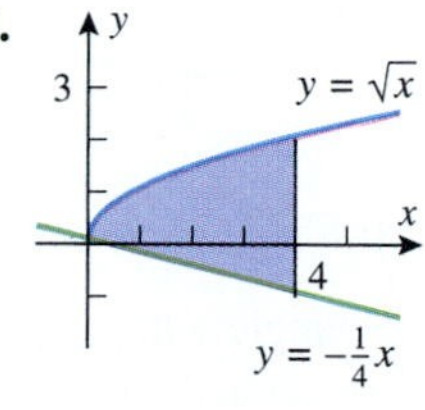

3.

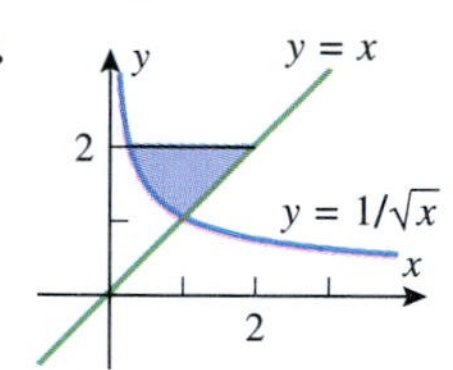

4.

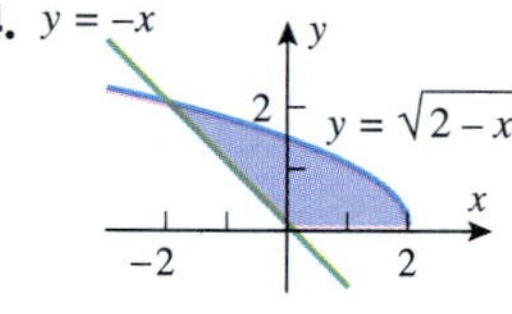

5. Find the area of the region enclosed by the curves $y = x^2$ and $y = 4x$ by integrating
(a) with respect to x (b) with respect to y.

6. Find the area of the region enclosed by the curves $y^2 = 4x$ and $y = 2x - 4$ by integrating
(a) with respect to x (b) with respect to y.

In Exercises 7–16, sketch the region enclosed by the curves, and find its area.

7. $y = x^2,\ y = \sqrt{x},\ x = 1/4,\ x = 1$

8. $y = x^3 - 4x,\ y = 0,\ x = 0,\ x = 2$

9. $y = \cos 2x,\ y = 0,\ x = \pi/4,\ x = \pi/2$

10. $y = \sec^2 x,\ y = 2,\ x = -\pi/4,\ x = \pi/4$

11. $x = \sin y,\ x = 0,\ y = \pi/4,\ y = 3\pi/4$

12. $x^2 = y,\ x = y - 2$

13. $y = e^x,\ y = e^{2x},\ x = 0,\ x = \ln 2$

14. $x = 1/y,\ x = 0,\ y = 1,\ y = e$

15. $y = 2 + |x - 1|,\ y = -\frac{1}{5}x + 7$

16. $y = x,\ y = 4x,\ y = -x + 2$

In Exercises 17–22, use a graphing utility, where helpful, to find the area of the region enclosed by the curves.

17. $y = x^3 - 4x^2 + 3x,\ y = 0,\ x = 0,\ x = 3$

18. $y = x^3 - 2x^2,\ y = 2x^2 - 3x,\ x = 0,\ x = 3$

19. $y = \sin x,\ y = \cos x,\ x = 0,\ x = 2\pi$

20. $y = x^3 - 4x,\ y = 0,\ x = -2,\ x = 2$

21. $x = y^3 - y,\ x = 0$

22. $x = y^3 - 4y^2 + 3y,\ x = y^2 - y$

23. Use a CAS to find the area enclosed by $y = 3 - 2x$ and $y = x^6 + 2x^5 - 3x^4 + x^2$.

24. Use a CAS to find the exact area enclosed by the curves $y = x^5 - 2x^3 - 3x$ and $y = x^3$.

25. Find a horizontal line $y = k$ that divides the area between $y = x^2$ and $y = 9$ into two equal parts.

26. Find a vertical line $x = k$ that divides the area enclosed by $x = \sqrt{y}$, $x = 2$, and $y = 0$ into two equal parts.

27. (a) Find the area of the region enclosed by the parabola $y = 2x - x^2$ and the x-axis.
(b) Find the value of m so that the line $y = mx$ divides the region in part (a) into two regions of equal area.

28. Find the area between the curve $y = \sin x$ and the line segment joining the points $(0, 0)$ and $(5\pi/6, 1/2)$ on the curve.

29. Suppose that f and g are integrable on $[a, b]$, but neither $f(x) \geq g(x)$ nor $g(x) \geq f(x)$ holds for all x in $[a, b]$ [i.e., the curves $y = f(x)$ and $y = g(x)$ are intertwined].
(a) What is the geometric significance of the integral

$$\int_a^b [f(x) - g(x)]\,dx?$$

(b) What is the geometric significance of the integral

$$\int_a^b |f(x) - g(x)|\,dx?$$

30. Let $A(n)$ be the area in the first quadrant enclosed by the curves $y = \sqrt[n]{x}$ and $y = x$.
(a) By considering how the graph of $y = \sqrt[n]{x}$ changes as n increases, make a conjecture about the limit of $A(n)$ as $n \to +\infty$.
(b) Confirm your conjecture by calculating the limit.

In Exercises 31 and 32, use Newton's Method (Section 6.4), where needed, to approximate the x-coordinates of the intersections of the curves to at least four decimal places; and then use those approximations to approximate the area of the region.

31. The region that lies below the curve $y = \sin x$ and above the line $y = 0.2x$, where $x \geq 0$.

32. The region enclosed by the graphs of $y = x^2$ and $y = \cos x$.

33. The accompanying figure shows velocity versus time curves for two cars that move along a straight track, accelerating from rest at a common starting line.
(a) How far apart are the cars after 60 seconds?
(b) How far apart are the cars after T seconds, where $0 \leq T \leq 60$?

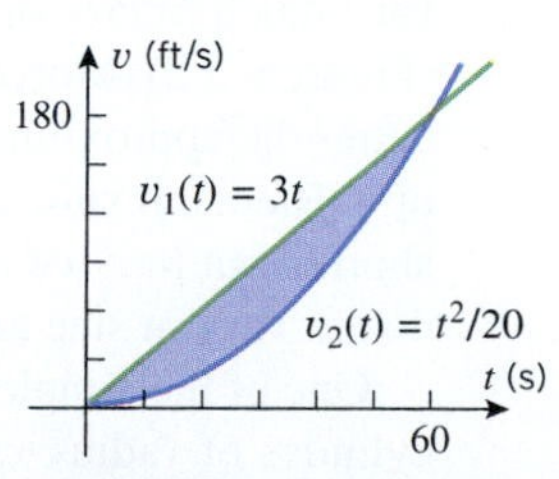

Figure Ex-33

34. The accompanying figure shows acceleration time curves for two cars that move along a straight track, accelerating from rest at the starting line. What does the area A between the curves over the interval $0 \leq t \leq T$ represent? Justify your answer.

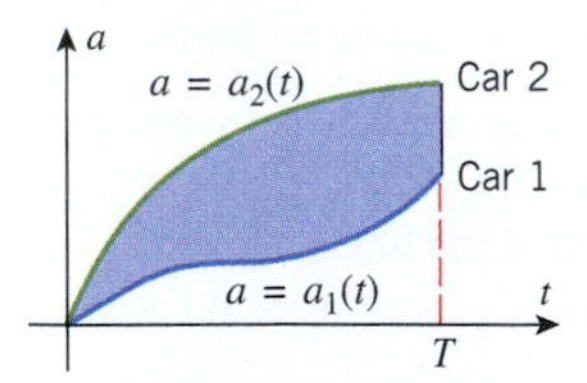

Figure Ex-34

35. Find the area of the region enclosed between the curve $x^{1/2} + y^{1/2} = a^{1/2}$ and the coordinate axes.

36. Show that the area of the ellipse in the accompanying figure is πab. [*Hint:* Use a formula from geometry.]

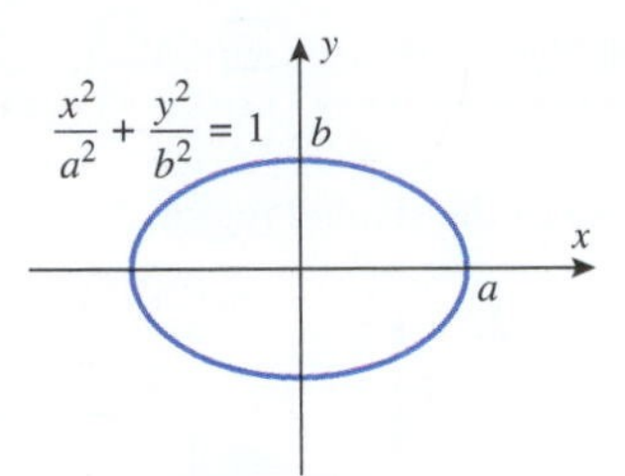

Figure Ex-36

37. A rectangle with edges parallel to the coordinate axes has one vertex at the origin and the diagonally opposite vertex on the curve $y = kx^m$ at the point where $x = b$ $(b > 0, k > 0,$ and $m \geq 0)$. Show that the fraction of the area of the rectangle that lies between the curve and the x-axis depends on m but not on k or b.

8.2 VOLUMES BY SLICING; DISKS AND WASHERS

In the last section we showed that the area of a plane region bounded by two curves can be obtained by integrating the length of a general cross section over an appropriate interval. In this section we will see that the same basic principle can be used to find volumes of certain three-dimensional solids.

VOLUMES BY SLICING

Recall that the underlying principle for finding the area of a plane region is to divide the region into thin strips, approximate the area of each strip by the area of a rectangle, add the approximations to form a Riemann sum, and take the limit of the Riemann sums to produce an integral for the area. Under appropriate conditions, the same strategy can be used to find the volume of a solid. The idea is to divide the solid into thin slabs, approximate the volume of each slab, add the approximations to form a Riemann sum, and take the limit of the Riemann sums to produce an integral for the volume (Figure 8.2.1).

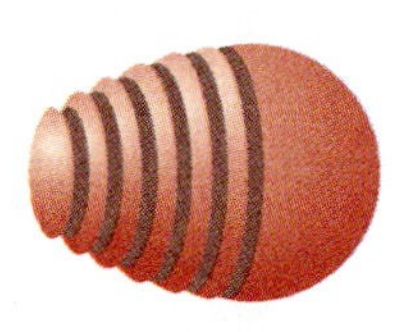

Figure 8.2.1

In a thin slab, the cross sections do not vary much in size and shape.

Figure 8.2.2

What makes this method work is the fact that a *thin* slab has cross sections that do not vary much in size or shape, which, as we will see, makes its volume easy to approximate (Figure 8.2.2). Moreover, the thinner the slab, the less variation in its cross sections and the better the approximation. Thus, once we approximate the volumes of the slabs, we can set up a Riemann sum whose limit is the volume of the entire solid. We will give the details shortly, but first we need to discuss how to find the volume of a solid whose cross sections do not vary in size and shape (i.e., are congruent).

One of the simplest examples of a solid with congruent cross sections is a right circular cylinder of radius r, since all cross sections taken perpendicular to the central axis are circular regions of radius r. The volume V of a right circular cylinder of radius r and height

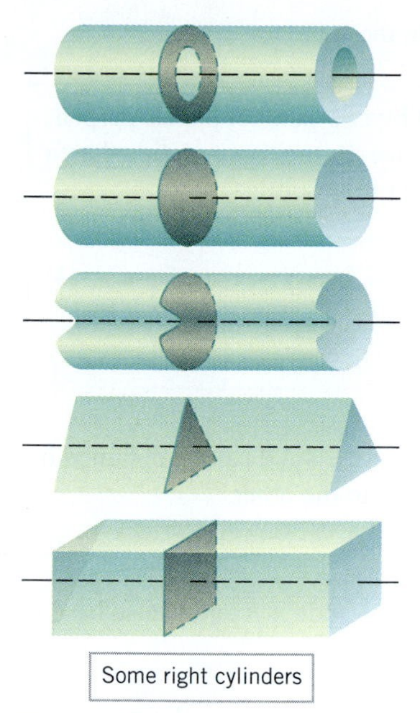

Figure 8.2.3

h can be expressed in terms of the height and the area of a cross section as

$$V = \pi r^2 h = [\text{area of a cross section}] \times [\text{height}] \tag{1}$$

This is a special case of a more general volume formula that applies to solids called *right cylinders*. A ***right cylinder*** is a solid that is generated when a plane region is translated along a line or ***axis*** that is perpendicular to the region (Figure 8.2.3). The distance h that the region is translated is called the ***height*** or sometimes the ***width*** of the cylinder, and each cross section is a duplicate of the translated region. We will assume that the volume V of a right cylinder with cross-sectional area A and height h is given by

$$V = A \cdot h = [\text{area of a cross section}] \times [\text{height}] \tag{2}$$

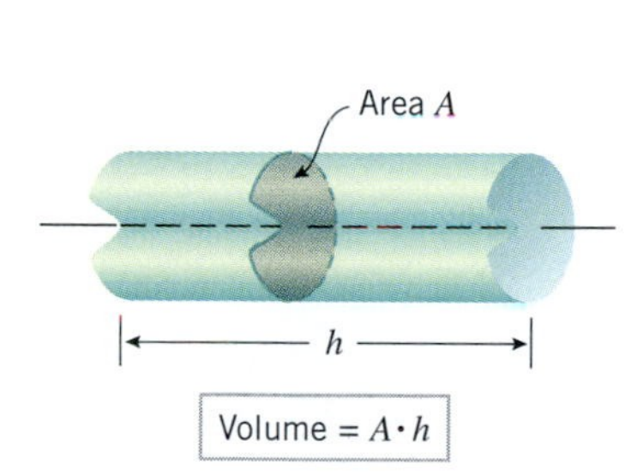

Figure 8.2.4

(Figure 8.2.4). Note that this is consistent with Formula (1) for the volume of a right circular cylinder. We now have all of the tools required to solve the following problem.

8.2.1 PROBLEM. Let S be a solid that extends along the x-axis and is bounded on the left and right, respectively, by the planes that are perpendicular to the x-axis at $x = a$ and $x = b$ (Figure 8.2.5*a*). Find the volume V of the solid, assuming that its cross-sectional area $A(x)$ is known at each point x in the interval $[a, b]$.

To solve this problem we divide the interval $[a, b]$ into n subintervals, which has the effect of dividing the solid into n slabs (Figure 8.2.5*b*).

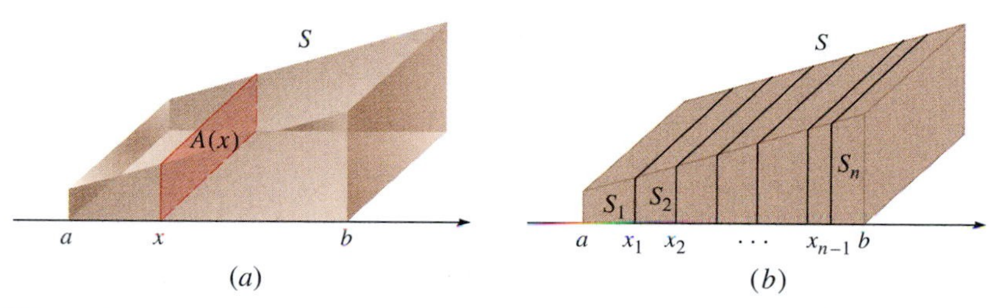

Figure 8.2.5

If we assume that the width of the kth slab is Δx_k, then the volume of the slab can be approximated by the volume of a right cylinder of width (height) Δx_k and cross-sectional area $A(x_k^*)$, where x_k^* is any point in the kth subinterval (Figure 8.2.6). Adding these approximations yields the following Riemann sum that approximates the volume V:

$$V \approx \sum_{k=1}^{n} A(x_k^*)\Delta x_k$$

Taking the limit as n increases and the widths of the subintervals approach zero yields the definite integral

$$V = \lim_{\max \Delta x_k \to 0} \sum_{k=1}^{n} A(x_k^*)\Delta x_k = \int_a^b A(x)\,dx$$

In summary, we have the following result:

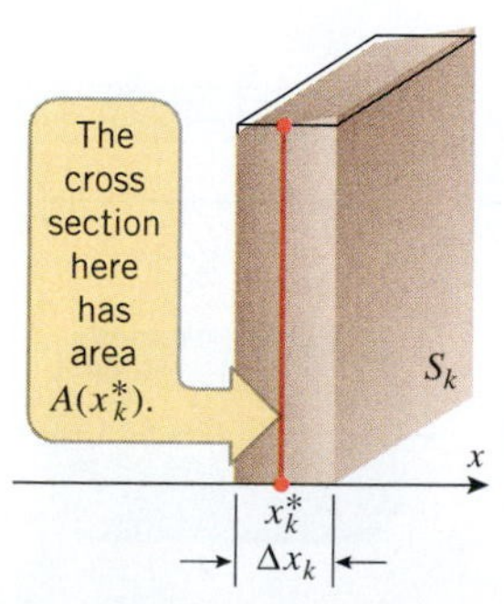

Figure 8.2.6

8.2.2 VOLUME FORMULA. Let S be a solid bounded by two parallel planes perpendicular to the x-axis at $x = a$ and $x = b$. If, for each x in $[a, b]$, the cross-sectional area of S perpendicular to the x-axis is $A(x)$, then the volume of the solid is

$$V = \int_a^b A(x)\,dx \tag{3}$$

provided $A(x)$ is integrable.

There is a similar result for cross sections perpendicular to the y-axis.

> **8.2.3 VOLUME FORMULA.** Let S be a solid bounded by two parallel planes perpendicular to the y-axis at $y = c$ and $y = d$. If, for each y in $[c, d]$, the cross-sectional area of S perpendicular to the y-axis is $A(y)$, then the volume of the solid is
>
> $$V = \int_c^d A(y)\,dy \tag{4}$$
>
> provided $A(y)$ is integrable.

REMARK. In words, these formulas state that the volume of the solid can be obtained by integrating the cross-sectional area from one end of the solid to the other.

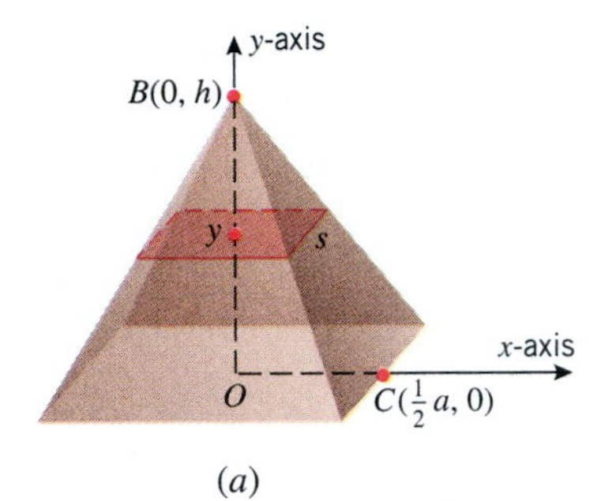

(a)

B
h − y
½s
y
h
O ½a C

(b)

Figure 8.2.7

Example 1

Derive the formula for the volume of a right pyramid whose altitude is h and whose base is a square with sides of length a.

Solution. As illustrated in Figure 8.2.7a, we introduce a rectangular coordinate system in which the y-axis passes through the apex and is perpendicular to the base, and the x-axis passes through the base and is parallel to a side of the base.

At any point y in the interval $[0, h]$ on the y-axis, the cross section perpendicular to the y-axis is a square. If s denotes the length of a side of this square, then by similar triangles (Figure 8.2.7b)

$$\frac{\frac{1}{2}s}{\frac{1}{2}a} = \frac{h-y}{h} \quad \text{or} \quad s = \frac{a}{h}(h-y)$$

Thus, the area $A(y)$ of the cross section at y is

$$A(y) = s^2 = \frac{a^2}{h^2}(h-y)^2$$

and by (4) the volume is

$$V = \int_0^h A(y)\,dy = \int_0^h \frac{a^2}{h^2}(h-y)^2\,dy = \frac{a^2}{h^2}\int_0^h (h-y)^2\,dy$$

$$= \frac{a^2}{h^2}\left[-\frac{1}{3}(h-y)^3\right]_{y=0}^h = \frac{a^2}{h^2}\left[0 + \frac{1}{3}h^3\right] = \frac{1}{3}a^2h$$

That is, the volume is $\frac{1}{3}$ of the area of the base times the altitude. ◀

SOLIDS OF REVOLUTION

A ***solid of revolution*** is a solid that is generated by revolving a plane region about a line that lies in the same plane as the region; the line is called the ***axis of revolution***. Many familiar solids are of this type (Figure 8.2.8).

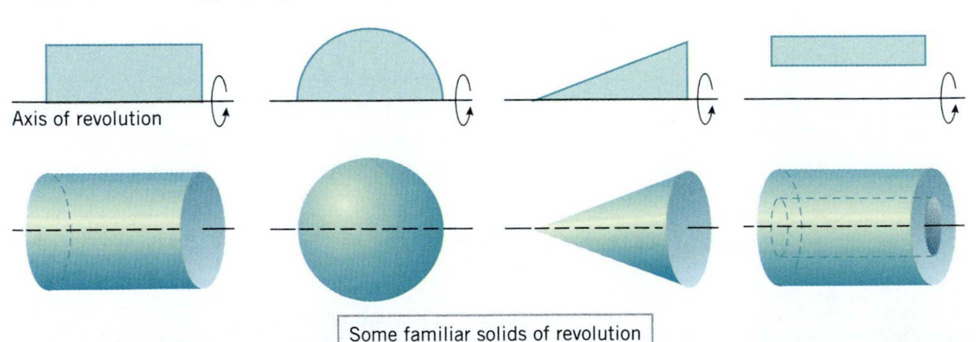

Figure 8.2.8

We will be interested in the following general problem:

> **8.2.4** PROBLEM. Let f be continuous and nonnegative on $[a, b]$, and let R be the region that is bounded above by $y = f(x)$, below by the x-axis, and on the sides by the lines $x = a$ and $x = b$ (Figure 8.2.9a). Find the volume of the solid of revolution that is generated by revolving the region R about the x-axis.

We can solve this problem by slicing. For this purpose, observe that the cross section of the solid taken perpendicular to the x-axis at the point x is a circular disk of radius $f(x)$ (Figure 8.2.9b). The area of this region is

$$A(x) = \pi[f(x)]^2$$

Thus, from (3) the volume of the solid is

$$V = \int_a^b \pi[f(x)]^2\,dx \tag{5}$$

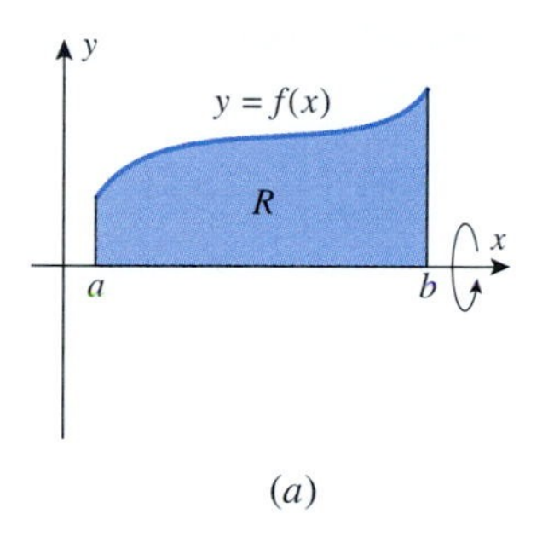

(a)

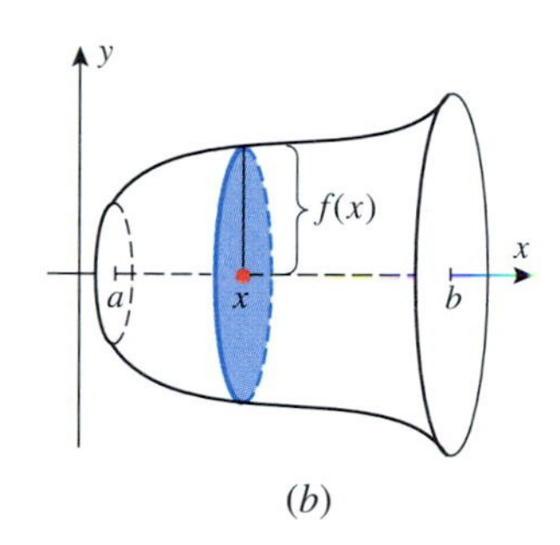

(b)

Figure 8.2.9

VOLUMES BY DISKS PERPENDICULAR TO THE x-AXIS

Because the cross sections are disk shaped, the application of this formula is called the ***method of disks***.

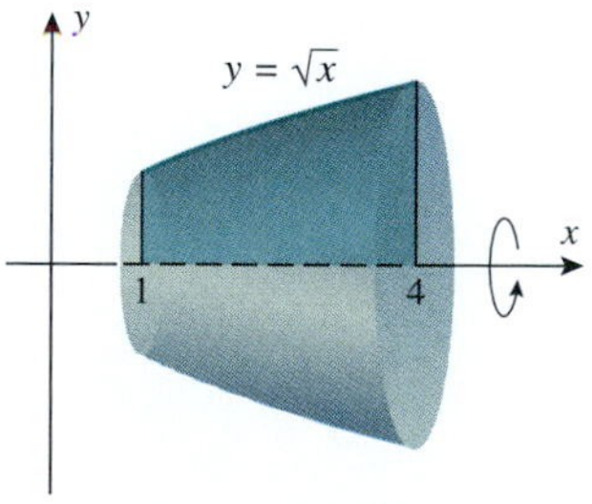

Example 2

Find the volume of the solid that is obtained when the region under the curve $y = \sqrt{x}$ over the interval $[1, 4]$ is revolved about the x-axis (Figure 8.2.10).

Figure 8.2.10

Solution. From (5), the volume is

$$V = \int_a^b \pi[f(x)]^2\,dx = \int_1^4 \pi x\,dx = \left.\frac{\pi x^2}{2}\right]_1^4 = 8\pi - \frac{\pi}{2} = \frac{15\pi}{2} \quad \blacktriangleleft$$

Example 3

Derive the formula for the volume of a sphere of radius r.

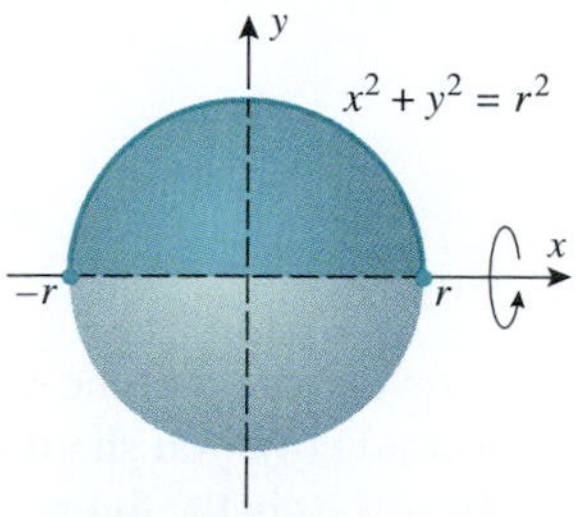

Figure 8.2.11

Solution. As indicated in Figure 8.2.11, a sphere of radius r can be generated by revolving the upper semicircular disk enclosed between the x-axis and

$$x^2 + y^2 = r^2$$

about the x-axis. Since the upper half of this circle is the graph of $y = f(x) = \sqrt{r^2 - x^2}$, it follows from (5) that the volume of the sphere is

$$V = \int_a^b \pi[f(x)]^2\,dx = \int_{-r}^{r} \pi(r^2 - x^2)\,dx = \pi\left[r^2 x - \frac{x^3}{3}\right]_{-r}^{r} = \frac{4}{3}\pi r^3 \quad \blacktriangleleft$$

VOLUMES BY WASHERS PERPENDICULAR TO THE x-AXIS

Not all solids of revolution have solid interiors; some have holes or channels that create interior surfaces, as in the last part of Figure 8.2.8. Thus, we will be interested in problems of the following type.

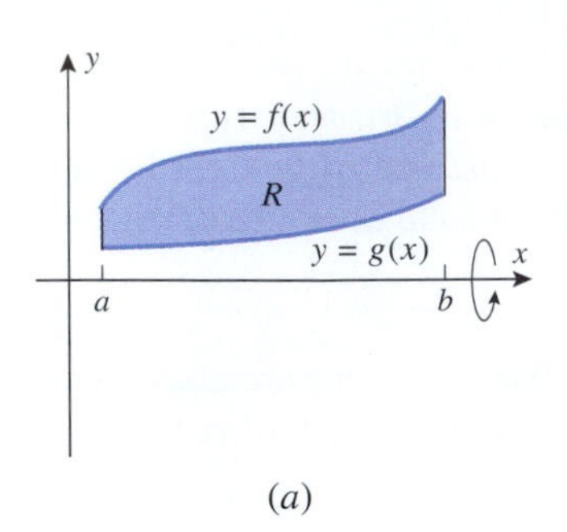

(a)

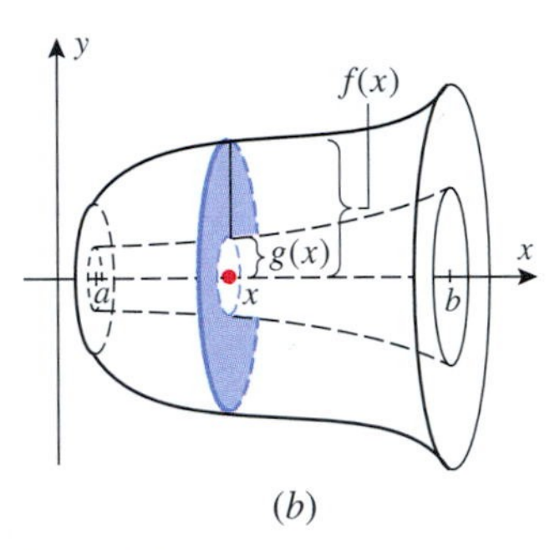

(b)

Figure 8.2.12

8.2.5 PROBLEM. Let f and g be continuous and nonnegative on $[a, b]$, and suppose that $f(x) \geq g(x)$ for all x in the interval $[a, b]$. Let R be the region that is bounded above by $y = f(x)$, below by $y = g(x)$, and on the sides by the lines $x = a$ and $x = b$ (Figure 8.2.12a). Find the volume of the solid of revolution that is generated by revolving the region R about the x-axis.

We can solve this problem by slicing. For this purpose, observe that the cross section of the solid taken perpendicular to the x-axis at the point x is the annular or "washer-shaped" region with inner radius $g(x)$ and outer radius $f(x)$ (Figure 8.2.12b); hence its area is

$$A(x) = \pi[f(x)]^2 - \pi[g(x)]^2 = \pi([f(x)]^2 - [g(x)]^2)$$

Thus, from (3) the volume of the solid is

$$V = \int_a^b \pi([f(x)]^2 - [g(x)]^2)\,dx \tag{6}$$

Because the cross sections are washer shaped, the application of this formula is called the ***method of washers***.

Example 4

Find the volume of the solid generated when the region between the graphs of the equations $f(x) = \frac{1}{2} + x^2$ and $g(x) = x$ over the interval $[0, 2]$ is revolved about the x-axis (Figure 8.2.13).

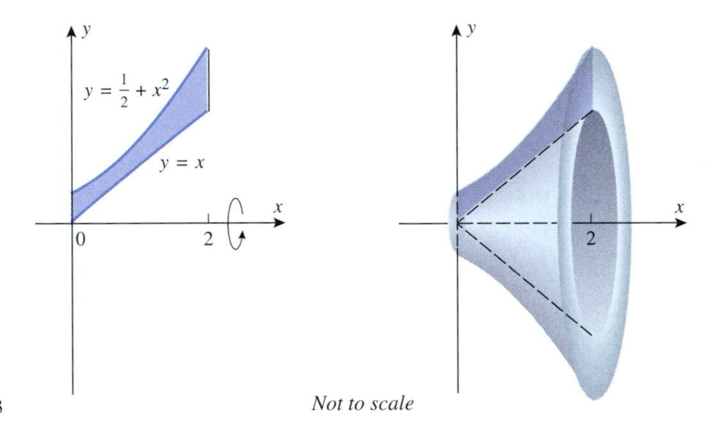

Figure 8.2.13

Solution. From (6) the volume is

$$V = \int_a^b \pi([f(x)]^2 - [g(x)]^2)\,dx = \int_0^2 \pi\left(\left[\tfrac{1}{2} + x^2\right]^2 - x^2\right)dx$$

$$= \int_0^2 \pi\left(\frac{1}{4} + x^4\right)dx = \pi\left[\frac{x}{4} + \frac{x^5}{5}\right]_0^2 = \frac{69\pi}{10}$$ ◀

VOLUMES BY DISKS AND WASHERS PERPENDICULAR TO THE y-AXIS

The methods of disks and washers have analogs for regions that are revolved about the y-axis (Figures 8.2.14 and 8.2.15). Using the method of slicing and Formula (4), you should have no trouble deducing the following formulas for the volumes of the solids in the figures.

$$V = \int_c^d \pi[u(y)]^2\,dy \qquad V = \int_c^d \pi([w(y)]^2 - [v(y)]^2)\,dy \tag{7–8}$$

Disks — Washers

Example 5

Find the volume of the solid generated when the region enclosed by $y = \sqrt{x}$, $y = 2$, and $x = 0$ is revolved about the y-axis (Figure 8.2.16).

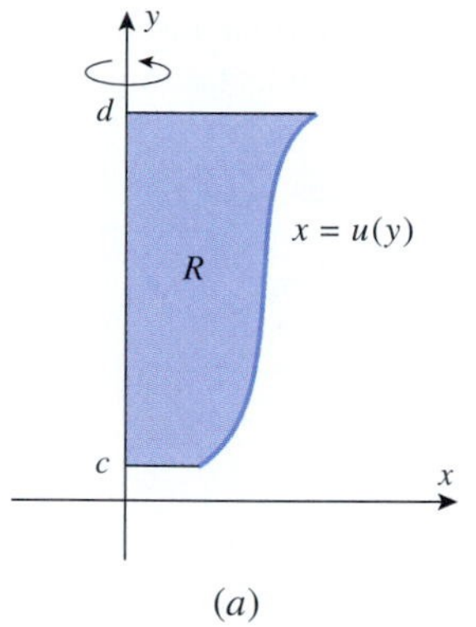

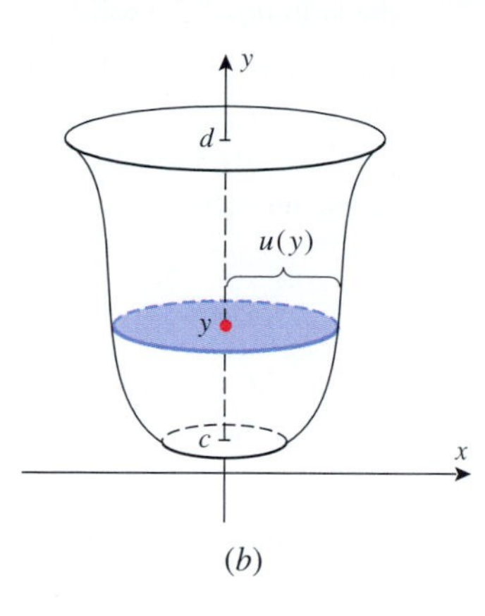

Figure 8.2.14

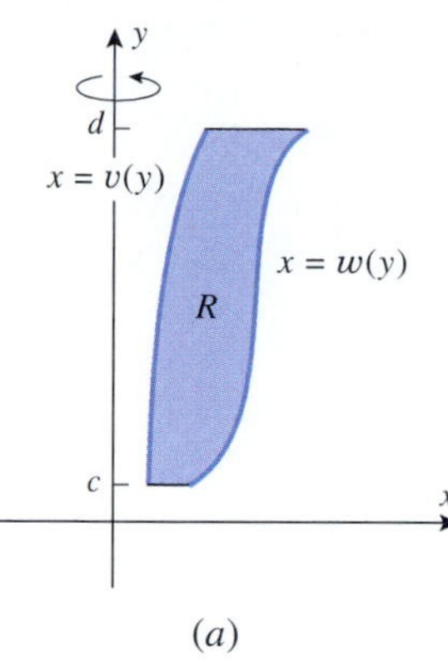

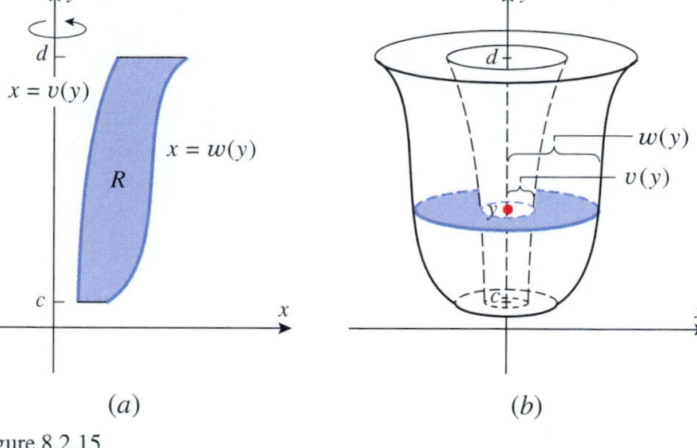

Figure 8.2.15

Solution. The cross sections taken perpendicular to the y-axis are disks, so we will apply (7). But first we must rewrite $y = \sqrt{x}$ as $x = y^2$. Thus, from (7) with $u(y) = y^2$, the volume is

$$V = \int_c^d \pi[u(y)]^2\, dy = \int_0^2 \pi y^4\, dy = \frac{\pi y^5}{5}\bigg]_0^2 = \frac{32\pi}{5}$$ ◀

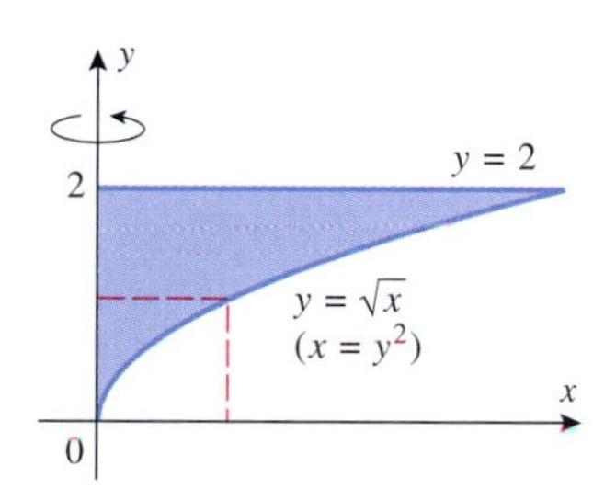

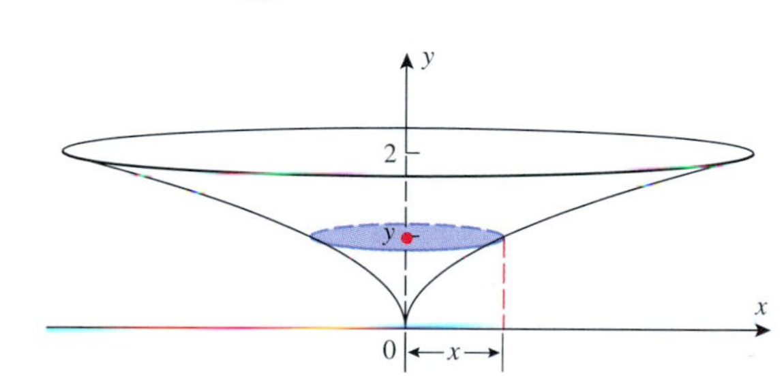

Figure 8.2.16

EXERCISE SET 8.2 C CAS

In Exercises 1–4, find the volume of the solid that results when the shaded region is revolved about the indicated axis.

1.

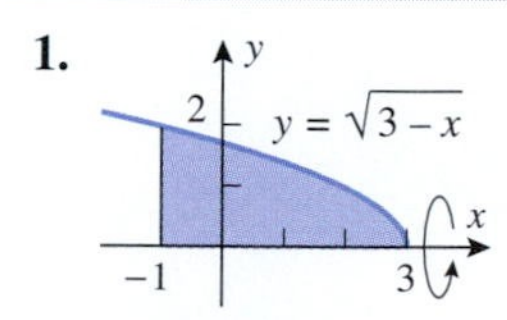

2.

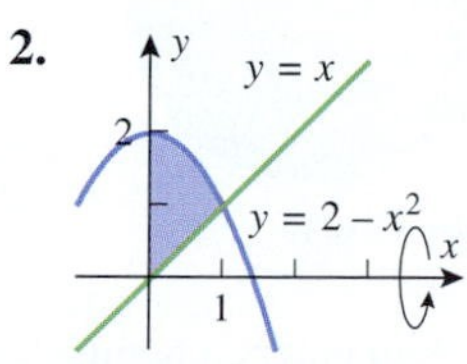

3.

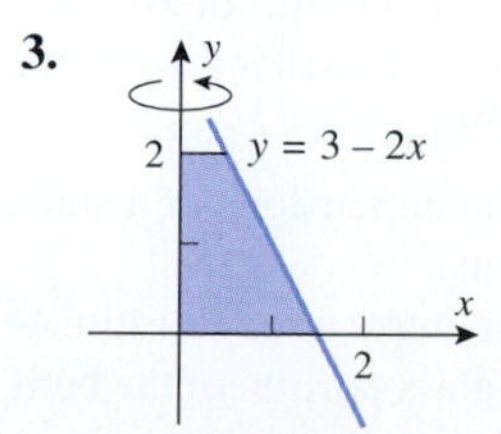

4.

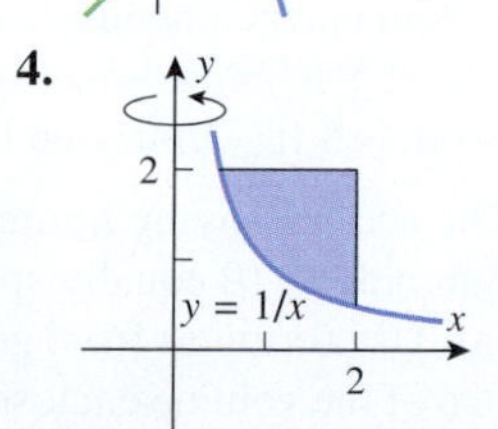

In Exercises 5–14, find the volume of the solid that results when the region enclosed by the given curves is revolved about the x-axis.

5. $y = x^2,\ x = 0,\ x = 2,\ y = 0$

6. $y = \sec x,\ x = \pi/4,\ x = \pi/3,\ y = 0$

7. $y = \sqrt{\cos x},\ x = \pi/4,\ x = \pi/2,\ y = 0$

8. $y = x^2,\ y = x^3$

9. $y = \sqrt{25 - x^2},\ y = 3$

10. $y = 9 - x^2,\ y = 0$

11. $y = e^x,\ y = 0,\ x = 0,\ x = \ln 3$

12. $y = e^{-2x},\ y = 0,\ x = 0,\ x = 1$

13. $x = \sqrt{y},\ x = y/4$

14. $y = \sin x,\ y = \cos x,\ x = 0,\ x = \pi/4$. [*Hint:* Use the identity $\cos 2x = \cos^2 x - \sin^2 x$.]

In Exercises 15–22, find the volume of the solid that results when the region enclosed by the given curves is revolved about the y-axis.

15. $y = x^3,\ x = 0,\ y = 1$

16. $x = 1 - y^2,\ x = 0$

17. $x = \sqrt{1+y},\ x = 0,\ y = 3$

18. $y = x^2 - 1,\ x = 2,\ y = 0$

19. $x = \csc y,\ y = \pi/4,\ y = 3\pi/4,\ x = 0$

20. $y = x^2,\ x = y^2$

21. $x = y^2,\ x = y + 2$

22. $x = 1 - y^2,\ x = 2 + y^2,\ y = -1,\ y = 1$

23. Find the volume of the solid that results when the region above the x-axis and below the ellipse

$$\frac{x^2}{a^2} + \frac{y^2}{b^2} = 1 \quad (a > 0, b > 0)$$

is revolved about the x-axis.

24. Let V be the volume of the solid that results when the region enclosed by $y = 1/x,\ y = 0,\ x = 2$, and $x = b$ $(0 < b < 2)$ is revolved about the x-axis. Find the value of b for which $V = 3$.

25. Find the volume of the solid generated when the region enclosed by $y = \sqrt{x+1},\ y = \sqrt{2x}$, and $y = 0$ is revolved about the x-axis. [*Hint:* Split the solid into two parts.]

26. Find the volume of the solid generated when the region enclosed by $y = \sqrt{x},\ y = 6 - x$, and $y = 0$ is revolved about the x-axis. [*Hint:* Split the solid into two parts.]

27. Find the volume of the solid that results when the region enclosed by $y = \sqrt{x},\ y = 0$, and $x = 9$ is revolved about the line $x = 9$.

28. Find the volume of the solid that results when the region in Exercise 27 is revolved about the line $y = 3$.

29. Find the volume of the solid that results when the region enclosed by $x = y^2$ and $x = y$ is revolved about the line $y = -1$.

30. Find the volume of the solid that results when the region in Exercise 29 is revolved about the line $x = -1$.

31. A nose cone for a space reentry vehicle is designed so that a cross section, taken x ft from the tip and perpendicular to the axis of symmetry, is a circle of radius $\frac{1}{4}x^2$ ft. Find the volume of the nose cone given that its length is 20 ft.

32. A certain solid is 1 ft high, and a horizontal cross section taken x ft above the bottom of the solid is an annulus of inner radius x^2 and outer radius $\sqrt{x}$. Find the volume of the solid.

33. Find the volume of the solid whose base is the region bounded between the curves $y = x$ and $y = x^2$, and whose cross sections perpendicular to the x-axis are squares.

34. The base of a certain solid is the region enclosed by $y = \sqrt{x}$, $y = 0$, and $x = 4$. Every cross section perpendicular to the x-axis is a semicircle with its diameter across the base. Find the volume of the solid.

35. Find the volume of the solid whose base is enclosed by the circle $x^2 + y^2 = 1$ and whose cross sections taken perpendicular to the base are

(a) semicircles

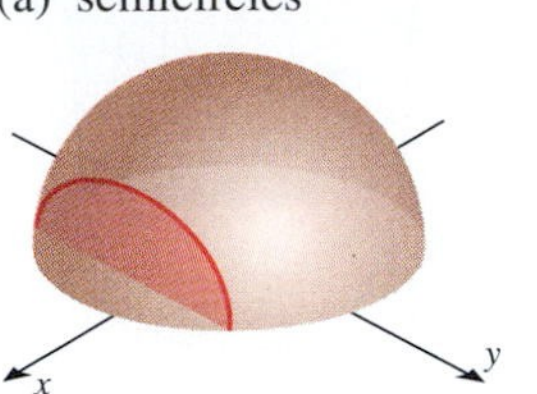

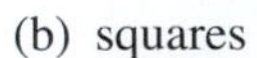
(b) squares

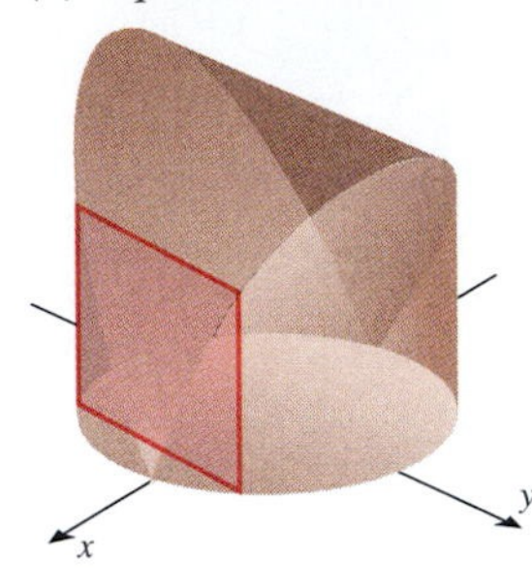

(c) equilateral triangles.

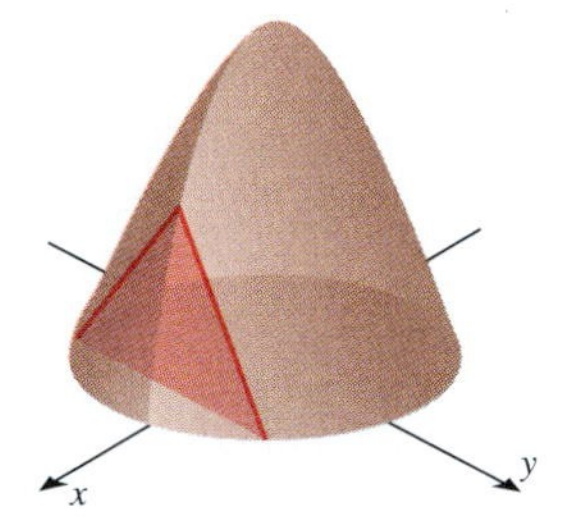

36. Derive the formula for the volume of a right circular cone with radius r and height h.

In Exercises 37 and 38, use a CAS to find the volume of the solid that results when the region enclosed by the curves is revolved about the stated axis.

C **37.** $y = \sin^8 x,\ y = 2x/\pi,\ x = 0,\ x = \pi/2$; x-axis

C **38.** $y = e^x,\ x = 1,\ y = 1$; y-axis

39. The accompanying figure shows a ***spherical cap*** of radius ρ and height h cut from a sphere of radius r. Show that the volume V of the spherical cap can be expressed as

(a) $V = \frac{1}{3}\pi h^2(3r - h)$ (b) $V = \frac{1}{6}\pi h(3\rho^2 + h^2)$

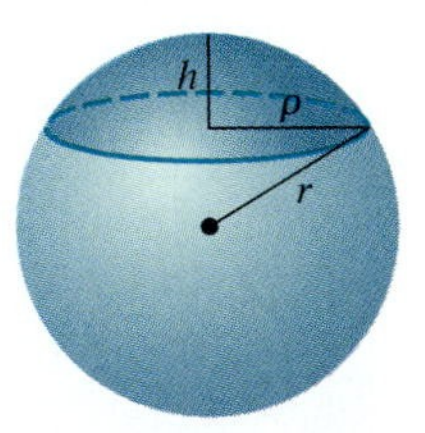

Figure Ex-39

40. If fluid enters a hemispherical vat with a radius of 10 ft at a rate of $\frac{1}{2}$ ft^3/min, how fast will the fluid be rising when the depth is 5 ft? [*Hint:* See Exercise 39.]

41. The accompanying figure shows the dimensions of a small lightbulb at 10 equally spaced points.

(a) Use formulas from geometry to make a rough estimate of the volume enclosed by the glass portion of the bulb.

(b) Use the average of left and right endpoint approximations to approximate the volume.

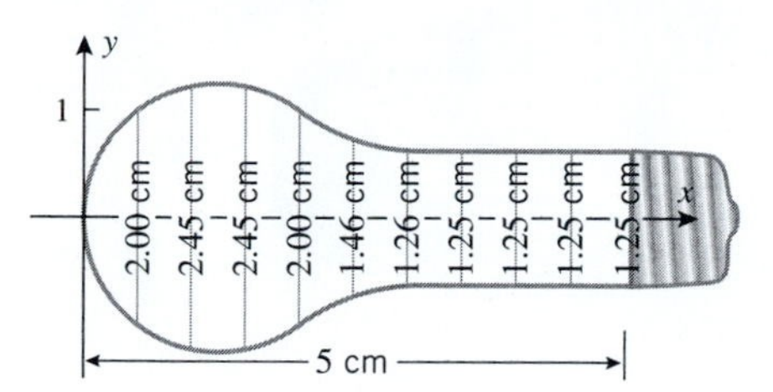

Figure Ex-41

42. Use the result in Exercise 39 to find the volume of the solid that remains when a hole of radius $r/2$ is drilled through the center of a sphere of radius r, and then check your answer by integrating.

43. As shown in the accompanying figure, a cocktail glass with a bowl shaped like a hemisphere of diameter 8 cm contains a cherry with a diameter of 2 cm. If the glass is filled to a depth of h cm, what is the volume of liquid it contains? [*Hint:* First consider the case where the cherry is partially submerged, then the case where it is totally submerged.]

44. Find the volume of the torus that results when the region enclosed by the circle of radius r with center at $(h, 0)$, $h > r$, is revolved about the y-axis. [*Hint:* Use an appropriate formula from plane geometry to help evaluate the definite integral.]

45. A wedge is cut from a right circular cylinder of radius r by two planes, one perpendicular to the axis of the cylinder and the other making an angle θ with the first. Find the volume of the wedge by slicing perpendicular to the y-axis as shown in the accompanying figure.

Figure Ex-43

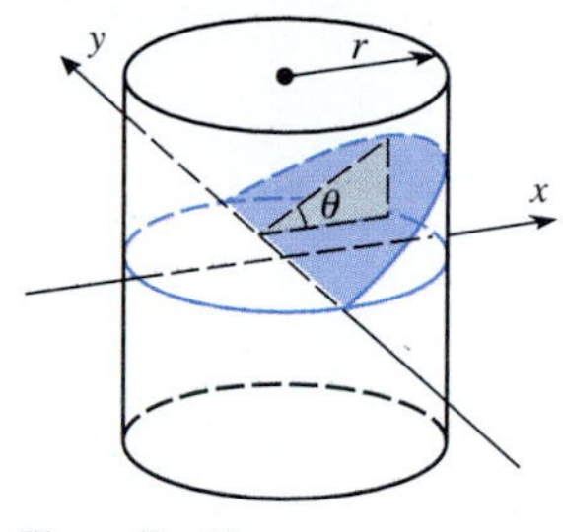

Figure Ex-45

46. Find the volume of the wedge described in Exercise 45 by slicing perpendicular to the x-axis.

47. Two right circular cylinders of radius r have axes that intersect at right angles. Find the volume of the solid common to the two cylinders. [*Hint:* One-eighth of the solid is sketched in the accompanying figure.]

48. In 1635 Bonaventura Cavalieri, a student of Galileo, stated the following result, called ***Cavalieri's principle***: *If two solids have the same height, and if the areas of their cross sections taken parallel to and at equal distances from their bases are always equal, then the solids have the same volume.* Use this result to find the volume of the oblique cylinder in the accompanying figure.

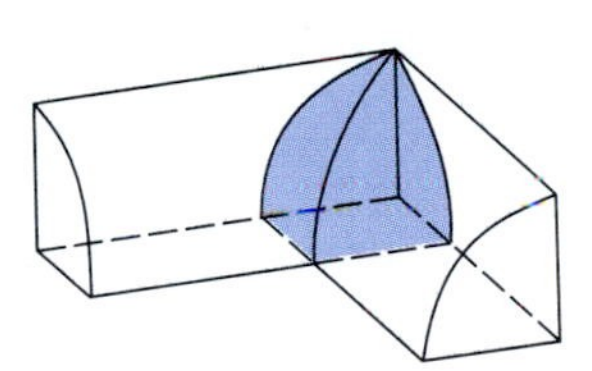

Figure Ex-47

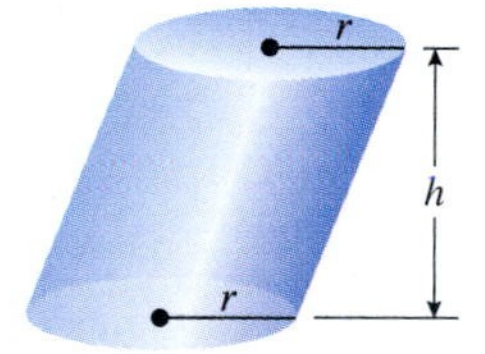

Figure Ex-48

8.3 VOLUMES BY CYLINDRICAL SHELLS

The methods for computing volumes that have been discussed so far depend on our ability to compute the cross-sectional area of the solid and to integrate that area across the solid. In this section we will develop another method for finding volumes that may be applicable when the cross-sectional area cannot be found or the integration is too difficult.

CYLINDRICAL SHELLS

In this section we will be interested in the following problem:

8.3.1 PROBLEM. Let f be continuous and nonnegative on $[a, b]$, and let R be the region that is bounded above by $y = f(x)$, below by the x-axis, and on the sides by the lines $x = a$ and $x = b$. Find the volume V of the solid of revolution S that is generated by revolving the region R about the y-axis (Figure 8.3.1).

Sometimes problems of this type can be solved by the method of disks or washers perpendicular to the y-axis, but when that method is not applicable or the resulting integral is difficult, the *method of cylindrical shells,* which we will discuss here, will often work.

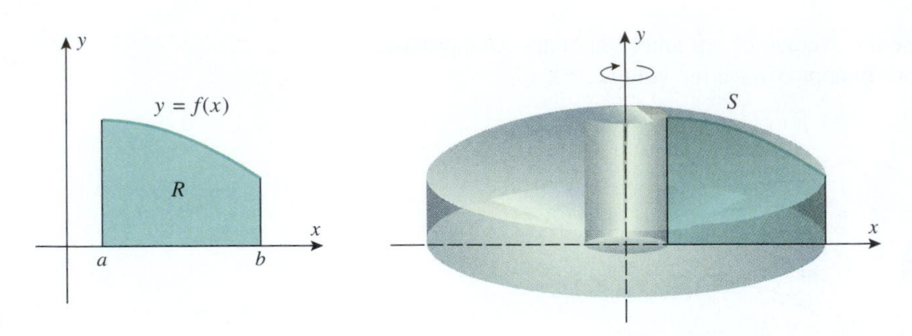

Figure 8.3.1

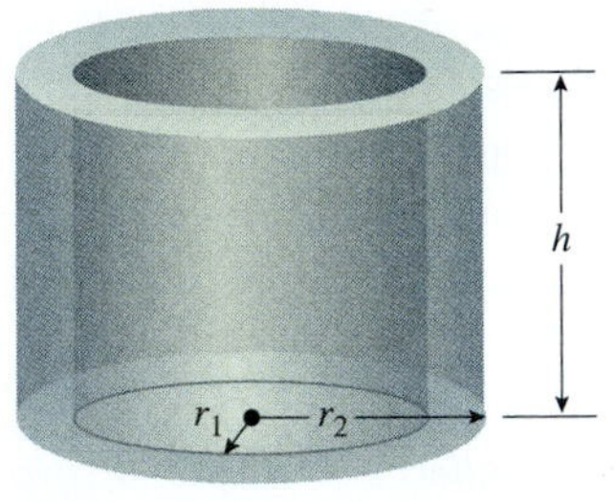

Figure 8.3.2

A ***cylindrical shell*** is a solid enclosed by two concentric right circular cylinders (Figure 8.3.2). The volume V of a cylindrical shell with inner radius r_1, outer radius r_2, and height h can be written as

$$\begin{aligned} V &= [\text{area of cross section}] \cdot [\text{height}] = (\pi r_2^2 - \pi r_1^2)h \\ &= \pi(r_2 + r_1)(r_2 - r_1)h = 2\pi \cdot \left[\tfrac{1}{2}(r_1 + r_2)\right] \cdot h \cdot (r_2 - r_1) \end{aligned}$$

But $\frac{1}{2}(r_1 + r_2)$ is the average radius of the shell and $r_2 - r_1$ is its thickness, so

$$V = 2\pi \cdot [\text{average radius}] \cdot [\text{height}] \cdot [\text{thickness}] \tag{1}$$

We will now show how this formula can be used to solve the problem posed above. The underlying idea is to divide the interval $[a, b]$ into n subintervals, thereby subdividing the region R into n strips, $R_1, R_2, \ldots, R_n$ (Figure 8.3.3a). When the region R is revolved about the y-axis, these strips generate "tube-like" solids $S_1, S_2, \ldots, S_n$ that are nested one inside the other and together comprise the entire solid S (Figure 8.3.3b). Thus, the volume V of the solid can be obtained by adding together the volumes of the tubes; that is,

$$V = V(S_1) + V(S_2) + \cdots + V(S_n)$$

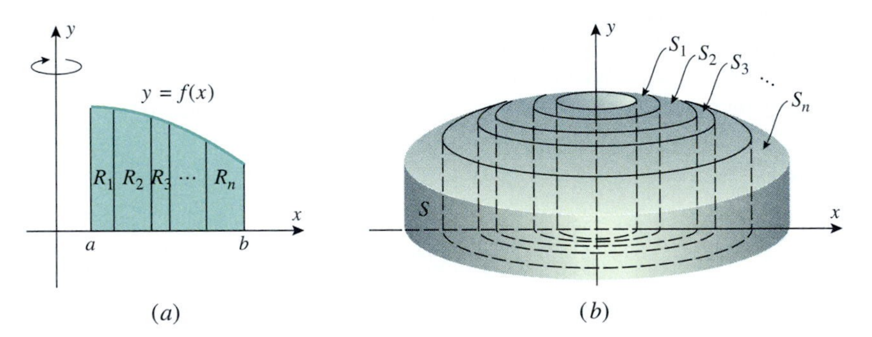

Figure 8.3.3

As a rule, the tubes will have curved upper surfaces, so there will be no simple formulas for their volumes. However, if the strips are thin, then we can approximate each strip by a rectangle (Figure 8.3.4a). These rectangles, when revolved about the y-axis, will produce cylindrical shells whose volumes closely approximate the volumes generated by the original strips (Figure 8.3.4b). We will show that by adding the volumes of the cylindrical shells we

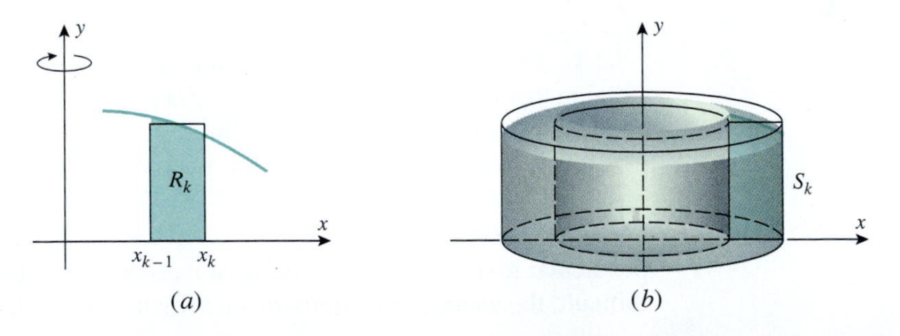

Figure 8.3.4

can obtain a Riemann sum that approximates the volume V, and by taking the limit of the Riemann sums we can obtain an integral for the exact volume V.

To implement this idea, suppose that the kth strip extends from the point x_{k-1} to the point x_k and that the width of this strip is

$$\Delta x_k = x_k - x_{k-1}$$

Figure 8.3.5

If we let x_k^* be the *midpoint* of the interval $[x_{k-1}, x_k]$, and if we construct a rectangle of height $f(x_k^*)$ over the interval, then revolving this rectangle about the y-axis produces a cylindrical shell of height $f(x_k^*)$, average radius x_k^*, and thickness Δx_k (Figure 8.3.5). From (1), the volume V_k of this cylindrical shell is

$$V_k = 2\pi x_k^* f(x_k^*)\Delta x_k$$

Adding the volumes of the n cylindrical shells yields the following Riemann sum that approximates the volume V:

$$V \approx \sum_{k=1}^{n} 2\pi x_k^* f(x_k^*)\Delta x_k$$

Taking the limit as n increases and the widths of the subintervals approach zero yields the definite integral

$$V = \lim_{\max \Delta x_k \to 0} \sum_{k=1}^{n} 2\pi x_k^* f(x_k^*)\Delta x_k = \int_a^b 2\pi x f(x)\,dx$$

In summary, we have the following result.

8.3.2 VOLUME BY CYLINDRICAL SHELLS ABOUT THE y-AXIS. Let f be continuous and nonnegative on $[a, b]$, and let R be the region that is bounded above by $y = f(x)$, below by the x-axis, and on the sides by the lines $x = a$ and $x = b$. Then the volume V of the solid of revolution that is generated by revolving the region R about the y-axis is given by

$$V = \int_a^b 2\pi x f(x)\,dx \qquad (2)$$

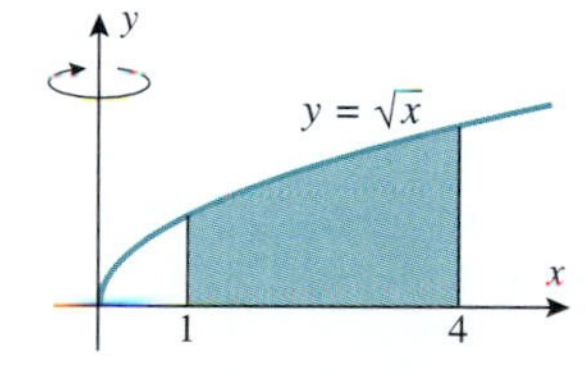

Cutaway view of the solid

Figure 8.3.6

Example 1

Use cylindrical shells to find the volume of the solid generated when the region enclosed between $y = \sqrt{x}$, $x = 1$, $x = 4$, and the x-axis is revolved about the y-axis (Figure 8.3.6).

Solution. Since $f(x) = \sqrt{x}$, $a = 1$, and $b = 4$, Formula (2) yields

$$V = \int_1^4 2\pi x\sqrt{x}\,dx = 2\pi\int_1^4 x^{3/2}\,dx = \left[2\pi\cdot\frac{2}{5}x^{5/2}\right]_1^4 = \frac{4\pi}{5}[32-1] = \frac{124\pi}{5} \blacktriangleleft$$

VARIATIONS OF THE METHOD OF CYLINDRICAL SHELLS

The method of cylindrical shells is applicable in a variety of situations that do not fit the conditions required by Formula (2). For example, the region may be enclosed between two curves, or the axis of revolution may be some line other than the y-axis. However, rather than develop a separate formula for every possible situation, we will give a general way of thinking about the method of cylindrical shells that can be adapted to each new situation as it arises.

For this purpose, we will need to reexamine the integrand in Formula (2): At each point x in the interval $[a, b]$, the vertical line segment from the x-axis to the curve $y = f(x)$ can be viewed as the cross section of the region R at x (Figure 8.3.7a). When the region R is revolved about the y-axis, the cross section at x sweeps out the *surface* of a right circular cylinder of height $f(x)$ and radius x (Figure 8.3.7*b*). The area of this surface is

$$2\pi x f(x)$$

(Figure 8.3.7*c*), which is the integrand in (2). Thus, Formula (2) can be viewed informally in the following way.

> **8.3.3 AN INFORMAL VIEWPOINT ABOUT CYLINDRICAL SHELLS.** The volume V of a solid of revolution that is generated by revolving a region R about an axis can be obtained by integrating the area of the surface generated by an arbitrary cross section of R taken parallel to the axis of revolution.

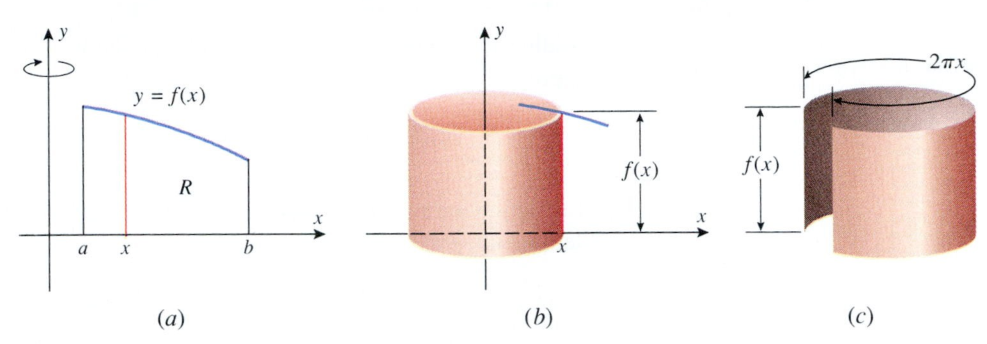

Figure 8.3.7

The following examples illustrate how to apply this result in situations where Formula (2) is not applicable:

Example 2

Use cylindrical shells to find the volume of the solid generated when the region R in the first quadrant enclosed between $y = x$ and $y = x^2$ is revolved about the y-axis (Figure 8.3.8).

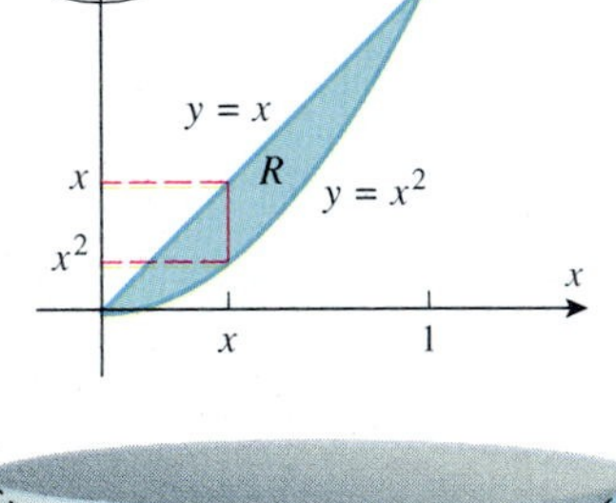

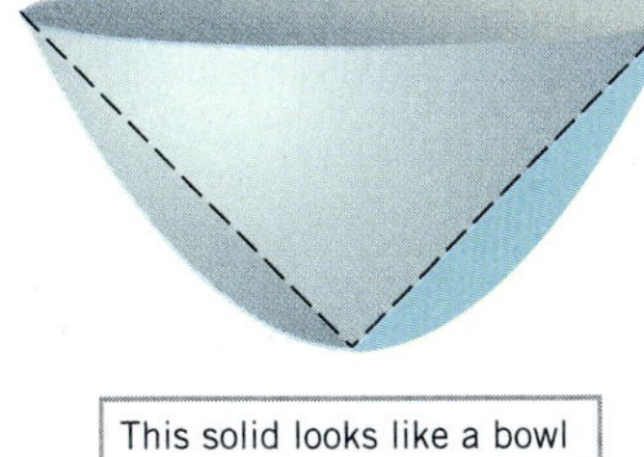

This solid looks like a bowl with a cone-shaped interior.

Figure 8.3.8

Solution. At each x in $[0, 1]$ the cross section of R parallel to the y-axis generates a cylindrical surface of height $x - x^2$ and radius x. Since the area of this surface is

$$2\pi x(x - x^2)$$

the volume of the solid is

$$V = \int_0^1 2\pi x(x-x^2)\,dx = 2\pi \int_0^1 (x^2-x^3)\,dx = 2\pi \left[\frac{x^3}{3} - \frac{x^4}{4}\right]_0^1 = 2\pi\left[\frac{1}{3} - \frac{1}{4}\right] = \frac{\pi}{6}$$

◀

FOR THE READER. The volume in this example can also be obtained by the method of washers. Confirm that the volume produced by that method agrees with the volume obtained by cylindrical shells.

Example 3

Use cylindrical shells to find the volume of the solid generated when the region R under $y = x^2$ over the interval $[0, 2]$ is revolved about the x-axis (Figure 8.3.9).

Solution. At each y in the interval $0 \le y \le 4$, the cross section of R parallel to the x-axis generates a cylindrical surface of height $2 - \sqrt{y}$ and radius y. Since the area of this surface is $2\pi y(2 - \sqrt{y})$, the volume of the solid is

$$V = \int_0^4 2\pi y(2 - \sqrt{y})\,dy = 2\pi \int_0^4 (2y - y^{3/2})\,dy = 2\pi \left[y^2 - \frac{2}{5}y^{5/2}\right]_0^4 = \frac{32\pi}{5}$$ ◀

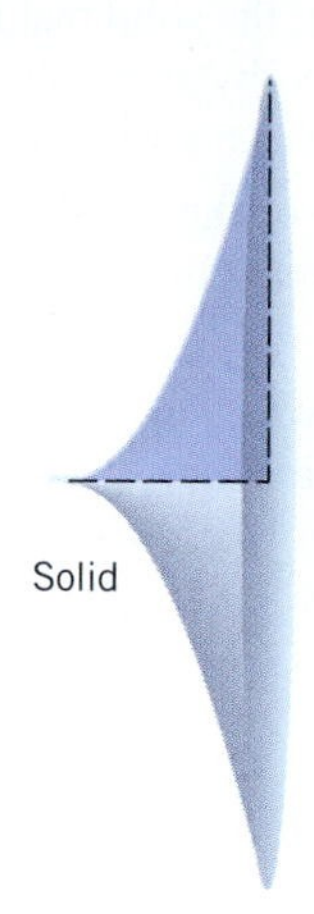

Figure 8.3.9

FOR THE READER. The volume in this example can also be obtained by the method of disks. Confirm that the volume produced by that method agrees with the volume obtained by cylindrical shells.

EXERCISE SET 8.3 [C] CAS

In Exercises 1–4, use cylindrical shells to find the volume of the solid generated when the shaded region is revolved about the indicated axis.

1.

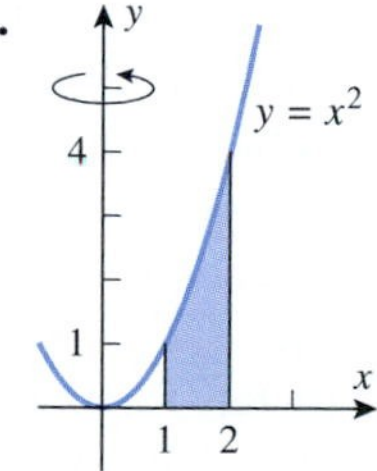

2.

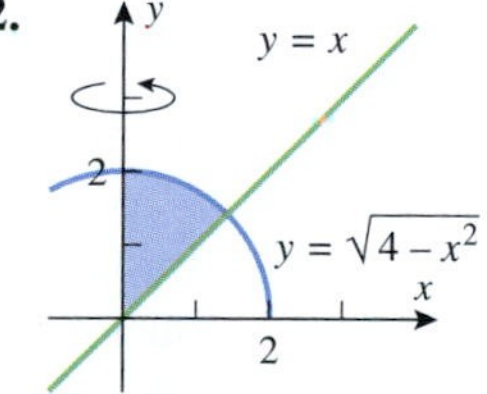

3.

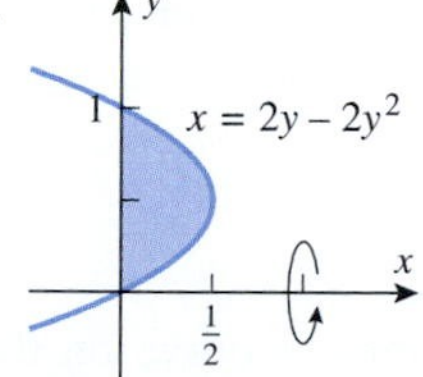

4.

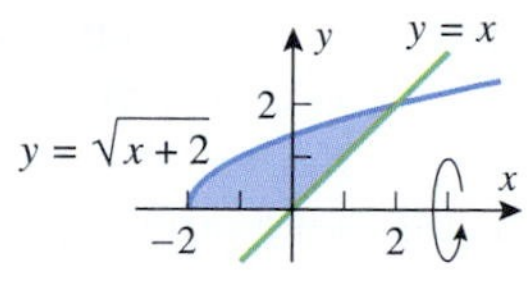

In Exercises 5–12, use cylindrical shells to find the volume of the solid generated when the region enclosed by the given curves is revolved about the y-axis.

5. $y = x^3, \ x = 1, \ y = 0$

6. $y = \sqrt{x}, \ x = 4, \ x = 9, \ y = 0$

7. $y = 1/x, \ y = 0, \ x = 1, \ x = 3$

8. $y = \cos(x^2), \ x = 0, \ x = \frac{1}{2}\sqrt{\pi}, \ y = 0$

9. $y = 2x - 1, \ y = -2x + 3, \ x = 2$

10. $y = \dfrac{1}{x^2 + 1}, \ x = 0, \ x = 1, \ y = 0$

11. $y = e^{x^2}, \ x = 1, \ x = \sqrt{3}, \ y = 0$

12. $y = 2x - x^2, \ y = 0$

In Exercises 13–16, use cylindrical shells to find the volume of the solid generated when the region enclosed by the given curves is revolved about the x-axis.

13. $y^2 = x, \ y = 1, \ x = 0$

14. $x = 2y, \ y = 2, \ y = 3, \ x = 0$

15. $y = x^2, \ x = 1, \ y = 0$

16. $xy = 4, \ x + y = 5$

[C] **17.** Use a CAS to find the volume of the solid generated when the region enclosed by $y = \sin x$ and $y = 0$ for $0 \le x \le \pi$ is revolved about the y-axis.

[C] **18.** Use a CAS to find the volume of the solid generated when the region enclosed by $y = \cos x$, $y = 0$, and $x = 0$ for $0 \le x \le \pi/2$ is revolved about the y-axis.

19. (a) Use cylindrical shells to find the volume of the solid that is generated when the region under the curve

$$y = x^3 - 3x^2 + 2x$$

over [0, 1] is revolved about the y-axis.

(b) For this problem, is the method of cylindrical shells easier or harder than the method of slicing discussed in the last section? Explain.

20. Use cylindrical shells to find the volume of the solid that is generated when the region that is enclosed by $y = 1/x^3$, $x = 1$, $x = 2$, $y = 0$ is revolved about the line $x = -1$.

21. Use cylindrical shells to find the volume of the solid that is generated when the region that is enclosed by $y = x^3$, $y = 1$, $x = 0$ is revolved about the line $y = 1$.

22. Let R_1 and R_2 be regions of the form shown in the accompanying figure. Use cylindrical shells to find a formula for the volume of the solid that results when
(a) region R_1 is revolved about the y-axis
(b) region R_2 is revolved about the x-axis.

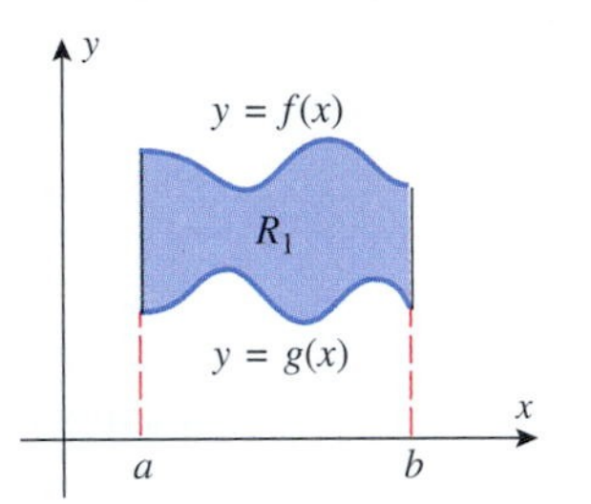

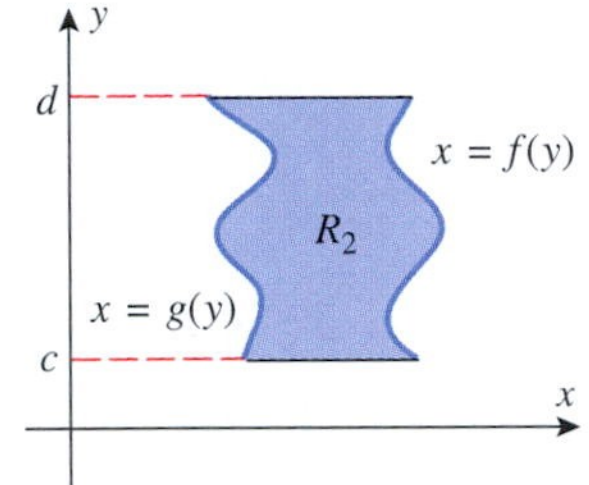

Figure Ex-22

23. Use cylindrical shells to find the volume of the cone generated when the triangle with vertices $(0, 0)$, $(0, r)$, $(h, 0)$, where $r > 0$ and $h > 0$, is revolved about the x-axis.

24. The region enclosed between the curve $y^2 = kx$ and the line $x = \frac{1}{4}k$ is revolved about the line $x = \frac{1}{2}k$. Use cylindrical shells to find the volume of the resulting solid. (Assume $k > 0$.)

25. A round hole of radius a is drilled through the center of a solid sphere of radius r. Use cylindrical shells to find the volume of the portion removed. (Assume $r > a$.)

26. Use cylindrical shells to find the volume of the torus obtained by revolving the circle $x^2 + y^2 = a^2$ about the line $x = b$, where $b > a > 0$. [*Hint:* It may help in the integration to think of an integral as an area.]

27. Let V_x and V_y be the volumes of the solids that result when the region enclosed by $y = 1/x$, $y = 0$, $x = \frac{1}{2}$, and $x = b$ $\left(b > \frac{1}{2}\right)$ is revolved about the x-axis and y-axis, respectively. Is there a value of b for which $V_x = V_y$?

8.4 LENGTH OF A PLANE CURVE

In this section we will consider the problem of finding the length of a plane curve.

ARC LENGTH

Although formulas for lengths of circular arcs appear in early historical records, very little was known about the lengths of more general curves until the mid-seventeenth century. About that time formulas were discovered for a few specific curves such as the length of an arch of a cycloid. However, such basic problems as finding the length of an ellipse defied the mathematicians of that period, and almost no progress was made on the general problem of finding lengths of curves until the advent of calculus in the next century.

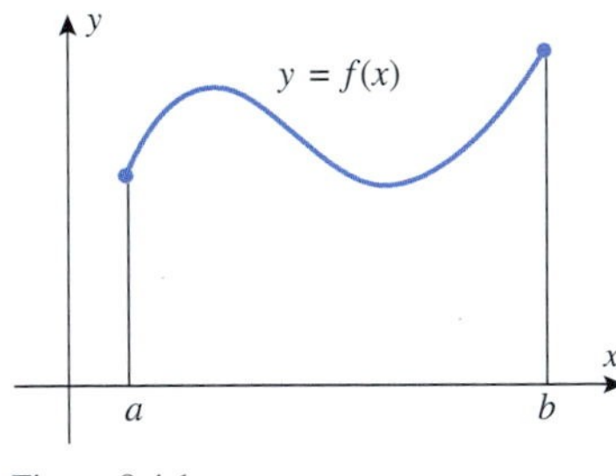

Figure 8.4.1

Our first objective in this section is to *define* what we mean by the length (also called the ***arc length***) of a plane curve $y = f(x)$ over an interval $[a, b]$ (Figure 8.4.1). Once that is done we will be able to focus on computational matters. To avoid some complications that would otherwise occur, we will impose the requirement that f' be continuous on $[a, b]$, in which case we will say that $y = f(x)$ is a ***smooth curve*** on $[a, b]$ (or that f is a ***smooth function*** on $[a, b]$).

We will be concerned with the following problem:

> **8.4.1** ARC LENGTH PROBLEM. Suppose that $y = f(x)$ is a smooth curve on the interval $[a, b]$. Define and find a formula for the arc length L of the curve $y = f(x)$ over the interval $[a, b]$.

The basic idea for defining arc length is to break up the curve into small segments, approximate the curve segments by line segments, add the lengths of the line segments to form a Riemann sum that approximates the arc length L, and take the limit of the Riemann sums to obtain an integral for L.

Figure 8.4.2

To implement this idea, divide the interval $[a, b]$ into n subintervals by inserting points $x_1, x_2, \ldots, x_{n-1}$ between a and b. As shown in Figure 8.4.2, let $P_0, P_1, \ldots, P_n$ be the points

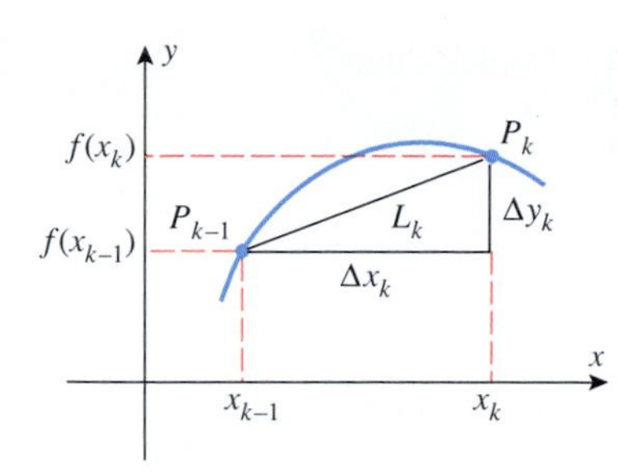

Figure 8.4.3

on the curve with x-coordinates $a, x_1, x_2, \dots, x_{n-1}, b$ and join these points with straight line segments. These line segments form a ***polygonal path*** that we can regard as an approximation to the curve $y = f(x)$. As suggested by Figure 8.4.3, the length L_k of the kth line segment in the polygonal path is

$$L_k = \sqrt{(\Delta x_k)^2 + (\Delta y_k)^2} = \sqrt{(\Delta x_k)^2 + [f(x_k) - f(x_{k-1})]^2} \tag{1}$$

If we now add the lengths of these line segments, we obtain the following approximation to the length L of the curve

$$L \approx \sum_{k=1}^{n} L_k = \sum_{k=1}^{n} \sqrt{(\Delta x_k)^2 + [f(x_k) - f(x_{k-1})]^2} \tag{2}$$

To put this in the form of a Riemann sum we will apply the Mean-Value Theorem (6.5.2). This theorem implies that there is a point x_k^* between x_{k-1} and x_k such that

$$\frac{f(x_k) - f(x_{k-1})}{x_k - x_{k-1}} = f'(x_k^*) \quad \text{or} \quad f(x_k) - f(x_{k-1}) = f'(x_k^*)\Delta x_k$$

and hence we can rewrite (2) as

$$L \approx \sum_{k=1}^{n} \sqrt{1 + [f'(x_k^*)]^2}\,\Delta x_k$$

Thus, taking the limit as n increases and the widths of the subintervals approach zero yields the following integral that defines the arc length L:

$$L = \lim_{\max \Delta x_k \to 0} \sum_{k=1}^{n} \sqrt{1 + [f'(x_k^*)]^2}\,\Delta x_k = \int_a^b \sqrt{1 + [f'(x)]^2}\, dx$$

In summary, we have the following definition:

8.4.2 DEFINITION. If $y = f(x)$ is a smooth curve on the interval $[a, b]$, then the arc length L of this curve over $[a, b]$ is defined as

$$L = \int_a^b \sqrt{1 + [f'(x)]^2}\, dx \tag{3}$$

This result provides both a definition and a formula for computing arc lengths. Where convenient, (3) can also be expressed as

$$L = \int_a^b \sqrt{1 + [f'(x)]^2}\, dx = \int_a^b \sqrt{1 + \left(\frac{dy}{dx}\right)^2}\, dx \tag{4}$$

Moreover, for a curve expressed in the form $x = g(y)$, where g' is continuous on $[c, d]$, the arc length L from $y = c$ to $y = d$ can be expressed as

$$L = \int_c^d \sqrt{1 + [g'(y)]^2}\, dy = \int_c^d \sqrt{1 + \left(\frac{dx}{dy}\right)^2}\, dy \tag{5}$$

Example 1

Find the arc length of the curve $y = x^{3/2}$ from $(1, 1)$ to $(2, 2\sqrt{2})$ (Figure 8.4.4) in two ways: (a) using Formula (4) and (b) using Formula (5).

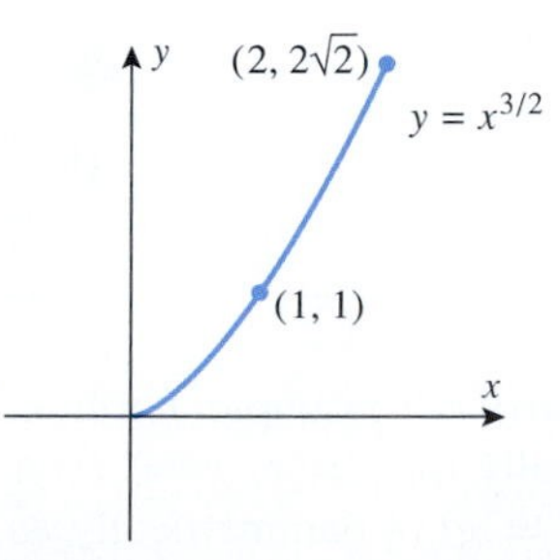

Figure 8.4.4

Solution (a). Since

$$\frac{dy}{dx} = \tfrac{3}{2}x^{1/2}$$

and since the curve extends from $x = 1$ to $x = 2$, it follows from (4) that

$$L = \int_1^2 \sqrt{1 + \frac{9}{4}x}\, dx$$

To evaluate this integral we make the u-substitution

$$u = 1 + \tfrac{9}{4}x, \quad du = \tfrac{9}{4}\, dx$$

and then change the x-limits of integration ($x = 1, x = 2$) to the corresponding u-limits $\left(u = \frac{13}{4}, u = \frac{22}{4}\right)$:

$$L = \frac{4}{9}\int_{13/4}^{22/4} u^{1/2}\, du = \frac{8}{27}u^{3/2}\bigg]_{13/4}^{22/4} = \frac{8}{27}\left[\left(\frac{22}{4}\right)^{3/2} - \left(\frac{13}{4}\right)^{3/2}\right]$$

$$= \frac{22\sqrt{22} - 13\sqrt{13}}{27} \approx 2.09$$

Solution (b). To apply Formula (5) we must first rewrite the equation $y = x^{3/2}$ so that x is expressed as a function of y. This yields $x = y^{2/3}$ and

$$\frac{dx}{dy} = \tfrac{2}{3}y^{-1/3}$$

Since the curve extends from $y = 1$ to $y = 2\sqrt{2}$, it follows from (5) that

$$L = \int_1^{2\sqrt{2}} \sqrt{1 + \frac{4}{9}y^{-2/3}}\, dy = \frac{1}{3}\int_1^{2\sqrt{2}} y^{-1/3}\sqrt{9y^{2/3} + 4}\, dy$$

To evaluate this integral we make the u-substitution

$$u = 9y^{2/3} + 4, \quad du = 6y^{-1/3}\, dy$$

and change the y-limits of integration ($y = 1, y = 2\sqrt{2}$) to the corresponding u-limits ($u = 13, u = 22$). This gives

$$L = \frac{1}{18}\int_{13}^{22} u^{1/2}\, du = \frac{1}{27}u^{3/2}\bigg]_{13}^{22} = \frac{1}{27}[(22)^{3/2} - (13)^{3/2}] = \frac{22\sqrt{22} - 13\sqrt{13}}{27}$$

This result agrees with that in part (a); however, the integration here is more tedious. In problems where there is a choice between using (4) or (5), it is often the case that one of the formulas leads to a simpler integral than the other. ◀

ARC LENGTH OF CURVES DEFINED PARAMETRICALLY

The following result provides a formula for finding the arc length of a curve from parametric equations for the curve. Its derivation is similar to that of Formula (3) and will be omitted.

8.4.3 ARC LENGTH FORMULA FOR PARAMETRIC CURVES. If no segment of the curve represented by the parametric equations

$$x = x(t), \quad y = y(t) \qquad (a \le t \le b)$$

is traced more than once as t increases from a to b, and if dx/dt and dy/dt are continuous functions for $a \le t \le b$, then the arc length L of the curve is given by

$$L = \int_a^b \sqrt{\left(\frac{dx}{dt}\right)^2 + \left(\frac{dy}{dt}\right)^2}\, dt \tag{6}$$

REMARK. Note that Formulas (4) and (5) are special cases of (6). For example, Formula (4) can be obtained from (6) by writing $y = f(x)$ parametrically as $x = t$, $y = f(t)$; similarly, Formula (5) can be obtained from (6) by writing $x = g(y)$ parametrically as $x = g(t)$, $y = t$. We leave the details as exercises.

Example 2

Use (6) to find the circumference of a circle of radius a from the parametric equations

$$x = a\cos t, \quad y = a\sin t \qquad (0 \le t \le 2\pi)$$

Solution.

$$L = \int_0^{2\pi} \sqrt{\left(\frac{dx}{dt}\right)^2 + \left(\frac{dy}{dt}\right)^2}\, dt = \int_0^{2\pi} \sqrt{(-a\sin t)^2 + (a\cos t)^2}\, dt$$

$$= \int_0^{2\pi} a\, dt = at\Big]_0^{2\pi} = 2\pi a$$

◀

FINDING ARC LENGTH BY NUMERICAL METHODS

As a rule, the integrals that arise in calculating arc length tend to be impossible to evaluate in terms of elementary functions, so it will often be necessary to approximate the integral using a numerical method such as the midpoint approximation (discussed in Section 7.5) or some other comparable method. Examples 1 and 2 are rare exceptions.

Example 3

From (4), the arc length of $y = \sin x$ from $x = 0$ to $x = \pi$ is given by the integral

$$L = \int_0^{\pi} \sqrt{1 + (\cos x)^2}\, dx$$

This integral cannot be evaluated in terms of elementary functions; however, using a calculating utility with a numerical integration capability yields the approximation $L \approx 3.8202$.

◀

FOR THE READER. In Figure 8.4.5, the scale on both axes is 2 centimeters per unit. Confirm that the result in Example 3 is reasonable by laying a piece of string as closely as possible along the curve in the figure and measuring its length in centimeters.

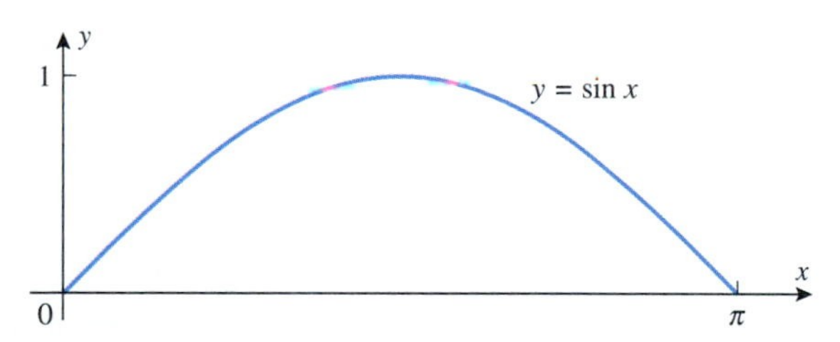

Figure 8.4.5

FOR THE READER. Computer algebra systems and some scientific calculators have commands for evaluating integrals numerically, and some scientific calculators have built-in commands for approximating arc lengths. If you have a scientific calculator with one of these capabilities or a CAS, read the documentation, and then use your calculator or CAS to check the result in Example 3.

EXERCISE SET 8.4 Graphing Calculator C CAS

1. Use the Theorem of Pythagoras to find the length of the line segment $y = 2x$ from $(1, 2)$ to $(2, 4)$, and confirm that the value is consistent with the length computed using
   (a) Formula (4) (b) Formula (5).

2. Use the Theorem of Pythagoras to find the length of the line segment $x = t$, $y = 5t$ $(0 \le t \le 1)$, and confirm that the value is consistent with the length computed using Formula (6).

In Exercises 3–8, find the exact arc length of the curve over the stated interval.

3. $y = 3x^{3/2} - 1$ from $x = 0$ to $x = 1$

4. $x = \frac{1}{3}(y^2 + 2)^{3/2}$ from $y = 0$ to $y = 1$

5. $y = x^{2/3}$ from $x = 1$ to $x = 8$

6. $y = (x^6 + 8)/16x^2$ from $x = 2$ to $x = 3$

7. $y = \frac{1}{2}(e^x + e^{-x})$ from $x = 0$ to $x = 3$

8. $x = \frac{1}{8}y^4 + \frac{1}{4}y^{-2}$ from $y = 1$ to $y = 4$

In Exercises 9–14, find the exact arc length of the parametric curve without eliminating the parameter.

9. $x = \frac{1}{3}t^3, \ y = \frac{1}{2}t^2 \quad (0 \le t \le 1)$

10. $x = (1 + t)^2, \ y = (1 + t)^3 \quad (0 \le t \le 1)$

11. $x = \cos 2t, \ y = \sin 2t \quad (0 \le t \le \pi/2)$

12. $x = \cos t + t \sin t, \ y = \sin t - t \cos t \quad (0 \le t \le \pi)$

13. $x = e^t \cos t, \ y = e^t \sin t \quad (0 \le t \le \pi/2)$

14. $x = e^t(\sin t + \cos t), \ y = e^t(\cos t - \sin t) \quad (1 \le t \le 4)$

In Exercises 15 and 16, express the exact arc length of the curve over the given interval as an integral that has been simplified to eliminate the radical, and then evaluate the integral using a CAS.

C **15.** $y = \ln(\sec x)$ from $x = 0$ to $x = \pi/4$

C **16.** $y = \ln(\sin x)$ from $x = \pi/4$ to $\pi/2$

C **17.** (a) Recall from Section 1.7 that a cycloid is the path traced by a point on the rim of a wheel that rolls along a line (Figure 1.7.13). Use the parametric equations in Formula (9) of that section to show that the length L of one arch of a cycloid is given by the integral

$$L = a \int_0^{2\pi} \sqrt{2(1 - \cos\theta)}\, d\theta$$

(b) Use a CAS to show that L is eight times the radius of the wheel (see the accompanying figure).

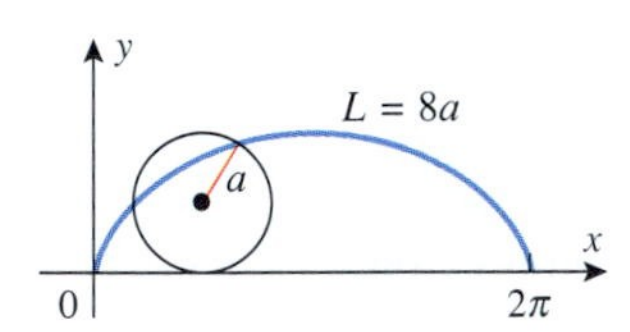

Figure Ex-17

18. It was stated in Exercise 41 of Section 1.7 that the curve given parametrically by the equations

$$x = a\cos^3\phi, \quad y = a\sin^3\phi$$

is called a *four-cusped hypocycloid* (also called an ***astroid***).

(a) Use a graphing utility to generate the graph in the case where $a = 1$, so that it is traced exactly once.

(b) Find the exact arc length of the curve in part (a).

19. Consider the curve $y = x^{2/3}$.

(a) Sketch the portion of the curve between $x = -1$ and $x = 8$.

(b) Explain why Formula (4) cannot be used to find the arc length of the curve sketched in part (a).

(c) Find the arc length of the curve sketched in part (a).

20. Derive Formulas (4) and (5) from Formula (6) by choosing appropriate parametrizations of the curves.

In Exercises 21 and 22, use the midpoint approximation with $n = 20$ subintervals to approximate the arc length of the curve over the given interval.

21. $y = x^2$ from $x = 0$ to $x = 2$

22. $x = \sin y$ from $y = 0$ to $y = \pi$

C **23.** Use a CAS or a scientific calculator with numerical integration capabilities to approximate the arc lengths in Exercises 21 and 22.

24. Let $y = f(x)$ be a smooth curve on the closed interval $[a, b]$. Prove that if there are nonnegative numbers m and M such that $m \le f'(x) \le M$ for all x in $[a, b]$, then the arc length L of $y = f(x)$ over the interval $[a, b]$ satisfies the inequalities

$$(b - a)\sqrt{1 + m^2} \le L \le (b - a)\sqrt{1 + M^2}$$

25. Use the result of Exercise 24 to show that the arc length L of $y = \sin x$ over the interval $0 \le x \le \pi/4$ satisfies

$$\frac{\pi}{4}\sqrt{\frac{3}{2}} \le L \le \frac{\pi}{4}\sqrt{2}$$

26. Show that the total arc length of the ellipse $x = a\cos t$, $y = b\sin t$, $0 \le t \le 2\pi$ for $a > b > 0$ is given by

$$4a \int_0^{\pi/2} \sqrt{1 - k^2\cos^2 t}\, dt$$

where $k = \sqrt{a^2 - b^2}/a$.

C **27.** (a) Show that the total arc length of the ellipse

$$x = 2\cos t, \quad y = \sin t \quad (0 \le t \le 2\pi)$$

is given by

$$4\int_0^{\pi/2} \sqrt{1 + 3\sin^2 t}\, dt$$

(b) Use a CAS or a scientific calculator with numerical integration capabilities to approximate the arc length in part (a). Round your answer to two decimal places.

(c) Suppose that the parametric equations in part (a) describe the path of a particle moving in the xy-plane, where t is time in seconds and x and y are in centimeters. Use a CAS or a scientific calculator with numerical integration capabilities to approximate the distance traveled by the particle from $t = 1.5$ s to $t = 4.8$ s. Round your answer to two decimal places.

C **28.** A basketball player makes a successful shot from the free throw line. Suppose that the path of the ball from the mo-

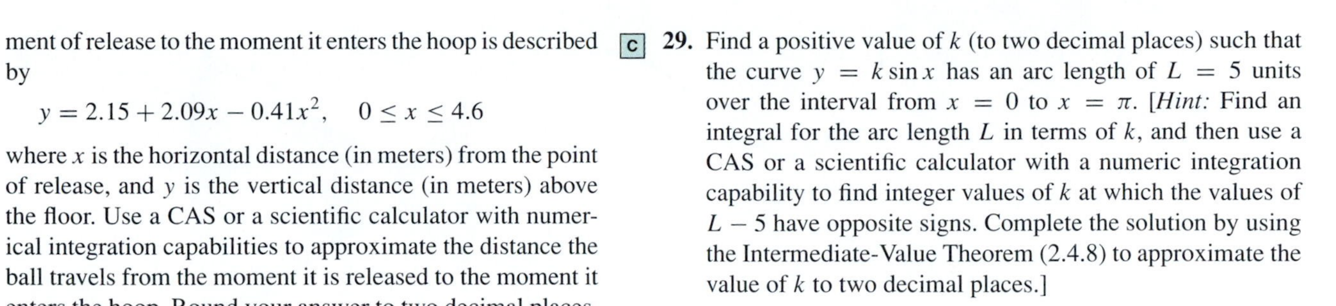

ment of release to the moment it enters the hoop is described by

$$y = 2.15 + 2.09x - 0.41x^2, \quad 0 \leq x \leq 4.6$$

where x is the horizontal distance (in meters) from the point of release, and y is the vertical distance (in meters) above the floor. Use a CAS or a scientific calculator with numerical integration capabilities to approximate the distance the ball travels from the moment it is released to the moment it enters the hoop. Round your answer to two decimal places.

C **29.** Find a positive value of k (to two decimal places) such that the curve $y = k \sin x$ has an arc length of $L = 5$ units over the interval from $x = 0$ to $x = \pi$. [*Hint:* Find an integral for the arc length L in terms of k, and then use a CAS or a scientific calculator with a numeric integration capability to find integer values of k at which the values of $L - 5$ have opposite signs. Complete the solution by using the Intermediate-Value Theorem (2.4.8) to approximate the value of k to two decimal places.]

8.5 AREA OF A SURFACE OF REVOLUTION

In this section we will consider the problem of finding the area of a surface that is generated by revolving a plane curve about a line.

SURFACE AREA

A ***surface of revolution*** is a surface that is generated by revolving a plane curve about an axis that lies in the same plane as the curve. For example, the surface of a sphere can be generated by revolving a semicircle about its diameter, and the lateral surface of a right circular cylinder can be generated by revolving a line segment about an axis that is parallel to it (Figure 8.5.1).

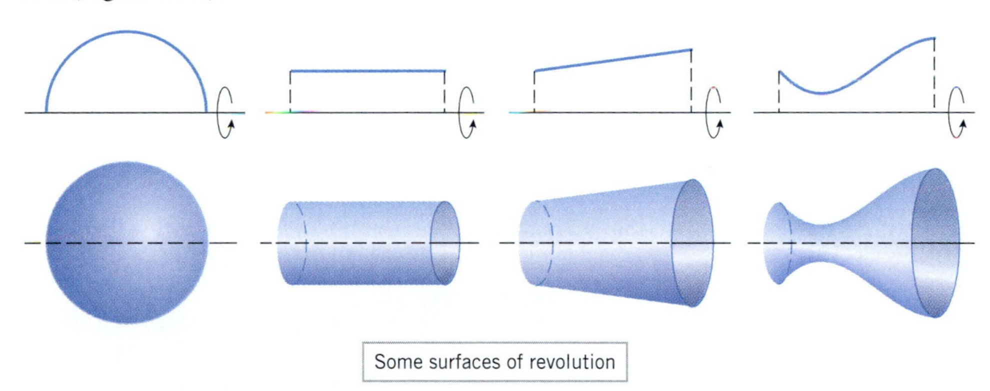

Figure 8.5.1

In this section we will be concerned with the following problem:

8.5.1 SURFACE AREA PROBLEM. Suppose that f is a smooth, nonnegative function on $[a, b]$ and that a surface of revolution is generated by revolving the portion of the curve $y = f(x)$ between $x = a$ and $x = b$ about the x-axis (Figure 8.5.2). Define what is meant by the *area* S of the surface, and find a formula for computing it.

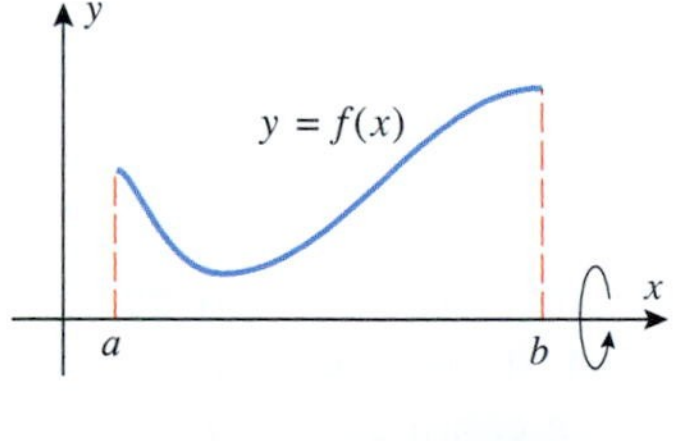

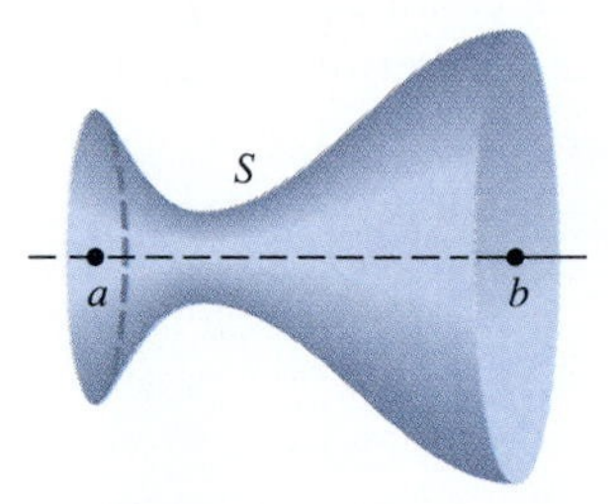

Figure 8.5.2

To motivate an appropriate definition for the area S of a surface of revolution, we will decompose the surface into small sections whose areas can be approximated by elementary formulas, add the approximations of the areas of the sections to form a Riemann sum that approximates S, and then take the limit of the Riemann sums to obtain an integral for the exact value of S.

To implement this idea, divide the interval $[a, b]$ into n subintervals by inserting points $x_1, x_2, \ldots, x_{n-1}$ between a and b. As illustrated in Figure 8.5.3a, these points define a polygonal path that approximates the curve $y = f(x)$ over the interval $[a, b]$. When this

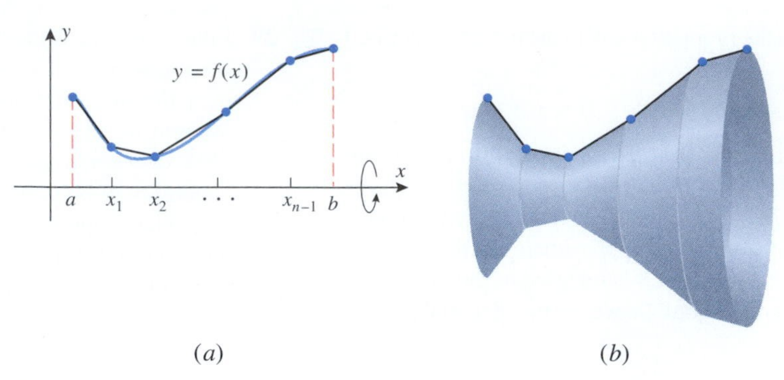

Figure 8.5.3

polygonal path is revolved about the x-axis, it generates a surface consisting of n parts, each of which is a frustum of a right circular cone (Figure 8.5.3b). Thus, the area of each part of the approximating surface can be obtained from the formula

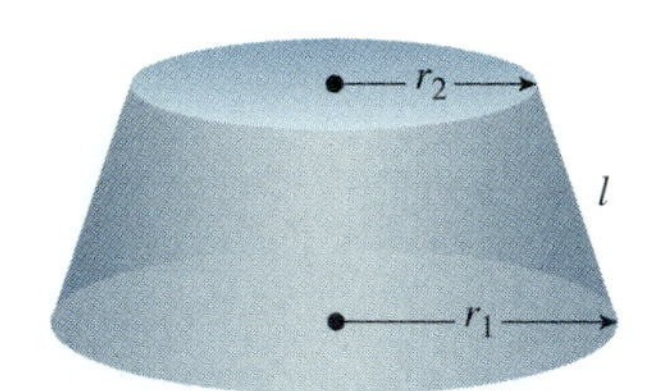

Figure 8.5.4

$$S = \pi(r_1 + r_2)l \tag{1}$$

for the lateral area S of a frustum of slant height l and base radii r_1 and r_2 (Figure 8.5.4). As suggested by Figure 8.5.5, the kth frustum has radii $f(x_{k-1})$ and $f(x_k)$ and height Δx_k. Its slant height is the length L_k of the kth line segment in the polygonal path, which from Formula (1) of Section 8.4 is

$$L_k = \sqrt{(\Delta x_k)^2 + [f(x_k) - f(x_{k-1})]^2}$$

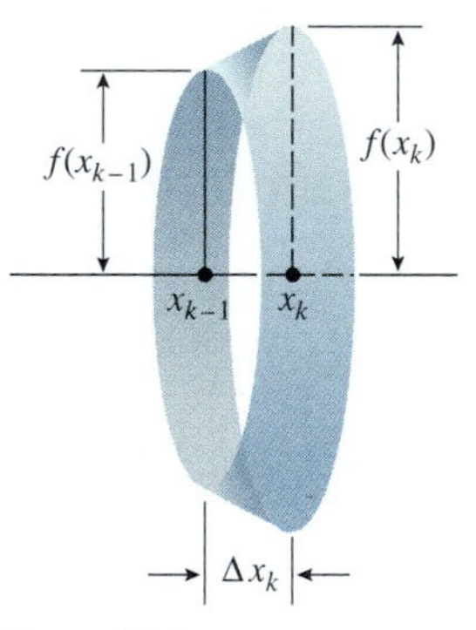

Figure 8.5.5

Thus, the lateral area S_k of the kth frustum is

$$S_k = \pi[f(x_{k-1}) + f(x_k)]\sqrt{(\Delta x_k)^2 + [f(x_k) - f(x_{k-1})]^2}$$

If we add these areas, we obtain the following approximation to the area S of the entire surface:

$$S \approx \sum_{k=1}^{n} \pi[f(x_{k-1}) + f(x_k)]\sqrt{(\Delta x_k)^2 + [f(x_k) - f(x_{k-1})]^2} \tag{2}$$

To put this in the form of a Riemann sum we will apply the Mean-Value Theorem (6.5.2). This theorem implies that there is a point x_k^* between x_{k-1} and x_k such that

$$\frac{f(x_k) - f(x_{k-1})}{x_k - x_{k-1}} = f'(x_k^*) \quad \text{or} \quad f(x_k) - f(x_{k-1}) = f'(x_k^*)\Delta x_k$$

and hence we can rewrite (2) as

$$S \approx \sum_{k=1}^{n} \pi[f(x_{k-1}) + f(x_k)]\sqrt{1 + [f'(x_k^*)]^2}\,\Delta x_k \tag{3}$$

However, this is not yet a Riemann sum because it involves the variables x_{k-1} and x_k. To eliminate these variables from the expression, observe that the average value of the numbers $f(x_{k-1})$ and $f(x_k)$ lies between these numbers, so the continuity of f and the Intermediate-Value Theorem (2.4.8) imply that there is a point x_k^{**} between x_{k-1} and x_k such that

$$\tfrac{1}{2}[f(x_{k-1}) + f(x_k)] = f(x_k^{**})$$

Thus, (2) can be expressed as

$$S \approx \sum_{k=1}^{n} 2\pi f(x_k^{**})\sqrt{1 + [f'(x_k^*)]^2}\,\Delta x_k$$

Although this expression is close to a Riemann sum in form, it is not a true Riemann sum because it involves two variables x_k^* and x_k^{**}, rather than x_k^* alone. However, it is proved in advanced calculus courses that this has no effect on the limit because of the continuity of f. Thus, we can assume that $x_k^{**} = x_k^*$ when taking the limit, and this suggests that S can be defined as

$$S = \lim_{\max \Delta x_k \to 0} \sum_{k=1}^{n} 2\pi f(x_k^{**})\sqrt{1 + [f'(x_k^*)]^2}\Delta x_k = \int_a^b 2\pi f(x)\sqrt{1 + [f'(x)]^2}\, dx$$

In summary, we have the following definition:

8.5.2 DEFINITION. If f is a smooth, nonnegative function on $[a, b]$, then the surface area S of the surface of revolution that is generated by revolving the portion of the curve $y = f(x)$ between $x = a$ and $x = b$ about the x-axis is defined as

$$S = \int_a^b 2\pi f(x)\sqrt{1 + [f'(x)]^2}\, dx$$

This result provides both a definition and formula for computing surface areas. Where convenient, this formula can also be expressed as

$$S = \int_a^b 2\pi f(x)\sqrt{1 + [f'(x)]^2}\, dx = \int_a^b 2\pi y\sqrt{1 + \left(\frac{dy}{dx}\right)^2}\, dx \tag{4}$$

Moreover, if g is nonnegative and $x = g(y)$ is a smooth curve on the interval $[c, d]$, then the area of the surface that is generated by revolving the portion of a curve $x = g(y)$ between $y = c$ and $y = d$ about the y-axis can be expressed as

$$S = \int_c^d 2\pi g(y)\sqrt{1 + [g'(y)]^2}\, dy = \int_c^d 2\pi x\sqrt{1 + \left(\frac{dx}{dy}\right)^2}\, dy \tag{5}$$

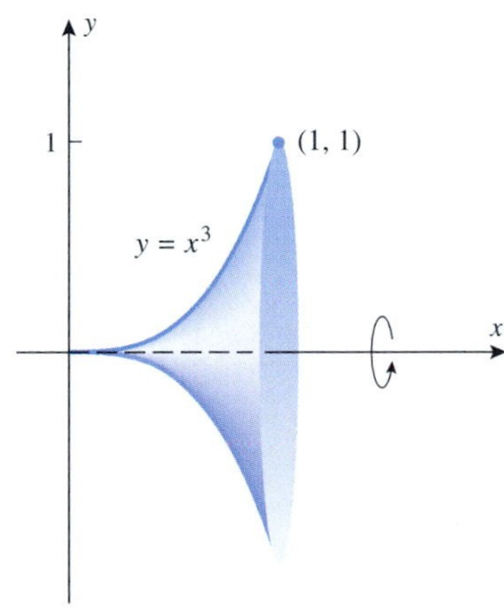

Figure 8.5.6

Example 1

Find the area of the surface that is generated by revolving the portion of the curve $y = x^3$ between $x = 0$ and $x = 1$ about the x-axis (Figure 8.5.6).

Solution. Since $y = x^3$, we have $dy/dx = 3x^2$, and hence from (4) the surface area S is

$$\begin{aligned}
S &= \int_0^1 2\pi y\sqrt{1 + \left(\frac{dy}{dx}\right)^2}\, dx \\
&= \int_0^1 2\pi x^3\sqrt{1 + (3x^2)^2}\, dx \\
&= 2\pi \int_0^1 x^3(1 + 9x^4)^{1/2}\, dx \\
&= \frac{2\pi}{36}\int_1^{10} u^{1/2}\, du \qquad \boxed{\begin{array}{c} u = 1 + 9x^4 \\ du = 36x^3\, dx \end{array}} \\
&= \frac{2\pi}{36} \cdot \frac{2}{3}u^{3/2}\Big]_{u=1}^{10} = \frac{\pi}{27}(10^{3/2} - 1) \approx 3.56
\end{aligned}$$

◀

Example 2

Find the area of the surface that is generated by revolving the portion of the curve $y = x^2$ between $x = 1$ and $x = 2$ about the y-axis (Figure 8.5.7).

Solution. Because the curve is revolved about the y-axis we will apply Formula (5). Toward this end, we rewrite $y = x^2$ as $x = \sqrt{y}$ and observe that the y-values corresponding to $x = 1$ and $x = 2$ are $y = 1$ and $y = 4$. Since $x = \sqrt{y}$, we have $dx/dy = 1/(2\sqrt{y})$, and hence from (5) the surface area S is

Figure 8.5.7

$$\begin{aligned}S &= \int_1^4 2\pi x\sqrt{1+\left(\frac{dx}{dy}\right)^2}\,dy\\ &= \int_1^4 2\pi\sqrt{y}\sqrt{1+\left(\frac{1}{2\sqrt{y}}\right)^2}\,dy\\ &= \pi\int_1^4 \sqrt{4y+1}\,dy\\ &= \frac{\pi}{4}\int_5^{17} u^{1/2}\,du \qquad \boxed{\begin{array}{l}u = 4y+1\\ du = 4\,dy\end{array}}\\ &= \frac{\pi}{4}\cdot\frac{2}{3}u^{3/2}\Big]_{u=5}^{17} = \frac{\pi}{6}(17^{3/2}-5^{3/2}) \approx 30.85\end{aligned}$$

EXERCISE SET 8.5 C CAS

In Exercises 1–4, find the area of the surface generated by revolving the given curve about the x-axis.

1. $y = 7x,\ 0 \le x \le 1$
2. $y = \sqrt{x},\ 1 \le x \le 4$
3. $y = \sqrt{4-x^2},\ -1 \le x \le 1$
4. $x = \sqrt[3]{y},\ 1 \le y \le 8$

In Exercises 5–8, find the area of the surface generated by revolving the given curve about the y-axis.

5. $x = 9y+1,\ 0 \le y \le 2$
6. $x = y^3,\ 0 \le y \le 1$
7. $x = \sqrt{9-y^2},\ -2 \le y \le 2$
8. $x = 2\sqrt{1-y},\ -1 \le y \le 0$

In Exercises 9–12, use a CAS to find the exact area of the surface generated by revolving the curve about the stated axis.

C 9. $y = \sqrt{x} - \frac{1}{3}x^{3/2},\ 1 \le x \le 3;$ x-axis

C 10. $y = \frac{1}{3}x^3 + \frac{1}{4}x^{-1},\ 1 \le x \le 2;$ x-axis

C 11. $8xy^2 = 2y^6 + 1,\ 1 \le y \le 2;$ y-axis

C 12. $x = \sqrt{16-y},\ 0 \le y \le 15;$ y-axis

In Exercises 13–16, use a CAS or a scientific calculator with numerical integration capabilities to approximate the area of the surface generated by revolving the curve about the stated axis. Round your answer to two decimal places.

C 13. $y = e^x,\ 0 \le x \le 1;$ x-axis

C 14. $y = \sin x,\ 0 \le x \le \pi;$ x-axis

C 15. $y = e^x,\ 1 \le y \le e;$ y-axis

C 16. $x = \tan y,\ 0 \le y \le \pi/4;$ y-axis

17. Use Formula (4) to show that the lateral area S of a right circular cone with height h and base radius r is

$$S = \pi r\sqrt{r^2+h^2}$$

18. Show that the area of the surface of a sphere of radius r is $4\pi r^2$. [*Hint:* Revolve the semicircle $y = \sqrt{r^2-x^2}$ about the x-axis.]

19. (a) The figure in Exercise 39 of Section 8.2 shows a spherical cap of height h cut from a sphere of radius r. Show that the surface area S of the cap is $S = 2\pi rh$. [*Hint:* Revolve an appropriate portion of the circle $x^2+y^2=r^2$ about the y-axis.]
    (b) The portion of a sphere that is cut by two parallel planes is called a ***zone***. Use the result in part (a) to show that the surface area of a zone depends on the radius of the sphere and the distance between the planes, but not on the location of the zone.

Exercises 20–26 require the formulas developed in the following discussion: If $x'(t)$ and $y'(t)$ are continuous functions and if no segment of the curve

$$x = x(t), \quad y = y(t) \qquad (a \le t \le b)$$

is traced more than once, then it can be shown that the area of the surface generated by revolving this curve about the x-axis is

$$S = \int_a^b 2\pi y(t)\sqrt{[x'(t)]^2 + [y'(t)]^2}\,dt \qquad \text{(A)}$$

and the area of the surface generated by revolving the curve about the y-axis is

$$S = \int_a^b 2\pi x(t)\sqrt{[x'(t)]^2 + [y'(t)]^2}\,dt \qquad \text{(B)}$$

20. Derive Formulas (4) and (5) from Formulas (A) and (B) above by choosing appropriate parametrizations for the curves $y = f(x)$ and $x = g(y)$.

21. Find the area of the surface generated by revolving the parametric curve $x = t^2$, $y = 2t$, $0 \le t \le 4$ about the x-axis.

C **22.** Use a CAS to find the area of the surface generated by revolving the parametric curve $x = e^t \cos t$, $y = e^t \sin t$, $0 \le t \le \pi/2$ about the x-axis.

23. Find the area of the surface generated by revolving the parametric curve $x = t$, $y = 2t^2$, $0 \le t \le 1$ about the y-axis.

24. Find the area of the surface generated by revolving the equations $x = \cos^2 t$, $y = \sin^2 t$, $0 \le t \le \pi/2$ about the y-axis.

25. By revolving the semicircle

$$x = r\cos t, \quad y = r\sin t \qquad (0 \le t \le \pi)$$

about the x-axis, show that the surface area of a sphere of radius r is $4\pi r^2$.

26. The equations

$$x = a\phi - a\sin\phi, \quad y = a - a\cos\phi \qquad (0 \le \phi \le 2\pi)$$

represent one arch of a cycloid. Show that the surface area generated by revolving this curve about the x-axis is $S = 64\pi a^2/3$. [*Hint:* Use the identities $\sin^2 \dfrac{\phi}{2} = \dfrac{1-\cos\phi}{2}$ and $\sin^3\phi = (1-\cos^2\phi)\sin\phi$ to help with the integration.]

27. (a) If a cone of slant height l and base radius r is cut along a lateral edge and laid flat, then as shown in the accompanying figure it becomes a sector of a circle of radius l. Use the formula $A = \frac{1}{2}l^2\theta$ for the area of a sector with radius l and central angle θ (in radians) to show that the lateral surface area of the cone is $\pi r l$.

(b) Use the result in part (a) to obtain Formula (1) for the lateral surface area of a frustum.

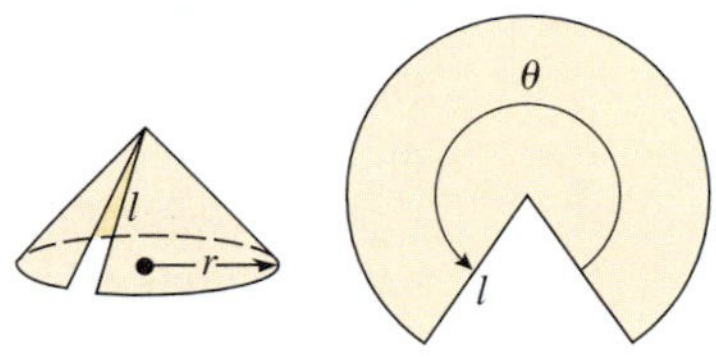

Figure Ex-27

28. Let $y = f(x)$ be a smooth curve on the interval $[a, b]$ and assume that $f(x) \ge 0$ for $a \le x \le b$. By the Extreme-Value Theorem 6.1.3, the function f has a maximum value K and a minimum value k on $[a, b]$. Prove: If L is the arc length of the curve $y = f(x)$ between $x = a$ and $x = b$ and if S is the area of the surface that is generated by revolving this curve about the x-axis, then

$$2\pi k L \le S \le 2\pi K L$$

29. Let $y = f(x)$ be a smooth curve on $[a, b]$ and assume that $f(x) \ge 0$ for $a \le x \le b$. Let A be the area under the curve $y = f(x)$ between $x = a$ and $x = b$ and let S be the area of the surface obtained when this section of curve is revolved about the x-axis.

(a) Prove that $2\pi A \le S$.

(b) For what functions f is $2\pi A = S$?

8.6 WORK

In this section we will use the integration tools developed in the preceding chapter to study some of the basic principles of "work," which is one of the fundamental concepts in physics and engineering.

THE ROLE OF WORK IN PHYSICS AND ENGINEERING

In this section we will be concerned with two related concepts, *work* and *energy*. To put these ideas in a familiar setting, when you push a stalled car for a certain distance you are performing work, and the effect of your work is to make the car move. The energy of motion caused by the work is called the *kinetic energy* of the car. The exact relationship between work and kinetic energy is governed by a principle of physics, called the *work–energy theorem*. Although we will touch on this idea in this section, a detailed study of the relationship between work and energy will be left for courses in physics and engineering. Our primary goal here will be to explain the role of integration in the study of work.

WORK DONE BY A CONSTANT FORCE APPLIED IN THE DIRECTION OF MOTION

When a stalled car is pushed, the speed that the car attains depends on the force F with which it is pushed and the distance d over which that force is applied (Figure 8.6.1). Thus, force and distance are the ingredients of work in the following definition.

8.6.1 DEFINITION. If a constant force of magnitude F is applied in the direction of motion of an object, and if that object moves a distance d, then we define the ***work*** W performed by the force on the object to be

$$W = F \cdot d \qquad (1)$$

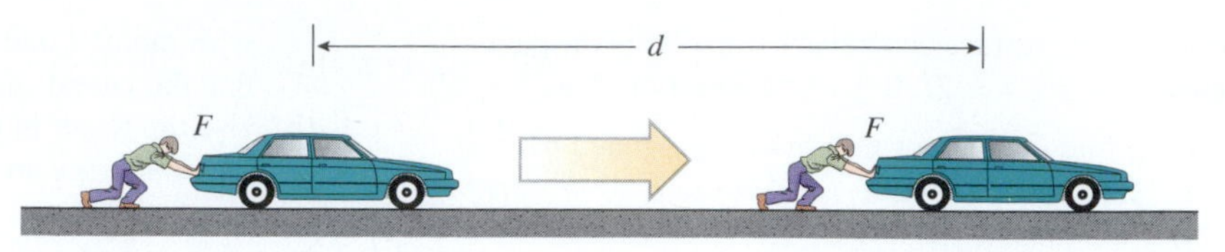

Figure 8.6.1

FOR THE READER. If you push against an immovable object, such as a brick wall, you may tire yourself out, but you will perform no work. Why?

Common units for measuring force are newtons (N) in the International System of Units (SI), dynes (dyn) in the CGS system, and pounds (lb) in the British Engineering system. One newton is the force required to give a mass of 1 kg an acceleration of 1 m/s^2, one dyne is the force required to give a mass of 1 g an acceleration of 1 cm/s^2, and one pound of force is the force required to give a mass of 1 slug an acceleration of 1 ft/s^2.

It follows from Definition 8.6.1 that work has units of force times distance. The most common units of work are newton-meters (N·m), dyne-centimeters (dyn·cm), and foot-pounds (ft·lb). As indicated in Table 8.6.1, one newton-meter is also called a ***joule*** (J), and one dyne-centimeter is also called an ***erg***. One foot-pound is approximately 1.36 J.

Table 8.6.1

SYSTEM	FORCE ×	DISTANCE =	WORK
SI	newton (N)	meter (m)	joule (J)
CGS	dyne (dyn)	centimeter (cm)	erg
BE	pound (lb)	foot (ft)	foot-pound (ft·lb)

CONVERSION FACTORS:

$1\text{ N} = 10^5\text{ dyn} \approx 0.225\text{ lb}$ $1\text{ lb} \approx 4.45\text{ N}$

$1\text{ J} = 10^7\text{ erg} \approx 0.738\text{ ft·lb}$ $1\text{ ft·lb} \approx 1.36\text{ J} = 1.36 \times 10^7\text{ erg}$

Example 1

An object moves 5 ft along a line while subjected to a constant force of 100 lb in its direction of motion. The work done is

$$W = F \cdot d = 100 \cdot 5 = 500 \text{ ft·lb}$$

An object moves 25 m along a line while subjected to a constant force of 4 N in its direction of motion. The work done is

$$W = F \cdot d = 4 \cdot 25 = 100 \text{ N·m} = 100 \text{ J}$$ ◀

Vasili Alexeev lifting a record-breaking 562 lb in the 1976 Olympics

Example 2

In the 1976 Olympics, Vasili Alexeev astounded the world by lifting a record-breaking 562 lb from the floor to above his head (about 2 m). Equally astounding was the feat of strongman Paul Anderson, who in 1957 braced himself on the floor and used his back to lift 6270 lb of lead and automobile parts a distance of 1 cm. Who did more work?

Solution. To lift an object one must apply sufficient force to overcome the gravitational force that the Earth exerts on that object. The force that the Earth exerts on an object is that object's weight; thus, in performing their feats, Alexeev applied a force of 562 lb over a distance of 2 m and Anderson applied a force of 6270 lb over a distance of 1 cm. Since pounds are units in the BE system, meters are units in SI, and centimeters are units in the CGS system, we will need to decide on the measurement system we want to use and be consistent. Let us agree to use SI and express the work of the two men in joules. Using the conversion factor in Table 8.6.1 we obtain

$$562 \text{ lb} \approx 562 \text{ lb} \times 4.45 \text{ N/lb} = 2500.9 \text{ N}$$
$$6270 \text{ lb} \approx 6270 \text{ lb} \times 4.45 \text{ N/lb} = 27{,}901.5 \text{ N}$$

Using these values and the fact that 1 cm = 0.01 m we obtain

$$\text{Alexeev's work} = (2500.9 \text{ N}) \times (2 \text{ m}) \approx 5002 \text{ J}$$
$$\text{Anderson's work} = (27{,}901.5 \text{ N}) \times (0.01 \text{ m}) \approx 279 \text{ J}$$

Therefore, even though Anderson's lift required a tremendous upward force, it was applied over such a short distance that Alexeev did more work. ◀

WORK DONE BY A VARIABLE FORCE APPLIED IN THE DIRECTION OF MOTION

Many important problems are concerned with finding the work done by a *variable* force that is applied in the direction of motion. For example, Figure 8.6.2*a* shows a spring in its natural state (neither compressed nor stretched). If we want to pull the block horizontally so that it moves with a uniform speed (Figure 8.6.2*b*), then we would have to apply more and more force to the block to overcome the increasing force of the stretching spring. Thus, our next objective is to define what is meant by the work performed by a variable force and to find a formula for computing it. This will require calculus.

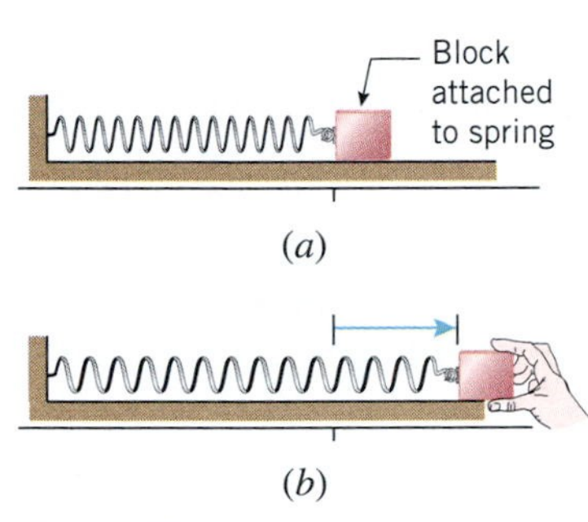

Figure 8.6.2

8.6.2 PROBLEM. Suppose that an object moves in the positive direction along a coordinate line while subjected to a variable force $F(x)$ that is applied in the direction of motion. Define what is meant by the *work* W performed by the force on the object as the object moves from $x = a$ to $x = b$, and find a formula for computing the work.

The basic idea for solving this problem is to break up the interval $[a, b]$ into subintervals that are sufficiently small that the force does not vary much on each subinterval. This will allow us to treat the force as constant on each subinterval and to approximate the work on each subinterval using Formula (1). By adding the approximations to the work on the subintervals, we will obtain a Riemann sum that approximates the work W over the entire interval, and by taking the limit of the Riemann sums we will obtain an integral for W.

To implement this idea, divide the interval $[a, b]$ into n subintervals by inserting points $x_1, x_2, \ldots, x_{n-1}$ between a and b. We can use Formula (1) to approximate the work W_k done in the kth subinterval by choosing any point x_k^* in this interval and regarding the force to have a constant value $F(x_k^*)$ throughout the interval. Since the width of the kth subinterval is $x_k - x_{k-1} = \Delta x_k$, this yields the approximation

$$W_k \approx F(x_k^*)\Delta x_k$$

Adding these approximations yields the following Riemann sum that approximates the work W done over the entire interval:

$$W \approx \sum_{k=1}^{n} F(x_k^*)\Delta x_k$$

Taking the limit as n increases and the widths of the subintervals approach zero yields the definite integral

$$W = \lim_{\max \Delta x_k \to 0} \sum_{k=1}^{n} F(x_k^*)\Delta x_k = \int_a^b F(x)\,dx$$

In summary, we have the following result:

8.6.3 DEFINITION. Suppose that an object moves in the positive direction along a coordinate line over the interval $[a, b]$ while subjected to a variable force $F(x)$ that is applied in the direction of motion. Then we define the ***work*** W performed by the force on the object to be

$$W = \int_a^b F(x)\,dx \tag{2}$$

Hooke's law [Robert Hooke (1635–1703), English physicist] states that under appropriate conditions a spring that is stretched x units beyond its natural length pulls back with a force

$$F(x) = kx$$

where k is a constant (called the ***spring constant*** or ***spring stiffness***). The value of k depends on such factors as the thickness of the spring and the material used in its composition. Since $k = F(x)/x$, the constant k has units of force per unit length.

Example 3

A spring exerts a force of 5 N when stretched 1 m beyond its natural length.

(a) Find the spring constant k.

(b) How much work is required to stretch the spring 1.8 m beyond its natural length?

Solution (a). From Hooke's law,

$$F(x) = kx$$

From the data, $F(x) = 5$ N when $x = 1$ m, so $5 = k \cdot 1$. Thus, the spring constant is $k = 5$ newtons per meter (N/m). This means that the force $F(x)$ required to stretch the spring x meters is

$$F(x) = 5x \tag{3}$$

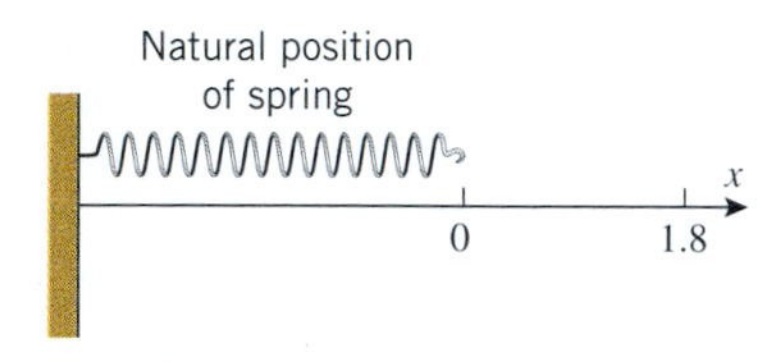

Figure 8.6.3

Solution (b). Place the spring along a coordinate line as shown in Figure 8.6.3. We want to find the work W required to stretch the spring over the interval from $x = 0$ to $x = 1.8$. From (2) and (3) the work W required is

$$W = \int_a^b F(x)\,dx = \int_0^{1.8} 5x\,dx = \frac{5x^2}{2}\Big]_0^{1.8} = 8.1 \text{ J}$$ ◀

Example 4

An astronaut's *weight* (or more precisely, *Earth weight*) is the force exerted on the astronaut by the Earth's gravity. As the astronaut moves upward into space, the gravitational pull of the Earth decreases, and hence so does his or her weight. We will show later in the text that if the Earth is assumed to be a sphere of radius 4000 mi, then an astronaut who weighs 150 lb on Earth will have a weight of

$$w(x) = \frac{2{,}400{,}000{,}000}{x^2} \text{ lb}$$

at a distance of x mi from the Earth's center. Use this formula to determine the work in foot-pounds required to lift the astronaut to a point that is 800 mi above the surface of the Earth (Figure 8.6.4).

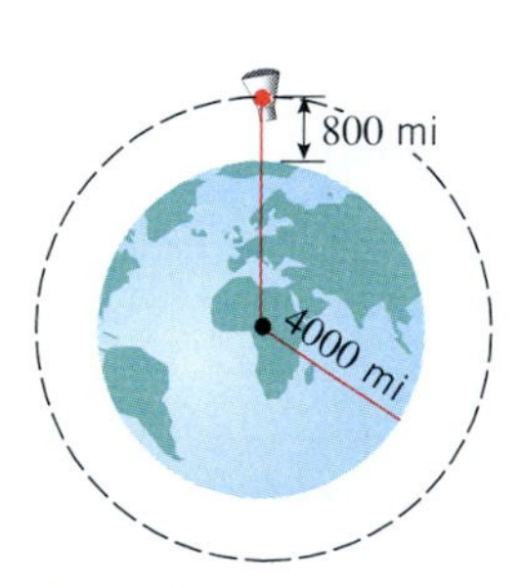

Figure 8.6.4

Solution. Since the Earth has a radius of 4000 mi, the astronaut is lifted from a point that is 4000 mi from the Earth's center to a point that is 4800 mi from the Earth's center. Thus, from (2), the work W required to lift the astronaut is

$$\begin{aligned} W &= \int_{4000}^{4800} \frac{2{,}400{,}000{,}000}{x^2}\,dx \\ &= -\frac{2{,}400{,}000{,}000}{x}\Big]_{4000}^{4800} \\ &= -500{,}000 + 600{,}000 \\ &= 100{,}000 \text{ mile-pounds} \\ &= (100{,}000 \text{ mi}\cdot\text{lb}) \times (5280 \text{ ft/mi}) \\ &= 5.28 \times 10^8 \text{ ft}\cdot\text{lb} \end{aligned}$$ ◀

CALCULATING WORK FROM BASIC PRINCIPLES

Some problems cannot be solved by mechanically substituting into formulas, and one must return to basic principles to obtain solutions. This is illustrated in the next example.

Example 5

A cylindrical water tank of radius 10 ft and height 30 ft is half filled with water. How much work is required to pump all of the water out through a hole in the top of the tank?

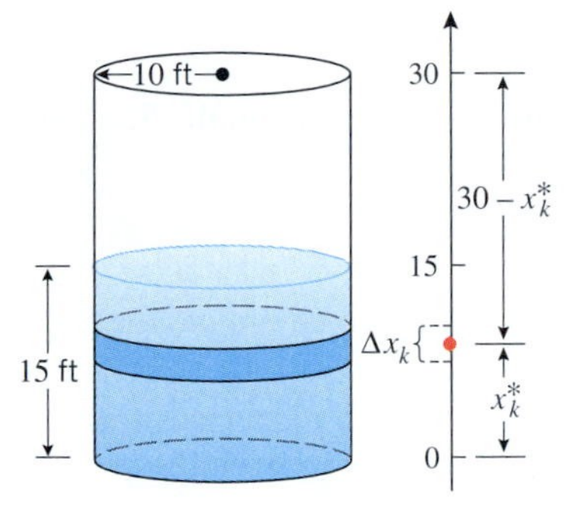

Figure 8.6.5

Solution. Our strategy will be to divide the water into thin layers, approximate the work required to move each layer to the top of the tank, add the approximations for the layers to obtain a Riemann sum that approximates the total work, and then take the limit of the Riemann sums to produce an integral for the total work.

To implement this idea, introduce an x-axis as shown in Figure 8.6.5, and divide the water into n layers with Δx_k denoting the thickness of the kth layer. Although the upper and lower surfaces of the kth layer are at different distances from the top, the difference will be small if the layer is thin, and we can reasonably assume that the entire layer is concentrated at a single point x_k^* (Figure 8.6.5). Thus, the work W_k required to move the kth layer to the top of the tank is approximately

$$W_k \approx F_k(30 - x_k^*) \tag{4}$$

where F_k is the force required to lift the kth layer. But the force required to lift the kth layer is the force needed to overcome gravity, and this is the same as the weight of the layer. To find the weight F_k of the kth layer we will multiply its volume by the weight density of water (62.4 lb/ft^3). This yields

$$F_k = (\pi(10)^2\Delta x_k)(62.4) = 6240\pi\Delta x_k$$

Thus, from (4)

$$W_k \approx \underbrace{(30 - x_k^*)}_{\text{Distance}} \cdot \underbrace{6240\pi\Delta x_k}_{\text{Force}}$$

and hence the work W required to move all n layers is approximately

$$W = \sum_{k=1}^{n} W_k \approx \sum_{k=1}^{n}(30 - x_k^*)(6240\pi)\Delta x_k$$

To find the *exact* value of the work we take the limit as $\max \Delta x_k \to 0$. This yields

$$W = \lim_{\max \Delta x_k \to 0} \sum_{k=1}^{n}(30 - x_k^*)(6240\pi)\Delta x_k = \int_0^{15} (30 - x)(6240\pi)\, dx$$

$$= 6240\pi\left(30x - \frac{x^2}{2}\right)\Bigg]_0^{15} = 2{,}106{,}000\pi \text{ ft·lb} \approx 6{,}616{,}194 \text{ ft·lb}$$ ◀

THE WORK–ENERGY THEOREM

When you see an object in motion, you can be certain that somehow work has been expended to create that motion. For example, when you drop a stone from a building the stone gathers speed because the force of the Earth's gravity is performing work on it, and when a hockey player strikes a puck with a hockey stick, the work performed on the puck during the brief period of contact with the stick creates the enormous speed of the puck across the ice.

The linkage between work and motion is based on the concept of *kinetic energy*, which we will now define. If a particle of mass m is moving with speed v at a certain instant, then its ***kinetic energy*** K at that instant is defined as

$$K = \tfrac{1}{2}mv^2 \tag{5}$$

It is the following fundamental principle of physics that relates work and kinetic energy.

8.6.4 WORK–ENERGY THEOREM. *When a force does work on an object, it causes a change in the kinetic energy of the object that is equal to the work performed; that is, if W is the work performed on an object of mass m, and if the initial and final speeds of the object are v_i and v_f, respectively, then*

$$W = \tfrac{1}{2}mv_f^2 - \tfrac{1}{2}mv_i^2 \tag{6}$$

The units of kinetic energy are the same as the units of work. For example, in the Standard International system kinetic energy is measured in joules (J).

Example 6

A space probe of mass $m = 5.00 \times 10^4$ kg travels in deep space subjected only to the force of its own engine. Starting at a time when the speed of the probe is $v = 1.10 \times 10^4$ m/s, the engine is fired continuously over a distance of 2.50×10^6 m with a constant force of 4.00×10^5 N in the direction of motion. What is the final speed of the probe?

Solution. Since the force applied by the engine is constant and in the direction of motion, the work W expended by the engine on the probe is

$$W = \text{force} \times \text{distance} = (4.00 \times 10^5 \text{ N}) \times (2.50 \times 10^6 \text{ m}) = 1.00 \times 10^{12} \text{ J}$$

From (6), the final kinetic energy $K_f = \frac{1}{2}mv_f^2$ of the probe can be expressed in terms of the work W and the initial kinetic energy $K_i = \frac{1}{2}mv_i^2$ as

$$K_f = W + K_i$$

Thus, from the known mass and initial speed we have

$$K_f = (1.00 \times 10^{12} \text{ J}) + \tfrac{1}{2}(5.00 \times 10^4 \text{ kg})(1.10 \times 10^4 \text{ m/s})^2 \approx 4.03 \times 10^{12} \text{ J}$$

The final kinetic energy is $K_f = \frac{1}{2}mv_f^2$, so the final speed of the probe is

$$v_f = \sqrt{\frac{2K_f}{m}} = \sqrt{\frac{2(4.03 \times 10^{12})}{5.00 \times 10^4}} \approx 1.27 \times 10^4 \text{ m/s}$$ ◀

EXERCISE SET 8.6

1. Find the work done when
  (a) a constant force of 30 lb in the positive x-direction moves an object from $x = -2$ to $x = 5$ ft
  (b) a variable force of $F(x) = 1/x^2$ lb in the positive x-direction moves an object from $x = 1$ to $x = 6$ ft.

2. A variable force $F(x)$ in the positive x-direction is graphed in the accompanying figure. Find the work done by the force on a particle that moves from $x = 0$ to $x = 5$.

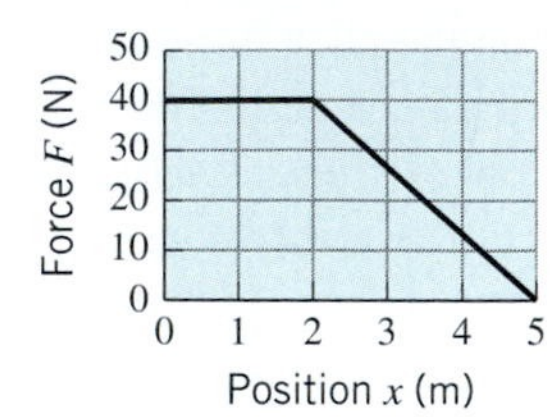

Figure Ex-2

3. A constant force of 10 lb in the positive x-direction is applied to a particle whose velocity versus time curve is shown in the accompanying figure. Find the work done by the force on the particle from time $t = 0$ to $t = 5$.

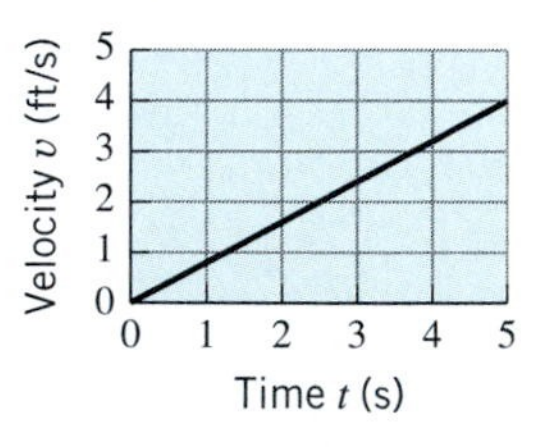

Figure Ex-3

4. A spring whose natural length is 15 cm exerts a force of 45 N when stretched to a length of 20 cm.
  (a) Find the spring constant (in newtons/meter).
  (b) Find the work that is done in stretching the spring 3 cm beyond its natural length.
  (c) Find the work done in stretching the spring from a length of 20 cm to a length of 25 cm.

5. A spring exerts a force of 100 N when it is stretched 0.2 m beyond its natural length. How much work is required to stretch the spring 0.8 m beyond its natural length?

6. Assume that a force of 6 N is required to compress a spring from a natural length of 4 m to a length of $3\frac{1}{2}$ m. Find the work required to compress the spring from its natural length to a length of 2 m. (Hooke's law applies to compression as well as extension.)

7. Assume that 10 ft·lb of work is required to stretch a spring 1 ft beyond its natural length. What is the spring constant?

8. A cylindrical tank of radius 5 ft and height 9 ft is two-thirds filled with water. Find the work required to pump all the water over the upper rim.

9. Solve Exercise 8 assuming that the tank is two-thirds filled with a liquid that weighs ρ lb/ft^3.

10. A cone-shaped water reservoir is 20 ft in diameter across the top and 15 ft deep. If the reservoir is filled to a depth of 10 ft, how much work is required to pump all the water to the top of the reservoir?

11. The vat shown in the accompanying figure contains water to a depth of 2 m. Find the work required to pump all the water to the top of the vat. [Use 9810 N/m^3 as the weight density of water.]

12. The cylindrical tank shown in the accompanying figure is filled with a liquid weighing 50 lb/ft^3. Find the work required to pump all the liquid to a level 1 ft above the top of the tank.

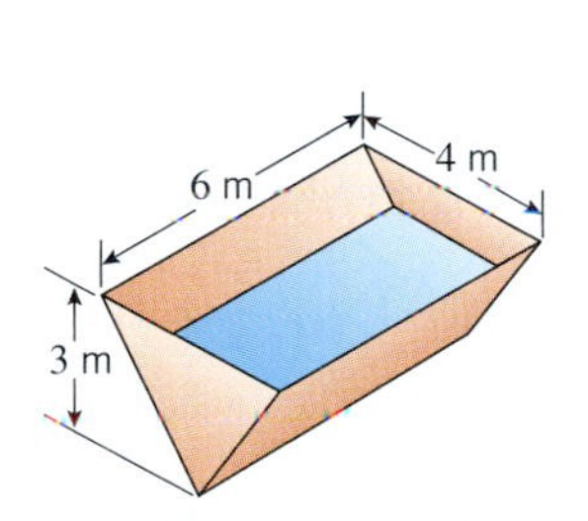

Figure Ex-11

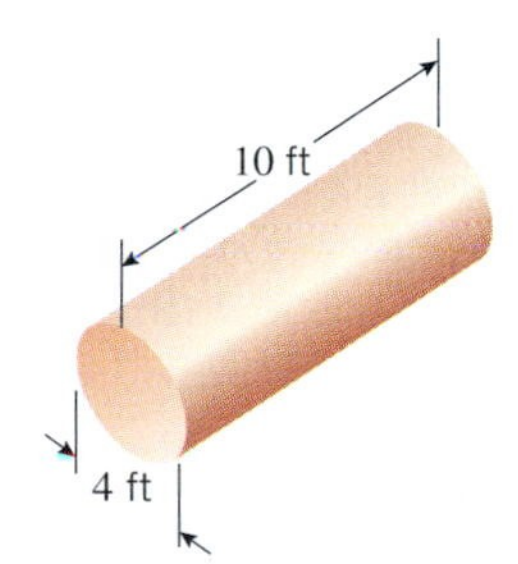

Figure Ex-12

13. A swimming pool is built in the shape of a rectangular parallelepiped 10 ft deep, 15 ft wide, and 20 ft long.
   (a) If the pool is filled to 1 ft below the top, how much work is required to pump all the water into a drain at the top edge of the pool?
   (b) A one-horsepower motor can do 550 ft·lb of work per second. What size motor is required to empty the pool in 1 hour?

14. A rocket weighing 3 tons is filled with 40 tons of liquid fuel. In the initial part of the flight, fuel is burned off at a constant rate of 2 tons per 1000 ft of vertical height. How much work is done in lifting the rocket to 3000 ft?

15. A 100-ft length of steel chain weighing 15 lb/ft is dangling from a pulley. How much work is required to wind the chain onto the pulley?

16. It follows from Coulomb's law in physics that two like electrostatic charges repel each other with a force inversely proportional to the square of the distance between them. Suppose that two charges A and B repel with a force of k newtons when they are positioned at points $A(-a, 0)$ and $B(a, 0)$, where a is measured in meters. Find the work W required to move charge A along the x-axis to the origin if charge B remains stationary.

17. It is a law of physics that the gravitational force exerted by the Earth on an object varies inversely as the square of its distance from the Earth's center. Thus, an object's weight $w(x)$ is related to its distance x from the Earth's center by a formula of the form

$$w(x) = \frac{k}{x^2}$$

where k is a constant of proportionality that depends on the mass of the object.
   (a) Use this fact and the assumption that the Earth is a sphere of radius 4000 mi to obtain the formula for $w(x)$ in Example 4.
   (b) Find a formula for the weight $w(x)$ of a satellite that is x mi from the Earth's surface if its weight is 6000 lb.
   (c) How much work is required to lift the satellite from the surface of the Earth to an orbital position that is 1000 mi high?

18. (a) The formula $w(x) = k/x^2$ in Exercise 17 is applicable to all celestial bodies. Assuming that the Moon is a sphere of radius 1080 mi, find the force that the Moon exerts on an astronaut who is x mi from the surface of the Moon if her weight on the Moon's surface is 20 lb.
   (b) How much work is required to lift the astronaut to a point that is 10.8 mi above the Moon's surface?

19. The Yamanashi Maglev Test Line in Japan that runs between Sakaigawa and Akiyama is currently testing magnetic levitation (MAGLEV) trains that are designed to levitate inches above powerful magnetic fields. Suppose that a MAGLEV train has a mass of $m = 4.00 \times 10^5$ kg and that starting at a time when the train has a speed of 20 m/s the engine applies a force of 6.40×10^5 N in the direction of motion over a distance of 3.00×10^3 m. Use the Work–Energy Theorem (8.6.4) to find the final speed of the train.

20. Assume that a Mars probe of mass $m = 2.00 \times 10^3$ kg is subjected only to the force of its own engine. Starting at a time when the speed of the probe is $v = 1.00 \times 10^4$ m/s, the engine is fired continuously over a distance of 2.00×10^5 m with a constant force of 2.00×10^5 N in the direction of motion. Use the Work–Energy Theorem (8.6.4) to find the final speed of the probe.

21. On August 10, 1972 a meteorite with an estimated mass of 4×10^6 kg and an estimated speed of 15 km/s skipped across the atmosphere above the western United States and Canada but fortunately did not hit the Earth.
   (a) Assuming that the meteorite had hit the Earth with a speed of 15 km/s, what would have been its change in kinetic energy in joules (J)?
   (b) Express the energy as a multiple of the explosive energy of 1 megaton of TNT, which is 4.2×10^{15} J.
   (c) The energy associated with the Hiroshima atomic bomb was 13 kilotons of TNT. To how many such bombs would the meteorite impact have been equivalent?

8.7 FLUID PRESSURE AND FORCE

In this section we will use the integration tools developed in the preceding chapter to study the pressures and forces exerted by fluids on submerged objects.

WHAT IS A FLUID?

A ***fluid*** is a substance that flows to conform to the boundaries of any container in which it is placed. Fluids include *liquids*, such as water, oil, and mercury, as well as *gases*, such as helium, oxygen, and air. The study of fluids falls into two categories: *fluid statics* (the study of fluids at rest) and *fluid dynamics* (the study of fluids in motion). In this section we will be concerned only with fluid statics; toward the end of this text we will investigate problems in fluid dynamics.

THE CONCEPT OF PRESSURE

The effect that a force has on an object depends on how that force is spread over the surface of the object. For example, when you walk on soft snow with boots, the weight of your body crushes the snow and you sink into it. However, if you put on a pair of skis to spread the weight of your body over a greater surface area, then the weight of your body has less of a crushing effect on the snow, and you are able to glide across the surface. The concept that accounts for both the magnitude of a force and the area over which it is applied is called *pressure*.

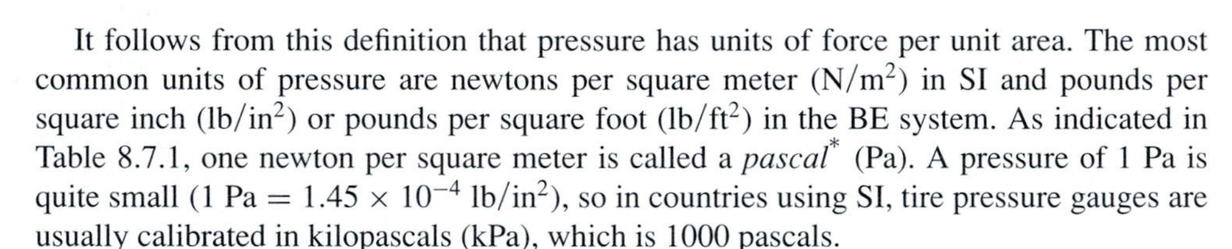

8.7.1 DEFINITION. If a force of magnitude F is applied to a surface of area A, then we define the ***pressure*** P exerted by the force on the surface to be

$$P = \frac{F}{A} \tag{1}$$

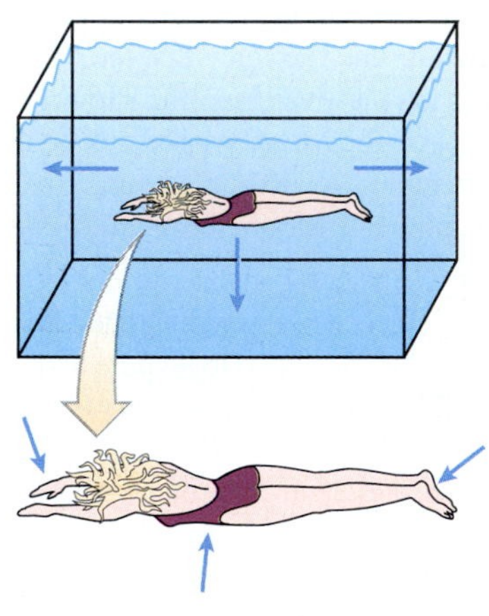

Fluid forces always act perpendicular to the surface of a submerged object.

Figure 8.7.1

It follows from this definition that pressure has units of force per unit area. The most common units of pressure are newtons per square meter (N/m^2) in SI and pounds per square inch (lb/in^2) or pounds per square foot (lb/ft^2) in the BE system. As indicated in Table 8.7.1, one newton per square meter is called a *pascal** (Pa). A pressure of 1 Pa is quite small ($1 \text{ Pa} = 1.45 \times 10^{-4} \text{ lb/in}^2$), so in countries using SI, tire pressure gauges are usually calibrated in kilopascals (kPa), which is 1000 pascals.

In this section we will be interested in pressures and forces on objects submerged in fluids. Pressures themselves have no directional characteristics, but the forces that they create always act perpendicular to the face of the submerged object. Thus, in Figure 8.7.1 the water pressure creates horizontal forces on the sides of the tank, vertical forces on the bottom of the tank, and forces that vary in direction, so as to be perpendicular to the different parts of the swimmer's body.

*BLAISE PASCAL (1623–1662). French mathematician and scientist. Pascal's mother died when he was three years old and his father, a highly educated magistrate, personally provided the boy's early education. Although Pascal showed an inclination for science and mathematics, his father refused to tutor him in those subjects until he mastered Latin and Greek. Pascal's sister and primary biographer claimed that he independently discovered the first thirty-two propositions of Euclid without ever reading a book on geometry. (However, it is generally agreed that the story is apocryphal.) Nevertheless, the precocious Pascal published a highly respected essay on conic sections by the time he was sixteen years old. Descartes, who read the essay, thought it so brilliant that he could not believe that it was written by such a young man. By age 18 his health began to fail and until his death he was in frequent pain. However, his creativity was unimpaired.

Pascal's contributions to physics include the discovery that air pressure decreases with altitude and the principle of fluid pressure that bears his name. However, the originality of his work is questioned by some historians. Pascal made major contributions to a branch of mathematics called "projective geometry," and he helped to develop probability theory through a series of letters with Fermat.

In 1646, Pascal's health problems resulted in a deep emotional crisis that led him to become increasingly concerned with religious matters. Although born a Catholic, he converted to a religious doctrine called Jansenism and spent most of his final years writing on religion and philosophy.

Table 8.7.1

SYSTEM	FORCE	÷	AREA	=	PRESSURE
SI	newton (N)		square meter (m^2)		pascal (Pa)
BE	pound (lb)		square foot (ft^2)		lb/ft^2
BE	pound (lb)		square inch (in^2)		lb/in^2 (psi)

CONVERSION FACTORS:
1 Pa $\approx 1.45 \times 10^{-4}$ lb/in^2 $\approx 2.09 \times 10^{-2}$ lb/ft^2
1 $lb/in^2 \approx 6.90 \times 10^3$ Pa 1 $lb/ft^2 \approx 47.9$ Pa

Example 1

Referring to Figure 8.7.1, suppose that the back of the swimmer's hand has a surface area of 8.4×10^{-3} m^2 and that the pressure acting on it is 1.2×10^5 Pa (a realistic value near the bottom of a deep diving pool). Find the force that acts on the swimmer's hand.

Solution. From (1), the force F is

$$F = PA = (1.2 \times 10^5 \text{ N/m}^2)(8.4 \times 10^{-3} \text{ m}^2) \approx 1.0 \times 10^3 \text{ N}$$

This is quite a large force (about 230 lb in the BE system). ◀

FLUID DENSITY

Scuba divers know that the deeper they dive, the greater the pressure and the forces that they feel on their bodies. This sense of pressure and force is caused by the weight of the water and air above—the deeper the diver goes, the greater the weight above and hence the greater the pressure and force that he or she feels.

To calculate pressures and forces on submerged objects, we need to know something about the characteristics of the fluids in which they are submerged. For simplicity, we will assume that the fluids under consideration are *homogeneous*, by which we mean that any two samples of the fluid with the same volume have the same mass. It follows from this assumption that the mass per unit volume is a constant δ that depends on the physical characteristics of the fluid but not on the size or location of the sample; we call

$$\delta = \frac{m}{V} \tag{2}$$

the ***mass density*** of the fluid. Sometimes it is more convenient to work with weight per unit volume than with mass per unit volume. Thus, we define the ***weight density*** ρ of a fluid to be

$$\rho = \frac{w}{V} \tag{3}$$

where w is the weight of a fluid sample of volume V. Thus, if the weight density of a fluid is known, then the weight w of a fluid sample of volume V can be computed from the formula $w = \rho V$. Table 8.7.2 shows some typical weight densities.

Table 8.7.2

WEIGHT DENSITIES	
SI	N/m^3
Machine oil	4,708
Gasoline	6,602
Fresh water	9,810
Seawater	10,045
Mercury	133,416
BE SYSTEM	lb/ft^3
Machine oil	30.0
Gasoline	42.0
Fresh water	62.4
Seawater	64.0
Mercury	849.0

All densities are affected by variations in temperature and pressure. Weight densities are affected by variations in g.

FLUID PRESSURE

To calculate fluid pressures and forces we will need ***Pascal's principle***, which states that *fluid pressure at a given depth is the same in all directions* (Figure 8.7.2). This implies, for example, that at the bottom corner of a swimming pool the pressure on the two side walls is the same as the pressure on the bottom.

It is a straightforward matter to calculate fluid force and pressure on a flat surface that is submerged *horizontally* because each point on the surface is at the same depth. If a flat surface of area A is submerged horizontally at a depth h in a container of fluid with weight density ρ, then the fluid exerts a force F that is perpendicular to the surface and is

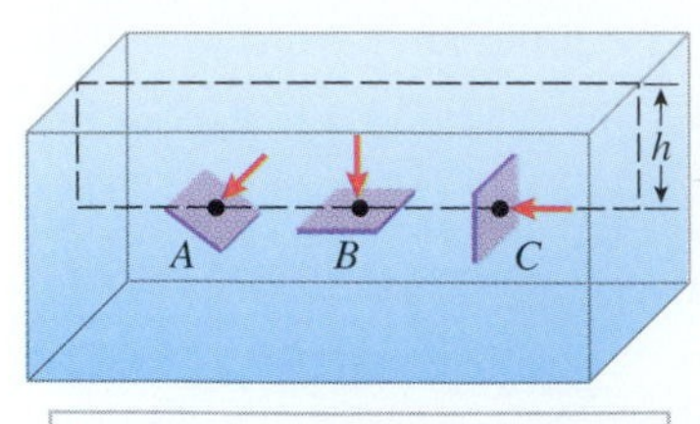

Figure 8.7.2

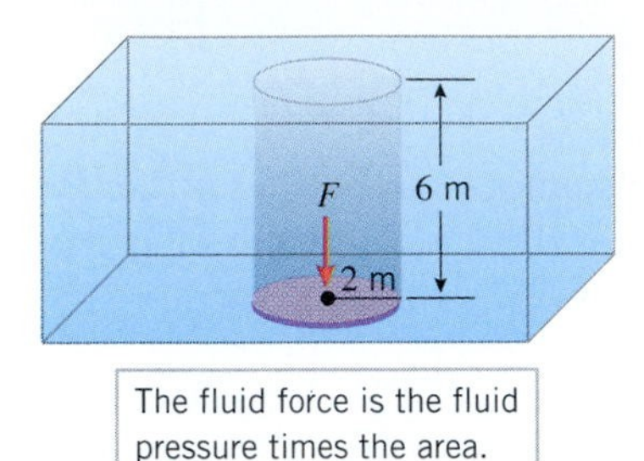

Figure 8.7.3

given by

$$F = \rho h A \tag{4}$$

Thus, it follows from (1) that the pressure P on the surface created by this force is

$$P = \frac{F}{A} = \rho h \tag{5}$$

Example 2

Find the fluid pressure and force on the top of a flat circular plate of radius 2 m that is submerged horizontally in water at a depth of 6 m (Figure 8.7.3).

Solution. Since the weight density of water is $\rho = 9810\ \text{N/m}^3$, it follows from (5) that the fluid pressure is

$$P = \rho h = (9810)(6) = 58{,}860\ \text{Pa}$$

and it follows from (4) that the fluid force is

$$F = \rho h A = \rho h(\pi r^2) = (9810)(6)(4\pi) = 235{,}440\pi \approx 739{,}657\ \text{N}$$ ◀

FLUID FORCE ON A VERTICAL SURFACE

It was easy to calculate the fluid force on the horizontal plate in Example 2 because each point on the plate was at the same depth. The problem of finding the fluid force on a vertical surface is more complicated because the depth, and hence the pressure, is not constant over the surface. To find the fluid force on a vertical surface we will need calculus.

8.7.2 PROBLEM. Suppose that a flat surface is immersed vertically in a fluid of weight density ρ and that the submerged portion of the surface extends from $x = a$ to $x = b$ along an x-axis whose positive direction is down (Figure 8.7.4a). For $a \le x \le b$, suppose that $w(x)$ is the width of the surface and that $h(x)$ is the depth of the point x. Define what is meant by the *fluid force* F on the surface, and find a formula for computing it.

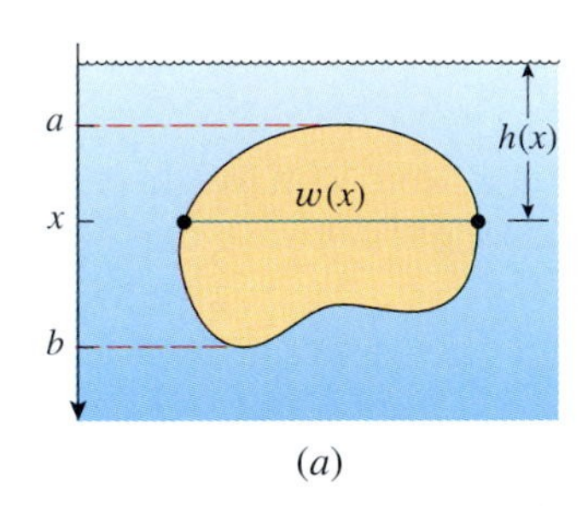

(a)

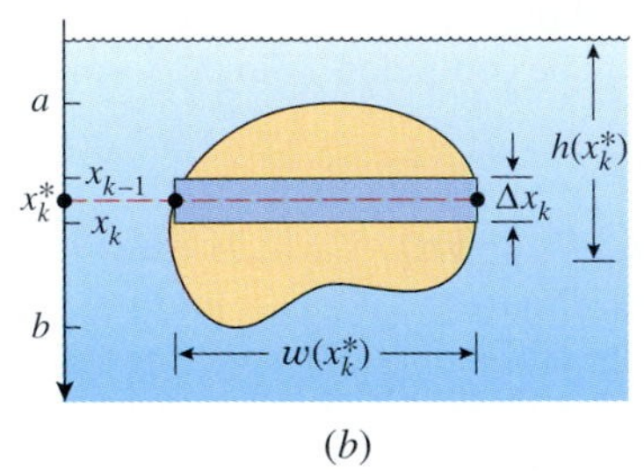

(b)

Figure 8.7.4

The basic idea for solving this problem is to break up the interval $[a, b]$ into subintervals that are sufficiently small that the depth does not vary much on each subinterval. This has the effect of dividing the plate into strips over each of which the depth can be treated as constant. This assumption will allow us to use Formula (4) to approximate the fluid force on each strip. By adding the approximations to the forces on the strips we will obtain a Riemann sum that approximates the total force F on the entire surface, and by taking the limit of the Riemann sums we will obtain an integral for F.

To implement this idea, divide the interval $[a, b]$ into n subintervals by inserting points $x_1, x_2, \ldots, x_{n-1}$ between a and b. To approximate the force on the kth strip we choose any point x_k^* in the kth interval and approximate the strip by a rectangle of length $w(x_k^*)$ and width $\Delta x_k = x_k - x_{k-1}$ (Figure 8.7.4b).

Although the top and bottom of the rectangle are at different depths, the difference will be small if the strip is thin and we can reasonably assume that the entire strip is at depth $h(x_k^*)$. Thus, from (4) we can approximate the force F_k on the kth strip as

$$F_k \approx \rho \underbrace{h(x_k^*)}_{\text{Depth}} \cdot \underbrace{w(x_k^*)\Delta x_k}_{\text{Area of rectangle}}$$

Adding these approximations yields the following Riemann sum that approximates the total force F on the surface:

$$F = \sum_{k=1}^{n} F_k \approx \sum_{k=1}^{n} \rho h(x_k^*) w(x_k^*) \Delta x_k$$

Taking the limit as n increases and the widths of the subintervals approach zero yields the definite integral

$$F = \lim_{\max \Delta x_k \to 0} \sum_{k=1}^{n} \rho h(x_k^*) w(x_k^*) \Delta x_k = \int_a^b \rho h(x) w(x)\, dx$$

In summary, we have the following result:

8.7.3 DEFINITION. Suppose that a flat surface is immersed vertically in a fluid of weight density ρ and that the submerged portion of the surface extends from $x = a$ to $x = b$ along an x-axis whose positive direction is down (Figure 8.7.4b). For $a \le x \le b$, suppose that $w(x)$ is the width of the surface and that $h(x)$ is the depth of the point x. Then we define the ***fluid force*** F on the surface to be

$$F = \int_a^b \rho h(x) w(x)\, dx \tag{6}$$

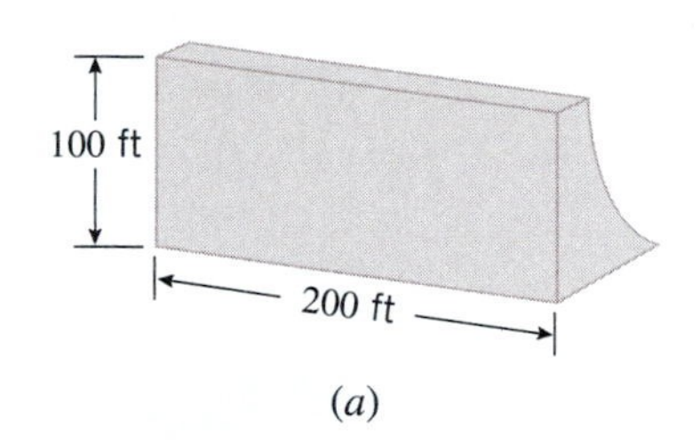

(a)

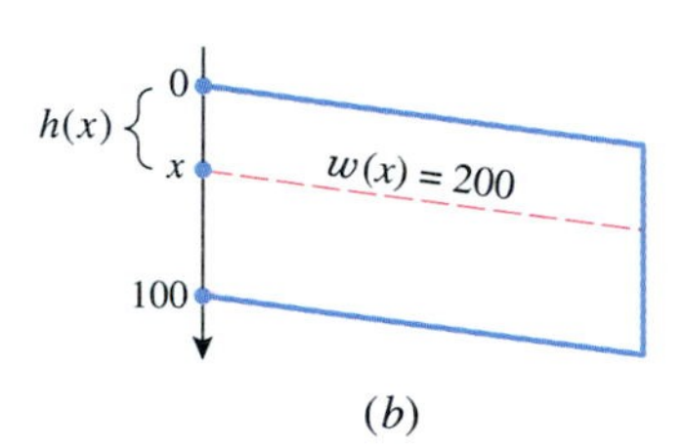

(b)

Figure 8.7.5

Example 3

The face of a dam is a vertical rectangle of height 100 ft and width 200 ft (Figure 8.7.5a). Find the total fluid force exerted on the face when the water surface is level with the top of the dam.

Solution. Introduce an x-axis with its origin at the water surface as shown in Figure 8.7.5b. At a point x on this axis, the width of the dam in feet is $w(x) = 200$ and the depth in feet is $h(x) = x$. Thus, from (6) with $\rho = 62.4\ \text{lb/ft}^3$ (the weight density of water) we obtain as the total force on the face

$$F = \int_0^{100} (62.4)(x)(200)\, dx = 12{,}480 \int_0^{100} x\, dx = 12{,}480\, \frac{x^2}{2}\bigg]_0^{100} = 62{,}400{,}000 \text{ lb}$$

◀

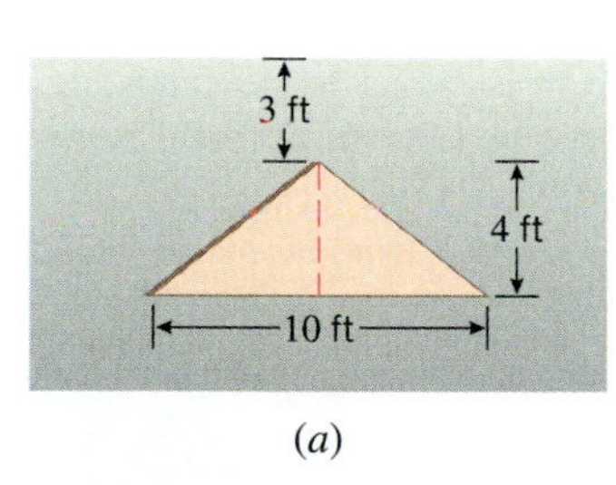

(a)

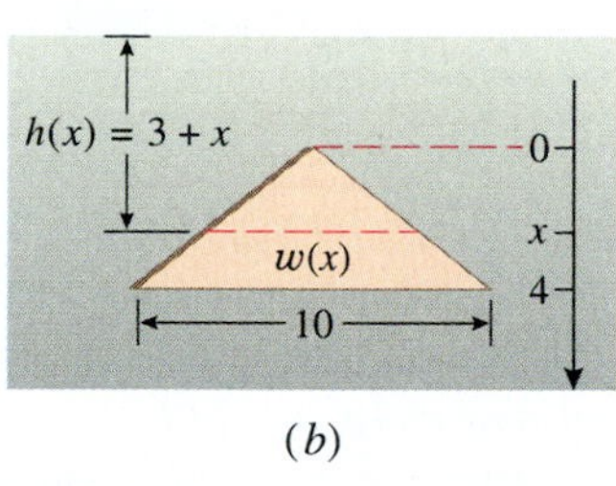

(b)

Figure 8.7.6

Example 4

A plate in the form of an isosceles triangle with base 10 ft and altitude 4 ft is submerged vertically in machine oil as shown in Figure 8.7.6a. Find the fluid force F against the plate surface if the oil has weight density $\rho = 30\ \text{lb/ft}^3$.

Solution. Introduce an x-axis as shown in Figure 8.7.6b. By similar triangles, the width of the plate, in feet, at a depth of $h(x) = (3 + x)$ ft satisfies

$$\frac{w(x)}{10} = \frac{x}{4}, \quad \text{so} \quad w(x) = \frac{5}{2}x$$

Thus, it follows from (6) that the force on the plate is

$$F = \int_a^b \rho h(x) w(x)\, dx = \int_0^4 (30)(3 + x)\left(\frac{5}{2}x\right) dx$$

$$= 75 \int_0^4 (3x + x^2)\, dx = 75 \left[\frac{3x^2}{2} + \frac{x^3}{3}\right]_0^4 = 3400 \text{ lb}$$

◀

EXERCISE SET 8.7

In this exercise set, refer to Table 8.7.2 for weight densities of fluids, when needed.

1. A flat rectangular plate is submerged horizontally in a liquid.

(a) Find the force (in lb) and the pressure (in lb/ft^2) on the top surface of the plate if its area is 100 ft^2, the liquid is water, and the surface is at a depth of 5 ft.

(b) Find the force (in N) and the pressure (in Pa) on the top surface of the plate if its area is 25 m^2, the liquid is water, and the surface is at a depth of 10 m.

2. (a) Find the force (in N) on the deck of a sunken ship if its area is 160 m^2 and the pressure acting on it is 6.0×10^5 Pa.

(b) Find the force (in lb) on a diver's face mask if its area is 60 in^2 and the pressure acting on it is 100 lb/in^2.

In Exercises 3–8, the flat surfaces shown are submerged vertically in water. Find the fluid force against the surface.

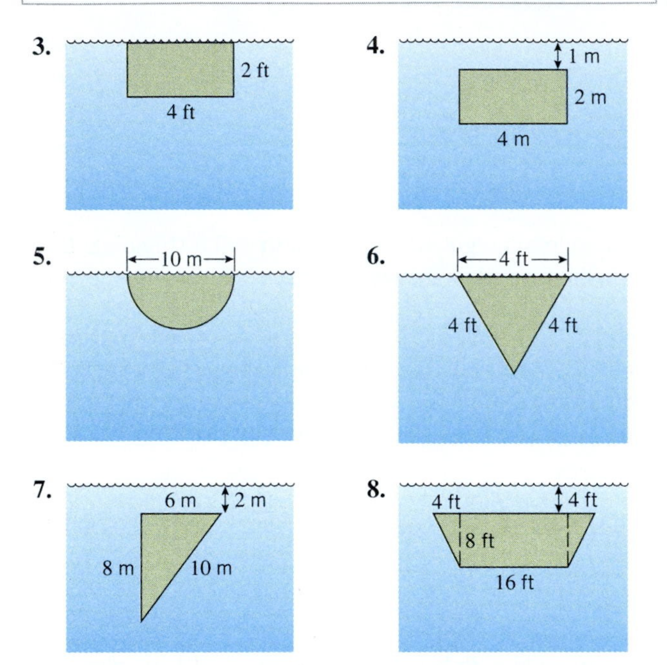

9. Suppose that a flat surface is immersed vertically in a fluid of weight density ρ. If ρ is doubled, is the force on the plate also doubled? Explain your reasoning.

10. An oil tank is shaped like a right circular cylinder of diameter 4 ft. Find the total fluid force against one end when the axis is horizontal and the tank is half filled with oil of weight density 50 lb/ft^3.

11. A square plate of side a feet is dipped in a liquid of weight density ρ lb/ft^3. Find the fluid force on the plate if a vertex is at the surface and a diagonal is perpendicular to the surface.

12. The accompanying figure shows a dam whose face is an inclined rectangle. Find the fluid force on the face when the water is level with the top of this dam.

13. The accompanying figure shows a rectangular swimming pool whose bottom is an inclined plane. Find the fluid force on the bottom when the pool is filled to the top.

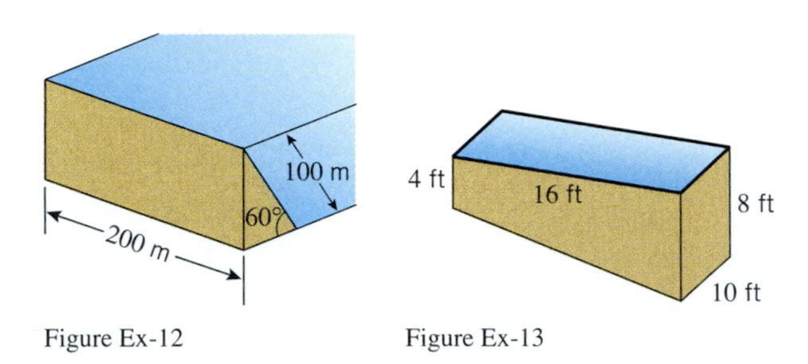

Figure Ex-12

Figure Ex-13

14. An observation window on a submarine is a square with 2-ft sides. Using ρ_0 for the weight density of seawater, find the fluid force on the window when the submarine has descended so that the window is vertical and its top is at a depth of h feet.

15. (a) Show: If the submarine in Exercise 14 descends vertically at a constant rate, then the fluid force on the window increases at a constant rate.

(b) At what rate is the force on the window increasing if the submarine is descending vertically at 20 ft/min?

8.8 HYPERBOLIC FUNCTIONS AND HANGING CABLES

In this section we will study certain combinations of e^x and e^{-x}, called "hyperbolic functions." These functions, which arise in various engineering applications, have many properties in common with the trigonometric functions. This similarity is somewhat surprising, since there is little on the surface to suggest that there should be any relationship between exponential and trigonometric functions. This is because the relationship occurs within the context of complex numbers, a topic which we will leave for more advanced courses.

DEFINITIONS OF HYPERBOLIC FUNCTIONS

To introduce the hyperbolic functions, observe that the function e^x can be expressed in the following way as the sum of an even function and an odd function:

$$e^x = \underbrace{\frac{e^x + e^{-x}}{2}}_{\text{Even}} + \underbrace{\frac{e^x - e^{-x}}{2}}_{\text{Odd}}$$

These functions are sufficiently important that there are names and notation associated with them: the odd function is called the *hyperbolic sine* of x and the even function is called the *hyperbolic cosine* of x. They are denoted by

$$\sinh x = \frac{e^x - e^{-x}}{2} \quad \text{and} \quad \cosh x = \frac{e^x + e^{-x}}{2}$$

where sinh is pronounced "cinch" and cosh rhymes with "gosh." From these two building blocks we can create four more functions to produce the following set of six ***hyperbolic functions***.

8.8.1 DEFINITION.

Hyperbolic sine $\quad \sinh x = \dfrac{e^x - e^{-x}}{2}$

Hyperbolic cosine $\quad \cosh x = \dfrac{e^x + e^{-x}}{2}$

Hyperbolic tangent $\quad \tanh x = \dfrac{\sinh x}{\cosh x} = \dfrac{e^x - e^{-x}}{e^x + e^{-x}}$

Hyperbolic cotangent $\quad \coth x = \dfrac{\cosh x}{\sinh x} = \dfrac{e^x + e^{-x}}{e^x - e^{-x}}$

Hyperbolic secant $\quad \operatorname{sech} x = \dfrac{1}{\cosh x} = \dfrac{2}{e^x + e^{-x}}$

Hyperbolic cosecant $\quad \operatorname{csch} x = \dfrac{1}{\sinh x} = \dfrac{2}{e^x - e^{-x}}$

REMARK. The terms "tanh," "sech," and "csch" are pronounced "tanch," "seech," and "coseech," respectively.

Example 1

$$\sinh 0 = \frac{e^0 - e^{-0}}{2} = \frac{1-1}{2} = 0$$

$$\cosh 0 = \frac{e^0 + e^{-0}}{2} = \frac{1+1}{2} = 1$$

$$\sinh 2 = \frac{e^2 - e^{-2}}{2} \approx 3.6269$$

◀

GRAPHS OF THE HYPERBOLIC FUNCTIONS

The graphs of the hyperbolic functions, which are shown in Figure 8.8.1, can be generated with a graphing utility, but it is worthwhile to observe that the general shape of the graph of $y = \cosh x$ can be obtained by sketching the graphs of $y = \frac{1}{2}e^x$ and $y = \frac{1}{2}e^{-x}$ separately and adding the corresponding y-coordinates [see part (a) of the figure]. Similarly, the general shape of the graph of $y = \sinh x$ can be obtained by sketching the graphs of $y = \frac{1}{2}e^x$ and $y = -\frac{1}{2}e^{-x}$ separately and adding corresponding y-coordinates [see part (b) of the figure].

Observe that $\sinh x$ has a domain of $(-\infty, +\infty)$ and a range of $(-\infty, +\infty)$, whereas $\cosh x$ has a domain of $(-\infty, +\infty)$ and a range of $[1, +\infty)$. Observe also that $y = \frac{1}{2}e^x$ and $y = \frac{1}{2}e^{-x}$ are ***curvilinear asymptotes*** for $y = \cosh x$ in the sense that the graph of $y = \cosh x$ gets closer and closer to the graph of $y = \frac{1}{2}e^x$ as $x \to +\infty$ and gets closer and closer to the graph of $y = \frac{1}{2}e^{-x}$ as $x \to -\infty$. Similarly, $y = \frac{1}{2}e^x$ is a curvilinear asymptote for $y = \sinh x$ as $x \to +\infty$ and $y = -\frac{1}{2}e^{-x}$ is a curvilinear asymptote as $x \to -\infty$. Other properties of the hyperbolic functions are explored in the exercises.

The design of the Gateway Arch near St. Louis is based on an inverted hyperbolic cosine curve.

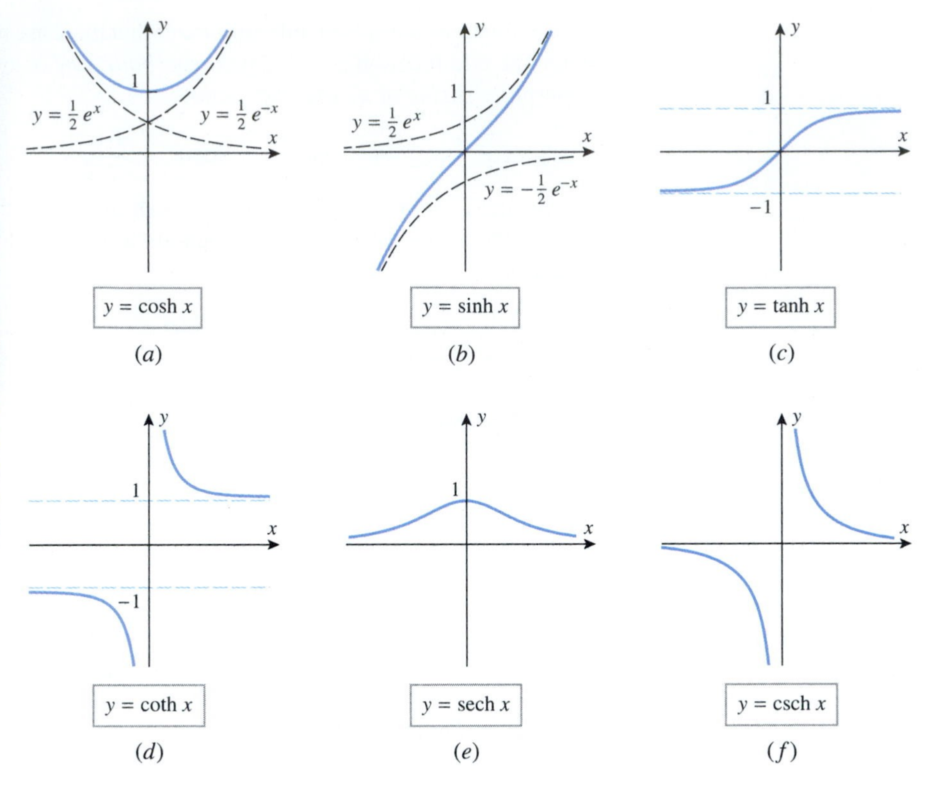

Figure 8.8.1

HANGING CABLES AND OTHER APPLICATIONS

Hyperbolic functions arise in vibratory motions inside elastic solids and more generally in many problems where mechanical energy is gradually absorbed by a surrounding medium. They also occur when a homogeneous, flexible cable is suspended between two points, as with a telephone line hanging between two poles. Such a cable forms a curve, called a ***catenary*** (from the Latin *catena*, meaning "chain"). If, as in Figure 8.8.2, a coordinate system is introduced so that the low point of the cable lies on the y-axis, then it can be shown using principles of physics that the cable has an equation of the form

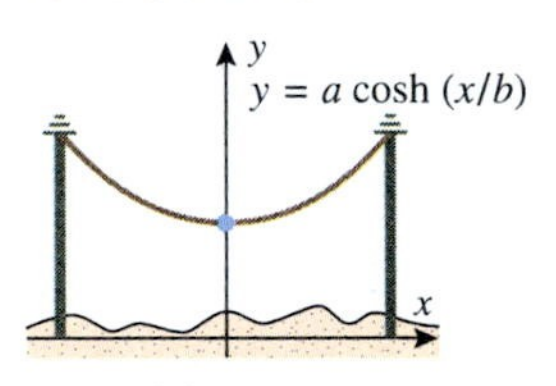

Figure 8.8.2

$$y = a\cosh\left(\frac{x}{b}\right)$$

HYPERBOLIC IDENTITIES

The hyperbolic functions satisfy various identities that are similar to identities for trigonometric functions. The most fundamental of these is

$$\cosh^2 x - \sinh^2 x = 1 \tag{1}$$

which can be proved by writing

$$\begin{aligned}\cosh^2 x - \sinh^2 x &= \left(\frac{e^x + e^{-x}}{2}\right)^2 - \left(\frac{e^x - e^{-x}}{2}\right)^2 \\ &= \tfrac{1}{4}(e^{2x} + 2e^0 + e^{-2x}) - \tfrac{1}{4}(e^{2x} - 2e^0 + e^{-2x}) \\ &= 1\end{aligned}$$

Other hyperbolic identities can be derived in a similar manner or, alternatively, by performing algebraic operations on known identities. For example, if we divide (1) by $\cosh^2 x$, we obtain

$$1 - \tanh^2 x = \operatorname{sech}^2 x$$

and if we divide (1) by $\sinh^2 x$, we obtain

$$\coth^2 x - 1 = \operatorname{csch}^2 x$$

The following theorem summarizes some of the more useful hyperbolic identities. The proofs of those not already obtained are left as exercises.

8.8.2 THEOREM.

$$\cosh x + \sinh x = e^x$$
$$\cosh x - \sinh x = e^{-x}$$
$$\cosh^2 x - \sinh^2 x = 1$$
$$1 - \tanh^2 x = \operatorname{sech}^2 x$$
$$\coth^2 x - 1 = \operatorname{csch}^2 x$$
$$\cosh(-x) = \cosh x$$
$$\sinh(-x) = -\sinh x$$

$$\sinh(x + y) = \sinh x \cosh y + \cosh x \sinh y$$
$$\cosh(x + y) = \cosh x \cosh y + \sinh x \sinh y$$
$$\sinh(x - y) = \sinh x \cosh y - \cosh x \sinh y$$
$$\cosh(x - y) = \cosh x \cosh y - \sinh x \sinh y$$
$$\sinh 2x = 2 \sinh x \cosh x$$
$$\cosh 2x = \cosh^2 x + \sinh^2 x$$
$$\cosh 2x = 2 \sinh^2 x + 1$$
$$\cosh 2x = 2 \cosh^2 x - 1$$

WHY THEY ARE CALLED HYPERBOLIC FUNCTIONS

Recall that the parametric equations

$$x = \cos t, \quad y = \sin t \qquad (0 \le t \le 2\pi)$$

represent the unit circle $x^2 + y^2 = 1$ (Figure 8.8.3*a*), as may be seen by writing

$$x^2 + y^2 = \cos^2 t + \sin^2 t = 1$$

If $0 \le t \le 2\pi$, then the parameter t can be interpreted as the angle in radians from the positive x-axis to the point $(\cos t, \sin t)$ or, alternatively, as twice the shaded area of the sector in Figure 8.8.3*a* (verify). Analogously, the parametric equations

$$x = \cosh t, \quad y = \sinh t \qquad (-\infty < t < +\infty)$$

represent a portion of the curve $x^2 - y^2 = 1$, as may be seen by writing

$$x^2 - y^2 = \cosh^2 t - \sinh^2 t = 1$$

and observing that $x = \cosh t > 0$. This curve, which is shown in Figure 8.8.3*b*, is the right half of a larger curve called the ***unit hyperbola***; this is the reason why the functions in this section are called *hyperbolic* functions. It can be shown that if $t \ge 0$, then the parameter t can be interpreted as twice the shaded area in Figure 8.8.3*b* (We omit the details.)

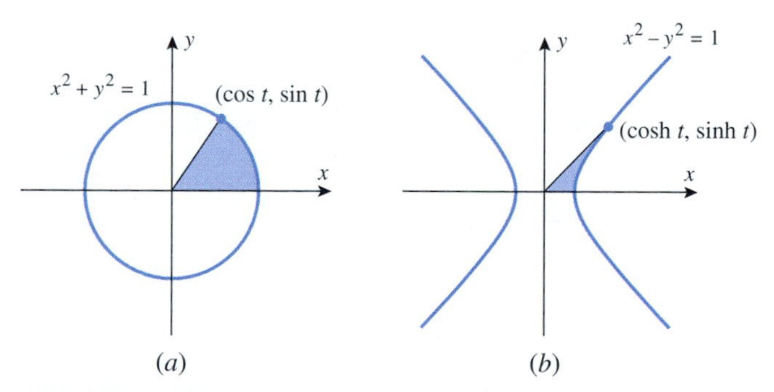

Figure 8.8.3

DERIVATIVE AND INTEGRAL FORMULAS

Derivative formulas for $\sinh x$ and $\cosh x$ can be obtained by expressing these functions in terms of e^x and e^{-x}:

$$\frac{d}{dx}[\sinh x] = \frac{d}{dx}\left[\frac{e^x - e^{-x}}{2}\right] = \frac{e^x + e^{-x}}{2} = \cosh x$$

$$\frac{d}{dx}[\cosh x] = \frac{d}{dx}\left[\frac{e^x + e^{-x}}{2}\right] = \frac{e^x - e^{-x}}{2} = \sinh x$$

Derivatives of the remaining hyperbolic functions can be obtained by expressing them in terms of sinh and cosh and applying appropriate identities. For example,

$$\frac{d}{dx}[\tanh x] = \frac{d}{dx}\left[\frac{\sinh x}{\cosh x}\right] = \frac{\cosh x \frac{d}{dx}[\sinh x] - \sinh x \frac{d}{dx}[\cosh x]}{\cosh^2 x}$$

$$= \frac{\cosh^2 x - \sinh^2 x}{\cosh^2 x} = \frac{1}{\cosh^2 x} = \operatorname{sech}^2 x$$

The following theorem provides a complete list of the generalized derivative formulas and corresponding integration formulas for the hyperbolic functions.

8.8.3 THEOREM.

$$\frac{d}{dx}[\sinh u] = \cosh u \frac{du}{dx} \qquad \int \sinh u\, du = \cosh u + C$$

$$\frac{d}{dx}[\cosh u] = \sinh u \frac{du}{dx} \qquad \int \cosh u\, du = \sinh u + C$$

$$\frac{d}{dx}[\tanh u] = \operatorname{sech}^2 u \frac{du}{dx} \qquad \int \operatorname{sech}^2 u\, du = \tanh u + C$$

$$\frac{d}{dx}[\coth u] = -\operatorname{csch}^2 u \frac{du}{dx} \qquad \int \operatorname{csch}^2 u\, du = -\coth u + C$$

$$\frac{d}{dx}[\operatorname{sech} u] = -\operatorname{sech} u \tanh u \frac{du}{dx} \qquad \int \operatorname{sech} u \tanh u\, du = -\operatorname{sech} u + C$$

$$\frac{d}{dx}[\operatorname{csch} u] = -\operatorname{csch} u \coth u \frac{du}{dx} \qquad \int \operatorname{csch} u \coth u\, du = -\operatorname{csch} u + C$$

Example 2

$$\frac{d}{dx}[\cosh(x^3)] = \sinh(x^3) \cdot \frac{d}{dx}[x^3] = 3x^2 \sinh(x^3)$$

$$\frac{d}{dx}[\ln(\tanh x)] = \frac{1}{\tanh x} \cdot \frac{d}{dx}[\tanh x] = \frac{\operatorname{sech}^2 x}{\tanh x}$$ ◀

Example 3

$$\int \sinh^5 x \cosh x\, dx = \tfrac{1}{6} \sinh^6 x + C$$

$u = \sinh x$
$du = \cosh x\, dx$

$$\int \tanh x\, dx = \int \frac{\sinh x}{\cosh x} dx$$

$$= \ln|\cosh x| + C$$

$u = \cosh x$
$du = \sinh x\, dx$

$$= \ln(\cosh x) + C$$

We were justified in dropping the absolute value signs since $\cosh x > 0$ for all x. ◀

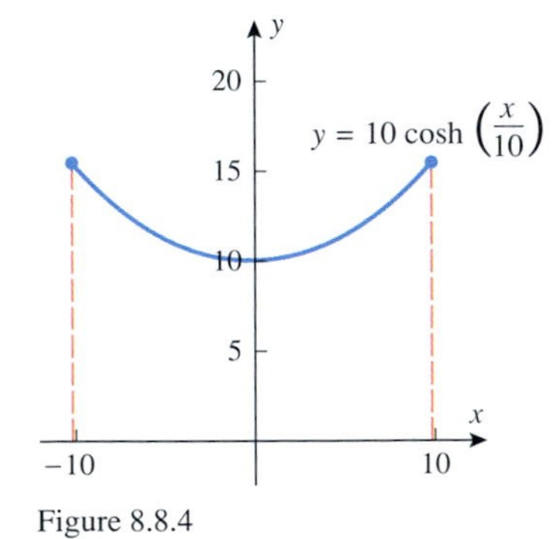

Figure 8.8.4

Example 4

Find the length of the catenary $y = 10\cosh(x/10)$ from $x = -10$ to $x = 10$ (Figure 8.8.4).

Solution. From Formula (4) of Section 8.4, the length L of the catenary is

$$L = \int_{-10}^{10} \sqrt{1 + \left(\frac{dy}{dx}\right)^2}\, dx$$

$$= 2\int_0^{10} \sqrt{1 + \left(\frac{dy}{dx}\right)^2}\, dx$$

By symmetry about the y-axis

$$= 2\int_0^{10} \sqrt{1 + \sinh^2\left(\frac{x}{10}\right)}\, dx$$

$$= 2\int_0^{10} \cosh\left(\frac{x}{10}\right) dx \qquad \text{By (1) and the fact that } \cosh x > 0$$

$$= 20 \sinh\left(\frac{x}{10}\right)\Big]_0^{10}$$

$$= 20[\sinh 1 - \sinh 0] = 20 \sinh 1 = 20\left(\frac{e - e^{-1}}{2}\right) \approx 23.50$$

REMARK. Computer algebra systems, such as *Mathematica*, *Maple*, and *Derive* have built-in capabilities for evaluating hyperbolic functions directly, but some calculators do not. However, if you need to evaluate a hyperbolic function on a calculator, you can do so by expressing it in terms of exponential functions, as in this example.

INVERSES OF HYPERBOLIC FUNCTIONS

Referring to Figure 8.8.1, it is evident that the graphs of $\sinh x$, $\tanh x$, $\coth x$, and $\operatorname{csch} x$ pass the horizontal line test, but the graphs of $\cosh x$ and $\operatorname{sech} x$ do not. In the latter case restricting x to be nonnegative makes the functions invertible (Figure 8.8.5). The graphs of the six inverse hyperbolic functions in Figure 8.8.6 were obtained by reflecting the graphs of the hyperbolic functions (with the appropriate restrictions) about the line $y = x$.

Table 8.8.1 summarizes the basic properties of the inverse hyperbolic functions. You should confirm that the domains and ranges listed in this table agree with the graphs in Figure 8.8.6.

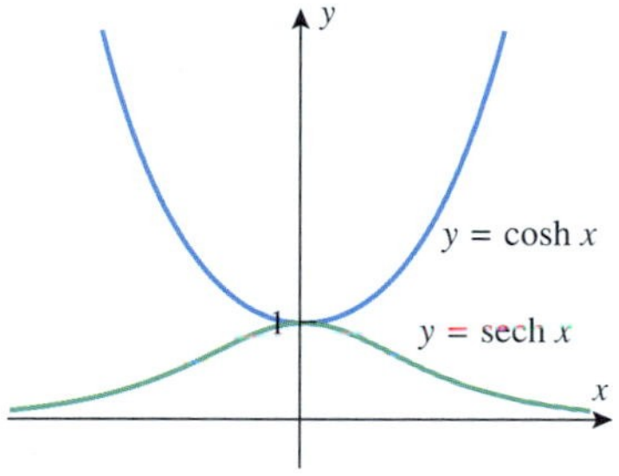

With the restriction that $x \geq 0$, the curves $y = \cosh x$ and $y = \operatorname{sech} x$ pass the horizontal line test.

Figure 8.8.5

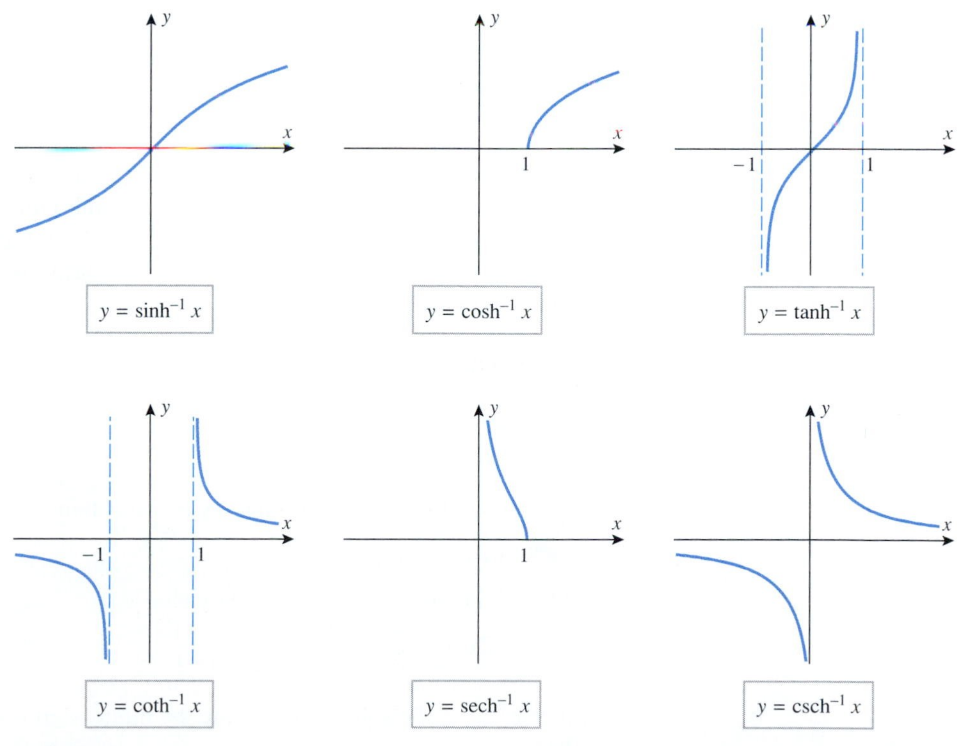

Figure 8.8.6

LOGARITHMIC FORMS OF INVERSE HYPERBOLIC FUNCTIONS

Because the hyperbolic functions are expressible in terms of e^x, it should not be surprising that the inverse hyperbolic functions are expressible in terms of natural logarithms; the next theorem shows that this is so.

Table 8.8.1

FUNCTION	DOMAIN	RANGE	BASIC RELATIONSHIPS
$\sinh^{-1} x$	$(-\infty, +\infty)$	$(-\infty, +\infty)$	$\sinh^{-1}(\sinh x) = x$ if $-\infty < x < +\infty$ $\sinh(\sinh^{-1} x) = x$ if $-\infty < x < +\infty$
$\cosh^{-1} x$	$[1, +\infty)$	$[0, +\infty)$	$\cosh^{-1}(\cosh x) = x$ if $x \geq 0$ $\cosh(\cosh^{-1} x) = x$ if $x \geq 1$
$\tanh^{-1} x$	$(-1, 1)$	$(-\infty, +\infty)$	$\tanh^{-1}(\tanh x) = x$ if $-\infty < x < +\infty$ $\tanh(\tanh^{-1} x) = x$ if $-1 < x < 1$
$\coth^{-1} x$	$(-\infty, -1) \cup (1, +\infty)$	$(-\infty, 0) \cup (0, +\infty)$	$\coth^{-1}(\coth x) = x$ if $x < 0$ or $x > 0$ $\coth(\coth^{-1} x) = x$ if $x < -1$ or $x > 1$
$\operatorname{sech}^{-1} x$	$(0, 1]$	$[0, +\infty)$	$\operatorname{sech}^{-1}(\operatorname{sech} x) = x$ if $x \geq 0$ $\operatorname{sech}(\operatorname{sech}^{-1} x) = x$ if $0 < x \leq 1$
$\operatorname{csch}^{-1} x$	$(-\infty, 0) \cup (0, +\infty)$	$(-\infty, 0) \cup (0, +\infty)$	$\operatorname{csch}^{-1}(\operatorname{csch} x) = x$ if $x < 0$ or $x > 0$ $\operatorname{csch}(\operatorname{csch}^{-1} x) = x$ if $x < 0$ or $x > 0$

8.8.4 THEOREM. *The following relationships hold for all x in the domain of the stated inverse hyperbolic function:*

$$\sinh^{-1} x = \ln(x + \sqrt{x^2+1}) \qquad \cosh^{-1} x = \ln(x + \sqrt{x^2-1})$$

$$\tanh^{-1} x = \frac{1}{2}\ln\left(\frac{1+x}{1-x}\right) \qquad \coth^{-1} x = \frac{1}{2}\ln\left(\frac{x+1}{x-1}\right)$$

$$\operatorname{sech}^{-1} x = \ln\left(\frac{1+\sqrt{1-x^2}}{x}\right) \qquad \operatorname{csch}^{-1} x = \ln\left(\frac{1}{x} + \frac{\sqrt{1+x^2}}{|x|}\right)$$

We will show how to derive the first formula in this theorem, and leave the rest as exercises. The basic idea is to write the equation $x = \sinh y$ in terms of exponential functions and solve this equation for y as a function of x. This will produce the equation $y = \sinh^{-1} x$ with $\sinh^{-1} x$ expressed in terms of natural logarithms. Expressing $x = \sinh y$ in terms of exponentials yields

$$x = \sinh y = \frac{e^y - e^{-y}}{2}$$

which can be rewritten as

$$e^y - 2x - e^{-y} = 0$$

Multiplying this equation through by e^y we obtain

$$e^{2y} - 2xe^y - 1 = 0$$

and applying the quadratic formula yields

$$e^y = \frac{2x \pm \sqrt{4x^2+4}}{2} = x \pm \sqrt{x^2+1}$$

Since $e^y > 0$, the solution involving the minus sign is extraneous and must be discarded. Thus,

$$e^y = x + \sqrt{x^2+1}$$

Taking natural logarithms yields

$$y = \ln(x + \sqrt{x^2+1}) \quad \text{or} \quad \sinh^{-1} x = \ln(x + \sqrt{x^2+1})$$

Example 5

$$\sinh^{-1} 1 = \ln(1 + \sqrt{1^2 + 1}) = \ln(1 + \sqrt{2}) \approx 0.8814$$

$$\tanh^{-1}\left(\frac{1}{2}\right) = \frac{1}{2}\ln\left(\frac{1 + \frac{1}{2}}{1 - \frac{1}{2}}\right) = \frac{1}{2}\ln 3 \approx 0.5493$$ ◀

DERIVATIVES AND INTEGRALS INVOLVING INVERSE HYPERBOLIC FUNCTIONS

Theorem 4.1.7 can be used to establish the differentiability of the inverse hyperbolic functions (we omit the details), and formulas for the derivatives can be obtained from Theorem 8.8.4. For example,

$$\frac{d}{dx}[\sinh^{-1} x] = \frac{d}{dx}[\ln(x + \sqrt{x^2 + 1})] = \frac{1}{x + \sqrt{x^2 + 1}}\left(1 + \frac{x}{\sqrt{x^2 + 1}}\right)$$

$$= \frac{\sqrt{x^2 + 1} + x}{(x + \sqrt{x^2 + 1})(\sqrt{x^2 + 1})} = \frac{1}{\sqrt{x^2 + 1}}$$

This computation leads to two integral formulas, a formula that involves $\sinh^{-1} x$ and an equivalent formula that involves logarithms:

$$\int \frac{dx}{\sqrt{x^2 + 1}} = \sinh^{-1} x + C = \ln(x + \sqrt{x^2 + 1}) + C$$

FOR THE READER. The derivative of $\sinh^{-1} x$ can also be obtained by letting $y = \sinh^{-1} x$ and differentiating the equation $x = \sinh y$ implicitly. Try it.

The following two theorems list the generalized derivative formulas and corresponding integration formulas for the inverse hyperbolic functions. Some of the proofs appear as exercises.

8.8.5 THEOREM.

$$\frac{d}{dx}(\sinh^{-1} u) = \frac{1}{\sqrt{1 + u^2}}\frac{du}{dx}$$

$$\frac{d}{dx}(\cosh^{-1} u) = \frac{1}{\sqrt{u^2 - 1}}\frac{du}{dx}, \quad u > 1$$

$$\frac{d}{dx}(\tanh^{-1} u) = \frac{1}{1 - u^2}\frac{du}{dx}, \quad |u| < 1$$

$$\frac{d}{dx}(\coth^{-1} u) = \frac{1}{1 - u^2}\frac{du}{dx}, \quad |u| > 1$$

$$\frac{d}{dx}(\operatorname{sech}^{-1} u) = -\frac{1}{u\sqrt{1 - u^2}}\frac{du}{dx}, \quad 0 < u < 1$$

$$\frac{d}{dx}(\operatorname{csch}^{-1} u) = -\frac{1}{|u|\sqrt{1 + u^2}}\frac{du}{dx}, \quad u \neq 0$$

8.8.6 THEOREM.

$$\int \frac{du}{\sqrt{1 + u^2}} = \sinh^{-1} u + C = \ln(u + \sqrt{u^2 + 1}) + C$$

$$\int \frac{du}{\sqrt{u^2 - 1}} = \cosh^{-1} u + C = \ln(u + \sqrt{u^2 - 1}) + C, \quad u > 1$$

$$\int \frac{du}{1 - u^2} = \begin{cases} \tanh^{-1} u + C, & |u| < 1 \\ \coth^{-1} u + C, & |u| > 1 \end{cases} = \frac{1}{2}\ln\left|\frac{1 + u}{1 - u}\right| + C$$

$$\int \frac{du}{u\sqrt{1 - u^2}} = -\operatorname{sech}^{-1}|u| + C = -\ln\left(\frac{1 + \sqrt{1 - u^2}}{|u|}\right) + C, \quad 0 < |u| < 1$$

$$\int \frac{du}{u\sqrt{1 + u^2}} = -\operatorname{csch}^{-1}|u| + C = -\ln\left(\frac{1 + \sqrt{1 + u^2}}{|u|}\right) + C, \quad u \neq 0$$

Example 6

Evaluate $\int \frac{dx}{\sqrt{4x^2-1}}, x > \frac{1}{2}$.

Solution. Let $u = 2x$. Thus, $du = 2\,dx$ and

$$\int \frac{dx}{\sqrt{4x^2-1}} = \frac{1}{2}\int \frac{2\,dx}{\sqrt{4x^2-1}} = \frac{1}{2}\int \frac{du}{\sqrt{u^2-1}}$$
$$= \frac{1}{2}\cosh^{-1} u + C = \frac{1}{2}\cosh^{-1}(2x) + C$$

Alternatively, we can use the logarithmic equivalent of $\cosh^{-1}(2x)$ and express the answer as

$$\int \frac{dx}{\sqrt{4x^2-1}} = \frac{1}{2}\ln(2x + \sqrt{4x^2-1}) + C$$

◀

EXERCISE SET 8.8 Graphing Calculator C CAS

In Exercises 1 and 2, approximate the expression to four decimal places.

1. (a) $\sinh 3$ (b) $\cosh(-2)$ (c) $\tanh(\ln 4)$ (d) $\sinh^{-1}(-2)$ (e) $\cosh^{-1} 3$ (f) $\tanh^{-1} \frac{3}{4}$
2. (a) $\operatorname{csch}(-1)$ (b) $\operatorname{sech}(\ln 2)$ (c) $\coth 1$ (d) $\operatorname{sech}^{-1}\frac{1}{2}$ (e) $\coth^{-1} 3$ (f) $\operatorname{csch}^{-1}(-\sqrt{3})$
3. In each part, find the exact numerical value of the expression.
   (a) $\sinh(\ln 3)$ (b) $\cosh(-\ln 2)$
   (c) $\tanh(2\ln 5)$ (d) $\sinh(-3\ln 2)$
4. In each part, rewrite the expression as a ratio of polynomials.
   (a) $\cosh(\ln x)$ (b) $\sinh(\ln x)$
   (c) $\tanh(2\ln x)$ (d) $\cosh(-\ln x)$
5. In each part, a value for one of the hyperbolic functions is given at an unspecified positive number x_0. Use appropriate identities to find the exact values of the remaining five hyperbolic functions at x_0.
   (a) $\sinh x_0 = 2$ (b) $\cosh x_0 = \frac{5}{4}$ (c) $\tanh x_0 = \frac{4}{5}$
6. Obtain the derivative formulas for $\operatorname{csch} x$, $\operatorname{sech} x$, and $\coth x$ from the derivative formulas for $\sinh x$, $\cosh x$, and $\tanh x$.
7. Find the derivatives of $\sinh^{-1} x$, $\cosh^{-1} x$, and $\tanh^{-1} x$ by differentiating the equations $x = \sinh y$, $x = \cosh y$, and $x = \tanh y$ implicitly.
8. C Use a CAS to find the derivatives of $\sinh^{-1} x$, $\cosh^{-1} x$, $\tanh^{-1} x$, $\coth^{-1} x$, $\operatorname{sech}^{-1} x$, and $\operatorname{csch}^{-1} x$, and confirm that your answers are consistent with those in Theorem 8.8.5.

In Exercises 9–28, find dy/dx.

9. $y = \sinh(4x-8)$
10. $y = \cosh(x^4)$
11. $y = \coth(\ln x)$
12. $y = \ln(\tanh 2x)$
13. $y = \operatorname{csch}(1/x)$
14. $y = \operatorname{sech}(e^{2x})$
15. $y = \sqrt{4x + \cosh^2(5x)}$
16. $y = \sinh^3(2x)$
17. $y = x^3 \tanh^2(\sqrt{x})$
18. $y = \sinh(\cos 3x)$
19. $y = \sinh^{-1}\left(\frac{1}{3}x\right)$
20. $y = \sinh^{-1}(1/x)$
21. $y = \ln(\cosh^{-1} x)$
22. $y = \cosh^{-1}(\sinh^{-1} x)$
23. $y = \frac{1}{\tanh^{-1} x}$
24. $y = (\coth^{-1} x)^2$
25. $y = \cosh^{-1}(\cosh x)$
26. $y = \sinh^{-1}(\tanh x)$
27. $y = e^x \operatorname{sech}^{-1}\sqrt{x}$
28. $y = (1 + x\operatorname{csch}^{-1} x)^{10}$
29. C Use a CAS to find the derivatives in Example 2. If the answers produced by the CAS do not match those in the text, then use appropriate identities to show that the answers are equivalent.
30. C For each of the derivatives you obtained in Exercises 9–28, use a CAS to check your answer. If the answer produced by the CAS does not match your own, show that the two answers are equivalent.

In Exercises 31–46, evaluate the integrals.

31. $\int \sinh^6 x \cosh x\,dx$
32. $\int \cosh(2x-3)\,dx$
33. $\int \sqrt{\tanh x}\operatorname{sech}^2 x\,dx$
34. $\int \operatorname{csch}^2(3x)\,dx$
35. $\int \tanh x\,dx$
36. $\int \coth^2 x \operatorname{csch}^2 x\,dx$
37. $\int_{\ln 2}^{\ln 3} \tanh x \operatorname{sech}^3 x\,dx$
38. $\int_0^{\ln 3} \frac{e^x - e^{-x}}{e^x + e^{-x}}\,dx$
39. $\int \frac{dx}{\sqrt{1+9x^2}}$
40. $\int \frac{dx}{\sqrt{x^2-2}}$ $(x > \sqrt{2})$

41. $\int \frac{dx}{\sqrt{1-e^{2x}}}$ $(x<0)$ **42.** $\int \frac{\sin\theta\, d\theta}{\sqrt{1+\cos^2\theta}}$

43. $\int \frac{dx}{x\sqrt{1+4x^2}}$ **44.** $\int \frac{dx}{\sqrt{9x^2-25}}$ $(x>5/3)$

45. $\int_0^{1/2} \frac{dx}{1-x^2}$ **46.** $\int_0^{\sqrt{3}} \frac{dt}{\sqrt{t^2+1}}$

C **47.** For each of the integrals you evaluated in Exercises 31–46, use a CAS to check your answer. If the answer produced by the CAS does not match your own, show that the two answers are equivalent.

48. Use a graphing utility to generate the graphs of $\sinh x$, $\cosh x$, and $\tanh x$ by expressing these functions in terms of e^x and e^{-x}. If your graphing utility can graph the hyperbolic functions directly, then generate the graphs that way as well.

49. Find the area enclosed by $y=\sinh 2x$, $y=0$, and $x=\ln 3$.

50. Find the volume of the solid that is generated when the region enclosed by $y=\operatorname{sech} x$, $y=0$, $x=0$, and $x=\ln 2$ is revolved about the x-axis.

51. Find the volume of the solid that is generated when the region enclosed by $y=\cosh 2x$, $y=\sinh 2x$, $x=0$, and $x=5$ is revolved about the x-axis.

52. Use Newton's Method to approximate the positive value of the constant a such that the area enclosed by $y=\cosh ax$, $y=0$, $x=0$, and $x=1$ is 2 square units. Express your answer to at least five decimal places.

53. Find the arc length of $y=\cosh x$ between $x=0$ and $x=\ln 2$.

54. Find the arc length of the catenary $y=a\cosh(x/a)$ between $x=0$ and $x=x_1$ $(x_1>0)$.

55. Prove that $\sinh x$ is an odd function of x and that $\cosh x$ is an even function of x, and check that this is consistent with the graphs in Figure 8.8.1.

In Exercises 56 and 57, prove the identities.

56. (a) $\cosh x+\sinh x=e^x$
(b) $\cosh x-\sinh x=e^{-x}$
(c) $\sinh(x+y)=\sinh x\cosh y+\cosh x\sinh y$
(d) $\sinh 2x=2\sinh x\cosh x$
(e) $\cosh(x+y)=\cosh x\cosh y+\sinh x\sinh y$
(f) $\cosh 2x=\cosh^2 x+\sinh^2 x$
(g) $\cosh 2x=2\sinh^2 x+1$
(h) $\cosh 2x=2\cosh^2 x-1$

57. (a) $1-\tanh^2 x=\operatorname{sech}^2 x$
(b) $\tanh(x+y)=\dfrac{\tanh x+\tanh y}{1+\tanh x\tanh y}$
(c) $\tanh 2x=\dfrac{2\tanh x}{1+\tanh^2 x}$

58. Prove:
(a) $\cosh^{-1}x=\ln(x+\sqrt{x^2-1}),\ x\geq 1$
(b) $\tanh^{-1}x=\dfrac{1}{2}\ln\left(\dfrac{1+x}{1-x}\right),\ -1<x<1.$

59. Use Exercise 58 to obtain the derivative formulas for $\cosh^{-1}x$ and $\tanh^{-1}x$.

60. Prove:

$$\operatorname{sech}^{-1}x=\cosh^{-1}(1/x),\quad 0<x\leq 1$$
$$\coth^{-1}x=\tanh^{-1}(1/x),\quad |x|>1$$
$$\operatorname{csch}^{-1}x=\sinh^{-1}(1/x),\quad x\neq 0$$

61. Use Exercise 60 to express the integral

$$\int \frac{du}{1-u^2}$$

entirely in terms of $\tanh^{-1}$.

62. Show that

(a) $\dfrac{d}{dx}[\operatorname{sech}^{-1}|x|]=-\dfrac{1}{x\sqrt{1-x^2}}$

(b) $\dfrac{d}{dx}[\operatorname{csch}^{-1}|x|]=-\dfrac{1}{x\sqrt{1+x^2}}$.

63. Find the limits, and confirm that they are consistent with the graphs in Figures 8.8.1 and 8.8.6.
(a) $\lim_{x\to+\infty}\sinh x$ (b) $\lim_{x\to-\infty}\sinh x$
(c) $\lim_{x\to+\infty}\tanh x$ (d) $\lim_{x\to-\infty}\tanh x$
(e) $\lim_{x\to+\infty}\sinh^{-1}x$ (f) $\lim_{x\to 1^-}\tanh^{-1}x$

64. In each part, find the limit.
(a) $\lim_{x\to+\infty}(\cosh^{-1}x-\ln x)$ (b) $\lim_{x\to+\infty}\dfrac{\cosh x}{e^x}$

65. Use the first and second derivatives to show that the graph of $y=\tanh^{-1}x$ is always increasing and has an inflection point at the origin.

66. The integration formulas for $1/\sqrt{u^2-1}$ in Theorem 8.8.6 are valid for $u>1$. Show that the following formula is valid for $u<-1$:

$$\int \frac{du}{\sqrt{u^2-1}}=-\cosh^{-1}(-u)+C=\ln|u+\sqrt{u^2-1}|+C$$

67. Show that $(\sinh x+\cosh x)^n=\sinh nx+\cosh nx$.

68. Show that

$$\int_{-a}^{a} e^{tx}\,dx=\frac{2\sinh at}{t}$$

69. A cable is suspended between two poles as shown in Figure 8.8.2. The equation of the curve formed by the cable is $y=a\cosh(x/a)$, where a is a positive constant. Suppose that the x-coordinates of the points of support are $x=-b$ and $x=b$, where $b>0$.
(a) Show that the length L of the cable is given by

$$L=2a\sinh\frac{b}{a}$$

(b) Show that the sag S (the vertical distance between the highest and lowest points on the cable) is given by

$$S=a\cosh\frac{b}{a}-a$$

Exercises 70 and 71 refer to the hanging cable described in Exercise 69.

70. Assuming that the cable is 120 ft long and the poles are 100 ft apart, approximate the sag in the cable by using Newton's Method to approximate a. Express your final answer to the nearest tenth of a foot. [*Hint:* First let $u = 50/a$.]

71. Assuming that the poles are 400 ft apart and the sag in the cable is 30 ft, approximate the length of the cable by using Newton's Method to approximate a. Express your final answer to the nearest tenth of a foot. [*Hint:* First let $u = 200/a$.]

72. The accompanying figure shows a person pulling a boat by holding a rope of length a attached to the bow and walking along the edge of a dock. If we assume that the rope is always tangent to the curve traced by the bow of the boat, then this curve, which is called a ***tractrix***, has the property that the segment of the tangent line between the curve and the y-axis has a constant length a. It can be proved that the equation of this tractrix is

$$y = a \operatorname{sech}^{-1}\frac{x}{a} - \sqrt{a^2 - x^2}$$

(a) Show that to move the bow of the boat to a point (x, y), the person must walk a distance

$$D = a \operatorname{sech}^{-1}\frac{x}{a}$$

from the origin.

(b) If the rope has a length of 15 m, how far must the person walk from the origin to bring the boat 10 m from the dock? Round your answer to two decimal places.

(c) Find the distance traveled by the bow along the tractrix as it moves from its initial position to the point where it is 5 m from the dock.

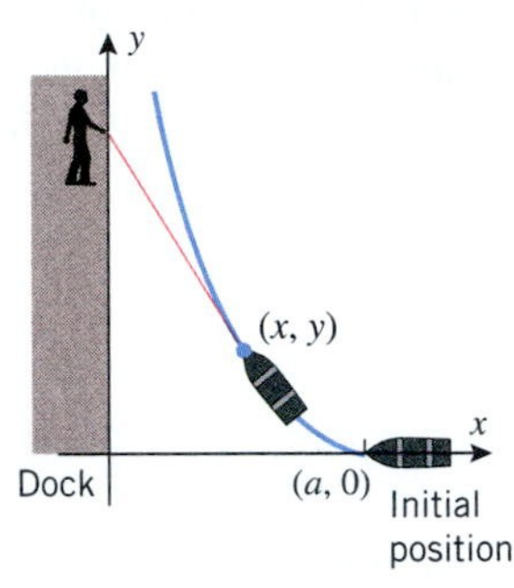

Figure Ex-72

SUPPLEMENTARY EXERCISES

1. State an integral formula for finding the arc length of a smooth curve $y = f(x)$ over an interval $[a, b]$, and use Riemann sums to derive the formula.

2. Describe the method of slicing for finding volumes, and use that method to derive an integral formula for finding volumes by the method of disks.

3. State an integral formula for finding a volume by the method of cylindrical shells, and use Riemann sums to derive the formula.

4. State an integral formula for the work W done by a variable force $F(x)$ applied in the direction of motion to an object moving from $x = a$ to $x = b$, and use Riemann sums to derive the formula.

5. State an integral formula for the fluid force F exerted on a vertical flat surface immersed in a fluid of weight density ρ, and use Riemann sums to derive the formula.

6. Let R be the region in the first quadrant enclosed by $y = x^2$, $y = 2 + x$, and $x = 0$. In each part, set up, but *do not evaluate*, an integral or a sum of integrals that will solve the problem.

(a) Find the area of R by integrating with respect to x.

(b) Find the area of R by integrating with respect to y.

(c) Find the volume of the solid generated by revolving R about the x-axis by integrating with respect to x.

(d) Find the volume of the solid generated by revolving R about the x-axis by integrating with respect to y.

(e) Find the volume of the solid generated by revolving R about the y-axis by integrating with respect to x.

(f) Find the volume of the solid generated by revolving R about the y-axis by integrating with respect to y.

7. (a) Set up a sum of definite integrals that represents the total shaded area between the curves $y = f(x)$ and $y = g(x)$ in the accompanying figure.

(b) Find the total area enclosed between $y = x^3$ and $y = x$ over the interval $[-1, 2]$.

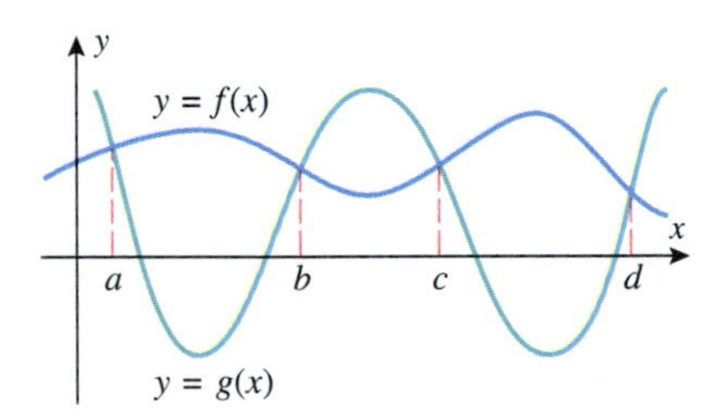

Figure Ex-7

8. Let C be the curve $27x - y^3 = 0$ between $y = 0$ and $y = 2$. In each part, set up, but *do not evaluate*, an integral or a sum of integrals that solves the problem.

(a) Find the area of the surface generated by revolving C about the y-axis by integrating with respect to x.

(b) Find the area of the surface generated by revolving C about the y-axis by integrating with respect to y.

(c) Find the area of the surface generated by revolving C about the line $y = -2$ by integrating with respect to y.

9. Find the arc length in the second quadrant of the curve $x^{2/3} + y^{2/3} = a^{2/3}$ from the point $x = -a$ to $x = -\frac{1}{8}a$ $(a > 0)$.

10. As shown in the accompanying figure, a cathedral dome is designed with six quarter-circular supports of radius r so that each horizontal cross section is a regular hexagon. Show that the volume of the dome is $r^3\sqrt{3}$.

11. As shown in the accompanying figure, a cylindrical hole is drilled all the way through the center of a sphere. Show that the volume of the remaining solid depends only on the length L of the hole, not on the size of the sphere.

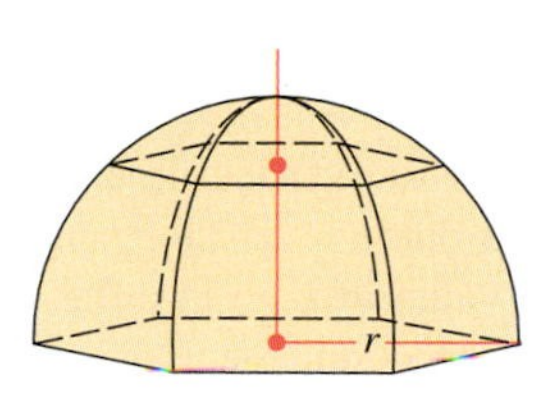

Figure Ex-10

Figure Ex-11

12. A football has the shape of the solid generated by revolving the region bounded between the x-axis and the parabola $y = 4R(x^2 - \frac{1}{4}L^2)/L^2$ about the x-axis. Find its volume.

13. The design of the Gateway Arch in St. Louis, Missouri, by architect Eero Saarinan was implemented using equations provided by Dr. Hannskarl Badel. The equation used for the centerline of the arch was

$$y = 693.8597 - 68.7672\cosh(0.0100333x) \text{ ft}$$

for x between -299.2239 and 299.2239.

(a) Use a graphing utility to graph the centerline of the arch.

(b) Find the length of the centerline to four decimal places.

(c) For what values of x is the height of the arch 100 ft? Round your answers to four decimal places.

(d) Approximate, to the nearest degree, the acute angle that the tangent line to the centerline makes with the ground at the ends of the arch.

14. A golfer makes a successful chip shot to the green. Suppose that the path of the ball from the moment it is struck to the moment it hits the green is described by

$$y = 12.54x - 0.41x^2$$

where x is the horizontal distance (in yards) from the point where the ball is struck, and y is the vertical distance (in yards) above the fairway. Find the distance the ball travels from the moment it is struck to the moment it hits the green. Assume that the fairway and green are at the same level and round your answer to two decimal places.

15. Derive integration formulas for

$$\int \frac{du}{\sqrt{a^2+u^2}}, \quad \int \frac{du}{\sqrt{u^2-a^2}}, \quad \int \frac{du}{a^2-u^2}$$

and use those formulas to evaluate

(a) $\displaystyle\int \frac{dx}{\sqrt{4+x^2}}$ (b) $\displaystyle\int \frac{dx}{\sqrt{x^2-9}}$

(c) $\displaystyle\int \frac{dx}{2-x^2}$ (d) $\displaystyle\int \frac{dx}{\sqrt{16+5x^2}}$.

16. In each part, prove the identity.

(a) $\cosh 3x = 4\cosh^3 x - 3\cosh x$

(b) $\cosh \frac{1}{2}x = \sqrt{\frac{1}{2}(\cosh x + 1)}$

(c) $\sinh \frac{1}{2}x = \pm\sqrt{\frac{1}{2}(\cosh x - 1)}$

17. (a) A spring exerts a force of 0.5 N when stretched 0.25 m beyond its natural length. Assuming that Hooke's law applies, how much work was performed in stretching the spring to this length?

(b) How far beyond its natural length can the spring be stretched with 25 J of work?

18. A boat is anchored so that the anchor is 150 ft below the surface of the water. In the water, the anchor weighs 2000 lb and the chain weighs 30 lb/ft. How much work is required to raise the anchor to the surface?

19. In each part, set up, but *do not evaluate*, an integral that solves the problem.

(a) Find the fluid force exerted on a side of a box that has a 3-m-square base and is filled to a depth of 1 m with a liquid of weight density ρ N/m^3.

(b) Find the fluid force exerted by a liquid of weight density ρ lb/ft^3 on a face of the vertical plate shown in part (*a*) of the accompanying figure.

(c) Find the fluid force exerted on the parabolic dam in part (*b*) of the accompanying figure by water that extends to the top of the dam.

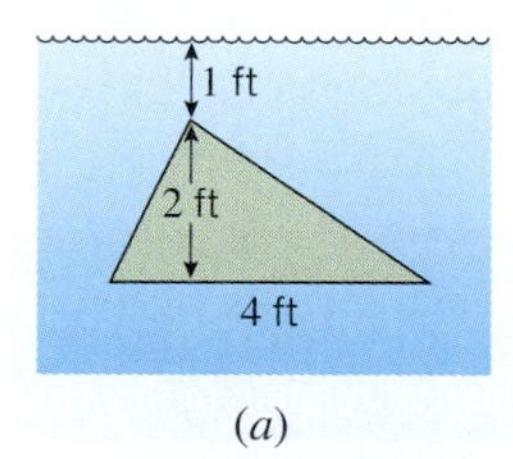

(*a*)

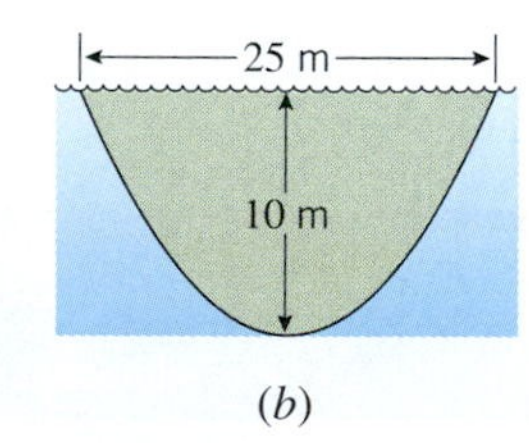

(*b*)

Figure Ex-19

20. Suppose that a hollow tube rotates with a constant angular velocity of ω rad/s about a horizontal axis at one end of the tube, as shown in the accompanying figure (next page). Assume that an object is free to slide without friction in the tube while the tube is rotating. Let r be the distance from the object to the pivot point at time $t \geq 0$, and assume that the object is at rest and $r = 0$ when $t = 0$. It can be shown that if the tube is horizontal at time $t = 0$, then

$$r = \frac{g}{2\omega^2}[\sinh(\omega t) - \sin(\omega t)]$$

during the period that the object is in the tube. Assume that t is in seconds and r is in meters, and use $g = 9.8\ \text{m/s}^2$ and $\omega = 2\ \text{rad/s}$.
(a) Graph r versus t for $0 \le t \le 1$.
(b) Assuming that the tube has a length of 1 m, approximately how long does it take for the object to reach the end of the tube?
(c) Use the result of part (b) to approximate dr/dt at the instant that the object reaches the end of the tube.

21. As shown in the accompanying figure, a horizontal beam with dimensions 2 in × 6 in × 16 ft is fixed at both ends and is subjected to a uniformly distributed load of 120 lb/ft. As a result of the load, the centerline of the beam undergoes a deflection that is described by

$$y = -1.67 \times 10^{-8}(x^4 - 2Lx^3 + L^2x^2)$$

$(0 \le x \le 192)$, where $L = 192$ inches is the length of the unloaded beam, x is the horizontal distance along the beam measured in inches from the left end, and y is the deflection of the centerline in inches.
(a) Graph y versus x for $0 \le x \le 192$.
(b) Find the maximum deflection of the centerline.
(c) Use a CAS or a calculator with a numerical integration capability to find the length of the centerline of the loaded beam. Round your answer to two decimal places.

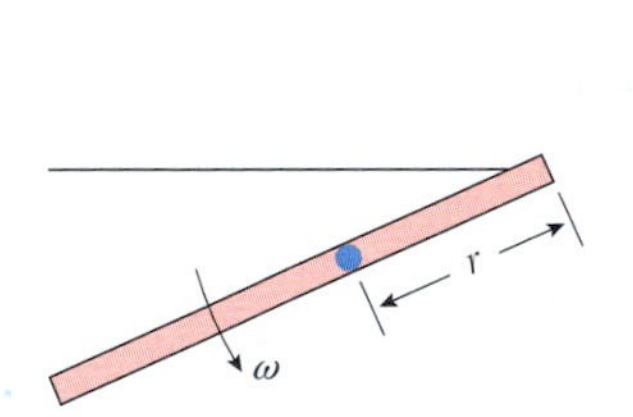

Figure Ex-20

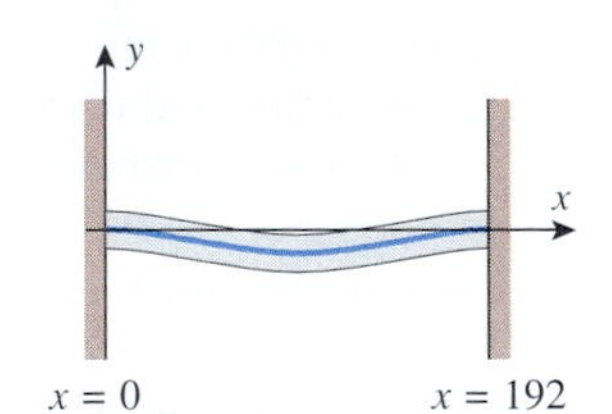

Figure Ex-21

Exercises 22–25 lead to equations that cannot be solved exactly. Use any method you want to approximate the solutions of those equations, and round your answers to two decimal places.

22. Find the area of the region enclosed by the curves $y = x^2 - 1$ and $y = 2\sin x$.

23. Referring to the accompanying figure, find the value of k so that the areas of the shaded regions are equal. [*Note:* This exercise is based on Problem A1 of the Fifty-Fourth Annual William Lowell Putnam Mathematical Competition.]

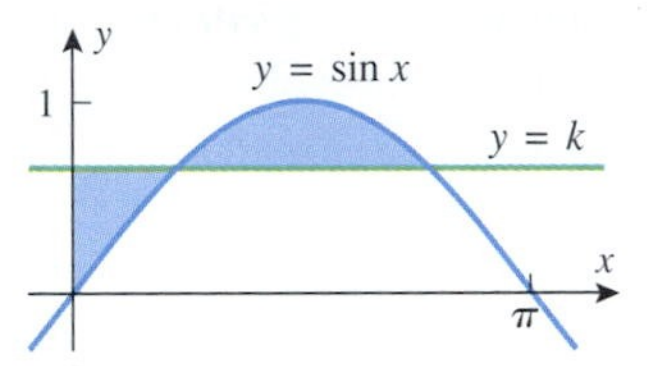

Figure Ex-23

C **24.** Consider the region to the left of the vertical line $x = k$ $(0 < k < \pi)$ and between the curve $y = \sin x$ and the x-axis. Use a CAS to find the value of k so that the solid generated by revolving the region about the y-axis has a volume of 8 cubic units.

25. Suppose that an object moves in the positive direction on an x-axis while subject to the force

$$F(x) = \frac{x}{\sqrt{1+x^3}}, \quad x \ge 0$$

where x is in meters and F is in newtons. The object moves 2 m from an unspecified starting point $x = a$ $(a \ge 0)$.
(a) Find a definite integral that gives the work done by F as a function of a.
(b) Find the value of a for which the work done by F is maximum. What is that maximum work? [*Hint:* See Exercise 24, Section 7.9.]

Expanding the Calculus Horizon

For additional material relating to this chapter, visit the Anton Website at http://www.wiley.com/college/anton

9

Augustin Cauchy

PRINCIPLES OF INTEGRAL EVALUATION

In earlier chapters we obtained many basic integration formulas from the corresponding differentiation formulas. For example, knowing that the derivative of $\sin x$ is $\cos x$ enabled us to deduce that the integral of $\cos x$ is $\sin x$. Subsequently, we expanded our integration repertoire by introducing the method of u-substitution. That method enabled us to integrate many functions by transforming the integrand of an unfamiliar integral into a familiar form. However, u-substitution alone is not adequate to handle the wide variety of integrals that arise in applications, so additional integration techniques are still needed. In this chapter we will discuss some of those techniques, and we will provide a more systematic procedure for attacking unfamiliar integrals. We will talk more about numerical approximations of definite integrals, and we will explore the idea of integrating over infinite intervals.

9.1 AN OVERVIEW OF INTEGRATION METHODS

In this section we will give a brief overview of methods for evaluating integrals, and we will review the integration formulas that were discussed in earlier sections.

METHODS FOR APPROACHING INTEGRATION PROBLEMS

There are three basic approaches for evaluating unfamiliar integrals:

- **Technology**—CAS programs such as *Mathematica*, *Maple*, and *Derive* are capable of evaluating extremely complicated integrals, and more and more modern research facilities are being equipped with such programs.
- **Tables**—Prior to the development of CAS programs, scientists relied heavily on tables to evaluate difficult integrals arising in applications. Such tables were compiled over many years, incorporating the skills and experience of many people. One such table appears in the endpapers of this text, but more comprehensive tables appear in various reference books such as the *CRC Standard Mathematical Tables and Formulae*, CRC Press, Inc., 1991.
- **Transformation Methods**—Transformation methods are methods for converting unfamiliar integrals into familiar integrals. These include u-substitution, algebraic manipulation of the integrand, and other methods that we will discuss in this chapter.

None of the three methods is perfect; for example, CAS programs often encounter integrals that they cannot evaluate and they sometimes produce answers that are excessively complicated, tables are not exhaustive and hence may not include a particular integral of interest, and transformation methods rely on human ingenuity that may prove to be inadequate in difficult problems.

In this chapter we will focus on transformation methods and tables, so it will *not be necessary* to have a CAS such as *Mathematica*, *Maple*, or *Derive*. However, if you have a CAS, then you can use it to confirm the results in the examples, and there are exercises that are designed to be solved with a CAS. If you have a CAS, keep in mind that many of the algorithms that it uses are based on the methods we will discuss here, so an understanding of these methods will help you to use your technology in a more informed way.

A REVIEW OF FAMILIAR INTEGRATION FORMULAS

The following is a list of basic integrals that we have encountered thus far:

CONSTANTS, POWERS, EXPONENTIALS

1. $\int du = u + C$ **2.** $\int a\,du = a\int du = au + C$

3. $\int u^r\,du = \dfrac{u^{r+1}}{r+1} + C,\ r \neq -1$ **4.** $\int \dfrac{du}{u} = \ln|u| + C$

5. $\int e^u\,du = e^u + C$ **6.** $\int b^u\,du = \dfrac{b^u}{\ln b} + C,\ b > 0, b \neq 1$

TRIGONOMETRIC FUNCTIONS

7. $\int \sin u\,du = -\cos u + C$ **8.** $\int \cos u\,du = \sin u + C$

9. $\int \sec^2 u\,du = \tan u + C$ **10.** $\int \csc^2 u\,du = -\cot u + C$

11. $\int \sec u \tan u\,du = \sec u + C$ **12.** $\int \csc u \cot u\,du = -\csc u + C$

13. $\int \tan u\,du = -\ln|\cos u| + C$ **14.** $\int \cot u\,du = \ln|\sin u| + C$

HYPERBOLIC FUNCTIONS

15. $\int \sinh u\, du = \cosh u + C$

16. $\int \cosh u\, du = \sinh u + C$

17. $\int \operatorname{sech}^2 u\, du = \tanh u + C$

18. $\int \operatorname{csch}^2 u\, du = -\coth u + C$

19. $\int \operatorname{sech} u \tanh u\, du = -\operatorname{sech} u + C$

20. $\int \operatorname{csch} u \coth u\, du = -\operatorname{csch} u + C$

ALGEBRAIC FUNCTIONS $(a > 0)$

21. $\int \frac{du}{\sqrt{a^2 - u^2}} = \sin^{-1} \frac{u}{a} + C \qquad (|u| < a)$

22. $\int \frac{du}{a^2 + u^2} = \frac{1}{a} \tan^{-1} \frac{u}{a} + C$

23. $\int \frac{du}{u\sqrt{u^2 - a^2}} = \frac{1}{a} \sec^{-1} \left|\frac{u}{a}\right| + C \qquad (0 < a < |u|)$

24. $\int \frac{du}{\sqrt{a^2 + u^2}} = \ln(u + \sqrt{u^2 + a^2}) + C$

25. $\int \frac{du}{\sqrt{u^2 - a^2}} = \ln\left|u + \sqrt{u^2 - a^2}\right| + C \qquad (0 < a < |u|)$

26. $\int \frac{du}{a^2 - u^2} = \frac{1}{2a} \ln\left|\frac{a + u}{a - u}\right| + C$

27. $\int \frac{du}{u\sqrt{a^2 - u^2}} = -\frac{1}{a} \ln\left|\frac{a + \sqrt{a^2 - u^2}}{u}\right| + C \qquad (0 < |u| < a)$

28. $\int \frac{du}{u\sqrt{a^2 + u^2}} = -\frac{1}{a} \ln\left|\frac{a + \sqrt{a^2 + u^2}}{u}\right| + C$

REMARK. Formulas 24–28 are generalizations of those in Theorem 8.8.6. Readers who did not cover that section can ignore those formulas for now, since we will develop other methods for obtaining them in this chapter.

EXERCISE SET 9.1

Review: Without looking at the text, complete the following integration formulas and then check your results by referring to the list of formulas at the beginning of this section.

Constants, Powers, Exponentials

$\int du =$ $\qquad \int a\, du =$

$\int u^r\, du =$ $\qquad \int \frac{du}{u} =$

$\int e^u\, du =$ $\qquad \int b^u\, du =$

Trigonometric Functions

$\int \sin u\, du =$ $\qquad \int \cos u\, du =$

$\int \sec^2 u\, du =$ $\qquad \int \csc^2 u\, du =$

$\int \sec u \tan u\, du =$ $\qquad \int \csc u \cot u\, du =$

$\int \tan u\, du =$ $\qquad \int \cot u\, du =$

Algebraic Functions

$\int \frac{du}{\sqrt{1-u^2}} =$ $\int \frac{du}{1+u^2} =$

$\int \frac{du}{u\sqrt{u^2-1}} =$ $\int \frac{du}{\sqrt{1+u^2}} =$

$\int \frac{du}{\sqrt{u^2-1}} =$ $\int \frac{du}{1-u^2} =$

$\int \frac{du}{u\sqrt{1-u^2}} =$ $\int \frac{du}{u\sqrt{1+u^2}} =$

Hyperbolic Functions

$\int \sinh u\, du =$ $\int \cosh u\, du =$

$\int \operatorname{sech}^2 u\, du =$ $\int \operatorname{csch}^2 u\, du =$

$\int \operatorname{sech} u \tanh u\, du =$

$\int \operatorname{csch} u \coth u\, du =$

In Exercises 1–30, evaluate the integrals by making appropriate u-substitutions and applying the formulas reviewed in this section.

1. $\int (3-2x)^3\, dx$ **2.** $\int \sqrt{4+9x}\, dx$

3. $\int x \sec^2(x^2)\, dx$ **4.** $\int 4x \tan(x^2)\, dx$

5. $\int \frac{\sin 3x}{2+\cos 3x}\, dx$ **6.** $\int \frac{1}{4+9x^2}\, dx$

7. $\int e^x \sinh(e^x)\, dx$ **8.** $\int \frac{\sec(\ln x)\tan(\ln x)}{x}\, dx$

9. $\int e^{\cot x} \csc^2 x\, dx$ **10.** $\int \frac{x}{\sqrt{1-x^4}}\, dx$

11. $\int \cos^5 7x \sin 7x\, dx$ **12.** $\int \frac{\cos x}{\sin x \sqrt{\sin^2 x + 1}}\, dx$

13. $\int \frac{e^x}{\sqrt{4+e^{2x}}}\, dx$ **14.** $\int \frac{e^{\tan^{-1} x}}{1+x^2}\, dx$

15. $\int \frac{e^{\sqrt{x-2}}}{\sqrt{x-2}}\, dx$

16. $\int (3x+1)\cot(3x^2+2x)\, dx$

17. $\int \frac{\cosh \sqrt{x}}{\sqrt{x}}\, dx$ **18.** $\int \frac{dx}{x \ln x}$

19. $\int \frac{dx}{\sqrt{x}\, 3^{\sqrt{x}}}$

20. $\int \sec(\sin\theta)\tan(\sin\theta)\cos\theta\, d\theta$

21. $\int \frac{\operatorname{csch}^2(2/x)}{x^2}\, dx$ **22.** $\int \frac{dx}{\sqrt{x^2-3}}$

23. $\int \frac{e^{-x}}{4-e^{-2x}}\, dx$ **24.** $\int \frac{\cos(\ln x)}{x}\, dx$

25. $\int \frac{e^x}{\sqrt{1-e^{2x}}}\, dx$ **26.** $\int \frac{\sinh(x^{-1/2})}{x^{3/2}}\, dx$

27. $\int \frac{x}{\sec(x^2)}\, dx$ **28.** $\int \frac{e^x}{\sqrt{4-e^{2x}}}\, dx$

29. $\int x 4^{-x^2}\, dx$ **30.** $\int 2^{\pi x}\, dx$

31. (a) Use Formulas (15), (17), and (19) of Section 4.5 to derive integration formulas for

$$\int \frac{dx}{\sqrt{1-x^2}}, \quad \int \frac{dx}{1+x^2}, \quad \int \frac{dx}{x\sqrt{x^2-1}}$$

(b) Use the integration formulas you obtained in part (a) to derive Formulas (21), (22), and (23) in this section.

9.2 INTEGRATION BY PARTS

In this section we will discuss an integration technique that is essentially the antiderivative formulation of the formula for differentiating a product of two functions.

DERIVATION OF THE FORMULA FOR INTEGRATION BY PARTS

If f and g are differentiable functions, then by the rule for differentiating products

$$\frac{d}{dx}[f(x)g(x)] = f(x)g'(x) + g(x)f'(x)$$

Integrating both sides we obtain

$$\int \frac{d}{dx}[f(x)g(x)]\, dx = \int f(x)g'(x)\, dx + \int g(x)f'(x)\, dx$$

or

$$f(x)g(x) + C = \int f(x)g'(x)\,dx + \int g(x)f'(x)\,dx$$

or

$$\int f(x)g'(x)\,dx = f(x)g(x) - \int g(x)f'(x)\,dx + C$$

Since the integral on the right will produce another constant of integration, there is no need to keep the C in this last equation; thus, we obtain

$$\int f(x)g'(x)\,dx = f(x)g(x) - \int g(x)f'(x)\,dx \tag{1}$$

which is called the formula for ***integration by parts***. By using this formula we can sometimes reduce a hard integration problem to an easier one.

In practice, it is usual to rewrite (1) by letting

$$u = f(x), \quad du = f'(x)\,dx$$
$$v = g(x), \quad dv = g'(x)\,dx$$

This yields the following alternative form for (1):

$$\int u\,dv = uv - \int v\,du \tag{2}$$

Example 1

Evaluate $\int xe^x\,dx$.

Solution. To apply (2) we must write the integral in the form

$$\int u\,dv$$

One way to do this is to let

$$u = x \quad \text{and} \quad dv = e^x\,dx$$

so that

$$du = dx \quad \text{and} \quad v = \int e^x\,dx = e^x$$

Thus, from (2)

$$\int xe^x\,dx = \int \underbrace{x}_{u}\,\underbrace{e^x\,dx}_{dv} = \underbrace{x}_{u}\,\underbrace{e^x}_{v} - \int \underbrace{e^x}_{v}\,\underbrace{dx}_{du} = xe^x - e^x + C$$ ◀

REMARK. In the calculation of v from dv above, we omitted the constant of integration and wrote $v = \int e^x\,dx = e^x$. Had we included a constant of integration and written $v = \int e^x\,dx = e^x + C_1$, the constant C_1 would have eventually canceled out [Exercise 58(a)]. This is always the case in integration by parts [Exercise 58(b)], so we will usually omit the constant when calculating v from dv.

To use integration by parts successfully, the choice of u and dv must be made so that the new integral is easier than the original. For example, had we decided above to let

$$u = e^x, \quad dv = x\,dx, \quad du = e^x\,dx, \quad v = \int x\,dx = \frac{x^2}{2}$$

then we would have obtained

$$\int xe^x\,dx = \int u\,dv = uv - \int v\,du = \frac{x^2}{2}e^x - \frac{1}{2}\int x^2e^x\,dx$$

For this choice of u and dv the new integral is actually more complicated than the original.

It is difficult to give hard and fast rules for choosing u and dv. It is a matter of experience that comes with lots of practice.

The next example shows that it is sometimes necessary to use integration by parts more than once in the same problem.

Example 2

Evaluate $\int x^2 e^{-x}\,dx$.

Solution. Let

$$u = x^2, \quad dv = e^{-x}\,dx, \quad du = 2x\,dx, \quad v = \int e^{-x}\,dx = -e^{-x}$$

so that

$$\int x^2 e^{-x}\,dx = \int u\,dv = uv - \int v\,du = -x^2 e^{-x} + 2\int x e^{-x}\,dx \tag{3}$$

The last integral is similar to the original except that we have replaced x^2 by x. Another integration by parts applied to $\int x e^{-x}\,dx$ will complete the problem. We let

$$u = x, \quad dv = e^{-x}\,dx, \quad du = dx, \quad v = \int e^{-x}\,dx = -e^{-x}$$

so that

$$\begin{aligned}\int x e^{-x}\,dx &= \int u\,dv = uv - \int v\,du \\ &= -xe^{-x} + \int e^{-x}\,dx \\ &= -xe^{-x} - e^{-x} + C_1\end{aligned}$$

Substituting in (3) we obtain

$$\begin{aligned}\int x^2 e^{-x}\,dx &= -x^2 e^{-x} + 2(-xe^{-x} - e^{-x} + C_1) \\ &= -x^2 e^{-x} - 2xe^{-x} - 2e^{-x} + 2C_1 \\ &= -(x^2 + 2x + 2)e^{-x} + C\end{aligned}$$

where $C = 2C_1$. ◀

Example 3

Evaluate $\int \ln x\,dx$.

Solution. Let

$$u = \ln x, \quad dv = dx, \quad du = \frac{1}{x}\,dx, \quad v = \int dx = x$$

so that

$$\begin{aligned}\int \ln x\,dx = \int u\,dv = uv - \int v\,du &= x\ln x - \int x\left(\frac{1}{x}\right)dx \\ &= x\ln x - \int dx = x\ln x - x + C\end{aligned}$$

◀

Example 4

Evaluate $\int e^x \cos x\,dx$.

Solution. Let

$$u = e^x, \quad dv = \cos x\,dx, \quad du = e^x\,dx, \quad v = \int \cos x\,dx = \sin x$$

Thus,

$$\int e^x \cos x\, dx = \int u\, dv = uv - \int v\, du = e^x \sin x - \int e^x \sin x\, dx \tag{4}$$

Since the integral $\int e^x \sin x\, dx$ is similar in form to the original integral $\int e^x \cos x\, dx$, it seems that nothing has been accomplished. However, let us integrate this new integral by parts. We let

$$u = e^x, \quad dv = \sin x\, dx, \quad du = e^x\, dx, \quad v = \int \sin x\, dx = -\cos x$$

Thus,

$$\int e^x \sin x\, dx = \int u\, dv = uv - \int v\, du = -e^x \cos x + \int e^x \cos x\, dx$$

Substituting in (4) yields

$$\int e^x \cos x\, dx = e^x \sin x - \left[-e^x \cos x + \int e^x \cos x\, dx\right]$$

or

$$\int e^x \cos x\, dx = e^x \sin x + e^x \cos x - \int e^x \cos x\, dx$$

which is an equation we can solve for the unknown integral. We obtain

$$2\int e^x \cos x\, dx = e^x \sin x + e^x \cos x$$

and hence

$$\int e^x \cos x\, dx = \tfrac{1}{2}e^x \sin x + \tfrac{1}{2}e^x \cos x + C$$ ◀

INTEGRATION BY PARTS FOR DEFINITE INTEGRALS

For definite integrals the formula corresponding to (2) is

$$\int_a^b u\, dv = uv\Big]_a^b - \int_a^b v\, du \tag{5}$$

REMARK. It is important to keep in mind that the variables u and v in this formula are functions of x and that the limits of integration in (5) are limits on the variable x. Sometimes it is helpful to emphasize this by writing (5) as

$$\int_{x=a}^{x=b} u\, dv = uv\Big]_{x=a}^{x=b} - \int_{x=a}^{x=b} v\, du \tag{6}$$

The next example illustrates how integration by parts can be used to integrate the inverse trigonometric functions.

Example 5

Evaluate $\displaystyle\int_0^1 \tan^{-1} x\, dx$.

Solution. Let

$$u = \tan^{-1} x, \quad dv = dx, \quad du = \frac{1}{1+x^2}\, dx, \quad v = \int dx = x$$

Thus,

$$\int_0^1 \tan^{-1} x\, dx = \int_0^1 u\, dv = uv\Big]_0^1 - \int_0^1 v\, du$$

$$= x \tan^{-1} x\Big]_0^1 - \int_0^1 \frac{x}{1+x^2}\, dx$$

The limits of integration refer to x; that is, $x = 0$ and $x = 1$.

But

$$\int_0^1 \frac{x}{1+x^2}\,dx = \frac{1}{2}\int_0^1 \frac{2x}{1+x^2}\,dx = \frac{1}{2}\ln(1+x^2)\Big]_0^1 = \frac{1}{2}\ln 2$$

so

$$\int_0^1 \tan^{-1} x\,dx = x\tan^{-1} x\Big]_0^1 - \frac{1}{2}\ln 2 = \left(\frac{\pi}{4} - 0\right) - \frac{1}{2}\ln 2 = \frac{\pi}{4} - \ln\sqrt{2}$$ ◀

REDUCTION FORMULAS

Integration by parts can be used to derive ***reduction formulas*** for integrals. These are formulas that express an integral involving a power of a function in terms of an integral that involves a *lower* power of that function. For example, if n is a positive integer and $n \geq 2$, then integration by parts can be used to obtain the reduction formulas

$$\int \sin^n x\,dx = -\frac{1}{n}\sin^{n-1} x\cos x + \frac{n-1}{n}\int \sin^{n-2} x\,dx \tag{7}$$

$$\int \cos^n x\,dx = \frac{1}{n}\cos^{n-1} x\sin x + \frac{n-1}{n}\int \cos^{n-2} x\,dx \tag{8}$$

To illustrate how such formulas can be obtained, let us derive (8). We begin by writing $\cos^n x$ as $\cos^{n-1} x \cdot \cos x$ and letting

$$u = \cos^{n-1} x \qquad dv = \cos x\,dx$$

$$du = (n-1)\cos^{n-2} x(-\sin x)\,dx \qquad v = \int \cos x\,dx = \sin x$$

$$= -(n-1)\cos^{n-2} x\sin x\,dx$$

so that

$$\int \cos^n x\,dx = \int \cos^{n-1} x\cos x\,dx = \int u\,dv = uv - \int v\,du$$

$$= \cos^{n-1} x\sin x + (n-1)\int \sin^2 x\cos^{n-2} x\,dx$$

$$= \cos^{n-1} x\sin x + (n-1)\int (1-\cos^2 x)\cos^{n-2} x\,dx$$

$$= \cos^{n-1} x\sin x + (n-1)\int \cos^{n-2} x\,dx - (n-1)\int \cos^n x\,dx$$

Transposing the last term on the right to the left side yields

$$n\int \cos^n x\,dx = \cos^{n-1} x\sin x + (n-1)\int \cos^{n-2} x\,dx$$

from which (8) follows.

Reduction formulas (7) and (8) reduce the exponent of sine (or cosine) by 2. Thus, if the formulas are applied repeatedly, the exponent can eventually be reduced to 0 if n is even or 1 if n is odd, at which point the integration can be completed. We will discuss this method in more detail in the next section, but for now, here is an example that illustrates how reduction formulas work.

Example 6

Evaluate $\int \cos^4 x\,dx$.

Solution. From (8) with $n = 4$

$$\begin{aligned}\int \cos^4 x\,dx &= \frac{1}{4}\cos^3 x \sin x + \frac{3}{4}\int \cos^2 x\,dx \\ &= \frac{1}{4}\cos^3 x \sin x + \frac{3}{4}\left(\frac{1}{2}\cos x \sin x + \frac{1}{2}\int dx\right) \\ &= \frac{1}{4}\cos^3 x \sin x + \frac{3}{8}\cos x \sin x + \frac{3}{8}x + C\end{aligned}$$

Now apply (8) with $n = 2$.

◀

EXERCISE SET 9.2 [C] CAS

In Exercises 1–40, evaluate the integral.

1. $\int xe^{-x}\,dx$
2. $\int xe^{3x}\,dx$
3. $\int x^2 e^x\,dx$
4. $\int x^2 e^{-2x}\,dx$
5. $\int x \sin 2x\,dx$
6. $\int x \cos 3x\,dx$
7. $\int x^2 \cos x\,dx$
8. $\int x^2 \sin x\,dx$
9. $\int \sqrt{x} \ln x\,dx$
10. $\int x \ln x\,dx$
11. $\int (\ln x)^2\,dx$
12. $\int \frac{\ln x}{\sqrt{x}}\,dx$
13. $\int \ln(2x+3)\,dx$
14. $\int \ln(x^2+4)\,dx$
15. $\int \sin^{-1} x\,dx$
16. $\int \cos^{-1}(2x)\,dx$
17. $\int \tan^{-1}(2x)\,dx$
18. $\int x \tan^{-1} x\,dx$
19. $\int e^x \sin x\,dx$
20. $\int e^{2x} \cos 3x\,dx$
21. $\int e^{ax} \sin bx\,dx$
22. $\int e^{-3\theta} \sin 5\theta\,d\theta$
23. $\int \sin(\ln x)\,dx$
24. $\int \cos(\ln x)\,dx$
25. $\int x \sec^2 x\,dx$
26. $\int x \tan^2 x\,dx$
27. $\int x^3 e^{x^2}\,dx$
28. $\int \frac{xe^x}{(x+1)^2}\,dx$
29. $\int_0^1 xe^{-5x}\,dx$
30. $\int_0^2 xe^{2x}\,dx$
31. $\int_1^e x^2 \ln x\,dx$
32. $\int_{\sqrt{e}}^e \frac{\ln x}{x^2}\,dx$
33. $\int_{-2}^2 \ln(x+3)\,dx$
34. $\int_0^{1/2} \sin^{-1} x\,dx$
35. $\int_2^4 \sec^{-1}\sqrt{\theta}\,d\theta$
36. $\int_1^2 x \sec^{-1} x\,dx$
37. $\int_0^{\pi/2} x \sin 4x\,dx$
38. $\int_0^{\pi} (x + x\cos x)\,dx$
39. $\int_1^3 \sqrt{x} \tan^{-1}\sqrt{x}\,dx$
40. $\int_0^2 \ln(x^2+1)\,dx$
41. In each part, evaluate the integral by making a u-substitution and then integrating by parts.
   (a) $\int e^{\sqrt{x}}\,dx$ (b) $\int \cos\sqrt{x}\,dx$
42. [C] For each of the integrals you evaluated in Exercises 1–41, use a CAS to check your answer. If the answer produced by the CAS does not match your own, show that the two answers are equivalent.
43. (a) Find the area of the region enclosed by $y = \ln x$, the line $x = e$, and the x-axis.
   (b) Find the volume of the solid generated when the region in part (a) is revolved about the x-axis.
44. Find the area of the region between $y = x\sin x$ and $y = x$ for $0 \le x \le \pi/2$.
45. Find the volume of the solid generated when the region between $y = \sin x$ and $y = 0$ for $0 \le x \le \pi$ is revolved about the y-axis.
46. Find the volume of the solid generated when the region enclosed between $y = \cos x$ and $y = 0$ for $0 \le x \le \pi/2$ is revolved about the y-axis.
47. A particle moving along the x-axis has velocity function $v(t) = t^2 e^{-t}$. How far does the particle travel from time $t = 0$ to $t = 5$?
48. The study of sawtooth waves in electrical engineering leads to integrals of the form
$$\int_{-\pi/\omega}^{\pi/\omega} t \sin(k\omega t)\,dt$$
where k is an integer and ω is a nonzero constant. Evaluate the integral.
49. Use reduction formula (7) to evaluate
   (a) $\int \sin^3 x\,dx$ (b) $\int_0^{\pi/4} \sin^4 x\,dx$.

50. Use reduction formula (8) to evaluate

(a) $\int \cos^5 x\,dx$ (b) $\int_0^{\pi/2} \cos^6 x\,dx$.

51. Derive reduction formula (7).

52. In each part, use integration by parts or other methods to derive the reduction formula.

(a) $\displaystyle\int \sec^n x\,dx = \frac{\sec^{n-2} x \tan x}{n-1} + \frac{n-2}{n-1}\int \sec^{n-2} x\,dx$

(b) $\displaystyle\int \tan^n x\,dx = \frac{\tan^{n-1} x}{n-1} - \int \tan^{n-2} x\,dx$

(c) $\displaystyle\int x^n e^x\,dx = x^n e^x - n\int x^{n-1} e^x\,dx$

In Exercises 53 and 54, use the reduction formulas in Exercise 52 to evaluate the integrals.

53. (a) $\int \tan^4 x\,dx$ (b) $\int \sec^4 x\,dx$ (c) $\int x^3 e^x\,dx$

54. (a) $\int x^2 e^{3x}\,dx$ (b) $\int_0^1 x e^{-\sqrt{x}}\,dx$

[*Hint:* First make a substitution.]

55. Let f be a function whose second derivative is continuous on $[-1, 1]$. Show that

$$\int_{-1}^{1} x f''(x)\,dx = f'(1) + f'(-1) - f(1) + f(-1)$$

56. Recall from Theorem 4.1.5 and the discussion preceding it that if $f'(x) > 0$, then the function f is increasing and has an inverse. The purpose of this problem is to show that if this condition is satisfied and if f' is continuous, then a definite integral of f^{-1} can be expressed in terms of a definite integral of f.

(a) Use integration by parts to show that

$$\int_a^b f(x)\,dx = bf(b) - af(a) - \int_a^b x f'(x)\,dx$$

(b) Use the result in part (a) to show that if $y = f(x)$, then

$$\int_a^b f(x)\,dx = bf(b) - af(a) - \int_{f(a)}^{f(b)} f^{-1}(y)\,dy$$

(c) Show that if we let $\alpha = f(a)$ and $\beta = f(b)$, then the result in part (b) can be written as

$$\int_\alpha^\beta f^{-1}(x)\,dx = \beta f^{-1}(\beta) - \alpha f^{-1}(\alpha) - \int_{f^{-1}(\alpha)}^{f^{-1}(\beta)} f(x)\,dx$$

57. In each part, use the result in Exercise 56 to obtain the equation, and then confirm that the equation is correct by performing the integrations.

(a) $\displaystyle\int_0^{1/2} \sin^{-1} x\,dx = \tfrac{1}{2}\sin^{-1}\left(\tfrac{1}{2}\right) - \int_0^{\pi/6} \sin x\,dx$

(b) $\displaystyle\int_e^{e^2} \ln x\,dx = (2e^2 - e) - \int_1^2 e^x\,dx$

58. (a) In Example 1, let

$$u = x, \quad dv = e^x\,dx,$$
$$du = dx, \quad v = \int e^x\,dx = e^x + C_1$$

and show that the constant C_1 cancels out, thus giving the same solution obtained by omitting C_1.

(b) Show that in general

$$uv - \int v\,du = u(v + C_1) - \int (v + C_1)\,du$$

thereby justifying the omission of the constant of integration when calculating v in integration by parts.

9.3 TRIGONOMETRIC INTEGRALS

In the last section we derived reduction formulas for integrating positive integer powers of sine, cosine, tangent, and secant. In this section we will show how to work with those reduction formulas, and we will discuss methods for integrating other kinds of integrals that involve trigonometric functions.

INTEGRATING POWERS OF SINE AND COSINE

In the preceding section we derived the reduction formulas

$$\int \sin^n x\,dx = -\frac{1}{n}\sin^{n-1} x \cos x + \frac{n-1}{n}\int \sin^{n-2} x\,dx \tag{1}$$

$$\int \cos^n x\,dx = \frac{1}{n}\cos^{n-1} x \sin x + \frac{n-1}{n}\int \cos^{n-2} x\,dx \tag{2}$$

In the case where $n = 2$, these formulas yield

$$\int \sin^2 x\,dx = -\frac{1}{2}\sin x \cos x + \frac{1}{2}\int dx = \frac{1}{2}x - \frac{1}{2}\sin x \cos x + C \tag{3}$$

$$\int \cos^2 x\,dx = \frac{1}{2}\cos x \sin x + \frac{1}{2}\int dx = \frac{1}{2}x + \frac{1}{2}\sin x \cos x + C \tag{4}$$

Alternative forms of these integration formulas can be derived from the trigonometric identities

$$\sin^2 x = \tfrac{1}{2}(1 - \cos 2x) \qquad \text{and} \qquad \cos^2 x = \tfrac{1}{2}(1 + \cos 2x) \tag{5–6}$$

which follow from the double-angle formulas

$$\cos 2x = 1 - 2\sin^2 x \quad \text{and} \quad \cos 2x = 2\cos^2 x - 1$$

These identities yield

$$\int \sin^2 x\,dx = \frac{1}{2}\int (1 - \cos 2x)\,dx = \frac{1}{2}x - \frac{1}{4}\sin 2x + C \tag{7}$$

$$\int \cos^2 x\,dx = \frac{1}{2}\int (1 + \cos 2x)\,dx = \frac{1}{2}x + \frac{1}{4}\sin 2x + C \tag{8}$$

Observe that the antiderivatives in Formulas (3) and (4) involve both sines and cosines, whereas those in (7) and (8) involve sines alone. However, the apparent discrepancy is easy to resolve by using the identity

$$\sin 2x = 2\sin x \cos x$$

to rewrite (7) and (8) in forms (3) and (4), or conversely.

In the case where $n = 3$, the reduction formulas for integrating $\sin^3 x$ and $\cos^3 x$ yield

$$\int \sin^3 x\,dx = -\frac{1}{3}\sin^2 x \cos x + \frac{2}{3}\int \sin x\,dx = -\frac{1}{3}\sin^2 x \cos x - \frac{2}{3}\cos x + C \tag{9}$$

$$\int \cos^3 x\,dx = \frac{1}{3}\cos^2 x \sin x + \frac{2}{3}\int \cos x\,dx = \frac{1}{3}\cos^2 x \sin x + \frac{2}{3}\sin x + C \tag{10}$$

If desired, Formula (9) can be expressed in terms of cosines alone by using the identity $\sin^2 x = 1 - \cos^2 x$, and Formula (10) can be expressed in terms of sines alone by using the identity $\cos^2 x = 1 - \sin^2 x$. We leave it for you to do this and confirm that

$$\int \sin^3 x\,dx = \tfrac{1}{3}\cos^3 x - \cos x + C \tag{11}$$

$$\int \cos^3 x\,dx = \sin x - \tfrac{1}{3}\sin^3 x + C \tag{12}$$

FOR THE READER. When asked to integrate $\sin^3 x$ and $\cos^3 x$, the *Maple* CAS produces forms (11) and (12). However, the *Mathematica* CAS produces

$$\int \sin^3 x\,dx = -\tfrac{3}{4}\cos x + \tfrac{1}{12}\cos 3x + C$$

$$\int \cos^3 x\,dx = \tfrac{3}{4}\sin x + \tfrac{1}{12}\sin 3x + C$$

See if you can reconcile *Mathematica*'s results with (11) and (12).

We leave it as an exercise to obtain the following formulas by first applying the reduction formulas, and then using appropriate trigonometric identities.

$$\int \sin^4 x\,dx = \tfrac{3}{8}x - \tfrac{1}{4}\sin 2x + \tfrac{1}{32}\sin 4x + C \tag{13}$$

$$\int \cos^4 x\,dx = \tfrac{3}{8}x + \tfrac{1}{4}\sin 2x + \tfrac{1}{32}\sin 4x + C \tag{14}$$

Example 1

Find the volume V of the solid that is obtained when the region under the curve $y = \sin^2 x$ over the interval $[0, \pi]$ is revolved about the x-axis (Figure 9.3.1).

Solution. Using the method of disks, Formula (5) of Section 8.2 yields

$$V = \int_0^{\pi} \pi \sin^4 x\, dx = \pi\left[\tfrac{3}{8}x - \tfrac{1}{4}\sin 2x + \tfrac{1}{32}\sin 4x\right]_0^{\pi} = \tfrac{3}{8}\pi^2 \quad \blacktriangleleft$$

INTEGRATING PRODUCTS OF SINES AND COSINES

If m and n are positive integers, then the integral

$$\int \sin^m x \cos^n x\, dx$$

can be evaluated by one of the three procedures stated in Table 9.3.1, depending on whether m and n are odd or even.

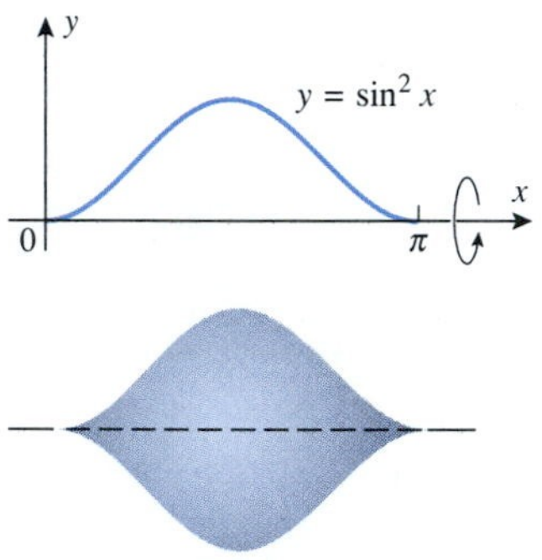

Figure 9.3.1

Example 2

Evaluate

(a) $\int \sin^4 x \cos^5 x\, dx$ (b) $\int \sin^4 x \cos^4 x\, dx$

Solution (a). Since $n = 5$ is odd, we will follow the first procedure in Table 9.3.1:

$$\begin{aligned}
\int \sin^4 x \cos^5 x\, dx &= \int \sin^4 x \cos^4 x \cos x\, dx \\
&= \int \sin^4 x (1 - \sin^2 x)^2 \cos x\, dx \\
&= \int u^4 (1 - u^2)^2\, du \\
&= \int (u^4 - 2u^6 + u^8)\, du \\
&= \tfrac{1}{5}u^5 - \tfrac{2}{7}u^7 + \tfrac{1}{9}u^9 + C \\
&= \tfrac{1}{5}\sin^5 x - \tfrac{2}{7}\sin^7 x + \tfrac{1}{9}\sin^9 x + C
\end{aligned}$$

Solution (b). Since $m = n = 4$, both exponents are even, so we will follow the third procedure in Table 9.3.1:

$$\begin{aligned}
\int \sin^4 x \cos^4 x\, dx &= \int (\sin^2 x)^2 (\cos^2 x)^2\, dx \\
&= \int \left(\frac{1}{2}[1 - \cos 2x]\right)^2 \left(\frac{1}{2}[1 + \cos 2x]\right)^2 dx \\
&= \frac{1}{16}\int (1 - \cos^2 2x)^2\, dx \\
&= \frac{1}{16}\int \sin^4 2x\, dx \\
&= \frac{1}{32}\int \sin^4 u\, du \\
&= \frac{1}{32}\left(\frac{3}{8}u - \frac{1}{4}\sin 2u + \frac{1}{32}\sin 4u\right) + C \\
&= \frac{3}{128}x - \frac{1}{128}\sin 4x + \frac{1}{1024}\sin 8x + C
\end{aligned}$$

Note that this can be obtained more directly from the original integral using the identity $\sin x \cos x = \frac{1}{2}\sin 2x$.

$u = 2x$
$du = 2\,dx$ or $dx = \frac{1}{2}\,du$

Formula (13) $\blacktriangleleft$

Table 9.3.1

$\int \sin^m x \cos^n x\,dx$	PROCEDURE	RELEVANT IDENTITIES
n odd	• Split off a factor of $\cos x$. • Apply the relevant identity. • Make the substitution $u = \sin x$.	$\cos^2 x = 1 - \sin^2 x$
m odd	• Split off a factor of $\sin x$. • Apply the relevant identity. • Make the substitution $u = \cos x$.	$\sin^2 x = 1 - \cos^2 x$
$\begin{cases} m \text{ even} \\ n \text{ even} \end{cases}$	• Use the relevant identities to reduce the powers on $\sin x$ and $\cos x$.	$\begin{cases} \sin^2 x = \frac{1}{2}(1 - \cos 2x) \\ \cos^2 x = \frac{1}{2}(1 + \cos 2x) \end{cases}$

Integrals of the form

$$\int \sin mx \cos nx\,dx, \quad \int \sin mx \sin nx\,dx, \quad \int \cos mx \cos nx\,dx \tag{15}$$

can be found by using the trigonometric identities

$$\sin\alpha\cos\beta = \tfrac{1}{2}[\sin(\alpha-\beta) + \sin(\alpha+\beta)] \tag{16}$$

$$\sin\alpha\sin\beta = \tfrac{1}{2}[\cos(\alpha-\beta) - \cos(\alpha+\beta)] \tag{17}$$

$$\cos\alpha\cos\beta = \tfrac{1}{2}[\cos(\alpha-\beta) + \cos(\alpha+\beta)] \tag{18}$$

to express the integrand as a sum or difference of sines and cosines.

Example 3

Evaluate $\displaystyle\int \sin 7x \cos 3x\,dx$.

Solution. Using (16) yields

$$\int \sin 7x \cos 3x\,dx = \frac{1}{2}\int (\sin 4x + \sin 10x)\,dx = -\frac{1}{8}\cos 4x - \frac{1}{20}\cos 10x + C \quad ◀$$

INTEGRATING POWERS OF TANGENT AND SECANT

The procedures for integrating powers of tangent and secant closely parallel those for sine and cosine. The idea is to use the following reduction formulas (which were derived in Exercise 52 of Section 9.2) to reduce the exponent in the integrand until the resulting integral can be evaluated:

$$\int \tan^n x\,dx = \frac{\tan^{n-1} x}{n-1} - \int \tan^{n-2} x\,dx \tag{19}$$

$$\int \sec^n x\,dx = \frac{\sec^{n-2} x \tan x}{n-1} + \frac{n-2}{n-1}\int \sec^{n-2} x\,dx \tag{20}$$

In the case where n is odd, the exponent can be reduced to 1, leaving us with the problem of integrating $\tan x$ or $\sec x$. These integrals are given by

$$\int \tan x\,dx = \ln|\sec x| + C \tag{21}$$

$$\int \sec x\,dx = \ln|\sec x + \tan x| + C \tag{22}$$

Formula (21) can be obtained by writing

$$\int \tan x\,dx = \int \frac{\sin x}{\cos x}\,dx$$

$$= -\ln|\cos x| + C \qquad \boxed{u = \cos x \\ du = -\sin x\,dx}$$

$$= \ln|\sec x| + C \qquad \boxed{\ln|\cos x| = -\ln\frac{1}{|\cos x|}}$$

Formula (22) requires a trick. We write

$$\int \sec x\,dx = \int \sec x\left(\frac{\sec x + \tan x}{\sec x + \tan x}\right)dx = \int \frac{\sec^2 x + \sec x\tan x}{\sec x + \tan x}\,dx$$

$$= \ln|\sec x + \tan x| + C \qquad \boxed{u = \sec x + \tan x \\ du = (\sec^2 x + \sec x\tan x)\,dx}$$

The following basic integrals occur frequently and are worth noting:

$$\int \tan^2 x\,dx = \tan x - x + C \tag{23}$$

$$\int \sec^2 x\,dx = \tan x + C \tag{24}$$

Formula (24) is already known to us, since the derivative of $\tan x$ is $\sec^2 x$. Formula (23) can be obtained by applying reduction formula (19) with $n = 2$ (verify) or, alternatively, by using the identity

$$1 + \tan^2 x = \sec^2 x$$

to write

$$\int \tan^2 x\,dx = \int (\sec^2 x - 1)\,dx = \tan x - x + C$$

The formulas

$$\int \tan^3 x\,dx = \tfrac{1}{2}\tan^2 x - \ln|\sec x| + C \tag{25}$$

$$\int \sec^3 x\,dx = \tfrac{1}{2}\sec x\tan x + \tfrac{1}{2}\ln|\sec x + \tan x| + C \tag{26}$$

can be deduced from (21), (22), and reduction formulas (19) and (20) as follows:

$$\int \tan^3 x\,dx = \frac{1}{2}\tan^2 x - \int \tan x\,dx = \frac{1}{2}\tan^2 x - \ln|\sec x| + C$$

$$\int \sec^3 x\,dx = \frac{1}{2}\sec x\tan x + \frac{1}{2}\int \sec x\,dx = \frac{1}{2}\sec x\tan x + \frac{1}{2}\ln|\sec x + \tan x| + C$$

INTEGRATING PRODUCTS OF TANGENTS AND SECANTS

If m and n are positive integers, then the integral

$$\int \tan^m x\sec^n x\,dx$$

can be evaluated by one of the three procedures stated in Table 9.3.2, depending on whether m and n are odd or even.

Example 4

Evaluate

(a) $\int \tan^2 x\sec^4 x\,dx$ (b) $\int \tan^3 x\sec^3 x\,dx$ (c) $\int \tan^2 x\sec x\,dx$

Table 9.3.2

$\int \tan^m x \sec^n x \, dx$	PROCEDURE	RELEVANT IDENTITIES
n even	• Split off a factor of $\sec^2 x$. • Apply the relevant identity. • Make the substitution $u = \tan x$.	$\sec^2 x = \tan^2 x + 1$
m odd	• Split off a factor of $\sec x \tan x$. • Apply the relevant identity. • Make the substitution $u = \sec x$.	$\tan^2 x = \sec^2 x - 1$
$\begin{cases} m \text{ even} \\ n \text{ odd} \end{cases}$	• Use the relevant identities to reduce the integrand to powers of $\sec x$ alone. • Then use the reduction formula for powers of $\sec x$.	$\tan^2 x = \sec^2 x - 1$

Solution (a). Since $n = 4$ is even, we will follow the first procedure in Table 9.3.2:

$$\begin{aligned}
\int \tan^2 x \sec^4 x \, dx &= \int \tan^2 x \sec^2 x \sec^2 x \, dx \\
&= \int \tan^2 x (\tan^2 x + 1) \sec^2 x \, dx \\
&= \int u^2(u^2 + 1) \, du \\
&= \tfrac{1}{5}u^5 + \tfrac{1}{3}u^3 + C = \tfrac{1}{5}\tan^5 x + \tfrac{1}{3}\tan^3 x + C
\end{aligned}$$

Solution (b). Since $m = 3$ is odd, we will follow the second procedure in Table 9.3.2:

$$\begin{aligned}
\int \tan^3 x \sec^3 x \, dx &= \int \tan^2 x \sec^2 x (\sec x \tan x) \, dx \\
&= \int (\sec^2 x - 1) \sec^2 x (\sec x \tan x) \, dx \\
&= \int (u^2 - 1) u^2 \, du \\
&= \tfrac{1}{5}u^5 - \tfrac{1}{3}u^3 + C = \tfrac{1}{5}\sec^5 x - \tfrac{1}{3}\sec^3 x + C
\end{aligned}$$

Solution (c). Since $m = 2$ is even and $n = 1$ is odd, we will follow the third procedure in Table 9.3.2:

$$\begin{aligned}
\int \tan^2 x \sec x \, dx &= \int (\sec^2 x - 1) \sec x \, dx \\
&= \int \sec^3 x \, dx - \int \sec x \, dx \qquad \text{See (26) and (22).} \\
&= \tfrac{1}{2}\sec x \tan x + \tfrac{1}{2}\ln|\sec x + \tan x| - \ln|\sec x + \tan x| + C \\
&= \tfrac{1}{2}\sec x \tan x - \tfrac{1}{2}\ln|\sec x + \tan x| + C
\end{aligned}$$

◀

AN ALTERNATIVE METHOD FOR INTEGRATING POWERS OF SINE, COSINE, TANGENT, AND SECANT

The methods in Tables 9.3.1 and 9.3.2 can sometimes be applied if $m = 0$ or $n = 0$ to integrate positive integer powers of sine, cosine, tangent, and secant without reduction formulas. For example, instead of using the reduction formula to integrate $\sin^3 x$, we can apply the second procedure in Table 9.3.1.

$$\begin{aligned}\int \sin^3 x\,dx &= \int (\sin^2 x)\sin x\,dx \\ &= \int (1-\cos^2 x)\sin x\,dx \qquad \boxed{\begin{array}{c} u=\cos x \\ du=-\sin x\,dx\end{array}} \\ &= -\int (1-u^2)\,du \\ &= \tfrac{1}{3}u^3 - u + C = \tfrac{1}{3}\cos^3 x - \cos x + C\end{aligned}$$

which agrees with (11).

REMARK. With the aid of the identity $1+\cot^2 x = \csc^2 x$ the techniques in Table 9.3.2 can be adapted to treat integrals of the form

$$\int \cot^m x \csc^n x\,dx$$

Also, there are reduction formulas for powers of cosecant and cotangent that are analogous to Formulas (19) and (20).

MERCATOR'S MAP OF THE WORLD

The integral of $\sec x$ plays an important role in the design of navigational maps for charting nautical and aeronautical courses. Sailors and pilots usually chart their courses along paths with constant compass headings; for example, the course might be $30°$ northeast or $135°$ southwest. Except for courses that are parallel to the equator or run due north or south, a course with constant compass heading spirals around the Earth toward one of the poles (as in Figure 9.3.2*a*). However, in 1569 the Flemish mathematician and geographer Gerhard Kramer (1512–1594) (better known by the Latin name Mercator) devised a world map, called the *Mercator projection*, in which spirals of constant compass headings appear as straight lines. This was extremely important because it enabled sailors to determine compass headings between two points by connecting them with a straight line on a map (Figure 9.3.2*b*).

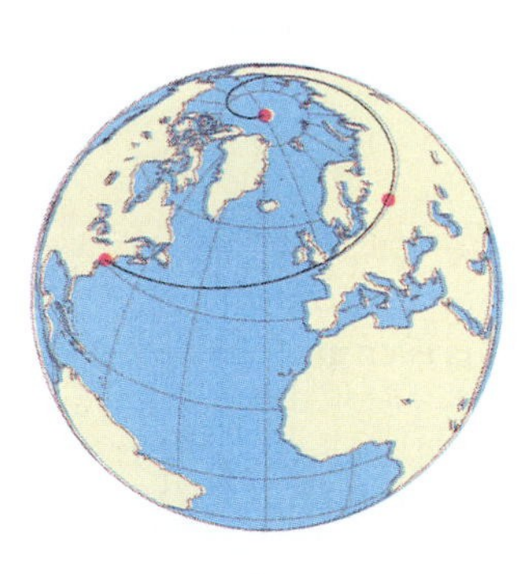

A flight with constant compass heading from New York City to Moscow as it appears on a globe

(*a*)

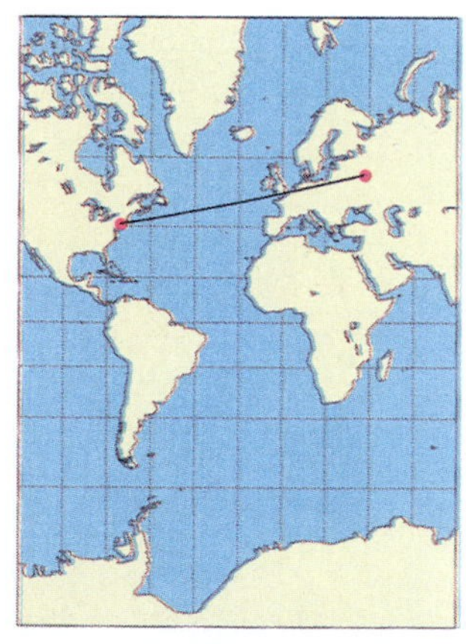

A flight with constant compass heading from New York City to Moscow as it appears on a Mercator projection

(*b*)

Figure 9.3.2

If the Earth is assumed to be a sphere of radius 4000 mi, then the lines of latitude at $1°$ increments are equally spaced about 70 mi apart (why?). However, in the Mercator projection, the lines of latitude become wider apart toward the poles, so that two widely spaced latitude lines near the poles may be actually the same distance apart on the Earth as two closely spaced latitude lines near the equator. It can be proved that on a Mercator map in which the equatorial line has length L, the vertical distance D_β on the map between the equator (latitude 0°) and the line of latitude β° is

$$D_\beta = \frac{L}{2\pi}\int_0^\beta \sec x\, dx \tag{27}$$

(see Exercises 59 and 60).

EXERCISE SET 9.3 C CAS

In Exercises 1–52, evaluate the integral.

1. $\int \cos^5 x \sin x\, dx$
2. $\int \sin^4 3x \cos 3x\, dx$
3. $\int \sin ax \cos ax\, dx$
4. $\int \cos^2 3x\, dx$
5. $\int \sin^2 5\theta\, d\theta$
6. $\int \cos^3 at\, dt$
7. $\int \cos^5 \theta\, d\theta$
8. $\int \sin^3 x \cos^3 x\, dx$
9. $\int \sin^2 2t \cos^3 2t\, dt$
10. $\int \sin^3 2x \cos^2 2x\, dx$
11. $\int \sin^2 x \cos^2 x\, dx$
12. $\int \sin^2 x \cos^4 x\, dx$
13. $\int \sin x \cos 2x\, dx$
14. $\int \sin 3\theta \cos 2\theta\, d\theta$
15. $\int \sin x \cos(x/2)\, dx$
16. $\int \cos^{1/5} x \sin x\, dx$
17. $\int_0^{\pi/4} \cos^3 x\, dx$
18. $\int_0^{\pi/2} \sin^2 \frac{x}{2} \cos^2 \frac{x}{2}\, dx$
19. $\int_0^{\pi/3} \sin^4 3x \cos^3 3x\, dx$
20. $\int_{-\pi}^{\pi} \cos^2 5\theta\, d\theta$
21. $\int_0^{\pi/6} \sin 2x \cos 4x\, dx$
22. $\int_0^{2\pi} \sin^2 kx\, dx$
23. $\int \sec^2(3x+1)\, dx$
24. $\int \tan 5x\, dx$
25. $\int e^{-2x} \tan(e^{-2x})\, dx$
26. $\int \cot 3x\, dx$
27. $\int \sec 2x\, dx$
28. $\int \frac{\sec(\sqrt{x})}{\sqrt{x}}\, dx$
29. $\int \tan^2 x \sec^2 x\, dx$
30. $\int \tan^5 x \sec^4 x\, dx$
31. $\int \tan^3 4x \sec^4 4x\, dx$
32. $\int \tan^4 \theta \sec^4 \theta\, d\theta$
33. $\int \sec^5 x \tan^3 x\, dx$
34. $\int \tan^5 \theta \sec \theta\, d\theta$
35. $\int \tan^4 x \sec x\, dx$
36. $\int \tan^2 \frac{x}{2} \sec^3 \frac{x}{2}\, dx$
37. $\int \tan 2t \sec^3 2t\, dt$
38. $\int \tan x \sec^5 x\, dx$
39. $\int \sec^4 x\, dx$
40. $\int \sec^5 x\, dx$
41. $\int \tan^4 x\, dx$
42. $\int \tan^3 4x\, dx$
43. $\int \sqrt{\tan x} \sec^4 x\, dx$
44. $\int \tan x \sec^{3/2} x\, dx$
45. $\int_0^{\pi/6} \tan^2 2x\, dx$
46. $\int_0^{\pi/6} \sec^3 \theta \tan \theta\, d\theta$
47. $\int_0^{\pi/2} \tan^5 \frac{x}{2}\, dx$
48. $\int_0^{1/4} \sec \pi x \tan \pi x\, dx$
49. $\int \cot^3 x \csc^3 x\, dx$
50. $\int \cot^2 3t \sec 3t\, dt$
51. $\int \cot^3 x\, dx$
52. $\int \csc^4 x\, dx$
53. Let m, n be distinct nonnegative integers. Use Formulas (16)–(18) to prove:
   (a) $\int_0^{2\pi} \sin mx \cos nx\, dx = 0$
   (b) $\int_0^{2\pi} \cos mx \cos nx\, dx = 0$
   (c) $\int_0^{2\pi} \sin mx \sin nx\, dx = 0.$
54. C For each of the integrals you evaluated in Exercises 1–52, use a CAS to check your answer. If the answer produced by the CAS does not match your own, show that the two answers are equivalent.
55. Find the arc length of the curve $y = \ln(\cos x)$ over the interval $[0, \pi/4]$.
56. Find the volume of the solid generated when the region enclosed by $y = \tan x$, $y = 1$, and $x = 0$ is revolved about the x-axis.
57. Find the volume of the solid that results when the region enclosed by $y = \cos x$, $y = \sin x$, $x = 0$, and $x = \pi/4$ is revolved about the x-axis.
58. The region bounded below by the x-axis and above by the portion of $y = \sin x$ from $x = 0$ to $x = \pi$ is revolved about the x-axis. Find the volume of the resulting solid.

59. Use Formula (27) to show that if the length of the equatorial line on a Mercator projection is L, then the vertical distance D between the latitude lines at $\alpha°$ and $\beta°$ on the same side of the equator (where $\alpha < \beta$) is

$$D = \frac{L}{2\pi} \ln \left| \frac{\sec\beta + \tan\beta}{\sec\alpha + \tan\alpha} \right|$$

60. Suppose that the equator has a length of 100 cm on a Mercator projection. In each part, use the result in Exercise 59 to answer the question.

(a) What is the vertical distance on the map between the equator and the line at 25° north latitude?

(b) What is the vertical distance on the map between New Orleans, Louisiana, at 30° north latitude and Winnepeg, Canada, at 50° north latitude?

61. (a) Show that

$$\int \csc x\, dx = -\ln|\csc x + \cot x| + C$$

(b) Show that the result in part (a) can also be written as

$$\int \csc x\, dx = \ln|\csc x - \cot x| + C$$

and

$$\int \csc x\, dx = \ln\left|\tan \tfrac{1}{2}x\right| + C$$

62. Rewrite $\sin x + \cos x$ in the form

$$A\sin(x + \phi)$$

and use your result together with Exercise 61 to evaluate

$$\int \frac{dx}{\sin x + \cos x}$$

63. Use the method of Exercise 62 to evaluate

$$\int \frac{dx}{a\sin x + b\cos x} \quad (a, b \text{ not both zero})$$

64. (a) Use Formula (7) in Section 9.2 to show that

$$\int_0^{\pi/2} \sin^n x\, dx = \frac{n-1}{n} \int_0^{\pi/2} \sin^{n-2} x\, dx$$

(b) Use this result to derive the ***Wallis sine formulas***:

$$\int_0^{\pi/2} \sin^n x\, dx = \frac{\pi}{2} \cdot \frac{1 \cdot 3 \cdot 5 \cdots (n-1)}{2 \cdot 4 \cdot 6 \cdots n} \quad \begin{pmatrix} n \text{ even} \\ \text{and} \geq 2 \end{pmatrix}$$

$$\int_0^{\pi/2} \sin^n x\, dx = \frac{2 \cdot 4 \cdot 6 \cdots (n-1)}{3 \cdot 5 \cdot 7 \cdots n} \quad \begin{pmatrix} n \text{ odd} \\ \text{and} \geq 3 \end{pmatrix}$$

65. Use the Wallis formulas in Exercise 64 to evaluate

(a) $\int_0^{\pi/2} \sin^3 x\, dx$ (b) $\int_0^{\pi/2} \sin^4 x\, dx$

(c) $\int_0^{\pi/2} \sin^5 x\, dx$ (d) $\int_0^{\pi/2} \sin^6 x\, dx$.

66. Use Formula (8) in Section 9.2 and the method of Exercise 64 to derive the ***Wallis cosine formulas***:

$$\int_0^{\pi/2} \cos^n x\, dx = \frac{\pi}{2} \cdot \frac{1 \cdot 3 \cdot 5 \cdots (n-1)}{2 \cdot 4 \cdot 6 \cdots n} \quad \begin{pmatrix} n \text{ even} \\ \text{and} \geq 2 \end{pmatrix}$$

$$\int_0^{\pi/2} \cos^n x\, dx = \frac{2 \cdot 4 \cdot 6 \cdots (n-1)}{3 \cdot 5 \cdot 7 \cdots n} \quad \begin{pmatrix} n \text{ odd} \\ \text{and} \geq 3 \end{pmatrix}$$

9.4 TRIGONOMETRIC SUBSTITUTIONS

In this section we will discuss a method for evaluating integrals containing radicals by making substitutions involving trigonometric functions. We will also show how integrals containing quadratic polynomials can sometimes be evaluated by completing the square.

THE METHOD OF TRIGONOMETRIC SUBSTITUTION

To start, we will be concerned with integrals that contain expressions of the form

$$\sqrt{a^2 - x^2}, \quad \sqrt{x^2 + a^2}, \quad \sqrt{x^2 - a^2}$$

in which a is a positive constant. The basic idea for evaluating such integrals is to make a substitution for x that will eliminate the radical. For example, to eliminate the radical in the expression $\sqrt{a^2 - x^2}$, we can make the substitution

$$x = a\sin\theta, \quad -\pi/2 \leq \theta \leq \pi/2 \tag{1}$$

which yields

$$\sqrt{a^2 - x^2} = \sqrt{a^2 - a^2\sin^2\theta} = \sqrt{a^2(1 - \sin^2\theta)}$$
$$= a\sqrt{\cos^2\theta} = a|\cos\theta| = a\cos\theta$$

$\cos\theta \geq 0$ since $-\pi/2 \leq \theta \leq \pi/2$

The restriction on θ in (1) serves two purposes—it enables us to replace $|\cos\theta|$ by $\cos\theta$ to simplify the calculations, and it also ensures that the substitutions can be rewritten as $\theta = \sin^{-1}(x/a)$, if needed.

Example 1

Evaluate $\displaystyle\int \frac{dx}{x^2\sqrt{4-x^2}}$.

Solution. To eliminate the radical we make the substitution

$$x = 2\sin\theta, \quad dx = 2\cos\theta\, d\theta$$

This yields

$$\begin{aligned}\int \frac{dx}{x^2\sqrt{4-x^2}} &= \int \frac{2\cos\theta\, d\theta}{(2\sin\theta)^2\sqrt{4-4\sin^2\theta}} \\ &= \int \frac{2\cos\theta\, d\theta}{(2\sin\theta)^2(2\cos\theta)} = \frac{1}{4}\int \frac{d\theta}{\sin^2\theta} \\ &= \frac{1}{4}\int \csc^2\theta\, d\theta = -\frac{1}{4}\cot\theta + C \end{aligned} \tag{2}$$

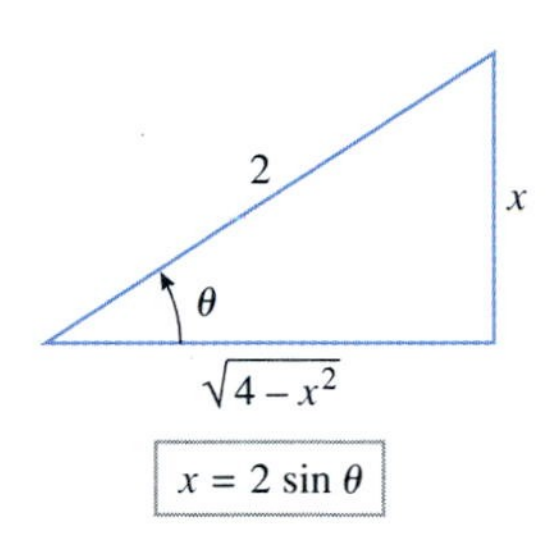

Figure 9.4.1

At this point we have completed the integration; however, because the original integral was expressed in terms of x, it is desirable to express $\cot\theta$ in terms of x as well. This can be done using trigonometric identities, but the expression can also be obtained by writing the substitution $x = 2\sin\theta$ as $\sin\theta = x/2$ and representing it geometrically as in Figure 9.4.1. From that figure we obtain

$$\cot\theta = \frac{\sqrt{4-x^2}}{x}$$

Substituting this in (2) yields

$$\int \frac{dx}{x^2\sqrt{4-x^2}} = -\frac{1}{4}\frac{\sqrt{4-x^2}}{x} + C$$

◀

Example 2

Evaluate $\displaystyle\int_1^{\sqrt{2}} \frac{dx}{x^2\sqrt{4-x^2}}$.

Solution. There are two possible approaches: we can make the substitution in the indefinite integral (as in Example 1) and then evaluate the definite integral using the x-limits of integration, or we can make the substitution in the definite integral and convert the x-limits to the corresponding θ-limits.

Method 1. Using the result from Example 1 with the x-limits of integration yields

$$\int_1^{\sqrt{2}} \frac{dx}{x^2\sqrt{4-x^2}} = -\frac{1}{4}\left[\frac{\sqrt{4-x^2}}{x}\right]_1^{\sqrt{2}} = -\frac{1}{4}[1-\sqrt{3}\,] = \frac{\sqrt{3}-1}{4}$$

Method 2. The substitution $x = 2\sin\theta$ can be expressed as $x/2 = \sin\theta$ or $\theta = \sin^{-1}(x/2)$, so the θ-limits that correspond to $x = 1$ and $x = \sqrt{2}$ are

$$\begin{aligned} x = 1&: \quad \theta = \sin^{-1}(1/2) = \pi/6 \\ x = \sqrt{2}&: \quad \theta = \sin^{-1}(\sqrt{2}/2) = \pi/4 \end{aligned}$$

Thus, from (2) in Example 1 we obtain

$$\int_1^{\sqrt{2}} \frac{dx}{x^2\sqrt{4-x^2}} = \int_{\pi/6}^{\pi/4} \frac{2\cos\theta\, d\theta}{(2\sin\theta)^2\sqrt{4-4\sin^2\theta}} = \frac{1}{4}\int_{\pi/6}^{\pi/4} \frac{d\theta}{\sin^2\theta}$$

$$= \frac{1}{4}\int_{\pi/6}^{\pi/4} \csc^2\theta\, d\theta = -\frac{1}{4}\left[\cot\theta\right]_{\pi/6}^{\pi/4}$$

$$= -\frac{1}{4}[1-\sqrt{3}] = \frac{\sqrt{3}-1}{4} \quad ◀$$

Example 3

Find the area of the ellipse

$$\frac{x^2}{a^2} + \frac{y^2}{b^2} = 1$$

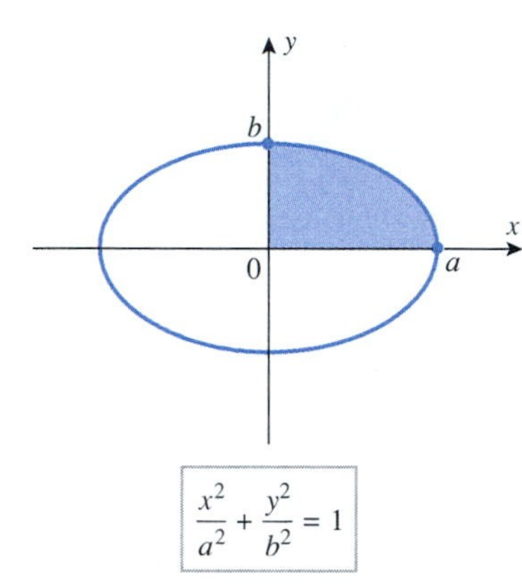

$$\frac{x^2}{a^2} + \frac{y^2}{b^2} = 1$$

Figure 9.4.2

Solution. Because the ellipse is symmetric about both axes, its area A is four times the area in the first quadrant (Figure 9.4.2). If we solve the equation of the ellipse for y in terms of x, we obtain

$$y = \pm\frac{b}{a}\sqrt{a^2 - x^2}$$

where the positive square root gives the equation of the upper half. Thus, the area A is given by

$$A = 4\int_0^a \frac{b}{a}\sqrt{a^2-x^2}\, dx = \frac{4b}{a}\int_0^a \sqrt{a^2-x^2}\, dx$$

To evaluate this integral, we will make the substitution $x = a\sin\theta$ ($dx = a\cos\theta\, d\theta$) and convert the x-limits of integration to θ-limits. Since the substitution can be expressed as $\theta = \sin^{-1}(x/a)$, the θ-limits of integration are

$$x = 0: \quad \theta = \sin^{-1}(0) = 0$$

$$x = a: \quad \theta = \sin^{-1}(1) = \pi/2$$

Thus, we obtain

$$A = \frac{4b}{a}\int_0^a \sqrt{a^2-x^2}\, dx = \frac{4b}{a}\int_0^{\pi/2} a\cos\theta \cdot a\cos\theta\, d\theta$$

$$= 4ab\int_0^{\pi/2} \cos^2\theta\, d\theta = 4ab\int_0^{\pi/2} \frac{1}{2}(1+\cos 2\theta)\, d\theta$$

$$= 2ab\left[\theta + \frac{1}{2}\sin 2\theta\right]_0^{\pi/2} = 2ab\left[\frac{\pi}{2} - 0\right] = \pi ab \quad ◀$$

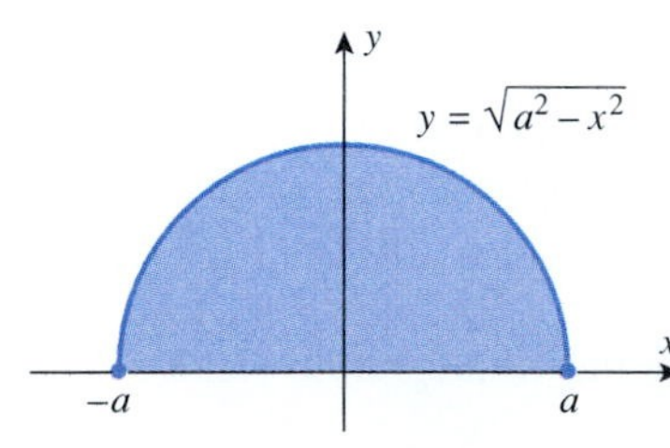

Figure 9.4.3

REMARK. In the special case where $a = b$, the ellipse becomes a circle of radius a, and the area formula becomes $A = \pi a^2$, as expected. It is worth noting that

$$\int_{-a}^{a} \sqrt{a^2-x^2}\, dx = \tfrac{1}{2}\pi a^2 \tag{3}$$

since this integral represents the area of the upper semicircle (Figure 9.4.3).

FOR THE READER. If you have a calculating utility with a numerical integration capability, use it and Formula (3) to approximate π to three decimal places.

Thus far, we have focused on using the substitution $x = a\sin\theta$ to evaluate integrals involving radicals of the form $\sqrt{a^2 - x^2}$. Table 9.4.1 summarizes this method and describes some other substitutions of this type.

Table 9.4.1

EXPRESSION IN THE INTEGRAND	SUBSTITUTION	RESTRICTION ON θ	SIMPLIFICATION
$\sqrt{a^2-x^2}$	$x = a\sin\theta$	$-\pi/2 \le \theta \le \pi/2$	$a^2-x^2 = a^2-a^2\sin^2\theta = a^2\cos^2\theta$
$\sqrt{a^2+x^2}$	$x = a\tan\theta$	$-\pi/2 < \theta < \pi/2$	$a^2+x^2 = a^2+a^2\tan^2\theta = a^2\sec^2\theta$
$\sqrt{x^2-a^2}$	$x = a\sec\theta$	$\begin{cases} 0 \le \theta < \pi/2 & (\text{if } x \ge a) \\ \pi/2 < \theta \le \pi & (\text{if } x \le -a) \end{cases}$	$x^2-a^2 = a^2\sec^2\theta - a^2 = a^2\tan^2\theta$

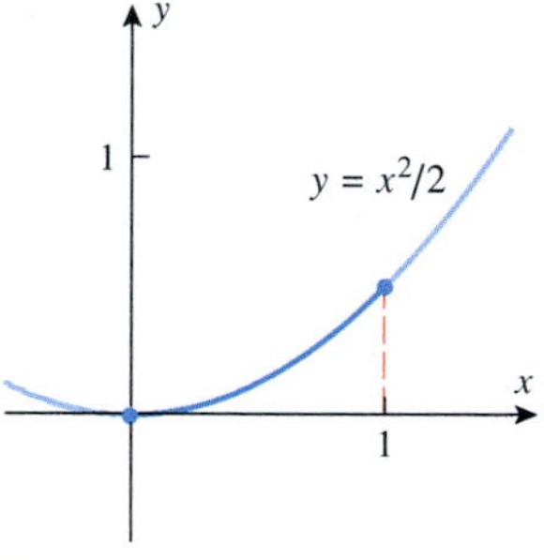

Figure 9.4.4

Example 4

Find the arc length of the curve $y = x^2/2$ from $x = 0$ to $x = 1$ (Figure 9.4.4).

Solution. From Formula (4) of Section 8.4 the arc length L of the curve is

$$L = \int_0^1 \sqrt{1+\left(\frac{dy}{dx}\right)^2}\,dx = \int_0^1 \sqrt{1+x^2}\,dx$$

The integrand involves a radical of the form $\sqrt{a^2+x^2}$ with $a = 1$, so from Table 9.4.1 we make the substitution

$$x = \tan\theta, \quad -\pi/2 < \theta < \pi/2$$

$$\frac{dx}{d\theta} = \sec^2\theta \quad \text{or} \quad dx = \sec^2\theta\,d\theta$$

Since this substitution can be expressed as $\theta = \tan^{-1}x$, the θ-limits of integration that correspond to the x-limits, $x = 0$ and $x = 1$, are

$$x = 0: \quad \theta = \tan^{-1}0 = 0$$

$$x = 1: \quad \theta = \tan^{-1}1 = \pi/4$$

Thus,

$$\begin{aligned} L = \int_0^1 \sqrt{1+x^2}\,dx &= \int_0^{\pi/4} \sqrt{1+\tan^2\theta}\,\sec^2\theta\,d\theta \\ &= \int_0^{\pi/4} \sqrt{\sec^2\theta}\,\sec^2\theta\,d\theta \\ &= \int_0^{\pi/4} |\sec\theta|\sec^2\theta\,d\theta \\ &= \int_0^{\pi/4} \sec^3\theta\,d\theta \qquad \boxed{\sec\theta > 0 \text{ since } -\pi/2 < \theta < \pi/2} \\ &= \left[\tfrac{1}{2}\sec\theta\tan\theta + \tfrac{1}{2}\ln|\sec\theta+\tan\theta|\right]_0^{\pi/4} \qquad \boxed{\text{Formula (26) of Section 9.3}} \\ &= \tfrac{1}{2}[\sqrt{2} + \ln(\sqrt{2}+1)] \approx 1.148 \end{aligned}$$

◀

Example 5

Evaluate $\displaystyle\int \frac{\sqrt{x^2-25}}{x}\,dx$, assuming that $x \ge 5$.

Solution. The integrand involves a radical of the form $\sqrt{x^2-a^2}$ with $a = 5$, so from Table 9.4.1 we make the substitution

$$x = 5\sec\theta, \quad 0 \le \theta < \pi/2$$

$$\frac{dx}{d\theta} = 5\sec\theta\tan\theta \quad \text{or} \quad dx = 5\sec\theta\tan\theta\,d\theta$$

Thus,

$$\int \frac{\sqrt{x^2-25}}{x}\,dx = \int \frac{\sqrt{25\sec^2\theta - 25}}{5\sec\theta}(5\sec\theta\tan\theta)\,d\theta$$

$$= \int \frac{5|\tan\theta|}{5\sec\theta}(5\sec\theta\tan\theta)\,d\theta$$

$$= 5\int \tan^2\theta\,d\theta \qquad \boxed{\tan\theta \geq 0 \text{ since } 0 \leq \theta < \pi/2}$$

$$= 5\int (\sec^2\theta - 1)\,d\theta = 5\tan\theta - 5\theta + C$$

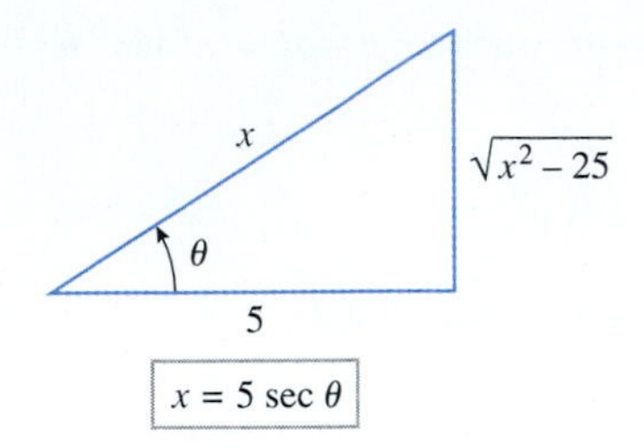

Figure 9.4.5

To express the solution in terms of x, we will represent the substitution $x = 5\sec\theta$ geometrically by the triangle in Figure 9.4.5, from which we obtain

$$\tan\theta = \frac{\sqrt{x^2-25}}{5}$$

From this and the fact that the substitution can be expressed as $\theta = \sec^{-1}(x/5)$, we obtain

$$\int \frac{\sqrt{x^2-25}}{x}\,dx = \sqrt{x^2-25} - 5\sec^{-1}\left(\frac{x}{5}\right) + C \qquad \blacktriangleleft$$

INTEGRALS INVOLVING $ax^2 + bx + c$

Integrals that involve a quadratic expression $ax^2 + bx + c$, where $a \neq 0$ and $b \neq 0$, can often be evaluated by first completing the square, then making an appropriate substitution. The following examples illustrate this idea:

Example 6

Evaluate $\displaystyle\int \frac{x}{x^2-4x+8}\,dx$.

Solution. Completing the square yields

$$x^2 - 4x + 8 = (x^2 - 4x + 4) + 8 - 4 = (x-2)^2 + 4$$

Thus, the substitution

$$u = x - 2, \quad du = dx$$

yields

$$\int \frac{x}{x^2-4x+8}\,dx = \int \frac{x}{(x-2)^2+4}\,dx = \int \frac{u+2}{u^2+4}\,du$$

$$= \int \frac{u}{u^2+4}\,du + 2\int \frac{du}{u^2+4}$$

$$= \frac{1}{2}\int \frac{2u}{u^2+4}\,du + 2\int \frac{du}{u^2+4}$$

$$= \frac{1}{2}\ln(u^2+4) + 2\left(\frac{1}{2}\right)\tan^{-1}\frac{u}{2} + C$$

$$= \frac{1}{2}\ln[(x-2)^2+4] + \tan^{-1}\left(\frac{x-2}{2}\right) + C \qquad \blacktriangleleft$$

Example 7

Evaluate $\displaystyle\int \frac{dx}{\sqrt{5-4x-2x^2}}$.

Solution. Completing the square yields

$$5 - 4x - 2x^2 = 5 - 2(x^2+2x) = 5 - 2(x^2+2x+1) + 2$$
$$= 5 - 2(x+1)^2 + 2 = 7 - 2(x+1)^2$$

Thus,

$$\begin{aligned}\int \frac{dx}{\sqrt{5-4x-2x^2}} &= \int \frac{dx}{\sqrt{7-2(x+1)^2}}\\ &= \int \frac{du}{\sqrt{7-2u^2}} && \begin{array}{l} u = x+1 \\ du = dx\end{array}\\ &= \frac{1}{\sqrt{2}}\int \frac{du}{\sqrt{(7/2)-u^2}}\\ &= \frac{1}{\sqrt{2}}\sin^{-1}\left(\frac{u}{\sqrt{7/2}}\right)+C && \text{Formula (21), Section 9.1 with } a=\sqrt{7/2}\\ &= \frac{1}{\sqrt{2}}\sin^{-1}(\sqrt{2/7}(x+1))+C\end{aligned}$$

◀

EXERCISE SET 9.4 [C] CAS

In Exercises 1–26, evaluate the integral.

1. $\int \sqrt{4-x^2}\,dx$
2. $\int \sqrt{1-4x^2}\,dx$
3. $\int \frac{x^2}{\sqrt{9-x^2}}\,dx$
4. $\int \frac{dx}{x^2\sqrt{16-x^2}}$
5. $\int \frac{dx}{(4+x^2)^2}$
6. $\int \frac{x^2}{\sqrt{5+x^2}}\,dx$
7. $\int \frac{\sqrt{x^2-9}}{x}\,dx$
8. $\int \frac{dx}{x^2\sqrt{x^2-16}}$
9. $\int \frac{x^3}{\sqrt{2-x^2}}\,dx$
10. $\int x^3\sqrt{5-x^2}\,dx$
11. $\int \frac{dx}{x^2\sqrt{4x^2-9}}$
12. $\int \frac{\sqrt{1+t^2}}{t}\,dt$
13. $\int \frac{dx}{(1-x^2)^{3/2}}$
14. $\int \frac{dx}{x^2\sqrt{x^2+25}}$
15. $\int \frac{dx}{\sqrt{x^2-1}}$
16. $\int \frac{dx}{1+2x^2+x^4}$
17. $\int \frac{dx}{(9x^2-1)^{3/2}}$
18. $\int \frac{x^2}{\sqrt{x^2-25}}\,dx$
19. $\int e^x\sqrt{1-e^{2x}}\,dx$
20. $\int \frac{\cos\theta}{\sqrt{2-\sin^2\theta}}\,d\theta$
21. $\int_0^4 x^3\sqrt{16-x^2}\,dx$
22. $\int_0^{1/3} \frac{dx}{(4-9x^2)^2}$
23. $\int_{\sqrt{2}}^2 \frac{dx}{x^2\sqrt{x^2-1}}$
24. $\int_{\sqrt{2}}^2 \frac{\sqrt{2x^2-4}}{x}\,dx$
25. $\int_1^3 \frac{dx}{x^4\sqrt{x^2+3}}$
26. $\int_0^3 \frac{x^3}{(3+x^2)^{5/2}}\,dx$

27. The integral

$$\int \frac{x}{x^2+4}\,dx$$

can be evaluated either by a trigonometric substitution or by the substitution $u = x^2+4$. Do it both ways and show that the results are equivalent.

[C] 28. For each of the integrals you evaluated in Exercises 1–27, use a CAS to check your answer. If the answer produced by the CAS does not match your own, show that the two answers are equivalent.

29. Find the arc length of the curve $y = \ln x$ from $x = 1$ to $x = 2$.

30. Find the arc length of the curve $y = x^2$ from $x = 0$ to $x = 1$.

31. Find the area of the surface generated when the curve in Exercise 30 is revolved about the x-axis.

32. Find the volume of the solid generated when the region enclosed by $x = y(1-y^2)^{1/4}$, $y = 0$, $y = 1$, and $x = 0$ is revolved about the y-axis.

In Exercise 33, the trigonometric substitutions $x = a\sec\theta$ and $x = a\tan\theta$ lead to difficult integrals; for such integrals it is sometimes possible to use the ***hyperbolic substitutions***

$$x = a\sinh u \text{ for integrals involving } \sqrt{x^2+a^2}$$

$$x = a\cosh u \text{ for integrals involving } \sqrt{x^2-a^2}$$

These substitutions are useful because in each case the hyperbolic identity

$$a^2\cosh^2 u - a^2\sinh^2 u = a^2$$

removes the radical.

33. (a) Evaluate

$$\int \frac{dx}{\sqrt{x^2+9}}$$

using the hyperbolic substitution that is suggested above.

(b) Evaluate the integral in part (a) by a trigonometric substitution and show that the results in parts (a) and (b) agree.

(c) Use a hyperbolic substitution to evaluate

$$\int \sqrt{x^2-1}\,dx, \quad x \geq 1$$

34. In Example 3 we found the area of an ellipse by making the substitution $x = a\sin\theta$ in the required integral. Find the area by making the substitution $x = a\cos\theta$, and discuss any restrictions on θ that are needed.

In Exercises 35–46, evaluate the integral.

35. $\int \frac{dx}{x^2-4x+13}$

36. $\int \frac{dx}{\sqrt{2x-x^2}}$

37. $\int \frac{dx}{\sqrt{8+2x-x^2}}$

38. $\int \frac{dx}{16x^2+16x+5}$

39. $\int \frac{dx}{\sqrt{x^2-6x+10}}$

40. $\int \frac{x}{x^2+6x+10}\,dx$

41. $\int \sqrt{3-2x-x^2}\,dx$

42. $\int \frac{e^x}{\sqrt{1+e^x+e^{2x}}}\,dx$

43. $\int \frac{dx}{2x^2+4x+7}$

44. $\int \frac{2x+3}{4x^2+4x+5}\,dx$

45. $\int_1^2 \frac{dx}{\sqrt{4x-x^2}}$

46. $\int_0^1 \sqrt{x(4-x)}\,dx$

C **47.** For each of the integrals you evaluated in Exercises 35–46, use a CAS to check your answer. If the answer produced by the CAS does not match your own, show that the two answers are equivalent.

In Exercises 48–50, there is a good chance that your CAS will not be able to evaluate the integral as stated. If this is so, make a substitution that converts the integral into one that your CAS can evaluate.

C **48.** $\int (x\cos x + \sin x)\sqrt{1+x^2\sin^2 x}\,dx$

C **49.** $\int \cos x \sin x \sqrt{1-\sin^4 x}\,dx$

C **50.** $\int_0^1 3^x\sqrt{9^x-1}\,dx$

9.5 INTEGRATING RATIONAL FUNCTIONS BY PARTIAL FRACTIONS

Recall that a rational function is a ratio of two polynomials. In this section we will give a general method for integrating rational functions that is based on the idea of decomposing a rational function into a sum of simple rational functions that can be integrated by the methods studied in earlier sections.

PARTIAL FRACTIONS

In algebra one learns to combine two or more fractions into a single fraction by finding a common denominator. For example,

$$\frac{2}{x-4}+\frac{3}{x+1}=\frac{2(x+1)+3(x-4)}{(x-4)(x+1)}=\frac{5x-10}{x^2-3x-4} \tag{1}$$

However, for purposes of integration, the left side of (1) is preferable to the right side since each of the terms is easy to integrate:

$$\int \frac{5x-10}{x^2-3x-4}\,dx = \int \frac{2}{x-4}\,dx + \int \frac{3}{x+1}\,dx = 2\ln|x-4| + 3\ln|x+1| + C$$

Thus, it is desirable to have some method that will enable us to obtain the left side of (1), starting with the right side. To illustrate how this can be done, we begin by noting that on the left side the numerators are constants and the denominators are the factors of the denominator on the right side. Thus, to find the left side of (1), starting from the right side, we could factor the denominator of the right side and look for constants A and B such that

$$\frac{5x-10}{(x-4)(x+1)}=\frac{A}{x-4}+\frac{B}{x+1} \tag{2}$$

One way to find the constants A and B is to multiply (2) through by $(x-4)(x+1)$ to clear fractions. This yields

$$5x-10 = A(x+1)+B(x-4) \tag{3}$$

This relationship holds for all x, so it holds in particular if $x=4$ or $x=-1$. Substituting

$x = 4$ in (3) makes the second term on the right drop out and yields the equation $10 = 5A$ or $A = 2$; and substituting $x = -1$ in (3) makes the first term on the right drop out and yields the equation $-15 = -5B$ or $B = 3$. Substituting these values in (2) we obtain

$$\frac{5x - 10}{(x - 4)(x + 1)} = \frac{2}{x - 4} + \frac{3}{x + 1} \tag{4}$$

which agrees with (1).

A second method for finding the constants A and B is to multiply out the right side of (3) and collect like powers of x to obtain

$$5x - 10 = (A + B)x + (A - 4B)$$

Since the polynomials on the two sides are identical, their corresponding coefficients must be the same. Equating the corresponding coefficients on the two sides yields the following system of equations in the unknowns A and B:

$$\begin{aligned} A + B &= 5 \\ A - 4B &= -10 \end{aligned}$$

Solving this system yields $A = 2$ and $B = 3$ as before (verify).

The terms on the right side of (4) are called ***partial fractions*** of the expression on the left side because they each constitute *part* of that expression. To find those partial fractions we first had to make a guess about their form, and then we had to find the unknown constants. Our next objective is to extend this idea to general rational functions. For this purpose, suppose that $P(x)/Q(x)$ is a ***proper rational function***, by which we mean that the degree of the numerator is less than the degree of the denominator. There is a theorem in advanced algebra which states that every proper rational function can be expressed as a sum

$$\frac{P(x)}{Q(x)} = F_1(x) + F_2(x) + \cdots + F_n(x)$$

where $F_1(x), F_2(x), \ldots, F_n(x)$ are rational functions of the form

$$\frac{A}{(ax + b)^k} \quad \text{or} \quad \frac{Ax + B}{(ax^2 + bx + c)^k}$$

in which the denominators are factors of $Q(x)$. The sum is called the ***partial fraction decomposition*** of $P(x)/Q(x)$, and the terms are called ***partial fractions***. As in our opening example, there are two parts to finding a partial fraction decomposition: determining the exact form of the decomposition and finding the unknown constants.

FINDING THE FORM OF A PARTIAL FRACTION DECOMPOSITION

The first step in finding the form of the partial fraction decomposition of a proper rational function $P(x)/Q(x)$ is to factor $Q(x)$ completely into linear and irreducible quadratic factors, and then collect all repeated factors so that $Q(x)$ is expressed as a product of *distinct* factors of the form

$$(ax + b)^m \quad \text{and} \quad (ax^2 + bx + c)^m$$

From these factors we can determine the form of the partial fraction decomposition using two rules that we will now discuss.

LINEAR FACTORS

If all of the factors of $Q(x)$ are linear, then the partial fraction decomposition of $P(x)/Q(x)$ can be determined by using the following rule:

LINEAR FACTOR RULE. For each factor of the form $(ax + b)^m$, the partial fraction decomposition contains the following sum of m partial fractions:

$$\frac{A_1}{ax + b} + \frac{A_2}{(ax + b)^2} + \cdots + \frac{A_m}{(ax + b)^m}$$

where $A_1, A_2, \ldots, A_m$ are constants to be determined. In the case where $m = 1$, only the first term in the sum appears.

Example 1

Evaluate $\displaystyle\int \frac{dx}{x^2+x-2}$.

Solution. The integrand is a proper rational function that can be written as

$$\frac{1}{x^2+x-2} = \frac{1}{(x-1)(x+2)}$$

The factors $x-1$ and $x+2$ are both linear and appear to the first power, so each contributes one term to the partial fraction decomposition by the linear factor rule. Thus, the decomposition has the form

$$\frac{1}{(x-1)(x+2)} = \frac{A}{x-1} + \frac{B}{x+2} \tag{5}$$

where A and B are constants to be determined. Multiplying this expression through by $(x-1)(x+2)$ yields

$$1 = A(x+2) + B(x-1) \tag{6}$$

As discussed earlier, there are two methods for finding A and B: we can substitute values of x that are chosen to make terms on the right drop out, or we can multiply out on the right and equate corresponding coefficients on the two sides to obtain a system of equations that can be solved for A and B. We will use the first approach.

Setting $x=1$ makes the second term in (6) drop out and yields $1 = 3A$ or $A = \frac{1}{3}$; and setting $x=-2$ makes the first term in (6) drop out and yields $1 = -3B$ or $B = -\frac{1}{3}$. Substituting these values in (5) yields the partial fraction decomposition

$$\frac{1}{(x-1)(x+2)} = \frac{\frac{1}{3}}{x-1} + \frac{-\frac{1}{3}}{x+2}$$

The integration can now be completed as follows:

$$\begin{aligned}\int \frac{dx}{(x-1)(x+2)} &= \frac{1}{3}\int \frac{dx}{x-1} - \frac{1}{3}\int \frac{dx}{x+2} \\ &= \frac{1}{3}\ln|x-1| - \frac{1}{3}\ln|x+2| + C = \frac{1}{3}\ln\left|\frac{x-1}{x+2}\right| + C\end{aligned}$$ ◀

If the factors of $Q(x)$ are linear and none are repeated, as in the last example, then the recommended method for finding the constants in the partial fraction decomposition is to substitute appropriate values of x to make terms drop out. However, if some of the linear factors are repeated, then it will not be possible to find all of the constants in this way. In this case the recommended procedure is to find as many constants as possible by substitution and then find the rest by equating coefficients. This is illustrated in the next example.

Example 2

Evaluate $\displaystyle\int \frac{2x+4}{x^3-2x^2}\,dx$.

Solution. The integrand can be rewritten as

$$\frac{2x+4}{x^3-2x^2} = \frac{2x+4}{x^2(x-2)}$$

Although x^2 is a quadratic factor, it is *not* irreducible since $x^2 = xx$. Thus, by the linear factor rule, x^2 introduces two terms (since $m=2$) of the form

$$\frac{A}{x} + \frac{B}{x^2}$$

and the factor $x - 2$ introduces one term (since $m = 1$) of the form

$$\frac{C}{x-2}$$

so the partial fraction decomposition is

$$\frac{2x+4}{x^2(x-2)} = \frac{A}{x} + \frac{B}{x^2} + \frac{C}{x-2} \tag{7}$$

Multiplying by $x^2(x-2)$ yields

$$2x + 4 = Ax(x-2) + B(x-2) + Cx^2 \tag{8}$$

which, after multiplying out and collecting like powers of x, becomes

$$2x + 4 = (A + C)x^2 + (-2A + B)x - 2B \tag{9}$$

Setting $x = 0$ in (8) makes the first and third terms drop out and yields $B = -2$, and setting $x = 2$ in (8) makes the first and second terms drop out and yields $C = 2$ (verify). However, there is no substitution in (8) that produces A directly, so we look to Equation (9) to find this value. This can be done by equating the coefficients of x^2 on the two sides to obtain

$$A + C = 0 \quad \text{or} \quad A = -C = -2$$

Substituting the values $A = -2$, $B = -2$, and $C = 2$ in (7) yields the partial fraction decomposition

$$\frac{2x+4}{x^2(x-2)} = \frac{-2}{x} + \frac{-2}{x^2} + \frac{2}{x-2}$$

Thus,

$$\begin{aligned}\int \frac{2x+4}{x^2(x-2)}\,dx &= -2\int \frac{dx}{x} - 2\int \frac{dx}{x^2} + 2\int \frac{dx}{x-2} \\ &= -2\ln|x| + \frac{2}{x} + 2\ln|x-2| + C = 2\ln\left|\frac{x-2}{x}\right| + \frac{2}{x} + C\end{aligned}$$ ◀

QUADRATIC FACTORS

If some of the factors of $Q(x)$ are irreducible quadratics, then the contribution of those factors to the partial fraction decomposition of $P(x)/Q(x)$ can be determined from the following rule:

QUADRATIC FACTOR RULE. For each factor of the form $(ax^2 + bx + c)^m$, the partial fraction decomposition contains the following sum of m partial fractions:

$$\frac{A_1x + B_1}{ax^2 + bx + c} + \frac{A_2x + B_2}{(ax^2 + bx + c)^2} + \cdots + \frac{A_mx + B_m}{(ax^2 + bx + c)^m}$$

where $A_1, A_2, \ldots, A_m, B_1, B_2, \ldots, B_m$ are constants to be determined. In the case where $m = 1$, only the first term in the sum appears.

Example 3

Evaluate $\displaystyle\int \frac{x^2 + x - 2}{3x^3 - x^2 + 3x - 1}\,dx$.

Solution. The denominator in the integrand can be factored by grouping:

$$\frac{x^2 + x - 2}{3x^3 - x^2 + 3x - 1} = \frac{x^2 + x - 2}{x^2(3x-1) + (3x-1)} = \frac{x^2 + x - 2}{(3x-1)(x^2+1)}$$

By the linear factor rule, the factor $3x - 1$ introduces one term; namely

$$\frac{A}{3x-1}$$

and by the quadratic factor rule, the factor $x^2 + 1$ introduces one term; namely

$$\frac{Bx+C}{x^2+1}$$

Thus, the partial fraction decomposition is

$$\frac{x^2+x-2}{(3x-1)(x^2+1)} = \frac{A}{3x-1} + \frac{Bx+C}{x^2+1} \tag{10}$$

Multiplying by $(3x - 1)(x^2 + 1)$ yields

$$x^2 + x - 2 = A(x^2+1) + (Bx+C)(3x-1) \tag{11}$$

We could find A by substituting $x = \frac{1}{3}$ to make the last term drop out, and then find the rest of the constants by equating corresponding coefficients. However, in this case it is just as easy to find *all* of the constants by equating coefficients and solving the resulting system. For this purpose we multiply out the right side of (11) and collect like terms:

$$x^2 + x - 2 = (A+3B)x^2 + (-B+3C)x + (A-C)$$

Equating corresponding coefficients gives

$$\begin{aligned} A + 3B \quad\quad &= 1 \\ -\ B + 3C &= 1 \\ A \quad\quad - C &= -2 \end{aligned}$$

To solve this system, subtract the third equation from the first to eliminate A. Then use the resulting equation together with the second equation to solve for B and C. Finally, determine A from the first or third equation. This yields (verify)

$$A = -\tfrac{7}{5}, \quad B = \tfrac{4}{5}, \quad C = \tfrac{3}{5}$$

Thus, (10) becomes

$$\frac{x^2+x-2}{(3x-1)(x^2+1)} = \frac{-\frac{7}{5}}{3x-1} + \frac{\frac{4}{5}x + \frac{3}{5}}{x^2+1}$$

and

$$\begin{aligned} \int \frac{x^2+x-2}{(3x-1)(x^2+1)}\,dx &= -\frac{7}{5}\int \frac{dx}{3x-1} + \frac{4}{5}\int \frac{x}{x^2+1}\,dx + \frac{3}{5}\int \frac{dx}{x^2+1} \\ &= -\frac{7}{15}\ln|3x-1| + \frac{2}{5}\ln(x^2+1) + \frac{3}{5}\tan^{-1} x + C \end{aligned}$$ ◀

FOR THE READER. Computer algebra systems have built-in capabilities for finding partial fraction decompositions. If you have a CAS, read the documentation on partial fraction decompositions, and use your CAS to find the decompositions in Examples 1, 2, and 3.

Example 4

Evaluate $\displaystyle\int \frac{3x^4 + 4x^3 + 16x^2 + 20x + 9}{(x+2)(x^2+3)^2}\,dx$.

Solution. Observe that the integrand is a proper rational function since the numerator has degree 4 and the denominator has degree 5. Thus, the method of partial fractions is applicable. By the linear factor rule, the factor $x + 2$ introduces the single term

$$\frac{A}{x+2}$$

and by the quadratic factor rule, the factor $(x^2 + 3)^2$ introduces two terms (since $m = 2$):

$$\frac{Bx + C}{x^2 + 3} + \frac{Dx + E}{(x^2 + 3)^2}$$

Thus, the partial fraction decomposition of the integrand is

$$\frac{3x^4 + 4x^3 + 16x^2 + 20x + 9}{(x + 2)(x^2 + 3)^2} = \frac{A}{x + 2} + \frac{Bx + C}{x^2 + 3} + \frac{Dx + E}{(x^2 + 3)^2} \tag{12}$$

Multiplying by $(x + 2)(x^2 + 3)^2$ yields

$$\begin{aligned} &3x^4 + 4x^3 + 16x^2 + 20x + 9 \\ &\qquad = A(x^2 + 3)^2 + (Bx + C)(x^2 + 3)(x + 2) + (Dx + E)(x + 2) \end{aligned} \tag{13}$$

which, after multiplying out and collecting like powers of x, becomes

$$\begin{aligned} &3x^4 + 4x^3 + 16x^2 + 20x + 9 \\ &\qquad = (A + B)x^4 + (2B + C)x^3 + (6A + 3B + 2C + D)x^2 \\ &\qquad\qquad + (6B + 3C + 2D + E)x + (9A + 6C + 2E) \end{aligned} \tag{14}$$

Equating corresponding coefficients in (14) yields the following system of five linear equations in five unknowns:

$$\begin{aligned} A + B &= 3 \\ 2B + C &= 4 \\ 6A + 3B + 2C + D &= 16 \\ 6B + 3C + 2D + E &= 20 \\ 9A + 6C + 2E &= 9 \end{aligned} \tag{15}$$

Efficient methods for solving systems of linear equations such as this are studied in a branch of mathematics called ***linear algebra***; those methods are outside the scope of this text. However, as a practical matter most linear systems of any size are solved by computer, and most computer algebra systems have commands that in many cases can solve linear systems exactly. In this particular case we can simplify the work by first substituting $x = -2$ in (13), which yields $A = 1$. Substituting this known value of A in (15) yields the simpler system

$$\begin{aligned} B &= 2 \\ 2B + C &= 4 \\ 3B + 2C + D &= 10 \\ 6B + 3C + 2D + E &= 20 \\ 6C + 2E &= 0 \end{aligned} \tag{16}$$

This system can be solved by starting at the top and working down, first substituting $B = 2$ in the second equation to get $C = 0$, then substituting the known values of B and C in the third equation to get $D = 4$, and so forth. This yields

$$A = 1, \quad B = 2, \quad C = 0, \quad D = 4, \quad E = 0$$

Thus, (12) becomes

$$\frac{3x^4 + 4x^3 + 16x^2 + 20x + 9}{(x + 2)(x^2 + 3)^2} = \frac{1}{x + 2} + \frac{2x}{x^2 + 3} + \frac{4x}{(x^2 + 3)^2}$$

and so

$$\begin{aligned} &\int \frac{3x^4 + 4x^3 + 16x^2 + 20x + 9}{(x + 2)(x^2 + 3)^2}\,dx \\ &\qquad = \int \frac{dx}{x + 2} + \int \frac{2x}{x^2 + 3}\,dx + 4\int \frac{x}{(x^2 + 3)^2}\,dx \\ &\qquad = \ln|x + 2| + \ln(x^2 + 3) - \frac{2}{x^2 + 3} + C \end{aligned}$$

◀

INTEGRATING IMPROPER RATIONAL FUNCTIONS

Although the method of partial fractions only applies to proper rational functions, an improper rational function can be integrated by performing a long division and expressing the function as the quotient plus the remainder over the divisor. The remainder over the divisor will be a proper rational function, which can then be decomposed into partial fractions. This idea is illustrated in the following example:

Example 5

Evaluate $\displaystyle\int \frac{3x^4+3x^3-5x^2+x-1}{x^2+x-2}\,dx$.

Solution. The integrand is an improper rational function since the numerator has degree 4 and the denominator has degree 2. Thus, we first perform the long division

$$
\begin{array}{r|l}
 & 3x^2+1 \\
x^2+x-2 & 3x^4+3x^3-5x^2+x-1 \\
 & \underline{3x^4+3x^3-6x^2} \\
 & \qquad\qquad\;\; x^2+x-1 \\
 & \qquad\qquad\;\; \underline{x^2+x-2} \\
 & \qquad\qquad\qquad\quad\;\; 1
\end{array}
$$

It follows that the integrand can be expressed as

$$\frac{3x^4+3x^3-5x^2+x-1}{x^2+x-2} = (3x^2+1) + \frac{1}{x^2+x-2}$$

and hence

$$\int \frac{3x^4+3x^3-5x^2+x-1}{x^2+x-2}\,dx = \int (3x^2+1)\,dx + \int \frac{dx}{x^2+x-2}$$

The second integral on the right now involves a proper rational function and can thus be evaluated by a partial fraction decomposition. Using the result of Example 1 we obtain

$$\int \frac{3x^4+3x^3-5x^2+x-1}{x^2+x-2}\,dx = x^3+x+\frac{1}{3}\ln\left|\frac{x-1}{x+2}\right|+C \qquad \blacktriangleleft$$

CONCLUDING REMARKS

There are some cases in which the method of partial fractions is inappropriate. For example, it would be illogical to use partial fractions to perform the integration

$$\int \frac{3x^2+2}{x^3+2x-8}\,dx = \ln|x^3+2x-8| + C$$

since the substitution $u = x^3+2x-8$ is more direct. Similarly, the integration

$$\int \frac{2x-1}{x^2+1}\,dx = \int \frac{2x}{x^2+1}\,dx - \int \frac{dx}{x^2+1} = \ln(x^2+1) - \tan^{-1}x + C$$

requires only a little algebra since the integrand is already in partial-fraction form.

EXERCISE SET 9.5 [C] CAS

In Exercises 1–8, write out the form of the partial fraction decomposition. (Do not find the numerical values of the coefficients.)

1. $\dfrac{3x-1}{(x-2)(x+5)}$

2. $\dfrac{5}{x(x^2-9)}$

3. $\dfrac{2x-3}{x^3-x^2}$

4. $\dfrac{x^2}{(x+2)^3}$

5. $\dfrac{1-5x^2}{x^3(x^2+1)}$

6. $\dfrac{2x}{(x-1)(x^2+5)}$

7. $\dfrac{4x^3-x}{(x^2+5)^2}$

8. $\dfrac{1-3x^4}{(x-2)(x^2+1)^2}$

In Exercises 9–32, evaluate the integral.

9. $\displaystyle\int \frac{dx}{x^2+3x-4}$

10. $\displaystyle\int \frac{dx}{x^2+8x+7}$

11. $\displaystyle\int \frac{11x+17}{2x^2+7x-4}\,dx$ **12.** $\displaystyle\int \frac{5x-5}{3x^2-8x-3}\,dx$

13. $\displaystyle\int \frac{2x^2-9x-9}{x^3-9x}\,dx$ **14.** $\displaystyle\int \frac{dx}{x(x^2-1)}$

15. $\displaystyle\int \frac{x^2+2}{x+2}\,dx$ **16.** $\displaystyle\int \frac{x^2-4}{x-1}\,dx$

17. $\displaystyle\int \frac{3x^2-10}{x^2-4x+4}\,dx$ **18.** $\displaystyle\int \frac{x^2}{x^2-3x+2}\,dx$

19. $\displaystyle\int \frac{x^5+2x^2+1}{x^3-x}\,dx$ **20.** $\displaystyle\int \frac{2x^5-x^3-1}{x^3-4x}\,dx$

21. $\displaystyle\int \frac{2x^2+3}{x(x-1)^2}\,dx$ **22.** $\displaystyle\int \frac{3x^2-x+1}{x^3-x^2}\,dx$

23. $\displaystyle\int \frac{x^2+x-16}{(x+1)(x-3)^2}\,dx$ **24.** $\displaystyle\int \frac{2x^2-2x-1}{x^3-x^2}\,dx$

25. $\displaystyle\int \frac{x^2}{(x+2)^3}\,dx$ **26.** $\displaystyle\int \frac{2x^2+3x+3}{(x+1)^3}\,dx$

27. $\displaystyle\int \frac{2x^2-1}{(4x-1)(x^2+1)}\,dx$ **28.** $\displaystyle\int \frac{dx}{x^3+x}$

29. $\displaystyle\int \frac{x^3+3x^2+x+9}{(x^2+1)(x^2+3)}\,dx$ **30.** $\displaystyle\int \frac{x^3+x^2+x+2}{(x^2+1)(x^2+2)}\,dx$

31. $\displaystyle\int \frac{x^3-3x^2+2x-3}{x^2+1}\,dx$

32. $\displaystyle\int \frac{x^4+6x^3+10x^2+x}{x^2+6x+10}\,dx$

In Exercises 33 and 34, evaluate the integral by making a substitution that converts the integrand to a rational function.

33. $\displaystyle\int \frac{\cos\theta}{\sin^2\theta+4\sin\theta-5}\,d\theta$ **34.** $\displaystyle\int \frac{e^t}{e^{2t}-4}\,dt$

35. Find the volume of the solid generated when the region enclosed by $y = x^2/(9-x^2)$, $y = 0$, $x = 0$, and $x = 2$ is revolved about the x-axis.

36. Find the area of the region under the curve $y = 1/(1+e^x)$, over the interval $[-\ln 5, \ln 5]$. [*Hint:* Make a substitution that converts the integrand to a rational function.]

In Exercises 37 and 38, use a CAS to evaluate the integral in two ways: (i) integrate directly; (ii) use the CAS to find the partial fraction decomposition and integrate the decomposition. Integrate by hand to check the results.

C **37.** $\displaystyle\int \frac{x^2+1}{(x^2+2x+3)^2}\,dx$

C **38.** $\displaystyle\int \frac{x^5+x^4+4x^3+4x^2+4x+4}{(x^2+2)^3}\,dx$

In Exercises 39 and 40, integrate by hand and check your answers using a CAS.

C **39.** $\displaystyle\int \frac{dx}{x^4-3x^3-7x^2+27x-18}$

C **40.** $\displaystyle\int \frac{dx}{16x^3-4x^2+4x-1}$

41. Show that

$$\int_0^1 \frac{x}{x^4+1}\,dx = \frac{\pi}{8}$$

42. Use partial fractions to derive the integration formula

$$\int \frac{1}{a^2-x^2}\,dx = \frac{1}{2a}\ln\left|\frac{a+x}{a-x}\right| + C$$

9.6 USING TABLES OF INTEGRALS AND COMPUTER ALGEBRA SYSTEMS

In this section we will discuss how to integrate using tables, and we will address some of the issues that relate to using computer algebra systems for integration. Readers who are not using computer algebra systems can skip that material with no problem.

INTEGRAL TABLES

Tables of integrals are useful for eliminating tedious hand computation. The endpapers of this text contain a relatively brief table of integrals that we will refer to as the ***Endpaper Integral Table***; more comprehensive tables are published in standard reference books such as the *CRC Standard Mathematical Tables and Formulae*, CRC Press, Inc., 1991.

All integral tables have their own scheme for classifying integrals according to the form of the integrand. For example, the Endpaper Integral Table classifies the integrals into 15 categories; *Basic Functions*, *Reciprocals of Basic Functions*, *Powers of Trigonometric Functions*, *Products of Trigonometric Functions*, and so forth. The first step in working with tables is to read through the classifications so that you understand the classification scheme and know where to look in the table for integrals of different types.

PERFECT MATCHES

If you are lucky, the integral you are attempting to evaluate will match up perfectly with one of the forms in the table. However, when looking for matches you may have to make an adjustment for the variable of integration. For example, the integral

$$\int x^2 \sin x \, dx$$

is a perfect match with Formula (46) in the Endpaper Integral Table, except for the letter used for the variable of integration. Thus, to apply Formula (46) to the given integral we need to change the variable of integration in the formula from u to x. With that minor modification we obtain

$$\int x^2 \sin x \, dx = 2x \sin x + (2 - x^2) \cos x + C$$

Here are some more examples of perfect matches:

Example 1

Use the Endpaper Integral Table to evaluate

(a) $\int \sin 7x \cos 2x \, dx$ (b) $\int x^2\sqrt{7+3x}\, dx$

(c) $\int \frac{\sqrt{2-x^2}}{x} dx$ (d) $\int (x^3 + 7x + 1) \sin \pi x \, dx$

Solution (a). The integrand can be classified as a product of trigonometric functions. Thus, from Formula (40) with $m = 7$ and $n = 2$ we obtain

$$\int \sin 7x \cos 2x \, dx = -\frac{\cos 9x}{18} - \frac{\cos 5x}{10} + C$$

Solution (b). The integrand can be classified as a power of x multiplying $\sqrt{a+bx}$. Thus, from Formula (103) with $a = 7$ and $b = 3$ we obtain

$$\int x^2\sqrt{7+3x}\, dx = \frac{2}{2835}(135x^2 - 252x + 392)(7+3x)^{3/2} + C$$

Solution (c). The integrand can be classified as a power of x dividing $\sqrt{a^2 - x^2}$. Thus, from Formula (79) with $a = \sqrt{2}$ we obtain

$$\int \frac{\sqrt{2-x^2}}{x} dx = \sqrt{2-x^2} - \sqrt{2} \ln \left| \frac{\sqrt{2} + \sqrt{2-x^2}}{x} \right| + C$$

Solution (d). The integrand can be classified as a polynomial multiplying a trigonometric function. Thus, we apply Formula (58) with $p(x) = x^3 + 7x + 1$ and $a = \pi$. The successive nonzero derivatives of $p(x)$ are

$$p'(x) = 3x^2 + 7, \quad p''(x) = 6x, \quad p'''(x) = 6$$

and hence

$$\int (x^3 + 7x + 1) \sin \pi x \, dx$$

$$= -\frac{x^3 + 7x + 1}{\pi} \cos \pi x + \frac{3x^2 + 7}{\pi^2} \sin \pi x + \frac{6x}{\pi^3} \cos \pi x - \frac{6}{\pi^4} \sin \pi x + C$$ ◀

MATCHES REQUIRING SUBSTITUTIONS

Sometimes an integral that does not match any table entry can be made to match by making an appropriate substitution. Here are some examples.

Example 2

Use the Endpaper Integral Table to evaluate $\int \sqrt{x - 4x^2}\, dx$.

Solution. The integrand does not match any of the forms in the table precisely. It comes closest to matching Formula (112), but it misses because of the factor of 4 multiplying x^2 inside the radical. However, if we make the substitution

$$u = 2x, \quad du = 2\,dx$$

then the $4x^2$ will become a u^2, and the transformed integral will be

$$\int \sqrt{x - 4x^2}\,dx = \frac{1}{2}\int \sqrt{\tfrac{1}{2}u - u^2}\,du$$

which matches Formula (112) with $a = \frac{1}{4}$. Thus, we obtain

$$\begin{aligned}
\int \sqrt{x - 4x^2}\,dx &= \frac{1}{2}\left[\frac{u - \frac{1}{4}}{2}\sqrt{\tfrac{1}{2}u - u^2} + \frac{1}{32}\sin^{-1}\left(\frac{u - \frac{1}{4}}{\frac{1}{4}}\right)\right] + C \\
&= \frac{1}{2}\left[\frac{2x - \frac{1}{4}}{2}\sqrt{x - 4x^2} + \frac{1}{32}\sin^{-1}\left(\frac{2x - \frac{1}{4}}{\frac{1}{4}}\right)\right] + C \\
&= \frac{8x - 1}{16}\sqrt{x - 4x^2} + \frac{1}{64}\sin^{-1}(8x - 1) + C
\end{aligned}$$

◀

Example 3

Use the Endpaper Integral Table to evaluate

$$\text{(a)}\ \int e^{\pi x}\sin^{-1}(e^{\pi x})\,dx \qquad \text{(b)}\ \int x\sqrt{x^2 - 4x + 5}\,dx$$

Solution (a). The integrand does not even come close to matching any of the forms in the table. However, a little thought suggests the substitution

$$u = e^{\pi x}, \quad du = \pi e^{\pi x}\,dx$$

from which we obtain

$$\int e^{\pi x}\sin^{-1}(e^{\pi x})\,dx = \frac{1}{\pi}\int \sin^{-1} u\,du$$

The integrand is now a basic function, and Formula (7) yields

$$\begin{aligned}
\int e^{\pi x}\sin^{-1}(e^{\pi x})\,dx &= \frac{1}{\pi}[u\sin^{-1} u + \sqrt{1 - u^2}] + C \\
&= \frac{1}{\pi}[e^{\pi x}\sin^{-1}(e^{\pi x}) + \sqrt{1 - e^{2\pi x}}] + C
\end{aligned}$$

Solution (b). Again, the integrand does not closely match any of the forms in the table. However, a little thought suggests that it may be possible to bring the integrand closer to the form $x\sqrt{x^2 + a^2}$ by completing the square to eliminate the term involving x inside the radical. Doing this yields

$$\int x\sqrt{x^2 - 4x + 5}\,dx = \int x\sqrt{(x^2 - 4x + 4) + 1}\,dx = \int x\sqrt{(x - 2)^2 + 1}\,dx \qquad (1)$$

At this point we are closer to the form $x\sqrt{x^2 + a^2}$, but we are not quite there because of the $(x - 2)^2$ rather than x^2 inside the radical. However, we can resolve that problem with the substitution

$$u = x - 2, \quad du = dx$$

With this substitution we have $x = u + 2$, so (1) can be expressed in terms of u as

$$\int x\sqrt{x^2 - 4x + 5}\,dx = \int (u + 2)\sqrt{u^2 + 1}\,du = \int u\sqrt{u^2 + 1}\,du + 2\int \sqrt{u^2 + 1}\,du$$

The first integral on the right is now a perfect match with Formula (84) with $a = 1$, and the second is a perfect match with Formula (72) with $a = 1$. Thus, applying these formulas and dropping the unnecessary absolute value signs we obtain

$$\int x\sqrt{x^2 - 4x + 5}\,dx = \left[\frac{1}{3}(u^2 + 1)^{3/2}\right] + 2\left[\frac{u}{2}\sqrt{u^2 + 1} + \frac{1}{2}\ln(u + \sqrt{u^2 + 1})\right] + C$$

If we now replace u by $x - 2$ (in which case $u^2 + 1 = x^2 - 4x + 5$), we obtain

$$\int x\sqrt{x^2 - 4x + 5}\,dx = \tfrac{1}{3}(x^2 - 4x + 5)^{3/2} + (x - 2)\sqrt{x^2 - 4x + 5} + \ln(x - 2 + \sqrt{x^2 - 4x + 5}) + C$$

Although correct, this form of the answer has an unnecessary mixture of radicals and fractional exponents. If desired, we can "clean up" the answer by writing

$$(x^2 - 4x + 5)^{3/2} = (x^2 - 4x + 5)\sqrt{x^2 - 4x + 5}$$

from which it follows that (verify)

$$\int x\sqrt{x^2 - 4x + 5}\,dx = \tfrac{1}{3}(x^2 - x - 1)\sqrt{x^2 - 4x + 1} + \ln(x - 2 + \sqrt{x^2 - 4x + 5}) + C \quad \blacktriangleleft$$

MATCHES REQUIRING REDUCTION FORMULAS

In cases where the entry in an integral table is a reduction formula, that formula will have to be applied first to reduce the given integral to a form in which it can be evaluated.

Example 4

Use the Endpaper Integral Table to evaluate $\displaystyle\int \frac{x^3}{\sqrt{1 + x}}\,dx$.

Solution. The integrand can be classified as a power of x multiplying the reciprocal of $\sqrt{a + bx}$. Thus, from reduction formula (107) with $a = 1$, $b = 1$, and $n = 3$, followed by Formula (106), we obtain

$$\begin{aligned}\int \frac{x^3}{\sqrt{1 + x}}\,dx &= \frac{2x^3\sqrt{1 + x}}{7} - \frac{6}{7}\int \frac{x^2}{\sqrt{1 + x}}\,dx \\ &= \frac{2x^3\sqrt{1 + x}}{7} - \frac{6}{7}\left[\frac{2}{15}(3x^2 - 4x + 8)\sqrt{1 + x}\right] + C \\ &= \left(\frac{2x^3}{7} - \frac{12x^2}{35} + \frac{16x}{35} - \frac{32}{35}\right)\sqrt{1 + x} + C \quad \blacktriangleleft\end{aligned}$$

MATCHES REQUIRING SPECIAL SUBSTITUTIONS

The Endpaper Integral Table has two entries involving an exponent of $3/2$ and numerous entries involving square roots (exponent $1/2$), but it has no entries with other fractional exponents. However, integrals involving fractional powers of x can often be simplified by making the substitution $u = x^{1/n}$ in which n is the least common multiple of the denominators of the exponents. Here are some examples.

Example 5

Evaluate

(a) $\displaystyle\int \frac{\sqrt{x}}{1 + \sqrt[3]{x}}\,dx$ (b) $\displaystyle\int \frac{dx}{2 + 2\sqrt{x}}$ (c) $\displaystyle\int \sqrt{1 + e^x}\,dx$

Solution (a). The integrand contains $x^{1/2}$ and $x^{1/3}$, so we make the substitution $u = x^{1/6}$, from which we obtain

$$x = u^6, \quad dx = 6u^5\,du$$

Thus,

$$\int \frac{\sqrt{x}}{1+\sqrt[3]{x}}\,dx = \int \frac{(u^6)^{1/2}}{1+(u^6)^{1/3}}(6u^5)\,du = 6\int \frac{u^8}{1+u^2}\,du$$

By long division

$$\frac{u^8}{1+u^2} = u^6 - u^4 + u^2 - 1 + \frac{1}{1+u^2}$$

from which it follows that

$$\begin{aligned}\int \frac{\sqrt{x}}{1+\sqrt[3]{x}}\,dx &= 6\int \left(u^6 - u^4 + u^2 - 1 + \frac{1}{1+u^2}\right) du\\ &= \tfrac{6}{7}u^7 - \tfrac{6}{5}u^5 + 2u^3 - 6u + 6\tan^{-1} u + C\\ &= \tfrac{6}{7}x^{7/6} - \tfrac{6}{5}x^{5/6} + 2x^{1/2} - 6x^{1/6} + 6\tan^{-1}(x^{1/6}) + C\end{aligned}$$

Solution (b). The integrand contains $x^{1/2}$ but does not match any of the forms in the Endpaper Integral Table. Thus, we make the substitution $u = x^{1/2}$, from which we obtain

$$x = u^2, \quad dx = 2u\,du$$

Making this substitution yields

$$\begin{aligned}\int \frac{dx}{2+2\sqrt{x}} &= \int \frac{2u}{2+2u}\,du\\ &= \int \left(1 - \frac{1}{1+u}\right) du && \text{Long division}\\ &= u - \ln|1+u| + C\\ &= \sqrt{x} - \ln(1+\sqrt{x}) + C && \text{Absolute value not needed}\end{aligned}$$

Solution (c). Again, the integral does not match any of the forms in the Endpaper Integral Table. However, the integrand contains $(1+e^x)^{1/2}$, which is analogous to the situation in part (b), except that here it is $1 + e^x$ rather than x that is raised to the $1/2$ power. This suggests the substitution $u = (1+e^x)^{1/2}$, from which we obtain (verify)

$$x = \ln(u^2 - 1), \quad dx = \frac{2u}{u^2-1}\,du$$

Thus,

$$\begin{aligned}\int \sqrt{1+e^x}\,dx &= \int u\left(\frac{2u}{u^2-1}\right) du\\ &= \int \frac{2u^2}{u^2-1}\,du\\ &= \int \left(2 + \frac{2}{u^2-1}\right) du && \text{Long division}\\ &= 2u + \int \left(\frac{1}{u-1} - \frac{1}{u+1}\right) du && \text{Partial fractions}\\ &= 2u + \ln|u-1| - \ln|u+1| + C\\ &= 2u + \ln\left|\frac{u-1}{u+1}\right| + C\\ &= 2\sqrt{1+e^x} + \ln\left[\frac{\sqrt{1+e^x}-1}{\sqrt{1+e^x}+1}\right] + C && \text{Absolute value not needed}\end{aligned}$$

◀

Functions that consist of finitely many sums, differences, quotients, and products of $\sin x$ and $\cos x$ are called ***rational functions of sin x and cos x***. Some examples are

$$\frac{\sin x + 3\cos^2 x}{\cos x + 4\sin x}, \quad \frac{\sin x}{1 + \cos x - \cos^2 x}, \quad \frac{3\sin^5 x}{1 + 4\sin x}$$

The Endpaper Integral Table gives a few formulas for integrating rational functions of $\sin x$ and $\cos x$ under the heading *Reciprocals of Basic Functions*. For example, it follows from Formula (18) that

$$\int \frac{1}{1 + \sin x}\, dx = \tan x - \sec x + C \tag{2}$$

However, since the integrand is a rational function of $\sin x$, it may be desirable in a particular application to express the value of the integral in terms of $\sin x$ and $\cos x$ and rewrite (2) as

$$\int \frac{1}{1 + \sin x}\, dx = \frac{\sin x - 1}{\cos x} + C$$

Many rational functions of $\sin x$ and $\cos x$ can be evaluated by an ingenious method that was discovered by the mathematician Karl Weierstrass (see p. 184). The idea is to make the substitution

$$u = \tan(x/2), \quad -\pi/2 < x/2 < \pi/2$$

from which it follows that

$$x = 2\tan^{-1} u, \quad dx = \frac{2}{1 + u^2}\, du$$

To implement this substitution we need to express $\sin x$ and $\cos x$ in terms of u. For this purpose we will use the identities

$$\sin x = 2\sin(x/2)\cos(x/2) \tag{3}$$

$$\cos x = \cos^2(x/2) - \sin^2(x/2) \tag{4}$$

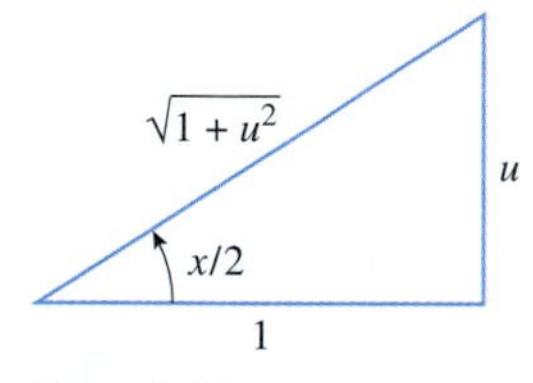

Figure 9.6.1

and the following relationships suggested by Figure 9.6.1:

$$\sin(x/2) = \frac{u}{\sqrt{1 + u^2}} \quad \text{and} \quad \cos(x/2) = \frac{1}{\sqrt{1 + u^2}}$$

Substituting these expressions in (3) and (4) yields

$$\sin x = 2\left(\frac{u}{\sqrt{1 + u^2}}\right)\left(\frac{1}{\sqrt{1 + u^2}}\right) = \frac{2u}{1 + u^2}$$

$$\cos x = \left(\frac{1}{\sqrt{1 + u^2}}\right)^2 - \left(\frac{u}{\sqrt{1 + u^2}}\right)^2 = \frac{1 - u^2}{1 + u^2}$$

In summary, we have shown that the substitution $u = \tan(x/2)$ can be implemented in a rational function of $\sin x$ and $\cos x$ by letting

$$\sin x = \frac{2u}{1 + u^2}, \quad \cos x = \frac{1 - u^2}{1 + u^2}, \quad dx = \frac{2}{1 + u^2}\, du \tag{5}$$

Example 6

Evaluate $\displaystyle\int \frac{dx}{1 - \sin x + \cos x}$.

Solution. The integrand is a rational function of $\sin x$ and $\cos x$ that does not match any of the formulas in the Endpaper Integral Table, so we make the substitution $u = \tan(x/2)$.

Thus, from (5) we obtain

$$\int \frac{dx}{1-\sin x+\cos x} = \int \frac{\dfrac{2\,du}{1+u^2}}{1-\left(\dfrac{2u}{1+u^2}\right)+\left(\dfrac{1-u^2}{1+u^2}\right)}$$

$$= \int \frac{2\,du}{(1+u^2)-2u+(1-u^2)}$$

$$= \int \frac{du}{1-u} = -\ln|1-u|+C = -\ln|1-\tan(x/2)|+C \quad ◀$$

REMARK. The substitution $u=\tan(x/2)$ will convert any rational function of $\sin x$ and $\cos x$ to an ordinary rational function of u. However, the method can lead to cumbersome partial fraction decompositions, so it may be worthwhile to explore the existence of simpler methods when hand computation is to be used.

INTEGRATING WITH COMPUTER ALGEBRA SYSTEMS

Integration tables are rapidly giving way to computerized integration using computer algebra systems. However, computerized integration is very much like computerized chess—the computer can sort through myriads of possibilities quickly, but the approach is sometimes mechanical and lacking in the imagination and judgment of human thought. As a result, answers produced by computer integration are sometimes less satisfactory than those in integral tables that have been refined over many years by many excellent mathematical minds.

Sometimes computer algebra systems do not produce the most general form of an indefinite integral. For example, the integral formula

$$\int \frac{dx}{x-1} = \ln|x-1|+C$$

which can be obtained by inspection or by using the substitution $u=x-1$ is valid for $x\neq 1$. However, *Mathematica*, *Maple*, and *Derive* evaluate this integral as*

$\ln(1-x)$,	$\ln(x-1)$,	$\ln(x-1)$
Mathematica	*Maple*	*Derive*

Observe that none of the systems put in the constant of integration—it is just assumed to be there. Observe also that none of the systems put in the absolute value signs; consequently, for *Maple* and *Derive* the resulting antiderivative is only valid if $x>1$, and for *Mathematica* it is only valid if $x<1$. Thus, although the computer algebra systems all produced a correct antiderivative, none of them produced the most general antiderivative.

Now let us examine how *Mathematica*, *Maple*, and *Derive* handle the integral

$$\int x\sqrt{x^2-4x+5}\,dx = \tfrac{1}{3}(x^2-x-1)\sqrt{x^2-4x+5} + \ln(x-2+\sqrt{x^2-4x+5}) \tag{6}$$

which we obtained in Example 3(b) (with the constant of integration included). *Derive* produces this result in a slightly different algebraic form, and *Maple* produces the result

$$\int x\sqrt{x^2-4x+5}\,dx = \tfrac{1}{3}(x^2-4x+5)^{3/2}+\tfrac{1}{2}(2x-4)\sqrt{x^2-4x+5}+\sinh^{-1}(x-2)$$

This can be rewritten as (6) by expressing the fractional exponent in radical form and expressing $\sinh^{-1}(x-2)$ in logarithmic form using Theorem 8.8.4 (verify). *Mathematica*

* Results produced by *Mathematica*, *Maple*, and *Derive* may vary depending on the version of the software that is used.

produces the result

$$\int x\sqrt{x^2-4x+5}\,dx = \tfrac{1}{3}(x^2-x-1)\sqrt{x^2-4x+5} - \sinh^{-1}(2-x)$$

which can be rewritten in form (6) by using Theorem 8.8.4 together with the identity $\sinh^{-1}(-x) = -\sinh^{-1} x$ (verify).

Computer algebra systems can sometimes produce inconvenient or unnatural answers to integration problems. For example, when *Mathematica*, *Maple*, and *Derive* are asked to integrate $(x+1)^7$, they produce the following results:

$$\underset{Maple}{\frac{(x+1)^8}{8}}, \quad \underset{Derive}{\frac{(x+1)^8}{8}}, \quad \underset{Mathematica}{x + \tfrac{7}{2}x^2 + 7x^3 + \tfrac{35}{4}x^4 + 7x^5 + \tfrac{7}{2}x^6 + x^7 + \tfrac{1}{8}x^8}$$

The answers produced by *Maple* and *Derive* are in keeping with the hand computation

$$\int (x+1)^7\,dx = \frac{(x+1)^8}{8} + C$$

that uses the substitution $u = x + 1$, $du = dx$, whereas the answer produced by *Mathematica* is based on expanding $(x+1)^7$ and integrating term by term.

FOR THE READER. If you expand the answers produced by *Maple* and *Derive*, you will discover that they contain the term $\frac{1}{8}$ that did not appear in the *Mathematica* result. What is the explanation?

In Example 2(a) of Section 9.3 we showed that

$$\int \sin^4 x \cos^5 x\,dx = \tfrac{1}{5}\sin^5 x - \tfrac{2}{7}\sin^7 x + \tfrac{1}{9}\sin^9 x + C$$

In contrast, *Mathematica* integrates this as

$$\tfrac{1}{80640}(1890\sin x - 420\sin 3x - 252\sin 5x + 45\sin 7x + 35\sin 9x)$$

and *Maple* and *Derive* essentially integrate it as

$$-\tfrac{1}{9}\sin^3 x\cos^6 x - \tfrac{1}{21}\sin x\cos^6 x + \tfrac{1}{105}\cos^4 x\sin x + \tfrac{4}{315}\cos^2 x\sin x + \tfrac{8}{315}\sin x$$

Although the three results look quite different, they can be obtained from one another using appropriate trigonometric identities.

COMPUTER ALGEBRA SYSTEMS CAN FAIL

Every computer algebra system has a library of functions that it can use to construct antiderivatives. Such libraries contain elementary functions, such as polynomials, rational functions, trigonometric functions, as well as various nonelementary functions that arise in engineering, physics, and other applied fields. If the result of an integration cannot be expressed in terms of the functions in the program's library, then the program will give some indication that it cannot evaluate the integral. For example, when asked to evaluate the integral

$$\int (1+\ln x)\sqrt{1+(x\ln x)^2}\,dx \tag{7}$$

Mathematica, *Maple*, and *Derive* all respond by displaying some form of the unevaluated integral as an answer to indicate that they could not perform the integration.

FOR THE READER. Sometimes integrals that cannot be evaluated by a CAS in their given form can be evaluated by first rewriting them in a different form or by making a substitution. Make a u-substitution in (7) that will enable you to evaluate the integral with your CAS.

Sometimes computer algebra systems respond by expressing an integral in terms of another integral. For example, if you try to integrate e^{x^2} using *Mathematica*, *Maple*, or

Derive, you will obtain an expression involving erf (which stands for ***error function***). This function is defined as

$$\operatorname{erf}(x) = \frac{2}{\sqrt{\pi}} \int_0^x e^{-t^2}\, dt$$

so that the three programs essentially do nothing but express the given integral in terms of a closely related integral.

Example 7

A particle moves along an x-axis in such a way that its velocity $v(t)$ at time t is

$$v(t) = 30 \cos^7 t \sin^4 t \quad (t \geq 0)$$

Graph the position versus time curve for the particle, given that the particle is at the point $x = 1$ when $t = 0$.

Solution. Since $dx/dt = v(t)$, the position function $x(t)$ is obtained by integrating the velocity function. We will perform the integration using *Mathematica*, but the procedure would be the same for *Maple* or *Derive*. Integrating $v(t)$ and adding the missing constant of integration yields

$$\begin{aligned} x &= \int 30 \cos^7 t \sin^4 t \, dt \\ &= \tfrac{1}{39424}(16170 \sin t - 2310 \sin 3t - 2541 \sin 5t - 165 \sin 7t + 385 \sin 9t \\ &\qquad + 105 \sin 11t) + C \end{aligned}$$

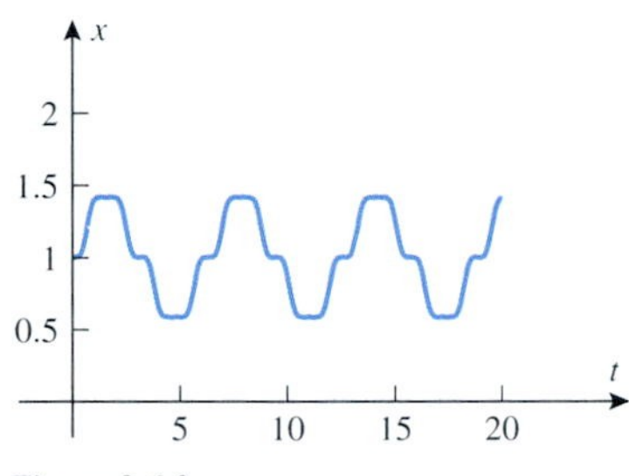

Figure 9.6.2

Since $x = 1$ if $t = 0$, it follows on substituting these values in this equation that $C = 1$. Thus, with this value for C the graph of x versus t is as shown in Figure 9.6.2. ◀

EXERCISE SET 9.6 C CAS

In Exercises 1–24:
(a) Use the Endpaper Integral Table to evaluate the integral.
(b) If you have a CAS, use it to evaluate the integral, and then confirm that the result is equivalent to the one that you found in part (a).

1. $\displaystyle\int \frac{3x}{4x-1}\, dx$ **2.** $\displaystyle\int \frac{x}{(2-3x)^2}\, dx$

3. $\displaystyle\int \frac{1}{x(2x+5)}\, dx$ **4.** $\displaystyle\int \frac{1}{x^2(1-5x)}\, dx$

5. $\displaystyle\int x\sqrt{2x-3}\, dx$ **6.** $\displaystyle\int \frac{x}{\sqrt{2-x}}\, dx$

7. $\displaystyle\int \frac{1}{x\sqrt{4-3x}}\, dx$ **8.** $\displaystyle\int \frac{1}{x\sqrt{3x-4}}\, dx$

9. $\displaystyle\int \frac{1}{5-x^2}\, dx$ **10.** $\displaystyle\int \frac{1}{x^2-9}\, dx$

11. $\displaystyle\int \sqrt{x^2-3}\, dx$ **12.** $\displaystyle\int \frac{\sqrt{x^2+5}}{x^2}\, dx$

13. $\displaystyle\int \frac{x^2}{\sqrt{x^2+4}}\, dx$ **14.** $\displaystyle\int \frac{1}{x^2\sqrt{x^2-2}}\, dx$

15. $\displaystyle\int \sqrt{9-x^2}\, dx$ **16.** $\displaystyle\int \frac{\sqrt{4-x^2}}{x^2}\, dx$

17. $\displaystyle\int \frac{\sqrt{3-x^2}}{x}\, dx$ **18.** $\displaystyle\int \frac{1}{x\sqrt{6x-x^2}}\, dx$

19. $\displaystyle\int \sin 3x \sin 2x\, dx$ **20.** $\displaystyle\int \sin 2x \cos 5x\, dx$

21. $\displaystyle\int x^3 \ln x\, dx$ **22.** $\displaystyle\int \frac{\ln x}{\sqrt{x}}\, dx$

23. $\displaystyle\int e^{-2x} \sin 3x\, dx$ **24.** $\displaystyle\int e^x \cos 2x\, dx$

In Exercises 25–36:
(a) Make the indicated u-substitution, and then use the Endpaper Integral Table to evaluate the integral.
(b) If you have a CAS, use it to evaluate the integral, and then confirm that the result is equivalent to the one that you found in part (a).

25. $\displaystyle\int \frac{e^{4x}}{(4-3e^{2x})^2}\, dx, \quad u = e^{2x}$

26. $\displaystyle\int \frac{\cos 2x}{(\sin 2x)(3-\sin 2x)}\, dx, \quad u = \sin 2x$

27. $\int \frac{1}{\sqrt{x}(9x+4)}\,dx,\ u = 3\sqrt{x}$

28. $\int \frac{\cos 4x}{9+\sin^2 4x}\,dx,\ u = \sin 4x$

29. $\int \frac{1}{\sqrt{9x^2-4}}\,dx,\ u = 3x$

30. $\int x\sqrt{2x^4+3}\,dx,\ u = \sqrt{2}x^2$

31. $\int \frac{x^5}{\sqrt{5-9x^4}}\,dx,\ u = 3x^2$

32. $\int \frac{1}{x^2\sqrt{3-4x^2}}\,dx,\ u = 2x$

33. $\int \frac{\sin^2(\ln x)}{x}\,dx,\ u = \ln x$

34. $\int e^{-2x}\cos^2(e^{-2x})\,dx,\ u = e^{-2x}$

35. $\int xe^{-2x}\,dx,\ u = -2x$

36. $\int \ln(5x-1)\,dx,\ u = 5x-1$

In Exercises 37–48:
(a) Make an appropriate u-substitution, and then use the Endpaper Integral Table to evaluate the integral.
(b) If you have a CAS, use it to evaluate the integral (no substitution), and then confirm that the result is equivalent to that in part (a).

37. $\int \frac{\sin 3x}{(\cos 3x)(\cos 3x+1)^2}\,dx$

38. $\int \frac{\ln x}{x\sqrt{4\ln x-1}}\,dx$

39. $\int \frac{x}{16x^4-1}\,dx$

40. $\int \frac{e^x}{3-4e^{2x}}\,dx$

41. $\int e^x\sqrt{3-4e^{2x}}\,dx$

42. $\int \frac{\sqrt{4-9x^2}}{x^2}\,dx$

43. $\int \sqrt{5x-9x^2}\,dx$

44. $\int \frac{1}{x\sqrt{x-5x^2}}\,dx$

45. $\int x\sin 3x\,dx$

46. $\int \cos\sqrt{x}\,dx$

47. $\int e^{-\sqrt{x}}\,dx$

48. $\int x\ln(2-3x^2)\,dx$

In Exercises 49–52:
(a) Complete the square, make an appropriate u-substitution, and then use the Endpaper Integral Table to evaluate the integral.
(b) If you have a CAS, use it to evaluate the integral (no substitution or square completion), and then confirm that the result is equivalent to that in part (a).

49. $\int \frac{1}{x^2+4x-5}\,dx$

50. $\int \sqrt{3-2x-x^2}\,dx$

51. $\int \frac{x}{\sqrt{5+4x-x^2}}\,dx$

52. $\int \frac{x}{x^2+6x+13}\,dx$

In Exercises 53–66:
(a) Make an appropriate u-substitution of the form $u = x^{1/n}$, $u = (x+a)^{1/n}$, or $u = x^n$, and then use the Endpaper Integral Table to evaluate the integral.
(b) If you have a CAS, use it to evaluate the integral, and then confirm that the result is equivalent to the one that you found in part (a).

53. $\int x\sqrt{x-2}\,dx$

54. $\int \frac{x}{\sqrt{x+1}}\,dx$

55. $\int x^5\sqrt{x^3+1}\,dx$

56. $\int \frac{1}{x\sqrt{x^3-1}}\,dx$

57. $\int \frac{dx}{\sqrt{x}+\sqrt[3]{x}}$

58. $\int \frac{dx}{x-x^{3/5}}$

59. $\int \frac{dx}{x(1-x^{1/4})}$

60. $\int \frac{x^{2/3}}{x+1}\,dx$

61. $\int \frac{dx}{x^{1/2}-x^{1/3}}$

62. $\int \frac{1+\sqrt{x}}{1-\sqrt{x}}\,dx$

63. $\int \frac{x^3}{\sqrt{1+x^2}}\,dx$

64. $\int \frac{x}{(x+3)^{1/5}}\,dx$

65. $\int \sin\sqrt{x}\,dx$

66. $\int e^{\sqrt{x}}\,dx$

In Exercises 67–72:
(a) Make u-substitution (5) to convert the integrand to a rational function of u, and then use the Endpaper Integral Table to evaluate the integral.
(b) If you have a CAS, use it to evaluate the integral (no substitution), and then confirm that the result is equivalent to that in part (a).

67. $\int \frac{dx}{1+\sin x+\cos x}$

68. $\int \frac{dx}{2+\sin x}$

69. $\int \frac{d\theta}{1-\cos\theta}$

70. $\int \frac{dx}{4\sin x-3\cos x}$

71. $\int \frac{\cos x}{2-\cos x}\,dx$

72. $\int \frac{dx}{\sin x+\tan x}$

In Exercises 73 and 74, use any method to solve for x.

73. $\int_2^x \frac{1}{t(4-t)}\,dt = 0.5,\ 2 < x < 4$

74. $\int_1^x \frac{1}{t\sqrt{2t-1}}\,dt = 1,\ x > \frac{1}{2}$

In Exercises 75–78, use any method to find the area of the region enclosed by the curves.

75. $y = \sqrt{25 - x^2}$, $y = 0$, $x = 0$, $x = 4$

76. $y = \sqrt{9x^2 - 4}$, $y = 0$, $x = 2$

77. $y = \dfrac{1}{25 - 16x^2}$, $y = 0$, $x = 0$, $x = 1$

78. $y = \sqrt{x}\ln x$, $y = 0$, $x = 4$

In Exercises 79–82, use any method to find the volume of the solid generated when the region enclosed by the curves is revolved about the y-axis.

79. $y = \cos x$, $y = 0$, $x = 0$, $x = \pi/2$

80. $y = \sqrt{x - 4}$, $y = 0$, $x = 8$

81. $y = e^{-x}$, $y = 0$, $x = 0$, $x = 3$

82. $y = \ln x$, $y = 0$, $x = 5$

In Exercises 83 and 84, use any method to find the arc length of the curve.

83. $y = 2x^2$, $0 \le x \le 2$

84. $y = 3\ln x$, $1 \le x \le 3$

In Exercises 85 and 86, use any method to find the area of the surface generated by revolving the curve about the x-axis.

85. $y = \sin x$, $0 \le x \le \pi$

86. $y = 1/x$, $1 \le x \le 4$

In Exercises 87 and 88, information is given about the motion of a particle moving along a coordinate line.
(a) Use a CAS to find the position function of the particle for $t \ge 0$. You may approximate the constants of integration, where necessary.
(b) Graph the position versus time curve.

C **87.** $v(t) = 20\cos^6 t \sin^3 t$, $s(0) = 2$

C **88.** $a(t) = e^{-t}\sin 2t \sin 4t$, $v(0) = 0$, $s(0) = 10$

89. (a) Use the substitution $u = \tan(x/2)$ to show that

$$\int \sec x\,dx = \ln\left|\frac{1 + \tan(x/2)}{1 - \tan(x/2)}\right| + C$$

and confirm that this is consistent with Formula (22) of Section 9.3.

(b) Use the result in part (a) to show that

$$\int \sec x\,dx = \ln\left|\tan\left(\frac{\pi}{4} + \frac{x}{2}\right)\right| + C$$

90. Use the substitution $u = \tan(x/2)$ to show that

$$\int \csc x\,dx = \frac{1}{2}\ln\left[\frac{1 - \cos x}{1 + \cos x}\right] + C$$

and confirm that this is consistent with the result in Exercise 61(a) of Section 9.3.

91. Find a substitution that can be used to integrate rational functions of $\sinh x$ and $\cosh x$ and use your substitution to evaluate

$$\int \frac{dx}{2\cosh x + \sinh x}$$

without expressing the integrand in terms of e^x and e^{-x}.

9.7 NUMERICAL INTEGRATION; SIMPSON'S RULE

The usual procedure for evaluating a definite integral is to find an antiderivative of the integrand and apply the Fundamental Theorem of Calculus. However, if an antiderivative of the integrand cannot be found, then we must settle for a numerical approximation of the integral. In earlier sections we discussed three procedures for approximating areas using Riemann sums—left endpoint approximation, right endpoint approximation, and midpoint approximation. In this section we will adapt those ideas to approximating general definite integrals, and we will discuss some new approximation methods that often provide more accuracy with less computation.

A REVIEW OF RIEMANN SUM APPROXIMATIONS

Recall from Formula (6) of Section 7.5 that the definite integral of a function f over an interval $[a, b]$ is defined as

$$\int_a^b f(x)\,dx = \lim_{n\to+\infty}\sum_{k=1}^{n} f(x_k^*)\Delta x$$

where the sum that appears on the right side is called a Riemann sum. In this formula, the interval $[a, b]$ is divided into n subintervals of width $\Delta x = (b - a)/n$, and x_k^* denotes an arbitrary point in the kth subinterval. It follows that as n increases the Riemann sum will eventually be a good approximation to the integral, which we denote by writing

$$\int_a^b f(x)\,dx \approx \sum_{k=1}^{n} f(x_k^*)\Delta x$$

or, equivalently,

$$\int_a^b f(x)\,dx \approx \Delta x\left[f(x_1^*) + f(x_2^*) + \cdots + f(x_n^*)\right]$$

In this section we will denote the values of f at the endpoints of the subintervals by

$$y_0 = f(a),\quad y_1 = f(x_1),\quad y_2 = f(x_2),\ldots,\quad y_{n-1} = f(x_{n-1}),\quad y_n = f(b)$$

and we will denote the values of f at the midpoints of the subintervals by

$$y_{m_1}, y_{m_2}, \ldots, y_{m_n}$$

(Figure 9.7.1).

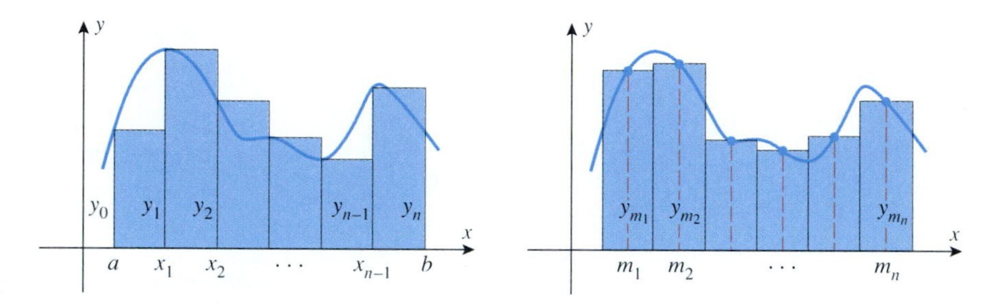

Figure 9.7.1

With this notation the left endpoint, right endpoint, and midpoint approximations discussed in Section 7.5 can be expressed as shown in Table 9.7.1.

Table 9.7.1

LEFT ENDPOINT APPROXIMATION	RIGHT ENDPOINT APPROXIMATION	MIDPOINT APPROXIMATION
$\int_a^b f(x)\,dx \approx \left(\frac{b-a}{n}\right)[y_0 + y_1 + \cdots + y_{n-1}]$	$\int_a^b f(x)\,dx \approx \left(\frac{b-a}{n}\right)[y_1 + y_2 + \cdots + y_n]$	$\int_a^b f(x)\,dx \approx \left(\frac{b-a}{n}\right)[y_{m_1} + y_{m_2} + \cdots + y_{m_n}]$

TRAPEZOIDAL APPROXIMATION

The left-hand and right-hand endpoint approximations are rarely used in applications; however, if we take the average of the left-hand and right-hand endpoint approximations, we obtain a result, called the ***trapezoidal approximation***, which is commonly used:

Trapezoidal Approximation

$$\int_a^b f(x)\,dx \approx \left(\frac{b-a}{2n}\right)[y_0 + 2y_1 + \cdots + 2y_{n-1} + y_n] \tag{1}$$

The name *trapezoidal approximation* can be explained by considering the case in which $f(x) \geq 0$ on $[a, b]$, so that $\int_a^b f(x)\,dx$ represents the area under $f(x)$ over $[a, b]$. Geometrically, the trapezoidal approximation formula results if we approximate this area by the sum of the trapezoidal areas shown in Figure 9.7.2 (Exercise 41).

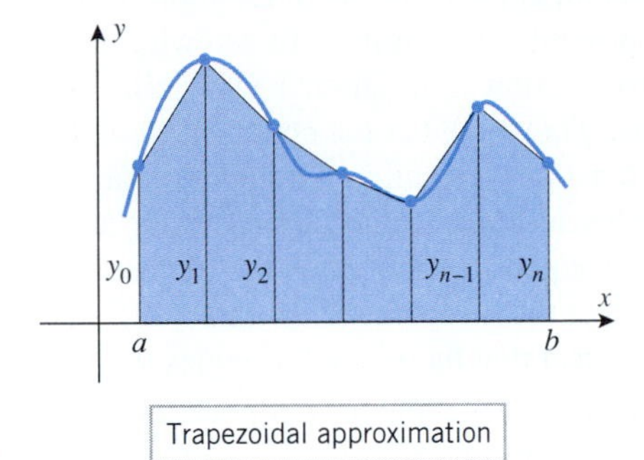

Figure 9.7.2

Example 1

In Table 9.7.2 we have approximated

$$\ln 2 = \int_1^2 \frac{1}{x}\,dx$$

using the midpoint approximation and the trapezoidal approximation. In each case we used $n = 10$ subdivisions of the interval $[1, 2]$, so that

$$\underbrace{\frac{b-a}{n} = \frac{2-1}{10} = 0.1}_{\text{Midpoint}} \quad \text{and} \quad \underbrace{\frac{b-a}{2n} = \frac{2-1}{20} = 0.05}_{\text{Trapezoidal}}$$

◀

REMARK. In Example 1 we rounded the numerical values to nine places to the right of the decimal point; we will follow this procedure throughout this section. If your calculator cannot produce this many places, then you will have to make the appropriate adjustments. What is important here is that you understand the principles involved.

Table 9.7.2

Midpoint Approximation

i	MIDPOINT m_i	$y_{m_i} = f(m_i) = 1/m_i$
1	1.05	0.952380952
2	1.15	0.869565217
3	1.25	0.800000000
4	1.35	0.740740741
5	1.45	0.689655172
6	1.55	0.645161290
7	1.65	0.606060606
8	1.75	0.571428571
9	1.85	0.540540541
10	1.95	0.512820513
		6.928353603

$$\int_1^2 \frac{1}{x}\,dx \approx (0.1)(6.928353603) = 0.692835360$$

Trapezoidal Approximation

i	ENDPOINT x_i	$y_i = f(x_i) = 1/x_i$	MULTIPLIER w_i	$w_i y_i$
0	1.0	1.000000000	1	1.000000000
1	1.1	0.909090909	2	1.818181818
2	1.2	0.833333333	2	1.666666667
3	1.3	0.769230769	2	1.538461538
4	1.4	0.714285714	2	1.428571429
5	1.5	0.666666667	2	1.333333333
6	1.6	0.625000000	2	1.250000000
7	1.7	0.588235294	2	1.176470588
8	1.8	0.555555556	2	1.111111111
9	1.9	0.526315789	2	1.052631579
10	2.0	0.500000000	1	0.500000000
				13.875428063

$$\int_1^2 \frac{1}{x}\,dx \approx (0.05)(13.875428063) = 0.693771403$$

COMPARISON OF THE MIDPOINT AND TRAPEZOIDAL APPROXIMATIONS

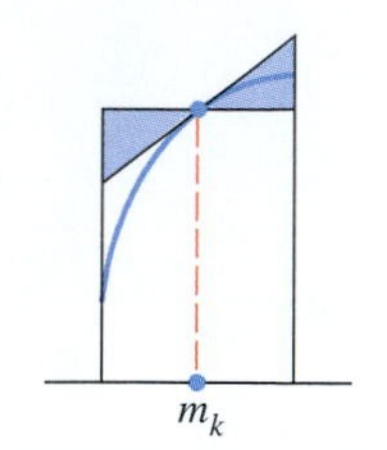

Figure 9.7.3

The value of ln 2 rounded to nine decimal places is

$$\ln 2 = \int_1^2 \frac{1}{x}\, dx \approx 0.693147181 \tag{2}$$

so that the midpoint approximation in Example 1 produced a more accurate result than the trapezoidal approximation (verify). To see why this should be so, we need to look at the midpoint approximation from another viewpoint. [For simplicity in the explanations, we will assume that $f(x) \geq 0$, but the conclusions will be true without this assumption.] For differentiable functions, the midpoint approximation is sometimes called the *tangent line approximation* because over each subinterval the area of the rectangle used in the midpoint approximation is equal to the area of the trapezoid whose upper boundary is the tangent line to $y = f(x)$ at the midpoint of the interval (Figure 9.7.3). The equality of these areas follows from the fact that the shaded triangles in Figure 9.7.3 are congruent.

In this section we will denote the midpoint and trapezoidal approximations of $\int_a^b f(x)\, dx$ with n subintervals by M_n and T_n, respectively, and we will denote the errors in these approximations by

$$|E_M| = \left|\int_a^b f(x)\, dx - M_n\right| \quad \text{and} \quad |E_T| = \left|\int_a^b f(x)\, dx - T_n\right|$$

In Figure 9.7.4*a* we have isolated a subinterval of $[a, b]$ on which the graph of a function f is concave down, and we have shaded the areas that represent the errors in the midpoint and trapezoidal approximations over the subinterval. In Figure 9.7.4*b* we show a succession of four illustrations which make it evident that the error from the midpoint approximation is less than that from the trapezoidal approximation. If the graph of f were concave up, analogous figures would lead to the same conclusion. (This argument, due to Frank Buck, appeared in *The College Mathematics Journal*, Vol. 16, No. 1, 1985.)

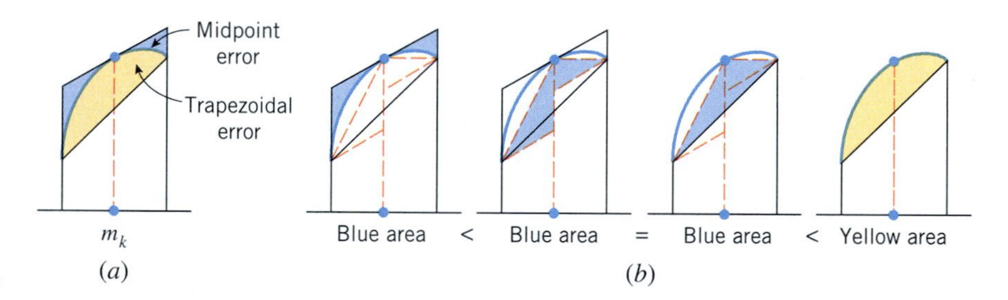

Figure 9.7.4

Figure 9.7.4*a* also suggests that on a subinterval where the graph is concave down, the midpoint approximation is larger than the value of the integral and the trapezoidal approximation is smaller. On an interval where the graph is concave up it is the other way around. In summary, we have the following result, which we state without formal proof:

> **9.7.1 THEOREM.** *Let f be continuous on $[a, b]$, and let $|E_M|$ and $|E_T|$ be the absolute errors that result from the midpoint and trapezoidal approximations of $\int_a^b f(x)\, dx$ using n subintervals.*
>
> (*a*) *If the graph of f is either concave up or concave down on (a, b), then $|E_M| < |E_T|$, that is, the error from the midpoint approximation is less than that from the trapezoidal approximation.*
>
> (*b*) *If the graph of f is concave down on (a, b), then*
>
> $$T_n < \int_a^b f(x)\, dx < M_n$$
>
> (*c*) *If the graph of f is concave up on (a, b), then*
>
> $$M_n < \int_a^b f(x)\, dx < T_n$$

Example 2

We observed earlier that the midpoint approximation of ln 2 obtained in Example 1 was more accurate than the trapezoidal approximation. This is consistent with part (*a*) of Theorem 9.7.1, since $f(x) = 1/x$ is continuous on $[1, 2]$ and concave up on $(1, 2)$. Moreover, a comparison of the two approximations to (2) shows that the midpoint approximation is smaller than ln 2 and the trapezoidal approximation is larger. This is consistent with part (*c*) of Theorem 9.7.1. ◀

REMARK. Do not erroneously conclude that the midpoint approximation is always better than the trapezoidal approximation; for functions with inflection points, the trapezoidal approximation can be more accurate.

SIMPSON'S RULE

Intuition suggests that we might improve on the midpoint and trapezoidal approximations by replacing the linear upper boundaries of the approximating strips in Figure 9.7.2 by curved upper boundaries chosen to fit the shape of the curve $y = f(x)$ more closely. This is the idea behind ***Simpson's** rule**, which uses parabolic curves of the form

$$y = ax^2 + bx + c \tag{3}$$

to approximate sections of the curve $y = f(x)$. [Recall from (7) in Appendix D that (3) is the equation of a parabola with axis of symmetry parallel to the y-axis.]

To simplify the description of Simpson's rule we will assume that $f(x) \geq 0$ on $[a, b]$ so that we can interpret $\int_a^b f(x)\,dx$ as an area. However, the method is valid without this assumption. The heart of Simpson's rule is the formula

$$A = \frac{h}{3}[Y_0 + 4Y_1 + Y_2] \tag{4}$$

which gives the area under the curve

$$y = ax^2 + bx + c$$

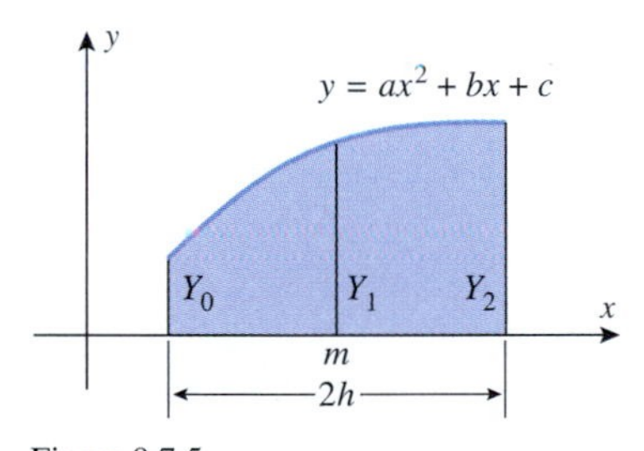

Figure 9.7.5

over an arbitrary interval of width $2h$. In this formula Y_0, Y_1, and Y_2 represent the y-values at the left-hand endpoint, the midpoint m, and the right-hand endpoint of the interval (Figure 9.7.5).

To derive (4), observe that the left-hand endpoint of the interval is $m-h$ and the right-hand endpoint is $m + h$, so the area A under $y = ax^2 + bx + c$ over this interval is

$$A = \int_{m-h}^{m+h} (ax^2 + bx + c)\,dx = \frac{a}{3}x^3 + \frac{b}{2}x^2 + cx\Big]_{m-h}^{m+h}$$

$$= \frac{a}{3}[(m+h)^3 - (m-h)^3] + \frac{b}{2}[(m+h)^2 - (m-h)^2] + c[(m+h) - (m-h)]$$

*THOMAS SIMPSON (1710–1761). English mathematician. Simpson was the son of a weaver. He was trained to follow in his father's footsteps and had little formal education in his early life. His interest in science and mathematics was aroused in 1724, when he witnessed an eclipse of the Sun and received two books from a peddler, one on astrology and the other on arithmetic. Simpson quickly absorbed their contents and soon became a successful local fortune teller. His improved financial situation enabled him to give up weaving and marry his landlady, an older woman. Then in 1733 some mysterious "unfortunate incident" forced him to move. He settled in Derby, where he taught in an evening school and worked at weaving during the day. In 1736 he moved to London and published his first mathematical work in a periodical called the *Ladies' Diary* (of which he later became the editor). In 1737 he published a successful calculus textbook that enabled him to give up weaving completely and concentrate on textbook writing and teaching. His fortunes improved further in 1740 when one Robert Heath accused him of plagiarism. The publicity was marvelous, and Simpson proceeded to dash off a succession of best-selling textbooks: *Algebra* (ten editions plus translations), *Geometry* (twelve editions plus translations), *Trigonometry* (five editions plus translations), and numerous others.

It is interesting to note that Simpson did not discover the rule that bears his name. It was a well-known result by Simpson's time.

or on simplifying,

$$A = \frac{h}{3}[a(6m^2 + 2h^2) + b(6m) + 6c] \tag{5}$$

But the values of $y = ax^2 + bx + c$ at the left-hand endpoint, the midpoint, and the right-hand endpoint are, respectively,

$$Y_0 = a(m-h)^2 + b(m-h) + c$$
$$Y_1 = am^2 + bm + c$$
$$Y_2 = a(m+h)^2 + b(m+h) + c$$

from which it follows that

$$Y_0 + 4Y_1 + Y_2 = a(6m^2 + 2h^2) + b(6m) + 6c \tag{6}$$

Thus, (4) follows from (5) and (6).

Simpson's rule is obtained by dividing the interval $[a, b]$ into an *even* number of subintervals of equal width h and applying Formula (4) to approximate the area under $y = f(x)$ over successive pairs of subintervals. The sum of these approximations then serves as an estimate of $\int_a^b f(x)\,dx$. More precisely, let $[a, b]$ be divided into n subintervals of width $h = (b-a)/n$ (n even) and let

$$y_0, y_1, \ldots, y_n$$

be the values of $y = f(x)$ at the subinterval endpoints

$$a = x_0, x_1, \ldots, x_n = b$$

By (4) the area under $y = f(x)$ over the first two subintervals is approximately

$$\frac{h}{3}[y_0 + 4y_1 + y_2]$$

and the area over the second pair of subintervals is approximately

$$\frac{h}{3}[y_2 + 4y_3 + y_4]$$

and the area over the last pair of subintervals is approximately

$$\frac{h}{3}[y_{n-2} + 4y_{n-1} + y_n]$$

Adding all the approximations, collecting terms, and replacing h by $(b-a)/n$ yields

Simpson's Rule

$$\int_a^b f(x)\,dx \approx \left(\frac{b-a}{3n}\right)[y_0 + 4y_1 + 2y_2 + 4y_3 + 2y_4 + \cdots + 2y_{n-2} + 4y_{n-1} + y_n]$$

We will denote the Simpson's rule approximation with n subintervals by S_n and the error in this approximation by

$$|E_S| = \left|\int_a^b f(x)\,dx - S_n\right|$$

Example 3

In Table 9.7.3 we have approximated

$$\ln 2 = \int_1^2 \frac{1}{x}\,dx$$

by Simpson's rule using $n = 10$ subdivisions so that

$$\frac{b-a}{3n} = \frac{2-1}{3(10)} = \frac{1}{30}$$

Observe, by comparing this result to (2), that Simpson's rule produced a more accurate approximation of ln 2 than either of the methods in Example 1. ◀

Table 9.7.3 Simpson's Rule

i	ENDPOINT x_i	$y_i = f(x_i) = 1/x_i$	MULTIPLIER w_i	$w_i y_i$
0	1.0	1.000000000	1	1.000000000
1	1.1	0.909090909	4	3.636363636
2	1.2	0.833333333	2	1.666666667
3	1.3	0.769230769	4	3.076923077
4	1.4	0.714285714	2	1.428571429
5	1.5	0.666666667	4	2.666666667
6	1.6	0.625000000	2	1.250000000
7	1.7	0.588235294	4	2.352941176
8	1.8	0.555555556	2	1.111111111
9	1.9	0.526315789	4	2.105263158
10	2.0	0.500000000	1	0.500000000
				20.794506921

$$\int_1^2 \frac{1}{x}\,dx \approx \left(\tfrac{1}{30}\right)(20.794506921) = 0.693150231$$

ERROR ESTIMATES

With all the methods studied in this section, there are two sources of error: the *intrinsic* or *truncation error* due to the approximation formula and the *roundoff* error introduced in the calculations. In general, increasing n reduces the truncation error but increases the roundoff error, since more computations are required for larger n. In practical applications, it is important to know how large n must be taken to ensure that a specified degree of accuracy is obtained. The analysis of roundoff error is complicated and will not be considered here. However, the following theorems, which are proved in books on ***numerical analysis***, provide upper bounds on the truncation errors in the midpoint, trapezoidal, and Simpson's rule approximations.

9.7.2 THEOREM (*Midpoint and Trapezoidal Error Estimates*). *If f'' is continuous on $[a, b]$ and if K_2 is the maximum value of $|f''(x)|$ on $[a, b]$, then for n subdivisions of $[a, b]$*

$$\text{(a)}\ |E_M| \le \frac{(b-a)^3 K_2}{24n^2} \qquad \text{(b)}\ |E_T| \le \frac{(b-a)^3 K_2}{12n^2} \tag{7–8}$$

9.7.3 THEOREM (*Simpson Error Estimate*). *If $f^{(4)}$ is continuous on $[a, b]$ and if K_4 is the maximum value of $|f^{(4)}(x)|$ on $[a, b]$, then for n subdivisions of $[a, b]$*

$$|E_S| \le \frac{(b-a)^5 K_4}{180n^4} \tag{9}$$

Example 4

Find an upper bound on the absolute error that results from approximating

$$\ln 2 = \int_1^2 \frac{1}{x}\,dx$$

using $n = 10$ subintervals by: (a) the trapezoidal approximation, (b) the midpoint approximation, and (c) Simpson's rule.

Solution. We will apply Formulas (7), (8), and (9) with

$$f(x) = \frac{1}{x}, \quad a = 1, \quad b = 2, \quad \text{and} \quad n = 10$$

We have

$$f'(x) = -\frac{1}{x^2}, \quad f''(x) = \frac{2}{x^3}, \quad f'''(x) = -\frac{6}{x^4}, \quad f^{(4)}(x) = \frac{24}{x^5}$$

Thus,

$$|f''(x)| = \left|\frac{2}{x^3}\right| = \frac{2}{x^3}, \quad |f^{(4)}(x)| = \left|\frac{24}{x^5}\right| = \frac{24}{x^5} \tag{10–11}$$

where we have dropped the absolute values because $f''(x)$ and $f^{(4)}(x)$ have positive values for $1 \le x \le 2$. Since (10) and (11) are continuous and decreasing on $[1, 2]$, both functions have their maximum values at $x = 1$; for (10) this maximum value is 2 and for (11) it is 24, so we can take $K_2 = 2$ in (7) and (8) and $K_4 = 24$ in (9). This yields

$$|E_T| \le \frac{(b-a)^3 K_2}{12n^2} = \frac{1^3 \cdot 2}{12 \cdot 10^2} \approx 0.001666667$$

$$|E_M| \le \frac{(b-a)^3 K_2}{24n^2} = \frac{1^3 \cdot 2}{24 \cdot 10^2} \approx 0.000833333$$

$$|E_S| \le \frac{(b-a)^5 K_4}{180n^4} = \frac{1^5 \cdot 24}{180 \cdot 10^4} \approx 0.000013333$$ ◀

Table 9.7.4 shows that the estimates in the preceding example are consistent with the computations in Examples 1 and 3. In the table we have obtained approximate values of $|E_T|$, $|E_M|$, and $|E_S|$ by computing the absolute value of the difference between the value of ln 2 (to nine decimal places) and the approximations obtained in Examples 1 and 3. Observe that these values of $|E_T|$, $|E_M|$, and $|E_S|$ satisfy the upper bounds obtained in Example 4; in fact, they are considerably smaller than the upper bounds. It is quite common that the actual errors in the approximations are substantially smaller than the upper bounds.

Table 9.7.4

ln 2 (NINE DECIMAL PLACES)	APPROXIMATION	ABSOLUTE VALUE OF THE DIFFERENCE
0.693147181	$T_{10} = 0.693771403$	$\lvert E_T\rvert \approx 0.000624222$
0.693147181	$M_{10} = 0.692835360$	$\lvert E_M\rvert \approx 0.000311821$
0.693147181	$S_{10} = 0.693150231$	$\lvert E_S\rvert \approx 0.000003050$

Example 5

How many subintervals should be used in approximating

$$\ln 2 = \int_1^2 \frac{1}{x}\,dx$$

by Simpson's rule for five decimal-place accuracy?

Solution. To obtain five decimal-place accuracy, we must choose the number of subintervals so that

$$|E_S| \le 0.000005 = 5 \times 10^{-6}$$

From (9), this can be achieved by taking n in Simpson's rule to satisfy

$$\frac{(b-a)^5 K_4}{180n^4} \le 5 \times 10^{-6}$$

Taking $a = 1, b = 2$, and $K_4 = 24$ (found in Example 4) in this inequality yields

$$\frac{24}{180n^4} \leq 5 \times 10^{-6}$$

which, on taking reciprocals, can be rewritten as

$$n^4 \geq \frac{2 \times 10^6}{75}$$

With the help of a calculating utility, and keeping in mind that n must be an even integer, you can verify that the smallest value of n that satisfies this requirement is $n = 14$. Thus, 14 subintervals will produce five decimal-place accuracy. ◀

REMARK. In cases where it is difficult to find the values of K_2 and K_4 required in Formulas (7), (8), and (9), these constants may be replaced by any larger constants if such constants are easier to find. For example, if $K_2 < K$, then

$$|E_T| \leq \frac{(b-a)^3 K_2}{12n^2} < \frac{(b-a)^3 K}{12n^2} \tag{12}$$

so the right side of (12) is also an upper bound on the value of $|E_T|$ (although it is larger and therefore less desirable than the upper bound using K_2).

Example 6

How many subintervals should be used in approximating

$$\int_0^1 \cos(x^2)\, dx$$

by the midpoint approximation for three decimal-place accuracy?

Solution. To obtain three decimal-place accuracy, we must choose n so that

$$|E_M| \leq 0.0005 = 5 \times 10^{-4} \tag{13}$$

However, from (7) with $f(x) = \cos(x^2)$, $a = 0$, and $b = 1$, an upper bound on the error $|E_M|$ is given by

$$|E_M| \leq \frac{K_2}{24n^2} \tag{14}$$

where K_2 is the maximum value of $|f''(x)|$ on the interval [0, 1]. But,

$$f'(x) = -2x\sin(x^2)$$

$$f''(x) = -4x^2\cos(x^2) - 2\sin(x^2) = -(4x^2\cos(x^2) + 2\sin(x^2))$$

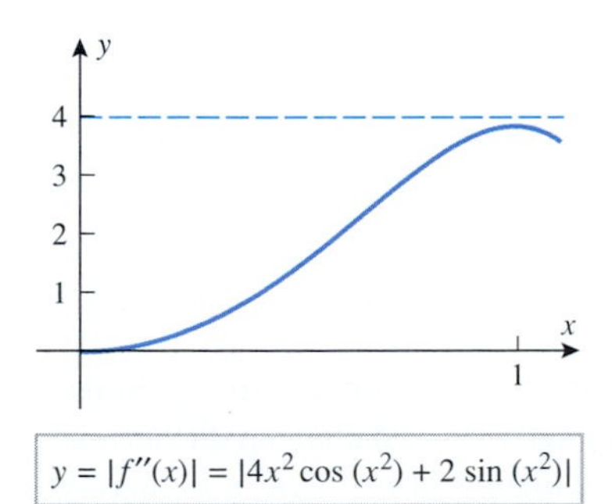

$y = |f''(x)| = |4x^2 \cos(x^2) + 2\sin(x^2)|$

Figure 9.7.6

so that

$$|f''(x)| = |4x^2\cos(x^2) + 2\sin(x^2)| \tag{15}$$

It would be tedious to look for the maximum value of this function on the interval [0, 1] analytically. However, it is evident from the graph of $|f''(x)|$ shown in Figure 9.7.6 that

$$|f''(x)| < 4 \qquad \text{for} \qquad 0 \leq x \leq 1$$

Thus, it follows from (14) that

$$|E_M| \leq \frac{K_2}{24n^2} < \frac{4}{24n^2} = \frac{1}{4n^2}$$

and hence we can satisfy (13) by choosing n so that

$$\frac{1}{6n^2} < 5 \times 10^{-4}$$

which, on taking reciprocals, can be written as

$$n^2 > \frac{10^4}{30}$$

With the help of a calculating utility, you can show that the smallest value of n that satisfies this requirement is $n = 19$. Thus, 19 subintervals will produce three decimal-place accuracy. ◀

A COMPARISON OF THE THREE METHODS

Of the three methods studied in this section, Simpson's rule generally produces more accurate results than the midpoint or trapezoidal approximations for the same amount of work. To make this plausible, let us express (7), (8), and (9) in terms of the subinterval width

$$\Delta x = \frac{b-a}{n}$$

We obtain

$$|E_M| \le \frac{1}{24}K_2(b-a)(\Delta x)^2 \tag{16}$$

$$|E_T| \le \frac{1}{12}K_2(b-a)(\Delta x)^2 \tag{17}$$

$$|E_S| \le \frac{1}{180}K_4(b-a)(\Delta x)^4 \tag{18}$$

(verify). Thus, for Simpson's rule the upper bound on the absolute error is proportional to $(\Delta x)^4$, whereas it is proportional to $(\Delta x)^2$ for the midpoint and trapezoidal approximations. Thus, reducing the interval width by a factor of 10, for example, reduces the error bound by a factor of 100 for the midpoint and trapezoidal approximations but reduces it by a factor of 10,000 for Simpson's rule. This suggests that the accuracy of Simpson's rule improves much more rapidly than that of the other approximations as n increases.

As a final note, observe that if $f(x)$ is a polynomial of degree 3 or less, then we have $f^{(4)}(x) = 0$ for all x, so $K_4 = 0$ in (9) and consequently $|E_S| = 0$. Thus, Simpson's rule gives exact results for polynomials of degree 3 or less. Similarly, the midpoint and trapezoidal approximations give exact results for polynomials of degree 1 or less. (You should also be able to see that this is so geometrically.)

EXERCISE SET 9.7 C CAS

In Exercises 1–6, use $n = 10$ subdivisions to approximate the integral by (a) the midpoint rule, (b) the trapezoidal rule, and (c) Simpson's rule. In each case find the exact value of the integral and approximate the absolute error. Express your answers to at least four decimal places.

1. $\int_0^3 \sqrt{x+1}\,dx$ **2.** $\int_1^4 \frac{1}{\sqrt{x}}\,dx$

3. $\int_0^{\pi} \sin x\,dx$ **4.** $\int_0^1 \cos x\,dx$

5. $\int_1^3 e^{-x}\,dx$ **6.** $\int_{-1}^1 \frac{1}{2x+3}\,dx$

In Exercises 7–12, use inequalities (7), (8), and (9) to find upper bounds on the errors in parts (a), (b), and (c) of the indicated exercise.

7. Exercise 1 **8.** Exercise 2

9. Exercise 3 **10.** Exercise 4

11. Exercise 5 **12.** Exercise 6

In Exercises 13–18, use inequalities (7), (8), and (9) to find a value for n to ensure that the absolute error will be less than the given value if n subdivisions are used to approximate the integral by (a) the midpoint rule, (b) the trapezoidal rule, and (c) Simpson's rule.

13. Exercise 1; 5×10^{-4} **14.** Exercise 2; 5×10^{-4}

15. Exercise 3; 10^{-3} **16.** Exercise 4; 10^{-3}

17. Exercise 5; 10^{-6} **18.** Exercise 6; 10^{-6}

In Exercises 19–24, approximate the integral using Simpson's rule with $n = 10$ subdivisions, and compare the answer to that produced by a calculating utility with a numerical integration capability. Express your answers to at least four decimal places.

19. $\int_0^1 e^{-x^2}\,dx$ **20.** $\int_0^2 \frac{x}{\sqrt{1+x^3}}\,dx$

21. $\int_1^2 \sqrt{1+x^3}\,dx$ **22.** $\int_0^{\pi} \frac{1}{2-\sin x}\,dx$

23. $\int_0^2 \sin(x^2)\,dx$ **24.** $\int_1^3 \sqrt{\ln x}\,dx$

In Exercises 25 and 26, the exact value of the integral is π (verify). Use $n = 10$ subdivisions to approximate the integral by (a) the midpoint rule, (b) the trapezoidal rule, and (c) Simpson's rule. Find an upper bound on the absolute error, and express your answers to at least four decimal places.

25. $\int_0^1 \frac{4}{1+x^2}\,dx$ **26.** $\int_0^2 \sqrt{4-x^2}\,dx$

27. In Example 5 we showed that taking $n = 14$ subdivisions ensures that the approximation of

$$\ln 2 = \int_1^2 \frac{1}{x}\,dx$$

by Simpson's rule is accurate to five decimal places. Confirm this by comparing the approximation of ln 2 produced by Simpson's rule with $n = 14$ to the value produced directly by your calculating utility.

28. In parts (a) and (b), determine whether an approximation of the integral by the trapezoidal rule would be less than or would be greater than the exact value of the integral.

(a) $\int_1^2 e^{-x^2}\,dx$ (b) $\int_0^{0.5} e^{-x^2}\,dx$

In Exercises 29 and 30, find a value for n to ensure that the absolute error in approximating the integral by the midpoint rule will be less than 10^{-4}.

29. $\int_0^2 x \sin x\,dx$ **30.** $\int_0^1 e^{\cos x}\,dx$

In Exercises 31 and 32, show that inequalities (7) and (8) are of no value in finding an upper bound on the absolute error that results from approximating the integral by either the midpoint rule or the trapezoidal rule.

31. $\int_0^1 \sqrt{x}\,dx$ **32.** $\int_0^1 \sin\sqrt{x}\,dx$

In Exercises 33 and 34, use Simpson's rule with $n = 10$ subdivisions to approximate the length of the curve. Express your answers to at least four decimal places.

33. $y = \sin x,\ 0 \le x \le \pi$ **34.** $y = 1/x,\ 1 \le x \le 3$

Numerical integration methods can be used in problems where only measured or experimentally determined values of the integrand are available. In Exercises 35–40, use Simpson's rule to estimate the value of the integral.

35. A graph of the speed v versus time t for a test run of an Infiniti G20 automobile is shown in the accompanying figure. Estimate the speeds at $t = 0, 5, 10, 15$, and 20 s from the graph, convert to ft/s using 1 mi/h = 22/15 ft/s, and use these speeds to approximate the number of feet traveled during the first 20 s. Round your answer to the nearest foot. [*Hint:* Distance traveled $= \int_0^{20} v(t)\,dt$.] [Data from *Road and Track*, October 1990.]

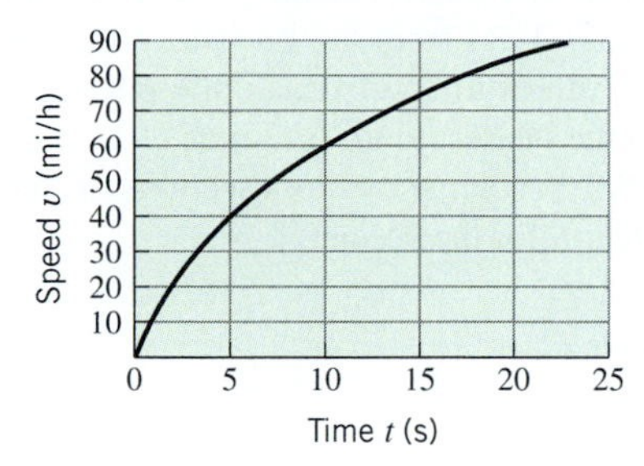

Figure Ex-35

36. A graph of the acceleration a versus time t for an object moving on a straight line is shown in the accompanying figure. Estimate the accelerations at $t = 0, 1, 2, \ldots, 8$ s from the graph and use them to approximate the change in velocity from $t = 0$ to $t = 8$ s. Round your answer to the nearest tenth cm/s. [*Hint:* Change in velocity $= \int_0^8 a(t)\,dt$.]

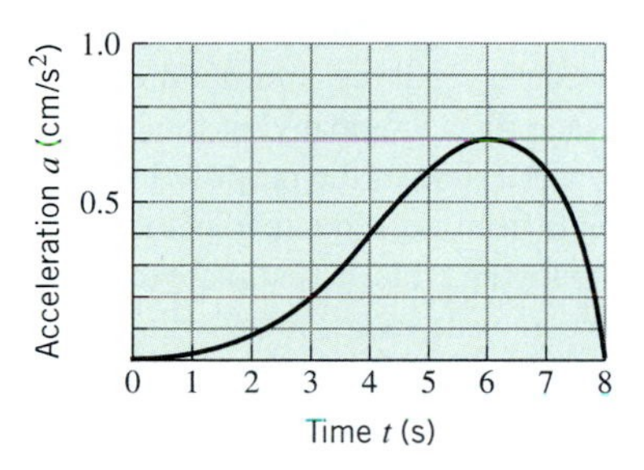

Figure Ex-36

37. The table in the accompanying figure gives the speeds, in miles per second, at various times for a test rocket that was fired upward from the surface of the Earth. Use these values to approximate the number of miles traveled during the first 180 s. Round your answer to the nearest tenth of a mile. [*Hint:* Distance traveled $= \int_0^{180} v(t)\,dt$.]

38. The table in the accompanying figure gives the speeds of a bullet at various distances from the muzzle of a rifle. Use these values to approximate the number of seconds for the bullet to travel 1800 ft. Express your answer to the nearest hundredth of a second. [*Hint:* If v is the speed of the bullet and x is the distance traveled, then $v = dx/dt$ so that $dt/dx = 1/v$ and $t = \int_0^{1800}(1/v)\,dx$.]

TIME t (s)	SPEED v (mi/s)
0	0.00
30	0.03
60	0.08
90	0.16
120	0.27
150	0.42
180	0.65

Figure Ex-37

DISTANCE x (ft)	SPEED v (ft/s)
0	3100
300	2908
600	2725
900	2549
1200	2379
1500	2216
1800	2059

Figure Ex-38

39. Measurements of a pottery shard recovered from an archaeological dig reveal that the shard came from a pot with a flat bottom and circular cross sections (see the accompanying figure). The figure shows interior radius measurements of the shard made every 4 cm from the bottom of the pot to the top. Use those values to approximate the interior volume of the pot to the nearest tenth of a liter (1 L = 1000 cm^3). [*Hint:* Use 8.2.3 (volume by cross sections) to set up an appropriate integral for the volume.]

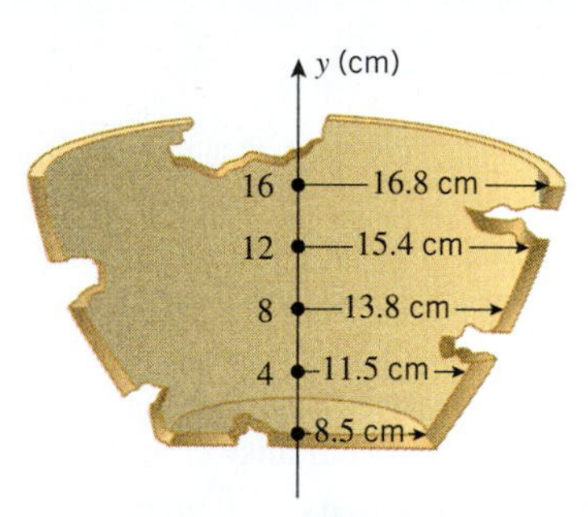

Figure Ex-39

40. Engineers want to construct a straight and level road 600 ft long and 75 ft wide by making a vertical cut through an intervening hill (see the accompanying figure). Heights of the hill above the centerline of the proposed road, as obtained at various points from a contour map of the region, are shown in the accompanying table. To estimate the construction costs, the engineers need to know the volume of earth that must be removed. Approximate this volume, rounded to the nearest cubic foot. [*Hint:* First, set up an integral for the cross-sectional area of the cut along the centerline of the road, then assume that the height of the hill does not vary between the centerline and edges of the road.]

HORIZONTAL DISTANCE x (ft)	HEIGHT h (ft)
0	0
100	7
200	16
300	24
400	25
500	16
600	0

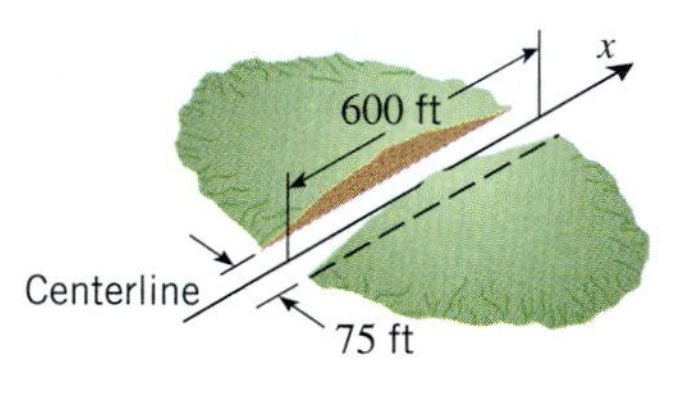

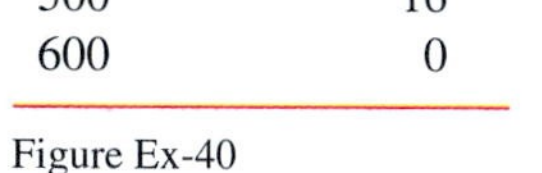

Figure Ex-40

41. Derive the trapezoidal rule by summing the areas of the trapezoids in Figure 9.7.2.

42. Let f be a function that is positive, continuous, decreasing, and concave down on the interval $[a, b]$. Assuming that $[a, b]$ is subdivided into n equal subintervals, arrange the following approximations of $\int_a^b f(x)\,dx$ in order of increasing value: left endpoint, right endpoint, midpoint, and trapezoidal.

C **43.** Let $f(x) = \cos(x^2)$.

(a) Use a CAS to approximate the maximum value of $|f''(x)|$ on the interval $[0, 1]$.

(b) How large must n be in the midpoint approximation of $\int_0^1 f(x)\,dx$ to ensure that the absolute error is less than 5×10^{-4}? Compare your result with that obtained in Example 6.

(c) Evaluate the integral using the midpoint approximation with the value of n obtained in part (b).

C **44.** Let $f(x) = \sqrt{1 + x^3}$.

(a) Use a CAS to approximate the maximum value of $|f''(x)|$ on the interval $[0, 1]$.

(b) How large must n be in the trapezoidal approximation of $\int_0^1 f(x)\,dx$ to ensure that the absolute error is less than 10^{-3}?

(c) Evaluate the integral using the trapezoidal approximation with the value of n obtained in part (b).

C **45.** Let $f(x) = \cos(x^2)$.

(a) Use a CAS to approximate the maximum value of $|f^{(4)}(x)|$ on the interval $[0, 1]$.

(b) How large must the value of n be in the approximation of $\int_0^1 f(x)\,dx$ by Simpson's rule to ensure that the absolute error is less than 10^{-4}?

(c) Evaluate the integral using Simpson's rule with the value of n obtained in part (b).

C **46.** Let $f(x) = \sqrt{1 + x^3}$.

(a) Use a CAS to approximate the maximum value of $|f^{(4)}(x)|$ on the interval $[0, 1]$.

(b) How large must the value of n be in the approximation of $\int_0^1 f(x)\,dx$ by Simpson's rule to ensure that the absolute error is less than 10^{-5}?

(c) Evaluate the integral using Simpson's rule with the value of n obtained in part (b).

9.8 IMPROPER INTEGRALS

Up to now we have focused on definite integrals with continuous integrands and finite intervals of integration. In this section we will extend the concept of a definite integral to include infinite intervals of integration and integrands that become infinite within the interval of integration.

IMPROPER INTEGRALS

It is assumed in the definition of the definite integral

$$\int_a^b f(x)\,dx$$

that $[a, b]$ is a finite interval and that the limit that defines the integral exists; that is, the function f is integrable. We observed in Theorem 7.5.8 that continuous functions are integrable, as are bounded functions with finitely many points of discontinuity. We also observed in that theorem that functions that are not bounded on the interval of integration are not integrable. Thus, for example, a function with a vertical asymptote within the interval of integration would not be integrable.

Our main objective in this section is to extend the concept of a definite integral to allow for infinite intervals of integration and integrands with vertical asymptotes within the interval of integration. We will call the vertical asymptotes ***infinite discontinuities***, and we will call integrals with infinite intervals of integration or infinite discontinuities within the interval of integration ***improper integrals***. Here are some examples:

- Improper integrals with infinite intervals of integration:

$$\int_1^{+\infty} \frac{dx}{x^2}, \quad \int_{-\infty}^{0} e^x\,dx, \quad \int_{-\infty}^{+\infty} \frac{dx}{1+x^2}$$

- Improper integrals with infinite discontinuities in the interval of integration:

$$\int_{-3}^{3} \frac{dx}{x^2}, \quad \int_1^2 \frac{dx}{x-1}, \quad \int_0^{\pi} \tan x\,dx$$

- Improper integrals with infinite discontinuities and infinite intervals of integration:

$$\int_0^{+\infty} \frac{dx}{\sqrt{x}}, \quad \int_{-\infty}^{+\infty} \frac{dx}{x^2-9}, \quad \int_1^{+\infty} \sec x\,dx$$

INTEGRALS OVER INFINITE INTERVALS

To motivate a reasonable definition for improper integrals of the form

$$\int_a^{+\infty} f(x)\,dx$$

let us begin with the case where f is continuous and nonnegative on $[a, +\infty)$, so we can think of the integral as the area under the curve $y = f(x)$ over the interval $[a, +\infty)$ (Figure 9.8.1). At first, you might be inclined to argue that this area is infinite because the region has infinite extent. However, such an argument would be based on vague intuition rather than precise mathematical logic, since the concept of area has only been defined over intervals of *finite extent*. Thus, before we can make any reasonable statements about the area of the region in Figure 9.8.1, we need to begin by defining what we mean by the area of this region. For that purpose, it will help to focus on a specific example.

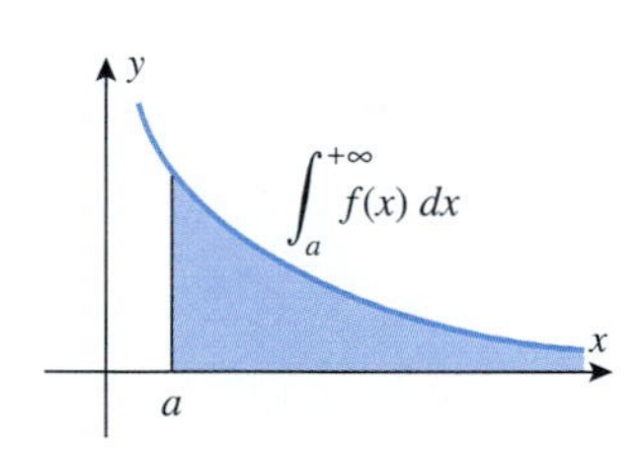

Figure 9.8.1

Suppose we are interested in the area A of the region that lies below the curve $y = 1/x^2$ and above the interval $[1, +\infty)$ on the x-axis. Instead of trying to find the entire area at once, let us begin by calculating the portion of the area that lies above a finite interval $[1, l]$, where $l > 1$ is arbitrary. That area is

$$\int_1^l \frac{dx}{x^2} = -\frac{1}{x}\bigg]_1^l = 1 - \frac{1}{l}$$

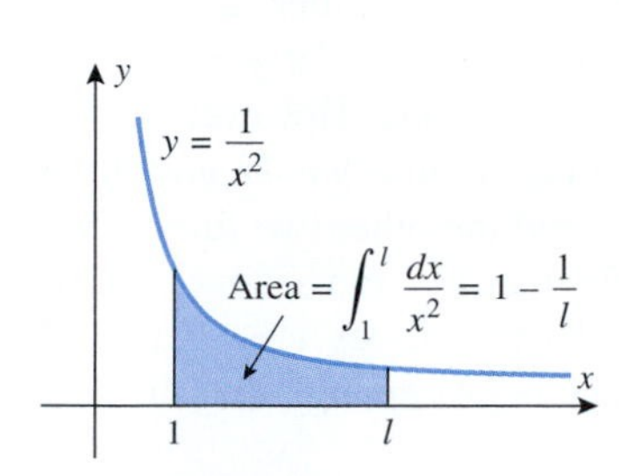

Figure 9.8.2

(Figure 9.8.2). If we now allow l to increase so that $l \to +\infty$, then the portion of the area over the interval $[1, l]$ will begin to fill out the area over the entire interval $[1, +\infty)$ (Figure 9.8.3), and hence we can reasonably define the area A under $y = 1/x^2$ over the interval $[1, +\infty)$ to be

$$A = \int_1^{+\infty} \frac{dx}{x^2} = \lim_{l\to+\infty} \int_1^l \frac{dx}{x^2} = \lim_{l\to+\infty}\left(1 - \frac{1}{l}\right) = 1 \tag{1}$$

Thus, the area has a finite value of 1 and is not infinite as we first conjectured.

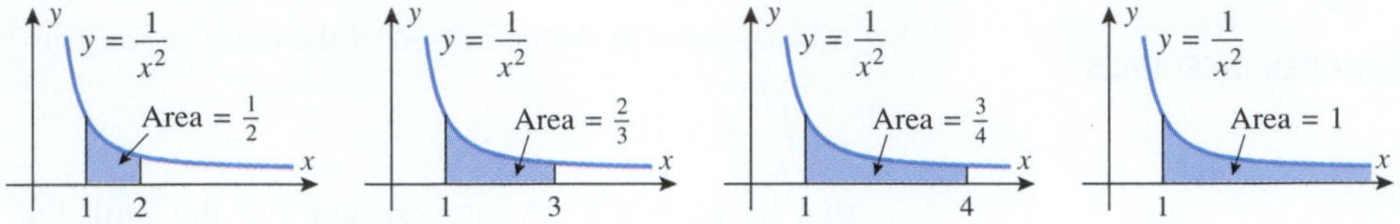

Figure 9.8.3

With the preceding discussion as our guide, we make the following definition (which is applicable to functions with both positive and negative values):

9.8.1 DEFINITION. The ***improper integral of f over the interval*** $[a, +\infty)$ is defined as

$$\int_a^{+\infty} f(x)\,dx = \lim_{l\to+\infty} \int_a^l f(x)\,dx$$

In the case where the limit exists, the improper integral is said to ***converge***, and the limit is defined to be the value of the integral. In the case where the limit does not exist, the improper integral is said to ***diverge***, and it is not assigned a value.

If f is nonnegative on $[a, +\infty)$ and the improper integral converges, then the value of the integral is regarded to be the area under the graph of f over the interval $[a, +\infty)$; and if the integral diverges, then the area under the graph of f over the interval $[a, +\infty)$ is regarded to be infinite.

Example 1

Evaluate

$$\text{(a)} \int_1^{+\infty} \frac{dx}{x^3} \qquad \text{(b)} \int_1^{+\infty} \frac{dx}{x}$$

Solution (a). Following the definition, we replace the infinite upper limit by a finite upper limit l, and then take the limit of the resulting integral. This yields

$$\int_1^{+\infty} \frac{dx}{x^3} = \lim_{l\to+\infty} \int_1^l \frac{dx}{x^3} = \lim_{l\to+\infty} \left[-\frac{1}{2x^2}\right]_1^l = \lim_{l\to+\infty} \left(\frac{1}{2} - \frac{1}{2l^2}\right) = \frac{1}{2}$$

Solution (b).

$$\int_1^{+\infty} \frac{dx}{x} = \lim_{l\to+\infty} \int_1^l \frac{dx}{x} = \lim_{l\to+\infty} \left[\ln x\right]_1^l = \lim_{l\to+\infty} \ln l = +\infty$$

In this case the integral diverges and hence has no value. ◀

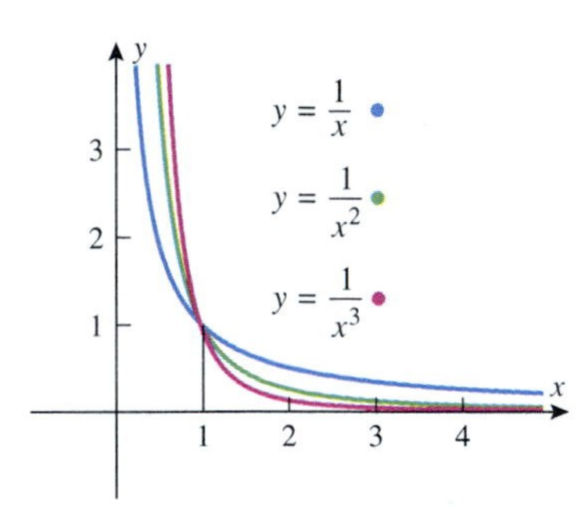

Figure 9.8.4

Because the functions $1/x^3$, $1/x^2$, and $1/x$ are nonnegative over the interval $[1, +\infty)$, it follows from (1) and the last example that over this interval the area under $y = 1/x^3$ is $\frac{1}{2}$, the area under $y = 1/x^2$ is 1, and the area under $y = 1/x$ is infinite. However, on the surface the graphs of the three functions seem very much alike (Figure 9.8.4), and there is nothing to suggest why one of the areas should be infinite and the other two finite. One explanation is that $1/x^3$ and $1/x^2$ approach zero more rapidly than $1/x$ as $x \to +\infty$, so that the area over the interval $[1, l]$ accumulates less rapidly under the curves $y = 1/x^3$ and $y = 1/x^2$ than under $y = 1/x$ as $l \to +\infty$, and the difference is just enough that the first two areas are finite and the third is infinite.

Example 2

For what values of p does the integral $\int_1^{+\infty} \frac{dx}{x^p}$ converge?

Solution. We know from the preceding example that the integral diverges if $p = 1$, so let us assume that $p \neq 1$. In this case we have

$$\int_1^{+\infty} \frac{dx}{x^p} = \lim_{l \to +\infty} \int_1^l x^{-p}\,dx = \lim_{l \to +\infty} \frac{x^{1-p}}{1-p}\bigg]_1^l = \lim_{l \to +\infty} \left[\frac{l^{1-p}}{1-p} - \frac{1}{1-p}\right]$$

If $p > 1$, then the exponent $1 - p$ is negative and $l^{1-p} \to 0$ as $l \to +\infty$; and if $p < 1$, then the exponent $1 - p$ is positive and $l^{1-p} \to +\infty$ as $l \to +\infty$. Thus, the integral converges if $p > 1$ and diverges otherwise. In the convergent case the value of the integral is

$$\int_1^{+\infty} \frac{dx}{x^p} = \left[0 - \frac{1}{1-p}\right] = \frac{1}{p-1} \quad (p > 1)$$ ◀

The following theorem summarizes this result:

9.8.2 THEOREM.

$$\int_1^{+\infty} \frac{dx}{x^p} = \begin{cases} \dfrac{1}{p-1} & \text{if} \quad p > 1 \\ \text{diverges} & \text{if} \quad p \leq 1 \end{cases}$$

Example 3

Evaluate $\displaystyle\int_0^{+\infty} (1 - x)e^{-x}\,dx$.

Solution. Integrating by parts with $u = 1 - x$ and $dv = e^{-x}\,dx$ yields

$$\int (1 - x)e^{-x}\,dx = -e^{-x}(1 - x) - \int e^{-x}\,dx = -e^{-x} + xe^{-x} + e^{-x} + C = xe^{-x} + C$$

Thus,

$$\int_0^{+\infty} (1 - x)e^{-x}\,dx = \lim_{l \to +\infty} \left[xe^{-x}\right]_0^l = \lim_{l \to +\infty} \frac{l}{e^l}$$

The limit is an indeterminate form of type ∞/∞, so we will apply L'Hôpital's rule by differentiating the numerator and denominator with respect to l. This yields

$$\int_0^{+\infty} (1 - x)e^{-x}\,dx = \lim_{l \to +\infty} \frac{1}{e^l} = 0$$

An explanation of why this integral is zero can be obtained by interpreting the integral as the net signed area between the graph of $y = (1 - x)e^{-x}$ and the interval $[0, +\infty)$ (Figure 9.8.5). ◀

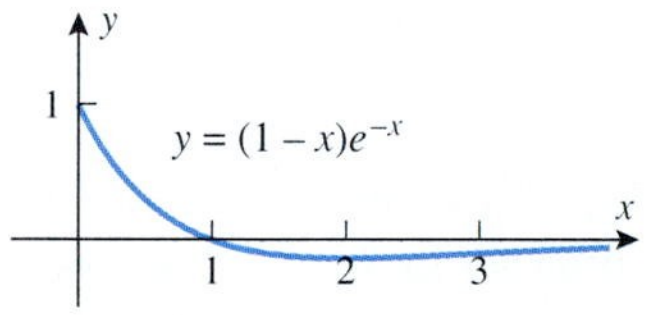

The net signed area between the graph and the interval $[0, +\infty)$ is zero.

Figure 9.8.5

We also make the following definition:

9.8.3 DEFINITION. The ***improper integral of f over the interval*** $(-\infty, b]$ is defined as

$$\int_{-\infty}^{b} f(x)\,dx = \lim_{l \to -\infty} \int_l^b f(x)\,dx \tag{2}$$

The integral is said to ***converge*** if the limit exists and ***diverge*** if it does not. The ***improper integral of f over the interval*** $(-\infty, +\infty)$ is defined as

$$\int_{-\infty}^{+\infty} f(x)\,dx = \int_{-\infty}^{c} f(x)\,dx + \int_c^{+\infty} f(x)\,dx \tag{3}$$

where c is any real number. The improper integral is said to ***converge*** if both terms converge and ***diverge*** if either term diverges.

REMARK. In this definition, if f is nonnegative on the interval of integration, then the improper integral is regarded to be the area under the graph of f over that interval; the area has a finite value if the integral converges and is infinite if it diverges. We also note that in (3) it is usual to choose $c = 0$, but the choice does not matter; it can be proved that neither the convergence nor the value of the integral depends on the choice of c.

Example 4

Evaluate $\displaystyle\int_{-\infty}^{+\infty} \frac{dx}{1+x^2}$.

Solution. We will evaluate the integral by choosing $c = 0$ in (3). With this value for c we obtain

$$\int_{0}^{+\infty} \frac{dx}{1+x^2} = \lim_{l\to+\infty} \int_{0}^{l} \frac{dx}{1+x^2} = \lim_{l\to+\infty} \left[\tan^{-1} x\right]_{0}^{l} = \lim_{l\to+\infty} (\tan^{-1} l) = \frac{\pi}{2}$$

$$\int_{-\infty}^{0} \frac{dx}{1+x^2} = \lim_{l\to-\infty} \int_{l}^{0} \frac{dx}{1+x^2} = \lim_{l\to-\infty} \left[\tan^{-1} x\right]_{l}^{0} = \lim_{l\to-\infty} (-\tan^{-1} l) = \frac{\pi}{2}$$

Thus, the integral converges and its value is

$$\int_{-\infty}^{+\infty} \frac{dx}{1+x^2} = \int_{-\infty}^{0} \frac{dx}{1+x^2} + \int_{0}^{+\infty} \frac{dx}{1+x^2} = \frac{\pi}{2} + \frac{\pi}{2} = \pi$$

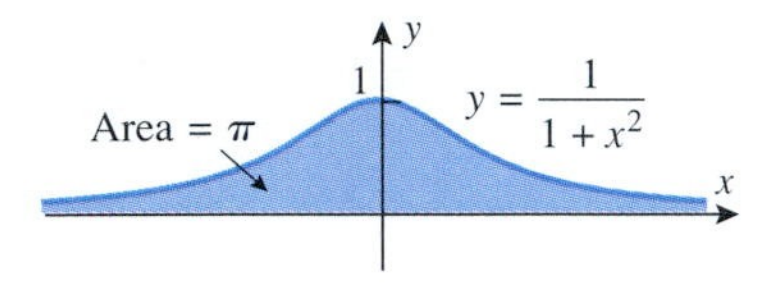

Figure 9.8.6

Since the integrand is nonnegative on the interval $(-\infty, +\infty)$, the integral represents the area of the region shown in Figure 9.8.6. ◀

INTEGRALS WHOSE INTEGRANDS HAVE INFINITE DISCONTINUITIES

Next we will consider improper integrals whose integrands have infinite discontinuities. We will start with the case where the interval of integration is a finite interval $[a, b]$ and the infinite discontinuity occurs at the right-hand endpoint.

To motivate an appropriate definition for such an integral let us consider the case where f is nonnegative on $[a, b]$, so we can interpret the improper integral $\int_a^b f(x)\,dx$ as the area of the region in Figure 9.8.7*a*. The problem of finding the area of this region is complicated by the fact that it extends indefinitely in the positive y-direction. However, instead of trying to find the entire area at once, we can proceed indirectly by calculating the portion of the area over the interval $[a, l]$ and then letting l approach b to fill out the area of the entire region (Figure 9.8.7*b*). Motivated by this idea, we make the following definition:

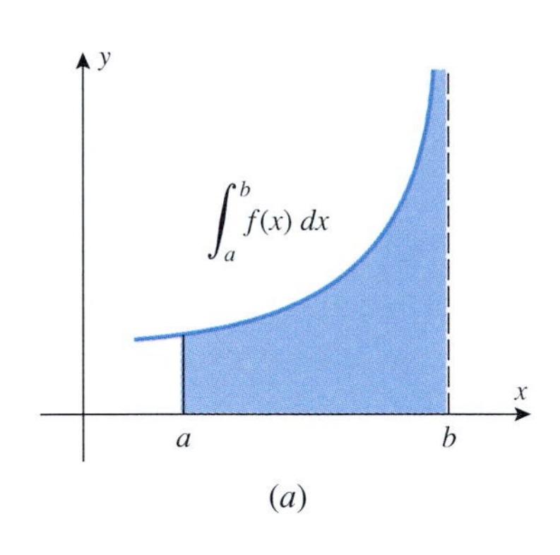

(*a*)

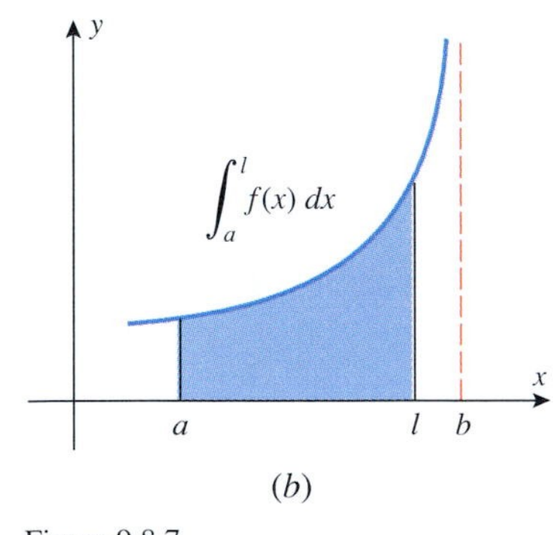

(*b*)

Figure 9.8.7

9.8.4 DEFINITION. If f is continuous on the interval $[a, b]$, except for an infinite discontinuity at b, then the ***improper integral of f over the interval [a, b]*** is defined as

$$\int_{a}^{b} f(x)\,dx = \lim_{l\to b^-} \int_{a}^{l} f(x)\,dx \tag{4}$$

In the case where the limit exists, the improper integral is said to ***converge***, and the limit is defined to be the value of the integral. In the case where the limit does not exist, the improper integral is said to ***diverge***, and it is not assigned a value.

Example 5

Evaluate $\displaystyle\int_{0}^{1} \frac{dx}{\sqrt{1-x}}$.

Solution. The integral is improper because the integrand approaches $+\infty$ as x approaches the upper limit 1 from the left. From (4),

$$\int_0^1 \frac{dx}{\sqrt{1-x}} = \lim_{l \to 1^-} \int_0^l \frac{dx}{\sqrt{1-x}} = \lim_{l \to 1^-} \left[-2\sqrt{1-x} \right]_0^l$$

$$= \lim_{l \to 1^-} [-2\sqrt{1-l} + 2] = 2 \qquad \blacktriangleleft$$

Improper integrals with an infinite discontinuity at the left-hand endpoint or inside the interval of integration are defined as follows.

9.8.5 DEFINITION. If f is continuous on the interval $[a, b]$, except for an infinite discontinuity at a, then the ***improper integral of f over the interval [a, b]*** is defined as

$$\int_a^b f(x)\,dx = \lim_{l \to a^+} \int_l^b f(x)\,dx \tag{5}$$

The integral is said to ***converge*** if the limit exists and ***diverge*** if it does not. If f is continuous on the interval $[a, b]$, except for an infinite discontinuity at a point c in (a, b), then the ***improper integral of f over the interval [a, b]*** is defined as

$$\int_a^b f(x)\,dx = \int_a^c f(x)\,dx + \int_c^b f(x)\,dx \tag{6}$$

The improper integral is said to ***converge*** if both terms converge and ***diverge*** if either term diverges (Figure 9.8.8).

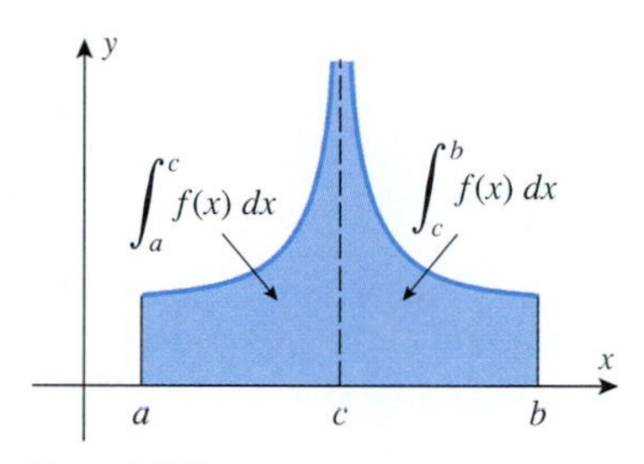

Figure 9.8.8

Example 6

Evaluate

(a) $\displaystyle\int_1^2 \frac{dx}{1-x}$ (b) $\displaystyle\int_1^4 \frac{dx}{(x-2)^{2/3}}$ (c) $\displaystyle\int_0^{+\infty} \frac{dx}{\sqrt{x}(x+1)}$

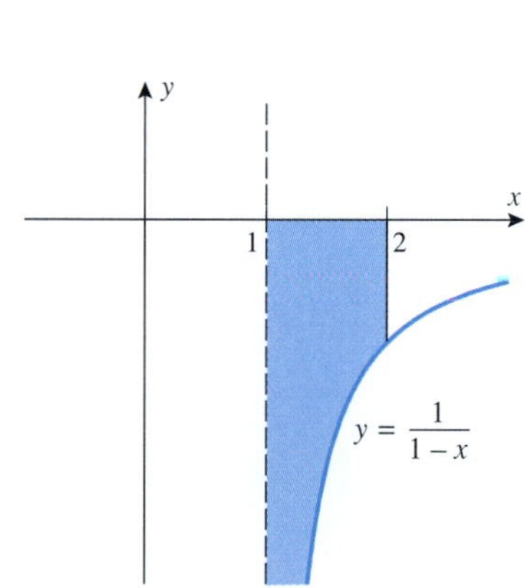

Figure 9.8.9

Solution (a). The integral is improper because the integrand approaches $-\infty$ as x approaches the lower limit 1 from the right (Figure 9.8.9). From Definition 9.8.5 we obtain

$$\int_1^2 \frac{dx}{1-x} = \lim_{l \to 1^+} \int_l^2 \frac{dx}{1-x} = \lim_{l \to 1^+} \Big[-\ln|1-x| \Big]_l^2$$

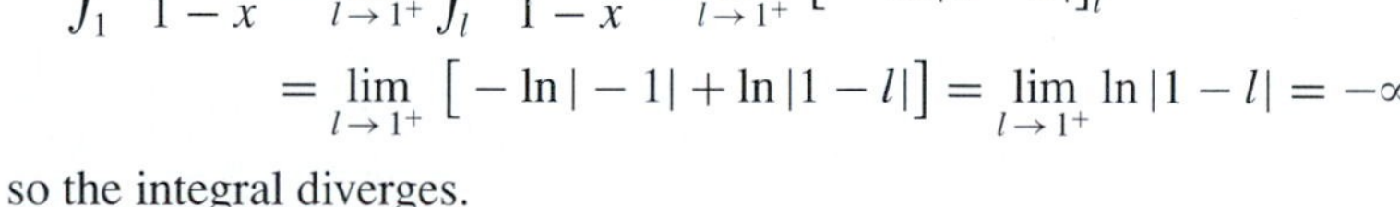

$$= \lim_{l \to 1^+} \Big[-\ln|-1| + \ln|1-l| \Big] = \lim_{l \to 1^+} \ln|1-l| = -\infty$$

so the integral diverges.

Solution (b). The integral is improper because the integrand approaches $+\infty$ at the point $x = 2$, which is inside the interval of integration. From Definition 9.8.5 we obtain

$$\int_1^4 \frac{dx}{(x-2)^{2/3}} = \int_1^2 \frac{dx}{(x-2)^{2/3}} + \int_2^4 \frac{dx}{(x-2)^{2/3}} \tag{7}$$

But

$$\int_1^2 \frac{dx}{(x-2)^{2/3}} = \lim_{l \to 2^-} \int_1^l \frac{dx}{(x-2)^{2/3}} = \lim_{l \to 2^-} [3(l-2)^{1/3} - 3(1-2)^{1/3}] = 3$$

$$\int_2^4 \frac{dx}{(x-2)^{2/3}} = \lim_{l \to 2^+} \int_l^4 \frac{dx}{(x-2)^{2/3}} = \lim_{l \to 2^+} [3(4-2)^{1/3} - 3(l-2)^{1/3}] = 3\sqrt[3]{2}$$

Thus, from (7)

$$\int_1^4 \frac{dx}{(x-2)^{2/3}} = 3 + 3\sqrt[3]{2}$$

Solution (c). This integral is improper for two reasons—the interval of integration is infinite, and there is an infinite discontinuity at $x = 0$. To evaluate this integral we will split the interval of integration at a convenient point, say $x = 1$, and write

$$\int_0^{+\infty} \frac{dx}{\sqrt{x}(x+1)} = \int_0^1 \frac{dx}{\sqrt{x}(x+1)} + \int_1^{+\infty} \frac{dx}{\sqrt{x}(x+1)}$$

The integrand in these two improper integrals does not match any of the forms in the Endpaper Integral Table, but the radical suggests the substitution $x = u^2$, $dx = 2u\,du$, from which we obtain

$$\int \frac{dx}{\sqrt{x}(x+1)} = \int \frac{2u\,du}{u(u^2+1)} = 2\int \frac{du}{u^2+1}$$
$$= 2\tan^{-1} u + C = 2\tan^{-1}\sqrt{x} + C$$

Thus,

$$\int_0^{+\infty} \frac{dx}{\sqrt{x}(x+1)} = 2\lim_{l\to 0^+}\left[\tan^{-1}\sqrt{x}\right]_l^1 + 2\lim_{l\to+\infty}\left[\tan^{-1}\sqrt{x}\right]_1^l$$
$$= 2\left[\frac{\pi}{4} - 0\right] + 2\left[\frac{\pi}{2} - \frac{\pi}{4}\right] = \pi$$ ◀

WARNING. It is sometimes tempting to apply the Fundamental Theorem of Calculus directly to an improper integral without taking the appropriate limits. To illustrate what can go wrong with this procedure, suppose we ignore the fact that the integral

$$\int_0^2 \frac{dx}{(x-1)^2} \tag{8}$$

is improper and write

$$\int_0^2 \frac{dx}{(x-1)^2} = -\frac{1}{x-1}\bigg]_0^2 = -1 - (1) = -2$$

This result is clearly nonsense because the integrand is never negative and consequently the integral cannot be negative! To evaluate (8) correctly we should write

$$\int_0^2 \frac{dx}{(x-1)^2} = \int_0^1 \frac{dx}{(x-1)^2} + \int_1^2 \frac{dx}{(x-1)^2}$$

But

$$\int_0^1 \frac{dx}{(x-1)^2} = \lim_{l\to 1^-}\int_0^l \frac{dx}{(x-1)^2} = \lim_{l\to 1^-}\left[-\frac{1}{l-1} - 1\right] = +\infty$$

so that (8) diverges.

THE APPLICATION OF IMPROPER INTEGRALS TO ARC LENGTH AND SURFACE AREA

In Definitions 8.4.2 and 8.5.2 for arc length and surface area we required the function f to be smooth (continuous first derivative) to ensure the integrability in the resulting formula. However, smoothness is overly restrictive since some of the most basic formulas in geometry involve functions that are not smooth but lead to convergent improper integrals. Accordingly, let us agree to extend the definitions of arc length and surface area to allow functions that are not smooth, but for which the resulting integral in the formula converges.

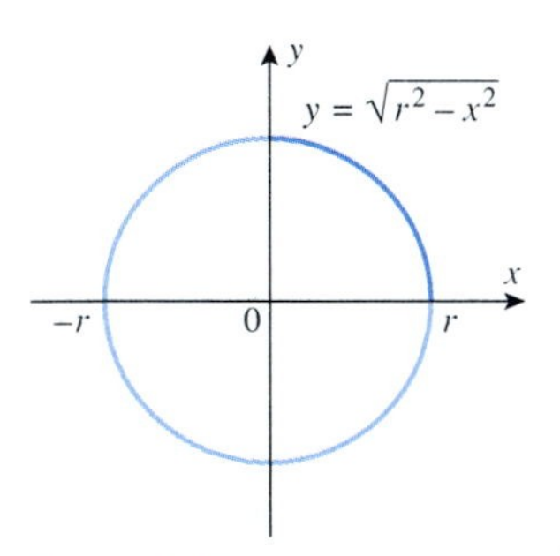

Figure 9.8.10

Example 7

Derive the formula for the circumference of a circle of radius r.

Solution. For convenience, let us assume that the circle is centered at the origin, in which case its equation is $x^2 + y^2 = r^2$. We will find the arc length of the portion of the circle that lies in the first quadrant and then multiply by 4 to obtain the total circumference (Figure 9.8.10).

Since the equation of the upper semicircle is $y = \sqrt{r^2 - x^2}$, it follows from Formula (4) of Section 8.4 that the circumference C is

$$C = 4\int_0^r \sqrt{1 + (dy/dx)^2}\, dx = 4\int_0^r \sqrt{1 + \left(-\frac{x}{\sqrt{r^2 - x^2}}\right)^2}\, dx$$

$$= 4r \int_0^r \frac{dx}{\sqrt{r^2 - x^2}}$$

This integral is improper because of the infinite discontinuity at $x = r$, and hence we evaluate it by writing

$$C = 4r \lim_{l \to r^-} \int_0^l \frac{dx}{\sqrt{r^2 - x^2}}$$

$$= 4r \lim_{l \to r^-} \left[\sin^{-1}\left(\frac{x}{r}\right)\right]_0^l$$

Formula (77) in the Endpaper Integral Table

$$= 4r \lim_{l \to r^-} \left[\sin^{-1}\left(\frac{l}{r}\right) - \sin^{-1} 0\right]$$

$$= 4r[\sin^{-1} 1 - \sin^{-1} 0] = 4r\left(\frac{\pi}{2} - 0\right) = 2\pi r$$ ◀

EXERCISE SET 9.8 Graphing Calculator CAS

1. In each part, determine whether the integral is improper, and if so, explain why.

(a) $\int_1^5 \frac{dx}{x-3}$ (b) $\int_1^5 \frac{dx}{x+3}$ (c) $\int_0^1 \ln x\, dx$

(d) $\int_1^{+\infty} e^{-x}\, dx$ (e) $\int_{-\infty}^{+\infty} \frac{dx}{\sqrt[3]{x-1}}$ (f) $\int_0^{\pi/4} \tan x\, dx$

2. In each part, determine all values of p for which the integral is improper.

(a) $\int_0^1 \frac{dx}{x^p}$ (b) $\int_1^2 \frac{dx}{x-p}$ (c) $\int_0^1 e^{-px}\, dx$

In Exercises 3–30, evaluate the integrals that converge.

3. $\int_0^{+\infty} e^{-x}\, dx$

4. $\int_{-1}^{+\infty} \frac{x}{1+x^2}\, dx$

5. $\int_4^{+\infty} \frac{2}{x^2-1}\, dx$

6. $\int_0^{+\infty} xe^{-x^2}\, dx$

7. $\int_e^{+\infty} \frac{1}{x \ln^3 x}\, dx$

8. $\int_2^{+\infty} \frac{1}{x\sqrt{\ln x}}\, dx$

9. $\int_{-\infty}^0 \frac{dx}{(2x-1)^3}$

10. $\int_{-\infty}^2 \frac{dx}{x^2+4}$

11. $\int_{-\infty}^0 e^{3x}\, dx$

12. $\int_{-\infty}^0 \frac{e^x\, dx}{3 - 2e^x}$

13. $\int_{-\infty}^{+\infty} x^3\, dx$

14. $\int_{-\infty}^{+\infty} \frac{x}{\sqrt{x^2+2}}\, dx$

15. $\int_{-\infty}^{+\infty} \frac{x}{(x^2+3)^2}\, dx$

16. $\int_{-\infty}^{+\infty} \frac{e^{-t}}{1+e^{-2t}}\, dt$

17. $\int_3^4 \frac{dx}{(x-3)^2}$

18. $\int_0^8 \frac{dx}{\sqrt[3]{x}}$

19. $\int_0^{\pi/2} \tan x\, dx$

20. $\int_0^9 \frac{dx}{\sqrt{9-x}}$

21. $\int_0^1 \frac{dx}{\sqrt{1-x^2}}$

22. $\int_{-3}^1 \frac{x\, dx}{\sqrt{9-x^2}}$

23. $\int_0^{\pi/6} \frac{\cos x}{\sqrt{1 - 2\sin x}}\, dx$

24. $\int_0^{\pi/4} \frac{\sec^2 x}{1 - \tan x}\, dx$

25. $\int_0^3 \frac{dx}{x-2}$

26. $\int_{-2}^2 \frac{dx}{x^2}$

27. $\int_{-1}^8 x^{-1/3}\, dx$

28. $\int_0^4 \frac{dx}{(x-2)^{2/3}}$

29. $\int_0^{+\infty} \frac{1}{x^2}\, dx$

30. $\int_1^{+\infty} \frac{dx}{x\sqrt{x^2-1}}$

In Exercises 31–34, make the u-substitution and evaluate the resulting definite integral.

31. $\int_0^{+\infty} \frac{e^{-\sqrt{x}}}{\sqrt{x}}\, dx;\ u = \sqrt{x}$ [*Note:* $u \to +\infty$ as $x \to +\infty$.]

32. $\int_0^{+\infty} \frac{dx}{\sqrt{x}(x+4)};\ u = \sqrt{x}$

33. $\int_0^{+\infty} \frac{e^{-x}}{\sqrt{1-e^{-x}}}\, dx;\ u = 1 - e^{-x}$

[*Note:* $u \to 1$ as $x \to +\infty$.]

34. $\int_0^{+\infty} \frac{e^{-x}}{\sqrt{1-e^{-2x}}}\, dx;\ u = e^{-x}$

C **35.** Read your CAS documentation to determine how to evaluate definite integrals with infinite limits of integration, and then for each of the integrals you evaluated in Exercises 1–34, check your answer with your CAS.

In Exercises 36 and 37, express the improper integral as a limit, and then evaluate that limit with a CAS. Confirm the answer by evaluating the integral directly with the CAS.

C **36.** $\int_0^{+\infty} xe^{-3x}\,dx$ C **37.** $\int_0^{+\infty} e^{-x}\cos x\,dx$

C **38.** In each part, confirm the result with a CAS.

(a) $\int_0^{+\infty} \frac{\sin x}{\sqrt{x}}\,dx = \sqrt{\frac{\pi}{2}}$ (b) $\int_{-\infty}^{+\infty} e^{-x^2}\,dx = \sqrt{\pi}$

(c) $\int_0^1 \frac{\ln x}{1+x}\,dx = -\frac{\pi^2}{12}$

C **39.** In each part, try to evaluate the integral exactly with a CAS. If your result is not a simple numerical answer, then use the CAS to find a numerical approximation of the integral.

(a) $\int_{-\infty}^{+\infty} \frac{1}{x^8+x+1}\,dx$ (b) $\int_0^{+\infty} \frac{1}{\sqrt{1+x^3}}\,dx$

(c) $\int_1^{+\infty} \frac{\ln x}{e^x}\,dx$ (d) $\int_1^{+\infty} \frac{\sin x}{x^2}\,dx$

40. Find the length of the curve $y = \sqrt{9-x^2}$ over the interval $[0, 3]$.

In Exercises 41 and 42, use L'Hôpital's rule to help evaluate the improper integral.

41. $\int_0^1 \ln x\,dx$ **42.** $\int_1^{+\infty} \frac{\ln x}{x^2}\,dx$

43. Find the area of the region between the x-axis and the curve $y = e^{-3x}$ for $x \geq 0$.

44. Find the area of the region between the x-axis and the curve $y = 8/(x^2-4)$ for $x \geq 3$.

45. Suppose that the region between the x-axis and the curve $y = e^{-x}$ for $x \geq 0$ is revolved about the x-axis.

(a) Find the volume of the solid that is generated.

(b) Find the surface area of the solid.

46. Suppose that f and g are continuous functions and that

$$0 \leq f(x) \leq g(x)$$

if $x \geq a$. Give a reasonable informal argument using areas to explain why the following results are true.

(a) If $\int_a^{+\infty} f(x)\,dx$ diverges, then $\int_a^{+\infty} g(x)\,dx$ diverges.

(b) If $\int_a^{+\infty} g(x)\,dx$ converges, then $\int_a^{+\infty} f(x)\,dx$ converges and $\int_a^{+\infty} f(x)\,dx \leq \int_a^{+\infty} g(x)\,dx$.

[*Note:* The results in this exercise are sometimes called ***comparison tests*** for improper integrals.]

In Exercises 47–51, use the results in Exercise 46.

47. (a) Confirm graphically and algebraically that $e^{-x^2} \leq e^{-x}$ if $x \geq 1$.

(b) Evaluate the integral

$$\int_1^{+\infty} e^{-x}\,dx$$

(c) What does the result obtained in part (b) tell you about the integral

$$\int_1^{+\infty} e^{-x^2}\,dx?$$

48. (a) Confirm graphically and algebraically that

$$\frac{1}{2x+1} \leq \frac{e^x}{2x+1} \quad (x \geq 0)$$

(b) Evaluate the integral

$$\int_0^{+\infty} \frac{dx}{2x+1}$$

(c) What does the result obtained in part (b) tell you about the integral

$$\int_0^{+\infty} \frac{e^x}{2x+1}\,dx?$$

49. Let R be the region to the right of $x = 1$ that is bounded by the x-axis and the curve $y = 1/x$. When this region is revolved about the x-axis it generates a solid whose surface is known as ***Gabriel's Horn*** (for reasons that should be clear from the accompanying figure). Show that the solid has a finite volume but its surface has an infinite area. [*Note:* It has been suggested that if one could saturate the interior of the solid with paint and allow it to seep through to the surface, then one could paint an infinite surface with a finite amount of paint! What do you think?]

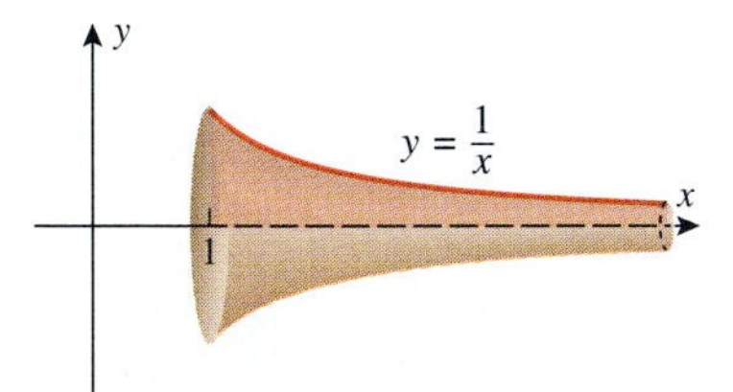

Figure Ex-49

50. In each part, use Exercise 46 to determine whether the integral converges or diverges. If it converges, then use part (b) of that exercise to find an upper bound on the value of the integral.

(a) $\int_2^{+\infty} \frac{\sqrt{x^3+1}}{x}\,dx$ (b) $\int_2^{+\infty} \frac{x}{x^5+1}\,dx$

(c) $\int_0^{+\infty} \frac{e^x}{2x+1}\,dx$

51. Show that

$$\lim_{x\to+\infty} \frac{\int_0^{2x} \sqrt{1+t^3}\,dt}{x^{5/2}}$$

is an indeterminate form of type ∞/∞, and then use L'Hôpital's rule to find the limit.

52. (a) Give a reasonable informal argument, based on areas, that explains why the integrals

$$\int_0^{+\infty} \sin x\,dx \quad \text{and} \quad \int_0^{+\infty} \cos x\,dx$$

diverge.

(b) Show that $\int_0^{+\infty} \frac{\cos\sqrt{x}}{\sqrt{x}}\,dx$ diverges.

53. In electromagnetic theory, the magnetic potential at a point on the axis of a circular coil is given by

$$u = \frac{2\pi NIr}{k} \int_a^{+\infty} \frac{dx}{(r^2+x^2)^{3/2}}$$

where N, I, r, k, and a are constants. Find u.

C **54.** The *average speed*, $\bar{v}$, of the molecules of an ideal gas is given by

$$\bar{v} = \frac{4}{\sqrt{\pi}}\left(\frac{M}{2RT}\right)^{3/2} \int_0^{+\infty} v^3 e^{-Mv^2/(2RT)}\,dv$$

and the *root-mean-square speed*, v_{rms}, by

$$v_{\text{rms}}^2 = \frac{4}{\sqrt{\pi}}\left(\frac{M}{2RT}\right)^{3/2} \int_0^{+\infty} v^4 e^{-Mv^2/(2RT)}\,dv$$

where v is the molecular speed, T is the gas temperature, M is the molecular weight of the gas, and R is the gas constant.

(a) Use a CAS to show that

$$\int_0^{+\infty} x^3 e^{-a^2x^2}\,dx = \frac{1}{2a^4}, \quad a > 0$$

and use this result to show that $\bar{v} = \sqrt{8RT/\pi M}$.

(b) Use a CAS to show that

$$\int_0^{+\infty} x^4 e^{-a^2x^2}\,dx = \frac{3\sqrt{\pi}}{8a^5}, \quad a > 0$$

and use this result to show that $v_{\text{rms}} = \sqrt{3RT/M}$.

55. In Exercise 17 of Section 8.6, we determined the work required to lift a 6000-lb satellite to an orbital position that is 1000 mi above the Earth's surface. The ideas discussed in that exercise will be needed here.

(a) Find a definite integral that represents the work required to lift a 6000-lb satellite to a position l miles above the Earth's surface.

(b) Find a definite integral that represents the work required to lift a 6000-lb satellite an "infinite distance" above the Earth's surface. Evaluate the integral. [*Note:* The result obtained here is sometimes called the work required to "escape" the Earth's gravity.]

A ***transform*** is a formula that converts or "transforms" one function into another. Transforms are used in applications to convert a difficult problem into an easier problem whose solution can then be used to solve the original difficult problem. The ***Laplace transform*** of a function $f(t)$, which plays an important role in the study of differential equations, is denoted by $\mathcal{L}\{f(t)\}$ and is defined by

$$\mathcal{L}\{f(t)\} = \int_0^{+\infty} e^{-st} f(t)\,dt$$

In this formula s is treated as a constant in the integration process; thus, the Laplace transform has the effect of transforming $f(t)$ into a function of s. Use this formula in Exercises 56 and 57.

56. Show that

(a) $\mathcal{L}\{1\} = \frac{1}{s}, \ s > 0$ (b) $\mathcal{L}\{e^{2t}\} = \frac{1}{s-2}, \ s > 2$

(c) $\mathcal{L}\{\sin t\} = \frac{1}{s^2+1}, \ s > 0$

(d) $\mathcal{L}\{\cos t\} = \frac{s}{s^2+1}, \ s > 0.$

57. In each part, find the Laplace transform.

(a) $f(t) = t, \ s > 0$ (b) $f(t) = t^2, \ s > 0$

(c) $f(t) = \begin{cases} 0, & t < 3 \\ 1, & t \geq 3 \end{cases}, \ s > 0$

58. Later in the text, we will show that

$$\int_0^{+\infty} e^{-x^2}\,dx = \tfrac{1}{2}\sqrt{\pi}$$

Confirm that this is reasonable by using a CAS or a calculator with a numerical integration capability.

59. Use the result in Exercise 58 to show that

(a) $\int_{-\infty}^{+\infty} e^{-ax^2}\,dx = \sqrt{\frac{\pi}{a}}, \ a > 0$

(b) $\frac{1}{\sqrt{2\pi}\sigma} \int_{-\infty}^{+\infty} e^{-x^2/2\sigma^2}\,dx = 1, \ \sigma > 0.$

A convergent improper integral over an infinite interval can be approximated by first replacing the infinite limit(s) of integration by finite limit(s), then using a numerical integration technique, such as Simpson's rule, to approximate the integral with finite limit(s). This technique is illustrated in Exercises 60 and 61.

60. Suppose that the integral in Exercise 58 is approximated by first writing it as

$$\int_0^{+\infty} e^{-x^2}\,dx = \int_0^K e^{-x^2}\,dx + \int_K^{+\infty} e^{-x^2}\,dx$$

then dropping the second term, and then applying Simpson's rule to the integral

$$\int_0^K e^{-x^2}\,dx$$

The resulting approximation has two sources of error: the error from Simpson's rule and the error

$$E = \int_K^{+\infty} e^{-x^2}\,dx$$

that results from discarding the second term. We call E the ***truncation error***.

(a) Approximate the integral in Exercise 58 by applying Simpson's rule with $n = 10$ subdivisions to the integral

$$\int_0^3 e^{-x^2}\,dx$$

Round your answer to four decimal places and compare it to $\frac{1}{2}\sqrt{\pi}$ rounded to four decimal places.

(b) Use the result that you obtained in Exercise 46 and the fact that $e^{-x^2} \le \frac{1}{3}xe^{-x^2}$ for $x \ge 3$ to show that the truncation error for the approximation in part (a) satisfies $0 < E < 2.1 \times 10^{-5}$.

61. (a) It can be shown that

$$\int_0^{+\infty} \frac{1}{x^6+1}\,dx = \frac{\pi}{3}$$

Approximate this integral by applying Simpson's rule with $n = 20$ subdivisions to the integral

$$\int_0^4 \frac{1}{x^6+1}\,dx$$

Round your answer to three decimal places and compare it to $\pi/3$ rounded to three decimal places.

(b) Use the result that you obtained in Exercise 46 and the fact that $1/(x^6+1) < 1/x^6$ for $x \ge 4$ to show that the truncation error for the approximation in part (a) satisfies $0 < E < 2 \times 10^{-4}$.

62. For what values of p does $\int_0^{+\infty} e^{px}\,dx$ converge?

63. Show that $\int_0^1 \frac{dx}{x^p}$ converges if $p < 1$ and diverges if $p \ge 1$.

C **64.** It is sometimes possible to convert an improper integral into a "proper" integral having the same value by making an appropriate substitution. Evaluate the following integral by making the indicated substitution, and investigate what happens if you evaluate the integral directly using a CAS.

$$\int_0^1 \sqrt{\frac{1+x}{1-x}}\,dx; \quad u = \sqrt{1-x}$$

In Exercises 65 and 66, transform the given improper integral into a proper integral by making the stated u-substitution, then approximate the proper integral by Simpson's rule with $n = 10$ subdivisions. Round your answer to three decimal places.

65. $\int_0^1 \frac{\cos x}{\sqrt{x}}\,dx; \quad u = \sqrt{x}$

66. $\int_0^1 \frac{\sin x}{\sqrt{1-x}}\,dx; \quad u = \sqrt{1-x}$

SUPPLEMENTARY EXERCISES

1. Consider the following methods for evaluating integrals: u-substitution, integration by parts, partial fractions, reduction formulas, and trigonometric substitutions. In each part, state the approach that you would try first to evaluate the integral. If none of them seems appropriate, then say so. You need not evaluate the integral.

(a) $\int x \sin x\,dx$ (b) $\int \cos x \sin x\,dx$

(c) $\int \tan^7 x\,dx$ (d) $\int \tan^7 x \sec^2 x\,dx$

(e) $\int \frac{3x^2}{x^3+1}\,dx$ (f) $\int \frac{3x^2}{(x+1)^3}\,dx$

(g) $\int \tan^{-1} x\,dx$ (h) $\int \sqrt{4-x^2}\,dx$

(i) $\int x\sqrt{4-x^2}\,dx$

2. Consider the following trigonometric substitutions:

$x = 3\sin\theta, \quad x = 3\tan\theta, \quad x = 3\sec\theta$

In each part, state the substitution that you would try first to evaluate the integral. If none seems appropriate, then state a trigonometric substitution that you would use. You need not evaluate the integral.

(a) $\int \sqrt{9+x^2}\,dx$ (b) $\int \sqrt{9-x^2}\,dx$

(c) $\int \sqrt{1-9x^2}\,dx$ (d) $\int \sqrt{x^2-9}\,dx$

(e) $\int \sqrt{9+3x^2}\,dx$ (f) $\int \sqrt{1+(9x)^2}\,dx$

3. (a) What condition must a rational function satisfy for the method of partial fractions to be applicable directly?

(b) If the condition in part (a) is not satisfied, what must you do if you want to use partial fractions?

4. What is an improper integral?

5. In each part, find the number of the formula in the Endpaper Integral Table that you would apply to evaluate the integral. You need not evaluate the integral.

(a) $\int \sin 7x \cos 9x\,dx$ (b) $\int (x^7 - x^5)e^{9x}\,dx$

(c) $\int x\sqrt{x-x^2}\,dx$ (d) $\int \frac{dx}{x\sqrt{4x+3}}$

(e) $\int x^9 \pi^x\,dx$ (f) $\int \frac{3x-1}{2+x^2}\,dx$

6. Evaluate the integral $\int_0^1 \frac{x^3}{\sqrt{x^2+1}}\,dx$ using
(a) integration by parts
(b) the substitution $u = \sqrt{x^2+1}$.

7. In each part, evaluate the integral by making an appropriate substitution and applying a reduction formula.
(a) $\int \sin^4 2x\,dx$ (b) $\int x\cos^5(x^2)\,dx$

8. Consider the integral $\int \frac{1}{x^3-x}\,dx$.
(a) Evaluate the integral using the substitution $x = \sec\theta$. For what values of x is your result valid?
(b) Evaluate the integral using the substitution $x = \sin\theta$. For what values of x is your result valid?
(c) Evaluate the integral using the method of partial fractions. For what values of x is your result valid?

9. (a) Evaluate the integral
$$\int \frac{1}{\sqrt{2x-x^2}}\,dx$$
three ways: using the substitution $u = \sqrt{x}$, using the substitution $u = \sqrt{2-x}$, and completing the square.
(b) Show that the answers in part (a) are equivalent.

10. Find the area of the region that is enclosed by the curves $y = (x-3)/(x^3+x^2)$, $y = 0$, $x = 1$, and $x = 2$.

11. Sketch the region whose area is $\int_0^{+\infty} \frac{dx}{1+x^2}$, and use your sketch to show that
$$\int_0^{+\infty} \frac{dx}{1+x^2} = \int_0^1 \sqrt{\frac{1-y}{y}}\,dy$$

12. Find the area that is enclosed between the x-axis and the curve $y = (\ln x - 1)/x^2$ for $x \ge e$.

13. Find the volume of the solid that is generated when the region between the x-axis and the curve $y = e^{-x}$ for $x \ge 0$ is revolved about the y-axis.

14. Find a positive value of a that satisfies the equation
$$\int_0^{+\infty} \frac{1}{x^2+a^2}\,dx = 1$$

In Exercises 15–30, evaluate the integral.

15. $\int \sqrt{\cos\theta}\sin\theta\,d\theta$

16. $\int_0^{\pi/4} \tan^7\theta\,d\theta$

17. $\int x\tan^2(x^2)\sec^2(x^2)\,dx$

18. $\int_{-1/\sqrt{2}}^{1/\sqrt{2}} (1-2x^2)^{3/2}\,dx$

19. $\int \frac{dx}{(3+x^2)^{3/2}}$

20. $\int \frac{\cos\theta}{\sin^2\theta - 6\sin\theta + 12}\,d\theta$

21. $\int \frac{x+3}{\sqrt{x^2+2x+2}}\,dx$

22. $\int \frac{\sec^2\theta}{\tan^3\theta - \tan^2\theta}\,d\theta$

23. $\int \frac{dx}{(x-1)(x+2)(x-3)}$

24. $\int \frac{dx}{x(x^2+x+1)}$

25. $\int_4^8 \frac{\sqrt{x-4}}{x}\,dx$

26. $\int_0^9 \frac{\sqrt{x}}{x+9}\,dx$

27. $\int \frac{1}{\sqrt{e^x+1}}\,dx$

28. $\int_0^{\ln 2} \sqrt{e^x-1}\,dx$

29. $\int_a^{+\infty} \frac{x\,dx}{(x^2+1)^2}$

30. $\int_0^{+\infty} \frac{dx}{a^2+b^2x^2}$, $a, b > 0$

Some integrals that can be evaluated by hand cannot be evaluated by all computer algebra systems. In Exercises 31–34, evaluate the integral by hand, and determine if it can be evaluated on your CAS.

C **31.** $\int \frac{x^3}{\sqrt{1-x^8}}\,dx$

C **32.** $\int (\cos^{32}x\sin^{30}x - \cos^{30}x\sin^{32}x)\,dx$

C **33.** $\int \sqrt{x - \sqrt{x^2-4}}\,dx$. [*Hint:* $\frac{1}{2}(\sqrt{x+2} - \sqrt{x-2})^2 = ?$]

C **34.** $\int \frac{1}{x^{10}+x}\,dx$. [*Hint:* Rewrite the denominator as $x^{10}(1+x^{-9})$.]

C **35.** Let
$$f(x) = \frac{-2x^5 + 26x^4 + 15x^3 + 6x^2 + 20x + 43}{x^6 - x^5 - 18x^4 - 2x^3 - 39x^2 - x - 20}$$
(a) Use a CAS to factor the denominator, and then write down the form of the partial fraction decomposition. You need not find the values of the constants.
(b) Check your answer in part (a) by using the CAS to find the partial fraction decomposition of f.
(c) Integrate f by hand, and then check your answer by integrating with the CAS.

36. The ***Gamma function***, $\Gamma(x)$, is defined as
$$\Gamma(x) = \int_0^{+\infty} t^{x-1}e^{-t}\,dt$$
It can be shown that this improper integral converges if and only if $x > 0$.
(a) Find $\Gamma(1)$.
(b) Prove: $\Gamma(x+1) = x\Gamma(x)$ for all $x > 0$. [*Hint:* Use integration by parts.]
(c) Use the results in parts (a) and (b) to find $\Gamma(2)$, $\Gamma(3)$, and $\Gamma(4)$; and then make a conjecture about $\Gamma(n)$ for positive integer values of n.
(d) Show that $\Gamma(\frac{1}{2}) = \sqrt{\pi}$. [*Hint:* See Exercise 58 of Section 9.8.]
(e) Use the results obtained in parts (b) and (d) to show that $\Gamma(\frac{3}{2}) = \frac{1}{2}\sqrt{\pi}$ and $\Gamma(\frac{5}{2}) = \frac{3}{4}\sqrt{\pi}$.

37. Refer to the Gamma function defined in Exercise 36 to show that

(a) $\displaystyle\int_0^1 (\ln x)^n\, dx = (-1)^n\Gamma(n+1), \quad n > 0.$

[*Hint:* Let $t = -\ln x$.]

(b) $\displaystyle\int_0^{+\infty} e^{-x^n}\, dx = \Gamma\left(\frac{n+1}{n}\right), \quad n > 0.$

[*Hint:* Let $t = x^n$. Use the result in Exercise 36(b).]

C **38.** A ***simple pendulum*** consists of a mass that swings in a vertical plane at the end of a massless rod of length L, as shown in the accompanying figure. Suppose that a simple pendulum is displaced through an angle θ_0 and released from rest. It can be shown that in the absence of friction, the time T required for the pendulum to make one complete back-and-forth swing, called the ***period***, is given by

$$T = \sqrt{\frac{8L}{g}}\int_0^{\theta_0} \frac{1}{\sqrt{\cos\theta - \cos\theta_0}}\, d\theta \tag{1}$$

where $\theta = \theta(t)$ is the angle the pendulum makes with the vertical at time t. The improper integral in (1) is difficult to evaluate numerically. By a substitution outlined below it can be shown that the period can be expressed as

$$T = 4\sqrt{\frac{L}{g}}\int_0^{\pi/2} \frac{1}{\sqrt{1 - k^2\sin^2\phi}}\, d\phi \tag{2}$$

where $k = \sin(\theta_0/2)$. The integral in (2) is called a ***complete elliptic integral of the first kind*** and is more easily evaluated by numerical methods.

(a) Obtain (2) from (1) by substituting

$$\cos\theta = 1 - 2\sin^2(\theta/2)$$
$$\cos\theta_0 = 1 - 2\sin^2(\theta_0/2)$$
$$k = \sin(\theta_0/2)$$

and then making the change of variable

$$\sin\phi = \sin(\theta/2)/\sin(\theta_0/2) = \sin(\theta/2)/k$$

(b) Use (2) and the numerical integration capability of your CAS to find the period of a simple pendulum for which $L = 1.5$ ft, $\theta_0 = 20°$, and $g = 32$ ft/s^2.

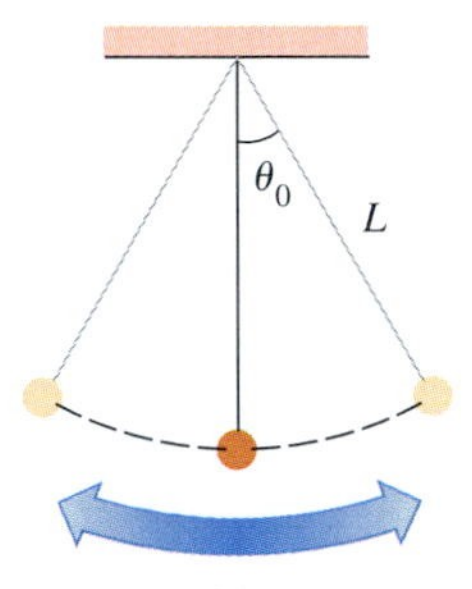

Figure Ex-38

EXPANDING THE CALCULUS HORIZON

Railroad Design

Your company has a contract to construct a track bed for a railroad line between towns A and B shown on the contour map in Figure 1. The bed can be created by cutting trenches through the surface or by using some combination of trenches and tunnels. As chief engineer, your assignment is to analyze the costs of trenches and tunnels and to propose a design strategy for minimizing the total construction cost.

Engineering Requirements

The Transportation Board submits the following engineering requirements to your company:

- The track bed is to be straight and 10 m wide. The grade is to increase at a constant rate from the existing elevation of 100 m at town A to an elevation of 110 m at point M and then decrease at a constant rate to the existing elevation of 88 m at town B.
- From town A to point M and from point N to town B the track bed is to be created by excavating a trench whose vertical cross sections are trapezoids with the dimensions shown in Figure 2.
- Between points M and N your company must decide whether to excavate a trench of the type in Figure 2 or to excavate a tunnel whose vertical cross sections have the dimensions shown in Figure 3.

CONTOUR MAP
ELEVATIONS IN METERS

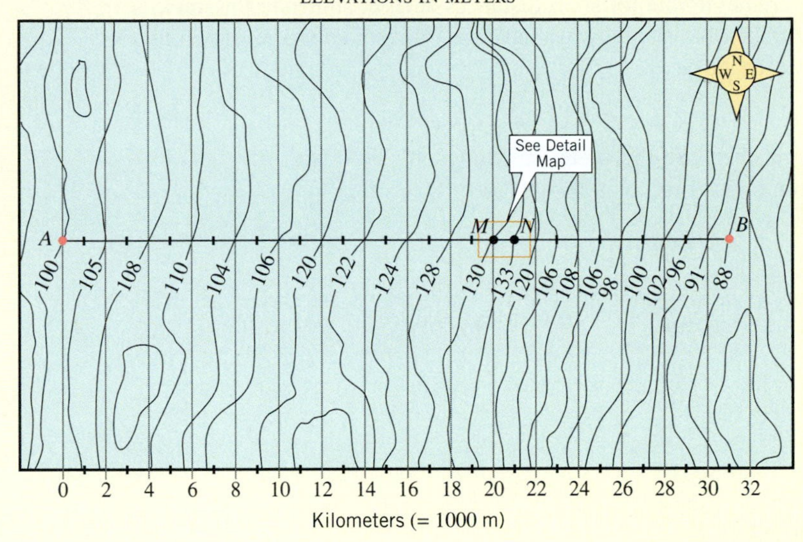

DETAIL MAP
PROPOSED TUNNEL CONSTRUCTION

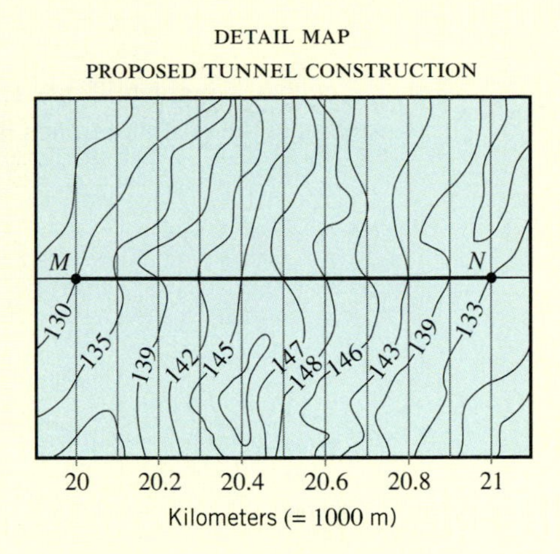

Figure 1

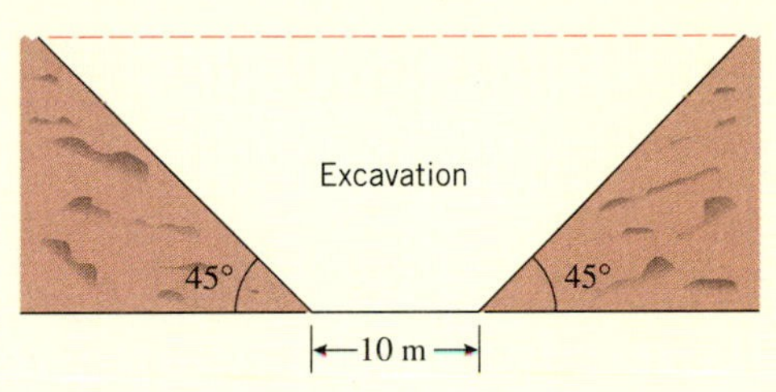

Figure 2

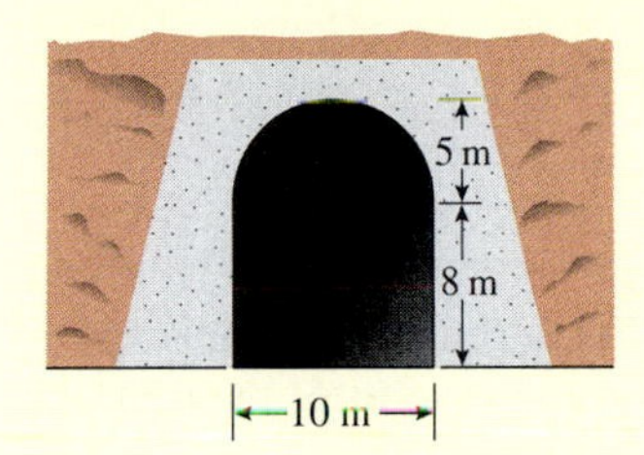

Figure 3

Cost Factors

Surface excavation of railbeds is performed using bulldozers, hydraulic excavators (backhoes), loading tractors, and other specialized equipment. Typically, the excavated dirt is piled at the side of the tracks to form sloped embankments, and the excavation cost is estimated from the volume of dirt to be removed and piled.

Tunnels in rock are often excavated by drilling shafts and inserting boring machines (called *moles*) to loosen and remove rock and dirt. Tunnels in soft ground are often excavated by starting at the tunnel face and using bucket or rotary excavators housed inside of shields. As the excavator progresses, tunnel liners are inserted behind it to support the earth and prevent cave-ins. Dirt removal is performed using conveyors or sometimes using railcars (called *muck cars*) that run on specially constructed tracks. Ventilation and air compression are other factors that add to the excavation cost of tunnels. In general, the excavation cost for a tunnel can be estimated from two components, the total volume of dirt to be removed and a cost that increases with the distance to the tunnel opening.

Make the following cost assumptions:

- The excavation and dirt-piling cost for a trench is \$4.00 per cubic meter.
- The drilling and dirt-piling cost for a tunnel is \$8.00 per cubic meter, and the costs involved in moving a load of dirt inside the tunnel a distance of 1 m toward the entrance along the track line is \$0.06 per cubic meter.

Cost Analysis of Trenches

Assume that variations in elevation are negligible for short distances at right angles to the track, so that the cross sections of the dirt to be excavated always have the trapezoidal shape shown in Figure 2 (straight horizontal edges at the surface).

Exercise 1 Complete Table 1, and then use the table and Simpson's rule with $n = 10$ to approximate the cost of a trench from town A to point M.

Table 1

DISTANCE x FROM TOWN A (m)	TERRAIN ELEVATION (m)	TRACK ELEVATION (m)	DEPTH OF CUT (m)	CROSS-SECTIONAL AREA $f(x)$ OF CUT (m^2)
0	100	100	0	0
2,000	105	101	4	56
4,000				
6,000				
8,000				
10,000				
12,000				
14,000				
16,000				
18,000				
20,000				

Exercise 2 As in Exercise 1, use Simpson's rule with $n = 10$ to approximate the cost of constructing a trench from (a) point M to point N, and (b) point N to town B.

Exercise 3 Find the total cost of the project if a trench is used along the entire line from town A to town B.

Cost Analysis of a Tunnel

Exercise 4

(a) Find the volume of dirt that must be removed from the tunnel, and calculate the drilling and dirt-piling cost.

(b) Find an integral for the cost of moving all of the dirt inside the tunnel to the tunnel entrance. [*Suggestion:* Use Riemann sums.]

(c) Find the total cost of excavating the tunnel.

Exercise 5 Find the total cost of the project using a trench from town A to point M, a tunnel from point M to point N, and a trench from point N to town B. Compare the cost to that obtained in Exercise 3 and state which method is cheaper.

Module by: *C. Lynn Kiaer, Rose-Hulman Institute of Technology*
David Ryeburn, Simon Fraser University
Howard Anton, Drexel University
Peter Dunn, Railroad Construction Company, Inc., Paterson, NJ

Additional material for this module can be found on the World Wide Web at http://www.wiley.com/college/anton

Colin Maclaurin

10

MATHEMATICAL MODELING WITH DIFFERENTIAL EQUATIONS

Many of the principles in science and engineering concern relationships between changing quantities. Since rates of change are represented mathematically by derivatives, it should not be surprising that such principles are often expressed in terms of differential equations. We introduced the concept of a differential equation in Section 7.2, but in this chapter we will go into more detail. We will discuss some important mathematical models that involve differential equations, and we will discuss some methods for solving and approximating solutions of some of the basic types of differential equations. However, we will only be able to touch the surface of this topic, leaving many important topics in differential equations to courses that are devoted completely to the subject.

10.1 FIRST-ORDER DIFFERENTIAL EQUATIONS AND APPLICATIONS

In this section we will introduce some basic terminology and concepts concerning differential equations. We will also discuss methods for solving certain basic types of differential equations, and we will give some applications of our work.

TERMINOLOGY

Recall from Section 7.2 that a ***differential equation*** is an equation involving one or more derivatives of an unknown function. In this section we will denote the unknown function by $y = y(x)$ unless the differential equation arises from an applied problem involving time, in which case we will denote it by $y = y(t)$. The ***order*** of a differential equation is the order of the highest derivative that it contains. Here are some examples:

DIFFERENTIAL EQUATION	ORDER
$\frac{dy}{dx} = 3y$	1
$\frac{d^2y}{dx^2} - 6\frac{dy}{dx} + 8y = 0$	2
$\frac{d^3y}{dx^3} - t\frac{dy}{dt} + (t^2 - 1)y = e^t$	3
$y' - y = e^{2x}$	1
$y'' + y' = \cos t$	2

In the last two equations the derivatives of y are expressed in "prime" notation. You will usually be able to tell from the equation itself or the context in which it arises whether to interpret y' as dy/dx or as dy/dt.

SOLUTIONS OF DIFFERENTIAL EQUATIONS

A function $y = y(x)$ is a ***solution*** of a differential equation on a given interval if the equation is satisfied for every x in that interval when y and its derivatives are substituted in the equation. For example, $y = e^{2x}$ is a solution of the differential equation

$$\frac{dy}{dx} - y = e^{2x} \tag{1}$$

on the interval $(-\infty, +\infty)$, since substituting y and its derivative into the left side of this equation yields

$$\frac{dy}{dx} - y = \frac{d}{dx}[e^{2x}] - e^{2x} = 2e^{2x} - e^{2x} = e^{2x}$$

for all real values of x. However, this is not the only solution on the interval $(-\infty, +\infty)$; for example, the function

$$y = Ce^x + e^{2x} \tag{2}$$

is also a solution for every real value of the constant C, since

$$\frac{dy}{dx} - y = \frac{d}{dx}[Ce^x + e^{2x}] - (Ce^x + e^{2x}) = (Ce^x + 2e^{2x}) - (Ce^x + e^{2x}) = e^{2x}$$

One can prove that *all* solutions of (1) on $(-\infty, +\infty)$ can be obtained by substituting values for the constant C in (2). On a given interval, a solution of a differential equation from which all solutions on that interval can be derived by substituting values for arbitrary constants is called the ***general solution*** of the equation on the interval. Thus, (2) is the general solution of (1) on the interval $(-\infty, +\infty)$.

REMARK. Usually, the general solution of an nth-order differential equation on an interval will contain n arbitrary constants. Although we will not prove this, it makes sense intuitively because n integrations are needed to recover a function from its nth derivative, and each integration introduces an arbitrary constant. For example, (2) has one arbitrary constant, which is consistent with the fact that it is the general solution of the *first-order* equation (1).

The graph of a solution of a differential equation is called an ***integral curve*** for the equation, so the general solution of a differential equation produces a family of integral curves corresponding to the different possible choices for the arbitrary constants. For example, Figure 10.1.1 shows some integral curves for (1), which were obtained by assigning values to the arbitrary constant in (2).

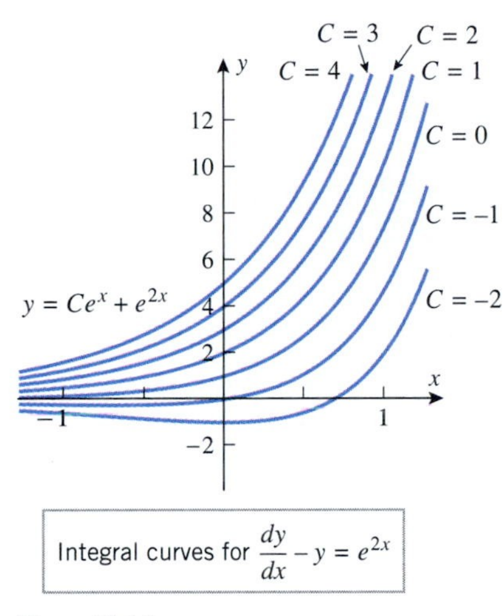

Integral curves for $\frac{dy}{dx} - y = e^{2x}$

Figure 10.1.1

INITIAL-VALUE PROBLEMS

When an applied problem leads to a differential equation, there are usually conditions in the problem that determine specific values for the arbitrary constants. As a rule of thumb, it requires n conditions to determine values for all n arbitrary constants in the general solution of an nth-order differential equation (one condition for each constant). For a first-order equation, the single arbitrary constant can be determined by specifying the value of the unknown function $y(x)$ at an arbitrary point x_0, say $y(x_0) = y_0$. This is called an ***initial condition***, and the problem of solving a first-order equation subject to an initial condition is called a ***first-order initial-value problem***. Geometrically, the initial condition $y(x_0) = y_0$ has the effect of isolating the integral curve that passes through the point (x_0, y_0) from the complete family of integral curves.

Example 1

The solution of the initial-value problem

$$\frac{dy}{dx} - y = e^{2x}, \quad y(0) = 3$$

can be obtained by substituting the initial condition $x = 0$, $y = 3$ in the general solution (2) to find C. We obtain

$$3 = Ce^0 + e^0 = C + 1$$

Thus, $C = 2$, and the solution of the initial-value problem, which is obtained by substituting this value of C in (2), is

$$y = 2e^x + e^{2x}$$

Geometrically, the graph of this solution is the integral curve in Figure 10.1.1 that passes through the point (0, 3). ◀

FIRST-ORDER EQUATIONS

The simplest first-order equations are those that can be written in the form

$$\frac{dy}{dx} = f(x) \tag{3}$$

Such equations can often be solved by integration. For example, if

$$\frac{dy}{dx} = x^3 \tag{4}$$

then

$$y = \int x^3\, dx = \frac{x^4}{4} + C$$

is the general solution of (4) on the interval $(-\infty, +\infty)$.

FIRST-ORDER SEPARABLE EQUATIONS

Equation (4) can be solved by integrating because the right side is a function of x. However, if the right side involves both x and y, as with

$$\frac{dy}{dx} = \sin(xy)$$

then direct integration is not possible and other methods must be used. In general, such equations can be complicated to solve exactly, and often one must settle for numerical approximations of solutions, as we will discuss in the next section. However, if the equation can be expressed in the form

$$h(y)\frac{dy}{dx} = g(x) \tag{5}$$

then we say that the equation is ***separable***, and we can often find the general solution by first rewriting the equation in the differential form

$$h(y)\, dy = g(x)\, dx \tag{6}$$

(all y's on one side and all x's on the other), and then integrating both sides to obtain

$$\int h(y)\, dy = \int g(x)\, dx \tag{7}$$

If the equation that results when these integrations are performed can be solved for y as a function of x, then this function provides an explicit formula for the general solution of (5). However, if the equation that results when these integrations are performed cannot be solved for y as a function of x, then the equation still defines solutions of (5), but it defines them implicitly.

The process of obtaining (6) from (5) is called ***separating variables***, and the method we have just discussed for solving (5) is called ***separation of variables***. A more detailed explanation of why this method works is given in the exercises.

Example 2

Solve the differential equation

$$\frac{dy}{dx} = -4xy^2$$

and then solve the initial-value problem

$$\frac{dy}{dx} = -4xy^2, \quad y(0) = 1$$

Solution. Separating variables and integrating yields

$$\frac{1}{y^2}\,dy = -4x\,dx$$

$$\int \frac{1}{y^2}\,dy = \int -4x\,dx$$

$$-\frac{1}{y} = -2x^2 + C$$

The integration on the left produces a constant c_1, and the integration on the right produces a constant c_2. We have combined these constants into the constant $C = c_2 - c_1$.

Solving for y as a function of x, we obtain

$$y = \frac{1}{2x^2 - C} \tag{8}$$

The initial condition $y(0) = 1$ requires that $y = 1$ when $x = 0$. Substituting these values in (8) yields $C = -1$ (verify). Thus, the solution of the initial-value problem is

$$y = \frac{1}{2x^2 + 1}$$

Some typical integral curves and the solution of the initial-value problem are graphed in Figure 10.1.2. ◀

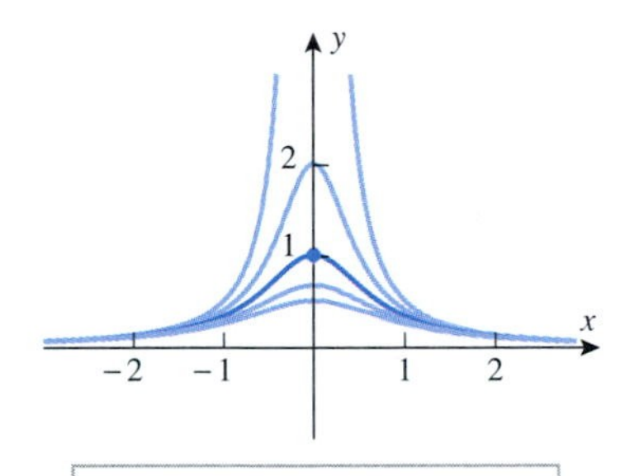

Integral curves for $\dfrac{dy}{dx} = -4xy^2$

Figure 10.1.2

Example 3

Solve the initial-value problem

$$(4y - \cos y)\frac{dy}{dx} - 3x^2 = 0, \quad y(0) = 0$$

Solution. First, we solve the differential equation. Separating variables and integrating yields

$$(4y - \cos y)\frac{dy}{dx} = 3x^2$$

$$(4y - \cos y)\,dy = 3x^2\,dx$$

$$\int (4y - \cos y)\,dy = \int 3x^2\,dx$$

$$2y^2 - \sin y = x^3 + C \tag{9}$$

Equation (9) defines the solutions of the differential equation implicitly; it cannot be solved explicitly for y as a function of x.

For the initial-value problem, the initial condition $y(0) = 0$ requires that $y = 0$ if $x = 0$. Substituting these values in (9) to determine the constant of integration yields $C = 0$ (verify). Thus, the solution of the initial-value problem is

$$2y^2 - \sin y = x^3$$ ◀

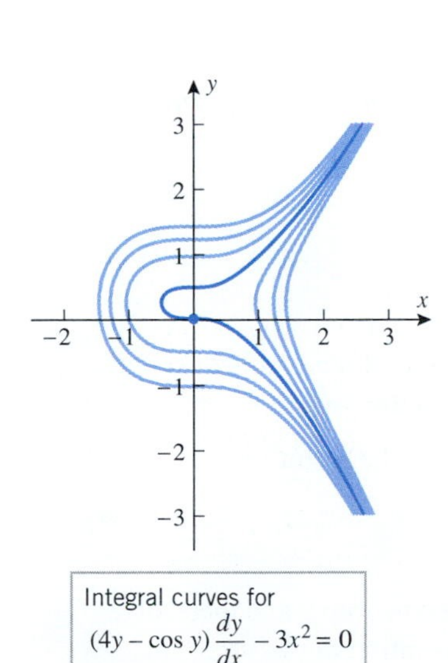

Integral curves for $(4y - \cos y)\dfrac{dy}{dx} - 3x^2 = 0$

Figure 10.1.3

FOR THE READER. Some computer algebra systems can graph implicit equations. For example, Figure 10.1.3 shows the graphs of (9) for $C = 0, \pm 1, \pm 2$, and ± 3, with emphasis on the solution of the initial-value problem. If you have a CAS that can graph implicit equations, read the documentation on graphing them and try to duplicate this figure. Also, try to determine which values of C produce which curves.

FIRST-ORDER LINEAR EQUATIONS

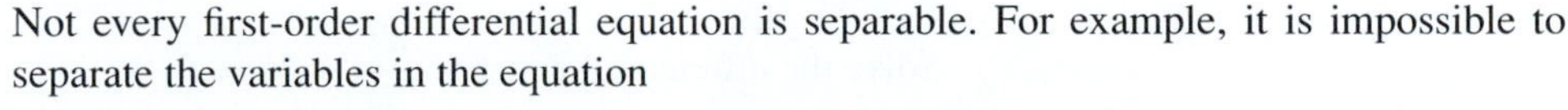

Not every first-order differential equation is separable. For example, it is impossible to separate the variables in the equation

$$\frac{dy}{dx} + 2xy = xe^{-x^2}$$

However, this equation can be solved by a different method that we will now consider.

A first-order differential equation is called ***linear*** if it is expressible in the form

$$\frac{dy}{dx} + p(x)y = q(x) \tag{10}$$

where the functions $p(x)$ and $q(x)$ are *continuous* and may or may not be constant. Some examples are

$$\frac{dy}{dx} + x^2y = e^x, \qquad \frac{dy}{dx} + (\sin x)y + x^3 = 0, \qquad \frac{dy}{dx} + 5y = 2$$

$p(x) = x^2, q(x) = e^x$ $\qquad$ $p(x) = \sin x, q(x) = -x^3$ $\qquad$ $p(x) = 5, q(x) = 2$

One procedure for solving (10) is based on the observation that if we define $\mu = \mu(x)$ by

$$\mu = e^{\int p(x)\,dx}$$

then

$$\frac{d\mu}{dx} = e^{\int p(x)\,dx} \cdot \frac{d}{dx}\int p(x)\,dx = \mu p(x)$$

Thus,

$$\frac{d}{dx}(\mu y) = \mu\frac{dy}{dx} + \frac{d\mu}{dx}y = \mu\frac{dy}{dx} + \mu p(x)y \tag{11}$$

If (10) is multiplied through by μ, and then simplified using (11), it becomes

$$\mu\frac{dy}{dx} + \mu p(x)y = \mu q(x)$$

$$\frac{d}{dx}(\mu y) = \mu q(x)$$

This equation can be solved by integrating both sides to obtain

$$\mu y = \int \mu q(x)\,dx + C \quad \text{or} \quad y = \frac{1}{\mu}\left[\int \mu q(x)\,dx + C\right]$$

To summarize, (10) can be solved in three steps, called ***the method of integrating factors***:

The Method of Integrating Factors

Step 1. Calculate

$$\mu = e^{\int p(x)\,dx}$$

This is called the ***integrating factor***. Since any μ will suffice, we can take the constant of integration to be zero in this step.

Step 2. Multiply both sides of (10) by μ and express the result as

$$\frac{d}{dx}[\mu y] = \mu q(x)$$

Step 3. Integrate both sides of the equation obtained in Step 2 and then solve for y. Be sure to include a constant of integration in this step.

Example 4

Solve the differential equation $\dfrac{dy}{dx} - 2y = 0$.

Solution. This equation is separable, but it is also linear, since it is of form (10) with $p(x) = -2$ and $q(x) = 0$; thus, we can solve it by separation of variables or by the method of integrating factors. We will solve it both ways, using the method of integrating factors first. The integrating factor is

$$\mu = e^{\int -2\,dx} = e^{-2x}$$

If we multiply the differential equation through by μ and follow Step 2 of the method of integrating factors, then we obtain

$$\frac{d}{dx}[e^{-2x}y] = 0$$

Integrating both sides of this equation yields

$$e^{-2x}y = C$$

which can be rewritten as

$$y = Ce^{2x}$$

Alternative Solution. Separating variables and integrating yields

$$\int \frac{dy}{y} = \int 2\,dx$$

$$\ln|y| = 2x + c$$

We have used c as the constant of integration here to reserve C for the constant in the final result.

$$|y| = e^{2x+c}$$

$$|y| = e^{c}e^{2x}$$

$$y = \pm e^{c}e^{2x}$$

$$y = Ce^{2x}$$

Letting $C = \pm e^{c}$

which agrees with the answer obtained above. ◀

REMARK. For first-order equations that are both linear and separable, the method of integrating factors is usually simpler than separation of variables, provided the integrating factor can be found easily. Moreover, the careful reader may have observed in the alternative solution of Example 4 that the constant $C = \pm e^{c}$ is not truly arbitrary, since $C = 0$ is not an allowable value. Thus, separation of variables missed the solution $y = 0$, which the method of integrating factors did not. This problem occurred because we had to divide by y to separate the variables.

Example 5

Solve the initial-value problem

$$x\frac{dy}{dx} - y = x, \quad y(1) = 2$$

Solution. The differential equation can be rewritten in form (10) by dividing through by x. This yields

$$\frac{dy}{dx} - \frac{1}{x}y = 1 \tag{12}$$

Comparing this to (10), we have $p(x) = -1/x$ and $q(x) = 1$. However, there is a difficulty here because the method of integrating factors requires that $p(x)$ and $q(x)$ be continuous, and $p(x)$ has a discontinuity at $x = 0$. Thus, the method of integrating factors can be applied if $x > 0$ or if $x < 0$, but not on an interval containing $x = 0$. However, the initial condition $y(1) = 2$ is imposed at $x = 1$, so we will assume that $x > 0$. With this assumption, the integrating factor is

$$\mu = e^{\int -(1/x)\,dx} = e^{-\ln|x|} = \frac{1}{|x|} = \frac{1}{x}$$

If we multiply (12) through by μ, then from Step 2 of the method of integrating factors we obtain

$$\frac{d}{dx}\left(\frac{1}{x}y\right) = \frac{1}{x}$$

Integrating both sides of this equation yields

$$\frac{1}{x}y = \ln x + C$$

or

$$y = x\ln x + Cx \tag{13}$$

The initial condition $y(1) = 2$ requires that $y = 2$ if $x = 1$. Substituting these values in (13) and solving for C yields $C = 2$ (verify), so the solution of the initial-value problem is

$$y = x\ln x + 2x$$ ◀

APPLICATIONS IN GEOMETRY

We conclude this section with some applications of first-order differential equations.

Example 6

Find a curve in the xy-plane that passes through (0, 3) and whose tangent line at a point (x, y) has slope $2x/y^2$.

Solution. Since the slope of the tangent line is dy/dx, we have

$$\frac{dy}{dx} = \frac{2x}{y^2} \tag{14}$$

and, since the curve passes through (0, 3), we have the initial condition

$$y(0) = 3 \tag{15}$$

Equation (14) is separable and can be written as

$$y^2\,dy = 2x\,dx$$

so

$$\int y^2\,dy = \int 2x\,dx \quad \text{or} \quad \tfrac{1}{3}y^3 = x^2 + C$$

It follows from the initial condition (15) that $y = 3$ if $x = 0$. Substituting these values in the last equation yields $C = 9$ (verify), so the equation of the desired curve is

$$\tfrac{1}{3}y^3 = x^2 + 9 \quad \text{or} \quad y = (3x^2 + 27)^{1/3}$$ ◀

MIXING PROBLEMS

In a typical mixing problem, a tank is filled to a specified level with a solution that contains a known amount of some soluble substance (say salt). The thoroughly stirred solution is allowed to drain from the tank at a known rate, and at the same time a solution with a known concentration of the soluble substance is added to the tank at a known rate that may or may not differ from the draining rate. As time progresses, the amount of the soluble substance in the tank will generally change, and the usual mixing problem seeks to determine the amount of the substance in the tank at a specified time. This type of problem serves as a model for many kinds of problems: discharge and filtration of pollutants in a river, injection and absorption of medication in the bloodstream, and migrations of species into and out of an ecological system, for example.

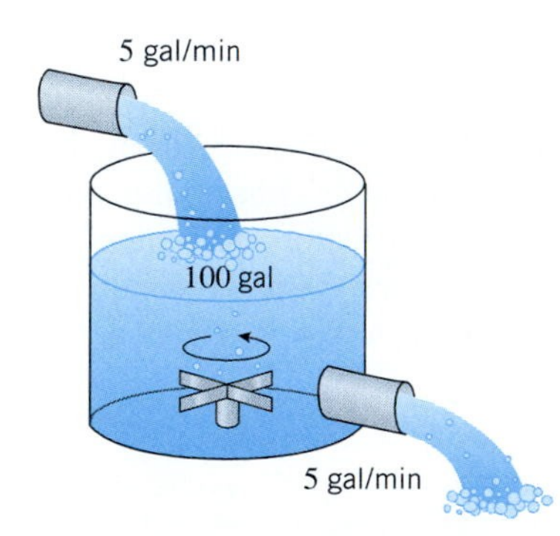

Figure 10.1.4

Example 7

At time $t = 0$, a tank contains 4 lb of salt dissolved in 100 gal of water. Suppose that brine containing 2 lb of salt per gallon of brine is allowed to enter the tank at a rate of 5 gal/min and that the mixed solution is drained from the tank at the same rate (Figure 10.1.4). Find the amount of salt in the tank after 10 minutes.

Solution. Let $y(t)$ be the amount of salt (in pounds) after t minutes. We are given that $y(0) = 4$, and we want to find $y(10)$. We will begin by finding a differential equation that is satisfied by $y(t)$. To do this, observe that dy/dt, which is the rate at which the amount of salt in the tank changes with time, can be expressed as

$$\frac{dy}{dt} = \text{rate in} - \text{rate out} \tag{16}$$

where *rate in* is the rate at which salt enters the tank and *rate out* is the rate at which salt leaves the tank. But the rate at which salt enters the tank is

$$\text{rate in} = (2\ \text{lb/gal}) \cdot (5\ \text{gal/min}) = 10\ \text{lb/min}$$

Since brine enters and drains from the tank at the same rate, the volume of brine in the tank stays constant at 100 gal. Thus, after t minutes have elapsed, the tank contains $y(t)$ lb of salt per 100 gal of brine, and hence the rate at which salt leaves the tank at that instant is

$$\text{rate out} = \left(\frac{y(t)}{100}\ \text{lb/gal}\right) \cdot (5\ \text{gal/min}) = \frac{y(t)}{20}\ \text{lb/min}$$

Therefore, (16) can be written as

$$\frac{dy}{dt} = 10 - \frac{y}{20} \quad \text{or} \quad \frac{dy}{dt} + \frac{y}{20} = 10$$

which is a first-order linear differential equation satisfied by $y(t)$. Since we are given that $y(0) = 4$, the function $y(t)$ can be obtained by solving the initial-value problem

$$\frac{dy}{dt} + \frac{y}{20} = 10, \quad y(0) = 4$$

The integrating factor for the differential equation is

$$\mu = e^{\int (1/20)\, dt} = e^{t/20}$$

If we multiply the differential equation through by μ, then from Step 2 of the method of integrating factors we obtain

$$\frac{d}{dt}(e^{t/20} y) = 10e^{t/20}$$

$$e^{t/20} y = \int 10e^{t/20}\, dt = 200e^{t/20} + C$$

$$y(t) = 200 + Ce^{-t/20} \tag{17}$$

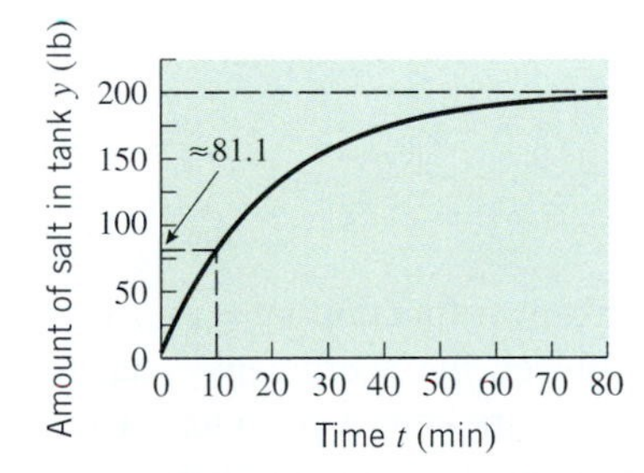

Figure 10.1.5

The initial condition states that $y = 4$ when $t = 0$. Substituting these values in (17) and solving for C yields $C = -196$ (verify), so

$$y(t) = 200 - 196e^{-t/20} \tag{18}$$

Thus, at time $t = 10$ the amount of salt in the tank is

$$y(10) = 200 - 196e^{-0.5} \approx 81.1\ \text{lb}$$ ◀

FOR THE READER. Figure 10.1.5 shows the graph of (18). Observe that $y(t) \to 200$ as $t \to +\infty$, which means that over an extended period of time the amount of salt in the tank tends toward 200 lb. Give an informal physical argument to explain why this result is to be expected.

A MODEL OF FREE-FALL MOTION RETARDED BY AIR RESISTANCE

In Section 6.3 we considered the free-fall model of an object moving along a vertical axis near the surface of the Earth. It was assumed in that model that there is no air resistance and that the only force acting on the object is the Earth's gravity. Our goal here is to find a model that takes air resistance into account. For this purpose we make the following assumptions:

- The object moves along a vertical s-axis whose origin is at the surface of the Earth and whose positive direction is up (Figure 6.3.7).

- At time $t = 0$ the height of the object is s_0 and the velocity is v_0.
- The only forces on the object are the force $F_G = -mg$ of the Earth's gravity acting down and the force F_R of air resistance acting opposite to the direction of motion. The force F_R is called the ***drag force***.

We will also need the following result from physics:

10.1.1 NEWTON'S SECOND LAW OF MOTION. If an object with mass m is subjected to a force F, then the object undergoes an acceleration a that satisfies the equation

$$F = ma \tag{19}$$

In the case of free-fall motion retarded by air resistance, the net force acting on the object is

$$F_G + F_R = -mg + F_R$$

and the acceleration is d^2s/dt^2, so Newton's second law implies that

$$-mg + F_R = m\frac{d^2s}{dt^2} \tag{20}$$

Experimentation has shown that the force F_R of air resistance depends on the shape of the object and its speed—the greater the speed, the greater the drag force. There are many possible models for air resistance, but one of the most basic assumes that the drag force F_R is proportional to the velocity of the object, that is,

$$F_R = -cv$$

where c is a positive constant that depends on the object's shape and properties of the air.* (The minus sign ensures that the drag force is opposite to the direction of motion.) Substituting this in (20) and writing d^2s/dt^2 as dv/dt, we obtain

$$-mg - cv = m\frac{dv}{dt}$$

or on dividing by m and rearranging we obtain

$$\frac{dv}{dt} + \frac{c}{m}v = -g$$

which is a first-order linear differential equation in the unknown function $v = v(t)$ with $p(t) = c/m$ and $q(t) = -g$ [see (10)]. For a specific object, the coefficient c can be determined experimentally, so we can assume that m, g, and c are known constants. Thus, the velocity function $v = v(t)$ can be obtained by solving the initial-value problem

$$\frac{dv}{dt} + \frac{c}{m}v = -g, \quad v(0) = v_0 \tag{21}$$

Once the velocity function is found, the position function $s = s(t)$ can be obtained by solving the initial-value problem

$$\frac{ds}{dt} = v(t), \quad s(0) = s_0 \tag{22}$$

In Exercise 47 we will ask you to solve (21) and show that

$$v(t) = e^{-ct/m}\left(v_0 + \frac{mg}{c}\right) - \frac{mg}{c} \tag{23}$$

---

*Other common models assume that $F_R = -cv^2$ or, more generally, $F_R = -cv^p$ for some value of p.

Note that

$$\lim_{t \to +\infty} v(t) = -\frac{mg}{c} \tag{24}$$

(verify). Thus, the speed $|v(t)|$ does not increase indefinitely, as in free fall; rather, because of the air resistance, it approaches a finite limiting speed v_τ given by

$$v_\tau = \left| -\frac{mg}{c} \right| = \frac{mg}{c} \tag{25}$$

This is called the ***terminal speed*** of the object, and (24) is called its ***terminal velocity***.

REMARK. Intuition suggests that near the limiting velocity, the velocity $v(t)$ changes very slowly; that is, $dv/dt \approx 0$. Thus, it should not be surprising that the limiting velocity can be obtained informally from (21) by setting $dv/dt = 0$ in the differential equation and solving for v. This yields

$$v = -\frac{mg}{c}$$

which agrees with (24).

EXERCISE SET 10.1 Graphing Calculator C CAS

1. Confirm that $y = 2e^{x^3/3}$ is a solution of the initial-value problem $y' = x^2 y$, $y(0) = 2$.

2. Confirm that $y = \frac{1}{4}x^4 + 2\cos x + 1$ is a solution of the initial-value problem $y' = x^3 - 2\sin x$, $y(0) = 3$.

In Exercises 3 and 4, state the order of the differential equation, and confirm that the functions in the given family are solutions.

3. (a) $(1+x)\dfrac{dy}{dx} = y$; $y = c(1+x)$

(b) $y'' + y = 0$; $y = c_1 \sin t + c_2 \cos t$

4. (a) $2\dfrac{dy}{dx} + y = x - 1$; $y = ce^{-x/2} + x - 3$

(b) $y'' - y = 0$; $y = c_1 e^t + c_2 e^{-t}$

In Exercises 5 and 6, use implicit differentiation to confirm that the equation defines implicit solutions of the differential equation.

5. $\ln y = xy + C$; $\dfrac{dy}{dx} = \dfrac{y^2}{1 - xy}$

6. $x^2 + xy^2 = C$; $2x + y^2 + 2xy\dfrac{dy}{dx} = 0$

In Exercises 7 and 8, solve the differential equation by the method of integrating factors and by separation of variables, and confirm that the two solutions are the same.

7. (a) $\dfrac{dy}{dx} + 3y = 0$ (b) $\dfrac{dy}{dt} - 2y = 0$

8. (a) $\dfrac{dy}{dx} - 4xy = 0$ (b) $\dfrac{dy}{dt} + y = 0$

In Exercises 9–18, solve the differential equation by separation of variables. Where reasonable, express the family of solutions as explicit functions of x.

9. $\dfrac{dy}{dx} = \dfrac{y}{x}$

10. $\dfrac{dy}{dx} = (1 + y^2)x^2$

11. $\dfrac{\sqrt{1+x^2}}{1+y}\dfrac{dy}{dx} = -x$

12. $(1 + x^4)\dfrac{dy}{dx} = \dfrac{x^3}{y}$

13. $(1 + y^2)y' = e^x y$

14. $y' = -xy$

15. $e^{-y}\sin x - y'\cos^2 x = 0$

16. $y' - (1+x)(1+y^2) = 0$

17. $\dfrac{dy}{dx} - \dfrac{y^2 - y}{\sin x} = 0$

18. $3\tan y - \dfrac{dy}{dx}\sec x = 0$

In Exercises 19–24, solve the differential equation by the method of integrating factors.

19. $\dfrac{dy}{dx} + 3y = e^{-2x}$

20. $\dfrac{dy}{dx} + 2xy = x$

21. $y' + y = \cos(e^x)$

22. $2\dfrac{dy}{dx} + 4y = 1$

23. $(x^2 + 1)\dfrac{dy}{dx} + xy = 0$

24. $\dfrac{dy}{dx} + y - \dfrac{1}{1 + e^x} = 0$

25. In each part, find the solution of the differential equation

$$x\frac{dy}{dx} + y = x$$

that satisfies the initial condition.

(a) $y(1) = 2$ (b) $y(-1) = 2$

26. In each part, find the solution of the differential equation

$$\frac{dy}{dx} = xy$$

that satisfies the initial condition.
(a) $y(0) = 1$ (b) $y(0) = \frac{1}{2}$

In Exercises 27–32, solve the initial-value problem by any method.

27. $\frac{dy}{dx} - xy = x, \quad y(0) = 3$

28. $\frac{dy}{dt} + y = 2, \quad y(0) = 1$

29. $y' = \frac{4x^2}{y + \cos y}, \quad y(1) = \pi$

30. $y' - xe^y = 2e^y, \quad y(0) = 0$

31. $\frac{dy}{dt} = \frac{2t + 1}{2y - 2}, \quad y(0) = -1$

32. $y' \cosh x + y \sinh x = \cosh^2 x, \quad y(0) = \frac{1}{4}$

33. (a) Sketch some typical integral curves of the differential equation $y' = y/2x$.
(b) Find an equation for the integral curve that passes through the point $(2, 1)$.

34. (a) Sketch some typical integral curves of the differential equation $y' = -x/y$.
(b) Find an equation for the integral curve that passes through the point $(3, 4)$.

In Exercises 35 and 36, solve the differential equation, and then use a graphing utility to generate the integral curves for $C = -2, -1, 0, 1, 2$.

35. $(x^2 + 4)\frac{dy}{dx} + xy = 0$ **36.** $y' + 2y - 3e^t = 0$

If you have a CAS that can graph implicit equations, solve the differential equations in Exercises 37 and 38, and then use the CAS to generate the integral curves for $C = -2, -1, 0, 1, 2$.

C **37.** $y' = \frac{x^2}{1 - y^2}$ C **38.** $y' = \frac{y}{1 + y^2}$

If you have a CAS, read the documentation on solving differential equations and initial-value problems, and then use the CAS in Exercises 39 and 40.

C **39.** Use a CAS to solve the differential equations in the odd-numbered Exercises 9–23, and confirm that the answers are consistent with those in the answer section of the text.

C **40.** Use a CAS to solve the initial-value problems in the odd-numbered Exercises 25–31, and confirm that the answers are consistent with those in the answer section of the text.

41. Find an equation of a curve with x-intercept 2 whose tangent line at any point (x, y) has slope xe^y.

42. Use a graphing utility to generate a curve that passes through the point $(1, 1)$ and whose tangent line at (x, y) is perpendicular to the line through (x, y) with slope $-2y/(3x^2)$.

43. At time $t = 0$, a tank contains 25 ounces of salt dissolved in 50 gal of water. Then brine containing 4 ounces of salt per gallon of brine is allowed to enter the tank at a rate of 2 gal/min and the mixed solution is drained from the tank at the same rate.
(a) How much salt is in the tank at an arbitrary time t?
(b) How much salt is in the tank after 25 min?

44. A tank initially contains 200 gal of pure water. Then at time $t = 0$ brine containing 5 lb of salt per gallon of brine is allowed to enter the tank at a rate of 10 gal/min and the mixed solution is drained from the tank at the same rate.
(a) How much salt is in the tank at an arbitrary time t?
(b) How much salt is in the tank after 30 min?

45. A tank with a 1000-gal capacity initially contains 500 gal of water that is polluted with 50 lb of particulate matter. At time $t = 0$, pure water is added at a rate of 20 gal/min and the mixed solution is drained off at a rate of 10 gal/min. How much particulate matter is in the tank when it reaches the point of overflowing?

46. The water in a polluted lake initially contains 1 lb of mercury salts per 100,000 gal of water. The lake is circular with diameter 30 m and uniform depth 3 m. Polluted water is pumped from the lake at a rate of 1000 gal/h and is replaced with fresh water at the same rate. Construct a table that shows the amount of mercury in the lake (in lbs) at the end of each hour over a 12-hour period. Discuss any assumptions you made. [Use 264 gal/m^3.]

47. (a) Use the method of integrating factors to confirm that (23) is the solution of initial-value problem (21). [*Note:* Keep in mind that c, m, and g are constants.]
(b) Show that (23) can be expressed in terms of the terminal speed (25) as

$$v(t) = e^{-gt/v_\tau}(v_0 + v_\tau) - v_\tau$$

(c) Show that if $s(0) = s_0$, then the position function of the object can be expressed as

$$s(t) = s_0 - v_\tau t + \frac{v_\tau}{g}(v_0 + v_\tau)(1 - e^{-gt/v_\tau})$$

48. Based on the air resistance model discussed in this section, a fully equipped sky diver weighing 240 lb would have a terminal speed of approximately 120 ft/s with a closed parachute and approximately 24 ft/s with an open parachute. Suppose that such a sky diver is dropped from an airplane at an altitude of 10,000 ft, falls for 25 s with a closed parachute, and then falls the rest of the way with an open parachute.
(a) Assuming that the sky diver's initial vertical velocity is zero, use Exercise 47 to find the sky diver's vertical velocity and height at the time the parachute opens. [Take $g = 32$ ft/s^2.]
(b) Approximate the total time, to the nearest second, that the sky diver is in the air. [*Hint:* You will not be able to solve for the time exactly, so consider using Newton's Method or the method of Example 6 of Section 2.4.]

49. The accompanying figure is a schematic diagram of a basic RL series electrical circuit that contains a power source with a time-dependent voltage of $V(t)$ volts (V), a resistor with a constant resistance of R ohms (Ω), and an inductor with a constant inductance of L henrys (H). If you don't know anything about electrical circuits, don't worry; all you need to know is that electrical theory states that a current of $I(t)$ amperes (A) flows through the circuit where $I(t)$ satisfies the differential equation

$$L\frac{dI}{dt} + RI = V(t)$$

(a) Find $I(t)$ if $R = 10\,\Omega$, $L = 4\,\text{H}$, V is a constant 12 V, and $I(0) = 0$ A.

(b) What happens to the current over a long period of time?

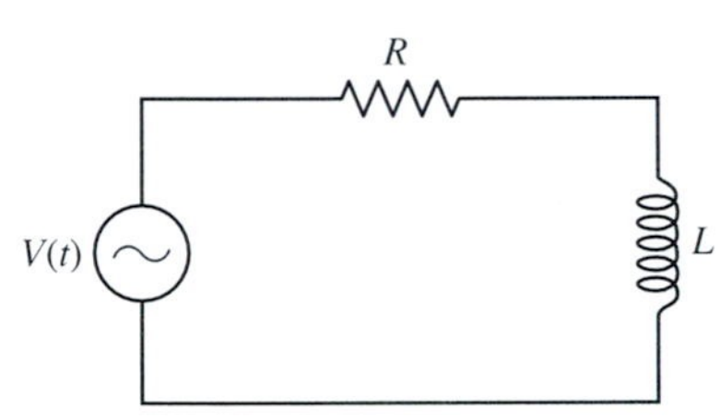

Figure Ex-49

50. Find $I(t)$ for the electrical circuit in Exercise 49 if $R = 6\,\Omega$, $L = 3$ H, $V(t) = 3\sin t$ V, and $I(0) = 15$ A.

51. A rocket, fired upward from rest at time $t = 0$, has an initial mass of m_0 (including its fuel). Assuming that the fuel is consumed at a constant rate k, the mass m of the rocket, while fuel is being burned, will be given by $m = m_0 - kt$. It can be shown that if air resistance is neglected and the fuel gases are expelled at a constant speed c relative to the rocket, then the velocity v of the rocket will satisfy the equation

$$m\frac{dv}{dt} = ck - mg$$

where g is the acceleration due to gravity.

(a) Find $v(t)$ keeping in mind that the mass m is a function of t.

(b) Suppose that the fuel accounts for 80% of the initial mass of the rocket and that all of the fuel is consumed in 100 s. Find the velocity of the rocket in meters per second at the instant the fuel is exhausted. [Take $g = 9.8$ m/s^2 and $c = 2500$ m/s.]

52. A bullet of mass m, fired straight up with an initial velocity of v_0, is slowed by the force of gravity and a drag force of air resistance kv^2, where g is the constant acceleration due to gravity and k is a positive constant. As the bullet moves upward, its velocity v satisfies the equation

$$m\frac{dv}{dt} = -(kv^2 + mg)$$

(a) Show that if $x = x(t)$ is the height of the bullet above the barrel opening at time t, then

$$mv\frac{dv}{dx} = -(kv^2 + mg)$$

(b) Express x in terms of v given that $x = 0$ when $v = v_0$.

(c) Assuming that

$$v_0 = 988 \text{ m/s}, \quad g = 9.8 \text{ m/s}^2$$
$$m = 3.56 \times 10^{-3} \text{ kg}, \quad k = 7.3 \times 10^{-6} \text{ kg/m}$$

use the result in part (b) to find out how high the bullet rises. [*Hint:* Find the velocity of the bullet at its highest point.]

The following discussion is needed for Exercises 53 and 54. Suppose that a tank containing a liquid is vented to the air at the top and has an outlet at the bottom through which the liquid can drain. It follows from ***Torricelli's law*** in physics that if the outlet is opened at time $t = 0$, then at each instant the depth of the liquid $h(t)$ and the area $A(h)$ of the liquid's surface are related by

$$A(h)\frac{dh}{dt} = -k\sqrt{h}$$

where k is a positive constant that depends on such factors as the viscosity of the liquid and the cross-sectional area of the outlet. Use this result in Exercises 53 and 54, assuming that h is in feet, $A(h)$ is in square feet, and t is in seconds. A calculator will be useful.

53. Suppose that the cylindrical tank in the accompanying figure is filled to a depth of 4 feet at time $t = 0$ and that the constant in Torricelli's law is $k = 0.025$.

(a) Find $h(t)$.

(b) How many minutes will it take for the tank to drain completely?

54. Follow the directions of Exercise 53 for the cylindrical tank in the accompanying figure, assuming that the tank is filled to a depth of 4 feet at time $t = 0$ and that the constant in Torricelli's law is $k = 0.025$.

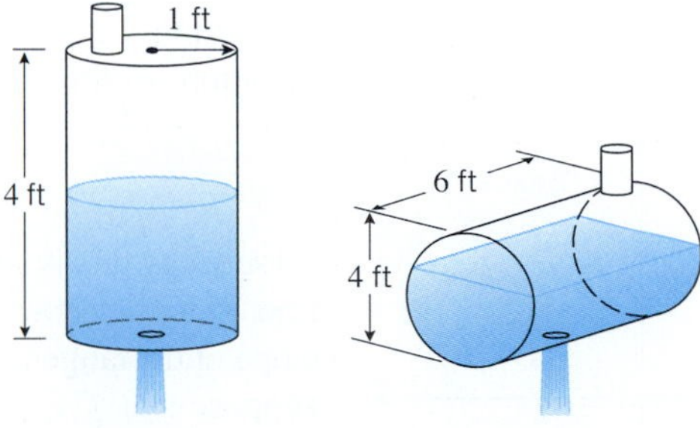

Figure Ex-53 Figure Ex-54

55. Suppose that a particle moving along the x-axis encounters a resisting force that results in an acceleration of $a = dv/dt = -0.04v^2$. Given that $x = 0$ cm and $v = 50$ cm/s at time $t = 0$, find the velocity v and position x as a function of t for $t \geq 0$.

56. Suppose that a particle moving along the x-axis encounters a resisting force that results in an acceleration of $a = dv/dt = -0.02\sqrt{v}$. Given that $x = 0$ cm and $v = 9$ cm/s at time $t = 0$, find the velocity v and position x as a function of t for $t \geq 0$.

57. Find an initial-value problem whose solution is

$$y = \cos x + \int_0^x e^{-t^2}\,dt$$

58. Derive Formula (7) for solving the separable differential equation

$$h(y)\frac{dy}{dx} = g(x)$$

by making the substitution $y = y(x)$, $dy = y'(x)\,dx$ in the integral

$$\int h(y)\,dy$$

10.2 DIRECTION FIELDS; EULER'S METHOD

In this section we will reexamine the concept of a direction field, and we will discuss a method for approximating solutions of first-order equations numerically. Numerical approximations are important in cases where the differential equation cannot be solved exactly.

FUNCTIONS OF TWO VARIABLES

We will be concerned here with first-order equations that are expressed with the derivative by itself on one side of the equation. For example,

$$y' = x^3 \quad \text{and} \quad y' = \sin(xy)$$

The first of these equations involves only x on the right side, so it has the form $y' = f(x)$. However, the second equation involves both x and y on the right side, so it has the form $y' = f(x, y)$, where the symbol $f(x, y)$ stands for a function of the two variables x and y. Later in the text we will study functions of two variables in more depth, but for now it will suffice to think of $f(x, y)$ as a formula that produces a unique output when values of x and y are given as inputs. For example, if

$$f(x, y) = x^2 + 3y$$

and if the inputs are $x = 2$ and $y = -4$, then the output is

$$f(2, -4) = 2^2 + 3(-4) = 4 - 12 = -8$$

REMARK. In applied problems involving time, it is usual to use t as the independent variable, in which case we would be concerned with equations of the form $y' = f(t, y)$, where $y' = dy/dt$.

DIRECTION FIELDS

In Section 7.2 we introduced the concept of a direction field in the context of differential equations of the form $y' = f(x)$; the same principles apply to differential equations of the form

$$y' = f(x, y)$$

To see why this is so, let us review the basic idea. If we interpret y' as the slope of a tangent line, then the differential equation states that at each point (x, y) on an integral curve, the slope of the tangent line is equal to the value of f at that point (Figure 10.2.1). For example, suppose that $f(x, y) = y - x$, in which case we have the differential equation

$$y' = y - x \tag{1}$$

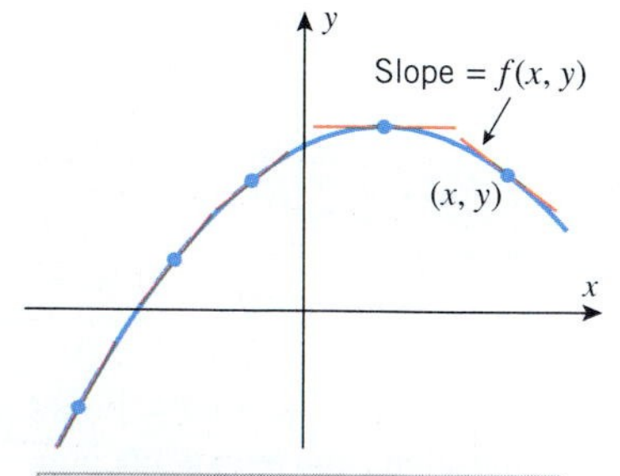

At each point (x, y) on an integral curve of $y' = f(x, y)$, the tangent line has slope $f(x, y)$.

Figure 10.2.1

A geometric description of the set of integral curves can be obtained by choosing a rectangular grid of points in the xy-plane, calculating the slopes of the tangent lines to the integral curves at the gridpoints, and drawing small segments of the tangent lines at those points. The resulting picture is called a ***direction field*** or a ***slope field*** for the differential equation because it shows the "direction" or "slope" of the integral curves at the gridpoints. The more gridpoints that are used, the better the description of the integral curves. For example, Figure 10.2.2 shows two direction fields for (1)—the first was obtained by hand calculation using the 49 gridpoints shown in the accompanying table, and the second, which gives a clearer picture of the integral curves, was obtained using 625 gridpoints and a CAS.

VALUES OF $f(x, y) = y - x$

	$y=-3$	$y=-2$	$y=-1$	$y=0$	$y=1$	$y=2$	$y=3$
$x=-3$	0	1	2	3	4	5	6
$x=-2$	−1	0	1	2	3	4	5
$x=-1$	−2	−1	0	1	2	3	4
$x=0$	−3	−2	−1	0	1	2	3
$x=1$	−4	−3	−2	−1	0	1	2
$x=2$	−5	−4	−3	−2	−1	0	1
$x=3$	−6	−5	−4	−3	−2	−1	0

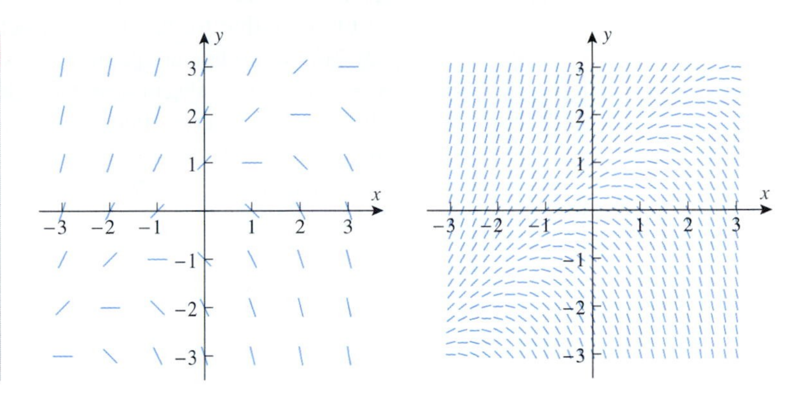

Figure 10.2.2

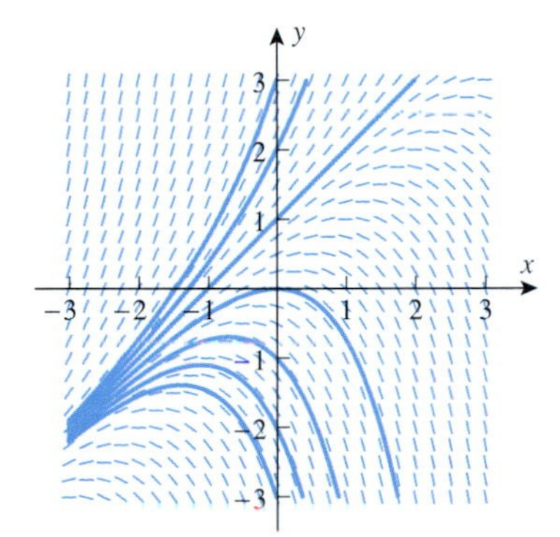

Figure 10.2.3

It so happens that Equation (1) can be solved exactly, since it can be written as

$$y' - y = -x$$

which, by comparison with Equation (10) in Section 10.1, is a first-order linear equation with $p(x) = -1$ and $q(x) = -x$. We leave it for you to use the method of integrating factors to show that the general solution of this equation is

$$y = x + 1 + Ce^x \tag{2}$$

Figure 10.2.3 shows some of the integral curves superimposed on the direction field. Observe, however, that it was not necessary to have the general solution to construct the direction field. Indeed, direction fields are important precisely because they can be constructed in cases where the differential equation cannot be solved exactly.

FOR THE READER. Confirm that the first direction field in Figure 10.2.2 is consistent with the values in the accompanying table.

Example 1

In Example 7 of Section 10.1 we considered a mixing problem in which the amount of salt $y(t)$ in a tank at time t was shown to satisfy the differential equation

$$\frac{dy}{dt} + \frac{y}{20} = 10$$

which can be rewritten as

$$y' = 10 - \frac{y}{20} \tag{3}$$

We subsequently found the general solution of this equation to be

$$y(t) = 200 + Ce^{-t/20} \tag{4}$$

and then we found the value of the arbitrary constant C from the initial condition in the problem [the known amount of salt $y(0)$ at time $t = 0$]. However, it follows from (4) that

$$\lim_{t \to +\infty} y(t) = 200$$

for all values of C, so regardless of the amount of salt that is present in the tank initially, the amount of salt in the tank will eventually begin to stabilize at 200 lb. This can also be seen geometrically from the direction field for (3) shown in Figure 10.2.4. This direction field suggests that if the amount of salt present in the tank is greater than 200 lb initially, then the amount of salt will decrease steadily over time toward a limiting value of 200 lb; and if it is less than 200 lb initially, then it will increase steadily toward a limiting value of 200 lb. The direction field also suggests that if the amount present initially is exactly 200 lb, then the amount of salt in the tank will stay constant at 200 lb. This can also be seen from (4), since $C = 0$ in this case (verify).

Observe that for the direction field shown in Figure 10.2.4 the tangent segments along any horizontal line are parallel. This occurs because the differential equation has the form $y' = f(y)$ with t absent from the right side [see (3)]. Thus, for a fixed y the slope y' does not change as time varies. Because of this time independence of slope, differential equations of the form $y' = f(y)$ are said to be ***autonomous*** (from the Greek word *autonomous*, meaning "independent"). ◀

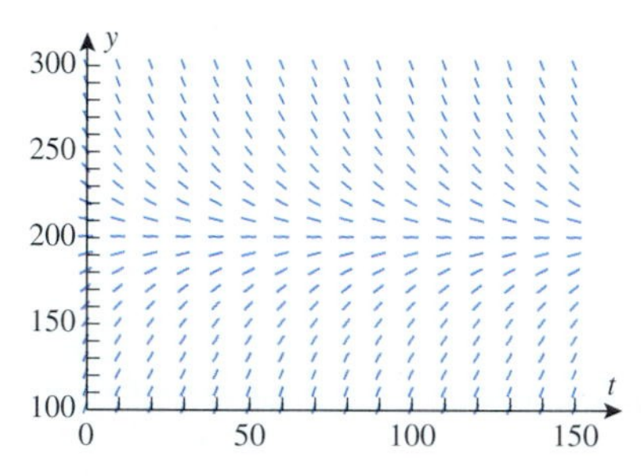

Figure 10.2.4

EULER'S METHOD

Our next objective is to develop a method for approximating the solution of an initial-value problem of the form

$$y' = f(x, y), \quad y(x_0) = y_0$$

We will not attempt to approximate $y(x)$ for all values of x; rather, we will choose some small increment h and focus on approximating the values of $y(x)$ at a succession of x-values spaced h units apart, starting from x_0. We will denote these x-values by

$$x_1 = x_0 + h, \quad x_2 = x_1 + h, \quad x_3 = x_2 + h, \quad x_4 = x_3 + h, \ldots$$

and we will denote the approximations of $y(x)$ at these points by

$$y_1 \approx y(x_1), \quad y_2 \approx y(x_2), \quad y_3 \approx y(x_3), \quad y_4 \approx y(x_4), \ldots$$

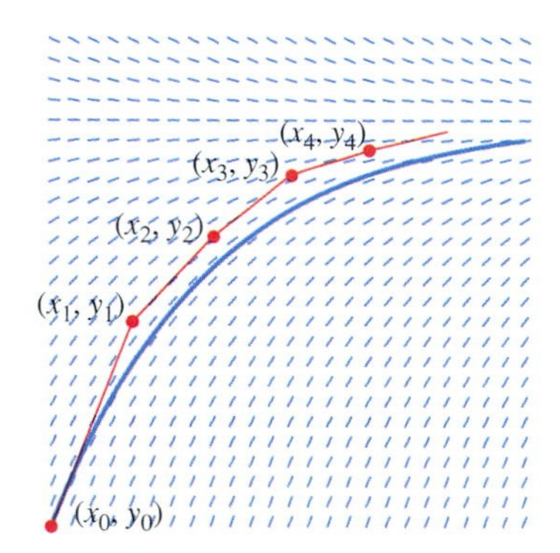

Figure 10.2.5

The technique that we will describe for obtaining these approximations is called ***Euler's Method***. Although there are better approximation methods available, many of them use Euler's Method as a starting point, so the underlying concepts are important to understand.

The basic idea behind Euler's Method is to start at the known initial point (x_0, y_0) and draw a line segment in the direction determined by the direction field until we reach the point (x_1, y_1) with x-coordinate $x_1 = x_0 + h$ (Figure 10.2.5). If h is small, then it is reasonable to expect that this line segment will not deviate much from the integral curve $y = y(x)$, and thus y_1 should closely approximate $y(x_1)$. To obtain the subsequent approximations, we repeat the process using the direction field as a guide at each step. Starting at the endpoint (x_1, y_1), we draw a line segment determined by the direction field until we reach the point (x_2, y_2) with x-coordinate $x_2 = x_1 + h$, and from that point we draw a line segment determined by the direction field to the point (x_3, y_3) with x-coordinate $x_3 = x_2 + h$, and so forth. As indicated in Figure 10.2.5, this procedure produces a polygonal path that tends to follow the integral curve closely, so it is reasonable to expect that the y-values $y_2, y_3, y_4, \ldots$ will closely approximate $y(x_2), y(x_3), y(x_4), \ldots$.

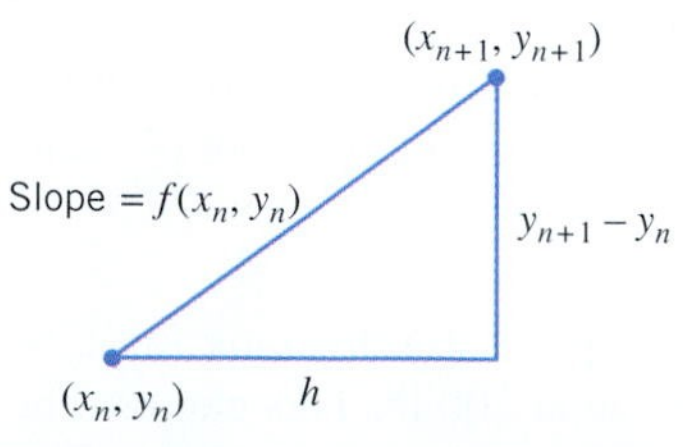

Figure 10.2.6

To explain how the approximations $y_1, y_2, y_3, \ldots$ can be computed, let us focus on a typical line segment. As indicated in Figure 10.2.6, assume that we have found the point (x_n, y_n), and we are trying to determine the next point (x_{n+1}, y_{n+1}), where $x_{n+1} = x_n + h$.

Since the slope of the line segment joining the points is determined by the direction field at the starting point, the slope is $f(x_n, y_n)$, and hence

$$\frac{y_{n+1} - y_n}{h} = f(x_n, y_n)$$

which we can rewrite as

$$y_{n+1} = y_n + f(x_n, y_n)h$$

This formula, which is the heart of Euler's Method, tells us how to use each approximation to compute the next approximation.

Euler's Method

To approximate the solution of the initial-value problem

$$y' = f(x, y), \quad y(x_0) = y_0$$

proceed as follows:

Step 1. Choose a nonzero number h to serve as an ***increment*** or ***step size*** along the x-axis, and let

$$x_1 = x_0 + h, \quad x_2 = x_1 + h, \quad x_3 = x_2 + h, \ldots$$

Step 2. Compute successively

$$\begin{aligned} y_1 &= y_0 + f(x_0, y_0)h \\ y_2 &= y_1 + f(x_1, y_1)h \\ y_3 &= y_2 + f(x_2, y_2)h \\ &\vdots \\ y_{n+1} &= y_n + f(x_n, y_n)h \end{aligned}$$

The numbers $y_1, y_2, y_3, \ldots$ in these equations are the approximations of $y(x_1), y(x_2), y(x_3), \ldots$.

Example 2

Use Euler's Method with a step size of 0.1 to make a table of approximate values of the solution of the initial-value problem

$$y' = y - x, \quad y(0) = 2 \tag{5}$$

over the interval $0 \le x \le 1$.

Solution. In this problem we have $f(x, y) = y - x$, $x_0 = 0$, and $y_0 = 2$. Moreover, since the step size is 0.1, the x-values at which the approximate values will be obtained are

$$x_1 = 0.1, \quad x_2 = 0.2, \quad x_3 = 0.3, \ldots, \quad x_9 = 0.9, \quad x_{10} = 1$$

The first three approximations are

$$y_1 = y_0 + f(x_0, y_0)h = 2 + (2 - 0)(0.1) = 2.2$$

$$y_2 = y_1 + f(x_1, y_1)h = 2.2 + (2.2 - 0.1)(0.1) = 2.41$$

$$y_3 = y_2 + f(x_2, y_2)h = 2.41 + (2.41 - 0.2)(0.1) = 2.631$$

Here is a way of organizing all 10 approximations rounded to five decimal places:

EULER'S METHOD FOR $y' = y - x$, $y(0) = 2$ WITH $h = 0.1$

n	x_n	y_n	$f(x_n, y_n)h$	$y_{n+1} = y_n + f(x_n, y_n)h$
0	0	2.00000	0.20000	2.20000
1	0.1	2.20000	0.21000	2.41000
2	0.2	2.41000	0.22100	2.63100
3	0.3	2.63100	0.23310	2.86410
4	0.4	2.86410	0.24641	3.11051
5	0.5	3.11051	0.26105	3.37156
6	0.6	3.37156	0.27716	3.64872
7	0.7	3.64872	0.29487	3.94359
8	0.8	3.94359	0.31436	4.25795
9	0.9	4.25795	0.33579	4.59374
10	1.0	4.59374	—	—

Observe that each entry in the last column becomes the next entry in the third column. ◀

ACCURACY OF EULER'S METHOD

It follows from (5) and the initial condition $y(0) = 2$ that the exact solution of the initial-value problem in Example 2 is

$$y = x + 1 + e^x$$

Thus, in this case we can compare the approximate values of $y(x)$ produced by Euler's Method with decimal approximations of the exact values (Table 10.2.1). In Table 10.2.1 the ***absolute error*** is calculated as

$$|\text{exact value} - \text{approximation}|$$

and the ***percentage error*** as

$$\frac{|\text{exact value} - \text{approximation}|}{|\text{exact value}|} \times 100\%$$

REMARK. As a rough rule of thumb, the absolute error in an approximation produced by Euler's Method is proportional to the step size; thus, reducing the step size by half reduces the absolute error (and hence the percentage error) by roughly half. However, reducing the step size also increases the amount of computation, thereby increasing the potential for round-off error. We will leave a detailed study of error issues for courses in differential equations or numerical analysis.

Table 10.2.1

x	EXACT SOLUTION	EULER APPROXIMATION	ABSOLUTE ERROR	PERCENTAGE ERROR
0	2.00000	2.00000	0.00000	0.00
0.1	2.20517	2.20000	0.00517	0.23
0.2	2.42140	2.41000	0.01140	0.47
0.3	2.64986	2.63100	0.01886	0.71
0.4	2.89182	2.86410	0.02772	0.96
0.5	3.14872	3.11051	0.03821	1.21
0.6	3.42212	3.37156	0.05056	1.48
0.7	3.71375	3.64872	0.06503	1.75
0.8	4.02554	3.94359	0.08195	2.04
0.9	4.35960	4.25795	0.10165	2.33
1.0	4.71828	4.59374	0.12454	2.64

EXERCISE SET 10.2 Graphing Calculator C CAS

1. Sketch the direction field for $y' = xy/8$ at the gridpoints (x, y), where $x = 0, 1, \ldots, 4$ and $y = 0, 1, \ldots, 4$.

2. Sketch the direction field for $y' + y = 2$ at the gridpoints (x, y), where $x = 0, 1, \ldots, 4$ and $y = 0, 1, \ldots, 4$.

3. A direction field for the differential equation $y' = 1 - y$ is shown in the accompanying figure. In each part, sketch the graph of the solution that satisfies the initial condition.
(a) $y(0) = -1$ (b) $y(0) = 1$ (c) $y(0) = 2$

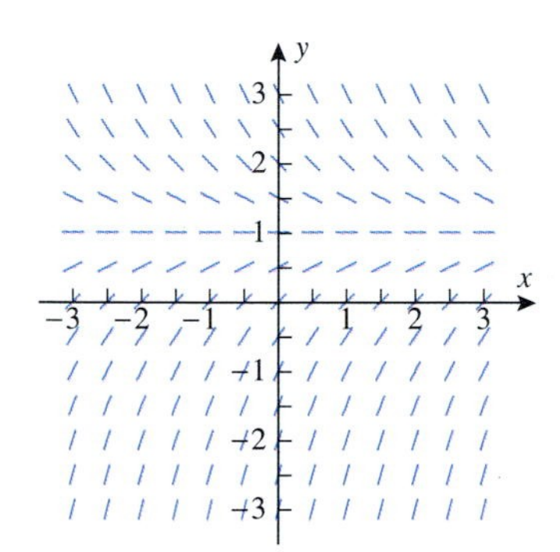

Figure Ex-3

4. Solve the initial-value problems in Exercise 3, and use a graphing utility to confirm that the integral curves for these solutions are consistent with the sketches you obtained from the direction field.

5. A direction field for the differential equation $y' = 2y - x$ is shown in the accompanying figure. In each part, sketch the graph of the solution that satisfies the initial condition.
(a) $y(1) = 1$ (b) $y(0) = -1$ (c) $y(-1) = 0$

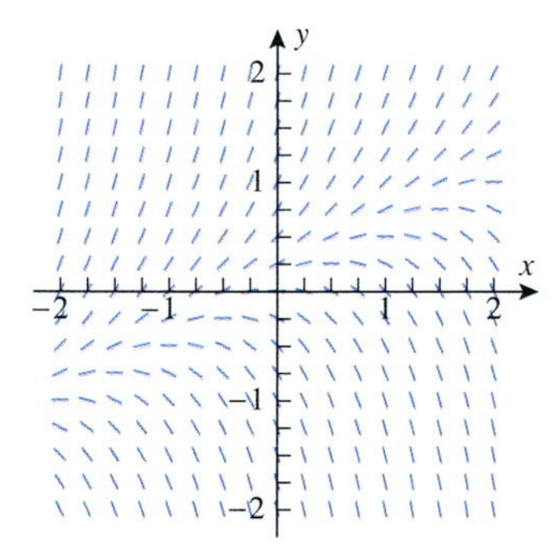

Figure Ex-5

6. Solve the initial-value problems in Exercise 5, and use a graphing utility to confirm that the integral curves for these solutions are consistent with the sketches you obtained from the direction field.

7. Use the direction field in Exercise 3 to make a conjecture about the behavior of the solutions of $y' = 1 - y$ as $x \to +\infty$, and confirm your conjecture by examining the general solution of the equation.

8. Use the direction field in Exercise 5 to make a conjecture about the effect of y_0 on the behavior of the solution of the initial-value problem $y' = 2y - x$, $y(0) = y_0$ as $x \to +\infty$, and check your conjecture by examining the solution of the initial-value problem.

9. In each part, match the differential equation with the direction field (see next page), and explain your reasoning.
(a) $y' = 1/x$ (b) $y' = 1/y$ (c) $y' = e^{-x^2}$
(d) $y' = y^2 - 1$ (e) $y' = \dfrac{x + y}{x - y}$
(f) $y' = (\sin x)(\sin y)$

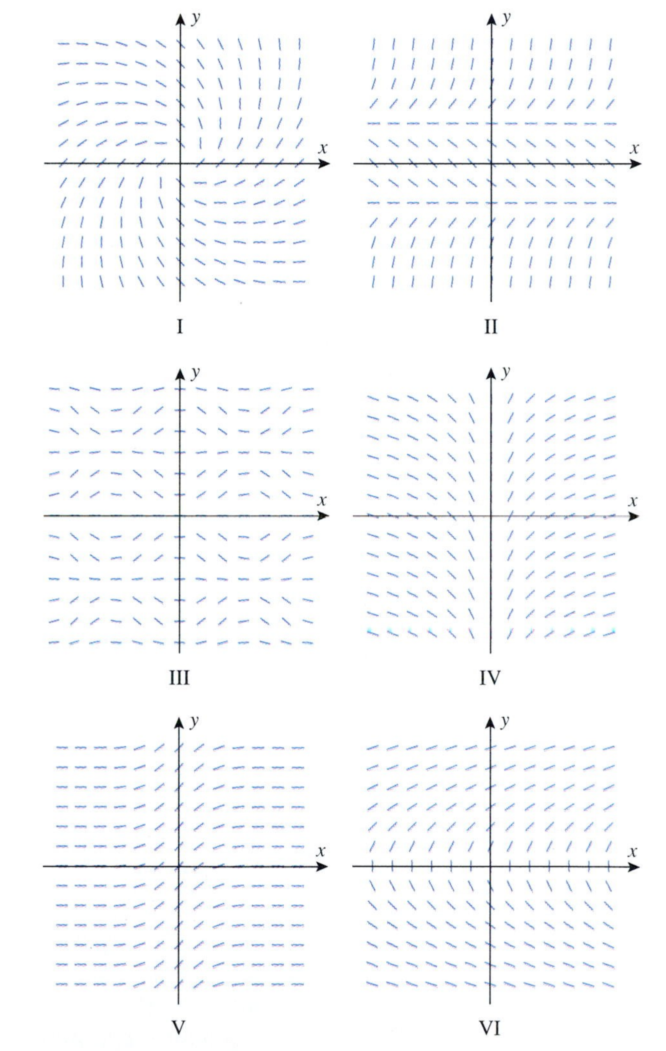

Figure Ex-9

10. If you have a CAS or a graphing utility that can generate direction fields, read the documentation on how to do it and check your answers in Exercise 9 by generating the direction fields for the differential equations.

11. (a) Use Euler's Method with a step size of $h = 0.2$ to approximate the solution of the initial-value problem

$$y' = x + y, \quad y(0) = 1$$

over the interval $0 \le x \le 1$.

(b) Solve the initial-value problem exactly, and calculate the error and the percentage error in each of the approximations in part (a).
(c) Sketch the exact solution and the approximate solution together.

12. It was stated at the end of this section that reducing the step size in Euler's Method by half reduces the error in each approximation by about half. Confirm that the error in $y(1)$ is reduced by about half if a step size of $h = 0.1$ is used in Exercise 11.

In Exercises 13–16, use Euler's Method with the given step size h to approximate the solution of the initial-value problem over the stated interval. Present your answer as a table and as a graph.

13. $dy/dx = \sqrt{y},\ y(0) = 1,\ 0 \le x \le 4,\ h = 0.5$

14. $dy/dx = x - y^2,\ y(0) = 1,\ 0 \le x \le 2,\ h = 0.25$

15. $dy/dt = \sin y,\ y(0) = 1,\ 0 \le t \le 2,\ h = 0.5$

16. $dy/dt = e^{-y},\ y(0) = 0,\ 0 \le t \le 1,\ h = 0.1$

17. Consider the initial-value problem

$$y' = \cos 2\pi t, \quad y(0) = 1$$

Use Euler's Method with five steps to approximate $y(1)$.

18. (a) Show that the solution of the initial-value problem $y' = e^{-x^2}$, $y(0) = 0$ is

$$y(x) = \int_0^x e^{-t^2}\,dt$$

(b) Use Euler's Method with $h = 0.05$ to approximate the value of

$$y(1) = \int_0^1 e^{-t^2}\,dt$$

and compare the answer to that produced by a calculating utility with a numerical integration capability.

19. The accompanying figure shows a direction field for the differential equation $y' = -x/y$.
(a) Use the direction field to estimate $y(\frac{1}{2})$ for the solution that satisfies the given initial condition $y(0) = 1$.
(b) Compare your estimate to the exact value of $y(\frac{1}{2})$.

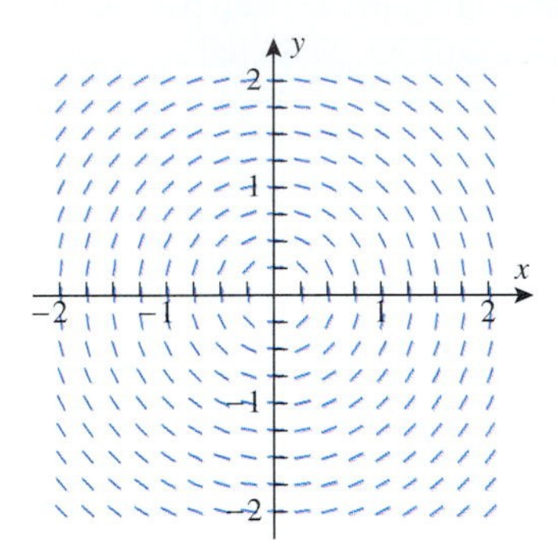

Figure Ex-19

20. Consider the initial-value problem

$$\frac{dy}{dx} = \frac{\sqrt{y}}{2}, \quad y(0) = 1$$

(a) Use Euler's Method with step sizes of $h = 0.2$, 0.1, and 0.05 to obtain three approximations of $y(1)$.
(b) Plot the three approximations versus h, and make a conjecture about the exact value of $y(1)$. Explain your reasoning.
(c) Check your conjecture by finding $y(1)$ exactly.

10.3 MODELING WITH DIFFERENTIAL EQUATIONS

Since many of the fundamental laws of the physical and social sciences involve rates of change, it should not be surprising that such laws are modeled by differential equations. In this section we will discuss the general idea of modeling with differential equations, and we will investigate some important models that can be applied to population growth, carbon dating, medicine, ecology, and the vibration of springs.

POPULATION GROWTH

One of the simplest models of population growth is based on the observation that when populations (people, plants, bacteria, and fruit flies, for example) are not constrained by environmental limitations, they tend to grow at a rate that is proportional to the size of the population—the larger the population, the more rapidly it grows.

To translate this principle into a mathematical model, suppose that $y = y(t)$ denotes the population at time t. At each point in time, the rate of increase of the population with respect to time is dy/dt, so the assumption that the rate of growth is proportional to the population is described by the differential equation

$$\frac{dy}{dt} = ky \tag{1}$$

where k is a positive constant of proportionality that can usually be determined experimentally. Thus, if the population is known at some point in time, say $y = y_0$ at time $t = 0$, then a general formula for the population $y(t)$ can be obtained by solving the initial-value problem

$$\frac{dy}{dt} = ky, \quad y(0) = y_0$$

PHARMACOLOGY

When a drug (say, penicillin or aspirin) is administered to an individual, it enters the bloodstream and then is absorbed by the body over time. Medical research has shown that the amount of a drug that is present in the bloodstream tends to decrease at a rate that is proportional to the amount of the drug present—the more of the drug that is present in the bloodstream, the more rapidly it is absorbed by the body.

To translate this principle into a mathematical model, suppose that $y = y(t)$ is the amount of the drug present in the bloodstream at time t. At each point in time, the rate of change in y with respect to t is dy/dt, so the assumption that the rate of decrease is proportional to the amount y in the bloodstream translates into the differential equation

$$\frac{dy}{dt} = -ky \tag{2}$$

where k is a positive constant of proportionality that depends on the drug and can be determined experimentally. The negative sign is required because y decreases with time. Thus, if the initial dosage of the drug is known, say $y = y_0$ at time $t = 0$, then a general formula for $y(t)$ can be obtained by solving the initial-value problem

$$\frac{dy}{dt} = -ky, \quad y(0) = y_0$$

SPREAD OF DISEASE

Suppose that a disease begins to spread in a population of L individuals. Logic suggests that at each point in time the rate at which the disease spreads will depend on how many individuals are already affected and how many are not—as more individuals are affected, the opportunity to spread the disease tends to increase, but at the same time there are fewer individuals who are not affected, so the opportunity to spread the disease tends to decrease. Thus, there are two conflicting influences on the rate at which the disease spreads.

To translate this into a mathematical model, suppose that $y = y(t)$ is the number of individuals who have the disease at time t, so of necessity the number of individuals who do not have the disease at time t is $L - y$. As the value of y increases, the value of $L - y$ decreases, so the conflicting influences of the two factors on the rate of spread dy/dt are taken into account by the differential equation

$$\frac{dy}{dt} = ky(L - y)$$

where k is a positive constant of proportionality that depends on the nature of the disease and the behavior patterns of the individuals and can be determined experimentally. Thus, if the number of affected individuals is known at some point in time, say $y = y_0$ at time $t = 0$, then a general formula for $y(t)$ can be obtained by solving the initial-value problem

$$\frac{dy}{dt} = ky(L - y), \quad y(0) = y_0 \tag{3}$$

INHIBITED POPULATION GROWTH

The population growth model that we discussed at the beginning of this section was predicated on the assumption that the population $y = y(t)$ is not constrained by the environment. For this reason, it is sometimes called the ***uninhibited growth model***. However, in the real world this assumption is usually not valid—populations generally grow within ecological systems that can only support a certain number of individuals; the number L of such individuals is called the ***carrying capacity*** of the system. Thus, when $y > L$, the population exceeds the capacity of the ecological system and tends to decrease toward L; when $y < L$, the population is below the capacity of the ecological system and tends to increase toward L; and when $y = L$, the population is in balance with the capacity of the ecological system and tends to remain stable.

To translate this into a mathematical model, we must look for a differential equation in which

$$\frac{dy}{dt} < 0 \quad \text{if} \quad \frac{y}{L} > 1$$

$$\frac{dy}{dt} > 0 \quad \text{if} \quad \frac{y}{L} < 1$$

$$\frac{dy}{dt} = 0 \quad \text{if} \quad \frac{y}{L} = 1$$

Moreover, logic suggests that when the population is far below the carrying capacity (i.e., $y/L \approx 0$), then the environmental constraints should have little effect, and the growth rate should behave very much like the uninhibited model. Thus, we want

$$\frac{dy}{dt} \approx ky \quad \text{if} \quad \frac{y}{L} \approx 0$$

A simple differential equation that meets all of these requirements is

$$\frac{dy}{dt} = k\left(1 - \frac{y}{L}\right)y$$

where k is a positive constant of proportionality. Thus, if k and L can be determined experimentally, and if the population is known at some point in time, say $y(0) = y_0$, then a general formula for the population $y(t)$ can be determined by solving the initial-value problem

$$\frac{dy}{dt} = k\left(1 - \frac{y}{L}\right)y, \quad y(0) = y_0 \tag{4}$$

This theory of population growth is due to the Belgian mathematician, P. F. Verhulst (1804–1849), who introduced it in 1838 and described it as "logistic growth."* Thus, the differential equation in (4) is called the ***logistic differential equation***, and the growth model described by (4) is called the ***logistic model*** or the ***inhibited growth model***.

REMARK. Observe that the differential equation in (3) can be expressed as

$$\frac{dy}{dt} = kL\left(1 - \frac{y}{L}\right)y$$

which is a logistic equation with kL rather than k as the constant of proportionality. Thus, this model for the spread of disease is also a logistic or inhibited growth model.

EXPONENTIAL GROWTH AND DECAY MODELS

Equations (1) and (2) are examples of a general class of models called *exponential models*. In general, exponential models arise in situations where a quantity increases or decreases at a rate that is proportional to the amount of the quantity present. More precisely, we make the following definition:

* Verhulst's model fell into obscurity for nearly a hundred years because he did not have sufficient census data to test its validity. However, interest in the model was revived in the 1930s when biologists used it successfully to describe the growth of fruit fly and flour beetle populations. Verhulst himself used the model to predict that an upper limit on Belgium's population would be approximately 9,400,000. In 1994 the population was about 10,118,000.

> **10.3.1 DEFINITION.** A quantity $y = y(t)$ is said to have an ***exponential growth model*** if it increases at a rate that is proportional to the amount of the quantity present, and it is said to have an ***exponential decay model*** if it decreases at a rate that is proportional to the amount of the quantity present. Thus, for an exponential growth model, the quantity $y(t)$ satisfies an equation of the form
>
> $$\frac{dy}{dt} = ky \quad (k > 0) \tag{5}$$
>
> and for an exponential decay model, the quantity $y(t)$ satisfies an equation of the form
>
> $$\frac{dy}{dt} = -ky \quad (k > 0) \tag{6}$$
>
> The constant k is called the ***growth constant*** or the ***decay constant***, as appropriate.

Equations (5) and (6) are first-order linear equations, since they can be rewritten as

$$\frac{dy}{dt} - ky = 0 \quad \text{and} \quad \frac{dy}{dt} + ky = 0$$

both of which have the form of Equation (10) in Section 10.1 (but with t rather than x as the independent variable); in the first equation we have $p(t) = -k$ and $q(t) = 0$, and in the second we have $p(t) = k$ and $q(t) = 0$.

To illustrate how these equations can be solved, suppose that a quantity $y = y(t)$ has an exponential growth model and we know the amount of the quantity at some point in time, say $y = y_0$ when $t = 0$. Thus, a general formula for $y(t)$ can be obtained by solving the initial-value problem

$$\frac{dy}{dt} - ky = 0, \quad y(0) = y_0$$

Multiplying the differential equation through by the integrating factor

$$\mu = e^{\int -k\,dt} = e^{-kt}$$

yields

$$\frac{d}{dt}(e^{-kt}y) = 0$$

and then integrating with respect to t yields

$$e^{-kt}y = C \quad \text{or} \quad y = Ce^{kt}$$

The initial condition implies that $y = y_0$ when $t = 0$, from which it follows that $C = y_0$ (verify). Thus, the solution of the initial-value problem is

$$y = y_0e^{kt} \tag{7}$$

We leave it for you to show that if $y = y(t)$ has an exponential decay model, and if $y(0) = y_0$, then

$$y = y_0e^{-kt} \tag{8}$$

INTERPRETING THE GROWTH AND DECAY CONSTANTS

The significance of the constant k in Formulas (7) and (8) can be understood by reexamining the differential equations that gave rise to these formulas. For example, in the case of the exponential growth model, Equation (5) can be rewritten as

$$k = \frac{dy/dt}{y}$$

which states that the growth rate as a fraction of the entire population remains constant over time, and this constant is k. For this reason, k is called the ***relative growth rate*** of the population. It is usual to express the relative growth rate as a percentage. Thus, a relative growth rate of 3% per unit of time in an exponential growth model means that $k = 0.03$. Similarly, the constant k in an exponential decay model is called the ***relative decay rate***.

REMARK. It is standard practice in applications to call the relative growth rate the *growth rate*, even though it is not really correct (the growth rate is dy/dt). However, the practice is so common that we will follow it here.

Example 1

According to United Nations data, the world population at the beginning of 1990 was approximately 5.3 billion and growing at a rate of about 2% per year. Assuming an exponential growth model, estimate the world population at the beginning of the year 2015.

Solution. Let

t = time elapsed from the beginning of 1990 (in years)

y = world population (in billions)

Since the beginning of 1990 corresponds to $t = 0$, it follows from the given data that

$$y_0 = y(0) = 5.3 \text{ (billion)}$$

Since the growth rate is 2% ($k = 0.02$), it follows from (7) that the world population at time t will be

$$y(t) = y_0 e^{kt} = 5.3e^{0.02t} \tag{9}$$

Since the beginning of the year 2015 corresponds to an elapsed time of $t = 25$ years ($2015 - 1990 = 25$), it follows from (9) that the world population by the year 2015 will be

$$y(25) = 5.3e^{0.02(25)} = 5.3e^{0.5} \approx 8.7$$

which is a population of approximately 8.7 billion. ◀

REMARK. In this example, the growth rate was given, so there was no need to calculate it. If the growth rate or decay rate in an exponential model is unknown, then it can be calculated using the initial condition and the value of y at one other point in time (Exercise 42).

DOUBLING TIME AND HALF-LIFE

If a quantity y has an exponential growth model, then the time required for the original size to double is called the ***doubling time***, and if y has an exponential decay model, then the time required for the original size to reduce by half is called the ***half-life***. As it turns out, doubling time and half-life depend only on the growth or decay rate and not on the amount present initially. To see why this is so, suppose that $y = y(t)$ has an exponential growth model

$$y = y_0 e^{kt} \tag{10}$$

and let T denote the amount of time required for y to double in size. Thus, at time $t = T$ the value of y will be $2y_0$, and hence from (10)

$$2y_0 = y_0 e^{kT} \quad \text{or} \quad e^{kT} = 2$$

Taking the natural logarithm of both sides yields $kT = \ln 2$, which implies that the doubling time is

$$T = \frac{1}{k}\ln 2 \tag{11}$$

We leave it as an exercise to show that Formula (11) also gives the half-life of an exponential decay model. Observe that this formula does not involve the initial amount y_0, so that in an exponential growth or decay model, the quantity y doubles (or reduces by half) every T units (Figure 10.3.1).

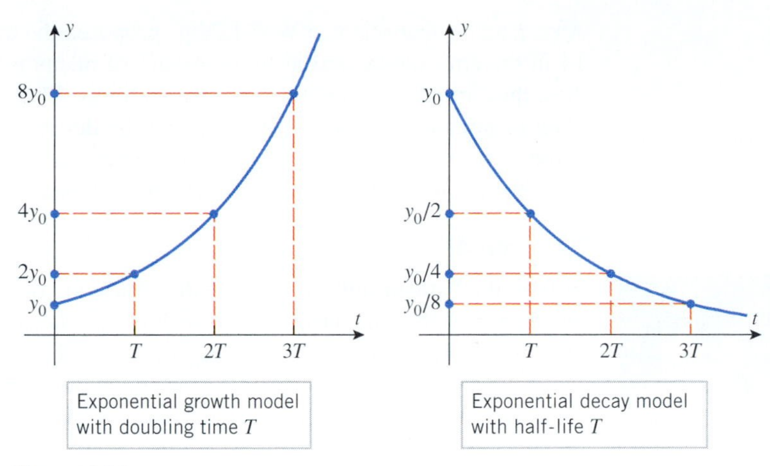

Figure 10.3.1

Example 2

It follows from (11) that with a continued growth rate of 2% per year, the doubling time for the world population will be

$$T = \frac{1}{0.02}\ln 2 \approx 34.657$$

or approximately 35 years. Thus, with a continued 2% annual growth rate the population of 5.3 billion in 1990 will double to 10.6 billion by the year 2025 and will double again to 21.2 billion by 2060. ◀

RADIOACTIVE DECAY

It is a fact of physics that radioactive elements disintegrate spontaneously in a process called ***radioactive decay***. Experimentation has shown that the rate of disintegration is proportional to the amount of the element present, which implies that the amount $y = y(t)$ of a radioactive element present as a function of time has an exponential decay model.

Every radioactive element has a specific half-life; for example, the half-life of radioactive carbon-14 is about 5730 years. Thus, from (11), the decay constant for this element is

$$k = \frac{1}{T}\ln 2 = \frac{\ln 2}{5730} \approx 0.000121$$

and this implies that if there are y_0 units of carbon-14 present at time $t = 0$, then the number of units present at a time t will be approximately

$$y(t) = y_0 e^{-0.000121t} \tag{12}$$

Example 3

If 100 grams of radioactive carbon-14 are stored in a cave for 1000 years, how many grams will be left at that time?

Solution. From (12) with $y_0 = 100$ and $t = 1000$, we obtain

$$y(1000) = 100e^{-0.000121(1000)} = 100e^{-0.121} \approx 88.6$$

Thus, about 88.6 grams will be left. ◀

CARBON DATING

When the nitrogen in the Earth's upper atmosphere is bombarded by cosmic radiation, the radioactive element carbon-14 is produced. This carbon-14 combines with oxygen to form carbon dioxide, which is ingested by plants, which in turn are eaten by animals. In this way all living plants and animals absorb quantities of radioactive carbon-14. In 1947 the

American nuclear scientist W. F. Libby* proposed the theory that the percentage of carbon-14 in the atmosphere and in living tissues of plants is the same. When a plant or animal dies, the carbon-14 in the tissue begins to decay. Thus, the age of an artifact that contains plant or animal material can be estimated by determining what percentage of its original carbon-14 content remains. Various procedures, called ***carbon dating*** or ***carbon-14 dating***, have been developed for measuring this percentage.

Example 4

The Shroud of Turin

In 1988 the Vatican authorized the British Museum to date a cloth relic known as the Shroud of Turin, possibly the burial shroud of Jesus of Nazareth. This cloth, which first surfaced in 1356, contains the negative image of a human body that was widely believed to be that of Jesus. The report of the British Museum showed that the fibers in the cloth contained between 92% and 93% of their original carbon-14. Use this information to estimate the age of the shroud.

Solution. From (12), the fraction of the original carbon-14 that remains after t years is

$$\frac{y(t)}{y_0} = e^{-0.000121t}$$

Taking the natural logarithm of both sides and solving for t, we obtain

$$t = -\frac{1}{0.000121}\ln\left(\frac{y(t)}{y_0}\right)$$

Thus, taking $y(t)/y_0$ to be 0.93 and 0.92, we obtain

$$t = -\frac{1}{0.000121}\ln(0.93) \approx 600$$

$$t = -\frac{1}{0.000121}\ln(0.92) \approx 689$$

This means that when the test was done in 1988, the shroud was between 600 and 689 years old, thereby placing its origin between 1299 A.D. and 1388 A.D. Thus, if one accepts the validity of carbon-14 dating, the Shroud of Turin cannot be the burial shroud of Jesus of Nazareth. ◀

LOGISTIC MODELS

Recall that the logistic model of population growth in an ecological system with carrying capacity L is determined by initial-value problem (4). To illustrate how this initial-value problem can be solved for $y(t)$, let us focus on the differential equation

$$\frac{dy}{dt} = k\left(1 - \frac{y}{L}\right)y \tag{13}$$

It will be convenient to rewrite Equation (13) as

$$\frac{dy}{dt} = \frac{k}{L}(L-y)y = \frac{k}{L}y(L-y)$$

This equation is separable, since it can be rewritten in differential form as

$$\frac{L}{y(L-y)}\,dy = k\,dt$$

Integrating both sides yields the equation

$$\int \frac{L}{y(L-y)}\,dy = \int k\,dt$$

*W. F. Libby, "Radiocarbon Dating," *American Scientist*, Vol. 44, 1956, pp. 98–112.

Using partial fractions on the left side, we can rewrite this equation as (verify)

$$\int \left(\frac{1}{y} + \frac{1}{L-y}\right) dy = \int k\,dt$$

Integrating and rearranging the form of the result, we obtain

$$\ln|y| - \ln|L-y| = kt + C$$

$$\ln\left|\frac{y}{L-y}\right| = kt + C$$

$$\left|\frac{y}{L-y}\right| = e^{kt+C}$$

$$\left|\frac{L-y}{y}\right| = e^{-kt-C} = e^{-C}e^{-kt}$$

$$\frac{L-y}{y} = \pm e^{-C}e^{-kt}$$

$$\frac{L}{y} - 1 = Ae^{-kt} \quad (\text{where } A = \pm e^{-C})$$

Solving this equation for y yields (verify)

$$y = \frac{L}{1 + Ae^{-kt}} \tag{14}$$

As the final step, we want to use the initial condition in (4) to determine the constant A. But the initial condition implies that $y = y_0$ if $t = 0$, so from (14)

$$y_0 = \frac{L}{1 + A}$$

from which we obtain

$$A = \frac{L - y_0}{y_0}$$

Thus, the solution of the initial-value problem (4) is

$$y = \frac{L}{1 + \left(\dfrac{L - y_0}{y_0}\right)e^{-kt}}$$

which can be rewritten more simply as

$$y = \frac{y_0 L}{y_0 + (L - y_0)e^{-kt}} \tag{15}$$

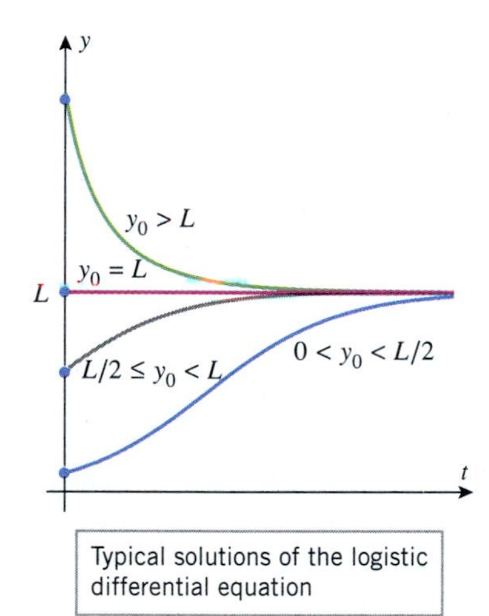

Figure 10.3.2

The graph of (15) has one of four general shapes, depending on the relationship between the initial population y_0 and the carrying capacity L (Figure 10.3.2).

Example 5

Figure 10.3.3 shows the graph of a population $y = y(t)$ with a logistic growth model. Estimate the values of y_0, L, and k, and use the estimates to deduce a formula for y as a function of t.

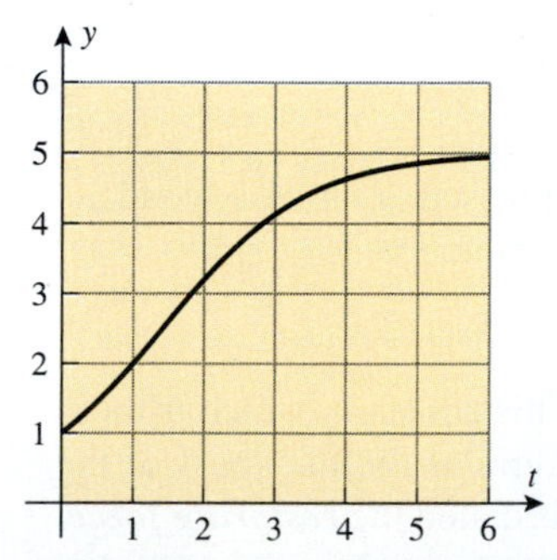

Figure 10.3.3

Solution. The graph suggests that the carrying capacity is $L = 5$, and the population at time $t = 0$ is $y_0 = 1$. Thus, from (15), the equation has the form

$$y = \frac{5}{1 + 4e^{-kt}} \tag{16}$$

where k must still be determined. However, the graph passes through the point $(1, 2)$, which tells us that $y = 2$ if $t = 1$. Substituting these values in (16) yields

$$2 = \frac{5}{1 + 4e^{-k}}$$

Solving for k we obtain (verify)

$$k = \log \tfrac{8}{3} \approx 0.98$$

and substituting this in (16) yields

$$y = \frac{5}{1 + 4e^{-0.98t}}$$

◀

VIBRATIONS OF SPRINGS

We conclude this section with an engineering model that leads to a second-order differential equation.

As shown in Figure 10.3.4, consider a block of mass m that is suspended from a vertical spring and allowed to settle into an ***equilibrium position***. Assume that the block is then set into vertical vibratory motion by pulling or pushing on it and releasing it at time $t = 0$. We will be interested in finding a mathematical model that describes the vibratory motion of the block over time.

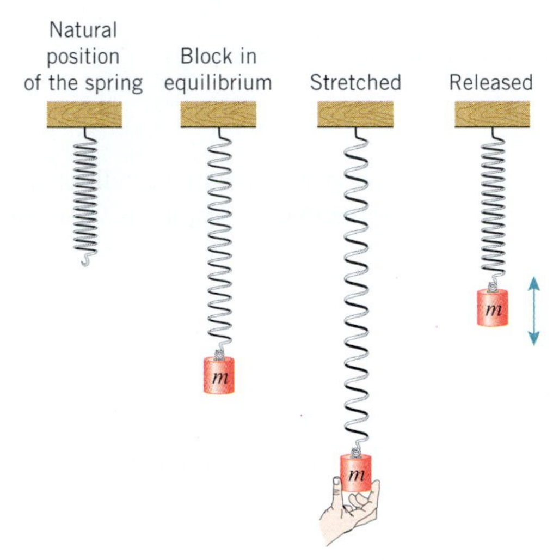

Figure 10.3.4

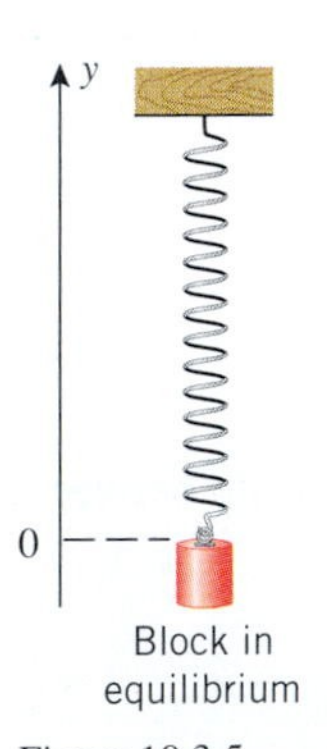

Figure 10.3.5

To translate this problem into mathematical form, we introduce a vertical y-axis whose positive direction is up and whose origin is at the connection of the spring to the block when the block is in equilibrium (Figure 10.3.5). Our goal is to find the coordinate $y = y(t)$ of the top of the block as a function of time. For this purpose we will need Newton's Second Law of Motion,

$$F = ma$$

[see (19) in Section 10.1], as well as the following two results from physics:

10.3.2 HOOKE'S LAW. If a spring is stretched (or compressed) l units beyond its natural position, then it pulls (or pushes) with a force of magnitude

$$F = kl$$

where k is a positive constant, called the ***spring constant***. This constant, which is measured in units of force per unit length, depends on such factors as the thickness of the spring and its composition. The force exerted by the spring is called the ***restoring force***.

> **10.3.3 WEIGHT.** The gravitational force exerted by the Earth on an object is called the object's ***weight*** (or more precisely, its ***Earth weight***). It follows from Newton's Second Law of Motion that an object with mass m has a weight w of magnitude mg, where g is the acceleration due to gravity. However, if the positive direction is up, as we are assuming here, then the force of the Earth's gravity is in the negative direction, so
>
> $$w = -mg$$
>
> The weight of an object is measured in units of force.

The motion of the block in Figure 10.3.4 will depend on how far it is stretched or compressed initially and the forces that act on it while it moves. In our model we will assume that there are only two such forces: its weight w and the restoring force F_s of the spring. In particular, we will ignore such forces as air resistance, internal frictional forces in the spring, forces due to movement of the spring support, and so forth. With these assumptions, the model is called the ***simple harmonic model*** and the motion of the block is called ***simple harmonic motion***.

Our goal is to produce a differential equation whose solution gives the position function $y(t)$ of the block as a function of time. We will do this by determining the net force $F(t)$ acting on the block at a general time t and then applying Newton's Second Law of Motion. Since the only forces acting on the block are its weight $w = -mg$ and the restoring force F_s of the spring, and since the acceleration of the block at time t is $y''(t)$, it follows from Newton's second law that

$$F_s(t) - mg = my''(t) \tag{17}$$

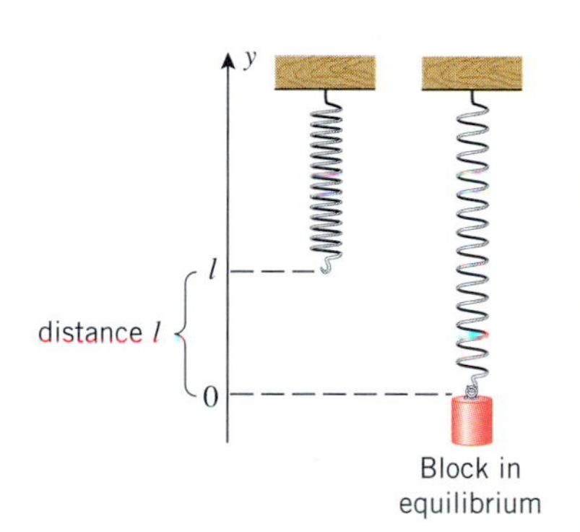

Figure 10.3.6

To express $F_s(t)$ in terms of $y(t)$, we will begin by examining the forces on the block when it is in its equilibrium position. In this position the downward force of the weight is perfectly balanced by the upward restoring force of the spring, so that the sum of these two forces must be zero. Thus, if we assume that the spring constant is k and that the spring is stretched a distance of l units beyond its natural length when the block is in equilibrium (Figure 10.3.6), then

$$kl - mg = 0 \tag{18}$$

Now let us examine the restoring force acting on the block when the connection point has coordinate $y(t)$. At this point the end of the spring is displaced $l - y(t)$ units from its natural position (Figure 10.3.7), so Hooke's law implies that the restoring force is

$$F_s(t) = k(l - y(t)) = kl - ky(t)$$

which from (18) can be rewritten as

$$F_s(t) = mg - ky(t)$$

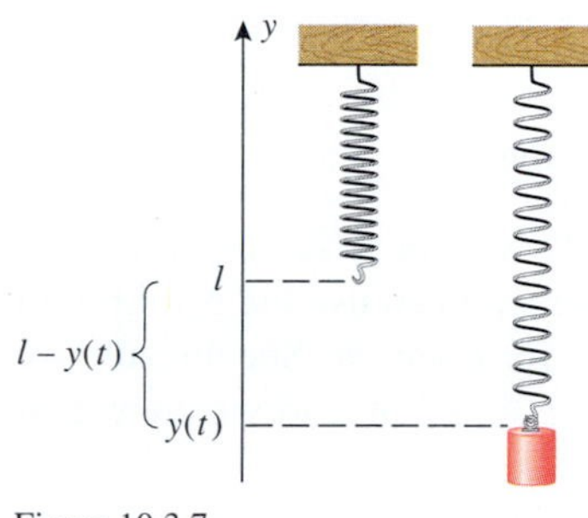

Figure 10.3.7

Substituting this in (17) and canceling the mg terms yields

$$-ky(t) = my''(t)$$

which we can rewrite as

$$y''(t) + \left(\frac{k}{m}\right) y(t) = 0 \tag{19}$$

This is a second-order differential equation whose solution is the position function of the block, and it is shown in courses on differential equations that the general solution of this equation is

$$y(t) = c_1 \cos\left(\sqrt{\frac{k}{m}}\, t\right) + c_2 \sin\left(\sqrt{\frac{k}{m}}\, t\right) \tag{20}$$

where c_1 and c_2 are arbitrary constants. Note that there are two arbitrary constants because (19) is a second-order differential equation.

FOR THE READER. Confirm that the functions in family (20) are solutions of (19).

Since (20) has two arbitrary constants, it requires two initial conditions to determine $y(t)$ uniquely. These can be obtained from the initial position and velocity of the block. Specifically, we will ask you to show in Exercise 40 that if the position of the block at time $t = 0$ is y_0, and if the initial velocity of the block is zero (i.e., it is *released* from rest), then

$$y(t) = y_0 \cos\left(\sqrt{\frac{k}{m}}\, t\right) \tag{21}$$

This formula describes a periodic vibration with an amplitude of $|y_0|$, a period T given by

$$T = \frac{2\pi}{\sqrt{k/m}} = 2\pi\sqrt{m/k} \tag{22}$$

and a frequency f given by

$$f = \frac{1}{T} = \frac{\sqrt{k/m}}{2\pi} \tag{23}$$

(Figure 10.3.8).

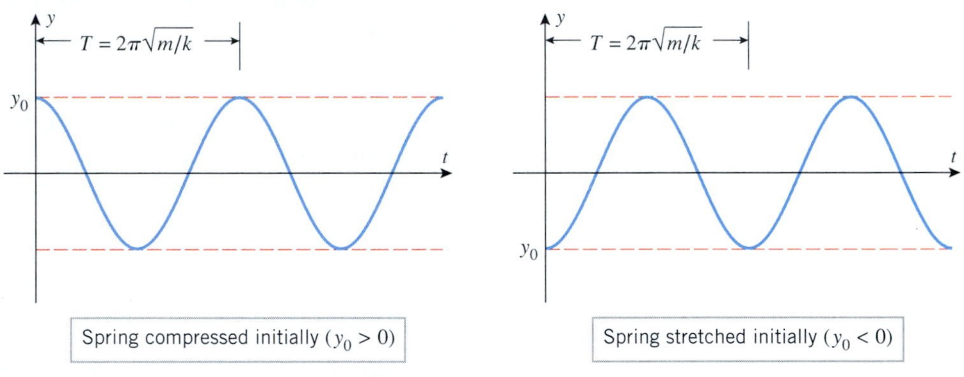

Figure 10.3.8

Example 6

Suppose that the block in Figure 10.3.4 stretches the spring 0.2 m in equilibrium. Suppose also that the block is pulled 0.5 m below its equilibrium position and released at time $t = 0$.

(a) Find the position function $y(t)$ of the block.

(b) Find the amplitude, period, and frequency of the vibration.

Solution (a). The appropriate formula is (21). Although we are not given the mass m of the block or the spring constant k, it does not matter because we can use the equilibrium condition (18) to find the ratio k/m without having values for k and m. Specifically, we are given that in equilibrium the block stretches the spring $l = 0.2$ m, and we know that $g = 9.8 \text{ m/s}^2$. Thus, (18) implies that

$$\frac{k}{m} = \frac{g}{l} = \frac{9.8}{0.2} = 49 \text{ s}^{-2} \tag{24}$$

Substituting this in (21) yields

$$y(t) = y_0 \cos 7t$$

where y_0 is the coordinate of the block at time $t = 0$. However, we are given that the block is initially 0.5 m *below* the equilibrium position, so $y_0 = -0.5$ and hence the position function of the block is $y(t) = -0.5 \cos 7t$

Solution (b). The amplitude of the vibration is

$$\text{amplitude} = |y_0| = |-0.5| = 0.5 \text{ m}$$

and from (22), (23), and (24) the period and frequency are

$$\text{period} = T = 2\pi\sqrt{\frac{m}{k}} = 2\pi\sqrt{\frac{1}{49}} = \frac{2\pi}{7} \text{ s} \qquad \text{frequency} = f = \frac{1}{T} = \frac{7}{2\pi} \text{ Hz}$$ ◄

EXERCISE SET 10.3 Graphing Calculator C CAS

1. (a) Suppose that a quantity $y = y(t)$ increases at a rate that is proportional to the square of the amount present, and suppose that at time $t = 0$, the amount present is y_0. Find an initial-value problem whose solution is $y(t)$.
   (b) Suppose that a quantity $y = y(t)$ decreases at a rate that is proportional to the square of the amount present, and suppose that at a time $t = 0$, the amount present is y_0. Find an initial-value problem whose solution is $y(t)$.

2. (a) Suppose that a quantity $y = y(t)$ changes in such a way that $dy/dt = k\sqrt{y}$, where $k > 0$. Describe how y changes in words.
   (b) Suppose that a quantity $y = y(t)$ changes in such a way that $dy/dt = -ky^3$, where $k > 0$. Describe how y changes in words.

3. (a) Suppose that a particle moves along an s-axis in such a way that its velocity $v(t)$ is always half of $s(t)$. Find a differential equation whose solution is $s(t)$.
   (b) Suppose that an object moves along an s-axis in such a way that its acceleration $a(t)$ is always twice the velocity. Find a differential equation whose solution is $s(t)$.

4. Suppose that a body moves along an s-axis through a resistive medium in such a way that the velocity $v = v(t)$ decreases at a rate that is twice the square of the velocity.
   (a) Find a differential equation whose solution is the velocity $v(t)$.
   (b) Find a differential equation whose solution is the position $s(t)$.

5. Suppose that an initial population of 10,000 bacteria grows exponentially at a rate of 1% per hour and that $y = y(t)$ is the number of bacteria present t hours later.
   (a) Find an initial-value problem whose solution is $y(t)$.
   (b) Find a formula for $y(t)$.
   (c) How long does it take for the initial population of bacteria to double?
   (d) How long does it take for the population of bacteria to reach 45,000?

6. A cell of the bacterium *E. coli* divides into two cells every 20 minutes when placed in a nutrient culture. Let $y = y(t)$ be the number of cells that are present t minutes after a single cell is placed in the culture. Assume that the growth of the bacteria is approximated by a continuous exponential growth model.
   (a) Find an initial-value problem whose solution is $y(t)$.
   (b) Find a formula for $y(t)$.
   (c) How many cells are present after 2 hours?
   (d) How long does it take for the number of cells to reach 1,000,000?

7. Radon-222 is a radioactive gas with a half-life of 3.83 days. This gas is a health hazard because it tends to get trapped in the basements of houses, and many health officials suggest that homeowners seal their basements to prevent entry of the gas. Assume that 5.0×10^7 radon atoms are trapped in a basement at the time it is sealed and that $y(t)$ is the number of atoms present t days later.
   (a) Find an initial-value problem whose solution is $y(t)$.
   (b) Find a formula for $y(t)$.
   (c) How many atoms will be present after 30 days?
   (d) How long will it take for 90% of the original quantity of gas to decay?

8. Polonium-210 is a radioactive element with a half-life of 140 days. Assume that 10 milligrams of the element are placed in a lead container and that $y(t)$ is the number of milligrams present t days later.
   (a) Find an initial-value problem whose solution is $y(t)$.
   (b) Find a formula for $y(t)$.
   (c) How many milligrams will be present after 10 weeks?
   (d) How long will it take for 70% of the original sample to decay?

9. Suppose that 100 fruit flies are placed in a breeding container that can support at most 5000 flies. Assuming that the population grows exponentially at a rate of 2% per day, how long will it take for the container to reach capacity?

10. Suppose that the town of Grayrock had a population of 10,000 in 1987 and a population of 12,000 in 1997. Assuming an exponential growth model, in what year will the population reach 20,000?

11. A scientist wants to determine the half-life of a certain radioactive substance. She determines that in exactly 5 days a 10.0-milligram sample of the substance decays to 3.5 milligrams. Based on these data, what is the half-life?

12. Suppose that 40% of a certain radioactive substance decays in 5 years.
(a) What is the half-life of the substance in years?
(b) Suppose that a certain quantity of this substance is stored in a cave. What percentage of it will remain after t years?

13. In each part, find an exponential growth model $y = y_0e^{kt}$ that satisfies the stated conditions.
(a) $y_0 = 2$; doubling time $T = 5$
(b) $y(0) = 5$; growth rate 1.5%
(c) $y(1) = 1$; $y(10) = 100$
(d) $y(1) = 1$; doubling time $T = 5$

14. In each part, find an exponential decay model $y = y_0e^{-kt}$ that satisfies the stated conditions.
(a) $y_0 = 10$; half-life $T = 5$
(b) $y(0) = 10$; decay rate 1.5%
(c) $y(1) = 100$; $y(10) = 1$
(d) $y(1) = 10$; half-life $T = 5$

15. (a) Make a conjecture about the effect on the graphs of $y = y_0e^{kt}$ and $y = y_0e^{-kt}$ of varying k and keeping y_0 fixed. Confirm your conjecture with a graphing utility.
(b) Make a conjecture about the effect on the graphs of $y = y_0e^{kt}$ and $y = y_0e^{-kt}$ of varying y_0 and keeping k fixed. Confirm your conjecture with a graphing utility.

16. (a) What effect does increasing y_0 and keeping k fixed have on the doubling time or half-life of an exponential model? Justify your answer.
(b) What effect does increasing k and keeping y_0 fixed have on the doubling time and half-life of an exponential model? Justify your answer.

17. (a) There is a trick, called the ***Rule of 70***, that can be used to get a quick estimate of the doubling time or half-life of an exponential model. According to this rule, the doubling time or half-life is roughly 70 divided by the percentage growth or decay rate. For example, we showed in Example 2 that with a continued growth rate of 2% per year the world population would double every 35 years. This result agrees with the Rule of 70, since $70/2 = 35$. Explain why this rule works.
(b) Use the Rule of 70 to estimate the doubling time of a population that grows exponentially at a rate of 1% per year.
(c) Use the Rule of 70 to estimate the half-life of a population that decreases exponentially at a rate of 3.5% per hour.
(d) Use the Rule of 70 to estimate the growth rate that would be required for a population growing exponentially to double every 10 years.

18. Find a formula for the tripling time of an exponential growth model.

19. In 1950, a research team digging near Folsom, New Mexico, found charred bison bones along with some leaf-shaped projectile points (called the "Folsom points") that had been made by a Paleo-Indian hunting culture. It was clear from the evidence that the bison had been cooked and eaten by the makers of the points, so that carbon-14 dating of the bones made it possible for the researchers to determine when the hunters roamed North America. Tests showed that the bones contained between 27% and 30% of their original carbon-14. Use this information to show that the hunters lived roughly between 9000 B.C. and 8000 B.C.

20. (a) Use a graphing utility to make a graph of p_{rem} versus t, where p_{rem} is the percentage of carbon-14 that remains in an artifact after t years.
(b) Use the graph to estimate the percentage of carbon-14 that would have to have been present in the 1988 test of the Shroud of Turin for it to have been the burial shroud of Jesus. [See Example 4.]

In Exercises 21 and 22, the graph of a logistic model

$$y = \frac{y_0L}{y_0 + (L - y_0)e^{-kt}}$$

is shown. Estimate y_0, L, and k.

21.

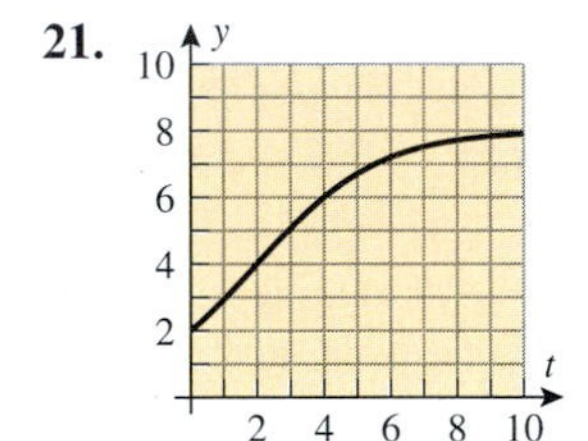

22.

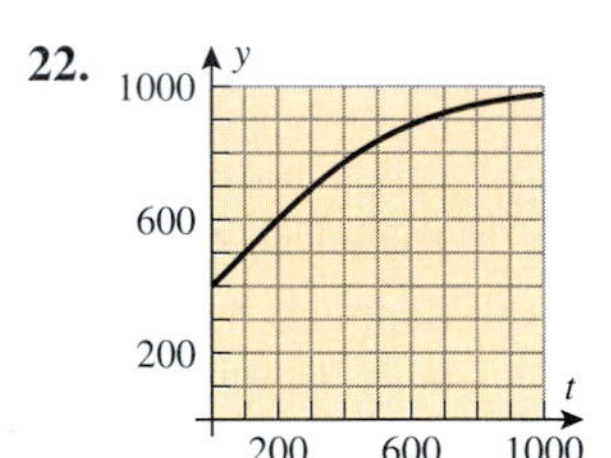

23. Suppose that the growth of a population $y = y(t)$ is given by the logistic equation

$$y = \frac{60}{5 + 7e^{-t}}$$

(a) What is the population at time $t = 0$?
(b) What is the carrying capacity L?
(c) What is the constant k?
(d) When does the population reach half of the carrying capacity?
(e) Find an initial-value problem whose solution is $y(t)$.

24. Suppose that the growth of a population $y = y(t)$ is given by the logistic equation

$$y = \frac{1000}{1 + 999e^{-0.9t}}$$

(a) What is the population at time $t = 0$?
(b) What is the carrying capacity L?
(c) What is the constant k?
(d) When does the population reach 75% of the carrying capacity?
(e) Find an initial-value problem whose solution is $y(t)$.

25. Suppose that a population $y(t)$ grows in accordance with the logistic model

$$\frac{dy}{dt} = 10(1 - 0.1y)y$$

(a) What is the carrying capacity?
(b) What is the value of k?
(c) For what value of y is the population growing most rapidly?

26. Suppose that a population $y(t)$ grows in accordance with the logistic model

$$\frac{dy}{dt} = 50y - 0.001y^2$$

(a) What is the carrying capacity?
(b) What is the value of k?
(c) For what value of y is the population growing most rapidly?

27. Suppose that a college residence hall houses 1000 students. Following the semester break, 20 students in the hall return with the flu, and 5 days later 35 students have the flu.
(a) Use model (4) to set up an initial-value problem whose solution is the number of students who will have had the flu t days after the return from the break. [*Note:* The differential equation in this case will involve a constant of proportionality.]
(b) Solve the initial-value problem, and use the given data to find the constant of proportionality.
(c) Make a table that illustrates how the flu spreads day to day over a 2-week period.
(d) Use a graphing utility to generate a graph that illustrates how the flu spreads over a 2-week period.

28. It has been observed experimentally that at a constant temperature the rate of change of the atmospheric pressure p with respect to the altitude h above sea level is proportional to the pressure.
(a) Assuming that the pressure at sea level is p_0, find an initial-value problem whose solution is $p(h)$. [*Note:* The differential equation in this case will involve a constant of proportionality.]
(b) Find a formula for $p(h)$ in atmospheres (atm) if the pressure at sea level is 1 atm and the pressure at 5000 ft above sea level is 0.83 atm.

Newton's Law of Cooling states that the rate at which the temperature of a cooling object decreases and the rate at which a warming object increases are proportional to the difference between the temperature of the object and the temperature of the surrounding medium. Use this result in Exercises 29–32.

29. A cup of water with a temperature of 95°C is placed in a room with a constant temperature 21°C.
(a) Assuming that Newton's Law of Cooling applies, set up and solve an initial-value problem whose solution is the temperature of the water t minutes after it is placed in the room. [*Note:* The differential equation will involve a constant of proportionality.]
(b) How many minutes will it take for the water to reach a temperature of 51°C if it cools to 85°C in 1 minute?

30. A glass of lemonade with a temperature of 40°F is placed in a room with a constant temperature of 70°F, and 1 hour later its temperature is 52°F. We stated in Example 8 of Section 4.4 that t hours after the lemonade is placed in the room its temperature is given by $T = 70 - 30e^{-0.5t}$. Confirm this using Newton's Law of Cooling and the method used in Exercise 29.

31. The great detective Sherlock Holmes and his assistant Dr. Watson are discussing the murder of actor Cornelius McHam. McHam was shot in the head, and his understudy, Barry Moore, was found standing over the body with the murder weapon in hand. Let's listen in.

Watson: Open-and-shut case Holmes—Moore is the murderer.
Holmes: Not so fast Watson—you are forgetting Newton's Law of Cooling!
Watson: Huh?
Holmes: Elementary my dear Watson—Moore was found standing over McHam at 10:06 P.M., at which time the coroner recorded a body temperature of 77.9°F and noted that the room thermostat was set to 72°F. At 11:06 P.M. the coroner took another reading and recorded a body temperature of 75.6°F. Since McHam's normal temperature is 98.6°F, and since Moore was on stage between 6:00 P.M. and 8:00 P.M., Moore is obviously innocent.
Watson: Huh?
Holmes: Sometimes you are so dull Watson. Ask any calculus student to figure it out for you.
Watson: Hrrumph....

32. Suppose that at time $t = 0$ an object with temperature T_0 is placed in a room with constant temperature T_a. If $T_0 < T_a$, then the temperature of the object will increase, and if $T_0 > T_a$, then the temperature will decrease. Assuming that Newton's Law of Cooling applies, show that in both cases the temperature $T(t)$ at time t is given by

$$T(t) = T_a + (T_0 - T_a)e^{-kt}$$

where k is a positive constant.

Exercises 33–38 involve vibrations of the block pictured in Figure 10.3.4. Assume that the y-axis is as shown in Figure 10.3.5 and that the simple harmonic model applies.

33. Suppose that the block has a mass of 1 kg, the spring constant is $k = 0.25$ N/m, and the block is pushed 0.3 m above its equilibrium position and released at time $t = 0$.
(a) Find the position function $y(t)$ of the block.

(b) Find the period and frequency of the vibration.
(c) Sketch the graph of $y(t)$.
(d) At what time does the block first pass through the equilibrium position?
(e) At what time does the block first reach its maximum distance below the equilibrium position?

34. Suppose that the block has a weight of 64 lb, the spring constant is $k = 0.25$ lb/ft, and the block is pushed 1 ft above its equilibrium position and released at time $t = 0$.
(a) Find the position function $y(t)$ of the block.
(b) Find the period and frequency of the vibration.
(c) Sketch the graph of $y(t)$.
(d) At what time does the block first pass through the equilibrium position?
(e) At what time does the block first reach its maximum distance below the equilibrium position?

35. Suppose that the block stretches the spring 0.05 m in equilibrium, and the block is pulled 0.12 m below the equilibrium position and released at time $t = 0$.
(a) Find the position function $y(t)$ of the block.
(b) Find the period and frequency of the vibration.
(c) Sketch the graph of $y(t)$.
(d) At what time does the block first pass through the equilibrium position?
(e) At what time does the block first reach its maximum distance above the equilibrium position?

36. Suppose that the block stretches the spring 0.5 ft in equilibrium, and is pulled 1.5 ft below the equilibrium position and released at time $t = 0$.
(a) Find the position function $y(t)$ of the block.
(b) Find the period and frequency of the vibration.
(c) Sketch the graph of $y(t)$.
(d) At what time does the block first pass through the equilibrium position?
(e) At what time does the block first reach its maximum distance above the equilibrium position?

37. (a) For what values of y would you expect the block in Exercise 36 to have its maximum speed? Confirm your answer to this question mathematically.
(b) For what values of y would you expect the block to have its minimum speed? Confirm your answer to this question mathematically.

38. Suppose that the block weighs w pounds and vibrates with a period of 3 s when it is pulled below the equilibrium position and released. Suppose also that if the process is repeated with an additional 4 lb of weight, then the period is 5 s.
(a) Find the spring constant.
(b) Find w.

39. As shown in the accompanying figure, suppose that a toy cart of mass m is attached to a wall by a spring with spring constant k, and let a horizontal x-axis be introduced with its origin at the connection point of the spring and cart when the cart is in equilibrium. Suppose that the cart is pulled or pushed horizontally to a point x_0 and then released at time $t = 0$. Find an initial-value problem whose solution is the position function of the cart, and state any assumptions you have made.

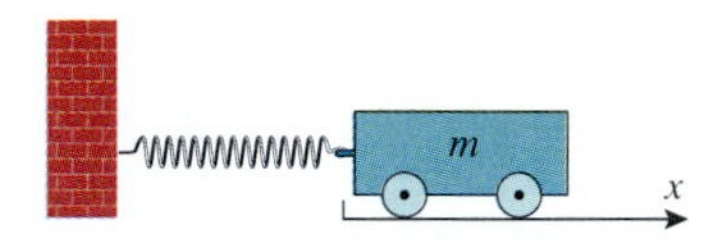

Figure Ex-39

40. Use the initial position $y(0) = y_0$ and the initial velocity $v(0) = 0$ to find the constants c_1 and c_2 in (20).

41. (a) Show that if $b > 1$, then the equation $y = y_0 b^t$ can be expressed as $y = y_0 e^{kt}$ for some positive constant k. [*Note:* This shows that if $b > 1$, and if y grows in accordance with the equation $y = y_0 b^t$, then y has an exponential growth model.]
(b) Show that if $0 < b < 1$, then the equation $y = y_0 b^t$ can be expressed as $y = y_0 e^{-kt}$ for some positive constant k. [*Note:* This shows that if $0 < b < 1$, and if y decays in accordance with the equation $y = y_0 b^t$, then y has an exponential decay model.]
(c) Express $y = 4(2^t)$ in the form $y = y_0 e^{kt}$.
(d) Express $y = 4(0.5^t)$ in the form $y = y_0 e^{-kt}$.

42. Suppose that a quantity y has an exponential growth model $y = y_0 e^{kt}$ or an exponential decay model $y = y_0 e^{-kt}$, and it is known that $y = y_1$ if $t = t_1$. In each case find a formula for k in terms of y_0, y_1, and t_1, assuming that $t_1 \neq 0$.

SUPPLEMENTARY EXERCISES

1. What is the relationship between the order of a differential equation and the number of arbitrary constants in its general solution? Give an informal explanation of why one would expect such a relationship.

2. Write a paragraph that describes Euler's Method.

3. (a) List the steps in the method of integrating factors for solving first-order linear differential equations.
(b) What would you do if you had to solve an important initial-value problem involving a first-order linear differential equation whose integrating factor could not be obtained because of the complexity of the integration?

4. Which of the following differential equations are separable?

(a) $\dfrac{dy}{dx} = f(x)g(y)$ (b) $\dfrac{dy}{dx} = \dfrac{f(x)}{g(y)}$

(c) $\dfrac{dy}{dx} = f(x) + g(y)$ (d) $\dfrac{dy}{dx} = \sqrt{f(x)g(y)}$

5. Classify the following first-order differential equations as separable, linear, both, or neither.

(a) $\dfrac{dy}{dx} - 3y = \sin x$ (b) $\dfrac{dy}{dx} + xy = x$

(c) $y\dfrac{dy}{dx} - x = 1$ (d) $\dfrac{dy}{dx} + xy^2 = \sin(xy)$

6. Confirm that the methods of integrating factors and separation of variables produce the same solution of the differential equation

$$\frac{dy}{dx} - 4xy = x$$

7. Consider the model $dy/dt = ky(L - y)$ for the spread of a disease, where $k > 0$ and $0 < y \leq L$. For what value of y is the disease spreading most rapidly, and at what rate is it spreading?

8. (a) Show that if a quantity $y = y(t)$ has an exponential model, and if $y(t_1) = y_1$ and $y(t_2) = y_2$, then the doubling time or the half-life T is

$$T = \left| \frac{(t_2 - t_1)\ln 2}{\ln(y_2/y_1)} \right|$$

(b) In a certain 1-hour period the number of bacteria in a colony increases by 25%. Assuming an exponential growth model, what is the doubling time for the colony?

9. Assume that a spherical meteoroid burns up at a rate that is proportional to its surface area. Given that the radius is originally 4 m and 1 min later its radius is 3 m, find a formula for the radius as a function of time.

10. A tank contains 1000 gal of fresh water. At time $t = 0$ brine containing 5 ounces of salt per gallon of brine is poured into the tank at a rate of 10 gal/min, and the mixed solution is drained from the tank at the same rate. After 15 min that process is stopped and fresh water is poured into the tank at the rate of 5 gal/min, and the mixed solution is drained from the tank at the same rate. Find the amount of salt in the tank at time $t = 30$.

11. Suppose that a room containing 1200 ft^3 of air is free of carbon monoxide. At time $t = 0$ cigarette smoke containing 4% carbon monoxide is introduced at the rate of 0.1 ft^3/min, and the well-circulated mixture is vented from the room at the same rate.

(a) Find a formula for the percentage of carbon monoxide in the room at time t.

(b) Extended exposure to air containing 0.012% carbon monoxide is considered dangerous. How long will it take to reach this level? [This is based on a problem from William E. Boyce and Richard C. DiPrima, *Elementary Differential Equations*, 6th ed., John Wiley & Sons, 1997.]

In Exercises 12–16, solve the initial-value problem.

12. $y' = 1 + y^2, \quad y(0) = 1$

13. $y' = \dfrac{y^5}{x(1 + y^4)}, \quad y(1) = 1$

14. $xy' + 2y = 4x^2, \quad y(1) = 2$

15. $y' = 4y^2 \sec^2 2x, \quad y(\pi/8) = 1$

16. $y' = 6 - 5y + y^2, \quad y(0) = \ln 2$

C **17.** (a) Solve the initial-value problem

$$y' - y = x \sin 3x, \quad y(0) = 1$$

by the method of integrating factors, using a CAS to perform any difficult integrations.

(b) Use the CAS to solve the initial-value problem directly, and confirm that the answer is consistent with that obtained in part (a).

(c) Graph the solution.

C **18.** Use a CAS to derive Formula (23) of Section 10.1 by solving initial-value problem (21).

19. (a) It is currently accepted that the half-life of carbon-14 might vary ± 40 years from its nominal value of 5730 years. Does this variation make it possible that the Shroud of Turin dates to the time of Jesus of Nazareth? [See Example 4 of Section 10.3.]

(b) Review the subsection of Section 3.6 entitled Error Propagation in Applications, and then estimate the percentage error that results in the computed age of an artifact from an $r\%$ error in the half-life of carbon-14.

20. (a) Use Euler's Method with a step-size of $h = 0.1$ to approximate the solution of the initial-value problem

$$y' = 1 + 5t - y, \quad y(1) = 5$$

over the interval $[1, 2]$.

(b) Find the percentage error in the values computed.

21. (a) Confirm that e^x and e^{-x} are solutions of the second-order differential equation $y'' - y = 0$.

(b) Find some more solutions.

(c) Find a solution $y(x)$ such that $y(0) = 1$ and $y'(0) = 1$.

22. (a) Sketch the integral curve of $2yy' = 1$ that passes through the point $(0, 1)$ and the integral curve that passes through the point $(0, -1)$.

(b) Sketch the integral curve of $y' = -2xy^2$ that passes through the point $(0, 1)$.

23. Suppose that a herd of 19 deer is moved to a small island whose estimated carrying capacity is 95 deer, and assume that the population has a logistic growth model.

(a) Given that 1 year later the population is 25, how long will it take for the deer population to reach 80% of the island's carrying capacity?

(b) Find an initial-value problem whose solution gives the deer population as a function of time.

C **24.** If the block in Figure 10.3.4 is displaced y_0 units from its equilibrium position and given an initial velocity of v_0, rather than being released with an initial velocity of 0, then its position function $y(t)$ given in Equation (20) of Section 10.3 must satisfy the initial conditions $y(0) = y_0$ and $y'(0) = v_0$.

(a) Show that

$$y(t) = y_0 \cos\left(\sqrt{\frac{k}{m}}t\right) + v_0\sqrt{\frac{m}{k}} \sin\left(\sqrt{\frac{k}{m}}t\right)$$

(b) Suppose that a block with a mass of 1 kg stretches the spring 0.5 m in equilibrium. Use a graphing utility to graph the position function of the block if it is set in motion by pulling it down 1 m and imparting it an initial upward velocity of 0.25 m/s.

(c) What is the maximum displacement of the block from the equilibrium position?

25. A block attached to a vertical spring is displaced from its equilibrium position and released, thereby causing it to vibrate with amplitude $|y_0|$ and period T.

(a) Show that the velocity of the block has maximum magnitude $2\pi|y_0|/T$ and that the maximum occurs when the block is at its equilibrium position.

(b) Show that the acceleration of the block has maximum magnitude $4\pi^2|y_0|/T^2$ and that the maximum occurs when the block is at a top or bottom point of its motion.

26. Suppose that P dollars is invested at an annual interest rate of $r \times 100\%$. If the accumulated interest is credited to the account at the end of the year, then the interest is said to be *compounded annually*; if it is credited at the end of each 6-month period, then it is said to be *compounded semiannually*; and if it is credited at the end of each 3-month period, then it is said to be *compounded quarterly*. The more frequently the interest is compounded, the better it is for the investor since more of the interest is itself earning interest.

(a) Show that if interest is compounded n times a year at equally spaced intervals, then the value A of the investment after t years is

$$A = P\left(1 + \frac{r}{n}\right)^{nt}$$

(b) One can imagine interest to be compounded each day, each hour, each minute, and so forth. Carried to the limit one can conceive of interest compounded at each instant of time; this is called ***continuous compounding***. Thus, from part (a), the value A of P dollars after t years when invested at an annual rate of $r \times 100\%$, compounded continuously, is

$$A = \lim_{n\to+\infty} P\left(1 + \frac{r}{n}\right)^{nt}$$

Use the fact that $\lim_{x\to 0}(1+x)^{1/x} = e$ to prove that $A = Pe^{rt}$.

(c) Use the result in part (b) to show that money invested at continuous compound interest increases at a rate proportional to the amount present.

27. (a) If \$1000 is invested at 8% per year compounded continuously (Exercise 26), what will the investment be worth after 5 years?

(b) If it is desired that an investment at 8% per year compounded continuously should have a value of \$10,000 after 10 years, how much should be invested now?

(c) How long does it take for an investment at 8% per year compounded continuously to double in value?

EXPANDING THE CALCULUS HORIZON

For additional material relating to this chapter, visit the Anton Website at http://www.wiley.com/college/anton

Brook Taylor

11

INFINITE SERIES

In this chapter we will be concerned with *infinite series*, which are sums that involve infinitely many terms. Infinite series play a fundamental role in both mathematics and science—they are used, for example, to approximate trigonometric functions and logarithms, to solve differential equations, to evaluate difficult integrals, to create new functions, and to construct mathematical models of physical laws. Since it is impossible to add up infinitely many numbers directly, our first goal will be to define exactly what we mean by the sum of an infinite series. However, unlike finite sums, it turns out that not all infinite series actually have a sum, so we will need to develop tools for determining which infinite series have sums and which do not. Once the basic ideas have been developed we will begin to apply our work; we will show how infinite series are used to evaluate such quantities as $\sin 17^\circ$ and $\ln 5$, how they are used to create functions, and finally, how they are used to model physical laws.

11.1 SEQUENCES

In everyday language, the term "sequence" means a succession of things in a definite order—chronological order, size order, or logical order, for example. In mathematics, the term "sequence" is commonly used to denote a succession of numbers whose order is determined by a rule or a function. In this section, we will develop some of the basic ideas concerning sequences of numbers.

DEFINITION OF A SEQUENCE

Stated informally, an ***infinite sequence***, or more simply a ***sequence***, is an unending succession of numbers, called ***terms***. It is understood that the terms have a definite order; that is, there is a first term a_1, a second term a_2, a third term a_3, a fourth term a_4, and so forth. Such a sequence would typically be written as

$$a_1, a_2, a_3, a_4, \dots$$

where the dots are used to indicate that the sequence continues indefinitely. Some specific examples are

$$1, 2, 3, 4, \dots, \qquad 1, \tfrac{1}{2}, \tfrac{1}{3}, \tfrac{1}{4}, \dots,$$
$$2, 4, 6, 8, \dots, \qquad 1, -1, 1, -1, \dots$$

Each of these sequences has a definite pattern that makes it easy to generate additional terms if we assume that those terms follow the same pattern as the displayed terms. However, such patterns can be deceiving, so it is better to have a rule or formula for generating the terms. One way of doing this is to look for a function that relates each term in the sequence to its term number. For example, in the sequence

$$2, 4, 6, 8, \dots$$

each term is twice the term number; that is, the nth term in the sequence is given by the formula $2n$. We denote this by writing the sequence as

$$2, 4, 6, 8, \dots, 2n, \dots$$

We call the function $f(n) = 2n$ the *general term* of this sequence. Now, if we want to know a specific term in the sequence, we need only substitute its term number in the formula for the general term. For example, the 37th term in the sequence is $2 \cdot 37 = 74$.

Example 1

In each part, find the general term of the sequence.

(a) $\tfrac{1}{2}, \tfrac{2}{3}, \tfrac{3}{4}, \tfrac{4}{5}, \dots$ (b) $\tfrac{1}{2}, \tfrac{1}{4}, \tfrac{1}{8}, \tfrac{1}{16}, \dots$

(c) $\tfrac{1}{2}, -\tfrac{2}{3}, \tfrac{3}{4}, -\tfrac{4}{5}, \dots$ (d) $1, 3, 5, 7, \dots$

Solution (a). In Table 11.1.1, the four known terms have been placed below their term numbers, from which we see that the numerator is the same as the term number and the denominator is one greater than the term number. This suggests that the nth term has numerator n and denominator $n + 1$, as indicated in the table. Thus, the sequence can be expressed as

$$\frac{1}{2}, \frac{2}{3}, \frac{3}{4}, \frac{4}{5}, \dots, \frac{n}{n+1}, \dots$$

Solution (b). In Table 11.1.2, the denominators of the four known terms have been expressed as powers of 2 and placed below their term numbers, from which we see that the exponent in the denominator is the same as the term number. This suggests that the denominator of the nth term is 2^n, as indicated in the table. Thus, the sequence can be expressed as

$$\frac{1}{2}, \frac{1}{4}, \frac{1}{8}, \frac{1}{16}, \dots, \frac{1}{2^n}, \dots$$

Table 11.1.1

TERM NUMBER	1	2	3	4	···	n	···
TERM	$\frac{1}{2}$	$\frac{2}{3}$	$\frac{3}{4}$	$\frac{4}{5}$	···	$\frac{n}{n+1}$	···

Table 11.1.2

TERM NUMBER	1	2	3	4	···	n	···
TERM	$\frac{1}{2}$	$\frac{1}{2^2}$	$\frac{1}{2^3}$	$\frac{1}{2^4}$	···	$\frac{1}{2^n}$	···

Solution (c). This sequence is identical to that in part (a), except for the alternating signs. Thus, the nth term in the sequence can be obtained by multiplying the nth term in part (a) by $(-1)^{n+1}$. This factor produces the correct alternating signs, since its successive values, starting with $n = 1$, are $1, -1, 1, -1, \ldots$. Thus, the sequence can be written as

$$\frac{1}{2}, -\frac{2}{3}, \frac{3}{4}, -\frac{4}{5}, \ldots, (-1)^{n+1}\frac{n}{n+1}, \ldots$$

Solution (d). In Table 11.1.3, the denominators of the four known terms have been placed below their term numbers, from which we see that each term is one less than twice its term number. This suggests that the nth term in the sequence is $2n - 1$, as indicated in the table. Thus, the sequence can be expressed as

$$1, 3, 5, 7, \ldots, 2n - 1, \ldots$$ ◀

Table 11.1.3

TERM NUMBER	1	2	3	4	···	n	···
TERM	1	3	5	7	···	$2n-1$	···

FOR THE READER. Consider the sequence whose general term is

$$f(n) = \tfrac{1}{3}(3 - 5n + 6n^2 - n^3)$$

Calculate the first three terms, and make a conjecture about the fourth term. Check your conjecture by calculating the fourth term. What message does this convey?

When the general term of a sequence

$$a_1, a_2, a_3, \ldots, a_n, \ldots \tag{1}$$

is known, there is no need to write out the initial terms, and it is common to write only the general term enclosed in braces. Thus, (1) might be written as

$$\{a_n\}_{n=1}^{+\infty}$$

For example, here are the four sequences in Example 1 expressed in brace notation.

SEQUENCE	BRACE NOTATION
$\frac{1}{2}, \frac{2}{3}, \frac{3}{4}, \frac{4}{5}, \ldots, \frac{n}{n+1}, \ldots$	$\left\{\frac{n}{n+1}\right\}_{n=1}^{+\infty}$
$\frac{1}{2}, \frac{1}{4}, \frac{1}{8}, \frac{1}{16}, \ldots, \frac{1}{2^n}, \ldots$	$\left\{\frac{1}{2^n}\right\}_{n=1}^{+\infty}$
$\frac{1}{2}, -\frac{2}{3}, \frac{3}{4}, -\frac{4}{5}, \ldots, (-1)^{n+1}\frac{n}{n+1}, \ldots$	$\left\{(-1)^{n+1}\frac{n}{n+1}\right\}_{n=1}^{+\infty}$
$1, 3, 5, 7, \ldots, 2n-1, \ldots$	$\{2n-1\}_{n=1}^{+\infty}$

The letter n in (1) is called the ***index*** for the sequence. It is not essential to use n for the index; any letter not reserved for another purpose can be used. For example, we might view the general term of the sequence $a_1, a_2, a_3, \ldots$ to be the kth term, in which case we would denote this sequence as $\{a_k\}_{k=1}^{+\infty}$. Moreover, it is not essential to start the index at

1; sometimes it is more convenient to start it at 0 (or some other integer). For example, consider the sequence

$$1, \frac{1}{2}, \frac{1}{2^2}, \frac{1}{2^3}, \ldots$$

One way to write this sequence is

$$\left\{\frac{1}{2^{n-1}}\right\}_{n=1}^{+\infty}$$

However, the general term will be simpler if we think of the initial term in the sequence as the zeroth term, in which case we can write the sequence as

$$\left\{\frac{1}{2^n}\right\}_{n=0}^{+\infty}$$

REMARK. In general discussions that involve sequences in which the specific terms and the starting point for the index are not important, it is common to write $\{a_n\}$ rather than $\{a_n\}_{n=1}^{+\infty}$ or $\{a_n\}_{n=0}^{+\infty}$. Moreover, we can distinguish between different sequences by using different letters for their general terms; thus, $\{a_n\}$, $\{b_n\}$, and $\{c_n\}$ denote three different sequences.

We began this section by describing a sequence as an unending succession of numbers. Although this conveys the general idea, it is not a satisfactory mathematical definition because it relies on the term "succession," which is itself an undefined term. To motivate a precise definition, consider the sequence

$$2, 4, 6, 8, \ldots, 2n, \ldots$$

If we denote the general term by $f(n) = 2n$, then we can write this sequence as

$$f(1), f(2), f(3), \ldots, f(n), \ldots$$

which is a "list" of values of the function

$$f(n) = 2n, \quad n = 1, 2, 3, \ldots$$

whose domain is the set of positive integers. This suggests the following definition.

11.1.1 DEFINITION. A ***sequence*** is a function whose domain is a set of integers. Specifically, we will regard the expression $\{a_n\}_{n=1}^{+\infty}$ to be an alternative notation for the function $f(n) = a_n$, $n = 1, 2, 3, \ldots$.

GRAPHS OF SEQUENCES

Since sequences are functions, it makes sense to talk about the graph of a sequence. For example, the graph of the sequence $\{1/n\}_{n=1}^{+\infty}$ is the graph of the equation

$$y = \frac{1}{n}, \quad n = 1, 2, 3, \ldots$$

Because the right side of this equation is defined only for positive integer values of n, the graph consists of a succession of isolated points (Figure 11.1.1a). This is in distinction to

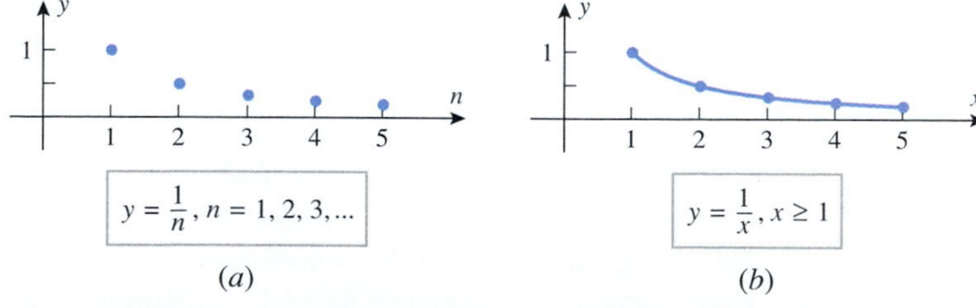

Figure 11.1.1

the graph of

$$y = \frac{1}{x}, \quad x \geq 1$$

which is a continuous curve (Figure 11.1.1*b*).

LIMIT OF A SEQUENCE

Since sequences are functions, we can inquire about their limits. However, because a sequence $\{a_n\}$ is only defined for integer values of n, the only limit that makes sense is the limit of a_n as $n \to +\infty$. In Figure 11.1.2 we have shown the graphs of four sequences, each of which behaves differently as $n \to +\infty$:

- The terms in the sequence $\{n+1\}$ increase without bound.
- The terms in the sequence $\{(-1)^{n+1}\}$ oscillate between -1 and 1.
- The terms in the sequence $\{n/(n+1)\}$ increase toward a "limiting value" of 1.
- The terms in the sequence $\left\{1 + \left(-\frac{1}{2}\right)^n\right\}$ also tend toward a "limiting value" of 1, but do so in an oscillatory fashion.

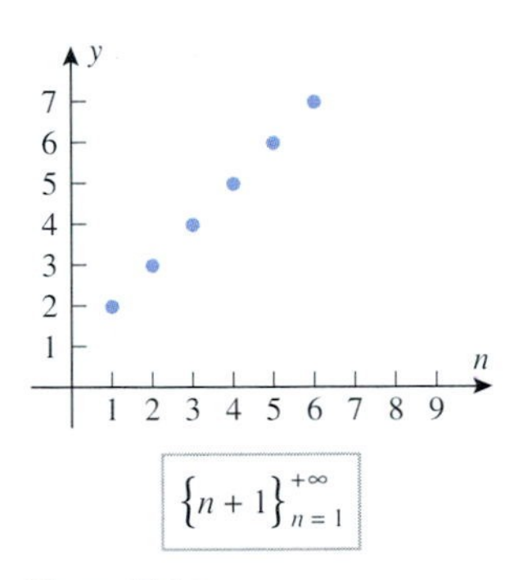

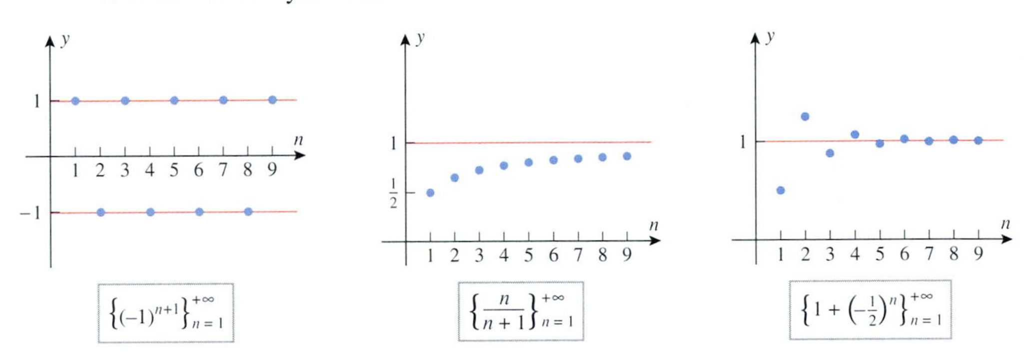

Figure 11.1.2

Informally speaking, the limit of a sequence $\{a_n\}$ is intended to describe how a_n behaves as $n \to +\infty$. To be more specific, we will say that *a sequence* $\{a_n\}$ *approaches a limit* L *if the terms in the sequence eventually become arbitrarily close to* L. Geometrically, this means that for any positive number ϵ there is a point in the sequence after which all terms lie between the lines $y = L - \epsilon$ and $y = L + \epsilon$ (Figure 11.1.3).

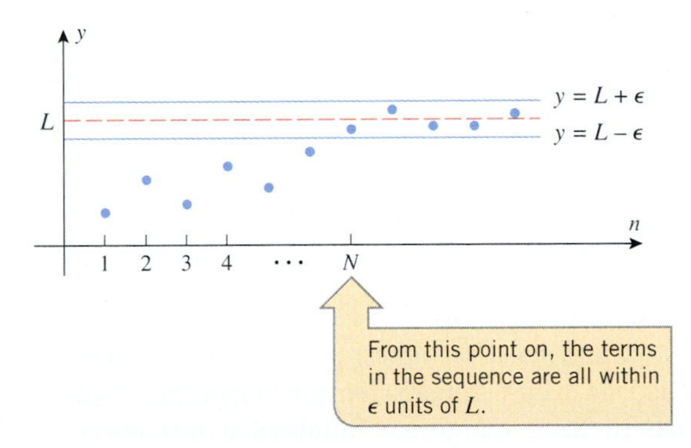

Figure 11.1.3

The following definition makes these ideas precise.

> **11.1.2 DEFINITION.** A sequence $\{a_n\}$ is said to ***converge*** to the ***limit*** L if given any $\epsilon > 0$, there is a positive integer N such that $|a_n - L| < \epsilon$ for $n \geq N$. In this case we write
>
> $$\lim_{n \to +\infty} a_n = L$$
>
> A sequence that does not converge to some finite limit is said to ***diverge***.

Example 2

The first two sequences in Figure 11.1.2 diverge, and the second two converge to 1; that is,

$$\lim_{n\to+\infty} \frac{n}{n+1} = 1 \quad \text{and} \quad \lim_{n\to+\infty} \left(1 + \left(-\tfrac{1}{2}\right)^n\right) = 1$$

◀

The following theorem, which we state without proof, shows that the familiar properties of limits apply to sequences. This theorem ensures that the algebraic techniques used to find limits of the form $\lim_{x\to+\infty}$ can also be used for limits of the form $\lim_{n\to+\infty}$.

11.1.3 THEOREM. *Suppose that the sequences $\{a_n\}$ and $\{b_n\}$ converge to limits L_1 and L_2, respectively, and c is a constant. Then*

(a) $\lim_{n\to+\infty} c = c$

(b) $\lim_{n\to+\infty} ca_n = c \lim_{n\to+\infty} a_n = cL_1$

(c) $\lim_{n\to+\infty} (a_n + b_n) = \lim_{n\to+\infty} a_n + \lim_{n\to+\infty} b_n = L_1 + L_2$

(d) $\lim_{n\to+\infty} (a_n - b_n) = \lim_{n\to+\infty} a_n - \lim_{n\to+\infty} b_n = L_1 - L_2$

(e) $\lim_{n\to+\infty} (a_n b_n) = \lim_{n\to+\infty} a_n \cdot \lim_{n\to+\infty} b_n = L_1 L_2$

(f) $\lim_{n\to+\infty} \left(\frac{a_n}{b_n}\right) = \dfrac{\lim_{n\to+\infty} a_n}{\lim_{n\to+\infty} b_n} = \dfrac{L_1}{L_2}$ (*if* $L_2 \neq 0$)

Example 3

In each part, determine whether the sequence converges or diverges. If it converges, find the limit.

(a) $\left\{\frac{n}{2n+1}\right\}_{n=1}^{+\infty}$ (b) $\left\{(-1)^{n+1}\frac{n}{2n+1}\right\}_{n=1}^{+\infty}$

(c) $\left\{(-1)^{n+1}\frac{1}{n}\right\}_{n=1}^{+\infty}$ (d) $\{8 - 2n\}_{n=1}^{+\infty}$

Solution (a). Dividing numerator and denominator by n yields

$$\lim_{n\to+\infty} \frac{n}{2n+1} = \lim_{n\to+\infty} \frac{1}{2 + 1/n} = \frac{\lim_{n\to+\infty} 1}{\lim_{n\to+\infty} (2 + 1/n)} = \frac{\lim_{n\to+\infty} 1}{\lim_{n\to+\infty} 2 + \lim_{n\to+\infty} 1/n}$$

$$= \frac{1}{2+0} = \frac{1}{2}$$

Thus, the sequence converges to $\frac{1}{2}$.

Solution (b). This sequence is the same as that in part (a), except for the factor of $(-1)^{n+1}$, which oscillates between $+1$ and -1. Thus, the terms in this sequence oscillate between positive and negative values, with the odd-numbered terms being identical to those in part (a) and the even-numbered terms being the negatives of those in part (a). Since the sequence in part (a) has a limit of $\frac{1}{2}$, it follows that the odd-numbered terms in this sequence approach $\frac{1}{2}$, while the even-numbered terms approach $-\frac{1}{2}$. Therefore, this sequence has no limit—it diverges.

Solution (c). Since $\lim_{n\to+\infty} 1/n = 0$, the product $(-1)^{n+1}(1/n)$ oscillates between positive and negative values, with the odd-numbered terms approaching 0 through positive values and the even-numbered terms approaching 0 through negative values. Thus,

$$\lim_{n\to+\infty} (-1)^{n+1}\frac{1}{n} = 0$$

so the sequence converges to 0.

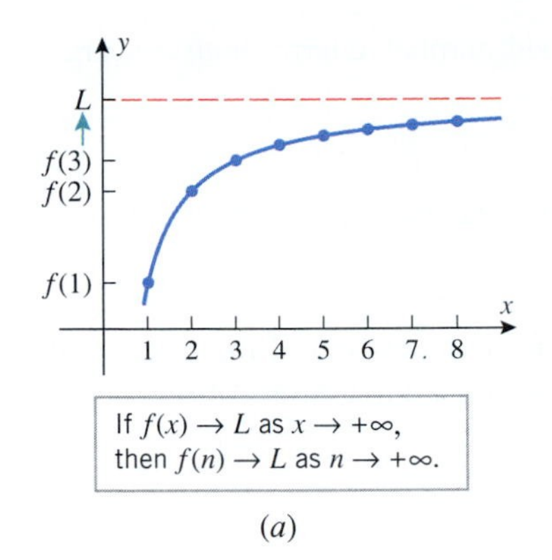

If $f(x) \to L$ as $x \to +\infty$, then $f(n) \to L$ as $n \to +\infty$.

(*a*)

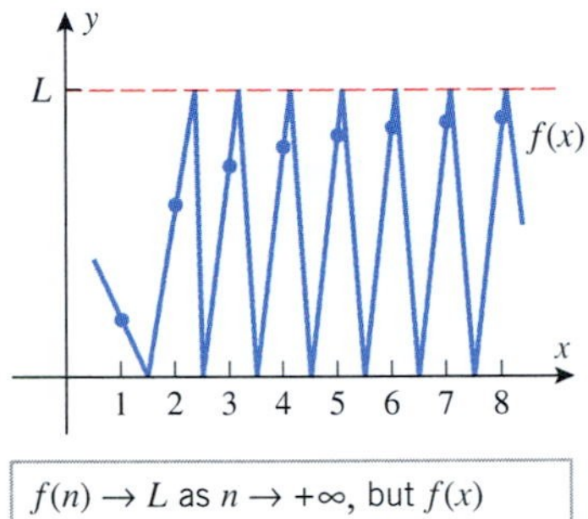

$f(n) \to L$ as $n \to +\infty$, but $f(x)$ diverges by oscillation as $x \to +\infty$.

(*b*)

Figure 11.1.4

Solution (d). $\lim_{n \to +\infty} (8 - 2n) = -\infty$, so the sequence $\{8 - 2n\}_{n=1}^{+\infty}$ diverges. ◀

If the general term of a sequence is $f(n)$, and if we replace n by x, where x can vary over the entire interval $[1, +\infty)$, then the values of $f(n)$ can be viewed as "sample values" of $f(x)$ taken at the positive integers. Thus, if $f(x) \to L$ as $x \to +\infty$, then it must also be true that $f(n) \to L$ as $n \to +\infty$ (Figure 11.1.4*a*). However, the converse is not true; that is, one cannot infer that $f(x) \to L$ as $x \to +\infty$ from the fact that $f(n) \to L$ as $n \to +\infty$ (Figure 11.1.4*b*).

Example 4

In each part, determine whether the sequence converges, and if so, find its limit.

(a) $1, \frac{1}{2}, \frac{1}{2^2}, \frac{1}{2^3}, \ldots, \frac{1}{2^n}, \ldots$ (b) $1, 2, 2^2, 2^3, \ldots, 2^n, \ldots$

Solution. Replacing n by x in the first sequence produces the power function $(1/2)^x$, and replacing n by x in the second sequence produces the power function 2^x. Now recall that if $0 < b < 1$, then $b^x \to 0$ as $x \to +\infty$, and if $b > 1$, then $b^x \to +\infty$ as $x \to +\infty$ (Figure 4.2.1). Thus,

$$\lim_{n \to +\infty} \frac{1}{2^n} = 0 \quad \text{and} \quad \lim_{n \to +\infty} 2^n = +\infty$$ ◀

Example 5

Find the limit of the sequence $\left\{\frac{n}{e^n}\right\}_{n=1}^{+\infty}$.

Solution. The expression n/e^n is an indeterminate form of type ∞/∞ as $n \to +\infty$, so L'Hôpital's rule is called for. However, we cannot apply this rule directly to n/e^n because the functions n and e^n are only defined at the positive integers, and hence are not differentiable functions. To circumvent this problem, we will replace n by x, and apply L'Hôpital's rule to the function x/e^x. This yields

$$\lim_{x \to +\infty} \frac{x}{e^x} = \lim_{x \to +\infty} \frac{1}{e^x} = 0$$

from which we can conclude that

$$\lim_{n \to +\infty} \frac{n}{e^n} = 0$$ ◀

Example 6

Show that $\lim_{n \to +\infty} \sqrt[n]{n} = 1$.

Solution.

$$\lim_{n \to +\infty} \sqrt[n]{n} = \lim_{n \to +\infty} n^{1/n} = \lim_{n \to +\infty} e^{(1/n)\ln n} = e^0 = 1$$ ◀

By L'Hôpital's rule applied to $(1/x)\ln x$

Sometimes the even-numbered and odd-numbered terms of a sequence behave sufficiently differently that it is desirable to investigate their convergence separately. The following theorem, whose proof is omitted, is helpful for that purpose.

11.1.4 THEOREM. *A sequence converges to a limit L if and only if the sequences of even-numbered terms and odd-numbered terms both converge to L.*

Example 7

The sequence

$$\frac{1}{2}, \frac{1}{3}, \frac{1}{2^2}, \frac{1}{3^2}, \frac{1}{2^3}, \frac{1}{3^3}, \ldots$$

converges to 0, since the even-numbered terms and the odd-numbered terms both converge to 0, and the sequence

$$1, \tfrac{1}{2}, 1, \tfrac{1}{3}, 1, \tfrac{1}{4}, \ldots$$

diverges, since the odd-numbered terms converge to 1 and the even-numbered terms converge to 0. ◀

THE SQUEEZING THEOREM FOR SEQUENCES

The following theorem, which we state without proof, is an adaptation of the Squeezing Theorem (2.5.2) to sequences. This theorem will be useful for finding limits of sequences that cannot be obtained directly.

11.1.5 THEOREM (*The Squeezing Theorem for Sequences*). *Let $\{a_n\}$, $\{b_n\}$, and $\{c_n\}$ be sequences such that*

$$a_n \le b_n \le c_n \quad (\textit{for all values of } n \textit{ beyond some index } N)$$

If the sequences $\{a_n\}$ and $\{c_n\}$ have a common limit L as $n \to +\infty$, then $\{b_n\}$ also has the limit L as $n \to +\infty$.

Table 11.1.4

n	$\dfrac{n!}{n^n}$
1	1.0000000000
2	0.5000000000
3	0.2222222222
4	0.0937500000
5	0.0384000000
6	0.0154320988
7	0.0061198990
8	0.0024032593
9	0.0009366567
10	0.0003628800
11	0.0001399059
12	0.0000537232

Example 8

Use numerical evidence to make a conjecture about the limit of the sequence*

$$\left\{\frac{n!}{n^n}\right\}_{n=1}^{+\infty}$$

and then confirm that your conjecture is correct.

Solution. Table 11.1.4, which was obtained with a calculating utility, suggests that the limit of the sequence may be 0. To confirm this we need to examine the limit of

$$a_n = \frac{n!}{n^n}$$

as $n \to +\infty$. Although this is an indeterminate form of type ∞/∞, L'Hôpital's rule is not helpful because we have no definition of $x!$ for values of x that are not integers. However, let us write out some of the initial terms and the general term in the sequence:

$$a_1 = 1, \quad a_2 = \frac{1 \cdot 2}{2 \cdot 2}, \quad a_3 = \frac{1 \cdot 2 \cdot 3}{3 \cdot 3 \cdot 3}, \ldots, \quad a_n = \frac{1 \cdot 2 \cdot 3 \cdots n}{n \cdot n \cdot n \cdots n}, \ldots$$

We can rewrite the general term as

$$a_n = \frac{1}{n}\left(\frac{2 \cdot 3 \cdots n}{n \cdot n \cdots n}\right)$$

from which it is evident that

$$0 \le a_n \le \frac{1}{n}$$

However, the two outside expressions have a limit of 0 as $n \to +\infty$; thus, the Squeezing Theorem for Sequences implies that $a_n \to 0$ as $n \to +\infty$, which confirms our conjecture. ◀

The following theorem is often useful for finding the limit of a sequence with both positive and negative terms—it states that if the sequence $\{|a_n|\}$ that is obtained by taking

---

*Recall that if n is a positive integer, then the symbol $n!$ (read "n factorial") denotes the product of the first n positive integers; that is,

$$n! = 1 \cdot 2 \cdot 3 \cdots n \quad \text{or equivalently,} \quad n! = n(n-1)(n-2)\cdots 1$$

Moreover, it is agreed by convention that $0! = 1$.

the absolute value of each term in the sequence $\{a_n\}$ converges to 0, then $\{a_n\}$ also converges to 0.

11.1.6 THEOREM. *If* $\lim_{n\to+\infty} |a_n| = 0$, *then* $\lim_{n\to+\infty} a_n = 0$.

Proof. Depending on the sign of a_n, either $a_n = |a_n|$ or $a_n = -|a_n|$. Thus, in all cases we have

$$-|a_n| \le a_n \le |a_n|$$

However, the limit of the two outside terms is 0, and hence the limit of a_n is 0 by the Squeezing Theorem for Sequences. ∎

Example 9

Consider the sequence

$$1, -\frac{1}{2}, \frac{1}{2^2}, -\frac{1}{2^3}, \dots, (-1)^n\frac{1}{2^n}, \dots$$

If we take the absolute value of each term, we obtain the sequence

$$1, \frac{1}{2}, \frac{1}{2^2}, \frac{1}{2^3}, \dots, \frac{1}{2^n}, \dots$$

which, as shown in Example 4, converges to 0. Thus, from Theorem 11.1.6 we have

$$\lim_{n\to+\infty}\left[(-1)^n\frac{1}{2^n}\right] = 0$$ ◀

SEQUENCES DEFINED RECURSIVELY

Some sequences do not arise from a formula for the general term, but rather from a formula or set of formulas that specify how to generate each term in the sequence from terms that precede it; such sequences are said to be defined ***recursively***, and the defining formulas are called ***recursion formulas***. A good example is the mechanic's rule for approximating square roots. In Formula (1) of the Introduction, we stated that the recursion formulas

$$y_0 = 1, \quad y_{n+1} = \frac{1}{2}\left(y_n + \frac{2}{y_n}\right), \qquad n = 0, 1, 2, \dots \tag{2}$$

generate a sequence $\{y_n\}$ that converges to $\sqrt{2}$, and in Table 1 of that section we used these recursion formulas to generate some of the terms in the sequence.

It would take us too far afield to investigate the convergence of sequences defined recursively, but we will conclude this section with a useful technique that can sometimes be used to compute limits of such sequences.

Example 10

Assuming that the sequence generated by (2) converges, show that the limit is $\sqrt{2}$.

Solution. Assume that $y_n \to L$, where L is to be determined. Since $n+1 \to +\infty$ as $n \to +\infty$, it is also true that $y_{n+1} \to L$ as $n \to +\infty$; thus, if we take the limit of the expression

$$y_{n+1} = \frac{1}{2}\left(y_n + \frac{2}{y_n}\right)$$

as $n \to +\infty$, we obtain

$$L = \frac{1}{2}\left(L + \frac{2}{L}\right)$$

which can be rewritten as $L^2 = 2$. The negative solution of this equation is extraneous because $y_n > 0$ for all n, so $L = \sqrt{2}$. ◀

EXERCISE SET 11.1 Graphing Calculator C CAS

1. In each part, find a formula for the general term of the sequence, starting with $n = 1$.
   (a) $1, \frac{1}{3}, \frac{1}{9}, \frac{1}{27}, \ldots$ (b) $1, -\frac{1}{3}, \frac{1}{9}, -\frac{1}{27}, \ldots$
   (c) $\frac{1}{2}, \frac{3}{4}, \frac{5}{6}, \frac{7}{8}, \ldots$ (d) $\frac{1}{\sqrt{\pi}}, \frac{4}{\sqrt[3]{\pi}}, \frac{9}{\sqrt[4]{\pi}}, \frac{16}{\sqrt[5]{\pi}}, \ldots$

2. In each part, find two formulas for the general term of the sequence, one starting with $n = 1$ and the other with $n = 0$.
   (a) $1, -r, r^2, -r^3, \ldots$ (b) $r, -r^2, r^3, -r^4, \ldots$

3. (a) Write out the first four terms of the sequence $\{1 + (-1)^n\}$, starting with $n = 0$.
   (b) Write out the first four terms of the sequence $\{\cos n\pi\}$, starting with $n = 0$.
   (c) Use the results in parts (a) and (b) to express the general term of the sequence $4, 0, 4, 0, \ldots$ in two different ways, starting with $n = 0$.

4. In each part, find a formula for the general term using factorials and starting with $n = 1$.
   (a) $1 \cdot 2, 1 \cdot 2 \cdot 3 \cdot 4, 1 \cdot 2 \cdot 3 \cdot 4 \cdot 5 \cdot 6,$ $1 \cdot 2 \cdot 3 \cdot 4 \cdot 5 \cdot 6 \cdot 7 \cdot 8, \ldots$
   (b) $1, 1 \cdot 2 \cdot 3, 1 \cdot 2 \cdot 3 \cdot 4 \cdot 5, 1 \cdot 2 \cdot 3 \cdot 4 \cdot 5 \cdot 6 \cdot 7, \ldots$

In Exercises 5–22, write out the first five terms of the sequence, determine whether the sequence converges, and if so find its limit.

5. $\left\{\frac{n}{n+2}\right\}_{n=1}^{+\infty}$
6. $\left\{\frac{n^2}{2n+1}\right\}_{n=1}^{+\infty}$
7. $\{2\}_{n=1}^{+\infty}$
8. $\left\{\ln\left(\frac{1}{n}\right)\right\}_{n=1}^{+\infty}$
9. $\left\{\frac{\ln n}{n}\right\}_{n=1}^{+\infty}$
10. $\left\{n \sin \frac{\pi}{n}\right\}_{n=1}^{+\infty}$
11. $\{1 + (-1)^n\}_{n=1}^{+\infty}$
12. $\left\{\frac{(-1)^{n+1}}{n^2}\right\}_{n=1}^{+\infty}$
13. $\left\{(-1)^n \frac{2n^3}{n^3+1}\right\}_{n=1}^{+\infty}$
14. $\left\{\frac{n}{2^n}\right\}_{n=1}^{+\infty}$
15. $\left\{\frac{(n+1)(n+2)}{2n^2}\right\}_{n=1}^{+\infty}$
16. $\left\{\frac{\pi^n}{4^n}\right\}_{n=1}^{+\infty}$
17. $\left\{\cos \frac{3}{n}\right\}_{n=1}^{+\infty}$
18. $\left\{\cos \frac{\pi n}{2}\right\}_{n=1}^{+\infty}$
19. $\{n^2 e^{-n}\}_{n=1}^{+\infty}$
20. $\{\sqrt{n^2+3n} - n\}_{n=1}^{+\infty}$
21. $\left\{\left(\frac{n+3}{n+1}\right)^n\right\}_{n=1}^{+\infty}$
22. $\left\{\left(1 - \frac{2}{n}\right)^n\right\}_{n=1}^{+\infty}$

In Exercises 23–30, find the general term of the sequence, starting with $n = 1$, determine whether the sequence converges, and if so find its limit.

23. $\frac{1}{2}, \frac{3}{4}, \frac{5}{6}, \frac{7}{8}, \ldots$
24. $0, \frac{1}{2^2}, \frac{2}{3^2}, \frac{3}{4^2}, \ldots$
25. $\frac{1}{3}, \frac{1}{9}, \frac{1}{27}, \frac{1}{81}, \ldots$
26. $-1, 2, -3, 4, -5, \ldots$
27. $\left(1 - \frac{1}{2}\right), \left(\frac{1}{2} - \frac{1}{3}\right), \left(\frac{1}{3} - \frac{1}{4}\right), \left(\frac{1}{4} - \frac{1}{5}\right), \ldots$
28. $3, \frac{3}{2}, \frac{3}{2^2}, \frac{3}{2^3}, \ldots$
29. $(\sqrt{2} - \sqrt{3}), (\sqrt{3} - \sqrt{4}), (\sqrt{4} - \sqrt{5}), \ldots$
30. $\frac{1}{3^5}, -\frac{1}{3^6}, \frac{1}{3^7}, -\frac{1}{3^8}, \ldots$

C 31. Read your CAS documentation to determine how to find limits approaching $+\infty$, and use the CAS to check the limits you calculated in Exercises 5–30.

C 32. (a) Use numerical evidence to make a conjecture about the limit of the sequence $\{\sqrt[n]{n^3}\}_{n=2}^{+\infty}$.
   (b) Use a CAS to confirm your conjecture.

33. (a) Starting with $n = 1$, write out the first six terms of the sequence $\{a_n\}$, where
$$a_n = \begin{cases} 1, & \text{if } n \text{ is odd} \\ n, & \text{if } n \text{ is even} \end{cases}$$
   (b) Starting with $n = 1$, and considering the even and odd terms separately, find a formula for the general term of the sequence
$$1, \frac{1}{2^2}, 3, \frac{1}{2^4}, 5, \frac{1}{2^6}, \ldots$$
   (c) Starting with $n = 1$, and considering the even and odd terms separately, find a formula for the general term of the sequence
$$1, \frac{1}{3}, \frac{1}{3}, \frac{1}{5}, \frac{1}{5}, \frac{1}{7}, \frac{1}{7}, \frac{1}{9}, \frac{1}{9}, \ldots$$
   (d) Determine whether the sequences in parts (a), (b), and (c) converge. For those that do, find the limit.

34. For what positive values of b does the sequence $b, 0, b^2, 0, b^3, 0, b^4, \ldots$ converge? Justify your answer.

35. In the discussion preceding Exercise 8 of the Introduction, we implied that the sequence defined recursively by
$$y_0 = 1, \quad y_{n+1} = \frac{1}{2}\left(y_n + \frac{p}{y_n}\right)$$
converges to $\sqrt{p}$. Assuming that this sequence converges, use the method of Example 10 to confirm that this is so.

36. Consider the sequence

$$a_1 = \sqrt{6}$$
$$a_2 = \sqrt{6 + \sqrt{6}}$$
$$a_3 = \sqrt{6 + \sqrt{6 + \sqrt{6}}}$$
$$a_4 = \sqrt{6 + \sqrt{6 + \sqrt{6 + \sqrt{6}}}}$$
$$\vdots$$

(a) Find a recursion formula for a_{n+1}.
(b) Assuming that the sequence converges, use the method of Example 10 to find the limit.

37. Consider the sequence $\{a_n\}_{n=1}^{+\infty}$, where

$$a_n = \frac{1}{n^2} + \frac{2}{n^2} + \cdots + \frac{n}{n^2}$$

(a) Find a_1, a_2, a_3, and a_4.
(b) Use numerical evidence to make a conjecture about the limit of the sequence.
(c) Confirm your conjecture by expressing a_n in closed form and calculating the limit.

38. Follow the directions in Exercise 37 with

$$a_n = \frac{1^2}{n^3} + \frac{2^2}{n^3} + \cdots + \frac{n^2}{n^3}$$

In Exercises 39 and 40, use numerical evidence to make a conjecture about the limit of the sequence, and then use the Squeezing Theorem for Sequences (Theorem 11.1.5) to confirm that your conjecture is correct.

39. $\displaystyle\lim_{n\to+\infty} \frac{\sin^2 n}{n}$

40. $\displaystyle\lim_{n\to+\infty} \left(\frac{1+n}{2n}\right)^n$

41. (a) A bored student enters the number 0.5 in a calculator display and then repeatedly computes the square of the number in the display. Taking $a_0 = 0.5$, find a formula for the general term of the sequence $\{a_n\}$ of numbers that appear in the display.
(b) Try this with a calculator and make a conjecture about the limit of a_n.
(c) Confirm your conjecture by finding the limit of a_n.
(d) For what values of a_0 will this procedure produce a convergent sequence?

42. Let

$$f(x) = \begin{cases} 2x, & 0 \le x < 0.5 \\ 2x - 1, & 0.5 \le x < 1 \end{cases}$$

Does the sequence $f(0.2)$, $f(f(0.2))$, $f(f(f(0.2))), \ldots$ converge? Justify your reasoning.

43. (a) Use a graphing utility to generate the graph of the equation $y = (2^x + 3^x)^{1/x}$, and then use the graph to make a conjecture about the limit of the sequence

$$\{(2^n + 3^n)^{1/n}\}_{n=1}^{+\infty}$$

(b) Confirm your conjecture by calculating the limit.

44. Consider the sequence $\{a_n\}_{n=1}^{+\infty}$ whose nth term is

$$a_n = \frac{1}{n}\sum_{k=1}^{n} \frac{1}{1 + (k/n)}$$

Show that $\displaystyle\lim_{n\to+\infty} a_n = \ln 2$ by interpreting a_n as the Riemann sum of a definite integral.

45. Let a_n be the average value of $f(x) = 1/x$ over the interval $[1, n]$. Determine whether the sequence $\{a_n\}$ converges, and if so find its limit.

46. The sequence whose terms are 1, 1, 2, 3, 5, 8, 13, 21, ... is called the ***Fibonacci sequence*** in honor of Leonardo ("Fibonacci") da Pisa (c. 1170–1250). This sequence has the property that after starting with two 1's, each term is the sum of the preceding two.
(a) Denoting the sequence by $\{a_n\}$ and starting with $a_1 = 1$ and $a_2 = 1$, show that

$$\frac{a_{n+2}}{a_{n+1}} = 1 + \frac{a_n}{a_{n+1}} \quad \text{if } n \ge 1$$

(b) Give a reasonable informal argument to show that if the sequence $\{a_{n+1}/a_n\}$ converges to some limit L, then the sequence $\{a_{n+2}/a_{n+1}\}$ must also converge to L.
(c) Assuming that the sequence $\{a_{n+1}/a_n\}$ converges, show that its limit is $(1 + \sqrt{5})/2$.

47. If we accept the fact that the sequence $\{1/n\}_{n=1}^{+\infty}$ converges to the limit $L = 0$, then according to Definition 11.1.2, for every $\epsilon > 0$, there exists an integer N such that $|a_n - L| = |(1/n) - 0| < \epsilon$ when $n \ge N$. In each part, find the smallest possible value of N for the given value of ϵ.
(a) $\epsilon = 0.5$ (b) $\epsilon = 0.1$ (c) $\epsilon = 0.001$

48. If we accept the fact that the sequence

$$\left\{\frac{n}{n+1}\right\}_{n=1}^{+\infty}$$

converges to the limit $L = 1$, then according to Definition 11.1.2, for every $\epsilon > 0$ there exists an integer N such that

$$|a_n - L| = \left|\frac{n}{n+1} - 1\right| < \epsilon$$

when $n \ge N$. In each part, find the smallest value of N for the given value of ϵ.
(a) $\epsilon = 0.25$ (b) $\epsilon = 0.1$ (c) $\epsilon = 0.001$

49. Use Definition 11.1.2 to prove that
(a) the sequence $\{1/n\}_{n=1}^{+\infty}$ converges to 0
(b) the sequence $\left\{\dfrac{n}{n+1}\right\}_{n=1}^{+\infty}$ converges to 1.

50. Find $\displaystyle\lim_{n\to+\infty} r^n$, where r is a real number. [*Hint:* Consider the cases $|r| < 1$, $|r| > 1$, $r = 1$, and $r = -1$ separately.]

11.2 MONOTONE SEQUENCES

There are many situations in which it is important to know whether a sequence converges, but the limit itself is not relevant to the problem at hand. In this section we will study several techniques that can be used to determine whether a sequence converges.

TERMINOLOGY

We begin with some terminology.

11.2.1 DEFINITION. A sequence $\{a_n\}_{n=1}^{+\infty}$ is called

strictly increasing if $a_1 < a_2 < a_3 < \cdots < a_n < \cdots$

increasing if $a_1 \le a_2 \le a_3 \le \cdots \le a_n \le \cdots$

strictly decreasing if $a_1 > a_2 > a_3 > \cdots > a_n > \cdots$

decreasing if $a_1 \ge a_2 \ge a_3 \ge \cdots \ge a_n \ge \cdots$

In words, a sequence is strictly increasing if each term is larger than its predecessor, increasing if each term is the same as or larger than its predecessor, strictly decreasing if each term is smaller than its predecessor, and decreasing if each term is the same as or smaller than its predecessor. It follows that every strictly increasing sequence is increasing (but not conversely), and every strictly decreasing sequence is decreasing (but not conversely). A sequence that is either strictly increasing or strictly decreasing is called ***strictly monotone***, and a sequence that is either increasing or decreasing is called ***monotone***.

Example 1

SEQUENCE	DESCRIPTION
$\frac{1}{2}, \frac{2}{3}, \frac{3}{4}, \ldots, \frac{n}{n+1}, \ldots$	Strictly increasing
$1, \frac{1}{2}, \frac{1}{3}, \ldots, \frac{1}{n}, \ldots$	Strictly decreasing
$1, 1, 2, 2, 3, 3, \ldots$	Increasing; not strictly increasing
$1, 1, \frac{1}{2}, \frac{1}{2}, \frac{1}{3}, \frac{1}{3}, \ldots$	Decreasing; not strictly decreasing
$1, -\frac{1}{2}, \frac{1}{3}, -\frac{1}{4}, \ldots, (-1)^{n+1}\frac{1}{n}, \ldots$	Neither increasing nor decreasing

The first and second sequences are strictly monotone, and the third and fourth sequences are monotone but not strictly monotone. The fifth sequence is not monotone. ◀

FOR THE READER. Can a sequence be both increasing and decreasing? Explain.

TESTING FOR MONOTONICITY

In order for a sequence to be strictly increasing, *all* pairs of successive terms, a_n and a_{n+1}, must satisfy $a_n < a_{n+1}$ or, equivalently, $a_{n+1} - a_n > 0$. More generally, monotone sequences can be classified as follows:

DIFFERENCE BETWEEN SUCCESSIVE TERMS	CLASSIFICATION
$a_{n+1} - a_n > 0$	Strictly increasing
$a_{n+1} - a_n < 0$	Strictly decreasing
$a_{n+1} - a_n \ge 0$	Increasing
$a_{n+1} - a_n \le 0$	Decreasing

Frequently, one can *guess* whether a sequence is monotone or strictly monotone by writing out some of the initial terms. However, to be certain that the guess is correct, one must give a precise mathematical argument. The following example illustrates one method for doing this.

Example 2

Show that

$$\frac{1}{2}, \frac{2}{3}, \frac{3}{4}, \ldots, \frac{n}{n+1}, \ldots$$

is a strictly increasing sequence.

Solution. The pattern of the initial terms suggests that the sequence is strictly increasing. To prove that this is so, let

$$a_n = \frac{n}{n+1}$$

We can obtain a_{n+1} by replacing n by $n + 1$ in this formula. This yields

$$a_{n+1} = \frac{n+1}{(n+1)+1} = \frac{n+1}{n+2}$$

Thus, for $n \geq 1$

$$a_{n+1} - a_n = \frac{n+1}{n+2} - \frac{n}{n+1} = \frac{n^2 + 2n + 1 - n^2 - 2n}{(n+1)(n+2)} = \frac{1}{(n+1)(n+2)} > 0$$

which proves that the sequence is strictly increasing. ◀

If a_n and a_{n+1} are any successive terms in a strictly increasing sequence, then $a_n < a_{n+1}$. If the terms in the sequence are all positive, then we can divide both sides of this inequality by a_n to obtain $1 < a_{n+1}/a_n$ or, equivalently, $a_{n+1}/a_n > 1$. More generally, monotone sequences with *positive* terms can be classified as follows:

RATIO OF SUCCESSIVE TERMS	CONCLUSION
$a_{n+1}/a_n > 1$	Strictly increasing
$a_{n+1}/a_n < 1$	Strictly decreasing
$a_{n+1}/a_n \geq 1$	Increasing
$a_{n+1}/a_n \leq 1$	Decreasing

Example 3

Show that the sequence in Example 2 is strictly increasing by examining the ratio of successive terms.

Solution. As shown in the solution of Example 2,

$$a_n = \frac{n}{n+1} \quad \text{and} \quad a_{n+1} = \frac{n+1}{n+2}$$

Thus,

$$\frac{a_{n+1}}{a_n} = \frac{(n+1)/(n+2)}{n/(n+1)} = \frac{n+1}{n+2} \cdot \frac{n+1}{n} = \frac{n^2 + 2n + 1}{n^2 + 2n} \tag{1}$$

Since the numerator in (1) exceeds the denominator, it follows that $a_{n+1}/a_n > 1$ for $n \geq 1$. This proves that the sequence is strictly increasing. ◀

The following example illustrates still a third technique for determining whether a sequence is strictly monotone.

Example 4

In Examples 2 and 3 we proved that the sequence

$$\frac{1}{2}, \frac{2}{3}, \frac{3}{4}, \ldots, \frac{n}{n+1}, \ldots$$

is strictly increasing by considering the difference and ratio of successive terms. Alternatively, we can proceed as follows. Let

$$f(x) = \frac{x}{x+1}$$

so that the nth term in the given sequence is $a_n = f(n)$. The function f is increasing for $x \geq 1$ since

$$f'(x) = \frac{(x+1)(1) - x(1)}{(x+1)^2} = \frac{1}{(x+1)^2} > 0$$

Thus,

$$a_n = f(n) < f(n+1) = a_{n+1}$$

which proves that the given sequence is strictly increasing. ◀

In general, if $f(n) = a_n$ is the nth term of a sequence, and if f is differentiable for $x \geq 1$, then we have the following results:

DERIVATIVE OF f FOR $x \geq 1$	CONCLUSION FOR THE SEQUENCE WITH $a_n = f(n)$
$f'(x) > 0$	Strictly increasing
$f'(x) < 0$	Strictly decreasing
$f'(x) \geq 0$	Increasing
$f'(x) \leq 0$	Decreasing

PROPERTIES THAT HOLD EVENTUALLY

Sometimes a sequence will behave erratically at first and then settle down into a definite pattern. For example, the sequence

$$9, -8, -17, 12, 1, 2, 3, 4, \ldots \tag{2}$$

is strictly increasing from the fifth term on, but the sequence as a whole cannot be classified as strictly increasing because of the erratic behavior of the first four terms. To describe such sequences, we introduce the following terminology.

> **11.2.2** DEFINITION. If discarding finitely many terms from the beginning of a sequence produces a sequence with a certain property, then the original sequence is said to have that property ***eventually***.

For example, although we cannot say that sequence (2) is strictly increasing, we can say that it is eventually strictly increasing.

Example 5

Show that the sequence $\left\{\dfrac{10^n}{n!}\right\}_{n=1}^{+\infty}$ is eventually strictly decreasing.

Solution. We have

$$a_n = \frac{10^n}{n!} \quad \text{and} \quad a_{n+1} = \frac{10^{n+1}}{(n+1)!}$$

so

$$\frac{a_{n+1}}{a_n} = \frac{10^{n+1}/(n+1)!}{10^n/n!} = \frac{10^{n+1}n!}{10^n(n+1)!} = 10\frac{n!}{(n+1)n!} = \frac{10}{n+1} \tag{3}$$

From (3), $a_{n+1}/a_n < 1$ for all $n \geq 10$, so the sequence is eventually strictly decreasing. ◀

AN INTUITIVE VIEW OF CONVERGENCE

Informally stated, the convergence or divergence of a sequence does not depend on the behavior of its *initial terms*, but rather on how the terms behave *eventually*. For example, the sequence

$$3,\ -9,\ -13,\ 17,\ 1,\ \frac{1}{2},\ \frac{1}{3},\ \frac{1}{4},\ldots$$

eventually behaves like the sequence

$$1,\ \frac{1}{2},\ \frac{1}{3},\ldots,\ \frac{1}{n},\ldots$$

and hence has a limit of 0.

CONVERGENCE OF MONOTONE SEQUENCES

The following two theorems, whose proofs are discussed at the end of this section, show that a monotone sequence either converges or becomes infinite—divergence by oscillation cannot occur.

11.2.3 THEOREM. *If a sequence $\{a_n\}$ is eventually increasing, then there are two possibilities:*

*(a) There is a constant M, called an **upper bound** for the sequence, such that $a_n \leq M$ for all n, in which case the sequence converges to a limit L satisfying $L \leq M$.*

(b) No upper bound exists, in which case $\lim_{n\to+\infty} a_n = +\infty$.

11.2.4 THEOREM. *If a sequence $\{a_n\}$ is eventually decreasing, then there are two possibilities:*

*(a) There is a constant M, called a **lower bound** for the sequence, such that $a_n \geq M$ for all n, in which case the sequence converges to a limit L satisfying $L \geq M$.*

(b) No lower bound exists, in which case $\lim_{n\to+\infty} a_n = -\infty$.

Note that these results do not give a method for obtaining limits; they tell us only whether a limit exists.

Example 6

Show that the sequence $\left\{\frac{10^n}{n!}\right\}_{n=1}^{+\infty}$ converges and find its limit.

Solution. We showed in Example 5 that the sequence is eventually strictly decreasing. Since all terms in the sequence are positive, it is bounded below by $M = 0$, and hence Theorem 11.2.4 guarantees that it converges to a nonnegative limit L. However, the limit is not evident directly from the formula $10^n/n!$ for the nth term, so we will need some ingenuity to obtain it.

Recall from Formula (3) of Example 5 that successive terms in the given sequence are related by the recursion formula

$$a_{n+1} = \frac{10}{n+1}a_n \tag{4}$$

where $a_n = 10^n/n!$. We will take the limit as $n \to +\infty$ of both sides of (4) and use the fact that

$$\lim_{n\to+\infty} a_{n+1} = \lim_{n\to+\infty} a_n = L$$

We obtain

$$L = \lim_{n\to+\infty} a_{n+1} = \lim_{n\to+\infty}\left(\frac{10}{n+1}a_n\right) = \lim_{n\to+\infty}\frac{10}{n+1}\lim_{n\to+\infty} a_n = 0\cdot L = 0$$

so that

$$L = \lim_{n\to+\infty}\frac{10^n}{n!} = 0$$

◀

REMARK. In the exercises we will show that the technique illustrated in this example can be adapted to obtain the limit

$$\lim_{n\to+\infty}\frac{x^n}{n!} = 0 \tag{5}$$

for any real value of x (Exercise 26). This result, which shows that $n!$ eventually increases more rapidly than any positive integer power of x, will be useful in our later work.

THE COMPLETENESS AXIOM

In this text we have accepted the familiar properties of real numbers without proof, and indeed, we have not even attempted to define the term *real number*. Although this is sufficient for many purposes, it was recognized by the late nineteenth century that the study of limits and functions in calculus requires a precise axiomatic formulation of the real numbers analogous to the axiomatic development of Euclidean geometry. Although we will not attempt to pursue this development, we will need to discuss one of the axioms about real numbers in order to prove Theorems 11.2.3 and 11.2.4. But first we will introduce some terminology.

If S is a nonempty set of real numbers, then we call u an ***upper bound*** for S if u is greater than or equal to every number in S, and we call l a ***lower bound*** for S if l is smaller than or equal to every number in S. For example, if S is the set of numbers in the interval $(1, 3)$, then $u = 4$, 10, and 100 are upper bounds for S and $l = -10$, 0, and $\frac{1}{2}$ are lower bounds for S. Observe also that $u = 3$ is the smallest of all upper bounds and $l = 1$ is the largest of all lower bounds. The existence of a smallest upper bound and a greatest lower bound for S is not accidental; it is a consequence of the following axiom.

11.2.5 AXIOM (***The Completeness Axiom***). *If a nonempty set S of real numbers has an upper bound, then it has a smallest upper bound (called the* ***least upper bound****), and if a nonempty set S of real numbers has a lower bound, then it has a largest lower bound (called the* ***greatest lower bound****).*

Proof of Theorem 11.2.3.

(*a*) Assume there exists a number M such that $a_n \le M$ for $n = 1, 2, \ldots$. Then M is an upper bound for the set of terms in the sequence. By the Completeness Axiom there is a least upper bound for the terms, call it L. Now let ϵ be any positive number. Since L is the least upper bound for the terms, $L - \epsilon$ is not an upper bound for the terms, which means that there is at least one term a_N such that

$$a_N > L - \epsilon$$

Moreover, since $\{a_n\}$ is an increasing sequence, we must have

$$a_n \ge a_N > L - \epsilon \tag{6}$$

when $n \geq N$. But a_n cannot exceed L since L is an upper bound for the terms. This observation together with (6) tells us that $L \geq a_n > L - \epsilon$ for $n \geq N$, so all terms from the Nth on are within ϵ units of L. This is exactly the requirement to have

$$\lim_{n \to +\infty} a_n = L$$

Finally, $L \leq M$ since M is an upper bound for the terms and L is the least upper bound. This proves part (*a*).

(*b*) If there is no number M such that $a_n \leq M$ for $n = 1, 2, \ldots$, then no matter how large we choose M, there is a term a_N such that

$$a_N > M$$

and, since the sequence is increasing,

$$a_n \geq a_N > M$$

when $n \geq N$. Thus, the terms in the sequence become arbitrarily large as n increases. That is,

$$\lim_{n \to +\infty} a_n = +\infty$$

The proof of Theorem 11.2.4 will be omitted since it is similar to that of 11.2.3.

EXERCISE SET 11.2

In Exercises 1–6, use $a_{n+1} - a_n$ to show that the given sequence $\{a_n\}$ is strictly increasing or strictly decreasing.

1. $\left\{\frac{1}{n}\right\}_{n=1}^{+\infty}$ 2. $\left\{1 - \frac{1}{n}\right\}_{n=1}^{+\infty}$ 3. $\left\{\frac{n}{2n+1}\right\}_{n=1}^{+\infty}$

4. $\left\{\frac{n}{4n-1}\right\}_{n=1}^{+\infty}$ 5. $\{n - 2^n\}_{n=1}^{+\infty}$ 6. $\{n - n^2\}_{n=1}^{+\infty}$

In Exercises 7–12, use a_{n+1}/a_n to show that the given sequence $\{a_n\}$ is strictly increasing or strictly decreasing.

7. $\left\{\frac{n}{2n+1}\right\}_{n=1}^{+\infty}$ 8. $\left\{\frac{2^n}{1+2^n}\right\}_{n=1}^{+\infty}$ 9. $\{ne^{-n}\}_{n=1}^{+\infty}$

10. $\left\{\frac{10^n}{(2n)!}\right\}_{n=1}^{+\infty}$ 11. $\left\{\frac{n^n}{n!}\right\}_{n=1}^{+\infty}$ 12. $\left\{\frac{5^n}{2^{(n^2)}}\right\}_{n=1}^{+\infty}$

In Exercises 13–18, use differentiation to show that the sequence is strictly increasing or strictly decreasing.

13. $\left\{\frac{n}{2n+1}\right\}_{n=1}^{+\infty}$ 14. $\left\{3 - \frac{1}{n}\right\}_{n=1}^{+\infty}$

15. $\left\{\frac{1}{n + \ln n}\right\}_{n=1}^{+\infty}$ 16. $\{ne^{-2n}\}_{n=1}^{+\infty}$

17. $\left\{\frac{\ln(n+2)}{n+2}\right\}_{n=1}^{+\infty}$ 18. $\{\tan^{-1} n\}_{n=1}^{+\infty}$

In Exercises 19–24, use any method to show that the given sequence is eventually strictly increasing or eventually strictly decreasing.

19. $\{2n^2 - 7n\}_{n=1}^{+\infty}$ 20. $\{n^3 - 4n^2\}_{n=1}^{+\infty}$

21. $\left\{\frac{n}{n^2+10}\right\}_{n=1}^{+\infty}$ 22. $\left\{n + \frac{17}{n}\right\}_{n=1}^{+\infty}$

23. $\left\{\frac{n!}{3^n}\right\}_{n=1}^{+\infty}$ 24. $\{n^5 e^{-n}\}_{n=1}^{+\infty}$

25. (a) Suppose that $\{a_n\}$ is a monotone sequence such that $1 \leq a_n \leq 2$. Must the sequence converge? If so, what can you say about the limit?
(b) Suppose that $\{a_n\}$ is a monotone sequence such that $a_n \leq 2$. Must the sequence converge? If so, what can you say about the limit?

26. The goal in this exercise is to prove Formula (5) in this section. The case where $x = 0$ is obvious, so we will focus on the case where $x \neq 0$.
(a) Let $a_n = |x|^n/n!$. Show that
$$a_{n+1} = \frac{|x|}{n+1} a_n$$
(b) Show that the sequence $\{a_n\}$ is eventually strictly decreasing.
(c) Show that the sequence $\{a_n\}$ converges.
(d) Use the results in parts (a) and (c) to show that $a_n \to 0$ as $n \to +\infty$.
(e) Obtain Formula (5) from the result in part (d).

27. Let $\{a_n\}$ be the sequence defined recursively by $a_1 = \sqrt{2}$ and $a_{n+1} = \sqrt{2 + a_n}$ for $n \geq 1$.
(a) List the first three terms of the sequence.
(b) Show that $a_n < 2$ for $n \geq 1$.
(c) Show that $a_{n+1}^2 - a_n^2 = (2 - a_n)(1 + a_n)$ for $n \geq 1$.
(d) Use the results in parts (b) and (c) to show that $\{a_n\}$ is a strictly increasing sequence. [*Hint:* If x and y are positive real numbers such that $x^2 - y^2 > 0$, then it follows by factoring that $x - y > 0$.]
(e) Show that $\{a_n\}$ converges and find its limit L.

28. Let $\{a_n\}$ be the sequence defined recursively by $a_1 = 1$ and $a_{n+1} = \frac{1}{2}[a_n + (3/a_n)]$ for $n \geq 1$.
(a) Show that $a_n \geq \sqrt{3}$ for $n \geq 2$. [*Hint:* What is the minimum value of $\frac{1}{2}[x + (3/x)]$ for $x > 0$?]
(b) Show that $\{a_n\}$ is eventually decreasing. [*Hint:* Examine $a_{n+1} - a_n$ or a_{n+1}/a_n and use the result in part (a).]
(c) Show that $\{a_n\}$ converges and find its limit L.

29. (a) Compare appropriate areas in the accompanying figure to deduce the following inequalities for $n \geq 2$:

$$\int_1^n \ln x \, dx < \ln n! < \int_1^{n+1} \ln x \, dx$$

(b) Use the result in part (a) to show that

$$\frac{n^n}{e^{n-1}} < n! < \frac{(n+1)^{n+1}}{e^n}, \quad n > 1$$

(c) Use the Squeezing Theorem for Sequences (Theorem 11.1.5) and the result in part (b) to show that

$$\lim_{n \to +\infty} \frac{\sqrt[n]{n!}}{n} = \frac{1}{e}$$

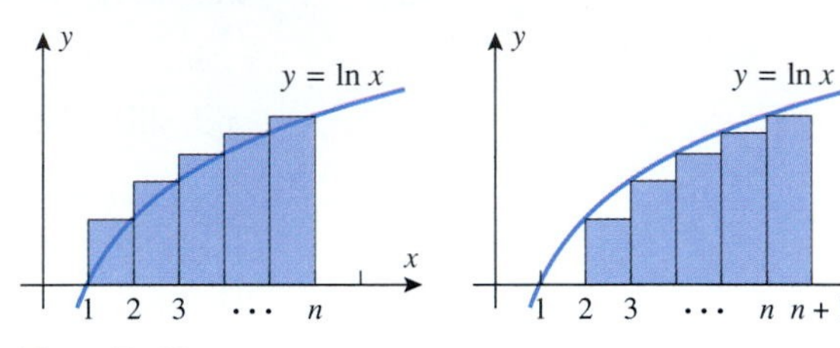

Figure Ex-29

30. Use the left inequality in Exercise 29(b) to show that

$$\lim_{n \to +\infty} \sqrt[n]{n!} = +\infty$$

11.3 INFINITE SERIES

The purpose of this section is to discuss sums that contain infinitely many terms. The most familiar examples of such sums occur in the decimal representation of real numbers. For example, when we write $\frac{1}{3}$ in the decimal form $\frac{1}{3} = 0.3333\ldots$, we mean

$$\frac{1}{3} = 0.3 + 0.03 + 0.003 + 0.0003 + \cdots$$

which suggests that the decimal representation of $\frac{1}{3}$ can be viewed as a sum of infinitely many real numbers.

SUMS OF INFINITE SERIES

Our first objective is to define what is meant by the "sum" of infinitely many real numbers. We begin with some terminology.

11.3.1 DEFINITION. An ***infinite series*** is an expression that can be written in the form

$$\sum_{k=1}^{\infty} u_k = u_1 + u_2 + u_3 + \cdots + u_k + \cdots$$

The numbers $u_1, u_2, u_3, \ldots$ are called the ***terms*** of the series.

Since it is impossible to add infinitely many numbers together directly, sums of infinite series are defined and computed by an indirect limiting process. To motivate the basic idea, consider the decimal

$$0.3333\ldots \tag{1}$$

This can be viewed as the infinite series

$$0.3 + 0.03 + 0.003 + 0.0003 + \cdots$$

or, equivalently,

$$\frac{3}{10} + \frac{3}{10^2} + \frac{3}{10^3} + \frac{3}{10^4} + \cdots \tag{2}$$

Since (1) is the decimal expansion of $\frac{1}{3}$, any reasonable definition for the sum of an infinite series should yield $\frac{1}{3}$ for the sum of (2). To obtain such a definition, consider the following sequence of (finite) sums:

$$\begin{aligned}
s_1 &= \frac{3}{10} = 0.3 \\
s_2 &= \frac{3}{10} + \frac{3}{10^2} = 0.33 \\
s_3 &= \frac{3}{10} + \frac{3}{10^2} + \frac{3}{10^3} = 0.333 \\
s_4 &= \frac{3}{10} + \frac{3}{10^2} + \frac{3}{10^3} + \frac{3}{10^4} = 0.3333 \\
&\vdots
\end{aligned}$$

The sequence of numbers $s_1, s_2, s_3, s_4, \ldots$ can be viewed as a succession of approximations to the "sum" of the infinite series, which we want to be $\frac{1}{3}$. As we progress through the sequence, more and more terms of the infinite series are used, and the approximations get better and better, suggesting that the desired sum of $\frac{1}{3}$ might be the *limit* of this sequence of approximations. To see that this is so, we must calculate the limit of the general term in the sequence of approximations, namely

$$s_n = \frac{3}{10} + \frac{3}{10^2} + \cdots + \frac{3}{10^n} \tag{3}$$

The problem of calculating

$$\lim_{n\to+\infty} s_n = \lim_{n\to+\infty} \left(\frac{3}{10} + \frac{3}{10^2} + \cdots + \frac{3}{10^n} \right)$$

is complicated by the fact that both the last term and the number of terms in the sum change with n. It is best to rewrite such limits in a closed form in which the number of terms does not vary, if possible. (See the remark following Example 3 in Section 7.4.) To do this, we multiply both sides of (3) by $\frac{1}{10}$ to obtain

$$\frac{1}{10} s_n = \frac{3}{10^2} + \frac{3}{10^3} + \cdots + \frac{3}{10^n} + \frac{3}{10^{n+1}} \tag{4}$$

and then subtract (4) from (3) to obtain

$$\begin{aligned}
s_n - \frac{1}{10} s_n &= \frac{3}{10} - \frac{3}{10^{n+1}} \\
\frac{9}{10} s_n &= \frac{3}{10}\left(1 - \frac{1}{10^n}\right) \\
s_n &= \frac{1}{3}\left(1 - \frac{1}{10^n}\right)
\end{aligned}$$

Since $1/10^n \to 0$ as $n \to +\infty$, it follows that

$$\lim_{n\to+\infty} s_n = \lim_{n\to+\infty} \frac{1}{3}\left(1 - \frac{1}{10^n}\right) = \frac{1}{3}$$

which we denote by writing

$$\frac{1}{3} = \frac{3}{10} + \frac{3}{10^2} + \frac{3}{10^3} + \cdots + \frac{3}{10^n} + \cdots$$

Motivated by the preceding example, we are now ready to define the general concept of the "sum" of an infinite series

$$u_1 + u_2 + u_3 + \cdots + u_k + \cdots$$

We begin with some terminology: Let s_n denote the sum of the first n terms of the series. Thus,

$$\begin{aligned} s_1 &= u_1 \\ s_2 &= u_1 + u_2 \\ s_3 &= u_1 + u_2 + u_3 \\ &\vdots \\ s_n &= u_1 + u_2 + u_3 + \cdots + u_n = \sum_{k=1}^{n} u_k \end{aligned}$$

The number s_n is called the ***nth partial sum*** of the series and the sequence $\{s_n\}_{n=1}^{+\infty}$ is called the ***sequence of partial sums***.

WARNING. In everyday language the words "sequence" and "series" are often used interchangeably. However, this is not so in mathematics—mathematically, a sequence is a *succession* and a series is a *sum*. It is essential that you keep this distinction in mind.

As n increases, the partial sum $s_n = u_1 + u_2 + \cdots + u_n$ includes more and more terms of the series. Thus, if s_n tends toward a limit as $n \to +\infty$, it is reasonable to view this limit as the sum of *all* the terms in the series. This suggests the following definition.

11.3.2 DEFINITION. Let $\{s_n\}$ be the sequence of partial sums of the series

$$u_1 + u_2 + u_3 + \cdots + u_k + \cdots$$

If the sequence $\{s_n\}$ converges to a limit S, then the series is said to ***converge*** to S, and S is called the ***sum*** of the series. We denote this by writing

$$S = \sum_{k=1}^{\infty} u_k$$

If the sequence of partial sums diverges, then the series is said to ***diverge***. A divergent series has no sum.

REMARK. Sometimes it will be desirable to start the summation index in an infinite series at $k = 0$ rather than $k = 1$, in which case we will view u_0 as the zeroth term and $s_0 = u_0$ as the zeroth partial sum. It can be proved that changing the starting value for the index has no effect on the convergence or divergence of an infinite series.

Example 1

Determine whether the series

$$1 - 1 + 1 - 1 + 1 - 1 + \cdots$$

converges or diverges. If it converges, find the sum.

Solution. It is tempting to conclude that the sum of the series is zero by arguing that the positive and negative terms cancel one another. However, this is *not correct*; the problem is that algebraic operations that hold for finite sums do not carry over to infinite series in all cases. Later, we will discuss conditions under which familiar algebraic operations can be applied to infinite series, but for this example we turn directly to Definition 11.3.2. The partial sums are

$$s_1 = 1$$
$$s_2 = 1 - 1 = 0$$
$$s_3 = 1 - 1 + 1 = 1$$
$$s_4 = 1 - 1 + 1 - 1 = 0$$

and so forth. Thus, the sequence of partial sums is

$$1, 0, 1, 0, 1, 0, \ldots$$

Since this is a divergent sequence, the given series diverges and consequently has no sum. ◀

GEOMETRIC SERIES

In many important geometric series, each term is obtained by multiplying the preceding term by some fixed constant. Thus, if the initial term of the series is a and each term is obtained by multiplying the preceding term by r, then the series has the form

$$\sum_{k=0}^{\infty} ar^k = a + ar + ar^2 + ar^3 + \cdots + ar^k + \cdots \quad (a \neq 0)$$

Such series are called ***geometric series***, and the number r is called the ***ratio*** for the series. Here are some examples:

$$1 + 2 + 4 + 8 + \cdots + 2^k + \cdots \qquad a = 1, r = 2$$

$$\frac{3}{10} + \frac{3}{10^2} + \frac{3}{10^3} + \cdots + \frac{3}{10^k} + \cdots \qquad a = \tfrac{3}{10}, r = \tfrac{1}{10}$$

$$\frac{1}{2} - \frac{1}{4} + \frac{1}{8} - \frac{1}{16} + \cdots + (-1)^{k+1}\frac{1}{2^k} + \cdots \qquad a = \tfrac{1}{2}, r = -\tfrac{1}{2}$$

$$1 + 1 + 1 + \cdots + 1 + \cdots \qquad a = 1, r = 1$$

$$1 - 1 + 1 - 1 + \cdots + (-1)^{k+1} + \cdots \qquad a = 1, r = -1$$

$$1 + x + x^2 + x^3 + \cdots + x^k + \cdots \qquad a = 1, r = x$$

REMARK. In some of these series we started the index of summation at $k = 0$ and in others at $k = 1$, depending on which choice produced the simpler general term.

The following theorem is the fundamental result on convergence of geometric series.

11.3.3 THEOREM. *A geometric series*

$$\sum_{k=0}^{\infty} ar^k = a + ar + ar^2 + \cdots + ar^k + \cdots \quad (a \neq 0)$$

converges if $|r| < 1$ *and diverges if* $|r| \geq 1$. *If the series converges, then the sum is*

$$\sum_{k=0}^{\infty} ar^k = \frac{a}{1 - r}$$

Proof. Let us treat the case $|r| = 1$ first. If $r = 1$, then the series is

$$a + a + a + a + \cdots$$

so the nth partial sum is $s_n = (n + 1)a$ and $\lim_{n\to+\infty} s_n = \lim_{n\to+\infty} (n + 1)a = \pm\infty$ (the sign depending on whether a is positive or negative). This proves divergence. If $r = -1$, the series is

$$a - a + a - a + \cdots$$

so the sequence of partial sums is

$$a, 0, a, 0, a, 0, \ldots$$

which diverges.

Now let us consider the case where $|r| \neq 1$. The nth partial sum of the series is

$$s_n = a + ar + ar^2 + \cdots + ar^n \tag{5}$$

Multiplying both sides of (5) by r yields

$$rs_n = ar + ar^2 + \cdots + ar^n + ar^{n+1} \tag{6}$$

and subtracting (6) from (5) gives

$$s_n - rs_n = a - ar^{n+1}$$

or

$$(1 - r)s_n = a - ar^{n+1} \tag{7}$$

Since $r \neq 1$ in the case we are considering, this can be rewritten as

$$s_n = \frac{a - ar^{n+1}}{1 - r} = \frac{a}{1 - r} - \frac{ar^{n+1}}{1 - r} \tag{8}$$

If $|r| < 1$, then $\lim_{n \to +\infty} r^{n+1} = 0$ (can you see why?), so $\{s_n\}$ converges. From (8)

$$\lim_{n \to +\infty} s_n = \frac{a}{1 - r}$$

If $|r| > 1$, then either $r > 1$ or $r < -1$. In the case $r > 1$, $\lim_{n \to +\infty} r^{n+1} = +\infty$, and in the case $r < -1$, r^{n+1} oscillates between positive and negative values that grow in magnitude, so $\{s_n\}$ diverges in both cases. ■

Example 2

The series

$$\sum_{k=0}^{\infty} \frac{5}{4^k} = 5 + \frac{5}{4} + \frac{5}{4^2} + \cdots + \frac{5}{4^k} + \cdots$$

is a geometric series with $a = 5$ and $r = \frac{1}{4}$. Since $|r| = \frac{1}{4} < 1$, the series converges and the sum is

$$\frac{a}{1 - r} = \frac{5}{1 - \frac{1}{4}} = \frac{20}{3}$$ ◀

Example 3

Find the rational number represented by the repeating decimal

$$0.784784784\ldots$$

Solution. We can write

$$0.784784784\ldots = 0.784 + 0.000784 + 0.000000784 + \cdots$$

so the given decimal is the sum of a geometric series with $a = 0.784$ and $r = 0.001$. Thus,

$$0.784784784\ldots = \frac{a}{1 - r} = \frac{0.784}{1 - 0.001} = \frac{0.784}{0.999} = \frac{784}{999}$$ ◀

Example 4

In each part, determine whether the series converges, and if so find its sum.

(a) $\sum_{k=1}^{\infty} 3^{2k} 5^{1-k}$ (b) $\sum_{k=0}^{\infty} x^k$

Solution (a). This is a geometric series in a concealed form, since we can rewrite it as

$$\sum_{k=1}^{\infty} 3^{2k} 5^{1-k} = \sum_{k=1}^{\infty} \frac{9^k}{5^{k-1}} = \sum_{k=1}^{\infty} 9\left(\frac{9}{5}\right)^{k-1}$$

Since $r = \frac{9}{5} > 1$, the series diverges.

Solution (b). The expanded form of the series is

$$\sum_{k=0}^{\infty} x^k = 1 + x + x^2 + \cdots + x^k + \cdots$$

The series is a geometric series with $a = 1$ and $r = x$, so it converges if $|x| < 1$ and diverges otherwise. When the series converges its sum is

$$\sum_{k=0}^{\infty} x^k = \frac{1}{1-x}$$

◀

Example 5

Determine whether the series

$$\sum_{k=1}^{\infty} \frac{1}{k(k+1)} = \frac{1}{1 \cdot 2} + \frac{1}{2 \cdot 3} + \frac{1}{3 \cdot 4} + \frac{1}{4 \cdot 5} + \cdots$$

converges or diverges. If it converges, find the sum.

Solution. The nth partial sum of the series is

$$s_n = \sum_{k=1}^{n} \frac{1}{k(k+1)} = \frac{1}{1 \cdot 2} + \frac{1}{2 \cdot 3} + \frac{1}{3 \cdot 4} + \cdots + \frac{1}{n(n+1)}$$

To calculate $\lim_{n \to +\infty} s_n$ we will rewrite s_n in closed form. This can be accomplished by using the method of partial fractions to obtain (verify)

$$\frac{1}{k(k+1)} = \frac{1}{k} - \frac{1}{k+1}$$

from which we obtain the telescoping sum

$$\begin{aligned} s_n &= \sum_{k=1}^{n} \left(\frac{1}{k} - \frac{1}{k+1}\right) \\ &= \left(1 - \frac{1}{2}\right) + \left(\frac{1}{2} - \frac{1}{3}\right) + \left(\frac{1}{3} - \frac{1}{4}\right) + \cdots + \left(\frac{1}{n} - \frac{1}{n+1}\right) \\ &= 1 + \left(-\frac{1}{2} + \frac{1}{2}\right) + \left(-\frac{1}{3} + \frac{1}{3}\right) + \cdots + \left(-\frac{1}{n} + \frac{1}{n}\right) - \frac{1}{n+1} \\ &= 1 - \frac{1}{n+1} \end{aligned}$$

so

$$\sum_{k=1}^{\infty} \frac{1}{k(k+1)} = \lim_{n \to +\infty} s_n = \lim_{n \to +\infty} \left(1 - \frac{1}{n+1}\right) = 1$$

◀

FOR THE READER. If you have a CAS, read the documentation to determine how to find sums of infinite series; then use the CAS to check the results in Example 5.

HARMONIC SERIES

One of the most important of all diverging series is the ***harmonic series***,

$$\sum_{k=1}^{\infty} \frac{1}{k} = 1 + \frac{1}{2} + \frac{1}{3} + \frac{1}{4} + \frac{1}{5} + \cdots$$

which arises in connection with the overtones produced by a vibrating musical string. It is not immediately evident that this series diverges. However, the divergence will become apparent when we examine the partial sums in detail. Because the terms in the series are all positive, the partial sums

$$s_1 = 1, \quad s_2 = 1 + \tfrac{1}{2}, \quad s_3 = 1 + \tfrac{1}{2} + \tfrac{1}{3}, \quad s_4 = 1 + \tfrac{1}{2} + \tfrac{1}{3} + \tfrac{1}{4}, \ldots$$

form a strictly increasing sequence

$$s_1 < s_2 < s_3 < \cdots < s_n < \cdots$$

Thus, by Theorem 11.2.3 we can prove divergence by demonstrating that there is no constant M that is greater than or equal to *every* partial sum. To this end, we will consider some selected partial sums, namely $s_2, s_4, s_8, s_{16}, s_{32}, \ldots$. Note that the subscripts are successive powers of 2, so that these are the partial sums of the form s_{2^n}. These partial sums satisfy the inequalities

$$\begin{aligned}
s_2 &= 1 + \tfrac{1}{2} > \tfrac{1}{2} + \tfrac{1}{2} = \tfrac{2}{2} \\
s_4 &= s_2 + \tfrac{1}{3} + \tfrac{1}{4} > s_2 + \left(\tfrac{1}{4} + \tfrac{1}{4}\right) = s_2 + \tfrac{1}{2} > \tfrac{3}{2} \\
s_8 &= s_4 + \tfrac{1}{5} + \tfrac{1}{6} + \tfrac{1}{7} + \tfrac{1}{8} > s_4 + \left(\tfrac{1}{8} + \tfrac{1}{8} + \tfrac{1}{8} + \tfrac{1}{8}\right) = s_4 + \tfrac{1}{2} > \tfrac{4}{2} \\
s_{16} &= s_8 + \tfrac{1}{9} + \tfrac{1}{10} + \tfrac{1}{11} + \tfrac{1}{12} + \tfrac{1}{13} + \tfrac{1}{14} + \tfrac{1}{15} + \tfrac{1}{16} \\
&> s_8 + \left(\tfrac{1}{16} + \tfrac{1}{16} + \tfrac{1}{16} + \tfrac{1}{16} + \tfrac{1}{16} + \tfrac{1}{16} + \tfrac{1}{16} + \tfrac{1}{16}\right) = s_8 + \tfrac{1}{2} > \tfrac{5}{2} \\
&\;\;\vdots \\
s_{2^n} &> \frac{n+1}{2}
\end{aligned}$$

If M is any constant, we can find a positive integer n such that $(n+1)/2 > M$. But for this n

$$s_{2^n} > \frac{n+1}{2} > M$$

so that no constant M is greater than or equal to *every* partial sum of the harmonic series. This proves divergence.

This divergence proof, which predates the discovery of calculus, is due to a French bishop and teacher, Nicole Oresme (1323–1382). This series eventually attracted the interest of Johann and Jakob Bernoulli (p. 99) and led them to begin thinking about the general concept of convergence, which was a new idea at that time.

XVI. *Summa seriei infinitæ harmonicè progressionalium*, $\frac{1}{1} + \frac{1}{2} + \frac{1}{3} + \frac{1}{4} + \frac{1}{5}$ *&c. est infinita.*

Id primus deprehendit Frater: inventa namque per præced. summa seriei $\frac{1}{2} + \frac{1}{6} + \frac{1}{12} + \frac{1}{20} + \frac{1}{30}$, &c. visurus porrò, quid emergeret ex ista serie, $\frac{1}{2} + \frac{2}{6} + \frac{3}{12} + \frac{4}{20} + \frac{5}{30}$, &c. si resolveretur methodo Prop. XIV. collegit propositionis veritatem ex absurditate manifesta, quæ sequeretur, si summa seriei harmonicæ finita statueretur. Animadvertit enim,

Seriem A, $\frac{1}{2} + \frac{1}{3} + \frac{1}{4} + \frac{1}{5} + \frac{1}{6} + \frac{1}{7}$, &c. ∞ (fractionibus singulis in alias, quarum numeratores sunt 1, 2, 3, 4, &c. transmutatis)
seriei B, $\frac{1}{2} + \frac{2}{6} + \frac{3}{12} + \frac{4}{20} + \frac{5}{30} + \frac{6}{42}$, &c. ∞ C + D + E + F, &c.

C. $\frac{1}{2} + \frac{1}{6} + \frac{1}{12} + \frac{1}{20} + \frac{1}{30} + \frac{1}{42}$, &c. ∞ per præc. $\frac{1}{1}$
D. . $+ \frac{1}{6} + \frac{1}{12} + \frac{1}{20} + \frac{1}{30} + \frac{1}{42}$, &c. ∞ C − $\frac{1}{2}$ ∞ $\frac{1}{2}$
E. . . $+ \frac{1}{12} + \frac{1}{20} + \frac{1}{30} + \frac{1}{42}$, &c. ∞ D − $\frac{1}{6}$ ∞ $\frac{1}{3}$
F. . . . . . $+ \frac{1}{20} + \frac{1}{30} + \frac{1}{42}$, &c. ∞ E − $\frac{1}{12}$ ∞ $\frac{1}{4}$
&c. ∞ &c. } ∞ G; unde sequitur, seriem G ∞ A, totum parti, si summa finita esset.

Ego

This is a proof of the divergence of the harmonic series, as it appeared in an appendix of Jakob Bernoulli's posthumous publication, *Ars Conjectandi*, which appeared in 1713.

EXERCISE SET 11.3 C CAS

1. In each part, find exact values for the first four partial sums, find a closed form for the nth partial sum, and determine whether the series converges by calculating the limit of the nth partial sum. If the series converges, then state its sum.
   (a) $2 + \dfrac{2}{5} + \dfrac{2}{5^2} + \cdots + \dfrac{2}{5^{k-1}} + \cdots$
   (b) $\dfrac{1}{4} + \dfrac{2}{4} + \dfrac{2^2}{4} + \cdots + \dfrac{2^{k-1}}{4} + \cdots$
   (c) $\dfrac{1}{2 \cdot 3} + \dfrac{1}{3 \cdot 4} + \dfrac{1}{4 \cdot 5} + \cdots + \dfrac{1}{(k+1)(k+2)} + \cdots$

2. In each part, find exact values for the first four partial sums, find a closed form for the nth partial sum, and determine whether the series converges by calculating the limit of the nth partial sum. If the series converges, then state its sum.
   (a) $\displaystyle\sum_{k=1}^{\infty} \left(\frac{1}{4}\right)^k$ (b) $\displaystyle\sum_{k=1}^{\infty} 4^{k-1}$ (c) $\displaystyle\sum_{k=1}^{\infty} \left(\frac{1}{k+3} - \frac{1}{k+4}\right)$

In Exercises 3–14, determine whether the series converges, and if so, find its sum.

3. $\sum_{k=1}^{\infty}\left(-\frac{3}{4}\right)^{k-1}$

4. $\sum_{k=1}^{\infty}\left(\frac{2}{3}\right)^{k+2}$

5. $\sum_{k=1}^{\infty}(-1)^{k-1}\frac{7}{6^{k-1}}$

6. $\sum_{k=1}^{\infty}\left(-\frac{3}{2}\right)^{k+1}$

7. $\sum_{k=1}^{\infty}\frac{1}{(k+2)(k+3)}$

8. $\sum_{k=1}^{\infty}\left(\frac{1}{2^k}-\frac{1}{2^{k+1}}\right)$

9. $\sum_{k=1}^{\infty}\frac{1}{9k^2+3k-2}$

10. $\sum_{k=2}^{\infty}\frac{1}{k^2-1}$

11. $\sum_{k=3}^{\infty}\frac{1}{k-2}$

12. $\sum_{k=5}^{\infty}\left(\frac{e}{\pi}\right)^{k-1}$

13. $\sum_{k=1}^{\infty}\frac{4^{k+2}}{7^{k-1}}$

14. $\sum_{k=1}^{\infty}5^{3k}7^{1-k}$

In Exercises 15–20, express the given repeating decimal as a fraction.

15. 0.4444...

16. 0.9999...

17. 5.373737...

18. 0.159159159...

19. 0.782178217821...

20. 0.451141414...

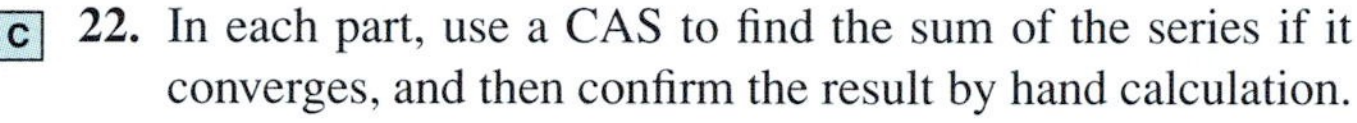

C **21.** Use a CAS to check your answers to Exercises 1–14.

C **22.** In each part, use a CAS to find the sum of the series if it converges, and then confirm the result by hand calculation.

(a) $\sum_{k=1}^{\infty}(-1)^{k+1}2^k 3^{2-k}$ (b) $\sum_{k=1}^{\infty}\frac{3^{3k}}{5^{k-1}}$ (c) $\sum_{k=1}^{\infty}\frac{1}{4k^2-1}$

23. A ball is dropped from a height of 10 m. Each time it strikes the ground it bounces vertically to a height that is $\frac{3}{4}$ of the preceding height. Find the total distance the ball will travel if it is assumed to bounce infinitely often.

24. The accompanying figure shows an "infinite staircase" constructed from cubes. Find the total volume of the staircase, given that the largest cube has a side of length 1 and each successive cube has a side whose length is half that of the preceding cube.

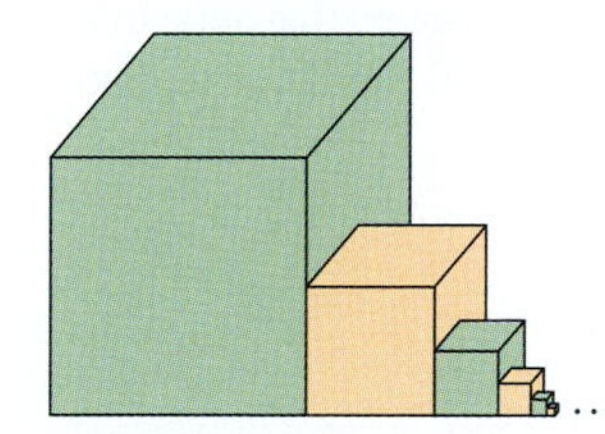

Figure Ex-24

25. In each part, find a closed form for the nth partial sum of the series, and determine whether the series converges. If so, find its sum.

(a) $\ln\frac{1}{2}+\ln\frac{2}{3}+\ln\frac{3}{4}+\cdots+\ln\frac{n}{n+1}+\cdots$

(b) $\ln\left(1-\frac{1}{4}\right)+\ln\left(1-\frac{1}{9}\right)+\ln\left(1-\frac{1}{16}\right)+\cdots+\ln\left(1-\frac{1}{(k+1)^2}\right)+\cdots$

26. Use geometric series to show that

(a) $\sum_{k=0}^{\infty}(-1)^k x^k=\frac{1}{1+x}$ if $-1<x<1$

(b) $\sum_{k=0}^{\infty}(x-3)^k=\frac{1}{4-x}$ if $2<x<4$

(c) $\sum_{k=0}^{\infty}(-1)^k x^{2k}=\frac{1}{1+x^2}$ if $-1<x<1$.

27. In each part, find all values of x for which the series converges, and find the sum of the series for those values of x.

(a) $x-x^3+x^5-x^7+x^9-\cdots$

(b) $\frac{1}{x^2}+\frac{2}{x^3}+\frac{4}{x^4}+\frac{8}{x^5}+\frac{16}{x^6}+\cdots$

(c) $e^{-x}+e^{-2x}+e^{-3x}+e^{-4x}+e^{-5x}+\cdots$

28. Show: $\sum_{k=1}^{\infty}\frac{\sqrt{k+1}-\sqrt{k}}{\sqrt{k^2+k}}=1$.

29. Show: $\sum_{k=1}^{\infty}\left(\frac{1}{k}-\frac{1}{k+2}\right)=\frac{3}{2}$.

30. Show: $\frac{1}{1\cdot 3}+\frac{1}{2\cdot 4}+\frac{1}{3\cdot 5}+\cdots=\frac{3}{4}$.

31. Show: $\frac{1}{1\cdot 3}+\frac{1}{3\cdot 5}+\frac{1}{5\cdot 7}+\cdots=\frac{1}{2}$.

32. Show that for all real values of x

$$\sin x-\frac{1}{2}\sin^2 x+\frac{1}{4}\sin^3 x-\frac{1}{8}\sin^4 x+\cdots=\frac{2\sin x}{2+\sin x}$$

33. Let a_1 be any real number, and let $\{a_n\}$ be the sequence defined recursively by

$$a_{n+1}=\tfrac{1}{2}(a_n+1)$$

Make a conjecture about the limit of the sequence, and confirm your conjecture by expressing a_n in terms of a_1 and taking the limit.

34. Recall that a *terminating decimal* is a decimal whose digits are all 0 from some point on ($0.5=0.50000\ldots$, for example). Show that a decimal of the form $0.a_1a_2\ldots a_n9999\ldots$, where $a_n\neq 9$, can be expressed as a terminating decimal.

35. The great Swiss mathematician Leonhard Euler (biography on p. 19) sometimes reached incorrect conclusions in his pioneering work on infinite series. For example, Euler deduced that

$$\tfrac{1}{2}=1-1+1-1+\cdots$$

and

$$-1=1+2+4+8+\cdots$$

by substituting $x=-1$ and $x=2$ in the formula

$$\frac{1}{1-x}=1+x+x^2+x^3+\cdots$$

What was the problem with his reasoning?

36. As shown in the accompanying figure, suppose that lines L_1 and L_2 form an angle θ, $0 < \theta < \pi/2$, at their point of intersection P. A point P_0 is chosen that is on L_1 and a units from P. Starting from P_0 a zig-zag path is constructed by successively going back and forth between L_1 and L_2 along a perpendicular from one line to the other. Find the following sums in terms of θ.

(a) $P_0P_1 + P_1P_2 + P_2P_3 + \cdots$

(b) $P_0P_1 + P_2P_3 + P_4P_5 + \cdots$

(c) $P_1P_2 + P_3P_4 + P_5P_6 + \cdots$

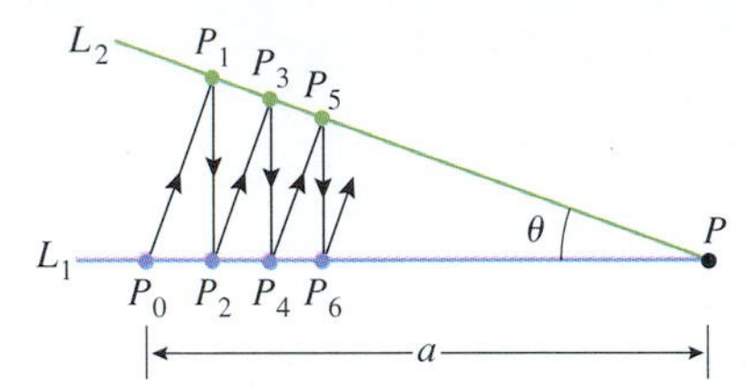

Figure Ex-36

37. As shown in the accompanying figure, suppose that an angle θ is bisected using a straightedge and compass to produce ray R_1, then the angle between R_1 and the initial side is bisected to produce ray R_2. Thereafter, rays R_3, R_4, $R_5, \ldots$ are constructed in succession by bisecting the angle between the preceding two rays. Show that the sequence of angles that these rays make with the initial side has a limit of $\theta/3$. [This problem is based on *Trisection of an Angle in an Infinite Number of Steps* by Eric Kincannon, which appeared in *The College Mathematics Journal*, Vol. 21, No. 5, November 1990.]

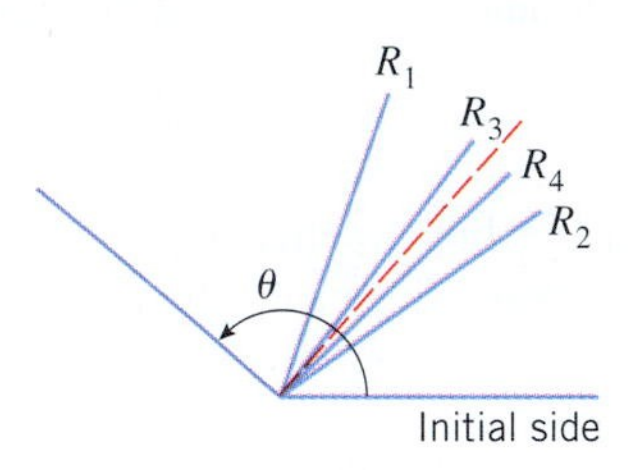

Figure Ex-37

38. In his *Treatise on the Configurations of Qualities and Motions* (written in the 1350s), the French Bishop of Lisieux, Nicole Oresme, used a geometric method to find the sum of the series

$$\sum_{k=1}^{\infty} \frac{k}{2^k} = \frac{1}{2} + \frac{2}{4} + \frac{3}{8} + \frac{4}{16} + \cdots$$

In part (a) of the accompanying figure, each term in the series is represented by the area of a rectangle, and in part (b) the configuration in part (a) has been divided into rectangles with areas $A_1, A_2, A_3, \ldots$. Find the sum $A_1 + A_2 + A_3 + \cdots$.

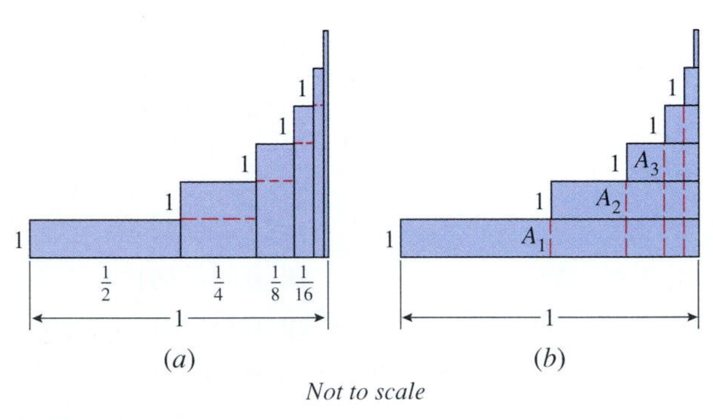

Figure Ex-38

39. C (a) See if your CAS can find the sum of the series

$$\sum_{k=1}^{\infty} \frac{6^k}{(3^{k+1} - 2^{k+1})(3^k - 2^k)}$$

(b) Find A and B such that

$$\frac{6^k}{(3^{k+1} - 2^{k+1})(3^k - 2^k)} = \frac{2^k A}{3^k - 2^k} + \frac{2^k B}{3^{k+1} - 2^{k+1}}$$

(c) Use the result in part (b) to find a closed form for the nth partial sum, and then find the sum of the series.

[This exercise is adapted from a problem that appeared in the Forty-Fifth Annual William Lowell Putnam Competition.]

11.4 CONVERGENCE TESTS

In the last section we showed how to find the sum of a series by finding a closed form for the nth partial sum and taking its limit. However, it is relatively rare that one can find a closed form for the nth partial sum of a series, so alternative methods are needed for finding sums of series. One possibility is to prove that the series converges, and then approximate the sum by a partial sum with sufficiently many terms to achieve the desired degree of accuracy. In this section we will develop various tests that can be used to determine whether a given series converges or diverges.

THE DIVERGENCE TEST

The kth term in an infinite series $\sum u_k$ is called the ***general term*** of the series. The following theorem establishes a relationship between the limit of the general term and the convergence properties of a series.

> **11.4.1** THEOREM (***The Divergence Test***).
> (*a*) *If* $\lim_{k\to+\infty} u_k \neq 0$, *then the series* $\sum u_k$ *diverges.*
> (*b*) *If* $\lim_{k\to+\infty} u_k = 0$, *then the series* $\sum u_k$ *may either converge or diverge.*

Proof (a). To prove this result, it suffices to show that if the series converges, then $\lim_{k\to+\infty} u_k = 0$ (why?). We will prove this alternative form of (*a*).

Let us assume that the series converges. The general term u_k can be written as

$$u_k = s_k - s_{k-1} \tag{1}$$

where s_k is the sum of the first k terms and s_{k-1} is the sum of the first $k-1$ terms. If S denotes the sum of the series, then $\lim_{k\to+\infty} s_k = S$, and since $(k-1)\to+\infty$ as $k\to+\infty$, we also have $\lim_{k\to+\infty} s_{k-1} = S$. Thus, from (1)

$$\lim_{k\to+\infty} u_k = \lim_{k\to+\infty} (s_k - s_{k-1}) = S - S = 0$$

Proof (b). To prove this result, it suffices to produce both a convergent series and a divergent series for which $\lim_{k\to+\infty} u_k = 0$. The following series both have this property:

$$\frac{1}{2} + \frac{1}{2^2} + \cdots + \frac{1}{2^k} + \cdots \quad \text{and} \quad 1 + \frac{1}{2} + \frac{1}{3} + \cdots + \frac{1}{k} + \cdots$$

The first is a convergent geometric series and the second is the divergent harmonic series. ∎

The alternative form of part (*a*) given in the preceding proof is sufficiently important that we state it separately for future reference.

> **11.4.2** THEOREM. *If the series* $\sum u_k$ *converges, then* $\lim_{k\to+\infty} u_k = 0$.

Example 1

The series

$$\sum_{k=1}^{\infty} \frac{k}{k+1} = \frac{1}{2} + \frac{2}{3} + \frac{3}{4} + \cdots + \frac{k}{k+1} + \cdots$$

diverges since

$$\lim_{k\to+\infty} \frac{k}{k+1} = \lim_{k\to+\infty} \frac{1}{1+1/k} = 1 \neq 0$$ ◀

WARNING. The converse of Theorem 11.4.2 is false. To prove that a series converges it does not suffice to show that $\lim_{k\to+\infty} u_k = 0$, since this property may hold for divergent as well as convergent series, as we saw in the proof of part (*b*) of Theorem 11.4.1.

ALGEBRAIC PROPERTIES OF INFINITE SERIES

For brevity, the proof of the following result is omitted.

11.4.3 THEOREM.

(*a*) *If* $\sum u_k$ *and* $\sum v_k$ *are convergent series, then* $\sum(u_k + v_k)$ *and* $\sum(u_k - v_k)$ *are convergent series and the sums of these series are related by*

$$\sum_{k=1}^{\infty}(u_k + v_k) = \sum_{k=1}^{\infty} u_k + \sum_{k=1}^{\infty} v_k$$

$$\sum_{k=1}^{\infty}(u_k - v_k) = \sum_{k=1}^{\infty} u_k - \sum_{k=1}^{\infty} v_k$$

(*b*) *If* c *is a nonzero constant, then the series* $\sum u_k$ *and* $\sum cu_k$ *both converge or both diverge. In the case of convergence, the sums are related by*

$$\sum_{k=1}^{\infty} cu_k = c\sum_{k=1}^{\infty} u_k$$

(*c*) *Convergence or divergence is unaffected by deleting a finite number of terms from a series; in particular, for any positive integer* K, *the series*

$$\sum_{k=1}^{\infty} u_k = u_1 + u_2 + u_3 + \cdots$$

$$\sum_{k=K}^{\infty} u_k = u_K + u_{K+1} + u_{K+2} + \cdots$$

both converge or both diverge.

REMARK. Do not read too much into part (*c*) of this theorem. Although the convergence is not affected when a finite number of terms is deleted from the beginning of a convergent series, the *sum* of a convergent series is changed by the removal of these terms.

Example 2

Find the sum of the series

$$\sum_{k=1}^{\infty}\left(\frac{3}{4^k} - \frac{2}{5^{k-1}}\right)$$

Solution. The series

$$\sum_{k=1}^{\infty}\frac{3}{4^k} = \frac{3}{4} + \frac{3}{4^2} + \frac{3}{4^3} + \cdots$$

is a convergent geometric series $\left(a = \frac{3}{4}, r = \frac{1}{4}\right)$, and the series

$$\sum_{k=1}^{\infty}\frac{2}{5^{k-1}} = 2 + \frac{2}{5} + \frac{2}{5^2} + \frac{2}{5^3} + \cdots$$

is also a convergent geometric series $\left(a = 2, r = \frac{1}{5}\right)$. Thus, from Theorems 11.4.3(*a*) and 11.3.3 the given series converges and

$$\sum_{k=1}^{\infty}\left(\frac{3}{4^k} - \frac{2}{5^{k-1}}\right) = \sum_{k=1}^{\infty}\frac{3}{4^k} - \sum_{k=1}^{\infty}\frac{2}{5^{k-1}} = \frac{\frac{3}{4}}{1-\frac{1}{4}} - \frac{2}{1-\frac{1}{5}} = -\frac{3}{2}$$ ◀

Example 3

Determine whether the following series converge or diverge.

(a) $\displaystyle\sum_{k=1}^{\infty}\frac{5}{k} = 5 + \frac{5}{2} + \frac{5}{3} + \cdots + \frac{5}{k} + \cdots$ (b) $\displaystyle\sum_{k=10}^{\infty}\frac{1}{k} = \frac{1}{10} + \frac{1}{11} + \frac{1}{12} + \cdots$

Solution. The first series is a constant times the divergent harmonic series, and hence diverges by part (b) of Theorem 11.4.3. The second series results by deleting the first nine terms from the divergent harmonic series, and hence diverges by part (c) of Theorem 11.4.3. ◀

THE INTEGRAL TEST

The expressions

$$\sum_{k=1}^{\infty} \frac{1}{k^2} \quad \text{and} \quad \int_1^{+\infty} \frac{1}{x^2}\,dx$$

are related in that the integrand in the improper integral results when the index k in the general term of the series is replaced by x and the limits of summation in the series are replaced by the corresponding limits of integration. The following theorem shows that there is a relationship between the convergence of the series and the integral.

11.4.4 THEOREM (*The Integral Test*). *Let $\sum u_k$ be a series with positive terms, and let $f(x)$ be the function that results when k is replaced by x in the general term of the series. If f is decreasing and continuous on the interval $[a, +\infty)$, then*

$$\sum_{k=1}^{\infty} u_k \quad \textit{and} \quad \int_a^{+\infty} f(x)\,dx$$

both converge or both diverge.

Example 4

Use the integral test to determine whether the following series converge or diverge.

(a) $\displaystyle\sum_{k=1}^{\infty} \frac{1}{k}$ (b) $\displaystyle\sum_{k=1}^{\infty} \frac{1}{k^2}$

Solution (a). We already know that this is the divergent harmonic series, so the integral test will simply provide another way of establishing the divergence. If we replace k by x in the general term $1/k$, we obtain the function $f(x) = 1/x$, which is decreasing and continuous for $x \geq 1$ (as required to apply the integral test with $a = 1$). Since

$$\int_1^{+\infty} \frac{1}{x}\,dx = \lim_{l\to+\infty} \int_1^{l} \frac{1}{x}\,dx = \lim_{l\to+\infty} [\ln l - \ln 1] = +\infty$$

the integral diverges and consequently so does the series.

Solution (b). If we replace k by x in the general term $1/k^2$, we obtain the function $f(x) = 1/x^2$, which is decreasing and continuous for $x \geq 1$. Since

$$\int_1^{+\infty} \frac{1}{x^2}\,dx = \lim_{l\to+\infty} \int_1^{l} \frac{dx}{x^2} = \lim_{l\to+\infty} \left[-\frac{1}{x}\right]_1^l = \lim_{l\to+\infty} \left[1 - \frac{1}{l}\right] = 1$$

the integral converges and consequently the series converges by the integral test with $a = 1$. ◀

REMARK. In part (b) of the last example, do *not* erroneously conclude that the sum of the series is 1 because the value of the corresponding integral is 1. It can be proved that the sum of the series is actually $\pi^2/6$ and, indeed, the sum of the first two terms alone exceeds 1.

***p*-SERIES**

The series in Example 4 are special cases of a class of series called ***p-series*** or ***hyperharmonic series***. A p-series is an infinite series of the form

$$\sum_{k=1}^{\infty} \frac{1}{k^p} = 1 + \frac{1}{2^p} + \frac{1}{3^p} + \cdots + \frac{1}{k^p} + \cdots$$

where $p > 0$. Examples of p-series are

$$\sum_{k=1}^{\infty} \frac{1}{k} = 1 + \frac{1}{2} + \frac{1}{3} + \cdots + \frac{1}{k} + \cdots \qquad p = 1$$

$$\sum_{k=1}^{\infty} \frac{1}{k^2} = 1 + \frac{1}{2^2} + \frac{1}{3^2} + \cdots + \frac{1}{k^2} + \cdots \qquad p = 2$$

$$\sum_{k=1}^{\infty} \frac{1}{\sqrt{k}} = 1 + \frac{1}{\sqrt{2}} + \frac{1}{\sqrt{3}} + \cdots + \frac{1}{\sqrt{k}} + \cdots \qquad p = \tfrac{1}{2}$$

The following theorem tells when a p-series converges.

11.4.5 THEOREM (*Convergence of p-Series*).

$$\sum_{k=1}^{\infty} \frac{1}{k^p} = 1 + \frac{1}{2^p} + \frac{1}{3^p} + \cdots + \frac{1}{k^p} + \cdots$$

converges if $p > 1$ and diverges if $0 < p \le 1$.

Proof. To establish this result when $p \neq 1$, we will use the integral test.

$$\int_1^{+\infty} \frac{1}{x^p}\,dx = \lim_{l\to+\infty} \int_1^l x^{-p}\,dx = \lim_{l\to+\infty} \left.\frac{x^{1-p}}{1-p}\right]_1^l = \lim_{l\to+\infty} \left[\frac{l^{1-p}}{1-p} - \frac{1}{1-p}\right]$$

If $p > 1$, then $1 - p < 0$, so $l^{1-p} \to 0$ as $l \to +\infty$. Thus, the integral converges [its value is $-1/(1-p)$] and consequently the series also converges. For $0 < p < 1$, it follows that $1 - p > 0$ and $l^{1-p} \to +\infty$ as $l \to +\infty$, so the integral and the series diverge. The case $p = 1$ is the harmonic series, which was previously shown to diverge. ∎

Example 5

$$1 + \frac{1}{\sqrt[3]{2}} + \frac{1}{\sqrt[3]{3}} + \cdots + \frac{1}{\sqrt[3]{k}} + \cdots$$

diverges since it is a p-series with $p = \frac{1}{3} < 1$. ◀

PROOF OF THE INTEGRAL TEST

Before we can prove the integral test, we need a basic result about convergence of series with *nonnegative* terms. If $u_1 + u_2 + u_3 + \cdots + u_k + \cdots$ is such a series, then its sequence of partial sums is increasing, that is,

$$s_1 \le s_2 \le s_3 \le \cdots \le s_n \le \cdots$$

Thus, from Theorem 11.2.3 the sequence of partial sums converges to a limit S if and only if it has some upper bound M, in which case $S \le M$. If no upper bound exists, then the sequence of partial sums diverges. Since convergence of the sequence of partial sums corresponds to convergence of the series, we have the following theorem.

11.4.6 THEOREM. *If $\sum u_k$ is a series with nonnegative terms, and if there is a constant M such that*

$$s_n = u_1 + u_2 + \cdots + u_n \le M$$

for every n, then the series converges and the sum S satisfies $S \le M$. If no such M exists, then the series diverges.

In words, this theorem implies that *a series with nonnegative terms converges if and only if its sequence of partial sums is bounded above.*

Proof of Theorem 11.4.4. We need only show that the series converges when the integral converges and that the series diverges when the integral diverges. For simplicity, we will limit the proof to the case where $a = 1$. Assume that $f(x)$ satisfies the hypotheses of the theorem for $x \geq 1$. Since

$$f(1) = u_1,\ f(2) = u_2, \ldots, f(n) = u_n, \ldots$$

the values of $u_1, u_2, \ldots, u_n, \ldots$ can be interpreted as the areas of the rectangles shown in Figure 11.4.1.

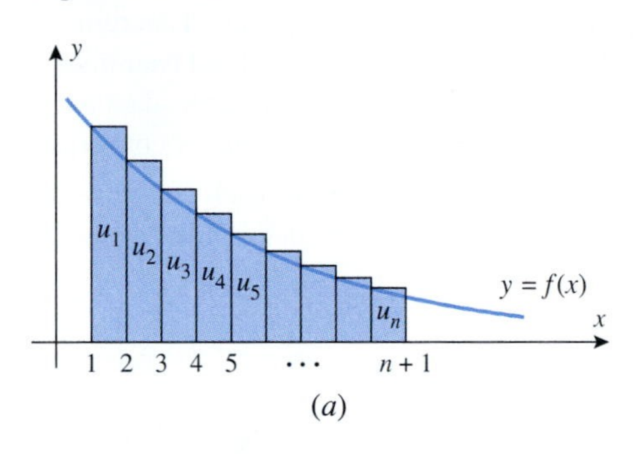

(a)

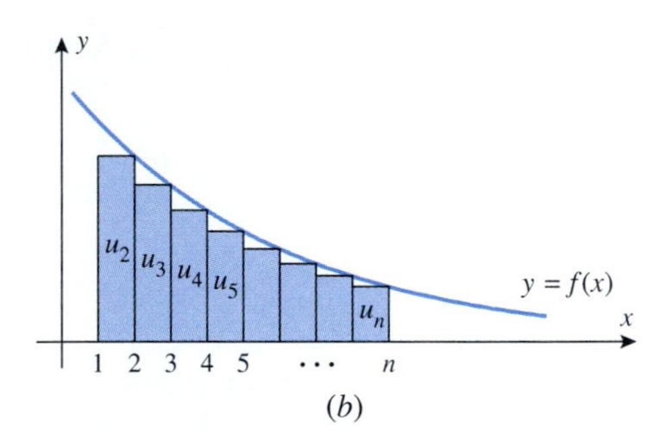

(b)

Figure 11.4.1

The following inequalities result by comparing the areas under the curve $y = f(x)$ to the areas of the rectangles in Figure 11.4.1 for $n > 1$:

$$\int_1^{n+1} f(x)\,dx < u_1 + u_2 + \cdots + u_n = s_n \qquad \text{Figure 11.4.1}a$$

$$s_n - u_1 = u_2 + u_3 + \cdots + u_n < \int_1^n f(x)\,dx \qquad \text{Figure 11.4.1}b$$

These inequalities can be combined as

$$\int_1^{n+1} f(x)\,dx < s_n < u_1 + \int_1^n f(x)\,dx \tag{2}$$

If the integral $\int_1^\infty f(x)\,dx$ converges to a finite value L, then from the right-hand inequality in (2)

$$s_n < u_1 + \int_1^n f(x)\,dx < u_1 + \int_1^\infty f(x)\,dx = u_1 + L$$

Thus, each partial sum is less than the finite constant $u_1 + L$, and the series converges by Theorem 11.4.6. On the other hand, if the integral $\int_1^\infty f(x)\,dx$ diverges, then

$$\lim_{n\to+\infty} \int_1^{n+1} f(x)\,dx = +\infty$$

so that from the left-hand inequality in (2), $\lim_{n\to+\infty} s_n = +\infty$. This implies that the series also diverges. ■

EXERCISE SET 11.4 Graphing Calculator C CAS

1. In each part, use Theorem 11.4.3 to find the sum of the series.

(a) $\left(\frac{1}{2} + \frac{1}{4}\right) + \left(\frac{1}{2^2} + \frac{1}{4^2}\right) + \cdots + \left(\frac{1}{2^k} + \frac{1}{4^k}\right) + \cdots$

(b) $\sum_{k=1}^{\infty} \left(\frac{1}{5^k} - \frac{1}{k(k+1)}\right)$

2. In each part, use Theorem 11.4.3 to find the sum of the series.

(a) $\sum_{k=2}^{\infty} \left[\frac{1}{k^2 - 1} - \frac{7}{10^{k-1}}\right]$ (b) $\sum_{k=1}^{\infty} \left[7^{-k}3^{k+1} - \frac{2^{k+1}}{5^k}\right]$

In Exercises 3 and 4, various p-series are given. In each case, find p and determine whether the series converges.

3. (a) $\sum_{k=1}^{\infty} \frac{1}{k^3}$ (b) $\sum_{k=1}^{\infty} \frac{1}{\sqrt{k}}$ (c) $\sum_{k=1}^{\infty} k^{-1}$ (d) $\sum_{k=1}^{\infty} k^{-2/3}$

4. (a) $\sum_{k=1}^{\infty} k^{-4/3}$ (b) $\sum_{k=1}^{\infty} \frac{1}{\sqrt[4]{k}}$ (c) $\sum_{k=1}^{\infty} \frac{1}{\sqrt[3]{k^5}}$ (d) $\sum_{k=1}^{\infty} \frac{1}{k^{\pi}}$

In Exercises 5 and 6, apply the divergence test, and state what it tells you about the series.

5. (a) $\sum_{k=1}^{\infty} \frac{k^2+k+3}{2k^2+1}$ (b) $\sum_{k=1}^{\infty} \left(1+\frac{1}{k}\right)^k$

(c) $\sum_{k=1}^{\infty} \cos k\pi$ (d) $\sum_{k=1}^{\infty} \frac{1}{k!}$

6. (a) $\sum_{k=1}^{\infty} \frac{k}{e^k}$ (b) $\sum_{k=1}^{\infty} \ln k$ (c) $\sum_{k=1}^{\infty} \frac{1}{\sqrt{k}}$ (d) $\sum_{k=1}^{\infty} \frac{\sqrt{k}}{\sqrt{k}+3}$

In Exercises 7 and 8, confirm that the integral test is applicable, and use it to determine whether the series converges.

7. (a) $\sum_{k=1}^{\infty} \frac{1}{5k+2}$ (b) $\sum_{k=1}^{\infty} \frac{1}{1+9k^2}$

8. (a) $\sum_{k=1}^{\infty} \frac{k}{1+k^2}$ (b) $\sum_{k=1}^{\infty} \frac{1}{(4+2k)^{3/2}}$

In Exercises 9–24, use any method to determine whether the series converges.

9. $\sum_{k=1}^{\infty} \frac{1}{k+6}$ **10.** $\sum_{k=1}^{\infty} \frac{3}{5k}$ **11.** $\sum_{k=1}^{\infty} \frac{1}{\sqrt{k+5}}$

12. $\sum_{k=1}^{\infty} \frac{1}{\sqrt[k]{e}}$ **13.** $\sum_{k=1}^{\infty} \frac{1}{\sqrt[3]{2k-1}}$ **14.** $\sum_{k=3}^{\infty} \frac{\ln k}{k}$

15. $\sum_{k=1}^{\infty} \frac{k}{\ln(k+1)}$ **16.** $\sum_{k=1}^{\infty} ke^{-k^2}$ **17.** $\sum_{k=1}^{\infty} \left(1+\frac{1}{k}\right)^{-k}$

18. $\sum_{k=1}^{\infty} \frac{k^2+1}{k^2+3}$ **19.** $\sum_{k=1}^{\infty} \frac{\tan^{-1} k}{1+k^2}$ **20.** $\sum_{k=1}^{\infty} \frac{1}{\sqrt{k^2+1}}$

21. $\sum_{k=1}^{\infty} k^2 \sin^2\left(\frac{1}{k}\right)$ **22.** $\sum_{k=1}^{\infty} k^2 e^{-k^3}$

23. $\sum_{k=5}^{\infty} 7k^{-1.01}$ **24.** $\sum_{k=1}^{\infty} \operatorname{sech}^2 k$

In Exercises 25 and 26, use the integral test to investigate the relationship between the value of p and the convergence of the series.

25. $\sum_{k=2}^{\infty} \frac{1}{k(\ln k)^p}$ **26.** $\sum_{k=3}^{\infty} \frac{1}{k(\ln k)[\ln(\ln k)]^p}$

C **27.** Use a CAS to confirm that

$$\sum_{k=1}^{\infty} \frac{1}{k^2} = \frac{\pi^2}{6} \quad \text{and} \quad \sum_{k=1}^{\infty} \frac{1}{k^4} = \frac{\pi^4}{90}$$

and then use these results in each part to find the sum of the series.

(a) $\sum_{k=1}^{\infty} \frac{3k^2-1}{k^4}$ (b) $\sum_{k=3}^{\infty} \frac{1}{k^2}$ (c) $\sum_{k=2}^{\infty} \frac{1}{(k-1)^4}$

28. Suppose that the series $\sum u_k$ converges and the series $\sum v_k$ diverges.

(a) Show that the series $\sum(u_k+v_k)$ and $\sum(u_k-v_k)$ both diverge. [*Hint:* Assume that each series converges and use Theorem 11.4.3 to obtain a contradiction.]

(b) Find examples to show that if $\sum u_k$ and $\sum v_k$ both diverge, then the series $\sum(u_k+v_k)$ and $\sum(u_k-v_k)$ may either converge or diverge.

29. In each part, use the results in Exercise 28, if needed, to determine whether the series diverges.

(a) $\sum_{k=1}^{\infty} \left[\left(\frac{2}{3}\right)^{k-1} + \frac{1}{k}\right]$ (b) $\sum_{k=1}^{\infty} \left[\frac{1}{3k+2} - \frac{1}{k^{3/2}}\right]$

(c) $\sum_{k=2}^{\infty} \left[\frac{1}{k(\ln k)^2} - \frac{1}{k^2}\right]$

Exercise 30 will show how a partial sum can be used to obtain upper and lower bounds on the sum of the series when the hypotheses of the integral test are satisfied. This result will be needed in Exercises 31–35.

30. (a) Let $\sum_{k=1}^{\infty} u_k$ be a convergent series with positive terms, let $f(x)$ be the function that results when k is replaced by x in the general term of the series, and suppose that f satisfies the hypotheses of the integral test for $x \geq n$ (Theorem 11.4.4). Use an area argument and the accompanying figure (following Exercise 35) to show that

$$\int_{n+1}^{+\infty} f(x)\,dx < \sum_{k=n+1}^{\infty} u_k < \int_{n}^{+\infty} f(x)\,dx$$

(b) Show that if S is the sum of the series $\sum_{k=1}^{\infty} u_k$ and s_n is the nth partial sum, then

$$s_n + \int_{n+1}^{+\infty} f(x)\,dx < S < s_n + \int_{n}^{+\infty} f(x)\,dx$$

31. (a) It was stated in Exercise 27 that

$$\sum_{k=1}^{\infty} \frac{1}{k^2} = \frac{\pi^2}{6}$$

Show that if s_n is the nth partial sum of this series, then

$$s_n + \frac{1}{n+1} < \frac{\pi^2}{6} < s_n + \frac{1}{n}$$

(b) Calculate s_3 exactly, and then use the result in part (a) to show that

$$\frac{29}{18} < \frac{\pi^2}{6} < \frac{61}{36}$$

(c) Use a calculating utility to confirm that the inequalities in part (b) are correct.

(d) Find upper and lower bounds on the error that results if the sum of the series is approximated by the 10th partial sum.

32. In each part, find upper and lower bounds on the error that results if the sum of the series is approximated by the 10th partial sum.

(a) $\sum_{k=1}^{\infty} \frac{1}{(2k+1)^2}$ (b) $\sum_{k=1}^{\infty} \frac{1}{k^2+1}$ (c) $\sum_{k=1}^{\infty} \frac{k}{e^k}$

33. Our objective in this problem is to approximate the sum of the series $\sum_{k=1}^{\infty} 1/k^3$ to two decimal-place accuracy.

(a) Show that if S is the sum of the series and s_n is the nth partial sum, then

$$s_n + \frac{1}{2(n+1)^2} < S < s_n + \frac{1}{2n^2}$$

(b) For two decimal-place accuracy, the error must be less than 0.005 (see Table 2.4.1 on p. 154). According to the Approximation Principle (2.4.10), we can achieve this by finding an interval of length 0.01 (or less) that contains S and approximating S by the midpoint of that interval. Find the smallest value of n such that the interval containing S in part (a) has a length of 0.01 or less.

(c) Approximate S to two decimal-place accuracy.

34. (a) Use the method of Exercise 33 to approximate the sum of the series $\sum_{k=1}^{\infty} 1/k^4$ to two decimal-place accuracy.

(b) It was stated in Exercise 27 that the sum of this series is $\pi^4/90$. Use a calculating utility to confirm that your answer in part (a) is accurate to two decimal places.

35. We showed in Section 11.3 that the harmonic series $\sum_{k=1}^{\infty} 1/k$ diverges. Our objective in this problem is to demonstrate that although the partial sums of this series approach $+\infty$, they increase extremely slowly.

(a) Use inequality (2) to show that for $n \geq 2$

$$\ln(n+1) < s_n < 1 + \ln n$$

(b) Use the inequalities in part (a) to find upper and lower bounds on the sum of the first million terms in the series.

(c) Show that the sum of the first billion terms in the series is less than 22.

(d) Find a value of n so that the sum of the first n terms is greater than 100.

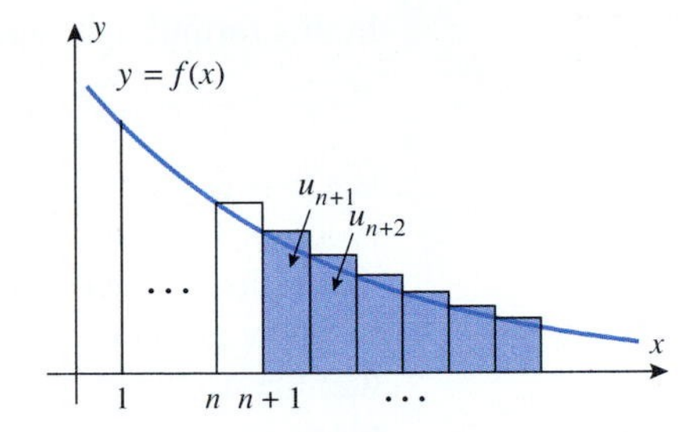

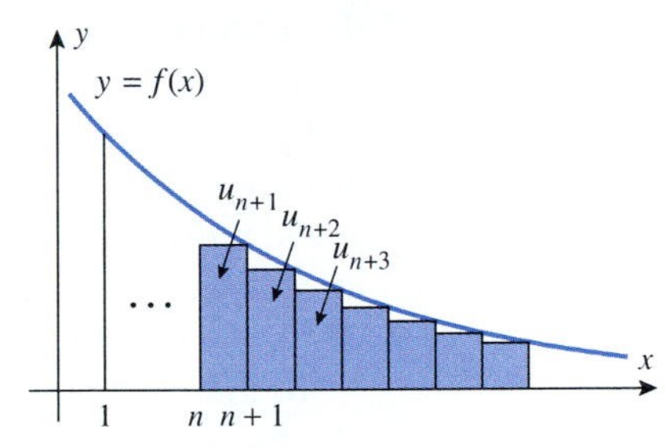

Figure Ex-30

36. Investigate the relationship between the value of a and the convergence of the series $\sum_{k=1}^{\infty} k^{-\ln a}$.

37. Use a graphing utility to confirm that the integral test applies to the series $\sum_{k=1}^{\infty} k^2 e^{-k}$, and then determine whether the series converges.

C **38.** (a) Show that the integral test applies to the series $\sum_{k=1}^{\infty} 1/(k^3+1)$.

(b) Use a CAS and the integral test to confirm that the series converges.

(c) Construct a table of partial sums for $n = 10, 20, 30, \ldots, 100$, showing at least six decimal places.

(d) Based on your table, make a conjecture about the sum of the series to three decimal-place accuracy.

(e) Use part (b) of Exercise 30 to check your conjecture.

11.5 TAYLOR AND MACLAURIN SERIES

In this section we will discuss methods for approximating values of trigonometric and logarithmic functions. This will lead us to the more general problem of approximating functions by polynomials and then to the problem of finding infinite series that converge to specific functions.

LOCAL QUADRATIC APPROXIMATIONS

Recall from Formula (6) in Section 3.6 that the local linear approximation of a function f at a point x_0 is

$$f(x) \approx f(x_0) + f'(x_0)(x - x_0) \tag{1}$$

In this formula, the approximating function

$$p(x) = f(x_0) + f'(x_0)(x - x_0)$$

is a first-degree polynomial whose value at x_0 is $f(x_0)$ and whose derivative at x_0 is $f'(x_0)$ (verify). Thus, the local linear approximation of f at x_0 has the property that its value and that of its first derivative match those of f at x_0.

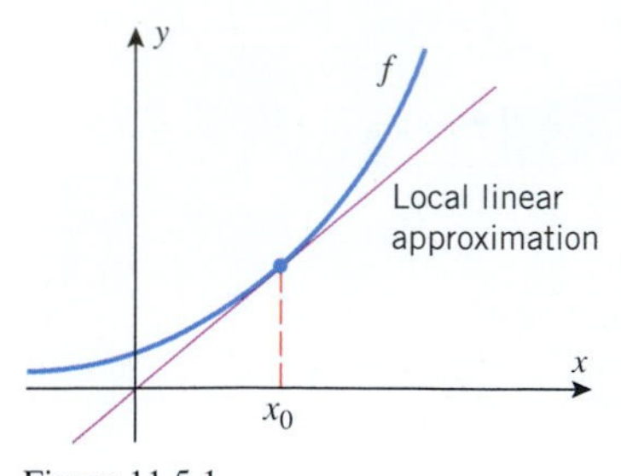

Figure 11.5.1

If the graph of a function f has a pronounced "bend" at a point x_0, then we can expect that the accuracy of the local linear approximation of f at x_0 will decrease rapidly as we progress away from x_0 (Figure 11.5.1). One way to deal with this problem is to approximate the function f at x_0 by a polynomial p of degree 2 with the property that the value of p and the value of its first two derivatives match those of f at x_0. This ensures that the graphs of f and p not only have the same tangent line at x_0, but they also bend in the same direction at that point (both concave up or concave down). As a result, we can expect that the graph of p will remain close to the graph of f over a larger interval around x_0 than the graph of the local linear approximation. The polynomial p is called the ***local quadratic approximation of f at the point $x = x_0$***.

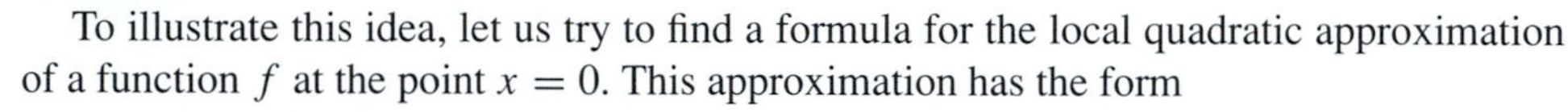

To illustrate this idea, let us try to find a formula for the local quadratic approximation of a function f at the point $x = 0$. This approximation has the form

$$f(x) \approx c_0 + c_1x + c_2x^2 \tag{2}$$

where c_0, c_1, and c_2 must be chosen so that the values of

$$p(x) = c_0 + c_1x + c_2x^2$$

and its first two derivatives match those of f at 0. Thus, we want

$$p(0) = f(0), \quad p'(0) = f'(0), \quad p''(0) = f''(0) \tag{3}$$

But the values of $p(0)$, $p'(0)$, and $p''(0)$ are as follows:

$$\begin{aligned} p(x) &= c_0 + c_1x + c_2x^2 & \qquad p(0) &= c_0 \\ p'(x) &= c_1 + 2c_2x & p'(0) &= c_1 \\ p''(x) &= 2c_2 & p''(0) &= 2c_2 \end{aligned}$$

Thus, it follows from (3) that

$$c_0 = f(0), \quad c_1 = f'(0), \quad c_2 = \frac{f''(0)}{2}$$

and substituting these in (2) yields the following formula for the local quadratic approximation of f at $x = 0$:

$$f(x) \approx f(0) + f'(0)x + \frac{f''(0)}{2}x^2 \tag{4}$$

REMARK. Observe that with $x_0 = 0$, Formula (1) becomes

$$f(x) \approx f(0) + f'(0)x \tag{5}$$

and hence the linear part of the local quadratic approximation of f at 0 is the local linear approximation of f at 0.

Example 1

Find the local linear and quadratic approximations of e^x at $x = 0$, and graph e^x and the two approximations together.

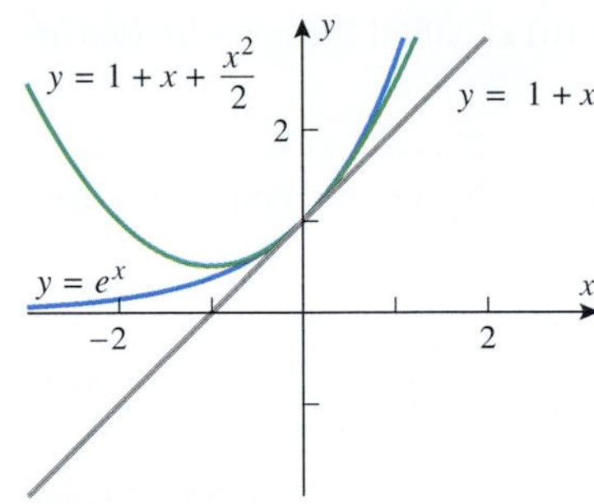

Figure 11.5.2

Solution. If we let $f(x) = e^x$, then $f'(x) = f''(x) = e^x$; and hence

$$f(0) = f'(0) = f''(0) = e^0 = 1$$

Thus, from (4) the local quadratic approximation of e^x at $x = 0$ is

$$e^x \approx 1 + x + \frac{x^2}{2}$$

and the local linear approximation (which is the linear part of the local quadratic approximation) is

$$e^x \approx 1 + x$$

The graphs of e^x and the two approximations are shown in Figure 11.5.2. As expected, the local quadratic approximation is more accurate than the local linear approximation near $x = 0$. ◀

MACLAURIN POLYNOMIALS

It is natural to ask whether one can improve on the accuracy of a local quadratic approximation by using a polynomial of degree 3. Specifically, one might look for a polynomial of degree 3 with the property that its value and the values of its first three derivatives match those of f at a point; and if this provides an improvement in accuracy, why not go on to polynomials of even higher degree? Thus, we are led to consider the following general problem.

> **11.5.1** PROBLEM. Given a function f that can be differentiated n times at a point x_0, find a polynomial p of degree n with the property that the value of p and the values of its first n derivatives match those of f at the point x_0.

We will begin by solving this problem in the case where $x_0 = 0$. Thus, we want a polynomial

$$p(x) = c_0 + c_1x + c_2x^2 + c_3x^3 + \cdots + c_nx^n \tag{6}$$

such that

$$f(0) = p(0), \quad f'(0) = p'(0), \quad f''(0) = p''(0), \quad \ldots, \quad f^{(n)}(0) = p^{(n)}(0) \tag{7}$$

But

$$\begin{aligned}
p(x) &= c_0 + c_1x + c_2x^2 + c_3x^3 + \cdots + c_nx^n \\
p'(x) &= c_1 + 2c_2x + 3c_3x^2 + \cdots + nc_nx^{n-1} \\
p''(x) &= 2c_2 + 3\cdot 2c_3x + \cdots + n(n-1)c_nx^{n-2} \\
p'''(x) &= 3\cdot 2c_3 + \cdots + n(n-1)(n-2)c_nx^{n-3} \\
&\ \vdots \\
p^{(n)}(x) &= n(n-1)(n-2)\cdots(1)c_n
\end{aligned}$$

Thus, to satisfy (7) we must have

$$\begin{aligned}
f(0) &= p(0) &&= c_0 \\
f'(0) &= p'(0) &&= c_1 \\
f''(0) &= p''(0) &&= 2c_2 = 2!c_2 \\
f'''(0) &= p'''(0) &&= 3\cdot 2c_3 = 3!c_3 \\
&\ \vdots \\
f^{(n)}(0) &= p^{(n)}(0) &&= n(n-1)(n-2)\cdots(1)c_n = n!c_n
\end{aligned}$$

which yields the following values for the coefficients of $p(x)$:

$$c_0 = f(0), \quad c_1 = f'(0), \quad c_2 = \frac{f''(0)}{2!}, \quad c_3 = \frac{f'''(0)}{3!}, \quad \ldots, \quad c_n = \frac{f^{(n)}(0)}{n!}$$

The polynomial that results by using these coefficients in (6) is called the *nth Maclaurin*[*] *polynomial for f*.

11.5.2 DEFINITION. If f can be differentiated n times at 0, then we define the ***nth Maclaurin polynomial for f*** to be

$$p_n(x) = f(0) + f'(0)x + \frac{f''(0)}{2!}x^2 + \frac{f'''(0)}{3!}x^3 + \cdots + \frac{f^{(n)}(0)}{n!}x^n \tag{8}$$

This polynomial has the property that its value and the values of its first n derivatives match the values of f and its first n derivatives at $x = 0$.

REMARK. Observe that $p_1(x)$ is the local linear approximation of f at 0 and $p_2(x)$ is the local quadratic approximation of f at $x = 0$.

Example 2

Find the Maclaurin polynomials p_0, p_1, p_2, p_3, and p_n for e^x.

Solution. Let $f(x) = e^x$. Thus,

$$f'(x) = f''(x) = f'''(x) = \cdots = f^{(n)}(x) = e^x$$

and

$$f(0) = f'(0) = f''(0) = f'''(0) = \cdots = f^{(n)}(0) = e^0 = 1$$

Therefore,

$$p_0(x) = f(0) = 1$$

$$p_1(x) = f(0) + f'(0)x = 1 + x$$

$$p_2(x) = f(0) + f'(0)x + \frac{f''(0)}{2!}x^2 = 1 + x + \frac{x^2}{2!} = 1 + x + \frac{1}{2}x^2$$

$$p_3(x) = f(0) + f'(0)x + \frac{f''(0)}{2!}x^2 + \frac{f'''(0)}{3!}x^3$$

$$= 1 + x + \frac{x^2}{2!} + \frac{x^3}{3!} = 1 + x + \frac{1}{2}x^2 + \frac{1}{6}x^3$$

$$p_n(x) = f(0) + f'(0)x + \frac{f''(0)}{2!}x^2 + \cdots + \frac{f^{(n)}(0)}{n!}x^n$$

$$= 1 + x + \frac{x^2}{2!} + \cdots + \frac{x^n}{n!}$$

◀

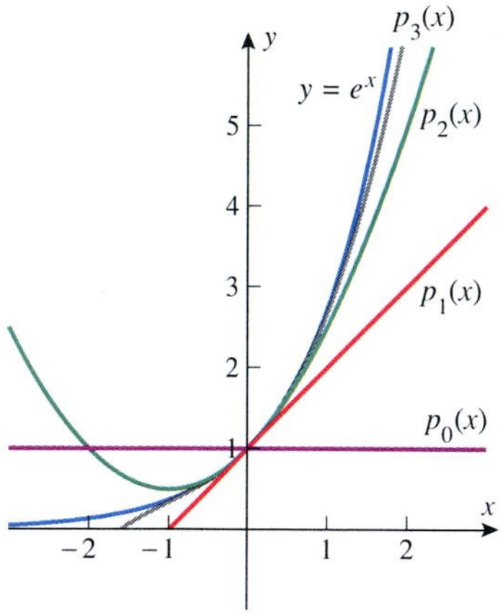

Figure 11.5.3

Figure 11.5.3 shows the graphs of e^x (in blue) and the graphs of the first four Maclaurin polynomials. Note that the graphs of $p_1(x)$, $p_2(x)$, and $p_3(x)$ are virtually indistinguishable from the graph of e^x near the origin, so that these polynomials are good approximations of e^x for x near 0. However, the farther x is from 0, the poorer these approximations become. This is typical of the Maclaurin polynomials for a function $f(x)$; they provide good

[*] COLIN MACLAURIN (1698–1746). Scottish mathematician. Maclaurin's father, a minister, died when the boy was only six months old, and his mother when he was nine years old. He was then raised by an uncle who was also a minister. Maclaurin entered Glasgow University as a divinity student, but transferred to mathematics after one year. He received his Master's degree at age 17 and, in spite of his youth, began teaching at Marischal College in Aberdeen, Scotland. He met Isaac Newton during a visit to London in 1719 and from that time on became Newton's disciple. During that era, some of Newton's analytic methods were bitterly attacked by major mathematicians and much of Maclaurin's important mathematical work resulted from his efforts to defend Newton's ideas geometrically. Maclaurin's work, *A Treatise of Fluxions* (1742), was the first systematic formulation of Newton's methods. The treatise was so carefully done that it was a standard of mathematical rigor in calculus until the work of Cauchy in 1821.

Maclaurin was an outstanding experimentalist. He devised numerous ingenious mechanical devices, made important astronomical observations, performed actuarial computations for insurance societies, and helped to improve maps of the islands around Scotland.

approximations of $f(x)$ near 0, but the accuracy diminishes as x progresses away from 0. However, it is usually the case that the higher the degree of the polynomial, the larger the interval on which it provides a specified accuracy. Accuracy issues will be investigated later.

TAYLOR POLYNOMIALS

Up to now we have focused on approximating a function f in the vicinity of the origin. Now we will consider the more general case of approximating f in the vicinity of an arbitrary point x_0. The basic idea is the same as before; we want to find an nth-degree polynomial p with the property that its value and the values of its first n derivatives match those of f at x_0. However, rather than expressing $p(x)$ in powers of x, it will simplify the computations if we express it in powers of $x - x_0$; that is,

$$p(x) = c_0 + c_1(x - x_0) + c_2(x - x_0)^2 + \cdots + c_n(x - x_0)^n \tag{9}$$

We will leave it as an exercise for you to imitate the computations used in the case where $x_0 = 0$ to show that

$$c_0 = f(x_0), \quad c_1 = f'(x_0), \quad c_2 = \frac{f''(x_0)}{2!}, \quad c_3 = \frac{f'''(x_0)}{3!}, \quad \ldots, \quad c_n = \frac{f^{(n)}(x_0)}{n!}$$

Substituting these values in (9) we obtain a polynomial called the *nth Taylor* polynomial about $x = x_0$ for f.*

11.5.3 DEFINITION. If f can be differentiated n times at x_0, then we define the ***nth Taylor polynomial for f about $x = x_0$*** to be

$$p_n(x) = f(x_0) + f'(x_0)(x - x_0) + \frac{f''(x_0)}{2!}(x - x_0)^2 + \frac{f'''(x_0)}{3!}(x - x_0)^3 + \cdots + \frac{f^{(n)}(x_0)}{n!}(x - x_0)^n \tag{10}$$

REMARK. Observe that the Maclaurin polynomials are special cases of the Taylor polynomials; that is, the nth-order Maclaurin polynomial is the nth-order Taylor polynomial about $x = 0$. Observe also that $p_1(x)$ is the local linear approximation of f at $x = x_0$ and $p_2(x)$ is the local quadratic approximation of f at $x = x_0$.

Example 3

Find the first four Taylor polynomials for $\ln x$ about $x = 2$.

Solution. Let $f(x) = \ln x$. Thus,

$$\begin{aligned} f(x) &= \ln x & f(2) &= \ln 2 \\ f'(x) &= 1/x & f'(2) &= 1/2 \\ f''(x) &= -1/x^2 & f''(2) &= -1/4 \\ f'''(x) &= 2/x^3 & f'''(2) &= 1/4 \end{aligned}$$

* BROOK TAYLOR (1685–1731). English mathematician. Taylor was born of well-to-do parents. Musicians and artists were entertained frequently in the Taylor home, which undoubtedly had a lasting influence on young Brook. In later years, Taylor published a definitive work on the mathematical theory of perspective and obtained major mathematical results about the vibrations of strings. There also exists an unpublished work, *On Musick*, that was intended to be part of a joint paper with Isaac Newton. Taylor's life was scarred with unhappiness, illness, and tragedy. Because his first wife was not rich enough to suit his father, the two men argued bitterly and parted ways. Subsequently, his wife died in childbirth. Then, after he remarried, his second wife also died in childbirth, though his daughter survived. Taylor's most productive period was from 1714 to 1719, during which time he wrote on a wide range of subjects—magnetism, capillary action, thermometers, perspective, and calculus. In his final years, Taylor devoted his writing efforts to religion and philosophy. According to Taylor, the results that bear his name were motivated by coffeehouse conversations about works of Newton on planetary motion and works of Halley ("Halley's comet") on roots of polynomials.

Taylor's writing style was so terse and hard to understand that he never received credit for many of his innovations.

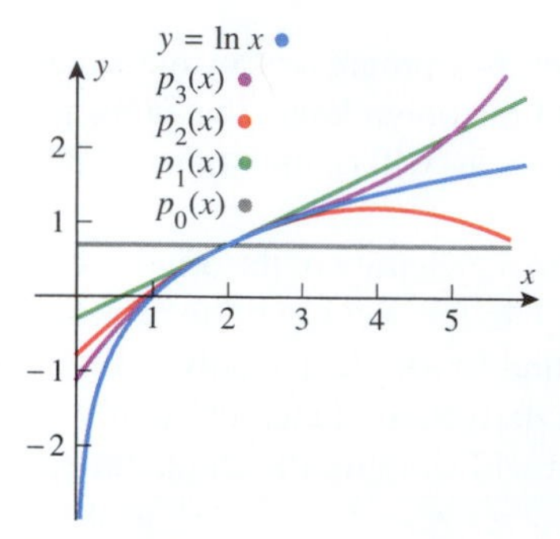

Figure 11.5.4

Substituting in (10) with $x_0 = 2$ yields

$$p_0(x) = f(2) = \ln 2$$
$$p_1(x) = f(2) + f'(2)(x-2) = \ln 2 + \tfrac{1}{2}(x-2)$$
$$p_2(x) = f(2) + f'(2)(x-2) + \frac{f''(2)}{2!}(x-2)^2 = \ln 2 + \tfrac{1}{2}(x-2) - \tfrac{1}{8}(x-2)^2$$
$$\begin{aligned} p_3(x) &= f(2) + f'(2)(x-2) + \frac{f''(2)}{2!}(x-2)^2 + \frac{f'''(2)}{3!}(x-2)^3 \\ &= \ln 2 + \tfrac{1}{2}(x-2) - \tfrac{1}{8}(x-2)^2 + \tfrac{1}{24}(x-2)^3 \end{aligned}$$

The graph of $\ln x$ (in blue) and its first four Taylor polynomials about $x = 2$ are shown in Figure 11.5.4. As expected, these polynomials produce their best approximations of $\ln x$ near 2. ◀

SIGMA NOTATION FOR TAYLOR AND MACLAURIN POLYNOMIALS

Frequently, we will want to express Formula (10) in sigma notation. To do this, we use the notation $f^{(k)}(x_0)$ to denote the kth derivative of f at $x = x_0$, and we make the convention that $f^{(0)}(x_0)$ denotes $f(x_0)$. This enables us to write

$$\begin{aligned} \sum_{k=0}^{n} \frac{f^{(k)}(x_0)}{k!}(x-x_0)^k &= f(x_0) + f'(x_0)(x-x_0) \\ &\quad + \frac{f''(x_0)}{2!}(x-x_0)^2 + \cdots + \frac{f^{(n)}(x_0)}{n!}(x-x_0)^n \end{aligned} \tag{11}$$

In particular, we can write the nth-order Maclaurin polynomial for $f(x)$ as

$$\sum_{k=0}^{n} \frac{f^{(k)}(0)}{k!}x^k = f(0) + f'(0)x + \frac{f''(0)}{2!}x^2 + \cdots + \frac{f^{(n)}(0)}{n!}x^n \tag{12}$$

TAYLOR AND MACLAURIN SERIES

For a fixed value of x near x_0, one would expect the approximation of $f(x)$ by its Taylor polynomial $p_n(x)$ about $x = x_0$ to improve as n increases, since increasing n has the effect of matching higher and higher derivatives of $f(x)$ with those of $p_n(x)$ at $x = x_0$. Indeed, it seems plausible that one might be able to achieve any desired degree of accuracy at a point x by choosing n sufficiently large; that is, the values of $p_n(x)$ might actually converge to $f(x)$ as $n \to +\infty$. Should this happen, we would have

$$f(x) = \lim_{n\to+\infty} \sum_{k=0}^{n} \frac{f^{(k)}(x_0)}{k!}(x-x_0)^k = \sum_{k=0}^{\infty} \frac{f^{(k)}(x_0)}{k!}(x-x_0)^k$$

Later we will study conditions under which the series on the right actually converges to $f(x)$. For the remainder of this section, we will focus on the computational aspects of finding these series. We make the following definition.

11.5.4 DEFINITION. If f has derivatives of all orders at x_0, then we call the series

$$\begin{aligned} \sum_{k=0}^{\infty} \frac{f^{(k)}(x_0)}{k!}(x-x_0)^k &= f(x_0) + f'(x_0)(x-x_0) + \frac{f''(x_0)}{2!}(x-x_0)^2 \\ &\quad + \cdots + \frac{f^{(k)}(x_0)}{k!}(x-x_0)^k + \cdots \end{aligned} \tag{13}$$

the ***Taylor series for f about x = x₀***. In the special case where $x_0 = 0$ this series becomes

$$\sum_{k=0}^{\infty} \frac{f^{(k)}(0)}{k!}x^k = f(0) + f'(0)x + \frac{f''(0)}{2!}x^2 + \cdots + \frac{f^{(k)}(0)}{k!}x^k + \cdots \tag{14}$$

in which case we call it the ***Maclaurin series for f***.

REMARK. Because the summation index in (13) starts at $k = 0$, it is convenient to call the initial term in this series the zeroth term. Thus, a Taylor series has a zeroth partial sum, a first partial sum, a second partial sum, and so forth. With this convention the nth partial sum of a Taylor series is the nth Taylor polynomial and the nth partial sum of a Maclaurin series is the nth Maclaurin polynomial [see (11) and (12)].

Example 4

Find the Maclaurin series for

(a) e^x (b) $\sin x$ (c) $\cos x$ (d) $\dfrac{1}{1-x}$

Solution (a). In Example 2 we found the nth Maclaurin polynomial for the function e^x to be

$$\sum_{k=0}^{n} \frac{x^k}{k!} = 1 + x + \frac{x^2}{2!} + \cdots + \frac{x^n}{n!}$$

Thus, the Maclaurin series for e^x is

$$\sum_{k=0}^{\infty} \frac{x^k}{k!} = 1 + x + \frac{x^2}{2!} + \frac{x^3}{3!} + \cdots + \frac{x^k}{k!} + \cdots$$

Solution (b). In the Maclaurin polynomials for $\sin x$, only the odd powers of x appear explicitly. To see this, let $f(x) = \sin x$; thus,

$$\begin{aligned}
f(x) &= \sin x & f(0) &= 0 \\
f'(x) &= \cos x & f'(0) &= 1 \\
f''(x) &= -\sin x & f''(0) &= 0 \\
f'''(x) &= -\cos x & f'''(0) &= -1
\end{aligned}$$

Since $f^{(4)}(x) = \sin x = f(x)$, the pattern $0, 1, 0, -1$ will repeat as we evaluate successive derivatives at 0. Therefore, the successive Maclaurin polynomials for $\sin x$ are

$$\begin{aligned}
p_0(x) &= 0 \\
p_1(x) &= 0 + x \\
p_2(x) &= 0 + x + 0 \\
p_3(x) &= 0 + x + 0 - \frac{x^3}{3!} \\
p_4(x) &= 0 + x + 0 - \frac{x^3}{3!} + 0 \\
p_5(x) &= 0 + x + 0 - \frac{x^3}{3!} + 0 + \frac{x^5}{5!} \\
p_6(x) &= 0 + x + 0 - \frac{x^3}{3!} + 0 + \frac{x^5}{5!} + 0 \\
p_7(x) &= 0 + x + 0 - \frac{x^3}{3!} + 0 + \frac{x^5}{5!} + 0 - \frac{x^7}{7!} \\
&\vdots
\end{aligned}$$

Because of the zero terms, each even-order Maclaurin polynomial [after $p_0(x)$] is the same as the preceding odd-order Maclaurin polynomial; that is,

$$p_{2k+1}(x) = p_{2k+2}(x) = x - \frac{x^3}{3!} + \frac{x^5}{5!} - \frac{x^7}{7!} + \cdots + (-1)^k \frac{x^{2k+1}}{(2k+1)!} \quad (k = 0, 1, 2, \ldots)$$

Thus, the Maclaurin series for $\sin x$ is

$$\sum_{k=0}^{\infty}(-1)^k \frac{x^{2k+1}}{(2k+1)!} = x - \frac{x^3}{3!} + \frac{x^5}{5!} - \frac{x^7}{7!} + \cdots + (-1)^k \frac{x^{2k+1}}{(2k+1)!} + \cdots$$

The graphs of $\sin x$, $p_1(x)$, $p_3(x)$, $p_5(x)$, and $p_7(x)$ are shown in Figure 11.5.5.

Solution (c). In the Maclaurin polynomials for $\cos x$, only the even powers of x appear explicitly; the computations are similar to those in part (b). The reader should be able to show that

$$p_0(x) = p_1(x) = 1$$
$$p_2(x) = p_3(x) = 1 - \frac{x^2}{2!}$$
$$p_4(x) = p_5(x) = 1 - \frac{x^2}{2!} + \frac{x^4}{4!}$$
$$p_6(x) = p_7(x) = 1 - \frac{x^2}{2!} + \frac{x^4}{4!} - \frac{x^6}{6!}$$

In general, the Maclaurin polynomials for $\cos x$ are

$$p_{2k}(x) = p_{2k+1}(x) = 1 - \frac{x^2}{2!} + \frac{x^4}{4!} - \cdots + (-1)^k \frac{x^{2k}}{(2k)!} \quad (k = 0, 1, 2, \ldots)$$

from which it follows that the Maclaurin series for $\cos x$ is

$$\sum_{k=0}^{\infty}(-1)^k \frac{x^{2k}}{(2k)!} = 1 - \frac{x^2}{2!} + \frac{x^4}{4!} - \frac{x^6}{6!} + \cdots + (-1)^k \frac{x^{2k}}{(2k)!} + \cdots$$

The graphs of $\cos x$, $p_0(x)$, $p_2(x)$, $p_4(x)$, and $p_6(x)$ are shown in Figure 11.5.6.

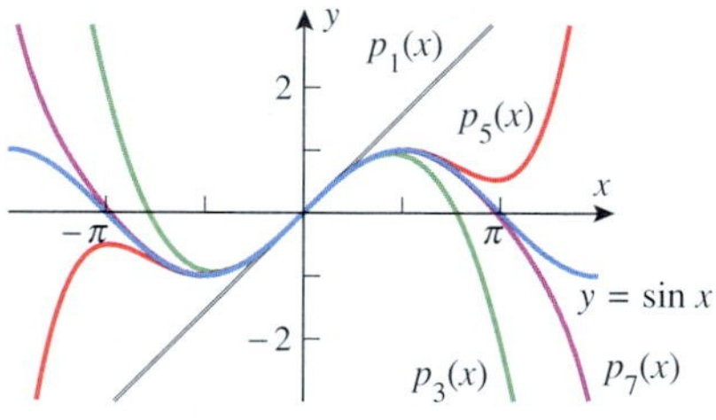

Figure 11.5.5

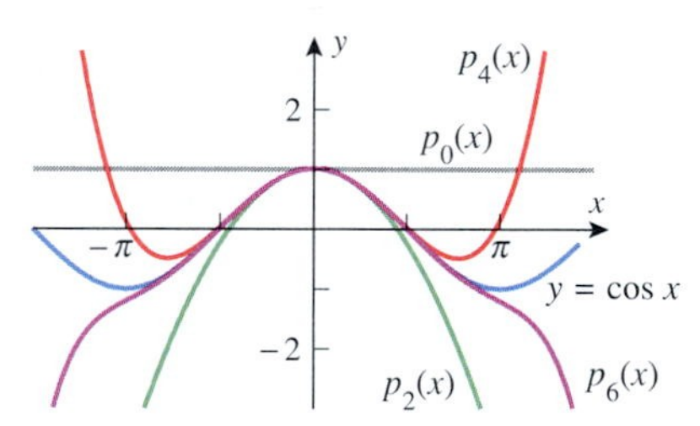

Figure 11.5.6

Solution (d). Let $f(x) = 1/(1-x)$. Thus, the values of f and its first k derivatives at $x = 0$ are as follows:

$$\begin{aligned}
f(x) &= \frac{1}{1-x} & f(0) &= 1 = 0! \\
f'(x) &= \frac{1}{(1-x)^2} & f'(0) &= 1 = 1! \\
f''(x) &= \frac{2}{(1-x)^3} & f''(0) &= 2 = 2! \\
f'''(x) &= \frac{3 \cdot 2}{(1-x)^4} & f'''(0) &= 3! \\
f^{(4)}(x) &= \frac{4 \cdot 3 \cdot 2}{(1-x)^5} & f^{(4)}(0) &= 4! \\
&\vdots & &\vdots \\
f^{(k)}(x) &= \frac{k!}{(1-x)^{k+1}} & f^{(k)}(0) &= k! \\
&\vdots & &\vdots
\end{aligned}$$

Substituting $f^{(k)}(0) = k!$ in Formula (14) yields the Maclaurin series

$$\sum_{k=0}^{\infty} x^k = 1 + x + x^2 + \cdots + x^k + \cdots$$

Thus, the Maclaurin series for $1/(1-x)$ happens to be the geometric series with initial term 1 and ratio x. ◀

Example 5

Find the Taylor series about $x = 1$ for $1/x$.

Solution. Let $f(x) = 1/x$. The computations are similar to those in part (d) of Example 4. We leave it for you to show that

$$f(1) = 1,\ f'(1) = -1,\ f''(1) = 2!,\ f'''(1) = -3!,$$
$$f^{(4)}(1) = 4!,\ \ldots,\ f^{(k)}(1) = (-1)^k k!$$

Thus, substituting $f^{(k)}(1) = (-1)^k k!$ into Formula (13) with $x_0 = 1$ yields the Taylor series

$$\sum_{k=0}^{\infty} (-1)^k (x-1)^k = 1 - (x-1) + (x-1)^2 - (x-1)^3 + \cdots$$ ◀

FOR THE READER. CAS programs have commands for generating Taylor polynomials of any specified degree. If you have a CAS, read the documentation to determine how this is done, and then use the CAS to confirm the computations in the examples in this section.

EXERCISE SET 11.5 Graphing Calculator C CAS

1. In each part, find the local quadratic approximation of f at $x = x_0$, and use that approximation to find the local linear approximation of f at x_0.
(a) $f(x) = e^{-x};\ x_0 = 0$
(b) $f(x) = \cos x;\ x_0 = 0$
(c) $f(x) = \sin x;\ x_0 = \pi/2$
(d) $f(x) = \sqrt{x};\ x_0 = 1$

C **2.** In each part, use a CAS to find the local quadratic approximation of f at $x = x_0$, and use that approximation to find the local linear approximation of f at $x = x_0$.
(a) $f(x) = e^{\sin x};\ x_0 = 0$
(b) $f(x) = \sqrt{x};\ x_0 = 9$
(c) $f(x) = \sec^{-1} x;\ x_0 = 2$
(d) $f(x) = \sin^{-1} x;\ x_0 = 0$

3. (a) Find the local quadratic approximation of $\sqrt{x}$ at $x_0 = 1$.
(b) Use the result obtained in part (a) to approximate $\sqrt{1.1}$, and compare your approximation to that produced directly by your calculating utility. [See Example 4 of Section 3.6.]

4. (a) Find the local quadratic approximation of $\cos x$ at $x_0 = 0$.
(b) Use the result obtained in part (a) to approximate $\cos 2^\circ$, and compare the approximation to that produced directly by your calculating utility.

5. Use an appropriate local quadratic approximation to approximate $\tan 61^\circ$, and compare the result to that produced directly by your calculating utility.

6. Use an appropriate local quadratic approximation to approximate $\sqrt{36.03}$, and compare the result to that produced directly by your calculating utility.

In Exercises 7–16, find the Maclaurin polynomials of orders $n = 0, 1, 2, 3$, and 4, and then find the Maclaurin series for the function in sigma notation.

7. e^{-x} **8.** e^{ax} **9.** $\cos \pi x$

10. $\sin \pi x$ **11.** $\ln(1+x)$ **12.** $\dfrac{1}{1+x}$

13. $\cosh x$ **14.** $\sinh x$ **15.** $x \sin x$

16. xe^x

17. (a) Find the Maclaurin series for the polynomial $f(x) = 1 + 2x - x^2 + x^3$.
(b) Find the Maclaurin series for the polynomial $f(x) = c_0 + c_1 x + c_2 x^2 + \cdots + c_n x^n$.

C **18.** For each of the Exercises 7–16 that you worked on, use a CAS to check the Maclaurin polynomial of order $n = 4$.

In Exercises 19–26, find the Taylor polynomials of orders $n = 0, 1, 2, 3$, and 4 about $x = x_0$, and then find the Taylor series for the function in sigma notation.

19. $e^x;\ x_0 = 1$ **20.** $e^{-x};\ x_0 = \ln 2$

21. $\dfrac{1}{x};\ x_0 = -1$ **22.** $\dfrac{1}{x+2};\ x_0 = 3$

23. $\sin \pi x;\ x_0 = \frac{1}{2}$

24. $\cos x;\ x_0 = \frac{\pi}{2}$

25. $\ln x;\ x_0 = 1$

26. $\ln x;\ x_0 = e$

27. (a) Find the Taylor series about $x = 1$ for the polynomial

$$f(x) = 1 + 2(x-1) - (x-1)^2 + (x-1)^3$$

(b) Find the Taylor series about $x = x_0$ for the polynomial

$$f(x) = c_0 + c_1(x - x_0) + c_2(x - x_0)^2 + \cdots + c_n(x - x_0)^n$$

C **28.** For each of the Exercises 19–26 that you worked on, use a CAS to check the Taylor polynomial of order $n = 4$.

In Exercises 29–32, find the first four distinct Taylor polynomials about $x = x_0$, and use a graphing utility to graph the given function and the Taylor polynomials on the same screen.

29. $f(x) = e^{-2x};\ x_0 = 0$

30. $f(x) = \sin x;\ x_0 = \pi/2$

31. $f(x) = \cos x;\ x_0 = \pi$

32. $\ln(x+1);\ x_0 = 0$

33. Which of the functions graphed in the following figure is most likely to have $p(x) = 1 - x + 2x^2$ as its second-order Maclaurin polynomial? Explain your reasoning.

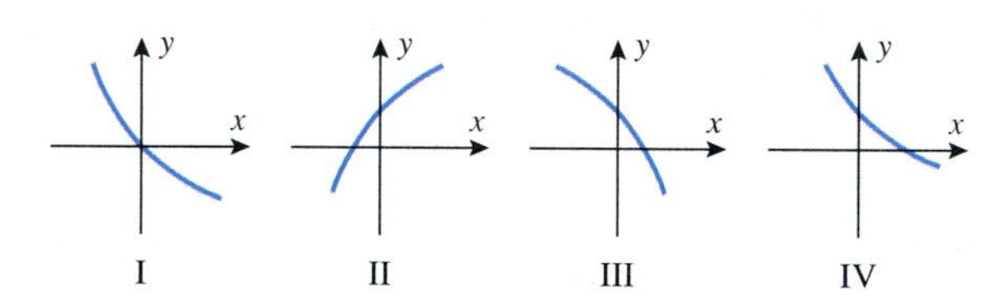

34. Suppose that the values of a function f and its first three derivatives at $x = 1$ are

$$f(1) = 2, \quad f'(1) = -3, \quad f''(1) = 0, \quad f'''(1) = 6$$

Find as many Taylor polynomials for f as you can about $x = 1$.

35. Show that the Taylor series for $\sinh x$ about $x = \ln 4$ is

$$\sum_{k=0}^{\infty} \frac{16 - (-1)^k}{8k!}(x - \ln 4)^k$$

36. (a) The accompanying figure shows a sector of radius r and central angle 2α. Assuming that the angle α is small, use the local quadratic approximation of $\cos \alpha$ at $\alpha = 0$ to show that $x \approx r\alpha^2/2$.

(b) Assuming that the Earth is a sphere of radius 4000 mi, use the result in part (a) to approximate the maximum amount by which a 100-mi arc along the equator will diverge from its chord.

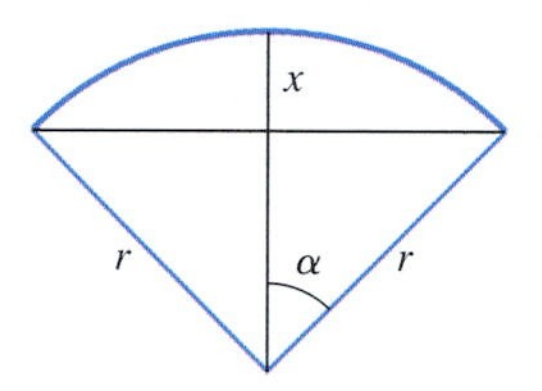

Figure Ex-36

37. Let $p_1(x)$ and $p_2(x)$ be the local linear and local quadratic approximations of $f(x) = e^{\sin x}$ at $x = 0$.

(a) Use a graphing utility to generate the graphs of $f(x)$, $p_1(x)$, and $p_2(x)$ on the same screen for $-1 \le x \le 1$.

(b) Construct a table of values of $f(x)$, $p_1(x)$, and $p_2(x)$ for $x = -1.00, -0.75, -0.50, -0.25, 0, 0.25, 0.50, 0.75, 1.00$. Round the values to three decimal places.

(c) Generate the graph of $|f(x) - p_1(x)|$, and use the graph to determine an interval on which $p_1(x)$ approximates $f(x)$ with an error of at most ± 0.01. [*Suggestion:* Review the discussion relating to Figure 3.6.9.]

(d) Generate the graph of $|f(x) - p_2(x)|$, and use the graph to determine an interval on which $p_2(x)$ approximates $f(x)$ with an error of at most ± 0.01.

11.6 THE COMPARISON, RATIO, AND ROOT TESTS

In this section we will develop some more basic convergence tests for series with nonnegative terms. Later, we will use some of these tests to study the convergence of Taylor series.

THE COMPARISON TEST

We will begin with a test that is useful in its own right and is also the building block for other important convergence tests. The underlying idea of this test is to use the known convergence or divergence of a series to deduce the convergence or divergence of another series.

11.6.1 THEOREM (*The Comparison Test*). *Let $\sum_{k=1}^{\infty} a_k$ and $\sum_{k=1}^{\infty} b_k$ be series with nonnegative terms and suppose that*

$$a_1 \le b_1,\ a_2 \le b_2,\ a_3 \le b_3, \ldots, a_k \le b_k, \ldots$$

(a) If the "bigger series" Σb_k converges, then the "smaller series" Σa_k also converges.

(b) If the "smaller series" Σa_k diverges, then the "bigger series" Σb_k also diverges.

We have left the proof of this theorem for the exercises; however, it is easy to visualize why the theorem is true by interpreting the terms in the series as areas of rectangles (Figure 11.6.1). The comparison test states that if the total area $\sum b_k$ is finite, then the total area $\sum a_k$ must also be finite; and if the total area $\sum a_k$ is infinite, then the total area $\sum b_k$ must also be infinite.

REMARK. As one would expect, it is not essential in Theorem 11.6.1 that the condition $a_k \leq b_k$ hold for all k, as stated; the conclusions of the theorem remain true if this condition is eventually true.

USING THE COMPARISON TEST

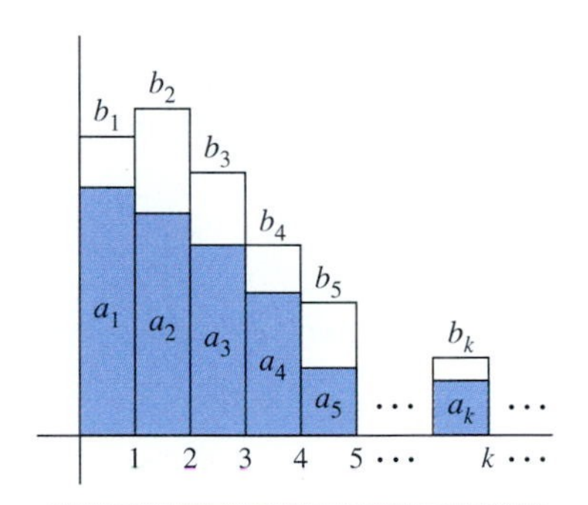

For each rectangle, b_k is the entire area and a_k is the area of the blue portion.

Figure 11.6.1

There are two steps required for using the comparison test to determine whether a series $\sum u_k$ with positive terms converges:

- Guess at whether the series $\sum u_k$ converges or diverges.
- Find a series that proves the guess to be correct. That is, if the guess is divergence, we must find a divergent series whose terms are "smaller" than the corresponding terms of $\sum u_k$, and if the guess is convergence, we must find a convergent series whose terms are "bigger" than the corresponding terms of $\sum u_k$.

To help with the guessing process in the first step, we have formulated two principles that sometimes *suggest* whether a series is likely to converge or diverge. We have called these "informal principles" because they are not intended as formal theorems. In fact, we will not guarantee that they *always* work. However, they work often enough to be useful.

11.6.2 INFORMAL PRINCIPLE. *Constant terms in the denominator of u_k can usually be deleted without affecting the convergence or divergence of the series.*

11.6.3 INFORMAL PRINCIPLE. *If a polynomial in k appears as a factor in the numerator or denominator of u_k, all but the leading term in the polynomial can usually be discarded without affecting the convergence or divergence of the series.*

Example 1

Use the comparison test to determine whether the following series converge or diverge.

(a) $\displaystyle\sum_{k=1}^{\infty} \frac{1}{\sqrt{k}-\frac{1}{2}}$ (b) $\displaystyle\sum_{k=1}^{\infty} \frac{1}{2k^2+k}$

Solution (a). According to Principle 11.6.2, we should be able to drop the constant in the denominator without affecting the convergence or divergence. Thus, the given series is likely to behave like

$$\sum_{k=1}^{\infty} \frac{1}{\sqrt{k}} \tag{1}$$

which is a divergent p-series $\left(p=\frac{1}{2}\right)$. Thus, we will guess that the given series diverges and try to prove this by finding a divergent series that is "smaller" than the given series. However, series (1) does the trick since

$$\frac{1}{\sqrt{k}-\frac{1}{2}} > \frac{1}{\sqrt{k}} \quad \text{for } k = 1, 2, \ldots$$

Thus, we have proved that the given series diverges.

Solution (b). According to Principle 11.6.3, we should be able to discard all but the leading term in the polynomial without affecting the convergence or divergence. Thus, the given series is likely to behave like

$$\sum_{k=1}^{\infty} \frac{1}{2k^2} = \frac{1}{2} \sum_{k=1}^{\infty} \frac{1}{k^2} \tag{2}$$

which converges since it is a constant times a convergent p-series ($p = 2$). Thus, we will guess that the given series converges and try to prove this by finding a convergent series that is "bigger" than the given series. However, series (2) does the trick since

$$\frac{1}{2k^2 + k} < \frac{1}{2k^2} \quad \text{for } k = 1, 2, \ldots$$

Thus, we have proved that the given series converges. ◀

THE LIMIT COMPARISON TEST

In the last example, Principles 11.6.2 and 11.6.3 provided the guess about convergence or divergence as well as the series needed to apply the comparison test. Unfortunately, it is not always so straightforward to find the series required for comparison, so we will now consider an alternative to the comparison test that is usually easier to apply. The proof is given in Appendix G.

> **11.6.4 THEOREM** (*The Limit Comparison Test*). *Let $\sum a_k$ and $\sum b_k$ be series with positive terms and suppose that*
>
> $$\rho = \lim_{k \to +\infty} \frac{a_k}{b_k}$$
>
> *If ρ is finite and $\rho > 0$, then the series both converge or both diverge.*

The cases where $\rho = 0$ or $\rho = +\infty$ are discussed in the exercises (Exercise 54).

Example 2

Use the limit comparison test to determine whether the following series converge or diverge.

$$\text{(a)} \sum_{k=2}^{\infty} \frac{1}{\sqrt{k} - 1} \qquad \text{(b)} \sum_{k=1}^{\infty} \frac{1}{2k^2 + k} \qquad \text{(c)} \sum_{k=1}^{\infty} \frac{3k^3 - 2k^2 + 4}{k^7 - k^3 + 2}$$

Solution (a). As in Example 1, Principle 11.6.2 suggests that the series is likely to behave like the divergent p-series (1). To prove that the given series diverges, we will apply the limit comparison test with

$$a_k = \frac{1}{\sqrt{k} - 1} \quad \text{and} \quad b_k = \frac{1}{\sqrt{k}}$$

We obtain

$$\rho = \lim_{k \to +\infty} \frac{a_k}{b_k} = \lim_{k \to +\infty} \frac{\sqrt{k}}{\sqrt{k} - 1} = \lim_{k \to +\infty} \frac{1}{1 - \dfrac{1}{\sqrt{k}}} = 1$$

Since ρ is finite and positive, it follows from Theorem 11.6.4 that the given series diverges, which agrees with the conclusion reached in Example 1 using the comparison test.

Solution (b). As in Example 1, Principle 11.6.3 suggests that the series is likely to behave like the convergent series (2). To prove that the given series converges, we will apply the limit comparison test with

$$a_k = \frac{1}{2k^2 + k} \quad \text{and} \quad b_k = \frac{1}{2k^2}$$

We obtain

$$\rho = \lim_{k \to +\infty} \frac{a_k}{b_k} = \lim_{k \to +\infty} \frac{2k^2}{2k^2 + k} = \lim_{k \to +\infty} \frac{2}{2 + \frac{1}{k}} = 1$$

Since ρ is finite and positive, it follows from Theorem 11.6.4 that the given series converges, which agrees with the conclusion reached in Example 1 using the comparison test.

Solution (c). From Principle 11.6.3, the series is likely to behave like

$$\sum_{k=1}^{\infty} \frac{3k^3}{k^7} = \sum_{k=1}^{\infty} \frac{3}{k^4} \tag{3}$$

which converges since it is a constant times a convergent p-series. Thus, the given series is likely to converge. To prove this, we will apply the limit comparison test to series (3) and the given series. We obtain

$$\rho = \lim_{k \to +\infty} \frac{\dfrac{3k^3 - 2k^2 + 4}{k^7 - k^3 + 2}}{\dfrac{3}{k^4}} = \lim_{k \to +\infty} \frac{3k^7 - 2k^6 + 4k^4}{3k^7 - 3k^3 + 6} = 1$$

Since ρ is finite and nonzero, it follows from Theorem 11.6.4 that the given series converges, since (3) converges. ◀

THE RATIO TEST

The comparison test and the limit comparison test hinge on first making a guess about convergence and then finding an appropriate series for comparison, both of which can be difficult tasks in cases where Principles 11.6.2 and 11.6.3 cannot be applied. In such cases the next test can often be used, since it works exclusively with the terms of the given series—it requires neither an initial guess about convergence nor the discovery of a series for comparison. Its proof is given in Appendix G.

11.6.5 THEOREM (*The Ratio Test*). *Let $\sum u_k$ be a series with positive terms and suppose that*

$$\rho = \lim_{k \to +\infty} \frac{u_{k+1}}{u_k}$$

(*a*) *If $\rho < 1$, the series converges.*

(*b*) *If $\rho > 1$ or $\rho = +\infty$, the series diverges.*

(*c*) *If $\rho = 1$, the series may converge or diverge, so that another test must be tried.*

Example 3

Use the ratio test to determine whether the following series converge or diverge.

(a) $\displaystyle\sum_{k=1}^{\infty} \frac{1}{k!}$ (b) $\displaystyle\sum_{k=1}^{\infty} \frac{k}{2^k}$ (c) $\displaystyle\sum_{k=1}^{\infty} \frac{k^k}{k!}$ (d) $\displaystyle\sum_{k=3}^{\infty} \frac{(2k)!}{4^k}$ (e) $\displaystyle\sum_{k=1}^{\infty} \frac{1}{2k-1}$

Solution (a). The series converges, since

$$\rho = \lim_{k \to +\infty} \frac{u_{k+1}}{u_k} = \lim_{k \to +\infty} \frac{1/(k+1)!}{1/k!} = \lim_{k \to +\infty} \frac{k!}{(k+1)!} = \lim_{k \to +\infty} \frac{1}{k+1} = 0 < 1$$

Solution (b). The series converges, since

$$\rho = \lim_{k \to +\infty} \frac{u_{k+1}}{u_k} = \lim_{k \to +\infty} \frac{k+1}{2^{k+1}} \cdot \frac{2^k}{k} = \frac{1}{2} \lim_{k \to +\infty} \frac{k+1}{k} = \frac{1}{2} < 1$$

Solution (c). The series diverges, since

$$\rho = \lim_{k\to+\infty} \frac{u_{k+1}}{u_k} = \lim_{k\to+\infty} \frac{(k+1)^{k+1}}{(k+1)!} \cdot \frac{k!}{k^k} = \lim_{k\to+\infty} \frac{(k+1)^k}{k^k} = \lim_{k\to+\infty} \left(1+\frac{1}{k}\right)^k = e > 1$$

See Theorem 7.9.2(*b*)

Solution (d). In the preceding parts, the index of summation started at 1. Although we could rewrite this series to make the index start at 1, it is not necessary to do so, since the requirements for the ratio test need only hold eventually. The series diverges, since

$$\rho = \lim_{k\to+\infty} \frac{u_{k+1}}{u_k} = \lim_{k\to+\infty} \frac{[2(k+1)]!}{4^{k+1}} \cdot \frac{4^k}{(2k)!} = \lim_{k\to+\infty} \left(\frac{(2k+2)!}{(2k)!} \cdot \frac{1}{4}\right)$$

$$= \frac{1}{4} \lim_{k\to+\infty} (2k+2)(2k+1) = +\infty$$

Solution (e). The ratio test is of no help since

$$\rho = \lim_{k\to+\infty} \frac{u_{k+1}}{u_k} = \lim_{k\to+\infty} \frac{1}{2(k+1)-1} \cdot \frac{2k-1}{1} = \lim_{k\to+\infty} \frac{2k-1}{2k+1} = 1$$

However, the integral test proves that the series diverges since

$$\int_1^{+\infty} \frac{dx}{2x-1} = \lim_{l\to+\infty} \int_1^l \frac{dx}{2x-1} = \lim_{l\to+\infty} \frac{1}{2}\ln(2x-1)\bigg]_1^l = +\infty$$

Both the comparison test and the limit comparison test would also have worked here (verify). ◀

THE ROOT TEST

In cases where it is difficult or inconvenient to find the limit required for the ratio test, the next test is sometimes useful. Since its proof is similar to the proof of the ratio test, we will omit it.

11.6.6 THEOREM (*The Root Test*). *Let $\sum u_k$ be a series with positive terms and suppose that*

$$\rho = \lim_{k\to+\infty} \sqrt[k]{u_k} = \lim_{k\to+\infty} (u_k)^{1/k}$$

(*a*) *If $\rho < 1$, the series converges.*

(*b*) *If $\rho > 1$, or $\rho = +\infty$, the series diverges.*

(*c*) *If $\rho = 1$, the series may converge or diverge, so that another test must be tried.*

Example 4

Use the root test to determine whether the following series converge or diverge.

(a) $\displaystyle\sum_{k=2}^{\infty} \left(\frac{4k-5}{2k+1}\right)^k$ (b) $\displaystyle\sum_{k=1}^{\infty} \frac{1}{(\ln(k+1))^k}$

Solution (a). The series diverges, since

$$\rho = \lim_{k\to+\infty} (u_k)^{1/k} = \lim_{k\to+\infty} \frac{4k-5}{2k+1} = 2 > 1$$

Solution (b). The series converges, since

$$\rho = \lim_{k\to+\infty} (u_k)^{1/k} = \lim_{k\to+\infty} \frac{1}{\ln(k+1)} = 0 < 1$$

◀

EXERCISE SET 11.6 [C] CAS

In Exercises 1 and 2, make a guess about the convergence or divergence of the series, and confirm your guess using the comparison test.

1. (a) $\sum_{k=1}^{\infty} \frac{1}{5k^2 - k}$ (b) $\sum_{k=1}^{\infty} \frac{3}{k - \frac{1}{4}}$

2. (a) $\sum_{k=2}^{\infty} \frac{k+1}{k^2 - k}$ (b) $\sum_{k=1}^{\infty} \frac{2}{k^4 + k}$

3. In each part, use the comparison test to show that the series converges.

(a) $\sum_{k=1}^{\infty} \frac{1}{3^k + 5}$ (b) $\sum_{k=1}^{\infty} \frac{5 \sin^2 k}{k!}$

4. In each part, use the comparison test to show that the series diverges.

(a) $\sum_{k=1}^{\infty} \frac{\ln k}{k}$ (b) $\sum_{k=1}^{\infty} \frac{k}{k^{3/2} - \frac{1}{2}}$

In Exercises 5–10, use the limit comparison test to determine whether the series converges.

5. $\sum_{k=1}^{\infty} \frac{4k^2 - 2k + 6}{8k^7 + k - 8}$ **6.** $\sum_{k=1}^{\infty} \frac{1}{9k + 6}$

7. $\sum_{k=1}^{\infty} \frac{5}{3^k + 1}$ **8.** $\sum_{k=1}^{\infty} \frac{k(k+3)}{(k+1)(k+2)(k+5)}$

9. $\sum_{k=1}^{\infty} \frac{1}{\sqrt[3]{8k^2 - 3k}}$ **10.** $\sum_{k=1}^{\infty} \frac{1}{(2k+3)^{17}}$

In Exercises 11–16, use the ratio test to determine whether the series converges. If the test is inconclusive, then say so.

11. $\sum_{k=1}^{\infty} \frac{3^k}{k!}$ **12.** $\sum_{k=1}^{\infty} \frac{4^k}{k^2}$ **13.** $\sum_{k=1}^{\infty} \frac{1}{5k}$

14. $\sum_{k=1}^{\infty} k \left(\frac{1}{2}\right)^k$ **15.** $\sum_{k=1}^{\infty} \frac{k!}{k^3}$ **16.** $\sum_{k=1}^{\infty} \frac{k}{k^2 + 1}$

In Exercises 17–20, use the root test to determine whether the series converges. If the test is inconclusive, then say so.

17. $\sum_{k=1}^{\infty} \left(\frac{3k+2}{2k-1}\right)^k$ **18.** $\sum_{k=1}^{\infty} \left(\frac{k}{100}\right)^k$

19. $\sum_{k=1}^{\infty} \frac{k}{5^k}$ **20.** $\sum_{k=1}^{\infty} (1 - e^{-k})^k$

In Exercises 21–44, use any method to determine whether the series converges.

21. $\sum_{k=0}^{\infty} \frac{7^k}{k!}$ **22.** $\sum_{k=1}^{\infty} \frac{1}{2k+1}$ **23.** $\sum_{k=1}^{\infty} \frac{k^2}{5^k}$

24. $\sum_{k=1}^{\infty} \frac{k!10^k}{3^k}$ **25.** $\sum_{k=1}^{\infty} k^{50} e^{-k}$ **26.** $\sum_{k=1}^{\infty} \frac{k^2}{k^3 + 1}$

27. $\sum_{k=1}^{\infty} \frac{\sqrt{k}}{k^3 + 1}$ **28.** $\sum_{k=1}^{\infty} \frac{4}{2 + 3^k k}$

29. $\sum_{k=1}^{\infty} \frac{1}{\sqrt{k(k+1)}}$ **30.** $\sum_{k=1}^{\infty} \frac{2 + (-1)^k}{5^k}$

31. $\sum_{k=1}^{\infty} \frac{2 + \sqrt{k}}{(k+1)^3 - 1}$ **32.** $\sum_{k=1}^{\infty} \frac{4 + |\cos k|}{k^3}$

33. $\sum_{k=1}^{\infty} \frac{1}{1 + \sqrt{k}}$ **34.** $\sum_{k=1}^{\infty} \frac{k!}{k^k}$ **35.** $\sum_{k=1}^{\infty} \frac{\ln k}{e^k}$

36. $\sum_{k=1}^{\infty} \frac{k!}{e^{k^2}}$ **37.** $\sum_{k=0}^{\infty} \frac{(k+4)!}{4!k!4^k}$ **38.** $\sum_{k=1}^{\infty} \left(\frac{k}{k+1}\right)^{k^2}$

39. $\sum_{k=1}^{\infty} \frac{1}{4 + 2^{-k}}$ **40.** $\sum_{k=1}^{\infty} \frac{\sqrt{k} \ln k}{k^3 + 1}$ **41.** $\sum_{k=1}^{\infty} \frac{\tan^{-1} k}{k^2}$

42. $\sum_{k=1}^{\infty} \frac{5^k + k}{k! + 3}$ **43.** $\sum_{k=0}^{\infty} \frac{(k!)^2}{(2k)!}$ **44.** $\sum_{k=1}^{\infty} \frac{(k!)^2 2^k}{(2k+2)!}$

In Exercises 45 and 46, find the general term of the series, and use the ratio test to show that the series converges.

45. $1 + \frac{1 \cdot 2}{1 \cdot 3} + \frac{1 \cdot 2 \cdot 3}{1 \cdot 3 \cdot 5} + \frac{1 \cdot 2 \cdot 3 \cdot 4}{1 \cdot 3 \cdot 5 \cdot 7} + \cdots$

46. $1 + \frac{1 \cdot 3}{3!} + \frac{1 \cdot 3 \cdot 5}{5!} + \frac{1 \cdot 3 \cdot 5 \cdot 7}{7!} + \cdots$

In Exercises 47 and 48, use a CAS to investigate the convergence of the series.

[C] **47.** $\sum_{k=1}^{\infty} \frac{\ln k}{3^k}$ [C] **48.** $\sum_{k=1}^{\infty} \frac{[\pi(k+1)]^k}{k^{k+1}}$

49. (a) Make a conjecture about the convergence of the series $\sum_{k=1}^{\infty} \sin(\pi/k)$ by considering the local linear approximation of $\sin x$ near $x = 0$.

(b) Try to confirm your conjecture using the limit comparison test.

50. (a) Make a conjecture about the convergence of the series

$$\sum_{k=1}^{\infty} \left[1 - \cos\left(\frac{1}{k}\right)\right]$$

by considering the local quadratic approximation of $\cos x$ near $x = 0$.

(b) Try to confirm your conjecture using the limit comparison test.

51. Show that $\ln x < \sqrt{x}$ if $x > 0$, and use this result to investigate the convergence of

(a) $\sum_{k=1}^{\infty} \frac{\ln k}{k^2}$ (b) $\sum_{k=2}^{\infty} \frac{1}{(\ln k)^2}$

52. For which positive values of α does the series $\sum_{k=1}^{\infty}(\alpha^k/k^\alpha)$ converge?

53. Use Theorem 11.4.6 to prove the comparison test (Theorem 11.6.1).

54. Let $\sum a_k$ and $\sum b_k$ be series with positive terms. Prove:

(a) If $\lim_{k\to+\infty}(a_k/b_k) = 0$ and $\sum b_k$ converges, then $\sum a_k$ converges.

(b) If $\lim_{k\to+\infty}(a_k/b_k) = +\infty$ and $\sum b_k$ diverges, then $\sum a_k$ diverges.

11.7 ALTERNATING SERIES; CONDITIONAL CONVERGENCE

Up to now we have focused exclusively on series with nonnegative terms. In this section we will discuss series that contain both positive and negative terms.

ALTERNATING SERIES

Series whose terms alternate between positive and negative, called ***alternating series***, are of special importance. Some examples are

$$\sum_{k=1}^{\infty}(-1)^{k+1}\frac{1}{k} = 1 - \frac{1}{2} + \frac{1}{3} - \frac{1}{4} + \frac{1}{5} - \cdots$$

$$\sum_{k=1}^{\infty}(-1)^{k}\frac{1}{k} = -1 + \frac{1}{2} - \frac{1}{3} + \frac{1}{4} - \frac{1}{5} + \cdots$$

In general, an alternating series has one of the following two forms:

$$\sum_{k=1}^{\infty}(-1)^{k+1}a_k = a_1 - a_2 + a_3 - a_4 + \cdots \tag{1}$$

$$\sum_{k=1}^{\infty}(-1)^{k}a_k = -a_1 + a_2 - a_3 + a_4 - \cdots \tag{2}$$

where the a_k's are assumed to be positive in both cases.

The following theorem is the key result on convergence of alternating series.

11.7.1 THEOREM (*Alternating Series Test*). *An alternating series of either form* (1) *or form* (2) *converges if the following two conditions are satisfied:*

(*a*) $a_1 > a_2 > a_3 > \cdots > a_k > \cdots$

(*b*) $\lim_{k\to+\infty} a_k = 0$

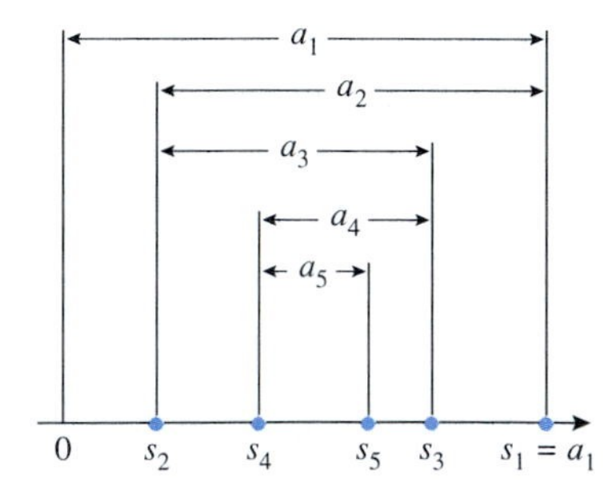

Figure 11.7.1

Proof. We will consider only alternating series of form (1). The idea of the proof is to show that if conditions (*a*) and (*b*) hold, then the sequences of even-numbered and odd-numbered partial sums converge to a common limit S. It will then follow from Theorem 11.1.4 that the entire sequence of partial sums converges to S.

Figure 11.7.1 shows how successive partial sums satisfying conditions (*a*) and (*b*) appear when plotted on a horizontal axis. The even-numbered partial sums

$$s_2, s_4, s_6, s_8, \ldots, s_{2n}, \ldots$$

form an increasing sequence bounded above by a_1, and the odd-numbered partial sums

$$s_1, s_3, s_5, \ldots, s_{2n-1}, \ldots$$

form a decreasing sequence bounded below by 0. Thus, by Theorems 11.2.3 and 11.2.4, the even-numbered partial sums converge to some limit S_E and the odd-numbered partial sums converge to some limit S_O. To complete the proof we must show that $S_E = S_O$. But the $(2n)$-th term in the series is $-a_{2n}$, so that $s_{2n} - s_{2n-1} = -a_{2n}$, which can be written as

$$s_{2n-1} = s_{2n} + a_{2n}$$

However, $2n \to +\infty$ and $2n - 1 \to +\infty$ as $n \to +\infty$, so that

$$S_O = \lim_{n\to+\infty} s_{2n-1} = \lim_{n\to+\infty} (s_{2n} + a_{2n}) = S_E + 0 = S_E$$

which completes the proof. ■

REMARK. As might be expected, it is not essential for condition (a) in the alternating series test to hold for all terms; an alternating series will converge if condition (b) is true and condition (a) holds eventually.

Example 1

Use the alternating series test to show that the following series converge.

$$\text{(a)} \sum_{k=1}^{\infty}(-1)^{k+1}\frac{1}{k} \qquad \text{(b)} \sum_{k=1}^{\infty}(-1)^{k+1}\frac{k+3}{k(k+1)}$$

Solution (a). The two conditions in the alternating series test are satisfied since

$$a_k = \frac{1}{k} > \frac{1}{k+1} = a_{k+1} \quad \text{and} \quad \lim_{k\to+\infty} a_k = \lim_{k\to+\infty} \frac{1}{k} = 0$$

Solution (b). The two conditions in the alternating series test are satisfied since

$$\frac{a_{k+1}}{a_k} = \frac{k+4}{(k+1)(k+2)} \cdot \frac{k(k+1)}{k+3} = \frac{k^2+4k}{k^2+5k+6} = \frac{k^2+4k}{(k^2+4k)+(k+6)} < 1$$

so

$$a_k > a_{k+1}$$

and

$$\lim_{k\to+\infty} a_k = \lim_{k\to+\infty} \frac{k+3}{k(k+1)} = \lim_{k\to+\infty} \frac{\dfrac{1}{k} + \dfrac{3}{k^2}}{1 + \dfrac{1}{k}} = 0$$

◀

REMARK. The series in part (a) of the last example is called the ***alternating harmonic series***. Observe that this series converges, whereas the harmonic series diverges.

REMARK. If an alternating series violates condition (b) of the alternating series test, then the series must diverge by the divergence test (Theorem 11.4.1). However, if condition (b) is satisfied, but condition (a) is not, the series can either converge or diverge.*

APPROXIMATING SUMS OF ALTERNATING SERIES

The following theorem is concerned with the error that results when the sum of an alternating series is approximated by a partial sum.

11.7.2 THEOREM. *If an alternating series satisfies the hypotheses of the alternating series test, and if S is the sum of the series, then:*

(a) *S lies between any two successive partial sums; that is, either*

$$s_n < S < s_{n+1} \quad \text{or} \quad s_{n+1} < S < s_n \tag{3}$$

depending on which partial sum is larger.

(b) *If S is approximated by s_n, then the absolute error $|S - s_n|$ satisfies*

$$|S - s_n| < a_{n+1} \tag{4}$$

Moreover, the sign of the error $S - s_n$ is the same as that of the coefficient of a_{n+1}.

---

*The interested reader will find some nice examples in an article by R. Lariviere, "On a Convergence Test for Alternating Series," *Mathematics Magazine*, Vol. 29, 1956, p. 88.

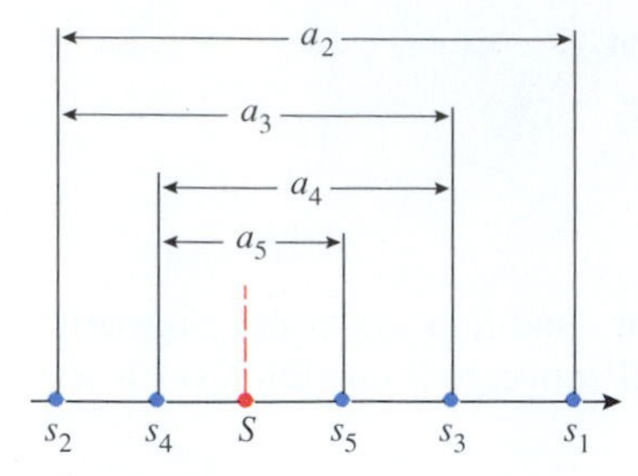

Figure 11.7.2

Proof. We will prove the theorem for series of form (1). Referring to Figure 11.7.2 and keeping in mind our observation in the proof of Theorem 11.7.1 that the odd-numbered partial sums form a decreasing sequence converging to S and the even-numbered partial sums form an increasing sequence converging to S, we see that successive partial sums oscillate from one side of S to the other in smaller and smaller steps with the odd-numbered partial sums being larger than S and the even-numbered partial sums being smaller than S. Thus, depending on whether n is even or odd, we have

$$s_n < S < s_{n+1} \quad \text{or} \quad s_{n+1} < S < s_n$$

which proves (3). Moreover, in either case we have

$$|S - s_n| < |s_{n+1} - s_n| \tag{5}$$

But $s_{n+1} - s_n = \pm a_{n+1}$ (the sign depending on whether n is even or odd). Thus, it follows from (5) that $|S - s_n| < a_{n+1}$, which proves (4). Finally, since the odd-numbered partial sums are larger than S and the even-numbered partial sums are smaller than S, it follows that $S - s_n$ has the same sign as the coefficient of a_{n+1} (verify). ■

REMARK. In words, inequality (4) states that for a series satisfying the hypotheses of the alternating series test, the magnitude of the error that results from approximating S by s_n is less than that of the first term that is *not* included in the partial sum.

Example 2

Later in this chapter we will show that the sum of the alternating harmonic series is

$$\ln 2 = 1 - \frac{1}{2} + \frac{1}{3} - \frac{1}{4} + \cdots + (-1)^{k+1}\frac{1}{k} + \cdots$$

(a) Accepting this to be so, find an upper bound on the magnitude of the error that results if ln 2 is approximated by the sum of the first eight terms in the series.

(b) Find a partial sum that approximates ln 2 to one decimal-place accuracy (the nearest tenth).

Solution (a). It follows from (4) that

$$|\ln 2 - s_8| < a_9 = \frac{1}{9} < 0.12 \tag{6}$$

As a check, let us compute s_8 exactly. We obtain

$$s_8 = 1 - \frac{1}{2} + \frac{1}{3} - \frac{1}{4} + \frac{1}{5} - \frac{1}{6} + \frac{1}{7} - \frac{1}{8} = \frac{533}{840}$$

Thus, with the help of a calculator

$$|\ln 2 - s_8| = \left|\ln 2 - \frac{533}{840}\right| \approx 0.059$$

This shows that the error is well under the estimate provided by upper bound (6).

Solution (b). For one decimal-place accuracy, we must choose n so that $|\ln 2 - s_n| \le 0.05$. However, it follows from (4) that

$$|\ln 2 - s_n| < a_{n+1}$$

so it suffices to choose n so that $a_{n+1} \le 0.05$.

One way to find n is to use a calculating utility to obtain numerical values for a_1, a_2, $a_3, \ldots$ until you encounter the first value that is less than or equal to 0.05. If you do this, you will find that it is $a_{20} = 0.05$; this tells us that partial sum s_{19} will provide the desired accuracy. Another way to find n is to solve the inequality

$$\frac{1}{n+1} \le 0.05$$

algebraically. We can do this by taking reciprocals, reversing the sense of the inequality, and then simplifying to obtain $n \ge 19$. Thus, s_{19} will provide the required accuracy, which is consistent with the previous result.

With the help of a calculating utility, the value of s_{19} is approximately $s_{19} \approx 0.7$ and the value of $\ln 2$ obtained directly is approximately $\ln 2 \approx 0.69$, which agrees with s_{19} when rounded to one decimal place. ◀

REMARK. As this example illustrates, the alternating harmonic series does not provide an efficient way to approximate $\ln 2$, since too much computation is required to achieve reasonable accuracy. Later, we will develop better ways to approximate logarithms.

ABSOLUTE CONVERGENCE

The series

$$1 - \frac{1}{2} - \frac{1}{2^2} + \frac{1}{2^3} + \frac{1}{2^4} - \frac{1}{2^5} - \frac{1}{2^6} + \cdots$$

does not fit in any of the categories studied so far—it has mixed signs, but is not alternating. We will now develop some convergence tests that can be applied to such series.

11.7.3 DEFINITION. A series

$$\sum_{k=1}^{\infty} u_k = u_1 + u_2 + \cdots + u_k + \cdots$$

is said to ***converge absolutely*** if the series of absolute values

$$\sum_{k=1}^{\infty} |u_k| = |u_1| + |u_2| + \cdots + |u_k| + \cdots$$

converges and is said to ***diverge absolutely*** if the series of absolute values diverges.

Example 3

Determine whether the following series converge absolutely.

(a) $1 - \frac{1}{2} - \frac{1}{2^2} + \frac{1}{2^3} + \frac{1}{2^4} - \frac{1}{2^5} - \cdots$ (b) $1 - \frac{1}{2} + \frac{1}{3} - \frac{1}{4} + \frac{1}{5} - \cdots$

Solution (a). The series of absolute values is the convergent geometric series

$$1 + \frac{1}{2} + \frac{1}{2^2} + \frac{1}{2^3} + \frac{1}{2^4} + \frac{1}{2^5} + \cdots$$

so the given series converges.

Solution (b). The series of absolute values is the divergent harmonic series

$$1 + \frac{1}{2} + \frac{1}{3} + \frac{1}{4} + \frac{1}{5} + \cdots$$

so the given series diverges absolutely. ◀

It is important to distinguish between the notions of convergence and absolute convergence. For example, the series in part (b) of Example 3 converges, since it is the alternating harmonic series, yet we demonstrated that it does not converge absolutely. However, the following theorem shows that *if a series converges absolutely, then it converges.*

11.7.4 THEOREM. *If the series*

$$\sum_{k=1}^{\infty} |u_k| = |u_1| + |u_2| + \cdots + |u_k| + \cdots$$

converges, then so does the series

$$\sum_{k=1}^{\infty} u_k = u_1 + u_2 + \cdots + u_k + \cdots$$

Proof. Our proof is based on a trick. We will write the series $\sum u_k$ as

$$\sum_{k=1}^{\infty} u_k = \sum_{k=1}^{\infty} [(u_k + |u_k|) - |u_k|] \tag{7}$$

We are assuming that $\sum |u_k|$ converges, so that if we can show that $\sum(u_k + |u_k|)$ converges, then it will follow from (7) and Theorem 11.4.3(a) that $\sum u_k$ converges. However, the value of $u_k + |u_k|$ is either 0 or $2|u_k|$, depending on the sign of u_k. Thus, in all cases it is true that

$$0 \le u_k + |u_k| \le 2|u_k|$$

But $\sum 2|u_k|$ converges, since it is a constant times the convergent series $\sum |u_k|$; hence $\sum(u_k + |u_k|)$ converges by the comparison test. ■

Theorem 11.7.4 is important because it provides a way of inferring convergence of a series with positive and negative terms from the convergence of a series with nonnegative terms (the series of absolute values). This is important because most of the convergence tests we have developed apply only to series with nonnegative terms.

Example 4

Show that the following series converge.

(a) $1 - \frac{1}{2} - \frac{1}{2^2} + \frac{1}{2^3} + \frac{1}{2^4} - \frac{1}{2^5} - \frac{1}{2^6} + \cdots$ (b) $\sum_{k=1}^{\infty} \frac{\cos k}{k^2}$

Solution (a). Observe that this is not an alternating series because the signs alternate in pairs after the first term. Thus, we have no convergence test that can be applied directly. However, we showed in Example 3(a) that the series converges absolutely, so Theorem 11.7.4 implies that it converges.

Solution (b). With the help of a calculating utility, you will be able to verify that the signs of the terms in this series vary irregularly. Thus, we will test for absolute convergence. The series of absolute values is

$$\sum_{k=1}^{\infty} \left| \frac{\cos k}{k^2} \right|$$

However,

$$\left| \frac{\cos k}{k^2} \right| \le \frac{1}{k^2}$$

But $\sum 1/k^2$ is a convergent p-series ($p = 2$), so the series of absolute values converges by the comparison test. Thus, the given series converges absolutely and hence converges. ◀

CONDITIONAL CONVERGENCE

Although Theorem 11.7.4 is a useful tool for series that converge absolutely, it provides no information about the convergence or divergence of a series that diverges absolutely. For example, consider the two series

$$1 - \frac{1}{2} + \frac{1}{3} - \frac{1}{4} + \cdots + (-1)^{k+1}\frac{1}{k} + \cdots \tag{8}$$

$$-1 - \frac{1}{2} - \frac{1}{3} - \frac{1}{4} - \cdots - \frac{1}{k} - \cdots \tag{9}$$

Both of these series diverge absolutely, since in each case the series of absolute values is the divergent harmonic series

$$1+\frac{1}{2}+\frac{1}{3}+\cdots+\frac{1}{k}+\cdots$$

However, series (8) converges, since it is the alternating harmonic series, and series (9) diverges, since it is a constant times the divergent harmonic series. As a matter of terminology, a series that converges but diverges absolutely is said to ***converge conditionally*** (or to be ***conditionally convergent***). Thus, (9) is a conditionally convergent series.

THE RATIO TEST FOR ABSOLUTE CONVERGENCE

Although one cannot generally infer convergence or divergence of a series from absolute divergence, the following variation of the ratio test provides a way of deducing divergence from absolute divergence in certain situations. We omit the proof.

11.7.5 THEOREM (***Ratio Test for Absolute Convergence***). *Let $\sum u_k$ be a series with nonzero terms and suppose that*

$$\rho=\lim_{k\to+\infty}\frac{|u_{k+1}|}{|u_k|}$$

(*a*) *If $\rho<1$, then the series $\sum u_k$ converges absolutely and therefore converges.*

(*b*) *If $\rho>1$ or if $\rho=+\infty$, then the series $\sum u_k$ diverges.*

(*c*) *If $\rho=1$, no conclusion about convergence or absolute convergence can be drawn from this test.*

Example 5

Use the ratio test for absolute convergence to determine whether the series converges.

$$\text{(a)}\ \sum_{k=1}^{\infty}(-1)^k\frac{2^k}{k!}\qquad\text{(b)}\ \sum_{k=1}^{\infty}(-1)^k\frac{(2k-1)!}{3^k}$$

Solution (a). Taking the absolute value of the general term u_k we obtain

$$|u_k|=\left|(-1)^k\frac{2^k}{k!}\right|=\frac{2^k}{k!}$$

Thus,

$$\rho=\lim_{k\to+\infty}\frac{|u_{k+1}|}{|u_k|}=\lim_{k\to+\infty}\frac{2^{k+1}}{(k+1)!}\cdot\frac{k!}{2^k}=\lim_{k\to+\infty}\frac{2}{k+1}=0<1$$

which implies that the series converges absolutely and therefore converges.

Solution (b). Taking the absolute value of the general term u_k we obtain

$$|u_k|=\left|(-1)^k\frac{(2k-1)!}{3^k}\right|=\frac{(2k-1)!}{3^k}$$

Thus,

$$\rho=\lim_{k\to+\infty}\frac{|u_{k+1}|}{|u_k|}=\lim_{k\to+\infty}\frac{[2(k+1)-1]!}{3^{k+1}}\cdot\frac{3^k}{(2k-1)!}$$

$$=\lim_{k\to+\infty}\frac{1}{3}\cdot\frac{(2k+1)!}{(2k-1)!}=\frac{1}{3}\lim_{k\to+\infty}(2k)(2k+1)=+\infty$$

which implies that the series diverges. ◀

SUMMARY OF CONVERGENCE TESTS

We conclude this section with a summary of convergence tests that can be used for reference.

Summary of Convergence Tests

NAME	STATEMENT	COMMENTS
Divergence Test (11.4.1)	If $\lim_{k\to+\infty} u_k \neq 0$, then $\sum u_k$ diverges.	If $\lim_{k\to+\infty} u_k = 0$, then $\sum u_k$ may or may not converge.
Integral Test (11.4.4)	Let $\sum u_k$ be a series with positive terms, and let $f(x)$ be the function that results when k is replaced by x in the general term of the series. If f is decreasing and continuous for $x \geq 1$, then $\sum_{k=1}^{\infty} u_k$ and $\int_1^{+\infty} f(x)\,dx$ both converge or both diverge.	This test only applies to series that have positive terms. Try this test when $f(x)$ is easy to integrate.
Comparison Test (11.6.1)	Let $\sum_{k=1}^{\infty} a_k$ and $\sum_{k=1}^{\infty} b_k$ be series with nonnegative terms such that $a_1 \leq b_1, a_2 \leq b_2, \ldots, a_k \leq b_k, \ldots$ If $\sum b_k$ converges, then $\sum a_k$ converges, and if $\sum a_k$ diverges, then $\sum b_k$ diverges.	This test only applies to series with nonnegative terms. Try this test as a last resort; other tests are often easier to apply.
Ratio Test (11.6.5)	Let $\sum u_k$ be a series with positive terms and suppose that $\rho = \lim_{k\to+\infty} \frac{u_{k+1}}{u_k}$ (a) Series converges if $\rho < 1$. (b) Series diverges if $\rho > 1$ or $\rho = +\infty$. (c) The test is inconclusive if $\rho = 1$.	Try this test when u_k involves factorials or kth powers.
Root Test (11.6.6)	Let $\sum u_k$ be a series with positive terms such that $\rho = \lim_{k\to+\infty} \sqrt[k]{u_k}$ (a) The series converges if $\rho < 1$. (b) The series diverges if $\rho > 1$ or $\rho = +\infty$. (c) The test is inconclusive if $\rho = 1$.	Try this test when u_k involves kth powers.
Limit Comparison Test (11.6.4)	Let $\sum a_k$ and $\sum b_k$ be series with positive terms such that $\rho = \lim_{k\to+\infty} \frac{a_k}{b_k}$ If $0 < \rho < +\infty$, then both series converge or both diverge.	This is easier to apply than the comparison test, but still requires some skill in choosing the series $\sum b_k$ for comparison.
Alternating Series Test (11.7.1)	If $a_k > 0$ for $k = 1, 2, 3, \ldots$, then the series $a_1 - a_2 + a_3 - a_4 + \cdots$ $-a_1 + a_2 - a_3 + a_4 - \cdots$ converge if the following conditions hold: (a) $a_1 > a_2 > a_3 > \cdots$ (b) $\lim_{k\to+\infty} a_k = 0$	This test applies only to alternating series.
Ratio Test for Absolute Convergence (11.7.5)	Let $\sum u_k$ be a series with nonzero terms such that $\rho = \lim_{k\to+\infty} \frac{\lvert u_{k+1}\rvert}{\lvert u_k\rvert}$ (a) The series converges absolutely if $\rho < 1$. (b) The series diverges absolutely if $\rho > 1$ or $\rho = +\infty$. (c) The test is inconclusive if $\rho = 1$.	The series need not have positive terms and need not be alternating to use this test.

Exercise Set 11.7 Graphing Calculator C CAS

In Exercises 1 and 2 show that the series converges by confirming that it satisfies the hypotheses of the alternating series test (Theorem 11.7.1).

1. $\sum_{k=1}^{\infty} \frac{(-1)^{k+1}}{2k+1}$ **2.** $\sum_{k=1}^{\infty} (-1)^{k+1} \frac{k}{3^k}$

In Exercises 3–6, determine whether the alternating series converges, and justify your answer.

3. $\sum_{k=1}^{\infty} (-1)^{k+1} \frac{k+1}{3k+1}$ **4.** $\sum_{k=1}^{\infty} (-1)^{k+1} \frac{k+1}{\sqrt{k}+1}$

5. $\sum_{k=1}^{\infty} (-1)^{k+1} e^{-k}$ **6.** $\sum_{k=3}^{\infty} (-1)^{k} \frac{\ln k}{k}$

In Exercises 7–12, use the ratio test for absolute convergence (Theorem 11.7.5) to determine whether the series converges or diverges. If the test is inconclusive, then say so.

7. $\sum_{k=1}^{\infty} \left(-\frac{3}{5}\right)^k$ **8.** $\sum_{k=1}^{\infty} (-1)^{k+1} \frac{2^k}{k!}$

9. $\sum_{k=1}^{\infty} (-1)^{k+1} \frac{3^k}{k^2}$ **10.** $\sum_{k=1}^{\infty} (-1)^{k} \frac{k}{5^k}$

11. $\sum_{k=1}^{\infty} (-1)^{k} \frac{k^3}{e^k}$ **12.** $\sum_{k=1}^{\infty} (-1)^{k+1} \frac{k^k}{k!}$

In Exercises 13–30, classify the series as absolutely convergent, conditionally convergent, or divergent.

13. $\sum_{k=1}^{\infty} \frac{(-1)^{k+1}}{3k}$ **14.** $\sum_{k=1}^{\infty} \frac{(-1)^{k+1}}{k^{4/3}}$ **15.** $\sum_{k=1}^{\infty} \frac{(-4)^k}{k^2}$

16. $\sum_{k=1}^{\infty} \frac{(-1)^{k+1}}{k!}$ **17.** $\sum_{k=1}^{\infty} \frac{\cos k\pi}{k}$ **18.** $\sum_{k=3}^{\infty} \frac{(-1)^k \ln k}{k}$

19. $\sum_{k=1}^{\infty} (-1)^{k+1} \frac{k+2}{k(k+3)}$ **20.** $\sum_{k=1}^{\infty} \frac{(-1)^{k+1} k^2}{k^3+1}$

21. $\sum_{k=1}^{\infty} \sin \frac{k\pi}{2}$ **22.** $\sum_{k=1}^{\infty} \frac{\sin k}{k^3}$

23. $\sum_{k=2}^{\infty} \frac{(-1)^k}{k \ln k}$ **24.** $\sum_{k=1}^{\infty} \frac{(-1)^k}{\sqrt{k(k+1)}}$

25. $\sum_{k=2}^{\infty} \left(-\frac{1}{\ln k}\right)^k$ **26.** $\sum_{k=1}^{\infty} \frac{(-1)^{k+1}}{\sqrt{k+1}+\sqrt{k}}$

27. $\sum_{k=2}^{\infty} \frac{(-1)^k (k^2+1)}{k^3+2}$ **28.** $\sum_{k=1}^{\infty} \frac{k \cos k\pi}{k^2+1}$

29. $\sum_{k=1}^{\infty} \frac{(-1)^{k+1} k!}{(2k-1)!}$ **30.** $\sum_{k=1}^{\infty} (-1)^{k+1} \frac{3^{2k-1}}{k^2+1}$

In Exercises 31–34, the series satisfies the hypotheses of the alternating series test. For the stated value of n, find an upper bound on the absolute error that results if the sum of the series is approximated by the nth partial sum.

31. $\sum_{k=1}^{\infty} \frac{(-1)^{k+1}}{k}$; $n = 7$ **32.** $\sum_{k=1}^{\infty} \frac{(-1)^{k+1}}{k!}$; $n = 5$

33. $\sum_{k=1}^{\infty} \frac{(-1)^{k+1}}{\sqrt{k}}$; $n = 99$

34. $\sum_{k=1}^{\infty} \frac{(-1)^{k+1}}{(k+1)\ln(k+1)}$; $n = 3$

In Exercises 35–38, the series satisfies the hypotheses of the alternating series test. Find a value of n for which the nth partial sum is ensured to approximate the sum of the series to the stated accuracy.

35. $\sum_{k=1}^{\infty} \frac{(-1)^{k+1}}{k}$; $|\text{error}| < 0.0001$

36. $\sum_{k=1}^{\infty} \frac{(-1)^{k+1}}{k!}$; $|\text{error}| < 0.00001$

37. $\sum_{k=1}^{\infty} \frac{(-1)^{k+1}}{\sqrt{k}}$; two decimal places

38. $\sum_{k=1}^{\infty} \frac{(-1)^{k+1}}{(k+1)\ln(k+1)}$; one decimal place

In Exercises 39 and 40, find an upper bound on the absolute error that results if s_{10} is used to approximate the sum of the given *geometric* series. Compute s_{10} rounded to four decimal places and compare this value with the exact sum of the series.

39. $\frac{3}{4} - \frac{3}{8} + \frac{3}{16} - \frac{3}{32} + \cdots$ **40.** $1 - \frac{2}{3} + \frac{4}{9} - \frac{8}{27} + \cdots$

In Exercises 41–44, the series satisfies the hypotheses of the alternating series test. Approximate the sum of the series to two decimal-place accuracy.

41. $1 - \frac{1}{3!} + \frac{1}{5!} - \frac{1}{7!} + \cdots$ **42.** $1 - \frac{1}{2!} + \frac{1}{4!} - \frac{1}{6!} + \cdots$

43. $\frac{1}{1 \cdot 2} - \frac{1}{2 \cdot 2^2} + \frac{1}{3 \cdot 2^3} - \frac{1}{4 \cdot 2^4} + \cdots$

44. $\frac{1}{1^5 + 4 \cdot 1} - \frac{1}{3^5 + 4 \cdot 3} + \frac{1}{5^5 + 4 \cdot 5} - \frac{1}{7^5 + 4 \cdot 7} + \cdots$

C **45.** The purpose of this exercise is to show that the error bound in part (*b*) of Theorem 11.7.2 can be overly conservative in certain cases.

(a) Use a CAS to confirm that
$$\frac{\pi}{4} = 1 - \frac{1}{3} + \frac{1}{5} - \frac{1}{7} + \cdots$$
(b) Use the CAS to show that $|(\pi/4) - s_{26}| < 10^{-2}$.
(c) According to the error bound in part (*b*) of Theorem 11.7.2, what value of n is required to ensure that $|(\pi/4) - s_n| < 10^{-2}$?

46. Show that the alternating p-series
$$1 - \frac{1}{2^p} + \frac{1}{3^p} - \frac{1}{4^p} + \cdots + (-1)^{k+1}\frac{1}{k^p} + \cdots$$
converges absolutely if $p > 1$, converges conditionally if $0 < p \leq 1$, and diverges if $p \leq 0$.

It can be proved that any series that is constructed from an absolutely convergent series by rearranging the terms is absolutely convergent and has the same sum as the original series. Use this fact together with parts (*a*) and (*b*) of Theorem 11.4.3 in Exercises 47 and 48.

47. It was stated in Exercise 27 of Section 11.4 that
$$\frac{\pi^2}{6} = 1 + \frac{1}{2^2} + \frac{1}{3^2} + \frac{1}{4^2} + \cdots$$
Use this to show that
$$\frac{\pi^2}{8} = 1 + \frac{1}{3^2} + \frac{1}{5^2} + \frac{1}{7^2} + \cdots$$

48. It was stated in Exercise 27 of Section 11.4 that
$$\frac{\pi^4}{90} = 1 + \frac{1}{2^4} + \frac{1}{3^4} + \frac{1}{4^4} + \cdots$$
Use this to show that
$$\frac{\pi^4}{96} = 1 + \frac{1}{3^4} + \frac{1}{5^4} + \frac{1}{7^4} + \cdots$$

49. It can be proved that the terms of any conditionally convergent series can be rearranged to give either a divergent series or a conditionally convergent series whose sum is any given number S. For example, we stated in Example 2 that
$$\ln 2 = 1 - \frac{1}{2} + \frac{1}{3} - \frac{1}{4} + \frac{1}{5} - \frac{1}{6} + \cdots$$
Show that we can rearrange this series so that its sum is $\frac{1}{2}\ln 2$ by rewriting it as
$$\left(1 - \frac{1}{2} - \frac{1}{4}\right) + \left(\frac{1}{3} - \frac{1}{6} - \frac{1}{8}\right) + \left(\frac{1}{5} - \frac{1}{10} - \frac{1}{12}\right) + \cdots$$
[*Hint:* Add the first two terms in each set of parentheses.]

50. (a) Use a graphing utility to graph
$$f(x) = \frac{4x-1}{4x^2-2x}, \quad x \geq 1$$
(b) Based on your graph, do think that the series
$$\sum_{k=1}^{\infty}(-1)^{k+1}\frac{4k-1}{4k^2-2k}$$
converges? Explain your reasoning.

51. As illustrated in the accompanying figure, a bug, starting at point A on a 180-cm wire, walks the length of the wire, stops and walks in the opposite direction for half the length of the wire, stops again and walks in the opposite direction for one-third the length of the wire, stops again and walks in the opposite direction for one-fourth the length of the wire, and so forth until it stops for the 1000th time.
(a) Give upper and lower bounds on the distance between the bug and point A when it finally stops. [*Hint:* As stated in Example 2, assume that the sum of the alternating harmonic series is ln 2.]
(b) Give upper and lower bounds on the total distance that the bug has traveled when it finally stops. [*Hint:* Use inequality (2) of Section 11.4.]

Figure Ex-51

52. (a) Prove that if $\sum a_k$ converges absolutely, then $\sum a_k^2$ converges.
(b) Show that the converse of part (a) is false by giving a counterexample.

11.8 POWER SERIES

In the last two sections we focused exclusively on series whose terms are numbers. In this section we will consider series whose terms are functions with the objective of developing the mathematical tools needed to investigate the convergence of Taylor and Maclaurin series.

POWER SERIES IN x

If $c_0, c_1, c_2, \ldots$ are constants and x is a variable, then a series of the form
$$\sum_{k=0}^{\infty} c_k x^k = c_0 + c_1 x + c_2 x^2 + \cdots + c_k x^k + \cdots \tag{1}$$

is called a ***power series in x***. Some examples are

$$\sum_{k=0}^{\infty} x^k = 1 + x + x^2 + x^3 + \cdots$$

$$\sum_{k=0}^{\infty} \frac{x^k}{k!} = 1 + x + \frac{x^2}{2!} + \frac{x^3}{3!} + \cdots$$

$$\sum_{k=0}^{\infty} (-1)^k \frac{x^{2k}}{(2k)!} = 1 - \frac{x^2}{2!} + \frac{x^4}{4!} - \frac{x^6}{6!} + \cdots$$

More generally, every Maclaurin series

$$\sum_{k=0}^{\infty} \frac{f^{(k)}(0)}{k!} x^k = f(0) + f'(0)x + \frac{f''(0)}{2!}x^2 + \cdots + \frac{f^{(k)}(0)}{k!}x^k + \cdots$$

is a power series in x.

RADIUS AND INTERVAL OF CONVERGENCE

If a numerical value is substituted for x in a power series $\sum c_k x^k$, then the resulting series of numbers may either converge or diverge. This leads to the problem of determining the set of x-values for which a given power series converges; this is called its ***convergence set***.

Observe that every power series in x converges at $x = 0$, since substituting this value in (1) produces the series

$$c_0 + 0 + 0 + 0 + \cdots + 0 + \cdots$$

whose sum is c_0. In rare cases $x = 0$ may be the only point in the convergence set, but more usually the convergence set is some finite or infinite interval containing $x = 0$. This is the content of the following theorem, whose proof will be omitted.

> **11.8.1** THEOREM. *For any power series in x, exactly one of the following is true:*
>
> (*a*) *The series converges only for $x = 0$.*
>
> (*b*) *The series converges absolutely (and hence converges) for all real values of x.*
>
> (*c*) *The series converges absolutely (and hence converges) for all x in some finite open interval $(-R, R)$, and diverges if $x < -R$ or $x > R$. At either of the points $x = R$ or $x = -R$, the series may converge absolutely, converge conditionally, or diverge, depending on the particular series.*

This theorem states that the convergence set for a power series in x is always an interval centered at $x = 0$ (possibly just the point $x = 0$ itself or possibly infinite). For this reason, the convergence set of a power series in x is called the ***interval of convergence***. In the case where the convergence set is the single point $x = 0$ we say that the series has ***radius of convergence* 0**, in the case where the convergence set is $(-\infty, +\infty)$ we say that the series has ***radius of convergence*** $+\infty$, and in the case where the convergence set extends between $-R$ and R we say that the series has ***radius of convergence R*** (Figure 11.8.1).

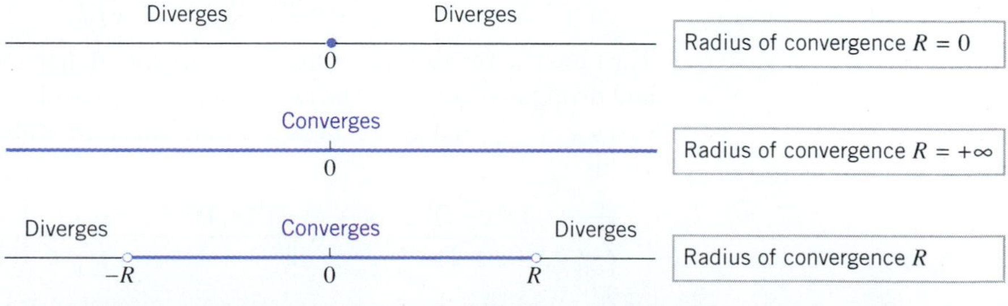

Figure 11.8.1

FINDING THE INTERVAL OF CONVERGENCE

The usual procedure for finding the interval of convergence of a power series is to apply the ratio test for absolute convergence (Theorem 11.7.5). The following example illustrates how this works.

Example 1

Find the interval of convergence and radius of convergence of the following power series.

$$\text{(a)} \sum_{k=0}^{\infty} x^k \qquad \text{(b)} \sum_{k=0}^{\infty} \frac{x^k}{k!} \qquad \text{(c)} \sum_{k=0}^{\infty} k!x^k \qquad \text{(d)} \sum_{k=0}^{\infty} \frac{(-1)^k x^k}{3^k(k+1)}$$

Solution (a). We apply the ratio test for absolute convergence. We have

$$\rho = \lim_{k\to+\infty} \left|\frac{u_{k+1}}{u_k}\right| = \lim_{k\to+\infty} \left|\frac{x^{k+1}}{x^k}\right| = \lim_{k\to+\infty} |x| = |x|$$

so the series converges absolutely if $\rho = |x| < 1$ and diverges if $\rho = |x| > 1$. The test is inconclusive if $|x| = 1$ (i.e., if $x = 1$ or $x = -1$), which means that we will have to investigate convergence at these points separately. At these points the series becomes

$$\sum_{k=0}^{\infty} 1^k = 1 + 1 + 1 + 1 + \cdots \qquad \boxed{x = 1}$$

$$\sum_{k=0}^{\infty} (-1)^k = 1 - 1 + 1 - 1 + \cdots \qquad \boxed{x = -1}$$

both of which diverge; thus, the interval of convergence for the given power series is $(-1, 1)$, and the radius of convergence is $R = 1$.

Solution (b). Applying the ratio test for absolute convergence, we obtain

$$\rho = \lim_{k\to+\infty} \left|\frac{u_{k+1}}{u_k}\right| = \lim_{k\to+\infty} \left|\frac{x^{k+1}}{(k+1)!} \cdot \frac{k!}{x^k}\right| = \lim_{k\to+\infty} \left|\frac{x}{k+1}\right| = 0$$

Since $\rho < 1$ for all x, the series converges absolutely for all x. Thus, the interval of convergence is $(-\infty, +\infty)$ and the radius of convergence is $R = +\infty$.

Solution (c). If $x \neq 0$, then the ratio test for absolute convergence yields

$$\rho = \lim_{k\to+\infty} \left|\frac{u_{k+1}}{u_k}\right| = \lim_{k\to+\infty} \left|\frac{(k+1)!x^{k+1}}{k!x^k}\right| = \lim_{k\to+\infty} |(k+1)x| = +\infty$$

Therefore, the series diverges for all nonzero values of x. Thus, the interval of convergence is the single point $x = 0$ and the radius of convergence is $R = 0$.

Solution (d). Since $|(-1)^k| = |(-1)^{k+1}| = 1$, we obtain

$$\begin{aligned}
\rho = \lim_{k\to+\infty} \left|\frac{u_{k+1}}{u_k}\right| &= \lim_{k\to+\infty} \left|\frac{x^{k+1}}{3^{k+1}(k+2)} \cdot \frac{3^k(k+1)}{x^k}\right| \\
&= \lim_{k\to+\infty} \left[\frac{|x|}{3} \cdot \left(\frac{k+1}{k+2}\right)\right] \\
&= \frac{|x|}{3} \lim_{k\to+\infty} \left(\frac{1 + (1/k)}{1 + (2/k)}\right) = \frac{|x|}{3}
\end{aligned}$$

The ratio test for absolute convergence implies that the series converges absolutely if $|x| < 3$ and diverges if $|x| > 3$. The ratio test fails to provide any information when $|x| = 3$, so the cases $x = -3$ and $x = 3$ need separate analyses. Substituting $x = -3$ in the given series yields

$$\sum_{k=0}^{\infty} \frac{(-1)^k(-3)^k}{3^k(k+1)} = \sum_{k=0}^{\infty} \frac{(-1)^k(-1)^k 3^k}{3^k(k+1)} = \sum_{k=0}^{\infty} \frac{1}{k+1}$$

which is the divergent harmonic series $1 + \frac{1}{2} + \frac{1}{3} + \frac{1}{4} + \cdots$. Substituting $x = 3$ in the given series yields

$$\sum_{k=0}^{\infty} \frac{(-1)^k 3^k}{3^k(k+1)} = \sum_{k=0}^{\infty} \frac{(-1)^k}{k+1} = 1 - \frac{1}{2} + \frac{1}{3} - \frac{1}{4} + \cdots$$

which is the conditionally convergent alternating harmonic series. Thus, the interval of convergence for the given series is $(-3, 3]$ and the radius of convergence is $R = 3$. ◀

POWER SERIES IN $x - x_0$

If x_0 is a constant, and if x is replaced by $x - x_0$ in (1), then the resulting series has the form

$$\sum_{k=0}^{\infty} c_k(x - x_0)^k = c_0 + c_1(x - x_0) + c_2(x - x_0)^2 + \cdots + c_k(x - x_0)^k + \cdots$$

This is called a ***power series in $x - x_0$***. Some examples are

$$\sum_{k=0}^{\infty} \frac{(x-1)^k}{k+1} = 1 + \frac{(x-1)}{2} + \frac{(x-1)^2}{3} + \frac{(x-1)^3}{4} + \cdots \qquad \boxed{x_0 = 1}$$

$$\sum_{k=0}^{\infty} \frac{(-1)^k(x+3)^k}{k!} = 1 - (x+3) + \frac{(x+3)^2}{2!} - \frac{(x+3)^3}{3!} + \cdots \qquad \boxed{x_0 = -3}$$

The first of these is a power series in $x - 1$ and the second is a power series in $x + 3$. Note that a power series in x is a power series in $x - x_0$ in which $x_0 = 0$. More generally, the Taylor series

$$\sum_{k=0}^{\infty} \frac{f^{(k)}(x_0)}{k!}(x - x_0)^k$$

is a power series in $x - x_0$.

The main result on convergence of a power series in $x - x_0$ can be obtained by substituting $x - x_0$ for x in Theorem 11.8.1. This leads to the following theorem.

> **11.8.2 THEOREM.** *For a power series $\sum c_k(x - x_0)^k$, exactly one of the following statements is true:*
>
> (*a*) *The series converges only for $x = x_0$.*
>
> (*b*) *The series converges absolutely (and hence converges) for all real values of x.*
>
> (*c*) *The series converges absolutely (and hence converges) for all x in some finite open interval $(x_0 - R, x_0 + R)$ and diverges if $x < x_0 - R$ or $x > x_0 + R$. At either of the points $x = x_0 - R$ or $x = x_0 + R$, the series may converge absolutely, converge conditionally, or diverge, depending on the particular series.*

It follows from this theorem that the set of values for which a power series in $x - x_0$ converges is always an interval centered at $x = x_0$; we call this the ***interval of convergence*** (Figure 11.8.2). In part (*a*) of Theorem 11.8.2 the interval of convergence reduces to the single point $x = x_0$, in which case we say that the series has ***radius of convergence $R = 0$***; in part (*b*) the interval of convergence is infinite (the entire real line), in which case we say that

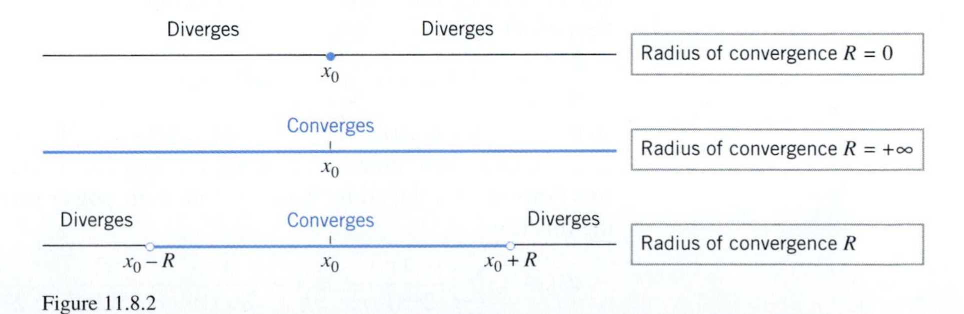

Figure 11.8.2

the series has ***radius of convergence*** $R = +\infty$; and in part (*c*) the interval extends between $x_0 - R$ and $x_0 + R$, in which case we say that the series has ***radius of convergence*** R.

Example 2

Find the interval of convergence and radius of convergence of the series

$$\sum_{k=1}^{\infty} \frac{(x-5)^k}{k^2}$$

Solution. We apply the ratio test for absolute convergence.

$$\rho = \lim_{k \to +\infty} \left| \frac{u_{k+1}}{u_k} \right| = \lim_{k \to +\infty} \left| \frac{(x-5)^{k+1}}{(k+1)^2} \cdot \frac{k^2}{(x-5)^k} \right|$$

$$= \lim_{k \to +\infty} \left[|x-5| \left(\frac{k}{k+1} \right)^2 \right]$$

$$= |x-5| \lim_{k \to +\infty} \left(\frac{1}{1+(1/k)} \right)^2 = |x-5|$$

Thus, the series converges absolutely if $|x-5| < 1$, or $-1 < x-5 < 1$, or $4 < x < 6$. The series diverges if $x < 4$ or $x > 6$.

To determine the convergence behavior at the endpoints $x = 4$ and $x = 6$, we substitute these values in the given series. If $x = 6$, the series becomes

$$\sum_{k=1}^{\infty} \frac{1^k}{k^2} = \sum_{k=1}^{\infty} \frac{1}{k^2} = 1 + \frac{1}{2^2} + \frac{1}{3^2} + \frac{1}{4^2} + \cdots$$

which is a convergent p-series ($p = 2$). If $x = 4$, the series becomes

$$\sum_{k=1}^{\infty} \frac{(-1)^k}{k^2} = -1 + \frac{1}{2^2} - \frac{1}{3^2} + \frac{1}{4^2} - \cdots$$

Since this series converges absolutely, the interval of convergence for the given series is $[4, 6]$. The radius of convergence is $R = 1$ (Figure 11.8.3). ◀

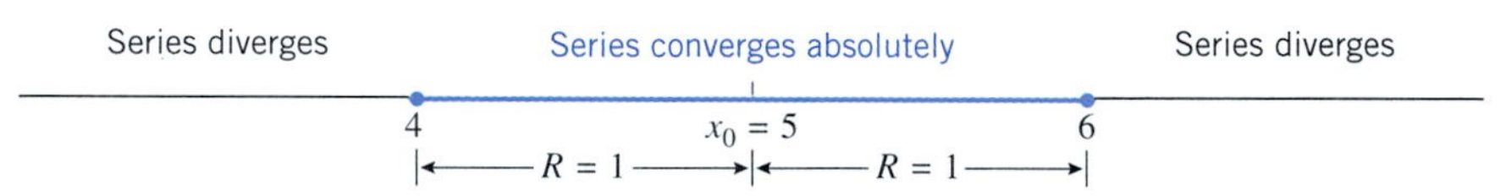

Figure 11.8.3

FOR THE READER. It will always be a waste of time to test for convergence at the endpoints of the interval of convergence using the ratio test, since ρ will always be 1 at those points if $\rho = \lim_{n \to +\infty} |a_{n+1}/a_n|$ exists. Explain why this must be so.

FUNCTIONS DEFINED BY POWER SERIES

If a function f is expressed as a power series on some interval, then we say that f is ***represented*** by the power series on that interval. For example, we saw in Example 4 of Section 11.3 that

$$\frac{1}{1-x} = 1 + x + x^2 + \cdots + x^k + \cdots$$

so that this power series represents the function $1/(1-x)$ on the interval $-1 < x < 1$.

Sometimes new functions actually originate as power series, and the properties of the functions are developed by working with their power series representations. For example, the functions

$$J_0(x) = \sum_{k=0}^{\infty} \frac{(-1)^k x^{2k}}{2^{2k}(k!)^2} = 1 - \frac{x^2}{2^2(1!)^2} + \frac{x^4}{2^4(2!)^2} - \frac{x^6}{2^6(3!)^2} + \cdots \qquad (2)$$

and

$$J_1(x) = \sum_{k=0}^{\infty} \frac{(-1)^k x^{2k+1}}{2^{2k+1}(k!)(k+1)!} = \frac{x}{2} - \frac{x^3}{2^3(1!)(2!)} + \frac{x^5}{2^5(2!)(3!)} - \cdots \tag{3}$$

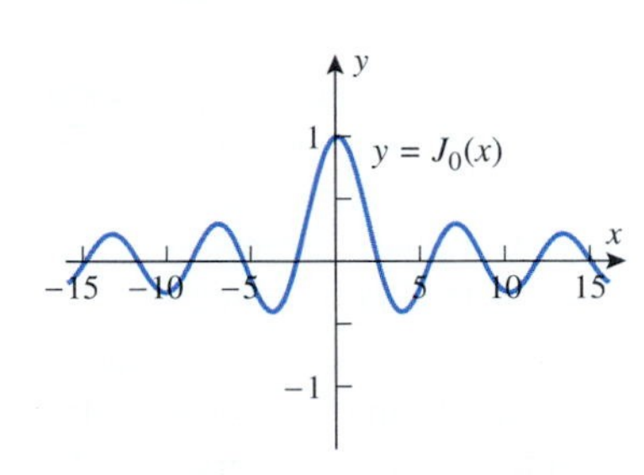

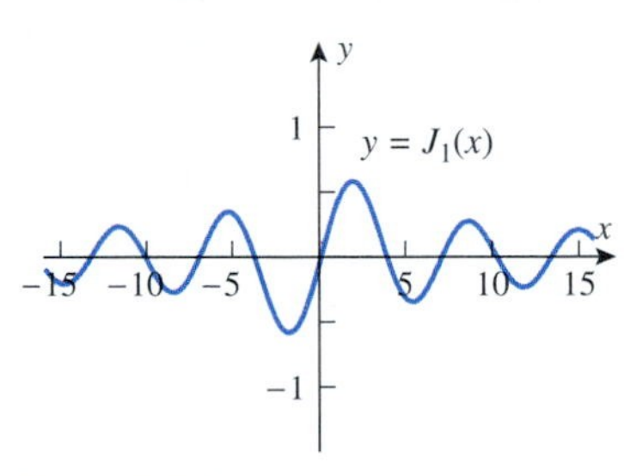

Figure 11.8.4

which are called ***Bessel functions*** in honor of the German mathematician and astronomer Friedrich Wilhelm Bessel (1784–1846), arise naturally in the study of planetary motion and in various problems that involve heat flow.

To find the domains of these functions, we must determine where their defining power series converge. For example, in the case $J_0(x)$ we have

$$\rho = \lim_{k\to+\infty}\left|\frac{u_{k+1}}{u_k}\right| = \lim_{k\to+\infty}\left|\frac{x^{2(k+1)}}{2^{2(k+1)}[(k+1)!]^2}\cdot\frac{2^{2k}(k!)^2}{x^{2k}}\right|$$

$$= \lim_{k\to+\infty}\left|\frac{x^2}{4(k+1)^2}\right| = 0 < 1$$

so that the series converges for all x; that is, the domain of $J_0(x)$ is $(-\infty, +\infty)$. We leave it as an exercise to show that the power series for $J_1(x)$ also converges for all x.

FOR THE READER. Many CAS programs have the Bessel functions as part of their libraries. If you have a CAS, read the documentation to determine whether it can graph $J_0(x)$ and $J_1(x)$; if so, generate the graphs shown in Figure 11.8.4.

EXERCISE SET 11.8 Graphing Calculator

In Exercises 1–4, find the interval of convergence of the power series, and find a familiar function that is represented by the power series on that interval.

1. $1 - x + x^2 - x^3 + \cdots + (-1)^k x^k + \cdots$

2. $1 + x^2 + x^4 + \cdots + x^{2k} + \cdots$

3. $1 + (x-2) + (x-2)^2 + \cdots + (x-2)^k + \cdots$

4. $1 - (x+3) + (x+3)^2 - (x+3)^3 + \cdots + (-1)^k(x+3)^k + \cdots$

5. Suppose that the function f is represented by the power series

$$f(x) = 1 - \frac{x}{2} + \frac{x^2}{4} - \frac{x^3}{8} + \cdots + (-1)^k\frac{x^k}{2^k} + \cdots$$

(a) Find the domain of f.
(b) Find $f(0)$ and $f(1)$.

6. Suppose that the function f is represented by the power series

$$f(x) = 1 - \frac{x-5}{3} + \frac{(x-5)^2}{3^2} - \frac{(x-5)^3}{3^3} + \cdots$$

(a) Find the domain of f.
(b) Find $f(3)$ and $f(6)$.

In Exercises 7–30, find the radius of convergence and the interval of convergence.

7. $\sum_{k=0}^{\infty} \frac{x^k}{k+1}$ **8.** $\sum_{k=0}^{\infty} 3^k x^k$ **9.** $\sum_{k=0}^{\infty} \frac{(-1)^k x^k}{k!}$

10. $\sum_{k=0}^{\infty} \frac{k!}{2^k} x^k$ **11.** $\sum_{k=1}^{\infty} \frac{5^k}{k^2} x^k$ **12.** $\sum_{k=2}^{\infty} \frac{x^k}{\ln k}$

13. $\sum_{k=1}^{\infty} \frac{x^k}{k(k+1)}$ **14.** $\sum_{k=0}^{\infty} \frac{(-2)^k x^{k+1}}{k+1}$

15. $\sum_{k=1}^{\infty} (-1)^{k-1}\frac{x^k}{\sqrt{k}}$ **16.** $\sum_{k=0}^{\infty} \frac{(-1)^k x^{2k}}{(2k)!}$

17. $\sum_{k=0}^{\infty} (-1)^k \frac{x^{2k+1}}{(2k+1)!}$ **18.** $\sum_{k=1}^{\infty} (-1)^k \frac{x^{3k}}{k^{3/2}}$

19. $\sum_{k=0}^{\infty} \frac{3^k}{k!} x^k$ **20.** $\sum_{k=2}^{\infty} (-1)^{k+1} \frac{x^k}{k(\ln k)^2}$

21. $\sum_{k=0}^{\infty} \frac{x^k}{1+k^2}$ **22.** $\sum_{k=0}^{\infty} \frac{(x-3)^k}{2^k}$

23. $\sum_{k=1}^{\infty} (-1)^{k+1} \frac{(x+1)^k}{k}$ **24.** $\sum_{k=0}^{\infty} (-1)^k \frac{(x-4)^k}{(k+1)^2}$

25. $\sum_{k=0}^{\infty} \left(\frac{3}{4}\right)^k (x+5)^k$ **26.** $\sum_{k=1}^{\infty} \frac{(2k+1)!}{k^3}(x-2)^k$

27. $\sum_{k=1}^{\infty} (-1)^k \frac{(x+1)^{2k+1}}{k^2+4}$ **28.** $\sum_{k=1}^{\infty} \frac{(\ln k)(x-3)^k}{k}$

29. $\sum_{k=0}^{\infty} \frac{\pi^k (x-1)^{2k}}{(2k+1)!}$ **30.** $\sum_{k=0}^{\infty} \frac{(2x-3)^k}{4^{2k}}$

31. Use the root test to find the interval of convergence of

$$\sum_{k=2}^{\infty} \frac{x^k}{(\ln k)^k}$$

32. Find the domain of the function

$$f(x) = \sum_{k=1}^{\infty} \frac{1\cdot 3\cdot 5\cdots(2k-1)}{(2k-2)!} x^k$$

33. If a function f is represented by a power series on an interval, then the graphs of the partial sums can be used as approximations to the graph of f.
(a) Use a graphing utility to generate the graph of $1/(1-x)$ together with the graphs of the first four partial sums of its Maclaurin series over the interval $(-1, 1)$.
(b) In general terms, where are the graphs of the partial sums the most accurate?

34. Show that the power series representation of the Bessel function $J_1(x)$ converges for all x [Formula (3)].

35. Show that if p is a positive integer, then the power series

$$\sum_{k=0}^{\infty} \frac{(pk)!}{(k!)^p} x^k$$

has a radius of convergence of $1/p^p$.

36. Show that if p and q are positive integers, then the power series

$$\sum_{k=0}^{\infty} \frac{(k+p)!}{k!(k+q)!} x^k$$

has a radius of convergence of $+\infty$.

37. (a) Suppose that the power series $\sum c_k(x-x_0)^k$ has radius of convergence R and p is a nonzero constant. What can you say about the radius of convergence of the power series $\sum pc_k(x-x_0)^k$? Explain your reasoning. [*Hint:* See Theorem 11.4.3.]
(b) Suppose that the power series $\sum c_k(x-x_0)^k$ has a finite radius of convergence R, and the power series $\sum d_k(x-x_0)^k$ has a radius of convergence of $+\infty$. What can you say about the radius of convergence of $\sum(c_k+d_k)(x-x_0)^k$? Explain your reasoning.
(c) Suppose that the power series $\sum c_k(x-x_0)^k$ has a finite radius of convergence R_1 and the power series $\sum d_k(x-x_0)^k$ has a finite radius of convergence R_2. What can you say about the radius of convergence of $\sum(c_k+d_k)(x-x_0)^k$? Explain your reasoning.

38. Prove: If $\lim_{k\to+\infty} |c_k|^{1/k} = L$, where $L \neq 0$, then $1/L$ is the radius of convergence of the power series $\sum_{k=0}^{\infty} c_k x^k$.

39. Prove: If the power series $\sum_{k=0}^{\infty} c_k x^k$ has radius of convergence R, then the series $\sum_{k=0}^{\infty} c_k x^{2k}$ has radius of convergence $\sqrt{R}$.

40. Prove: If the interval of convergence of the series $\sum_{k=0}^{\infty} c_k(x-x_0)^k$ is $(x_0-R, x_0+R]$, then the series converges conditionally at x_0+R.

11.9 CONVERGENCE OF TAYLOR SERIES; COMPUTATIONAL METHODS

In Section 11.5 we anticipated the possibility that a Taylor series for a function might actually converge to the function on some interval. In this section we will study the convergence of Taylor series, and we will show how they can be used to approximate trigonometric, exponential, and logarithmic functions.

THE nTH REMAINDER

Recall that the nth Taylor polynomial for a function f about $x = x_0$ has the property that its value and the values of its first n derivatives match those of f at x_0. As n increases, more and more derivatives match up, so it is reasonable to hope that for values of x near x_0 the values of the Taylor polynomials might converge to the value of $f(x)$; that is,

$$\sum_{k=0}^{n} \frac{f^{(k)}(x_0)}{k!}(x-x_0)^k \to f(x) \quad \text{as } n\to+\infty \tag{1}$$

However, the nth Taylor polynomial for f is the nth partial sum of the Taylor series for f, so (1) is equivalent to stating that the Taylor series for f converges at the point x, and its sum is $f(x)$. Thus, we are led to consider the following problem.

11.9.1 PROBLEM. Given a function f that has derivatives of all orders at a point x_0, determine whether there is an open interval containing x_0 such that $f(x)$ is the sum of its Taylor series about $x = x_0$ at each point in the interval; that is,

$$f(x) = \sum_{k=0}^{\infty} \frac{f^{(k)}(x_0)}{k!}(x-x_0)^k \tag{2}$$

FOR THE READER. Show that (2) holds at $x = x_0$, regardless of the function f.

To determine whether (2) holds on some open interval containing x_0, it will be convenient to consider the difference between $f(x)$ and its nth Taylor polynomial about $x = x_0$. This difference is called the ***nth remainder for f about x = x₀***, and is denoted by

$$R_n(x) = f(x) - \sum_{k=0}^{n} \frac{f^{(k)}(x_0)}{k!}(x - x_0)^k \tag{3}$$

This can also be written as

$$f(x) = \sum_{k=0}^{n} \frac{f^{(k)}(x_0)}{k!}(x - x_0)^k + R_n(x) \tag{4}$$

which is called ***Taylor's formula with remainder***.

One can think of $R_n(x)$ as the error that results at the point x when f is approximated by its nth Taylor polynomial. Thus, for the Taylor polynomials about x_0 to converge to f at a point x as $n \to +\infty$, the error $R_n(x)$ must approach 0; conversely, if $R_n(x) \to 0$ as $n \to +\infty$, then the Taylor polynomials converge to f at the point x. More precisely, we have the following theorem.

11.9.2 THEOREM. *The equality*

$$f(x) = \sum_{k=0}^{\infty} \frac{f^{(k)}(x_0)}{k!}(x - x_0)^k$$

holds at a point x if and only if $\lim_{n \to +\infty} R_n(x) = 0$.

ESTIMATING THE nTH REMAINDER

It is relatively rare that one can prove directly that $R_n(x) \to 0$ as $n \to +\infty$. Usually, this is proved indirectly by finding appropriate bounds on $|R_n(x)|$ and applying the Squeezing Theorem for Sequences. The following theorem, which is proved in Appendix G, provides a bound that can be used for this purpose.

11.9.3 THEOREM (*The Remainder Estimation Theorem*). *If the function f can be differentiated $n+1$ times on an interval I containing the point x_0, and if $|f^{(n+1)}(x)| \le M$ for all x in I, then*

$$|R_n(x)| \le \frac{M}{(n+1)!}|x - x_0|^{n+1} \tag{5}$$

for all x in I.

The following example illustrates how this theorem is applied.

Example 1

Show that the Maclaurin series for $\cos x$ converges to $\cos x$ for all x; that is,

$$\cos x = \sum_{k=0}^{\infty} (-1)^k \frac{x^{2k}}{(2k)!} = 1 - \frac{x^2}{2!} + \frac{x^4}{4!} - \frac{x^6}{6!} + \cdots \qquad (-\infty < x < +\infty)$$

Solution. From Theorem 11.9.2 we must show that $R_n(x) \to 0$ for all x as $n \to +\infty$. For this purpose let $f(x) = \cos x$, so that for all x we have

$$f^{(n+1)}(x) = \pm \cos x \quad \text{or} \quad f^{(n+1)}(x) = \pm \sin x$$

In all cases we have

$$|f^{(n+1)}(x)| \leq 1$$

so we can apply Theorem 11.9.3 with $M = 1$ and $x_0 = 0$ to conclude that

$$0 \leq |R_n(x)| \leq \frac{|x|^{n+1}}{(n+1)!} \tag{6}$$

However, it follows from Formula (5) of Section 11.2 with $n + 1$ in place of n and $|x|$ in place of x that

$$\lim_{n\to+\infty} \frac{|x|^{n+1}}{(n+1)!} = 0 \tag{7}$$

Thus, it follows from (6) and the Squeezing Theorem for Sequences (Theorem 11.1.5) that $|R_n(x)| \to 0$ as $n \to +\infty$; this implies that $R_n(x) \to 0$ as $n \to +\infty$ by Theorem 11.1.6. Since this is true for all x, we have proved that the Maclaurin series for $\cos x$ converges to $\cos x$ for all x. This is illustrated in Figure 11.9.1, where we can see how successive partial sums approximate the cosine curve more and more closely. ◀

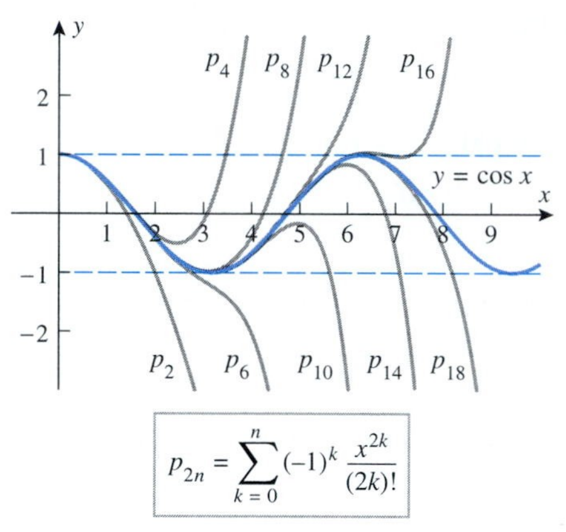

Figure 11.9.1

REMARK. The method used in Example 1 can be easily modified to prove that the Taylor series for $\cos x$ about any point $x = x_0$ converges to $\cos x$ for all x, and similarly that the Taylor series for $\sin x$ about any point $x = x_0$ converges to $\sin x$ for all x (Exercises 25 and 26). For reference, there is a list of some of the most important Maclaurin series in Table 11.9.1 at the end of this section.

APPROXIMATING TRIGONOMETRIC FUNCTIONS

In general, to approximate the value of a function f at a point x using a Taylor series, there are two basic questions that must be answered:

- About what point x_0 should the Taylor series be expanded?
- How many terms in the series should be used to achieve the desired accuracy?

In response to the first question, x_0 needs to be a point where the derivatives of f can be evaluated easily, since these values are needed for the coefficients in the Taylor series. Furthermore, if the function f is being evaluated at the point x, then x_0 should be chosen as close as possible to x, since Taylor series tend to converge more rapidly near x_0. For example, to approximate $\sin 3°$ $(= \pi/60$ radians), it would be reasonable to take $x_0 = 0$, since $\pi/60$ is close to 0 and the derivatives of $\sin x$ are easy to evaluate at 0. On the other hand, to approximate $\sin 85°$ $(= 17\pi/36$ radians), it would be more natural to take $x_0 = \pi/2$, since $17\pi/36$ is close to $\pi/2$ and the derivatives of $\sin x$ are easy to evaluate at $\pi/2$.

In response to the second question posed above, the number of terms required to achieve a specific accuracy needs to be determined on a problem-by-problem basis. The next example gives two methods for doing this.

Example 2

Use the Maclaurin series for $\sin x$ to approximate $\sin 3^\circ$ to five decimal-place accuracy.

Solution. In the Maclaurin series

$$\sin x = \sum_{k=0}^{\infty}(-1)^k \frac{x^{2k+1}}{(2k+1)!} = x - \frac{x^3}{3!} + \frac{x^5}{5!} - \frac{x^7}{7!} + \cdots \tag{8}$$

the angle x is assumed to be in radians (because the differentiation formulas for the trigonometric functions were derived with this assumption). Since $3^\circ = \pi/60$ radians, it follows from (8) that

$$\sin 3^\circ = \sin\frac{\pi}{60} = \left(\frac{\pi}{60}\right) - \frac{(\pi/60)^3}{3!} + \frac{(\pi/60)^5}{5!} - \frac{(\pi/60)^7}{7!} + \cdots \tag{9}$$

We must now determine how many terms in the series are required to achieve five decimal-place accuracy. We will consider two possible approaches, one using the Remainder Estimation Theorem (Theorem 11.9.3) and the other using the fact that (9) satisfies the hypotheses of the alternating series test (Theorem 11.7.1).

Method 1 (The Remainder Estimation Theorem). Since we want to achieve five decimal-place accuracy, our goal is to choose n so that the absolute value of the nth remainder at $x = \pi/60$ does not exceed $0.000005 = 5 \times 10^{-5}$; that is,

$$\left|R_n\left(\frac{\pi}{60}\right)\right| \le 0.000005 \tag{10}$$

However, if we let $f(x) = \sin x$, then $f^{(n+1)}(x)$ is either $\pm\sin x$ or $\pm\cos x$, and in either case $|f^{(n+1)}(x)| \le 1$ for all x. Thus, it follows from the Remainder Estimation Theorem with $M = 1$, $x_0 = 0$, and $x = \pi/60$ that

$$\left|R_n\left(\frac{\pi}{60}\right)\right| \le \frac{|\pi/60|^{n+1}}{(n+1)!}$$

Thus, we can satisfy (10) by choosing n so that

$$\frac{|\pi/60|^{n+1}}{(n+1)!} \le 0.000005$$

With the help of a calculating utility you can verify that the smallest value of n that meets this criterion is $n = 3$. Thus, to achieve five decimal-place accuracy we need only keep terms up to the third power in (9). This yields

$$\sin 3^\circ \approx \left(\frac{\pi}{60}\right) - \frac{(\pi/60)^3}{3!} \approx 0.05234 \tag{11}$$

(verify). As a check, the author's calculator gives $\sin 3^\circ \approx 0.05233595624$, which agrees with (11) when rounded to five decimal places.

Method 2 (The Alternating Series Test). We leave it for you to check that (9) satisfies the hypotheses of the alternating series test (Theorem 11.7.1).

Let s_n denote the sum of the terms in (9) up to and including the nth power of $\pi/60$. Since the exponents in the series are odd integers, the integer n must be odd, and the exponent of the first term *not* included in the sum s_n must be $n + 2$. Thus, it follows from part (b) of Theorem 11.7.2 that

$$|\sin 3^\circ - s_n| < \frac{(\pi/60)^{n+2}}{(n+2)!}$$

This means that for five decimal-place accuracy we must look for the first positive odd integer n such that

$$\frac{(\pi/60)^{n+2}}{(n+2)!} \leq 0.000005$$

With the help of a calculating utility you can verify that the smallest value of n that meets this criterion is $n = 3$. This agrees with the result obtained above using the Remainder Estimation Theorem and hence leads to approximation (11) as before. ◀

ROUNDOFF AND TRUNCATION ERROR

There are two types of errors that occur when computing with series. The first, called ***truncation error***, is the error that results when a series is approximated by a partial sum; and the second, called ***roundoff error***, is the error that arises from approximations in numerical computations. For example, in our derivation of (11) we took $n = 3$ to keep the truncation error below 0.000005. However, to evaluate the partial sum we had to approximate π, thereby introducing roundoff error. Had we not exercised some care in choosing this approximation, the roundoff error could easily have degraded the final result.

Methods for estimating and controlling roundoff error are studied in a branch of mathematics called ***numerical analysis***. However, as a rule of thumb, to achieve n decimal-place accuracy in a final result, all intermediate calculations must be accurate to at least $n + 1$ decimal places. Thus, in (11) at least six decimal-place accuracy in π is required to achieve the five decimal-place accuracy in the final numerical result. As a practical matter, a good working procedure is to perform all intermediate computations with the maximum number of digits that your calculating utility can handle and then round at the end.

APPROXIMATING EXPONENTIAL FUNCTIONS

Example 3

Show that the Maclaurin series for e^x converges to e^x for all x; that is,

$$e^x = \sum_{k=0}^{\infty} \frac{x^k}{k!} = 1 + x + \frac{x^2}{2!} + \frac{x^3}{3!} + \cdots + \frac{x^k}{k!} + \cdots \qquad (-\infty < x < +\infty)$$

Solution. Let $f(x) = e^x$, so that

$$f^{(n+1)}(x) = e^x$$

We want to show that $R_n(x) \to 0$ as $n \to +\infty$ for all x in the interval $-\infty < x < +\infty$. However, it will be helpful here to consider the cases $x \leq 0$ and $x > 0$ separately. If $x \leq 0$, then we will take the interval I in Theorem 11.9.3 to be $[x, 0]$, and if $x > 0$, then we will take it to be $[0, x]$. Since $f^{(n+1)}(x) = e^x$ is an increasing function, it follows that if c is in the interval $[x, 0]$, then

$$|f^{(n+1)}(c)| \leq |f^{(n+1)}(0)| = e^0 = 1$$

and if c is in the interval $[0, x]$, then

$$|f^{(n+1)}(c)| \leq |f^{(n+1)}(x)| = e^x$$

Thus, we can apply Theorem 11.9.3 with $M = 1$ in the case where $x \leq 0$ and with $M = e^x$ in the case where $x > 0$. This yields

$$0 \leq |R_n(x)| \leq \frac{|x|^{n+1}}{(n+1)!} \qquad \text{if } x \leq 0$$

$$0 \leq |R_n(x)| \leq e^x \frac{|x|^{n+1}}{(n+1)!} \qquad \text{if } x > 0$$

Thus, in both cases it follows from (7) and the Squeezing Theorem for Sequences that $|R_n(x)| \to 0$ as $n \to +\infty$, which in turn implies that $R_n(x) \to 0$ as $n \to +\infty$. Since this is true for all x, we have proved that the Maclaurin series for e^x converges to e^x for all x. ◀

Example 4

Use the Maclaurin series for e^x to approximate e to five decimal-place accuracy.

Solution. If we substitute $x = 1$ in the Maclaurin series

$$e^x = 1 + x + \frac{x^2}{2!} + \frac{x^3}{3!} + \cdots + \frac{x^k}{k!} + \cdots$$

we obtain

$$e = 1 + 1 + \frac{1}{2!} + \frac{1}{3!} + \cdots + \frac{1}{k!} + \cdots \tag{12}$$

and hence we can approximate e to any degree of accuracy using an appropriate partial sum

$$e \approx 1 + 1 + \frac{1}{2!} + \frac{1}{3!} + \cdots + \frac{1}{n!}$$

Thus, our problem is to determine how many terms in this partial sum are required to achieve five decimal-place accuracy; that is, we want to choose n so that the absolute value of the nth remainder at $x = 1$ in the Maclaurin series satisfies

$$|R_n(1)| \leq 0.000005$$

To determine n we will apply the Remainder Estimation Theorem with $f(x) = e^x$, $x = 1$, $x_0 = 0$, and I being the interval $[0, 1]$. In this case it follows from Formula (5) that

$$|R_n(1)| \leq \frac{M}{(n+1)!} \tag{13}$$

where M is an upper bound on the value of $f^{(n+1)}(x) = e^x$ for x in the interval $[0, 1]$. However, e^x is an increasing function, so its maximum value on the interval $[0, 1]$ occurs at $x = 1$; that is, $e^x \leq e$ on this interval. Thus, we can take $M = e$ in (13) to obtain

$$|R_n(1)| \leq \frac{e}{(n+1)!} \tag{14}$$

Unfortunately, this inequality is not very useful because it involves e, which is the very quantity we are trying to approximate. However, if we accept that $e < 3$, then we can replace (14) with the following less precise, but more useful, inequality:

$$|R_n(1)| < \frac{3}{(n+1)!}$$

Thus, we can achieve five decimal-place accuracy by choosing n so that

$$\frac{3}{(n+1)!} \leq 0.000005$$

With the help of a calculating utility you can verify that the smallest value of n that meets this criterion is $n = 9$. Thus, to five decimal-place accuracy

$$e \approx 1 + 1 + \frac{1}{2!} + \frac{1}{3!} + \frac{1}{4!} + \frac{1}{5!} + \frac{1}{6!} + \frac{1}{7!} + \frac{1}{8!} + \frac{1}{9!} \approx 2.71828$$

(verify). As a check, the author's calculator gives $e \approx 2.71828182846$, which agrees with the preceding approximation when rounded to five decimal places. ◀

APPROXIMATING LOGARITHMS

The Maclaurin series

$$\ln(1 + x) = x - \frac{x^2}{2} + \frac{x^3}{3} - \frac{x^4}{4} + \cdots \qquad (-1 < x \leq 1) \tag{15}$$

is the starting point for the approximation of natural logarithms. Unfortunately, the usefulness of this series is limited because of its slow convergence and the restriction $-1 < x \leq 1$. However, if we replace x by $-x$ in this series, we obtain

$$\ln(1 - x) = -x - \frac{x^2}{2} - \frac{x^3}{3} - \frac{x^4}{4} - \cdots \qquad (-1 \leq x < 1) \tag{16}$$

and on subtracting (16) from (15) we obtain

$$\ln\left(\frac{1+x}{1-x}\right) = 2\left(x + \frac{x^3}{3} + \frac{x^5}{5} + \frac{x^7}{7} + \cdots\right) \qquad (-1 < x < 1) \tag{17}$$

Series (17), first obtained by James Gregory* in 1668, can be used to compute the natural logarithm of any positive number y by letting

$$y = \frac{1+x}{1-x}$$

or, equivalently,

$$x = \frac{y-1}{y+1} \tag{18}$$

and noting that $-1 < x < 1$. For example, to compute $\ln 2$ we let $y = 2$ in (18), which yields $x = \frac{1}{3}$. Substituting this value in (17) gives

$$\ln 2 = 2\left[\frac{1}{3} + \frac{\left(\frac{1}{3}\right)^3}{3} + \frac{\left(\frac{1}{3}\right)^5}{5} + \frac{\left(\frac{1}{3}\right)^7}{7} + \cdots\right] \tag{19}$$

In Exercise 23 we will ask you to show that five decimal-place accuracy can be achieved using the partial sum with terms up to and including the 13th power of $\frac{1}{3}$. Thus, to five decimal-place accuracy

$$\ln 2 \approx 2\left[\frac{1}{3} + \frac{\left(\frac{1}{3}\right)^3}{3} + \frac{\left(\frac{1}{3}\right)^5}{5} + \frac{\left(\frac{1}{3}\right)^7}{7} + \cdots + \frac{\left(\frac{1}{3}\right)^{13}}{13}\right] \approx 0.69315$$

(verify). As a check, the author's calculator gives $\ln 2 \approx 0.69314718056$, which agrees with the preceding approximation when rounded to five decimal places.

REMARK. In Example 2 of Section 11.7, we stated without proof that

$$\ln 2 = 1 - \frac{1}{2} + \frac{1}{3} - \frac{1}{4} + \frac{1}{5} - \cdots$$

This result can be obtained letting $x = 1$ in (15). However, this series converges too slowly to be of practical value.

APPROXIMATING π

In the next section we will show that

$$\tan^{-1} x = x - \frac{x^3}{3} + \frac{x^5}{5} - \frac{x^7}{7} + \cdots \qquad (-1 \le x \le 1) \tag{20}$$

Letting $x = 1$, we obtain

$$\frac{\pi}{4} = \tan^{-1} 1 = 1 - \frac{1}{3} + \frac{1}{5} - \frac{1}{7} + \cdots$$

or

$$\pi = 4\left[1 - \frac{1}{3} + \frac{1}{5} - \frac{1}{7} + \cdots\right]$$

This famous series, obtained by Leibniz in 1674, converges too slowly to be of computational value. A more practical procedure for approximating π uses the identity

$$\frac{\pi}{4} = \tan^{-1}\frac{1}{2} + \tan^{-1}\frac{1}{3} \tag{21}$$

which was derived in Exercise 47 of Section 4.5. By using this identity and series (20)

* JAMES GREGORY (1638–1675). Scottish mathematician and astronomer. Gregory, the son of a minister, was famous in his time as the inventor of the Gregorian reflecting telescope, so named in his honor. Although he is not generally ranked with the great mathematicians, much of his work relating to calculus was studied by Leibniz and Newton and undoubtedly influenced some of their discoveries. There is a manuscript, discovered posthumously, which shows that Gregory had anticipated Taylor series well before Taylor.

to approximate $\tan^{-1}\frac{1}{2}$ and $\tan^{-1}\frac{1}{3}$, the value of π can be approximated efficiently to any degree of accuracy.

BINOMIAL SERIES

If m is a real number, then the Maclaurin series for $(1+x)^m$ is called the ***binomial series***; it is given by (verify)

$$1 + mx + \frac{m(m-1)}{2!}x^2 + \frac{m(m-1)(m-2)}{3!}x^3 + \cdots + \frac{m(m-1)\cdots(m-k+1)}{k!} + \cdots$$

In the case where m is a nonnegative integer, the function $f(x) = (1+x)^m$ is a polynomial of degree m, so

$$f^{(m+1)}(0) = f^{(m+2)}(0) = f^{(m+3)}(0) = \cdots = 0$$

and the binomial series reduces to the familiar binomial expansion

$$(1+x)^m = 1 + mx + \frac{m(m-1)}{2!}x^2 + \frac{m(m-1)(m-2)}{3!}x^3 + \cdots + x^m$$

which is valid for $-\infty < x < +\infty$.

It can be proved that if m is not a nonnegative integer, then the binomial series converges to $(1+x)^m$ if $|x| < 1$. Thus, for such values of x

$$(1+x)^m = 1 + mx + \frac{m(m-1)}{2!}x^2 + \cdots + \frac{m(m-1)\cdots(m-k+1)}{k!}x^k + \cdots \tag{22}$$

or in sigma notation,

$$(1+x)^m = 1 + \sum_{k=1}^{\infty} \frac{m(m-1)\cdots(m-k+1)}{k!}x^k \quad \text{if } |x| < 1 \tag{23}$$

Example 5

Find binomial series for

(a) $\dfrac{1}{(1+x)^2}$ (b) $\dfrac{1}{\sqrt{1+x}}$

Solution (a). Since the general term of the binomial series is complicated, you may find it helpful to write out some of the beginning terms of the series, as in Formula (22), to see developing patterns. Substituting $m = -2$ in this formula yields

$$\begin{aligned}
\frac{1}{(1+x)^2} = (1+x)^{-2} &= 1 + (-2)x + \frac{(-2)(-3)}{2!}x^2 \\
&\quad + \frac{(-2)(-3)(-4)}{3!}x^3 + \frac{(-2)(-3)(-4)(-5)}{4!}x^4 + \cdots \\
&= 1 - 2x + \frac{3!}{2!}x^2 - \frac{4!}{3!}x^3 + \frac{5!}{4!}x^4 + \cdots \\
&= 1 - 2x + 3x^2 - 4x^3 + 5x^4 + \cdots \\
&= \sum_{k=0}^{\infty}(-1)^k(k+1)x^k
\end{aligned}$$

Solution (b). Substituting $m = -\frac{1}{2}$ in (22) yields

$$\begin{aligned}
\frac{1}{\sqrt{1+x}} &= 1 - \frac{1}{2}x + \frac{\left(-\frac{1}{2}\right)\left(-\frac{1}{2}-1\right)}{2!}x^2 + \frac{\left(-\frac{1}{2}\right)\left(-\frac{1}{2}-1\right)\left(-\frac{1}{2}-2\right)}{3!}x^3 - \cdots \\
&= 1 - \frac{1}{2}x + \frac{1\cdot 3}{2^2\cdot 2!}x^2 - \frac{1\cdot 3\cdot 5}{2^3\cdot 3!}x^3 + \cdots \\
&= 1 + \sum_{k=1}^{\infty}(-1)^k\frac{1\cdot 3\cdot 5\cdots(2k-1)}{2^k k!}x^k
\end{aligned}$$

◀

For reference, Table 11.9.1 lists the Maclaurin series for some of the most important functions, together with a specification of the intervals over which the Maclaurin series converge to those functions. Some of these results are derived in the exercises and others will be derived in the next section using some special techniques that we will develop.

Table 11.9.1

MACLAURIN SERIES	INTERVAL OF CONVERGENCE
$\dfrac{1}{1-x}=\sum_{k=0}^{\infty} x^k = 1+x+x^2+x^3+\cdots$	$-1<x<1$
$\dfrac{1}{1+x^2}=\sum_{k=0}^{\infty} (-1)^k x^{2k} = 1-x^2+x^4-x^6+\cdots$	$-1<x<1$
$e^x=\sum_{k=0}^{\infty} \dfrac{x^k}{k!} = 1+x+\dfrac{x^2}{2!}+\dfrac{x^3}{3!}+\dfrac{x^4}{4!}+\cdots$	$-\infty<x<+\infty$
$\sin x=\sum_{k=0}^{\infty} (-1)^k \dfrac{x^{2k+1}}{(2k+1)!} = x-\dfrac{x^3}{3!}+\dfrac{x^5}{5!}-\dfrac{x^7}{7!}+\cdots$	$-\infty<x<+\infty$
$\cos x=\sum_{k=0}^{\infty} (-1)^k \dfrac{x^{2k}}{(2k)!} = 1-\dfrac{x^2}{2!}+\dfrac{x^4}{4!}-\dfrac{x^6}{6!}+\cdots$	$-\infty<x<+\infty$
$\ln(1+x)=\sum_{k=1}^{\infty} (-1)^{k+1} \dfrac{x^k}{k} = x-\dfrac{x^2}{2}+\dfrac{x^3}{3}-\dfrac{x^4}{4}+\cdots$	$-1<x\le 1$
$\tan^{-1} x=\sum_{k=0}^{\infty} (-1)^k \dfrac{x^{2k+1}}{2k+1} = x-\dfrac{x^3}{3}+\dfrac{x^5}{5}-\dfrac{x^7}{7}+\cdots$	$-1\le x\le 1$
$\sinh x=\sum_{k=0}^{\infty} \dfrac{x^{2k+1}}{(2k+1)!} = x+\dfrac{x^3}{3!}+\dfrac{x^5}{5!}+\dfrac{x^7}{7!}+\cdots$	$-\infty<x<+\infty$
$\cosh x=\sum_{k=0}^{\infty} \dfrac{x^{2k}}{(2k)!} = 1+\dfrac{x^2}{2!}+\dfrac{x^4}{4!}+\dfrac{x^6}{6!}+\cdots$	$-\infty<x<+\infty$
$(1+x)^m = 1+\sum_{k=1}^{\infty} \dfrac{m(m-1)\cdots(m-k+1)}{k!} x^k$	$-1<x<1$* $(m \ne 0, 1, 2, \ldots)$

*The behavior at the endpoints depends on m: For $m>0$ the series converges absolutely at both endpoints; for $m \le -1$ the series diverges at both endpoints; and for $-1<m<0$ the series converges conditionally at $x=1$ and diverges at $x=-1$.

EXERCISE SET 11.9 Graphing Calculator C CAS

1. Use both of the methods given in Example 2 to approximate $\sin 4^\circ$ to five decimal-place accuracy, and check your work by comparing your answer to that produced directly by your calculating utility.

2. Use both of the methods given in Example 2 to approximate $\cos 3^\circ$ to three decimal-place accuracy, and check your work by comparing your answer to that produced directly by your calculating utility.

3. Use the method of Example 4 to approximate $\sqrt{e}$ to four decimal-place accuracy, and check your work by comparing your answer to that produced directly by your calculating utility. [*Suggestion:* Write $\sqrt{e}$ as $e^{0.5}$.]

4. Use the method of Example 4 to approximate $1/e$ to three decimal-place accuracy, and check your work by comparing your answer to that produced directly by your calculating utility.

5. Use the Maclaurin series for $\cos x$ to approximate $\cos 0.1$ to five decimal-place accuracy, and check your work by comparing your answer to that produced directly by your calculating utility.

6. Use the Maclaurin series for $\tan^{-1} x$ to approximate $\tan^{-1} 0.1$ to three decimal-place accuracy, and check your work by comparing your answer to that produced directly by your calculating utility.

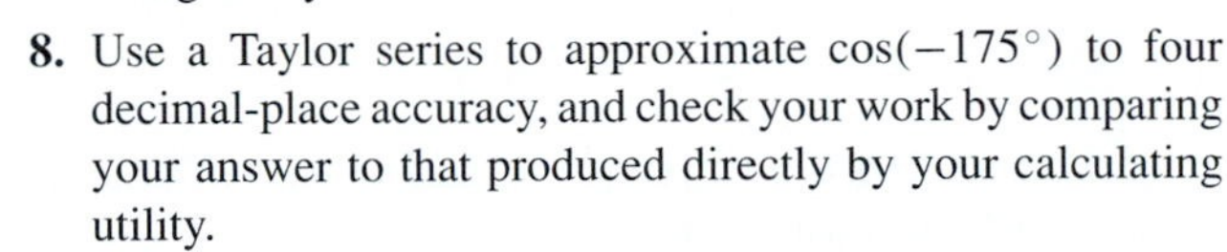

7. Use an appropriate Taylor series to approximate $\sin 85°$ to four decimal-place accuracy, and check your work by comparing your answer to that produced directly by your calculating utility.

8. Use a Taylor series to approximate $\cos(-175°)$ to four decimal-place accuracy, and check your work by comparing your answer to that produced directly by your calculating utility.

9. Use the Maclaurin series for $\sinh x$ to approximate $\sinh 0.5$ to three decimal-place accuracy. Check your work by computing $\sinh 0.5$ with a calculating utility.

10. Use the Maclaurin series for $\cosh x$ to approximate $\cosh 0.1$ to three decimal-place accuracy. Check your work by computing $\cosh 0.1$ with a calculating utility.

11. Use the Remainder Estimation Theorem and the method of Example 1 to prove that the Taylor series for $\sin x$ about $x = \pi/4$ converges to $\sin x$ for all x.

12. Use the Remainder Estimation Theorem and the method of Example 3 to prove that the Taylor series for e^x about $x = 1$ converges to e^x for all x.

13. (a) Use Formula (17) in the text to find a series that converges to $\ln 1.25$.
(b) Approximate $\ln 1.25$ using the first two terms of the series. Round your answer to three decimal places, and compare the result to that produced directly by your calculating utility.

14. (a) Use Formula (17) to find a series that converges to $\ln 3$.
(b) Approximate $\ln 3$ using the first two terms of the series. Round your answer to three decimal places, and compare the result to that produced directly by your calculating utility.

15. (a) Use the Maclaurin series for $\tan^{-1} x$ to approximate $\tan^{-1}\frac{1}{2}$ and $\tan^{-1}\frac{1}{3}$ to three decimal-place accuracy.
(b) Use the results in part (a) and Formula (21) to approximate π.
(c) Would you be willing to guarantee that your answer in part (b) is accurate to three decimal places? Explain your reasoning.
(d) Compare your answer in part (b) to that produced by your calculating utility.

16. Use an appropriate Taylor series for $\sqrt[3]{x}$ to approximate $\sqrt[3]{28}$ to three decimal-place accuracy, and check your answer by comparing it to that produced directly by your calculating utility.

17. (a) Use the Remainder Estimation Theorem to find an interval containing $x = 0$ over which $\sin x$ can be approximated by $x - (x^3/3!)$ to three decimal-place accuracy throughout the interval.
(b) Check your answer in part (a) by graphing

$$\left|\sin x - \left(x - \frac{x^3}{3!}\right)\right|$$

over the interval you obtained.

18. (a) Find an interval $[0, b]$ over which e^x can be approximated by $1 + x + (x^2/2!)$ to three decimal-place accuracy throughout the interval.
(b) Check your answer in part (a) by graphing

$$\left|e^x - \left(1 + x + \frac{x^2}{2!}\right)\right|$$

over the interval you obtained.

19. (a) Find an upper bound on the error that can result if $\cos x$ is approximated by $1 - (x^2/2!) + (x^4/4!)$ over the interval $[-0.2, 0.2]$.
(b) Check your answer in part (a) by graphing

$$\left|\cos x - \left(1 - \frac{x^2}{2!} + \frac{x^4}{4!}\right)\right|$$

over the interval.

20. (a) Find an upper bound on the error that can result if $\ln(1 + x)$ is approximated by x over the interval $[-0.01, 0.01]$.
(b) Check your answer in part (a) by graphing

$$|\ln(1 + x) - x|$$

over the interval.

21. Use Formula (22) for the binomial series to obtain the Maclaurin series for

(a) $\dfrac{1}{1+x}$ (b) $\sqrt[3]{1+x}$ (c) $\dfrac{1}{(1+x)^3}$.

22. If m is any real number, and k is a nonnegative integer, then we define the ***binomial coefficient***

$$\binom{m}{k} \text{ by the formulas } \binom{m}{0} = 1 \text{ and}$$

$$\binom{m}{k} = \frac{m(m-1)(m-2)\cdots(m-k+1)}{k!}$$

for $k \geq 1$. Express Formula (22) in the text in terms of binomial coefficients.

23. In this exercise we will use the Remainder Estimation Theorem to determine the number of terms that are required in Formula (19) to approximate $\ln 2$ to five decimal-place accuracy. For this purpose let

$$f(x) = \ln\frac{1+x}{1-x} = \ln(1+x) - \ln(1-x) \quad (-1 < x < 1)$$

(a) Show that

$$f^{(n+1)}(x) = n!\left[\frac{(-1)^n}{(1+x)^{n+1}} + \frac{1}{(1-x)^{n+1}}\right]$$

(b) Use the triangle inequality [Theorem 1.2.2(*d*)] to show that

$$|f^{(n+1)}(x)| \leq n!\left[\frac{1}{(1+x)^{n+1}} + \frac{1}{(1-x)^{n+1}}\right]$$

(c) Since we want to achieve five decimal-place accuracy, our goal is to choose n so that the absolute value of the nth remainder at $x = \frac{1}{3}$ does not exceed the value

$0.000005 = 0.5 \times 10^{-5}$; that is, $|R_n(\frac{1}{3})| \le 0.000005$. Use the Remainder Estimation Theorem to show that this condition will be satisfied if n is chosen so that

$$\frac{M}{(n+1)!}\left(\frac{1}{3}\right)^{n+1} \le 0.000005$$

where $|f^{(n+1)}(x)| \le M$ on the interval $[0, \frac{1}{3}]$.

(d) Use the result in part (b) to show that M can be taken as

$$M = n!\left[1 + \frac{1}{(\frac{2}{3})^{n+1}}\right]$$

(e) Use the results in parts (c) and (d) to show that five decimal-place accuracy will be achieved if n satisfies

$$\frac{1}{n+1}\left[\left(\frac{1}{3}\right)^{n+1} + \left(\frac{1}{2}\right)^{n+1}\right] \le 0.000005$$

and then show that the smallest value of n that satisfies this condition is $n = 13$.

24. Use Formula (17) and the method of Exercise 23 to approximate $\ln(\frac{5}{3})$ to five decimal-place accuracy. Then check your work by comparing your answer to that produced directly by your calculating utility.

25. Prove: The Taylor series for $\cos x$ about any point $x = x_0$ converges to $\cos x$ for all x.

26. Prove: The Taylor series for $\sin x$ about any point $x = x_0$ converges to $\sin x$ for all x.

C **27.** (a) In 1706 the British astronomer and mathematician John Machin discovered the following formula for $\pi/4$, called ***Machin's formula***:

$$\frac{\pi}{4} = 4\tan^{-1}\frac{1}{5} - \tan^{-1}\frac{1}{239}$$

Use a CAS to approximate $\pi/4$ using Machin's formula to 25 decimal places.

(b) In 1914 the brilliant Indian mathematician Srinivasa Ramanujan (1887–1920) showed that

$$\frac{1}{\pi} = \frac{\sqrt{8}}{9801}\sum_{k=0}^{\infty}\frac{(4k)!(1103 + 26{,}390k)}{(k!)^4 396^{4k}}$$

Use a CAS to compute the first four partial sums in ***Ramanujan's formula***.

28. The purpose of this exercise is to show that the Taylor series of a function f may possibly converge to a value different from $f(x)$ for certain x. Let

$$f(x) = \begin{cases} e^{-1/x^2}, & x \ne 0 \\ 0, & x = 0 \end{cases}$$

(a) Use the definition of a derivative to show that $f'(0) = 0$.

(b) With some difficulty it can be shown that $f^{(n)}(0) = 0$ for $n \ge 2$. Accepting this fact, show that the Maclaurin series of f converges for all x, but converges to $f(x)$ only at the point $x = 0$.

11.10 DIFFERENTIATING AND INTEGRATING POWER SERIES; MODELING WITH TAYLOR SERIES

In this section we will discuss methods for finding power series for derivatives and integrals of functions, and we will discuss some practical methods for finding Taylor series that can be used in situations where it is difficult or impossible to find the series directly.

DIFFERENTIATING POWER SERIES

We begin by considering the following problem:

11.10.1 PROBLEM. Suppose that a function f is represented by a power series on an open interval. How can we use the power series to find the derivative of f on that interval?

The solution to this problem can be motivated by considering the Maclaurin series for $\sin x$:

$$\sin x = x - \frac{x^3}{3!} + \frac{x^5}{5!} - \frac{x^7}{7!} + \cdots \qquad (-\infty < x < +\infty)$$

Of course, we already know that the derivative of $\sin x$ is $\cos x$; however, we are concerned here with using the Maclaurin series to deduce this. The solution is easy—all we need to do is differentiate the Maclaurin series term by term and observe that the resulting series is the Maclaurin series for $\cos x$:

$$\frac{d}{dx}\left[x - \frac{x^3}{3!} + \frac{x^5}{5!} - \frac{x^7}{7!} + \cdots\right] = 1 - 3\frac{x^2}{3!} + 5\frac{x^4}{5!} - 7\frac{x^6}{7!} + \cdots$$

$$= 1 - \frac{x^2}{2!} + \frac{x^4}{4!} - \frac{x^6}{6!} + \cdots = \cos x$$

Here is another example.

$$\frac{d}{dx}[e^x] = \frac{d}{dx}\left[1 + x + \frac{x^2}{2!} + \frac{x^3}{3!} + \frac{x^4}{4!} + \cdots\right]$$

$$= 1 + 2\frac{x}{2!} + 3\frac{x^2}{3!} + 4\frac{x^3}{4!} + \cdots = 1 + x + \frac{x^2}{2!} + \frac{x^3}{3!} + \cdots = e^x$$

FOR THE READER. See whether you can use this method to find the derivative of $\cos x$.

The preceding computations suggest that if a function f is represented by a power series on an open interval, then a power series representation of f' on that interval can be obtained by differentiating the power series for f term by term. This is stated more precisely in the following theorem, which we give without proof.

11.10.2 THEOREM (***Differentiation of Power Series***). *Suppose that a function f is represented by a power series in $x - x_0$ that has a nonzero radius of convergence R; that is,*

$$f(x) = \sum_{k=0}^{\infty} c_k(x - x_0)^k \qquad (x_0 - R < x < x_0 + R)$$

Then:

(*a*) *The function f is differentiable on the interval $(x_0 - R, x_0 + R)$.*

(*b*) *If the power series representation for f is differentiated term by term, then the resulting series has radius of convergence R and converges to f' on the interval $(x_0 - R, x_0 + R)$; that is,*

$$f'(x) = \sum_{k=0}^{\infty} \frac{d}{dx}[c_k(x - x_0)^k] \qquad (x_0 - R < x < x_0 + R)$$

This theorem has an important implication about the differentiability of functions that are represented by power series. According to the theorem, the power series for f' has the same radius of convergence as the power series for f, and this means that the theorem can be applied to f' as well as f. However, if we do this, then we conclude that f' is differentiable on the interval $(x_0 - R, x_0 + R)$, and the power series for f'' has the same radius of convergence as the power series for f and f'. We can now repeat this process ad infinitum, applying the theorem successively to $f'', f''', \ldots, f^{(n)}, \ldots$ to conclude that f has derivatives of all orders on the interval $(x_0 - R, x_0 + R)$. Thus, we have established the following result.

11.10.3 THEOREM. *If a function f can be represented by a power series in $x - x_0$ with a nonzero radius of convergence R, then f has derivatives of all orders on the interval $(x_0 - R, x_0 + R)$.*

In short, it is only the most "well-behaved" functions that can be represented by power series; that is, if a function f does not possess derivatives of all orders on an interval $(x_0 - R, x_0 + R)$, then it cannot be represented by a power series in $x - x_0$ on that interval.

Example 1

In Section 11.8, we showed that the Bessel function $J_0(x)$ is represented by the power series

$$J_0(x) = \sum_{k=0}^{\infty} \frac{(-1)^k x^{2k}}{2^{2k}(k!)^2} \tag{1}$$

with radius of convergence $+\infty$ [see Formula (2) of that section and the related discussion]. Thus, $J_0(x)$ has derivatives of all orders on the interval $(-\infty, +\infty)$, and these can be obtained by differentiating the series term by term. For example, if we write (1) as

$$J_0(x) = 1 + \sum_{k=1}^{\infty} \frac{(-1)^k x^{2k}}{2^{2k}(k!)^2}$$

and differentiate term by term, we obtain

$$J_0'(x) = \sum_{k=1}^{\infty} \frac{(-1)^k (2k) x^{2k-1}}{2^{2k}(k!)^2} = \sum_{k=1}^{\infty} \frac{(-1)^k x^{2k-1}}{2^{2k-1} k!(k-1)!}$$ ◀

REMARK. The computations in this example use some techniques that are worth noting. First, when a power series is expressed in sigma notation, the formula for the general term of the series will often not be of a form that can be used for differentiating the constant term. Thus, if the series has a nonzero constant term, as here, it is usually a good idea to split it off from the summation before differentiating. Second, observe how we simplified the final formula by canceling the factor k from one of the factorials in the denominator. This is a standard simplification technique.

INTEGRATING POWER SERIES

Since the derivative of a function that is represented by a power series can be obtained by differentiating the series term by term, it should not be surprising that an antiderivative of a function represented by a power series can be obtained by integrating the series term by term. For example, we know that $\sin x$ is an antiderivative of $\cos x$. Here is how this result can be obtained by integrating the Maclaurin series for $\cos x$ term by term:

$$\begin{aligned} \int \cos x\, dx &= \int \left[1 - \frac{x^2}{2!} + \frac{x^4}{4!} - \frac{x^6}{6!} + \cdots\right] dx \\ &= \left[x - \frac{x^3}{3(2!)} + \frac{x^5}{5(4!)} - \frac{x^7}{(6!)} + \cdots\right] + C \\ &= \left[x - \frac{x^3}{3!} + \frac{x^5}{5!} - \frac{x^7}{7!} + \cdots\right] + C = \sin x + C \end{aligned}$$

The same idea applies to definite integrals. For example, by direct integration we have

$$\int_0^1 \frac{dx}{1+x^2} = \tan^{-1} x \Big]_0^1 = \tan^{-1} 1 - \tan 0 = \frac{\pi}{4} - 0 = \frac{\pi}{4}$$

and we will show later in this section that

$$\frac{\pi}{4} = 1 - \frac{1}{3} + \frac{1}{5} - \frac{1}{7} + \cdots \tag{2}$$

Thus,

$$\int_0^1 \frac{dx}{1+x^2} = 1 - \frac{1}{3} + \frac{1}{5} - \frac{1}{7} + \cdots$$

Here is how this result can be obtained by integrating the Maclaurin series for $1/(1+x^2)$ term by term (see Table 11.9.1):

$$\begin{aligned} \int_0^1 \frac{dx}{1+x^2} &= \int_0^1 [1 - x^2 + x^4 - x^6 + \cdots]\, dx \\ &= x - \frac{x^3}{3} + \frac{x^5}{5} - \frac{x^7}{7} + \cdots \Big]_0^1 = 1 - \frac{1}{3} + \frac{1}{5} - \frac{1}{7} + \cdots \end{aligned}$$

The preceding computations are justified by the following theorem, which we give without proof.

11.10.4 THEOREM (*Integration of Power Series*). *Suppose that a function f is represented by a power series in $x - x_0$ that has a nonzero radius of convergence R; that is,*

$$f(x) = \sum_{k=0}^{\infty} c_k(x - x_0)^k \qquad (x_0 - R < x < x_0 + R)$$

(*a*) *If the power series representation of f is integrated term by term using an indefinite integral, then the resulting series has radius of convergence R and converges to $\int f(x)\,dx$ on the interval $(x_0 - R, x_0 + R)$; that is,*

$$\int f(x)\,dx = \sum_{k=0}^{\infty}\left[\int c_k(x - x_0)^k\,dx\right] + C \qquad (x_0 - R < x < x_0 + R)$$

(*b*) *If α and β are points in the interval $(x_0 - R, x_0 + R)$, and if the power series representation of f is integrated term by term from α to β, then the resulting series of numbers converges absolutely on the interval $(x_0 - R, x_0 + R)$ and*

$$\int_{\alpha}^{\beta} f(x)\,dx = \sum_{k=0}^{\infty}\left[\int_{\alpha}^{\beta} c_k(x - x_0)^k\,dx\right]$$

POWER SERIES REPRESENTATIONS MUST BE TAYLOR SERIES

For many functions it is difficult or impossible to find the derivatives that are required to obtain a Taylor series. For example, to find the Maclaurin series for $1/(1 + x^2)$ directly would require some tedious derivative computations (try it). A more practical approach is to substitute $-x^2$ for x in the geometric series

$$\frac{1}{1-x} = 1 + x + x^2 + x^3 + x^4 + \cdots \qquad (-1 < x < 1)$$

to obtain

$$\frac{1}{1+x^2} = 1 - x^2 + x^4 - x^6 + x^8 - \cdots$$

However, there are two questions of concern with this procedure:

- Where does the power series that we obtained for $1/(1 + x^2)$ actually converge to $1/(1 + x^2)$?
- How do we know that the power series we have obtained is actually the Maclaurin series for $1/(1 + x^2)$?

The first question is easy to resolve. Since the geometric series converges to $1/(1 - x)$ if $|x| < 1$, the second series will converge to $1/(1 + x^2)$ if $|-x^2| < 1$ or $|x^2| < 1$. However, this is true if and only if $|x| < 1$, so the power series we obtained for the function $1/(1 + x^2)$ converges to this function if $-1 < x < 1$.

The second question is more difficult to answer and leads us to the following general problem.

11.10.5 PROBLEM. Suppose that a function f is represented by a power series in $x - x_0$ that has a nonzero radius of convergence. What relationship exists between the given power series and the Taylor series for f about $x = x_0$?

The answer is that they are the same; and here is the theorem that proves it.

> **11.10.6 THEOREM.** *If a function f is represented by a power series in $x - x_0$ on some open interval containing x_0, then that power series is the Taylor series for f about $x = x_0$.*

Proof. Suppose that

$$f(x) = c_0 + c_1(x - x_0) + c_2(x - x_0)^2 + \cdots + c_k(x - x_0)^k + \cdots$$

for all x in some open interval containing x_0. To prove that this is the Taylor series for f about $x = x_0$, we must show that

$$c_k = \frac{f^{(k)}(x_0)}{k!} \quad \text{for} \quad k = 0, 1, 2, 3, \ldots$$

However, the assumption that the series converges to $f(x)$ on an open interval containing x_0 ensures that it has a nonzero radius of convergence R; hence we can differentiate term by term in accordance with Theorem 11.10.2. Thus,

$$\begin{aligned}
f(x) &= c_0 + c_1(x - x_0) + c_2(x - x_0)^2 + c_3(x - x_0)^3 + c_4(x - x_0)^4 + \cdots \\
f'(x) &= c_1 + 2c_2(x - x_0) + 3c_3(x - x_0)^2 + 4c_4(x - x_0)^3 + \cdots \\
f''(x) &= 2!c_2 + (3 \cdot 2)c_3(x - x_0) + (4 \cdot 3)c_4(x - x_0)^2 + \cdots \\
f'''(x) &= 3!c_3 + (4 \cdot 3 \cdot 2)c_4(x - x_0) + \cdots \\
&\vdots
\end{aligned}$$

On substituting $x = x_0$, all the powers of $x - x_0$ drop out, leaving

$$f(x_0) = c_0, \quad f'(x_0) = c_1, \quad f''(x_0) = 2!c_2, \quad f'''(x_0) = 3!c_3, \quad \ldots$$

from which we obtain

$$c_0 = f(x_0), \quad c_1 = f'(x_0), \quad c_2 = \frac{f''(x_0)}{2!}, \quad c_3 = \frac{f'''(x_0)}{3!}, \quad \ldots$$

which shows that the coefficients $c_0, c_1, c_2, c_3, \ldots$ are precisely the coefficients in the Taylor series about x_0 for $f(x)$. ∎

REMARK. This theorem tells us that no matter how we arrive at a power series representation of a function f, be it by substitution, by differentiation, by integration, or by some sort of algebraic manipulation, that series will be the Taylor series for f about $x = x_0$, provided that it converges to f on some open interval containing x_0.

SOME PRACTICAL WAYS TO FIND TAYLOR SERIES

Example 2

Find the Maclaurin series for $\tan^{-1} x$.

Solution. It would be tedious to find the Maclaurin series directly. A better approach is to start with the formula

$$\int \frac{1}{1 + x^2}\, dx = \tan^{-1} x + C$$

and integrate the Maclaurin series

$$\frac{1}{1 + x^2} = 1 - x^2 + x^4 - x^6 + x^8 - \cdots \qquad (-1 < x < 1)$$

term by term. This yields

$$\tan^{-1} x + C = \int \frac{1}{1 + x^2}\, dx = \int [1 - x^2 + x^4 - x^6 + x^8 - \cdots]\, dx$$

or

$$\tan^{-1} x = \left[x - \frac{x^3}{3} + \frac{x^5}{5} - \frac{x^7}{7} + \frac{x^9}{9} - \cdots\right] - C$$

The constant of integration can be evaluated by substituting $x = 0$ and using the condition $\tan^{-1} 0 = 0$. This gives $C = 0$, so that

$$\tan^{-1} x = x - \frac{x^3}{3} + \frac{x^5}{5} - \frac{x^7}{7} + \frac{x^9}{9} - \cdots \qquad (-1 < x < 1) \tag{3}$$

◀

REMARK. Observe that neither Theorem 11.10.2 nor Theorem 11.10.3 addresses what happens at the endpoints of the interval of convergence. However, it can be proved that if the Taylor series for f about $x = x_0$ converges to $f(x)$ for all x in the interval $(x_0 - R, x_0 + R)$, and if the Taylor series converges at the right endpoint $x_0 + R$, then the value that it converges to at that point is the limit of $f(x)$ as $x \to x_0 + R$ from the left; and if the Taylor series converges at the left endpoint $x_0 - R$, then the value that it converges to at that point is the limit of $f(x)$ as $x \to x_0 - R$ from the right.

For example, the Maclaurin series for $\tan^{-1} x$ given in (3) converges at both $x = -1$ and $x = 1$, since the hypotheses of the alternating series test (Theorem 11.7.1) are satisfied at those points. Thus, the continuity of $\tan^{-1} x$ on the interval $[-1, 1]$ implies that at $x = 1$ the Maclaurin series converges to

$$\lim_{x \to 1^-} \tan^{-1} x = \tan^{-1} 1 = \frac{\pi}{4}$$

and at $x = -1$ it converges to

$$\lim_{x \to -1^+} \tan^{-1} x = \tan^{-1}(-1) = -\frac{\pi}{4}$$

This shows that the Maclaurin series for $\tan^{-1} x$ actually converges to $\tan^{-1} x$ on the interval $-1 \leq x \leq 1$. Moreover, the convergence at $x = 1$ establishes Formula (2).

Taylor series provide an alternative to Simpson's rule and other numerical methods for approximating definite integrals.

Example 3

Approximate the integral

$$\int_0^1 e^{-x^2}\, dx$$

to three decimal-place accuracy by expanding the integrand in a Maclaurin series and integrating term by term.

Solution. The simplest way to obtain the Maclaurin series for e^{-x^2} is to replace x by $-x^2$ in the Maclaurin series

$$e^x = 1 + x + \frac{x^2}{2!} + \frac{x^3}{3!} + \frac{x^4}{4!} + \cdots$$

to obtain

$$e^{-x^2} = 1 - x^2 + \frac{x^4}{2!} - \frac{x^6}{3!} + \frac{x^8}{4!} - \cdots$$

Therefore,

$$\begin{aligned}
\int_0^1 e^{-x^2}\, dx &= \int_0^1 \left[1 - x^2 + \frac{x^4}{2!} - \frac{x^6}{3!} + \frac{x^8}{4!} - \cdots\right] dx \\
&= \left[x - \frac{x^3}{3} + \frac{x^5}{5(2!)} - \frac{x^7}{7(3!)} + \frac{x^9}{9(4!)} - \cdots\right]_0^1 \\
&= 1 - \frac{1}{3} + \frac{1}{5 \cdot 2!} - \frac{1}{7 \cdot 3!} + \frac{1}{9 \cdot 4!} - \cdots \\
&= \sum_{k=0}^{\infty} \frac{(-1)^k}{(2k+1)k!}
\end{aligned}$$

Since this series clearly satisfies the hypotheses of the alternating series test (Theorem 11.7.1), it follows from Theorem 11.7.2 that if we approximate the integral by s_n (the nth partial sum of the series), then

$$\left|\int_0^1 e^{-x^2}\,dx - s_n\right| < \frac{1}{[2(n+1)+1](n+1)!} = \frac{1}{(2n+3)(n+1)!}$$

Thus, for three decimal-place accuracy we must choose n such that

$$\frac{1}{(2n+3)(n+1)!} \le 0.0005 = 5 \times 10^{-4}$$

With the help of a calculating utility you can show that the smallest value of n that satisfies this condition is $n = 5$. Thus, the value of the integral to three decimal-place accuracy is

$$\int_0^1 e^{-x^2}\,dx \approx 1 - \frac{1}{3} + \frac{1}{5\cdot 2!} - \frac{1}{7\cdot 3!} + \frac{1}{9\cdot 4!} \approx 0.747$$

As a check, the author's CAS produced the approximation 0.746824, which agrees with our result when rounded to three decimal places. ◀

FOR THE READER. What advantages does the method in this example have over Simpson's rule? What are its disadvantages?

FINDING MACLAURIN SERIES BY MULTIPLICATION AND DIVISION

The following examples illustrate some algebraic techniques that are sometimes useful for finding Taylor series.

Example 4

Find the first three nonzero terms in the Maclaurin series for the function $f(x) = e^{-x^2}\tan^{-1} x$.

$$\begin{array}{r} 1 - x^2 + \frac{x^4}{2} - \cdots \\ \times \quad x - \frac{x^3}{3} + \frac{x^5}{5} - \cdots \\ \hline x - x^3 + \frac{x^5}{2} - \cdots \\ -\frac{x^3}{3} + \frac{x^5}{3} - \frac{x^7}{6} + \cdots \\ \frac{x^5}{5} - \frac{x^7}{5} + \cdots \\ \hline x - \frac{4}{3}x^3 + \frac{31}{30}x^5 - \cdots \end{array}$$

Solution. Using the series for e^{-x^2} and $\tan^{-1} x$ obtained in Examples 2 and 3 gives

$$e^{-x^2}\tan^{-1} x = \left(1 - x^2 + \frac{x^4}{2} - \cdots\right)\left(x - \frac{x^3}{3} + \frac{x^5}{5} - \cdots\right)$$

Multiplying, as shown in the margin, we obtain

$$e^{-x^2}\tan^{-1} x = x - \frac{4}{3}x^3 + \frac{31}{30}x^5 - \cdots$$

More terms in the series can be obtained by including more terms in the factors. Moreover, one can prove that a series obtained by this method converges at each point in the intersection of the intervals of convergence of the factors (and possibly on a larger interval). Thus, we can be certain that the series we have obtained converges for all x in the interval $-1 \le x \le 1$ (why?). ◀

$$\begin{array}{r|l} & x + \frac{x^3}{3} + \frac{2x^5}{15} + \cdots \\ \hline 1 - \frac{x^2}{2} + \frac{x^4}{24} - \cdots & x - \frac{x^3}{6} + \frac{x^5}{120} - \cdots \\ & x - \frac{x^3}{2} + \frac{x^5}{24} - \cdots \\ \hline & \frac{x^3}{3} - \frac{x^5}{30} + \cdots \\ & \frac{x^3}{3} - \frac{x^5}{6} + \cdots \\ \hline & \frac{2x^5}{15} + \cdots \end{array}$$

FOR THE READER. If you have a CAS, read the documentation about multiplying polynomials, and then use the CAS to duplicate the result in the last example.

Example 5

Find the first three nonzero terms in the Maclaurin series for $\tan x$.

Solution. Using the first three terms in the Maclaurin series for $\sin x$ and $\cos x$, we can express $\tan x$ as

$$\tan x = \frac{\sin x}{\cos x} = \frac{x - \dfrac{x^3}{3!} + \dfrac{x^5}{5!} - \cdots}{1 - \dfrac{x^2}{2!} + \dfrac{x^4}{4!} - \cdots}$$

Dividing, as shown in the margin, we obtain

$$\tan x = x + \frac{x^3}{3} + \frac{2x^5}{15} + \cdots$$ ◀

MODELING PHYSICAL LAWS WITH TAYLOR SERIES

Taylor series provide an important way of modeling physical laws. To illustrate the idea we will consider the problem of modeling the period of a simple pendulum (Figure 11.10.1). As explained in Exercise 38 of the supplementary exercises to Chapter 9, the period T of such a pendulum is given by

$$T = 4\sqrt{\frac{L}{g}}\int_0^{\pi/2} \frac{1}{\sqrt{1 - k^2 \sin^2\phi}}\, d\phi \tag{4}$$

where

L = length of the supporting rod

g = acceleration due to gravity

$k = \sin(\theta_0/2)$, where θ_0 is the initial angle of displacement from the vertical

$k \sin\phi = \sin(\theta/2)$, where θ is the displacement from the vertical at time t

The integral, which is called a ***complete elliptic integral of the first kind***, cannot be expressed in terms of elementary functions and is often approximated by numerical methods. Unfortunately, numerical values are so specific that they often give little insight into general physical principles. However, if we expand the integrand of (4) in a Maclaurin series and integrate term by term, then we can generate an infinite series that can be used to construct various mathematical models for the period T that give a deeper understanding of the behavior of the pendulum.

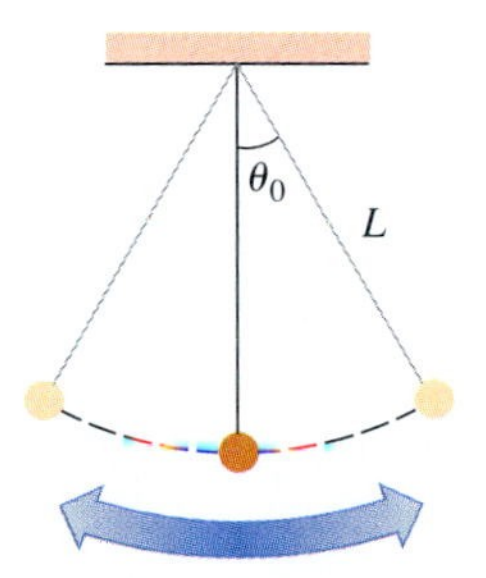

Figure 11.10.1

To obtain the Maclaurin series for the integrand, we will substitute $-k^2\sin^2\phi$ for x in the binomial series for $1/\sqrt{1+x}$ that we derived in Example 5 of Section 11.9. If we do this, then we can rewrite (4) as

$$T = 4\sqrt{\frac{L}{g}}\int_0^{\pi/2} \left[1 + \frac{1}{2}k^2\sin^2\phi + \frac{1\cdot 3}{2^2 2!}k^4\sin^4\phi + \frac{1\cdot 3\cdot 5}{2^3 3!}k^6\sin^6\phi + \cdots\right] d\phi \tag{5}$$

If we integrate term by term, then we can produce a Maclaurin series that converges to the period T. However, one of the most important cases of pendulum motion occurs when the initial displacement is small, in which case all subsequent displacements are small, and we can assume that $\phi \approx 0$. In this case we expect the convergence of the Maclaurin series for T to be rapid, and we can approximate the sum of the series by dropping all but the constant term in (5). This yields

$$T = 2\pi\sqrt{\frac{L}{g}} \tag{6}$$

which is called the ***first-order model*** of T or the model for ***small vibrations***. This model can be improved on by using more terms in the series. For example, if we use the first two terms in the Maclaurin series, we obtain the ***second-order model***

$$T = 2\pi\sqrt{\frac{L}{g}}\left(1 + \frac{k^2}{4}\right) \tag{7}$$

(verify).

EXERCISE SET 11.10 [C] CAS

1. In each part, obtain the Maclaurin series for the function by making an appropriate substitution in the Maclaurin series for $1/(1-x)$. Include the general term in your answer, and state the radius of convergence of the series.

(a) $\dfrac{1}{1+x}$ (b) $\dfrac{1}{1-x^2}$ (c) $\dfrac{1}{1-2x}$ (d) $\dfrac{1}{2-x}$

2. In each part, obtain the Maclaurin series for the function by making an appropriate substitution in the Maclaurin series for $\ln(1+x)$. Include the general term in your answer, and state the radius of convergence of the series.
(a) $\ln(1-x)$ (b) $\ln(1+x^2)$
(c) $\ln(1+2x)$ (d) $\ln(2+x)$

3. In each part, obtain the first four nonzero terms of the Maclaurin series for the function by making an appropriate substitution in one of the binomial series obtained in Example 5 of Section 11.9.
(a) $(2+x)^{-1/2}$ (b) $(1-x^2)^{-2}$

4. (a) Use the Maclaurin series for $1/(1-x)$ to find the Maclaurin series for $1/(a-x)$, where $a \neq 0$, and state the radius of convergence of the series.
(b) Use the binomial series for $1/(1+x)^2$ obtained in Example 5 of Section 11.9 to find the first four nonzero terms in the Maclaurin series for $1/(a+x)^2$, where $a \neq 0$, and state the radius of convergence of the series.

In Exercises 5–8, obtain the first four nonzero terms of the Maclaurin series for the function by making an appropriate substitution in a known Maclaurin series and performing any algebraic operations that are required. State the radius of convergence of the series.

5. (a) $\sin 2x$ (b) e^{-2x} (c) e^{x^2} (d) $x^2 \cos \pi x$

6. (a) $\cos 2x$ (b) $x^2 e^x$ (c) xe^{-x} (d) $\sin(x^2)$

7. (a) $\dfrac{x^2}{1+3x}$ (b) $x \sinh 2x$ (c) $x(1-x^2)^{3/2}$

8. (a) $\dfrac{x}{x-1}$ (b) $3\cosh(x^2)$ (c) $\dfrac{x}{(1+2x)^3}$

In Exercises 9 and 10, find the first four nonzero terms of the Maclaurin series for the function by using an appropriate trigonometric identity or property of logarithms and then substituting in a known Maclaurin series.

9. (a) $\sin^2 x$ (b) $\ln[(1+x^3)^{12}]$

10. (a) $\cos^2 x$ (b) $\ln\left(\dfrac{1-x}{1+x}\right)$

11. (a) Use a known Maclaurin series to find the Taylor series of $1/x$ about $x = 1$ by expressing this function as

$$\frac{1}{x} = \frac{1}{1-(1-x)}$$

(b) Find the interval of convergence of the Taylor series.

12. Use the method of Exercise 11 to find the Taylor series of $1/x$ about $x = x_0$, and state the interval of convergence of the Taylor series.

In Exercises 13 and 14, find the first four nonzero terms of the Maclaurin series for the function by multiplying the Maclaurin series of the factors.

13. (a) $e^x \sin x$ (b) $\sqrt{1+x}\ln(1+x)$

14. (a) $e^{-x^2}\cos x$ (b) $(1+x^2)^{4/3}(1+x)^{1/3}$

In Exercises 15 and 16, find the first four nonzero terms of the Maclaurin series for the function by dividing appropriate Maclaurin series.

15. (a) $\sec x \quad \left(= \dfrac{1}{\cos x}\right)$ (b) $\dfrac{\sin x}{e^x}$

16. (a) $\dfrac{\tan^{-1} x}{1+x}$ (b) $\dfrac{\ln(1+x)}{1-x}$

17. Use the Maclaurin series for e^x and e^{-x} to derive the Maclaurin series for $\sinh x$ and $\cosh x$. Include the general terms in your answers and state the radius of convergence of each series.

18. Use the Maclaurin series for $\sinh x$ and $\cosh x$ to obtain the first four nonzero terms in the Maclaurin series for $\tanh x$.

In Exercises 19 and 20, find the first five nonzero terms of the Maclaurin series for the function by using partial fractions and a known Maclaurin series.

19. $\dfrac{4x-2}{x^2-1}$

20. $\dfrac{x^3+x^2+2x-2}{x^2-1}$

In Exercises 21 and 22, confirm the derivative formula by differentiating the appropriate Maclaurin series term by term.

21. (a) $\dfrac{d}{dx}[\cos x] = -\sin x$ (b) $\dfrac{d}{dx}[\ln(1+x)] = \dfrac{1}{1+x}$

22. (a) $\dfrac{d}{dx}[\sinh x] = \cosh x$ (b) $\dfrac{d}{dx}[\tan^{-1} x] = \dfrac{1}{1+x^2}$

In Exercises 23 and 24, confirm the integration formula by integrating the appropriate Maclaurin series term by term.

23. (a) $\displaystyle\int e^x\,dx = e^x + C$ (b) $\displaystyle\int \sinh x\,dx = \cosh x + C$

24. (a) $\displaystyle\int \sin x\,dx = -\cos x + C$
(b) $\displaystyle\int \frac{1}{1+x}\,dx = \ln(1+x) + C$

25. (a) Use the Maclaurin series for $1/(1-x)$ to find the Maclaurin series for

$$f(x) = \frac{x}{1-x^2}$$

(b) Use the Maclaurin series obtained in part (a) to find $f^{(5)}(0)$ and $f^{(6)}(0)$.
(c) What can you say about the value of $f^{(n)}(0)$?

26. Let $f(x) = x^2 \cos 2x$. Use the method of Exercise 25 to find $f^{(99)}(0)$.

The limit of an indeterminate form as $x \to x_0$ can sometimes be found without using L'Hôpital's rule by expanding the functions involved in Taylor series about $x = x_0$ and taking the limit of the series term by term. Use this method to find the limits in Exercises 27 and 28.

27. (a) $\lim_{x \to 0} \frac{\sin x}{x}$ (b) $\lim_{x \to 0} \frac{\tan^{-1} x - x}{x^3}$

28. (a) $\lim_{x \to 0} \frac{1 - \cos x}{\sin x}$ (b) $\lim_{x \to 0} \frac{\ln \sqrt{1+x} - \sin 2x}{x}$

In Exercises 29–32, use Maclaurin series to approximate the integral to three decimal-place accuracy.

29. $\int_0^1 \sin(x^2)\, dx$ **30.** $\int_0^{1/2} \tan^{-1}(2x^2)\, dx$

31. $\int_0^{0.2} \sqrt[3]{1+x^4}\, dx$ **32.** $\int_0^{1/2} \frac{dx}{\sqrt[4]{x^2+1}}$

33. (a) Differentiate the Maclaurin series for $1/(1-x)$, and use the result to show that

$$\sum_{k=1}^{\infty} kx^k = \frac{x}{(1-x)^2} \quad \text{for } -1 < x < 1$$

(b) Integrate the Maclaurin series for $1/(1-x)$, and use the result to show that

$$\sum_{k=1}^{\infty} \frac{x^k}{k} = -\ln(1-x) \quad \text{for } -1 < x < 1$$

(c) Use the result in part (b) to show that

$$\sum_{k=1}^{\infty} (-1)^{k+1} \frac{x^k}{k} = \ln(1+x) \quad \text{for } -1 < x < 1$$

(d) Show that the series in part (c) converges if $x = 1$.

(e) Use the remark following Example 2 to show that

$$\sum_{k=1}^{\infty} (-1)^{k+1} \frac{x^k}{k} = \ln(1+x) \quad \text{for } -1 < x \le 1$$

34. In each part, use the results in Exercise 33 to find the sum of the series.

(a) $\sum_{k=1}^{\infty} \frac{k}{3^k} = \frac{1}{3} + \frac{2}{3^2} + \frac{3}{3^3} + \frac{4}{3^4} + \cdots$

(b) $\sum_{k=1}^{\infty} \frac{1}{k(4^k)} = \frac{1}{4} + \frac{1}{2(4^2)} + \frac{1}{3(4^3)} + \frac{1}{4(4^4)} + \cdots$

(c) $\sum_{k=1}^{\infty} (-1)^{k+1} \frac{1}{k} = 1 - \frac{1}{2} + \frac{1}{3} - \frac{1}{4} + \cdots$

35. (a) Use the relationship

$$\int \frac{1}{\sqrt{1+x^2}}\, dx = \sinh^{-1} x + C$$

to find the first four nonzero terms in the Maclaurin series for $\sinh^{-1} x$.

(b) Express the series in sigma notation.

(c) What is the radius of convergence?

36. (a) Use the relationship

$$\int \frac{1}{\sqrt{1-x^2}}\, dx = \sin^{-1} x + C$$

to find the first four nonzero terms in the Maclaurin series for $\sin^{-1} x$.

(b) Express the series in sigma notation.

(c) What is the radius of convergence?

37. We showed by Formula (12) of Section 10.3 that if there are y_0 units of radioactive carbon-14 present at time $t = 0$, then the number of units present t years later is

$$y(t) = y_0 e^{-0.000121t}$$

(a) Express $y(t)$ as a Maclaurin series.

(b) Use the first two terms in the series to show that the number of units present after 1 year is approximately $(0.999879)y_0$.

(c) Compare this to the value produced by the formula for $y(t)$.

38. In Section 10.1 we studied the motion of a falling object that has mass m and is retarded by air resistance. We showed that if the initial velocity is v_0 and the drag force F_R is proportional to the velocity, that is, $F_R = -cv$, then the velocity of the object at time t is

$$v(t) = e^{-ct/m}\left(v_0 + \frac{mg}{c}\right) - \frac{mg}{c}$$

where g is the acceleration due to gravity [see Formula (23) of Section 10.1].

(a) Use a Maclaurin series to show that if $ct/m \approx 0$, then the velocity can be approximated as

$$v(t) \approx v_0 - \left(\frac{cv_0}{m} + g\right)t$$

(b) Improve on the approximation in part (a).

C **39.** Suppose that a simple pendulum with a length of $L = 1$ meter is given an initial displacement of $\theta_0 = 5^\circ$ from the vertical.

(a) Approximate the period of the pendulum using Formula (6) for the first-order model. [Take $g = 9.8\ \text{m/s}^2$.]

(b) Approximate the period of the pendulum using Formula (7) for the second-order model.

(c) Use the numerical integration capability of a CAS to approximate the period of the pendulum from Formula (4), and compare it to the values obtained in parts (a) and (b).

40. Use the first three nonzero terms in Formula (5) and the Wallis sine formula in the Endpaper Integral Table (Formula 122) to obtain a model for the period of a simple pendulum.

41. Recall that the gravitational force exerted by the Earth on an object is called the object's *weight* (or more precisely, its *Earth weight*). We noted in statement 10.3.3 that if an object has mass m, then the magnitude of its weight is mg. However, this result presumes that the object is on the surface of the Earth (mean sea level). A more general formula for the magnitude of the gravitational force that the Earth exerts on an object of mass m is

$$F = \frac{mgR^2}{(R+h)^2}$$

where R is the radius of the Earth and h is the height of the object above the Earth's surface.

(a) Use the binomial series for $1/(1+x)^2$ obtained in Example 5 of Section 11.9 to express F as a Maclaurin series in powers of h/R.

(b) Show that if $h = 0$, then $F = mg$.

(c) Show that if $h/R \approx 0$, then $F \approx mg - (2mgh/R)$. [*Note:* The quantity $2mgh/R$ can be thought of as a "correction term" for the weight that takes the object's height above the Earth's surface into account.]

(d) If we assume that the Earth is a sphere of radius $R = 4000$ mi at mean sea level, by approximately what percentage does a person's weight change in going from mean sea level to the top of Mt. Everest (29,028 ft)?

42. (a) Show that the Bessel function $J_0(x)$ given by Formula (2) of Section 11.8 satisfies the differential equation $xy'' + y' + xy = 0$. (This is called the ***Bessel equation of order zero***.)

(b) Show that the Bessel function $J_1(x)$ given by Formula (3) of Section 11.8 satisfies the differential equation $x^2y'' + xy' + (x^2 - 1)y = 0$. (This is called the ***Bessel equation of order one***.)

(c) Show that $J_0'(x) = -J_1(x)$.

43. Prove: If the power series $\sum_{k=0}^{\infty} a_k x^k$ and $\sum_{k=0}^{\infty} b_k x^k$ have the same sum on an interval $(-r, r)$, then $a_k = b_k$ for all values of k.

SUPPLEMENTARY EXERCISES

1. What is the difference between an infinite sequence and an infinite series?

2. What is meant by the sum of an infinite series?

3. (a) What is a geometric series? Give some examples of convergent and divergent geometric series.

(b) What is a p-series? Give some examples of convergent and divergent p-series.

4. (a) Write down the formula for the Maclaurin series for f in sigma notation.

(b) Write down the formula for the Taylor series for f about $x = x_0$ in sigma notation.

5. State conditions under which an alternating series is guaranteed to converge.

6. (a) What does it mean to say that an infinite series converges absolutely?

(b) What relationship exists between convergence and absolute convergence of an infinite series?

7. If a power series in $x - x_0$ has radius of convergence R, what can you say about the set of x-values at which it converges?

8. State the Remainder Estimation Theorem, and describe some of its uses.

9. Are the following statements true or false? If true, state a theorem to justify your conclusion; if false, then give a counterexample.

(a) If $\sum u_k$ converges, then $u_k \to 0$ as $k \to +\infty$.

(b) If $u_k \to 0$ as $k \to +\infty$, then $\sum u_k$ converges.

(c) If $f(n) = a_n$ for $n = 1, 2, 3, \ldots$, and if $a_n \to L$ as $n \to +\infty$, then $f(x) \to L$ as $x \to +\infty$.

(d) If $f(n) = a_n$ for $n = 1, 2, 3, \ldots$, and if $f(x) \to L$ as $x \to +\infty$, then $a_n \to L$ as $n \to +\infty$.

(e) If $0 < a_n < 1$, then $\{a_n\}$ converges.

(f) If $0 < u_k < 1$, then $\sum u_k$ converges.

(g) If $\sum u_k$ and $\sum v_k$ converge, then $\sum(u_k + v_k)$ diverges.

(h) If $\sum u_k$ and $\sum v_k$ diverge, then $\sum(u_k - v_k)$ converges.

(i) If $0 \le u_k \le v_k$ and $\sum v_k$ converges, then $\sum u_k$ converges.

(j) If $0 \le u_k \le v_k$ and $\sum u_k$ diverges, then $\sum v_k$ diverges.

(k) If an infinite series converges, then it converges absolutely.

(l) If an infinite series diverges absolutely, then it diverges.

10. State whether each of the following is true or false. Justify your answers.

(a) The function $f(x) = x^{1/3}$ has a Maclaurin series.

(b) $1 + \frac{1}{2} - \frac{1}{2} + \frac{1}{3} - \frac{1}{3} + \frac{1}{4} - \frac{1}{4} + \cdots = 1$

(c) $1 + \frac{1}{2} - \frac{1}{2} + \frac{1}{2} - \frac{1}{2} + \frac{1}{2} - \frac{1}{2} + \cdots = 1$

In Exercises 11–14, use any method to determine whether the series converge.

11. (a) $\displaystyle\sum_{k=1}^{\infty} \frac{1}{5^k}$ (b) $\displaystyle\sum_{k=1}^{\infty} \frac{1}{5^k + 1}$ (c) $\displaystyle\sum_{k=1}^{\infty} \frac{9}{\sqrt{k} + 1}$

12. (a) $\displaystyle\sum_{k=1}^{\infty} (-1)^{k+1} \frac{k+4}{k^2+k}$ (b) $\displaystyle\sum_{k=1}^{\infty} (-1)^{k+1} \left(\frac{k+2}{3k-1}\right)^k$

(c) $\displaystyle\sum_{k=1}^{\infty} \frac{k^{-1/2}}{2 + \sin^2 k}$

13. (a) $\displaystyle\sum_{k=1}^{\infty} \frac{1}{k^3 + 2k + 1}$ (b) $\displaystyle\sum_{k=1}^{\infty} \frac{1}{(3+k)^{2/5}}$

(c) $\displaystyle\sum_{k=1}^{\infty} \frac{\cos(1/k)}{k^2}$

14. (a) $\sum_{k=1}^{\infty} \frac{\ln k}{k\sqrt{k}}$ (b) $\sum_{k=1}^{\infty} \frac{k^{4/3}}{8k^2+5k+1}$ (c) $\sum_{k=1}^{\infty} \frac{(-1)^{k+1}}{k^2+1}$

15. Find a formula for the exact error that results when the sum of the geometric series $\sum_{k=0}^{\infty}(1/5)^k$ is approximated by the sum of the first 100 terms in the series.

16. Does the series $1 - \frac{2}{3} + \frac{3}{5} - \frac{4}{7} + \frac{5}{9} + \cdots$ converge? Justify your answer.

17. (a) Find the first five Maclaurin polynomials of the function $p(x) = 1 - 7x + 5x^2 + 4x^3$.

(b) Make a general statement about the Maclaurin polynomials of a polynomial of degree n.

18. Use a Maclaurin series and properties of alternating series to show that $|\ln(1+x) - x| \le x^2/2$ if $0 < x < 1$.

19. Show that the approximation

$$\sin x \approx x - \frac{x^3}{3!} + \frac{x^5}{5!}$$

is accurate to four decimal places if $0 \le x \le \pi/4$.

20. Use Maclaurin series to approximate the integral

$$\int_0^1 \frac{1-\cos x}{x}\,dx$$

to three decimal-place accuracy.

21. It can be proved that

$$\lim_{n\to+\infty} \sqrt[n]{n!} = +\infty \quad \text{and} \quad \lim_{n\to+\infty} \frac{\sqrt[n]{n!}}{n} = \frac{1}{e}$$

In each part, use these limits and the root test to determine whether the series converges.

(a) $\sum_{k=0}^{\infty} \frac{2^k}{k!}$ (b) $\sum_{k=0}^{\infty} \frac{k^k}{k!}$

22. (a) Show that $k^k \ge k!$.

(b) Use the comparison test to show that $\sum_{k=1}^{\infty} k^{-k}$ converges.

(c) Use the root test to show that the series converges.

23. Suppose that $\sum_{k=1}^{n} u_k = 2 - \frac{1}{n}$. Find

(a) u_{100} (b) $\lim_{k\to+\infty} u_k$ (c) $\sum_{k=1}^{\infty} u_k$.

24. In each part, determine whether the series converges; if so, find its sum.

(a) $\sum_{k=1}^{\infty} \left(\frac{3}{2^k} - \frac{2}{3^k}\right)$ (b) $\sum_{k=1}^{\infty} [\ln(k+1) - \ln k]$

(c) $\sum_{k=1}^{\infty} \frac{1}{k(k+2)}$ (d) $\sum_{k=1}^{\infty} [\tan^{-1}(k+1) - \tan^{-1} k]$

25. In each part, find the sum of the series by associating it with some Maclaurin series.

(a) $2 + \frac{4}{2!} + \frac{8}{3!} + \frac{16}{4!} + \cdots$

(b) $\pi - \frac{\pi^3}{3!} + \frac{\pi^5}{5!} - \frac{\pi^7}{7!} + \cdots$

(c) $1 - \frac{e^2}{2!} + \frac{e^4}{4!} - \frac{e^6}{6!} + \cdots$

(d) $1 - \ln 3 + \frac{(\ln 3)^2}{2!} - \frac{(\ln 3)^3}{3!} + \cdots$

26. Suppose that the sequence $\{a_k\}$ is defined recursively by

$$a_0 = c, \quad a_{k+1} = \sqrt{a_k}$$

Assuming that the sequence converges, find its limit if

(a) $c = \frac{1}{2}$ (b) $c = \frac{3}{2}$.

27. Research has shown that the proportion p of the population with IQs (intelligence quotients) between α and β is approximately

$$p = \frac{1}{16\sqrt{2\pi}} \int_{\alpha}^{\beta} e^{-\frac{1}{2}\left(\frac{x-100}{16}\right)^2}\,dx$$

Use the first three terms of an appropriate Maclaurin series to estimate the proportion of the population that has IQs between 100 and 110.

28. Differentiate the Maclaurin series for xe^x and use the result to show that

$$\sum_{k=0}^{\infty} \frac{k+1}{k!} = 2e$$

29. Given: $\frac{\pi^2}{6} = 1 + \frac{1}{2^2} + \frac{1}{3^2} + \frac{1}{4^2} + \cdots$.

Show: $\frac{\pi^2}{12} = 1 - \frac{1}{2^2} + \frac{1}{3^2} - \frac{1}{4^2} + \cdots$.

30. Let a, b, and p be positive constants. For which values of p does the series $\sum_{k=1}^{\infty} \frac{1}{(a+bk)^p}$ converge?

31. In each part, write out the first four terms of the series, and then find the radius of convergence.

(a) $\sum_{k=1}^{\infty} \frac{1\cdot 2\cdot 3\cdots k}{1\cdot 4\cdot 7\cdots(3k-2)} x^k$

(b) $\sum_{k=1}^{\infty} (-1)^k \frac{1\cdot 2\cdot 3\cdots k}{1\cdot 3\cdot 5\cdots(2k-1)} x^{2k+1}$

32. Find the interval of convergence of

$$\sum_{k=0}^{\infty} \frac{(x-x_0)^k}{b^k} \quad (b > 0)$$

33. Show that the series

$$1 - \frac{x}{2!} + \frac{x^2}{4!} - \frac{x^3}{6!} + \cdots$$

converges to the function

$$f(x) = \begin{cases} \cos\sqrt{x}, & x \ge 0 \\ \cosh\sqrt{-x}, & x < 0 \end{cases}$$

[*Hint:* Use the Maclaurin series for $\cos x$ and $\cosh x$ to obtain series for $\cos\sqrt{x}$, where $x \ge 0$, and $\cosh\sqrt{-x}$, where $x \le 0$.]

34. Prove:

(a) If f is an even function, then all odd powers of x in its Maclaurin series have coefficient 0.

(b) If f is an odd function, then all even powers of x in its Maclaurin series have coefficient 0.

35. In Section 8.6 we defined the kinetic energy K of a particle with mass m and velocity v to be $K = \frac{1}{2}mv^2$ [see Formula (5) of that section]. In this formula the mass m is assumed to be constant, and K is called the ***Newtonian Kinetic Energy***. However, in Albert Einstein's relativity theory the mass m increases with the velocity and the kinetic energy K is given by the formula

$$K = m_0c^2\left[\frac{1}{\sqrt{1-(v/c)^2}} - 1\right]$$

in which m_0 is the mass of the particle when its velocity is zero, and c is the speed of light. This is called the ***relativistic kinetic energy***. Use an appropriate binomial series to show that if the velocity is small compared to the speed of light (i.e., $v/c \approx 0$), then the Newtonian and relativistic kinetic energies are in close agreement.

C **36.** If the constant p in the general p-series is replaced by a variable x for $x > 1$, then the resulting function is called the ***Riemann zeta function*** and is denoted by

$$\zeta(x) = \sum_{k=1}^{\infty} \frac{1}{k^x}$$

(a) Let s_n be the nth partial sum of the series for $\zeta(3.7)$. Find n such that s_n approximates $\zeta(3.7)$ to two decimal-place accuracy, and calculate s_n using this value of n. [*Hint:* Use the right inequality in Exercise 30(b) of Section 11.4 with $f(x) = 1/x^{3.7}$.]

(b) Determine whether your CAS can evaluate the Riemann zeta function directly. If so, compare the value produced by the CAS to the value of s_n obtained in part (a).

EXPANDING THE CALCULUS HORIZON

For additional material relating to this chapter, visit the Anton Website at http://www.wiley.com/college/anton

Johannes Kepler

12

ANALYTIC GEOMETRY IN CALCULUS

In this chapter we will study aspects of analytic geometry that are important in applications of calculus. We will begin by introducing *polar coordinate systems*, which are used, for example, in tracking the motion of planets and satellites, in identifying the location of objects from information on radar screens, and in the design of antennas. We will then discuss relationships between curves in polar coordinates and parametric curves in rectangular coordinates, and we will discuss methods for finding areas in polar coordinates and tangent lines to curves given in polar coordinates or parametrically in rectangular coordinates. We will then review the basic properties of parabolas, ellipses, and hyperbolas and discuss these curves in the context of polar coordinates. Finally, we will give some basic applications of our work in astronomy.

12.1 POLAR COORDINATES

Up to now we have specified the location of a point in the plane by means of coordinates relative to two perpendicular coordinate axes. However, sometimes a moving point has a special affinity for some fixed point, such as a planet moving in an orbit under the central attraction of the Sun. In such cases, the path of the particle is best described by its angular direction and its distance from the fixed point. In this section we will discuss a new kind of coordinate system that is based on this idea.

POLAR COORDINATE SYSTEMS

A ***polar coordinate system*** in a plane consists of a fixed point O, called the ***pole*** (or ***origin***), and a ray emanating from the pole, called the ***polar axis***. In such a coordinate system we can associate with each point P in the plane a pair of ***polar coordinates*** (r, θ), where r is the distance from P to the pole and θ is an angle from the polar axis to the ray OP (Figure 12.1.1). The number r is called the ***radial coordinate*** of P and the number θ the ***angular coordinate*** (or ***polar angle***) of P. In Figure 12.1.2, the points $(6, 45^\circ)$, $(5, 120^\circ)$, $(3, 225^\circ)$, and $(4, 330^\circ)$ are plotted in polar coordinate systems. If P is the pole, then $r = 0$, but there is no clearly defined polar angle. We will agree that an arbitrary angle can be used in this case; that is, $(0, \theta)$ are polar coordinates of the pole for all choices of θ.

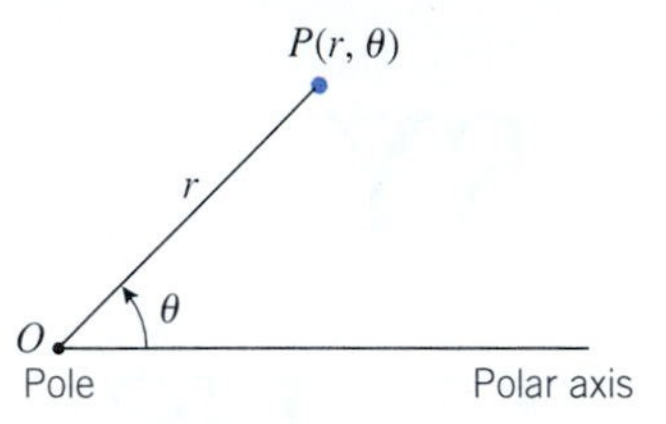

Figure 12.1.1

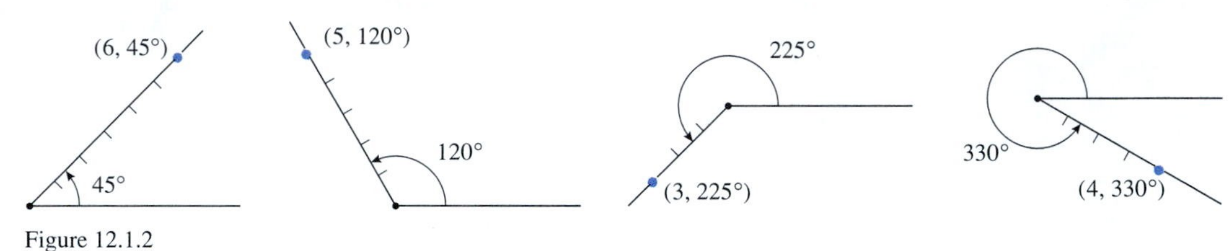

Figure 12.1.2

The polar coordinates of a point are not unique. For example, the polar coordinates

$$(1, 315^\circ), \quad (1, -45^\circ), \quad \text{and} \quad (1, 675^\circ)$$

all represent the same point (Figure 12.1.3). In general, if a point P has polar coordinates (r, θ), then

$$(r, \theta + n \cdot 360^\circ) \quad \text{and} \quad (r, \theta - n \cdot 360^\circ)$$

are also polar coordinates of P for any nonnegative integer n. Thus, every point has infinitely many pairs of polar coordinates.

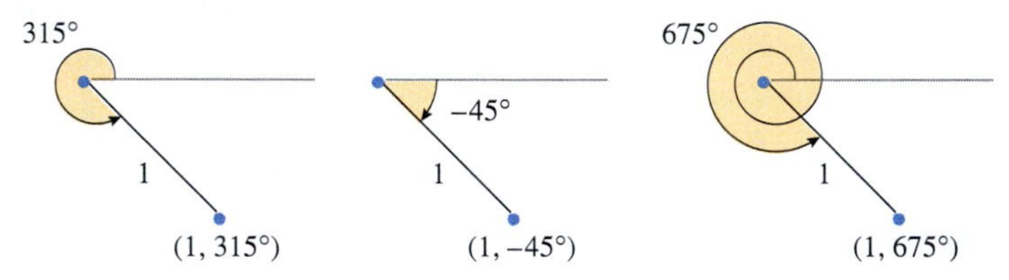

Figure 12.1.3

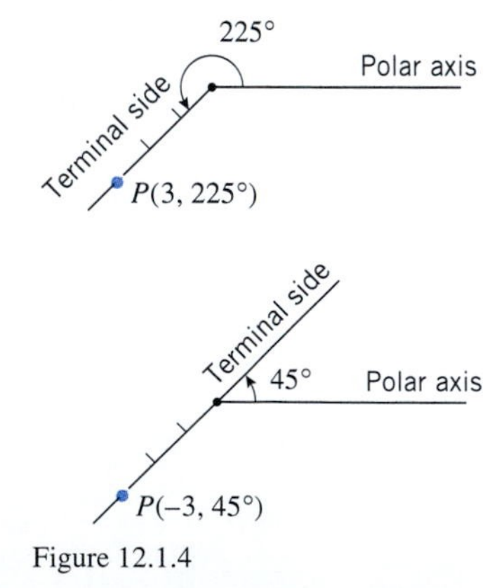

Figure 12.1.4

As defined above, the radial coordinate r of a point P is nonnegative, since it represents the distance from P to the pole. However, it will be convenient to allow for negative values of r as well. To motivate an appropriate definition, consider the point P with polar coordinates $(3, 225^\circ)$. As shown in Figure 12.1.4, we can reach this point by rotating the polar axis through an angle of 225° and then moving 3 units from the pole along the terminal side of the angle, or we can reach the point P by rotating the polar axis through an angle of 45° and then moving 3 units from the pole along the extension of the terminal side. This suggests that the point $(3, 225^\circ)$ might also be denoted by $(-3, 45^\circ)$, with the minus sign serving to indicate that the point is on the *extension* of the angle's terminal side rather than on the terminal side itself.

In general, the terminal side of the angle $\theta + 180°$ is the extension of the terminal side of θ, so we define negative radial coordinates by agreeing that

$$(-r, \theta) \quad \text{and} \quad (r, \theta + 180°)$$

are polar coordinates of the same point.

FOR THE READER. For many purposes it does not matter whether polar angles are measured in degrees or radians. However, in problems that involve derivatives or integrals they must be measured in radians, since the derivatives of the trigonometric functions were derived under this assumption. Henceforth, we will use radian measure for polar angles, except in certain applications where it is not required and degree measure is more convenient.

RELATIONSHIP BETWEEN POLAR AND RECTANGULAR COORDINATES

Frequently, it will be useful to superimpose a rectangular xy-coordinate system on top of a polar coordinate system, making the positive x-axis coincide with the polar axis. If this is done, then every point P will have both rectangular coordinates (x, y) and polar coordinates (r, θ). As suggested by Figure 12.1.5, these coordinates are related by the equations

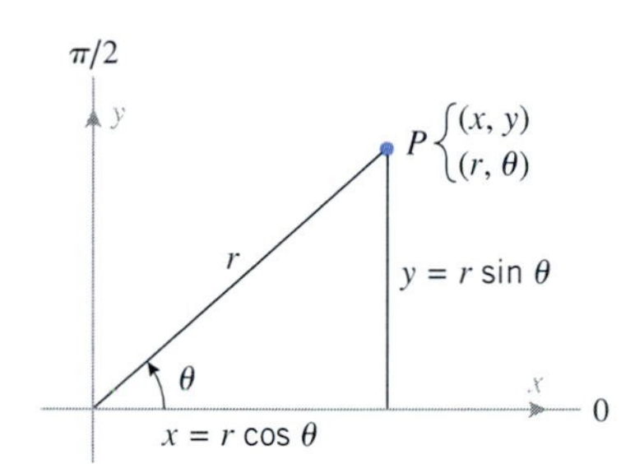

Figure 12.1.5

$$x = r\cos\theta, \quad y = r\sin\theta \tag{1}$$

These equations are well suited for finding x and y when r and θ are known. However, to find r and θ when x and y are known, it is preferable to use the identities $\sin^2\theta + \cos^2\theta = 1$ and $\tan\theta = \sin\theta/\cos\theta$ to rewrite (1) as

$$r^2 = x^2 + y^2, \quad \tan\theta = \frac{y}{x} \tag{2}$$

Example 1

Find the rectangular coordinates of the point P whose polar coordinates are $(6, 2\pi/3)$.

Solution. Substituting the polar coordinates $r = 6$ and $\theta = 2\pi/3$ in (1) yields

$$x = 6\cos\frac{2\pi}{3} = 6\left(-\frac{1}{2}\right) = -3$$

$$y = 6\sin\frac{2\pi}{3} = 6\left(\frac{\sqrt{3}}{2}\right) = 3\sqrt{3}$$

Thus, the rectangular coordinates of P are $(-3, 3\sqrt{3})$ (Figure 12.1.6). ◀

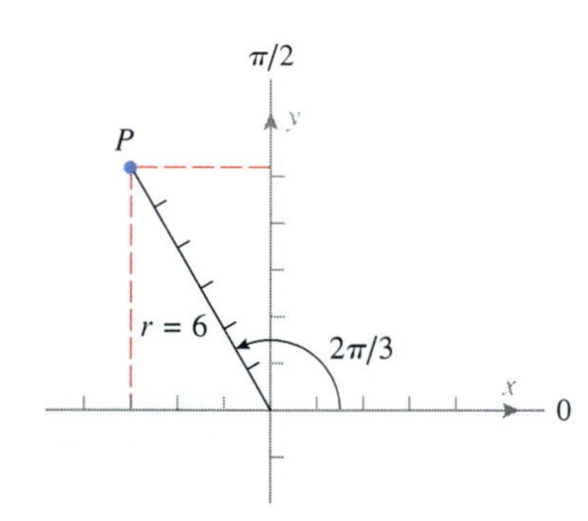

Figure 12.1.6

Example 2

Find polar coordinates of the point P whose rectangular coordinates are $(-2, 2\sqrt{3})$.

Solution. We will find the polar coordinates (r, θ) of P that satisfy the conditions $r > 0$ and $0 \le \theta < 2\pi$. From the first equation in (2),

$$r^2 = x^2 + y^2 = (-2)^2 + (2\sqrt{3})^2 = 4 + 12 = 16$$

so $r = 4$. From the second equation in (2),

$$\tan\theta = \frac{y}{x} = \frac{2\sqrt{3}}{-2} = -\sqrt{3}$$

From this and the fact that $(-2, 2\sqrt{3})$ lies in the second quadrant, it follows that the angle satisfying the requirement $0 \le \theta < 2\pi$ is $\theta = 2\pi/3$. Thus, $(4, 2\pi/3)$ are polar coordinates of P. All other polar coordinates of P are expressible in the form

$$\left(4, \frac{2\pi}{3} + 2n\pi\right) \quad \text{or} \quad \left(-4, \frac{5\pi}{3} + 2n\pi\right)$$

where n is an integer. ◀

GRAPHS IN POLAR COORDINATES

We will now consider the problem of graphing equations of the form $r = f(\theta)$ in polar coordinates, where θ is assumed to be measured in radians. Some examples of such equations are

$$r = 2\cos\theta, \quad r = \frac{4}{1 - 3\sin\theta}, \quad r = \theta$$

In a rectangular coordinate system the graph of an equation $y = f(x)$ consists of all points whose coordinates (x, y) satisfy the equation. However, in a polar coordinate system, points have infinitely many different pairs of polar coordinates, so that a given point may have some polar coordinates that satisfy the equation $r = f(\theta)$ and others that do not. Taking this into account, we define the ***graph of r = f(θ) in polar coordinates*** to consist of all points with *at least one* pair of coordinates (r, θ) that satisfy the equation.

The most elementary way to graph an equation $r = f(\theta)$ in polar coordinates is to plot points. The idea is to choose some typical values of θ, calculate the corresponding values of r, and then plot the resulting pairs (r, θ) in a polar coordinate system. Here are some examples.

Example 3

Sketch the graph of the equation $r = \sin\theta$ in polar coordinates by plotting points.

Solution. Table 12.1.1 shows the coordinates of points on the graph at increments of $\pi/6$ $(= 30^\circ)$.

Table 12.1.1

θ (RADIANS)	0	$\frac{\pi}{6}$	$\frac{\pi}{3}$	$\frac{\pi}{2}$	$\frac{2\pi}{3}$	$\frac{5\pi}{6}$	π	$\frac{7\pi}{6}$	$\frac{4\pi}{3}$	$\frac{3\pi}{2}$	$\frac{5\pi}{3}$	$\frac{11\pi}{6}$	2π
$r = \sin\theta$	0	$\frac{1}{2}$	$\frac{\sqrt{3}}{2}$	1	$\frac{\sqrt{3}}{2}$	$\frac{1}{2}$	0	$-\frac{1}{2}$	$-\frac{\sqrt{3}}{2}$	-1	$-\frac{\sqrt{3}}{2}$	$-\frac{1}{2}$	0
(r, θ)	$(0, 0)$	$\left(\frac{1}{2}, \frac{\pi}{6}\right)$	$\left(\frac{\sqrt{3}}{2}, \frac{\pi}{3}\right)$	$\left(1, \frac{\pi}{2}\right)$	$\left(\frac{\sqrt{3}}{2}, \frac{2\pi}{3}\right)$	$\left(\frac{1}{2}, \frac{5\pi}{6}\right)$	$(0, \pi)$	$\left(-\frac{1}{2}, \frac{7\pi}{6}\right)$	$\left(-\frac{\sqrt{3}}{2}, \frac{4\pi}{3}\right)$	$\left(-1, \frac{3\pi}{2}\right)$	$\left(-\frac{\sqrt{3}}{2}, \frac{5\pi}{3}\right)$	$\left(-\frac{1}{2}, \frac{11\pi}{6}\right)$	$(0, 2\pi)$

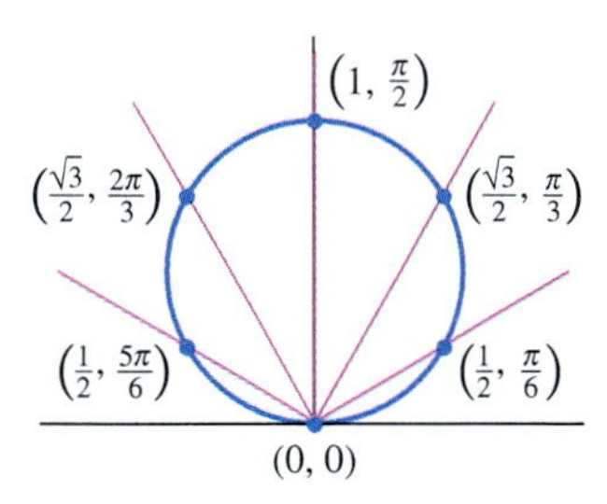

Figure 12.1.7

These points are plotted in Figure 12.1.7. Note, however, that there are 13 points listed in the table but only 6 distinct plotted points. This is because the pairs from $\theta = \pi$ on yield duplicates of the preceding points. For example, $(-1/2, 7\pi/6)$ and $(1/2, \pi/6)$ represent the same point. ◀

Observe that the points in Figure 12.1.7 appear to lie on a circle. We can confirm that this is so by expressing the polar equation $r = \sin\theta$ in terms of x and y. To do this, we multiply the equation through by r to obtain

$$r^2 = r\sin\theta$$

which now allows us to apply Formulas (1) and (2) to rewrite the equation as

$$x^2 + y^2 = y$$

Rewriting this equation as $x^2 + y^2 - y = 0$ and then completing the square yields

$$x^2 + \left(y - \tfrac{1}{2}\right)^2 = \tfrac{1}{4}$$

which is a circle of radius $\frac{1}{2}$ centered at the point $\left(0, \frac{1}{2}\right)$ in the xy-plane.

Just because an equation $r = f(\theta)$ involves the variables r and θ does not mean that it has to be graphed in a polar coordinate system. When useful, this equation can also be graphed in a rectangular coordinate system. For example, Figure 12.1.8 shows the graph of $r = \sin\theta$ in a rectangular θr-coordinate system. This graph can actually help to visualize how the polar graph in Figure 12.1.7 is generated:

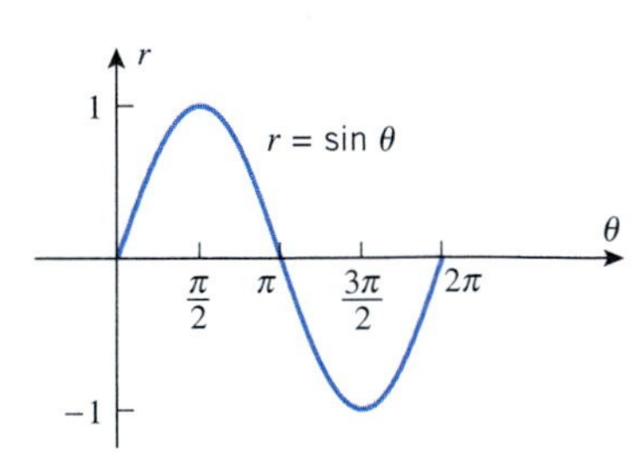

Figure 12.1.8

- At $\theta = 0$ we have $r = 0$, which corresponds to the pole $(0, 0)$ on the polar graph.
- As θ varies from 0 to $\pi/2$, the value of r increases from 0 to 1, so the point (r, θ) moves along the circle from the pole to the high point at $(1, \pi/2)$.
- As θ varies from $\pi/2$ to π, the value of r decreases from 1 back to 0, so the point (r, θ) moves along the circle from the high point back to the pole.
- As θ varies from π to $3\pi/2$, the values of r are negative, varying from 0 to -1. Thus, the point (r, θ) moves along the circle from the pole to the high point at $(1, \pi/2)$, which is the same as the point $(-1, 3\pi/2)$. This duplicates the motion that occurred for $0 \le \theta \le \pi/2$.
- As θ varies from $3\pi/2$ to 2π, the value of r varies from -1 to 0. Thus, the point (r, θ) moves along the circle from the high point back to the pole, duplicating the motion that occurred for $\pi/2 \le \theta \le \pi$.

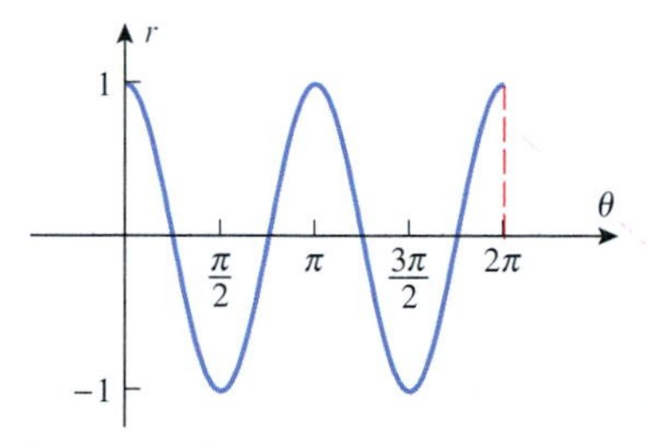

Figure 12.1.9

Example 4

Sketch the graph of $r = \cos 2\theta$ in polar coordinates.

Solution. Instead of plotting points, we will use the graph of $r = \cos 2\theta$ in rectangular coordinates (Figure 12.1.9) to visualize how the polar graph of this equation is generated. The analysis and the resulting polar graph are shown in Figure 12.1.10. This curve is called a *four-petal rose*. ◀

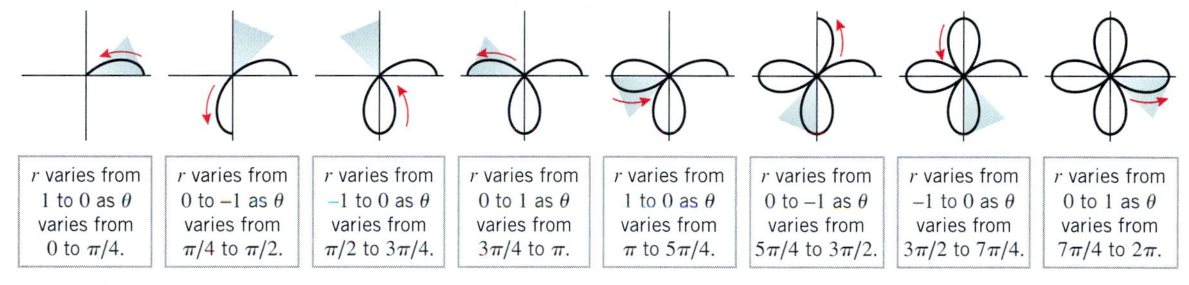

r varies from 1 to 0 as θ varies from 0 to $\pi/4$.	r varies from 0 to -1 as θ varies from $\pi/4$ to $\pi/2$.	r varies from -1 to 0 as θ varies from $\pi/2$ to $3\pi/4$.	r varies from 0 to 1 as θ varies from $3\pi/4$ to π.	r varies from 1 to 0 as θ varies from π to $5\pi/4$.	r varies from 0 to -1 as θ varies from $5\pi/4$ to $3\pi/2$.	r varies from -1 to 0 as θ varies from $3\pi/2$ to $7\pi/4$.	r varies from 0 to 1 as θ varies from $7\pi/4$ to 2π.

Figure 12.1.10

SYMMETRY TESTS

Observe that the polar graph of $r = \cos 2\theta$ in Figure 12.1.10 is symmetric about the x-axis and the y-axis. This symmetry could have been predicted from the following theorem, which is suggested by Figure 12.1.11 (we omit the proof).

> **12.1.1** THEOREM (***Symmetry Tests***).
>
> (*a*) *A curve in polar coordinates is symmetric about the x-axis if replacing θ by $-\theta$ in its equation produces an equivalent equation* (Figure 12.1.11*a*).
>
> (*b*) *A curve in polar coordinates is symmetric about the y-axis if replacing θ by $\pi - \theta$ in its equation produces an equivalent equation* (Figure 12.1.11*b*).
>
> (*c*) *A curve in polar coordinates is symmetric about the origin if replacing r by $-r$ in its equation produces an equivalent equation* (Figure 12.1.11*c*).

Example 5

Use Theorem 12.1.1 to confirm that the graph of $r = \cos 2\theta$ in Figure 12.1.10 is symmetric about the x-axis and y-axis.

Solution. To test for symmetry about the x-axis, we replace θ by $-\theta$. This yields

$$r = \cos(-2\theta) = \cos 2\theta$$

Thus, replacing θ by $-\theta$ does not alter the equation.

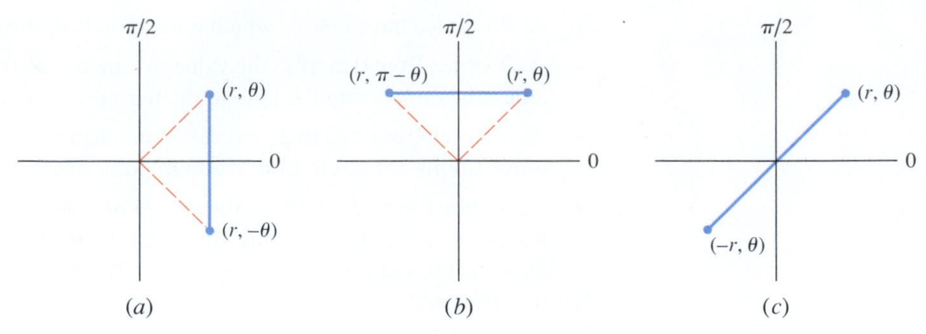

Figure 12.1.11

To test for symmetry about the y-axis, we replace θ by $\pi - \theta$. This yields

$$r = \cos 2(\pi - \theta) = \cos(2\pi - 2\theta) = \cos(-2\theta) = \cos 2\theta$$

Thus, replacing θ by $\pi - \theta$ does not alter the equation. ◀

Example 6

Sketch the graph of $r = a(1 - \cos\theta)$ in polar coordinates, assuming a to be a positive constant.

Solution. Observe first that replacing θ by $-\theta$ does not alter the equation, so we know in advance that the graph is symmetric about the polar axis. Thus, if we graph the upper half of the curve, then we can obtain the lower half by reflection about the polar axis.

As in our previous examples, we will first graph the equation in rectangular coordinates. This graph, which is shown in Figure 12.1.12a, can be obtained by rewriting the given equation as $r = a - a\cos\theta$, from which we see that the graph in rectangular coordinates can be obtained by first reflecting the graph of $r = a\cos\theta$ about the x-axis to obtain the graph of $r = -a\cos\theta$, and then translating that graph up a units to obtain the graph of $r = a - a\cos\theta$. Now we can see that:

- As θ varies from 0 to $\pi/3$, r increases from 0 to $a/2$.
- As θ varies from $\pi/3$ to $\pi/2$, r increases from $a/2$ to a.
- As θ varies from $\pi/2$ to $2\pi/3$, r increases from a to $3a/2$.
- As θ varies from $2\pi/3$ to π, r increases from $3a/2$ to $2a$.

This produces the polar curve shown in Figure 12.1.12b. The rest of the curve can be obtained by continuing the preceding analysis from π to 2π or, as noted above, by reflecting the portion already graphed about the x-axis (Figure 12.1.12c). This heart-shaped curve is called a ***cardioid*** (from the Greek word "kardia" for heart). ◀

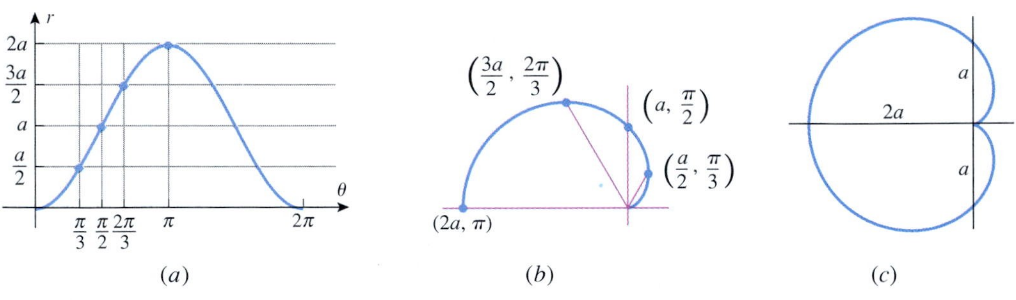

Figure 12.1.12

Example 7

Sketch the curves

$$\text{(a) } r = 1 \qquad \text{(b) } \theta = \frac{\pi}{4} \qquad \text{(c) } r = \theta \quad (\theta \geq 0)$$

in polar coordinates.

Solution (a). For all values of θ, the point $(1, \theta)$ is 1 unit away from the pole. Thus, the graph is the circle of radius 1 centered at the pole (Figure 12.1.13*a*).

Solution (b). For all values of r, the point $(r, \pi/4)$ lies on a line that makes an angle of $\pi/4$ with the polar axis (Figure 12.1.13*b*). Positive values of r correspond to points on the line in the first quadrant and negative values of r to points on the line in the third quadrant. Thus, in absence of any restriction on r, the graph is the entire line. Observe, however, that had we imposed the restriction $r \geq 0$, the graph would have been just the ray in the first quadrant.

Solution (c). Observe that as θ increases, so does r; thus, the graph is a curve that spirals out from the pole as θ increases. A reasonably accurate sketch of the spiral can be obtained by plotting the intersections with the x- and y-axes for values of θ that are multiples of $\pi/2$, keeping in mind that the value of r is always equal to the value of θ (Figure 12.1.13*c*). ◀

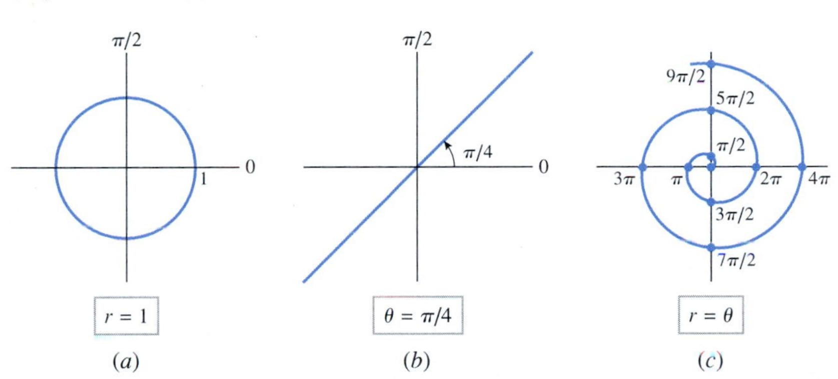

Figure 12.1.13

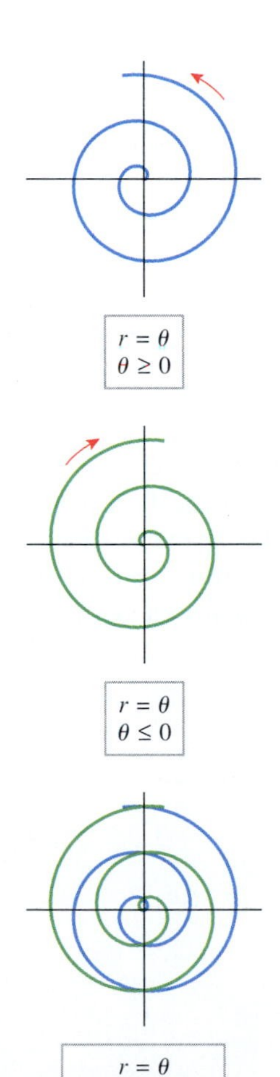

Figure 12.1.14

REMARK. The spiral in Figure 12.1.13*c*, which belongs to the family of ***Archimedean spirals*** $r = a\theta$, coils counterclockwise around the pole because of the restriction $\theta \geq 0$. Had we made the restriction $\theta \leq 0$, the spiral would have coiled clockwise, and had we allowed both positive and negative values of θ, the clockwise and counterclockwise spirals would have been superimposed to form a double Archimedean spiral (Figure 12.1.14).

Example 8

Sketch the graph of $r^2 = 4\cos 2\theta$ in polar coordinates.

Solution. This equation does not express r as a function of θ, since solving for r in terms of θ yields two functions:

$$r = 2\sqrt{\cos 2\theta} \quad \text{and} \quad r = -2\sqrt{\cos 2\theta}$$

Thus, to graph the equation $r^2 = 4\cos 2\theta$ we will have to graph the two functions separately and then combine those graphs.

We will start with the graph of $r = 2\sqrt{\cos 2\theta}$. Observe first that this equation is not changed if we replace θ by $-\theta$ or if we replace θ by $\pi - \theta$. Thus, the graph is symmetric about the x-axis and the y-axis. This means that the entire graph can be obtained by graphing the portion in the first quadrant, reflecting that portion about the y-axis to obtain the portion in the second quadrant and then reflecting those two portions about the x-axis to obtain the portions in the third and fourth quadrants.

To begin the analysis, we will graph the equation $r = 2\sqrt{\cos 2\theta}$ in rectangular coordinates (see Figure 12.1.15*a*). Note that there are gaps in that graph over the intervals $\pi/4 < \theta < 3\pi/4$ and $5\pi/4 < \theta < 7\pi/4$ because $\cos 2\theta$ is negative for those values of θ. From this graph we can see that:

- As θ varies from 0 to $\pi/4$, r decreases from 2 to 0.
- As θ varies from $\pi/4$ to $\pi/2$, no points are generated on the polar graph.

This produces the portion of the graph shown in Figure 12.1.15*b*. As noted above, we can complete the graph by a reflection about the y-axis followed by a reflection about the x-axis (12.1.15*c*). The resulting propeller-shaped graph is called a ***lemniscate*** (from the Greek word "lemniscos" for a looped ribbon resembling the number 8). We leave it for you to verify that the equation $r = 2\sqrt{\cos 2\theta}$ has the same graph as $r = -2\sqrt{\cos 2\theta}$, but traced in a diagonally opposite manner. Thus, the graph of the equation $r^2 = 4\cos 2\theta$ consists of two identical superimposed lemniscates. ◀

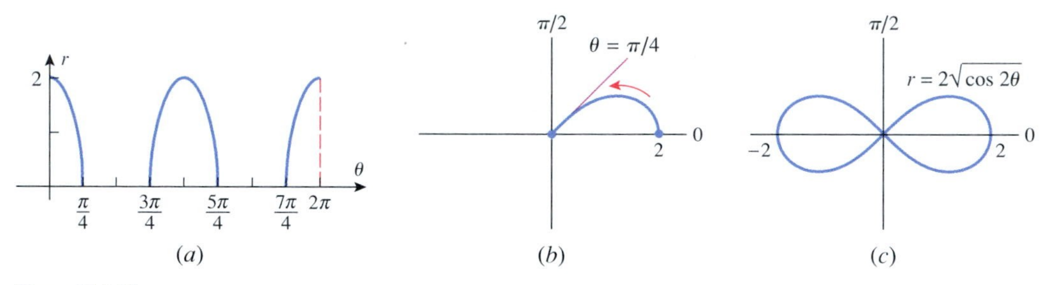

Figure 12.1.15

FAMILIES OF LINES AND RAYS THROUGH THE POLE

If θ_0 is a fixed angle, then for all values of r the point (r, θ_0) lies on the line that makes an angle of $\theta = \theta_0$ with the polar axis; and, conversely, every point on this line has a pair of polar coordinates of the form (r, θ_0). Thus, the equation $\theta = \theta_0$ represents the line that passes through the pole and makes an angle of θ_0 with the polar axis (Figure 12.1.16*a*). If r is restricted to be nonnegative, then the graph of the equation $\theta = \theta_0$ is the ray that emanates from the pole and makes an angle of θ_0 with the polar axis (Figure 12.1.16*b*). Thus, as θ_0 varies, the equation $\theta = \theta_0$ produces either a family of lines through the pole or a family of rays through the pole, depending on the restrictions on r.

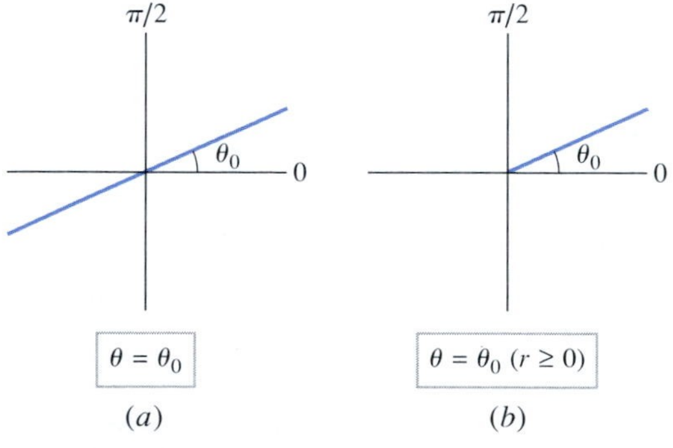

Figure 12.1.16

FAMILIES OF CIRCLES

We will consider three families of circles in which a is assumed to be a positive constant:

$$r = a \qquad r = 2a\cos\theta \qquad r = 2a\sin\theta \tag{3–5}$$

The equation $r = a$ represents a circle of radius a centered at the pole (Figure 12.1.17a). Thus, as a varies, this equation produces a family of circles centered at the pole. For families (4) and (5), recall from plane geometry that a triangle that is inscribed in a circle with a diameter of the circle for a side must be a right triangle. Thus, as indicated in Figures 12.1.17b and 12.1.17c, the equation $r = 2a\cos\theta$ represents a circle of radius a, centered on the x-axis and tangent to the y-axis at the origin; similarly, the equation $r = 2a\sin\theta$ represents a circle of radius a, centered on the y-axis and tangent to the x-axis at the origin. Thus, as a varies, Equations (4) and (5) produce the families illustrated in Figures 12.1.17d and 12.1.17e.

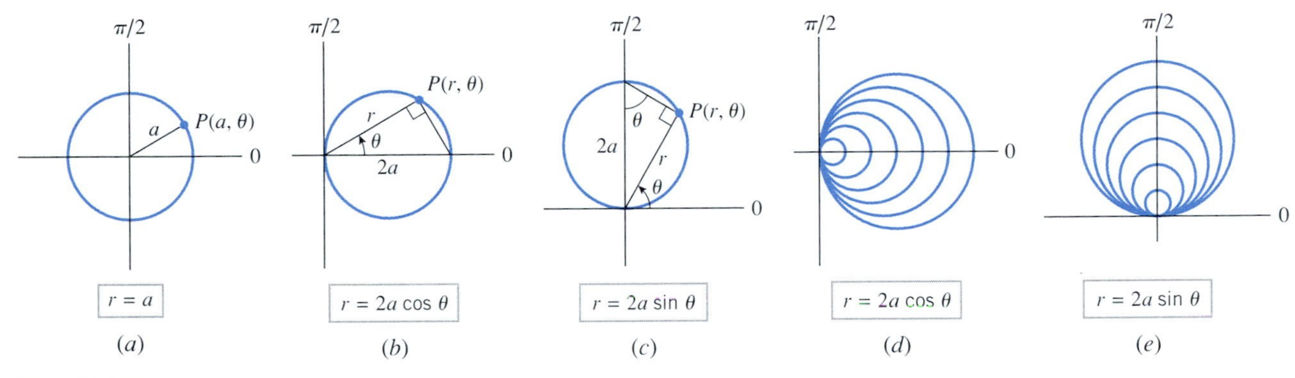

Figure 12.1.17

REMARK. Observe that replacing θ by $-\theta$ does not change the equation $r = 2a\cos\theta$, and replacing θ by $\pi - \theta$ does not change the equation $r = 2a\sin\theta$. This explains why the circles in Figure 12.1.17d are symmetric about the x-axis and those in Figure 12.1.17e are symmetric about the y-axis.

FAMILIES OF ROSE CURVES

In polar coordinates, equations of the form

$$r = a\sin n\theta \qquad r = a\cos n\theta \tag{6–7}$$

in which $a > 0$ and n is a positive integer represent families of flower-shaped curves called ***roses*** (Figure 12.1.18). The rose consists of n equally spaced petals of radius a if n is odd

ROSE CURVES

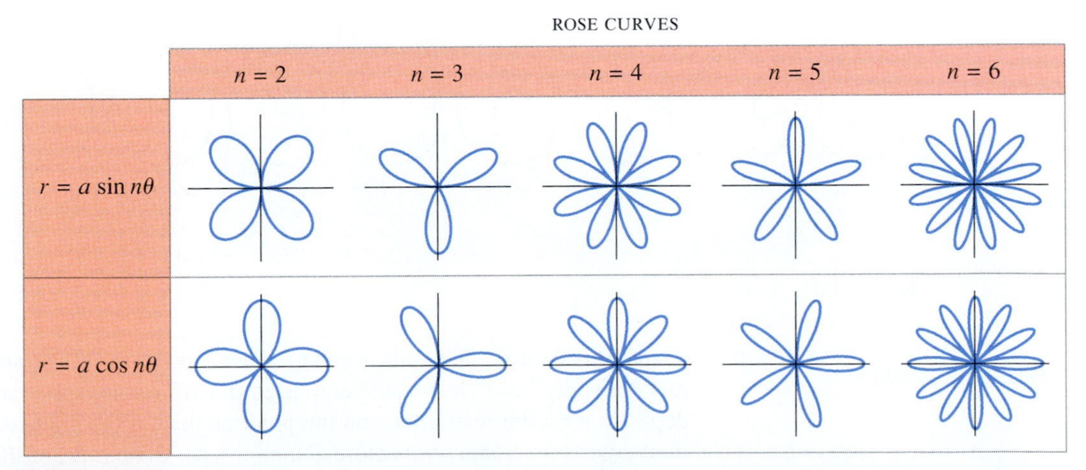

Figure 12.1.18

and $2n$ equally spaced petals of radius a if n is positive and even. It can be shown that a rose with an even number of petals is traced out exactly once as θ varies over the interval $0 \le \theta < 2\pi$ and a rose with an odd number of petals is traced out exactly once as θ varies over the interval $0 \le \theta < \pi$ (Exercise 73). A four-petal rose of radius 1 was graphed in Example 4.

FOR THE READER. What do the graphs of the one-petal roses look like?

FAMILIES OF CARDIOIDS AND LIMAÇONS

Equations with any of the four forms

$$r = a \pm b \sin\theta \qquad r = a \pm b \cos\theta \tag{8–9}$$

in which $a > 0$ and $b > 0$ represent polar curves called ***limaçons*** (from the Latin word "limax" for a snail-like creature that is commonly called a slug). There are four possible shapes for a limaçon that are determined by the ratio a/b (Figure 12.1.19). If $a = b$ (the case $a/b = 1$), then the limaçon is called a ***cardioid*** because of its heart-shaped appearance, as noted in Example 6.

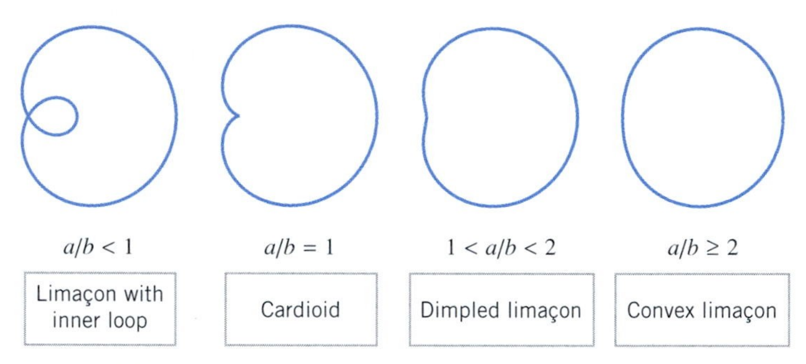

Figure 12.1.19

Example 9

Figure 12.1.20 shows the family of limaçons $r = a + \cos\theta$ with the constant a varying from 0.25 to 2.50 in steps of 0.25. In keeping with Figure 12.1.19, the limaçons evolve from the loop type to the convex type. As a increases from the starting value of 0.25, the loops get smaller and smaller until the cardioid is reached at $a = 1$. As a increases further, the limaçons evolve through the dimpled type into the convex type. ◀

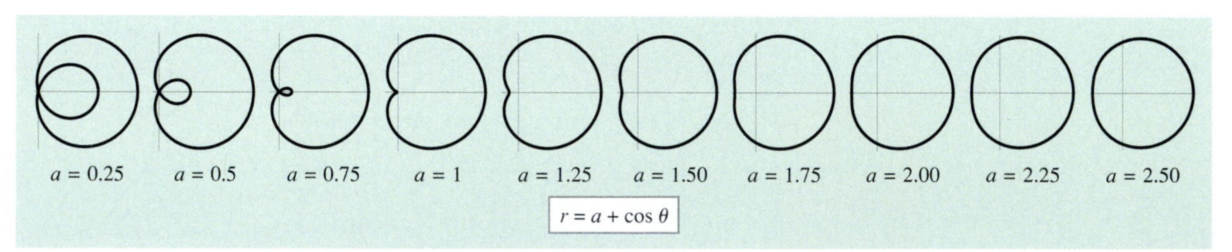

Figure 12.1.20

FAMILIES OF SPIRALS

A ***spiral*** is a curve that coils around a central point. As illustrated in Figure 12.1.14, spirals generally have "left-hand" and "right-hand" versions that coil in opposite directions, depending on the restrictions on the polar angle and the signs of constants that appear in their equations. Some of the more common types of spirals are shown in Figure 12.1.21 for nonnegative values of θ, a, and b.

Archimedean spiral
$r = a\theta$

Parabolic spiral
$r = a\sqrt{\theta}$

Logarithmic spiral
$r = ae^{b\theta}$

Lituus spiral
$r = a/\sqrt{\theta}$

Hyperbolic spiral
$r = a/\theta$

Figure 12.1.21

SPIRALS IN NATURE

Spirals of many kinds occur in nature. For example, the shell of the chambered nautilus (*below*) forms a logarithmic spiral, and a coiled sailor's rope forms an Archimedean spiral. Spirals also occur in flowers, the tusks of certain animals, and in the shapes of galaxies.

The shell of the chambered nautilus reveals a logarithmic spiral. The animal lives in the outermost chamber.

A sailor's coiled rope forms an Archimedean spiral.

GENERATING POLAR CURVES WITH GRAPHING UTILITIES

For polar curves that are too complicated for hand computation, graphing utilities must be used. Although many graphing utilities are capable of graphing polar curves directly, some are not. However, if a graphing utility is capable of graphing parametric equations, then it can be used to graph a polar curve $r = f(\theta)$ by converting this equation to parametric form. This can be done by substituting $f(\theta)$ for r in (1). This yields

$$x = f(\theta)\cos\theta, \quad y = f(\theta)\sin\theta \tag{10}$$

which is a pair of parametric equations for the polar curve in terms of the parameter θ.

Example 10

Express the polar equation

$$r = 2 + \cos\frac{5\theta}{2}$$

parametrically, and generate the polar graph from the parametric equations using a graphing utility.

Solution. Substituting the given expression for r in $x = r\cos\theta$ and $y = r\sin\theta$ yields the parametric equations

$$x = \left[2 + \cos\frac{5\theta}{2}\right]\cos\theta, \quad y = \left[2 + \cos\frac{5\theta}{2}\right]\sin\theta$$

Next, we need to find an interval over which to vary θ to produce the entire graph. To find such an interval, we will look for the smallest number of complete revolutions that must occur until the value of r begins to repeat. Algebraically, this amounts to finding the smallest positive integer n such that

$$2+\cos\left(\frac{5(\theta+2n\pi)}{2}\right)=2+\cos\frac{5\theta}{2}$$

or

$$\cos\left(\frac{5\theta}{2}+5n\pi\right)=\cos\frac{5\theta}{2}$$

For this equality to hold, the quantity $5n\pi$ must be an even multiple of π; the smallest n for which this occurs is $n=2$. Thus, the entire graph will be traced in two revolutions, which means it can be generated from the parametric equations

$$x=\left[2+\cos\frac{5\theta}{2}\right]\cos\theta,\quad y=\left[2+\cos\frac{5\theta}{2}\right]\sin\theta\qquad(0\le\theta\le 4\pi)$$

This yields the graph in Figure 12.1.22. ◀

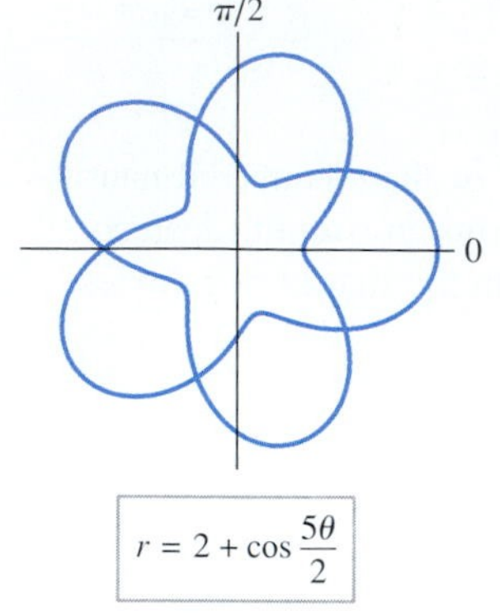

$r=2+\cos\dfrac{5\theta}{2}$

Figure 12.1.22

FOR THE READER. Some graphing utilities require that t be used for the parameter. If this is true of your graphing utility, then you will have to replace θ by t in (10) to generate graphs in polar coordinates. Use a graphing utility to duplicate the curve in Figure 12.1.22.

EXERCISE SET 12.1 Graphing Calculator

In Exercises 1 and 2, plot the points in polar coordinates.

1. (a) $(3, \pi/4)$ (b) $(5, 2\pi/3)$ (c) $(1, \pi/2)$
(d) $(4, 7\pi/6)$ (e) $(-6, -\pi)$ (f) $(-1, 9\pi/4)$

2. (a) $(2, -\pi/3)$ (b) $(3/2, -7\pi/4)$ (c) $(-3, 3\pi/2)$
(d) $(-5, -\pi/6)$ (e) $(2, 4\pi/3)$ (f) $(0, \pi)$

In Exercises 3 and 4, find the rectangular coordinates of the points whose polar coordinates are given.

3. (a) $(6, \pi/6)$ (b) $(7, 2\pi/3)$ (c) $(-6, -5\pi/6)$
(d) $(0, -\pi)$ (e) $(7, 17\pi/6)$ (f) $(-5, 0)$

4. (a) $(-8, \pi/4)$ (b) $(7, -\pi/4)$ (c) $(8, 9\pi/4)$
(d) $(5, 0)$ (e) $(-2, -3\pi/2)$ (f) $(0, \pi)$

5. In each part, a point is given in rectangular coordinates. Find two pairs of polar coordinates for the point, one pair satisfying $r\ge 0$ and $0\le\theta<2\pi$, and the second pair satisfying $r\ge 0$ and $-\pi<\theta\le\pi$.
(a) $(-5, 0)$ (b) $(2\sqrt{3}, -2)$ (c) $(0, -2)$
(d) $(-8, -8)$ (e) $(-3, 3\sqrt{3})$ (f) $(1, 1)$

6. In each part find polar coordinates satisfying the stated conditions for the point whose rectangular coordinates are $(-\sqrt{3}, 1)$.
(a) $r\ge 0$ and $0\le\theta<2\pi$
(b) $r\le 0$ and $0\le\theta<2\pi$
(c) $r\ge 0$ and $-2\pi<\theta\le 0$
(d) $r\le 0$ and $-\pi<\theta\le\pi$

In Exercises 7 and 8, use a calculating utility, where needed, to approximate the polar coordinates of the points whose rectangular coordinates are given.

7. (a) $(4, 3)$ (b) $(2, -5)$ (c) $(1, \tan^{-1} 1)$

8. (a) $(-3, 4)$ (b) $(-3, 1.7)$ (c) $(2, \sin^{-1}\frac{1}{2})$

In Exercises 9 and 10, identify the curve by transforming the given polar equation to rectangular coordinates.

9. (a) $r=2$ (b) $r\sin\theta=4$
(c) $r=3\cos\theta$ (d) $r=\dfrac{6}{3\cos\theta+2\sin\theta}$

10. (a) $r=5\sec\theta$ (b) $r=2\sin\theta$
(c) $r=4\cos\theta+4\sin\theta$ (d) $r=\sec\theta\tan\theta$

In Exercises 11 and 12, express the given equations in polar coordinates.

11. (a) $x=7$ (b) $x^2+y^2=9$
(c) $x^2+y^2-6y=0$ (d) $4xy=9$

12. (a) $y=-3$ (b) $x^2+y^2=5$
(c) $x^2+y^2+4x=0$ (d) $x^2(x^2+y^2)=y^2$

In Exercises 13–16, a graph is given in a rectangular θr-coordinate system. Sketch the corresponding graph in polar coordinates.

13.

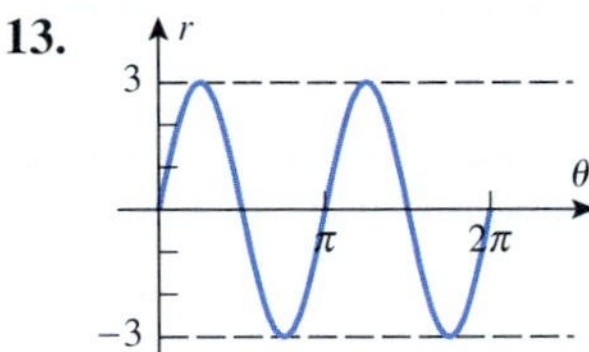

14.

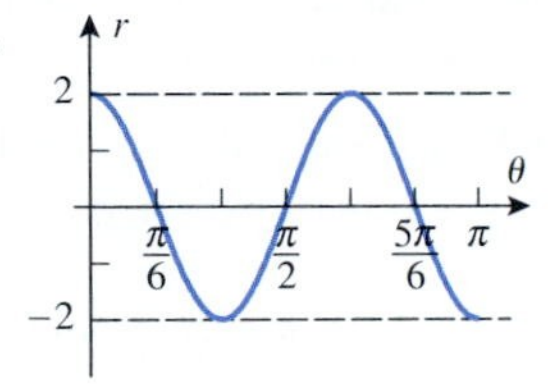

15.

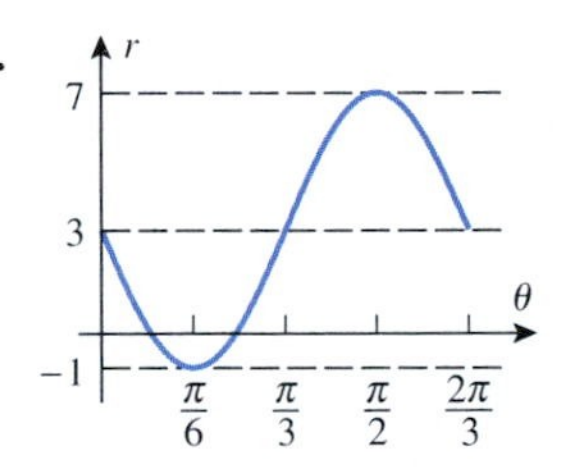

16.

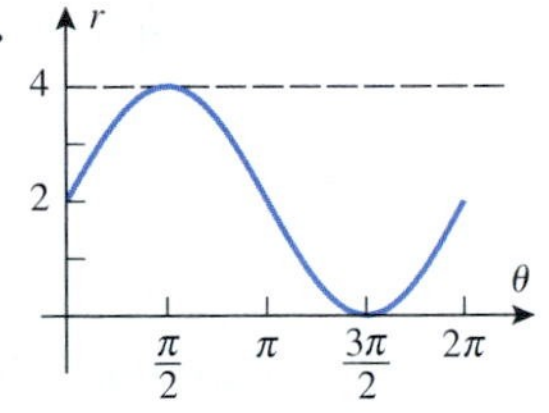

In Exercises 17–20, find an equation for the given polar graph.

17. (a)

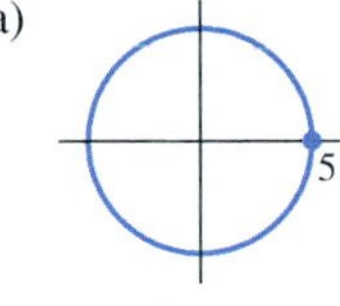

Circle

(b)

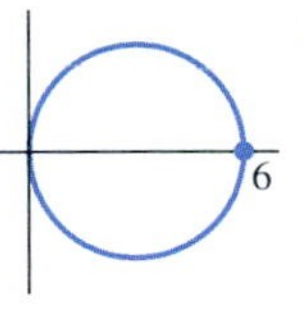

Circle

(c)

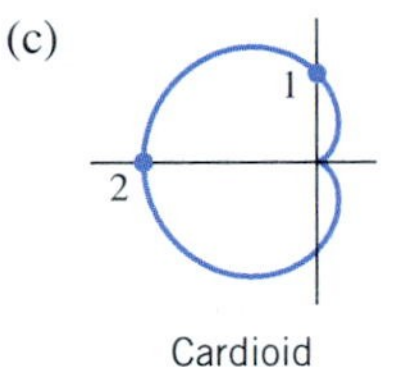

Cardioid

18. (a)

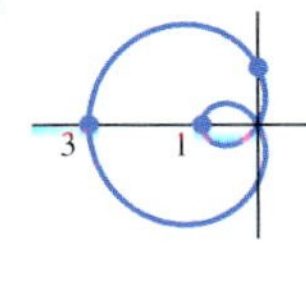

Limaçon

(b)

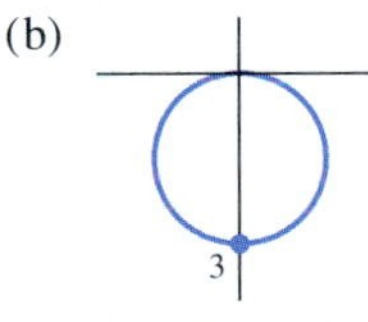

Circle

(c)

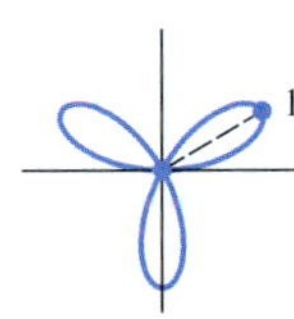

Three-petal rose

19. (a)

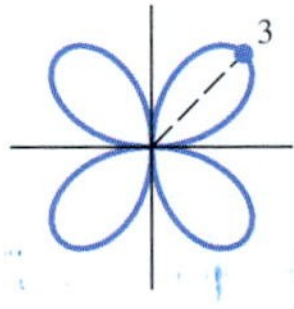

Four-petal rose

(b)

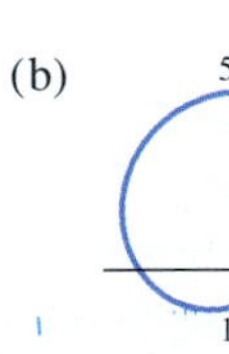

Limaçon

(c)

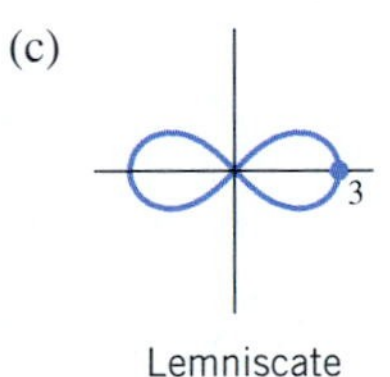

Lemniscate

20. (a)

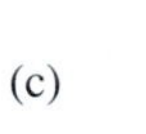

Cardioid

(b)

Five-petal rose

(c)

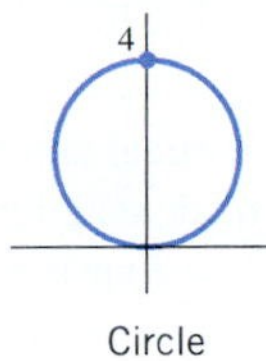

Circle

In Exercises 21–50, sketch the curve in polar coordinates.

21. $\theta = \dfrac{\pi}{6}$

22. $\theta = -\dfrac{3\pi}{4}$

23. $r = 3$

24. $r = 4\sin\theta$

25. $r = 6\cos\theta$

26. $r = 1 + \sin\theta$

27. $2r = \cos\theta$

28. $r - 2 = 2\cos\theta$

29. $r = 3(1 - \sin\theta)$

30. $r = -5 + 5\sin\theta$

31. $r = 4 - 4\cos\theta$

32. $r = 1 + 2\sin\theta$

33. $r = -1 - \cos\theta$

34. $r = 4 + 3\cos\theta$

35. $r = 2 + \sin\theta$

36. $r = 3 - \cos\theta$

37. $r = 3 + 4\cos\theta$

38. $r - 5 = 3\sin\theta$

39. $r = 5 - 2\cos\theta$

40. $r = -3 - 4\sin\theta$

41. $r^2 = 9\cos 2\theta$

42. $r^2 = \sin 2\theta$

43. $r^2 = 16\sin 2\theta$

44. $r = 4\theta \quad (\theta \ge 0)$

45. $r = 4\theta \quad (\theta \le 0)$

46. $r = 4\theta$

47. $r = \cos 2\theta$

48. $r = 3\sin 2\theta$

49. $r = 9\sin 4\theta$

50. $r = 2\cos 3\theta$

51. For each of the curves you sketched in Exercises 21–50, check your work with a graphing utility.

In Exercises 52–55, use a graphing utility to generate the polar graph. Be sure to choose the parameter interval so that a complete graph is generated.

52. $r = \sin\dfrac{\theta}{2}$

53. $r = 1 + 2\cos\dfrac{\theta}{4}$

54. $r = 0.5 + \cos\dfrac{\theta}{3}$

55. $r = \cos\dfrac{\theta}{5}$

56. The accompanying figure shows the graph of the "butterfly curve"

$$r = e^{\cos\theta} - 2\cos 4\theta + \sin^3\frac{\theta}{4}$$

Generate the complete butterfly with a graphing utility, and state the parameter interval you used.

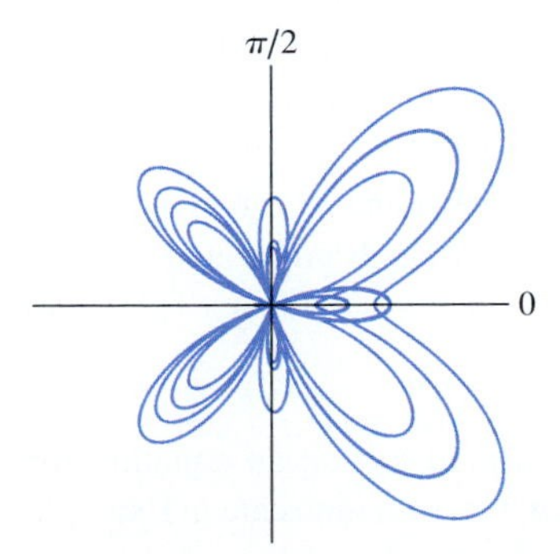

Figure Ex-56

57. Figure Ex-57 (next page) shows the Archimedean spiral $r = \theta/2$ produced with a graphing calculator.

(a) What interval of values for θ do you think was used to generate the graph?

(b) Duplicate the graph with your own graphing utility.

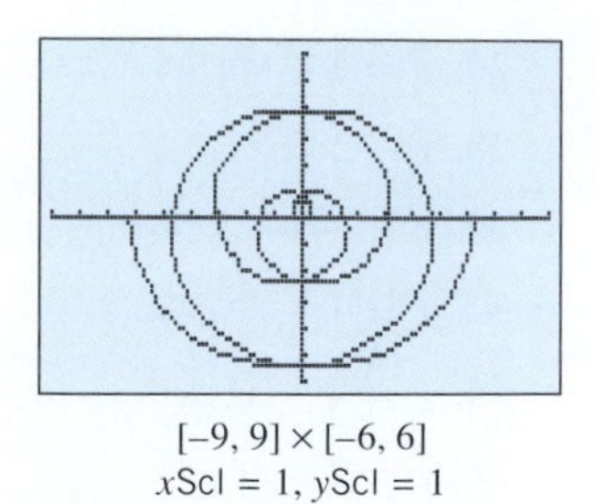

$[-9, 9] \times [-6, 6]$
$x\text{Scl} = 1$, $y\text{Scl} = 1$

Figure Ex-57

58. The accompanying figure shows graphs of the Archimedean spiral $r = \theta$ and the parabolic spiral $r = \sqrt{\theta}$. Which is which? Explain your reasoning.

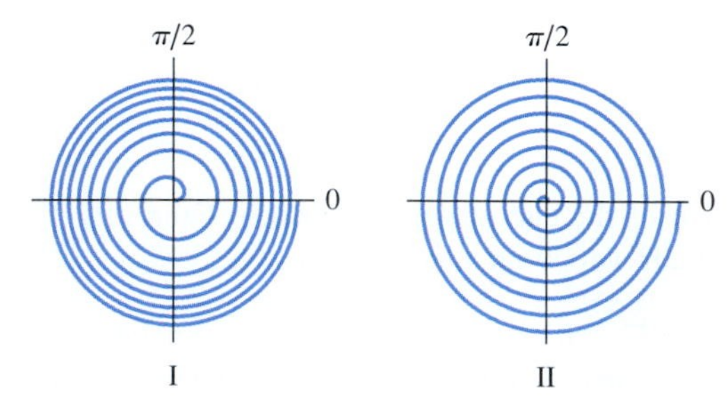

Figure Ex-58

59. (a) Show that if a varies, then the polar equation

$$r = a\sec\theta \quad (-\pi/2 < \theta < \pi/2)$$

describes a family of lines perpendicular to the polar axis.

(b) Show that if b varies, then the polar equation

$$r = b\csc\theta \quad (0 < \theta < \pi)$$

describes a family of lines parallel to the polar axis.

60. Show that if the polar graph of $r = f(\theta)$ is rotated counterclockwise around the origin through an angle α, then $r = f(\theta - \alpha)$ is an equation for the rotated curve. [*Hint:* If (r_0, θ_0) is any point on the original graph, then $(r_0, \theta_0 + \alpha)$ is a point on the rotated graph.]

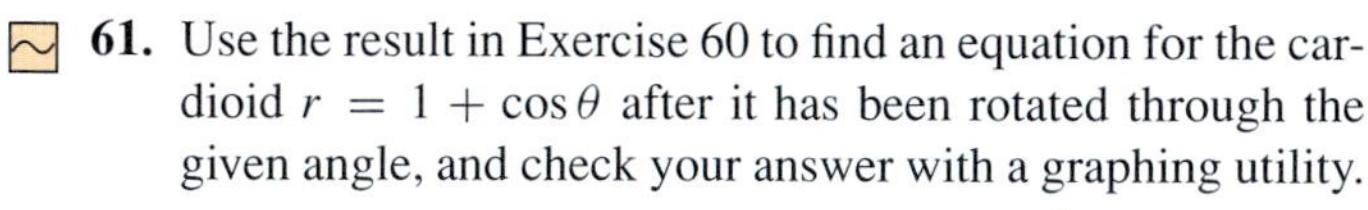

61. Use the result in Exercise 60 to find an equation for the cardioid $r = 1 + \cos\theta$ after it has been rotated through the given angle, and check your answer with a graphing utility.

(a) $\dfrac{\pi}{4}$ (b) $\dfrac{\pi}{2}$ (c) π (d) $\dfrac{5\pi}{4}$

62. Use the result in Exercise 60 to find an equation for the lemniscate that results when the lemniscate in Example 8 is rotated counterclockwise through an angle of $\pi/2$.

63. Sketch the polar graph of the equation $(r - 1)(\theta - 1) = 0$.

64. (a) Show that if A and B are not both zero, then the graph of the polar equation

$$r = A\sin\theta + B\cos\theta$$

is a circle. Find its radius.

(b) Derive Formulas (4) and (5) from the formula given in part (a).

65. Find the highest point on the cardioid $r = 1 + \cos\theta$.

66. Find the leftmost point on the upper half of the cardioid $r = 1 + \cos\theta$.

67. (a) Show that in a polar coordinate system the distance d between the points (r_1, θ_1) and (r_2, θ_2) is

$$d = \sqrt{r_1^2 + r_1^2 - 2r_1r_2\cos(\theta_1 - \theta_2)}$$

(b) Show that if $0 \le \theta_1 < \theta_2 \le \pi$ and if r_1 and r_2 are positive, then the area A of the triangle with vertices $(0, 0)$, (r_1, θ_1), and (r_2, θ_2) is

$$A = \tfrac{1}{2}r_1r_2\sin(\theta_2 - \theta_1)$$

(c) Find the distance between the points whose polar coordinates are $(3, \pi/6)$ and $(2, \pi/3)$.

(d) Find the area of the triangle whose vertices in polar coordinates are $(0, 0)$, $(1, 5\pi/6)$, and $(2, \pi/3)$.

68. In the late seventeenth century the Italian astronomer Giovanni Domenico Cassini (1625–1712) introduced the family of curves

$$(x^2 + y^2 + a^2)^2 - b^4 - 4a^2x^2 = 0 \quad (a > 0, b > 0)$$

in his studies of the relative motions of the Earth and the Sun. These curves, which are called ***Cassini ovals***, have one of the three basic shapes shown in the accompanying figure.

(a) Show that if $a = b$, then the polar equation of the Cassini oval is $r^2 = 2a^2\cos 2\theta$, which is a lemniscate.

(b) Use the formula in Exercise 67(a) to show that the lemniscate in part (a) is the curve traced by a point that moves in such a way that the product of its distances from the polar points $(a, 0)$ and (a, π) is a^2.

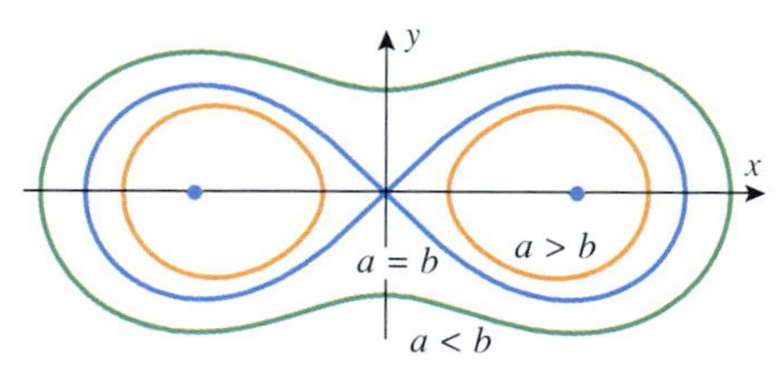

Figure Ex-68

Vertical and horizontal asymptotes of polar curves can often be detected by investigating the behavior of $x = r\cos\theta$ and $y = r\sin\theta$ as θ varies. This idea is used in Exercises 69–72.

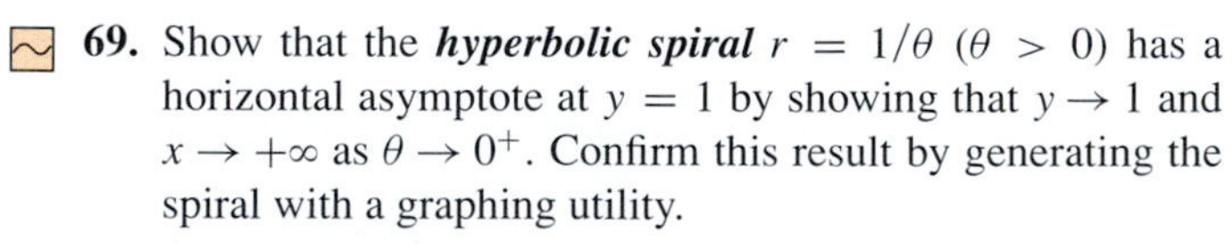

69. Show that the ***hyperbolic spiral*** $r = 1/\theta$ $(\theta > 0)$ has a horizontal asymptote at $y = 1$ by showing that $y \to 1$ and $x \to +\infty$ as $\theta \to 0^+$. Confirm this result by generating the spiral with a graphing utility.

70. Show that the spiral $r = 1/\theta^2$ does not have any horizontal asymptotes.

71. (a) Show that the ***kappa curve*** $r = 4\tan\theta$ $(0 \le \theta \le 2\pi)$ has a vertical asymptote at $x = 4$ by showing that $x \to 4$ and $y \to +\infty$ as $\theta \to \pi/2^-$ and that $x \to 4$ and $y \to -\infty$ as $\theta \to \pi/2^+$.

(b) Use the method in part (a) to show that the kappa curve also has a vertical asymptote at $x = -4$.

(c) Confirm the results in parts (a) and (b) by generating the kappa curve with a graphing utility.

72. Use a graphing utility to make a conjecture about the existence of asymptotes for the ***cissoid*** $r = 2\sin\theta\tan\theta$, and then confirm your conjecture by calculating appropriate limits.

73. Prove that a rose with an even number of petals is traced out exactly once as θ varies over the interval $0 \le \theta < 2\pi$ and a rose with an odd number of petals is traced out exactly once as θ varies over the interval $0 \le \theta < \pi$.

12.2 TANGENT LINES AND ARC LENGTH FOR PARAMETRIC AND POLAR CURVES

In this section we will derive the formulas required to find slopes, tangent lines, and arc lengths of parametric and polar curves.

TANGENT LINES TO PARAMETRIC CURVES

We will be concerned in this section with curves that are given by parametric equations

$$x = f(t), \quad y = g(t)$$

in which $f(t)$ and $g(t)$ have continuous first derivatives with respect to t. It can be proved that if $dx/dt \ne 0$, then y is a differentiable function of x, in which case the chain rule implies that

$$\frac{dy}{dx} = \frac{dy/dt}{dx/dt} \tag{1}$$

This formula makes it possible to find dy/dx directly from the parametric equations without eliminating the parameter.

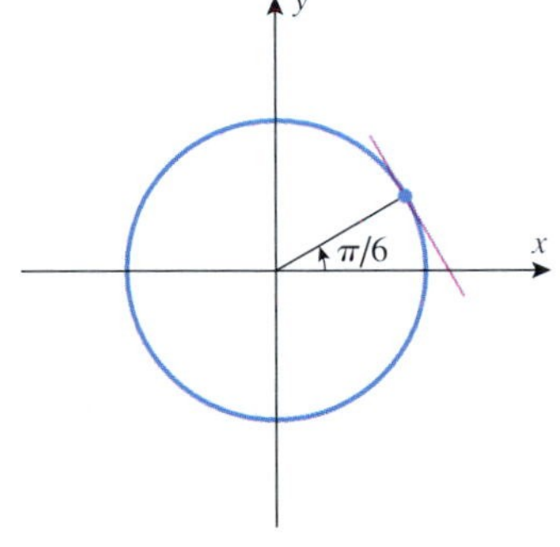

Figure 12.2.1

Example 1

Find the slope of the tangent line to the unit circle

$$x = \cos t, \quad y = \sin t \qquad (0 \le t \le 2\pi)$$

at the point where $t = \pi/6$ (Figure 12.2.1).

Solution. From (1), the slope at a general point on the circle is

$$\frac{dy}{dx} = \frac{dy/dt}{dx/dt} = \frac{\cos t}{-\sin t} = -\cot t \tag{2}$$

Thus, the slope at $t = \pi/6$ is

$$\left.\frac{dy}{dx}\right|_{t=\pi/6} = -\cot\frac{\pi}{6} = -\sqrt{3}$$ ◀

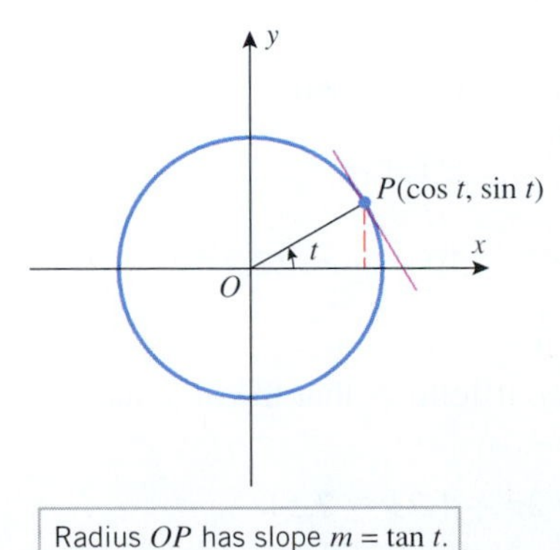

Radius OP has slope $m = \tan t$.

Figure 12.2.2

REMARK. Note that Formula (2) makes sense geometrically because the radius to the point $P(\cos t, \sin t)$ has slope $m = \tan t$; hence, the tangent line at P, being perpendicular to the radius, has slope $-1/m = -1/\tan t = -\cot t$ (Figure 12.2.2).

It follows from Formula (1) that the tangent line to a parametric curve will be horizontal at those points where $dy/dt = 0$ and $dx/dt \ne 0$, since $dy/dx = 0$ at such points. Two different situations occur when $dx/dt = 0$. At points where $dx/dt = 0$ and $dy/dt \ne 0$, the right side of (1) has a nonzero numerator and a zero denominator; we will agree that the curve has ***infinite slope*** and a ***vertical tangent line*** at such points. At points where dx/dt and dy/dt are both zero, the right side of (1) becomes an indeterminate form; we call such points ***singular points***. No general statement can be made about the behavior of parametric curves at singular points; they must be analyzed case by case.

Example 2

In a disastrous first flight, an experimental paper airplane follows the trajectory

$$x = t - 3\sin t, \quad y = 4 - 3\cos t \qquad (t \geq 0)$$

but crashes into a wall at time $t = 10$ (Figure 12.2.3).

(a) At what times was the airplane flying horizontally?

(b) At what times was it flying vertically?

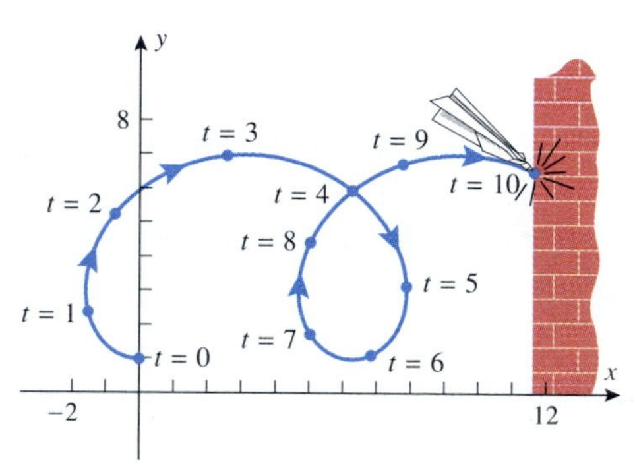

Figure 12.2.3

Solution (a). The airplane was flying horizontally at those times when $dy/dt = 0$ and $dx/dt \neq 0$. From the given trajectory we have

$$\frac{dy}{dt} = 3\sin t \quad \text{and} \quad \frac{dx}{dt} = 1 - 3\cos t \tag{3}$$

Setting $dy/dt = 0$ yields the equation $3\sin t = 0$, or, more simply, $\sin t = 0$. This equation has four solutions in the time interval $0 \leq t \leq 10$:

$$t = 0, \quad t = \pi, \quad t = 2\pi, \quad t = 3\pi$$

Since $dx/dt = 1 - 3\cos t \neq 0$ for these values of t (verify), the airplane was flying horizontally at times

$$t = 0, \quad t = \pi \approx 3.14, \quad t = 2\pi \approx 6.28, \quad \text{and} \quad t = 3\pi \approx 9.42$$

which is consistent with Figure 12.2.3.

Solution (b). The airplane was flying vertically at those times when $dx/dt = 0$ and $dy/dt \neq 0$. Setting $dx/dt = 0$ in (3) yields the equation

$$1 - 3\cos t = 0 \quad \text{or} \quad \cos t = \tfrac{1}{3}$$

This equation has three solutions in the time interval $0 \leq t \leq 10$ (Figure 12.2.4):

$$t = \cos^{-1}\tfrac{1}{3}, \quad t = 2\pi - \cos^{-1}\tfrac{1}{3}, \quad t = 2\pi + \cos^{-1}\tfrac{1}{3}$$

Since $dy/dt = 3\sin t$ is not zero at these points (why?), it follows that the airplane was flying vertically at times

$$t = \cos^{-1}\tfrac{1}{3} \approx 1.23, \quad t \approx 2\pi - 1.23 \approx 5.05, \quad t \approx 2\pi + 1.23 \approx 7.51$$

which again is consistent with Figure 12.2.3. ◀

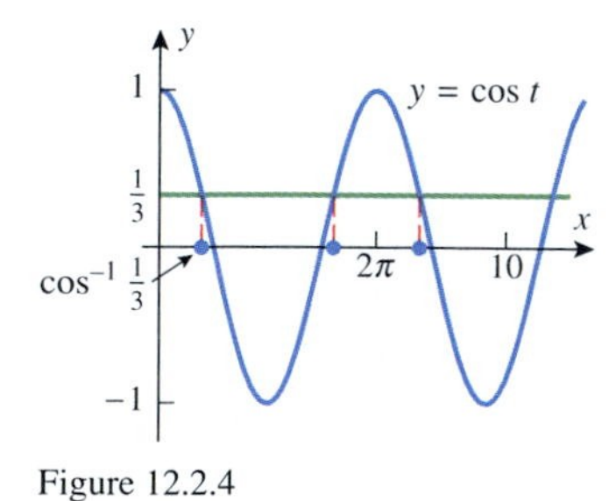

Figure 12.2.4

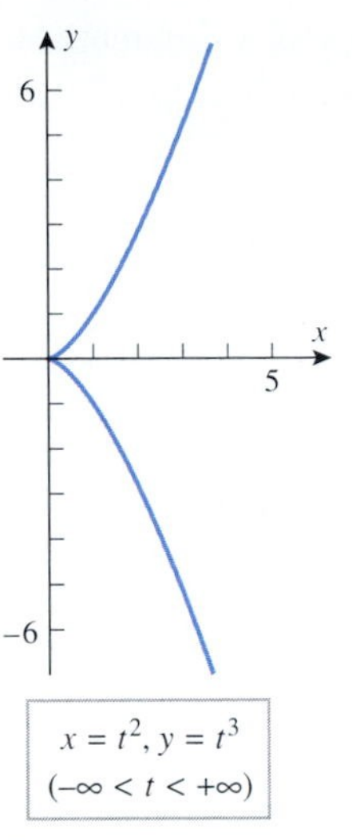

Figure 12.2.5

Example 3

The curve represented by the parametric equations

$$x = t^2, \quad y = t^3 \qquad (-\infty < t < +\infty)$$

is called a ***semicubical parabola***. The parameter t can be eliminated by cubing x and squaring y, from which it follows that $y^2 = x^3$. The graph of this equation, shown in Figure 12.2.5, consists of two branches: an upper branch obtained by graphing $y = x^{3/2}$ and a lower branch obtained by graphing $y = -x^{3/2}$. The two branches meet at the origin, which corresponds to $t = 0$ in the parametric equations. This is a singular point because the derivatives $dx/dt = 2t$ and $dy/dt = 3t^2$ are both zero there. ◀

Example 4

Without eliminating the parameter, find dy/dx and d^2y/dx^2 at the points $(1, 1)$ and $(1, -1)$ on the semicubical parabola given by the parametric equations in Example 3.

Solution. From (1) we have

$$\frac{dy}{dx} = \frac{dy/dt}{dx/dt} = \frac{3t^2}{2t} = \frac{3}{2}t \quad (t \neq 0) \tag{4}$$

and from (1) applied to $y' = dy/dx$ we have

$$\frac{d^2y}{dx^2} = \frac{dy'}{dx} = \frac{dy'/dt}{dx/dt} = \frac{3/2}{2t} = \frac{3}{4t} \tag{5}$$

Since the point $(1, 1)$ on the curve corresponds to $t = 1$ in the parametric equations, it follows from (4) and (5) that

$$\left.\frac{dy}{dx}\right|_{t=1} = \frac{3}{2} \quad \text{and} \quad \left.\frac{d^2y}{dx^2}\right|_{t=1} = \frac{3}{4}$$

Similarly, the point $(1, -1)$ corresponds to $t = -1$ in the parametric equations, so applying (4) and (5) again yields

$$\left.\frac{dy}{dx}\right|_{t=-1} = -\frac{3}{2} \quad \text{and} \quad \left.\frac{d^2y}{dx^2}\right|_{t=-1} = -\frac{3}{4}$$

Note that the values we obtained for the first and second derivatives are consistent with the graph in Figure 12.2.5, since at $(1, 1)$ on the upper branch the tangent line has positive slope and the curve is concave up, and at $(1, -1)$ on the lower branch the tangent line has negative slope and the curve is concave down.

Finally, observe that we were able to apply Formulas (4) and (5) for both $t = 1$ and $t = -1$, even though the points $(1, 1)$ and $(1, -1)$ lie on different branches. In contrast, had we chosen to perform the same computations by eliminating the parameter, we would have had to obtain separate derivative formulas for $y = x^{3/2}$ and $y = -x^{3/2}$. ◀

TANGENT LINES TO POLAR CURVES

Our next objective is to find a method for obtaining slopes of tangent lines to polar curves of the form $r = f(\theta)$ in which r is a differentiable function of θ. We showed in the last section that a curve of this form can be expressed parametrically in terms of the parameter θ by substituting $f(\theta)$ for r in the equations $x = r\cos\theta$ and $y = r\sin\theta$. This yields

$$x = f(\theta)\cos\theta, \quad y = f(\theta)\sin\theta$$

from which we obtain

$$\begin{aligned} \frac{dx}{d\theta} &= -f(\theta)\sin\theta + f'(\theta)\cos\theta = -r\sin\theta + \frac{dr}{d\theta}\cos\theta \\ \frac{dy}{d\theta} &= f(\theta)\cos\theta + f'(\theta)\sin\theta = r\cos\theta + \frac{dr}{d\theta}\sin\theta \end{aligned} \tag{6}$$

Thus, if $dx/d\theta$ and $dy/d\theta$ are continuous and if $dx/d\theta \neq 0$, then y is a differentiable function of x, and Formula (1) with θ in place of t yields

$$\frac{dy}{dx} = \frac{dy/d\theta}{dx/d\theta} = \frac{r\cos\theta + \sin\theta \dfrac{dr}{d\theta}}{-r\sin\theta + \cos\theta \dfrac{dr}{d\theta}} \tag{7}$$

Example 5

Find the slope of the tangent line to the circle $r = 4\cos\theta$ at the point where $\theta = \pi/4$.

Solution. From (7) with $r = 4\cos\theta$ we obtain (verify)

$$\frac{dy}{dx} = \frac{4\cos^2\theta - 4\sin^2\theta}{-8\sin\theta\cos\theta} = \frac{4\cos 2\theta}{-4\sin 2\theta} = -\cot 2\theta$$

Thus, at the point where $\theta = \pi/4$ the slope of the tangent line is

$$m = \left.\frac{dy}{dx}\right|_{\theta=\pi/4} = -\cot\frac{\pi}{2} = 0$$

which implies that the circle has a horizontal tangent line at the point where $\theta = \pi/4$ (Figure 12.2.6). ◀

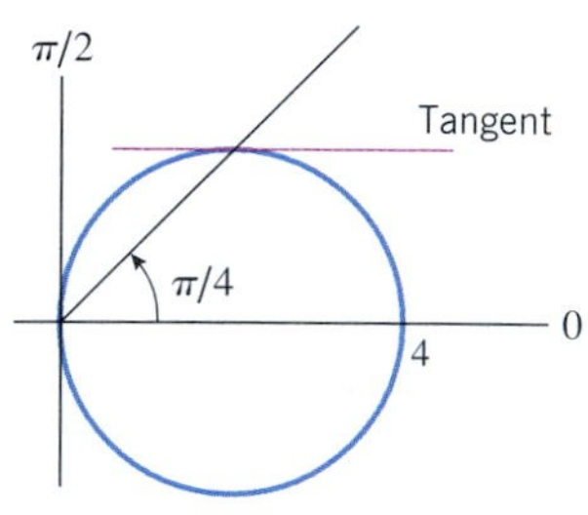

Figure 12.2.6

Example 6

Find the points on the cardioid $r = 1 - \cos\theta$ at which there is a horizontal tangent line, a vertical tangent line, or a singular point.

Solution. A horizontal tangent line will occur where $dy/d\theta = 0$ and $dx/d\theta \neq 0$, a vertical tangent line where $dy/d\theta \neq 0$ and $dx/d\theta = 0$, and a singular point where $dy/d\theta = 0$ and $dx/d\theta = 0$. We could find these derivatives from the formulas in (6). However, an alternative approach is go back to basic principles and express the cardioid parametrically by substituting $r = 1 - \cos\theta$ in the conversion formulas $x = r\cos\theta$ and $y = r\sin\theta$. This yields

$$x = (1 - \cos\theta)\cos\theta, \quad y = (1 - \cos\theta)\sin\theta \qquad (0 \le \theta \le 2\pi)$$

Differentiating these equations with respect to θ and then simplifying yields (verify)

$$\frac{dx}{d\theta} = \sin\theta(2\cos\theta - 1), \qquad \frac{dy}{d\theta} = (1 - \cos\theta)(1 + 2\cos\theta)$$

Thus, $dx/d\theta = 0$ if $\sin\theta = 0$ or $\cos\theta = \frac{1}{2}$, and $dy/d\theta = 0$ if $\cos\theta = 1$ or $\cos\theta = -\frac{1}{2}$. We leave it for you to solve these equations and show that the solutions of $dx/d\theta = 0$ on the interval $0 \le \theta \le 2\pi$ are

$$\frac{dx}{d\theta} = 0: \quad \theta = 0,\ \frac{\pi}{3},\ \pi,\ \frac{5\pi}{3},\ 2\pi$$

and the solutions of $dy/d\theta = 0$ on the interval $0 \le \theta \le 2\pi$ are

$$\frac{dy}{d\theta} = 0: \quad \theta = 0,\ \frac{2\pi}{3},\ \frac{4\pi}{3},\ 2\pi$$

Thus, horizontal tangent lines occur at $\theta = 2\pi/3$ and $\theta = 4\pi/3$; vertical tangent lines occur at $\theta = \pi/3$, π, and $5\pi/3$; and singular points occur at $\theta = 0$ and $\theta = 2\pi$ (Figure 12.2.7). Note, however, that $r = 0$ at both singular points, so there is really only one singular point on the cardioid—the pole. ◀

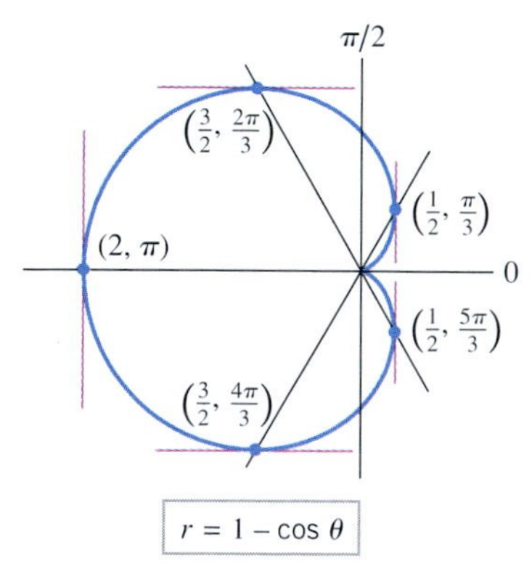

Figure 12.2.7

TANGENT LINES TO POLAR CURVES AT THE ORIGIN

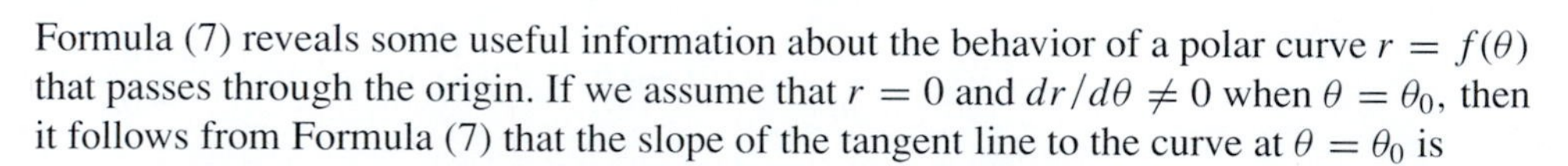

Formula (7) reveals some useful information about the behavior of a polar curve $r = f(\theta)$ that passes through the origin. If we assume that $r = 0$ and $dr/d\theta \neq 0$ when $\theta = \theta_0$, then it follows from Formula (7) that the slope of the tangent line to the curve at $\theta = \theta_0$ is

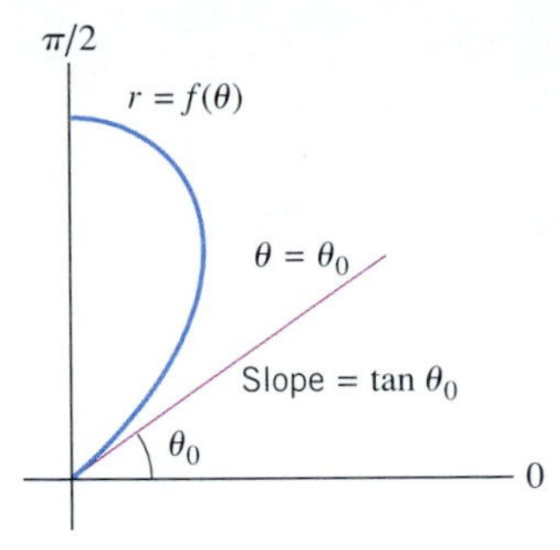

Figure 12.2.8

$$\frac{dy}{dx} = \frac{0 + \sin\theta_0 \dfrac{dr}{d\theta}}{0 + \cos\theta_0 \dfrac{dr}{d\theta}} = \frac{\sin\theta_0}{\cos\theta_0} = \tan\theta_0$$

(Figure 12.2.8). However, $\tan\theta_0$ is also the slope of the line $\theta = \theta_0$, so we can conclude that this line is tangent to the curve at the origin. Thus, we have established the following result.

> **12.2.1 THEOREM.** *If the polar curve $r = f(\theta)$ passes through the origin at $\theta = \theta_0$, and if $dr/d\theta \neq 0$ at $\theta = \theta_0$, then the line $\theta = \theta_0$ is tangent to the curve at the origin.*

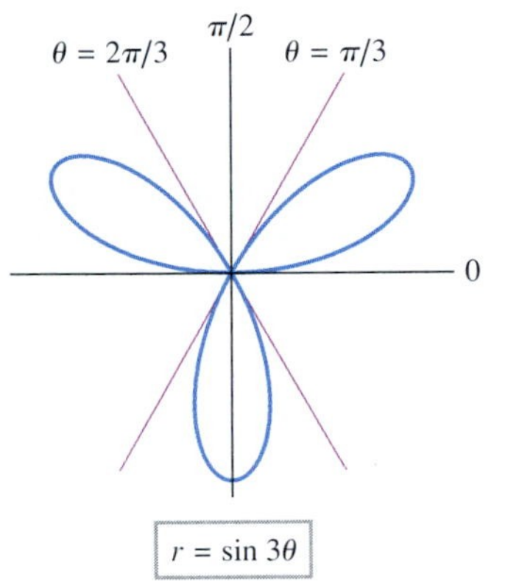

Figure 12.2.9

This theorem tells us that equations of the tangent lines at the origin to the curve $r = f(\theta)$ can be obtained by solving the equation $f(\theta) = 0$. It is important to keep in mind, however, that $r = f(\theta)$ may be zero for more than one value of θ, so there may be more than one tangent line at the origin. This is illustrated in the next example.

Example 7

The three-petal rose $r = \sin 3\theta$ in Figure 12.2.9 has three tangent lines at the origin, which can be found by solving the equation

$$\sin 3\theta = 0$$

It was shown in Exercise 73 of Section 12.1 that the complete rose is traced once as θ varies over the interval $0 \leq \theta < \pi$, so we need only look for solutions in this interval. We leave it for you to confirm that these solutions are

$$\theta = 0, \quad \theta = \frac{\pi}{3}, \quad \text{and} \quad \theta = \frac{2\pi}{3}$$

Since $dr/d\theta = 3\cos 3\theta \neq 0$ for these values of θ, these three lines are tangent to the rose at the origin, which is consistent with the figure. ◀

ARC LENGTH OF A POLAR CURVE

A formula for the arc length of a polar curve $r = f(\theta)$ can be derived by expressing the curve in parametric form and applying Formula (6) of Section 8.4 for the arc length of a parametric curve. We leave it as an exercise to show the following.

> **12.2.2 ARC LENGTH FORMULA FOR POLAR CURVES.** If no segment of the polar curve $r = f(\theta)$ is traced more than once as θ increases from α to β, and if $dr/d\theta$ is continuous for $\alpha \leq \theta \leq \beta$, then the arc length L from $\theta = \alpha$ to $\theta = \beta$ is
>
> $$L = \int_{\alpha}^{\beta} \sqrt{r^2 + \left(\frac{dr}{d\theta}\right)^2}\, d\theta \tag{8}$$

Example 8

Find the arc length of the spiral $r = e^{\theta}$ in Figure 12.2.10 between $\theta = 0$ and $\theta = \pi$.

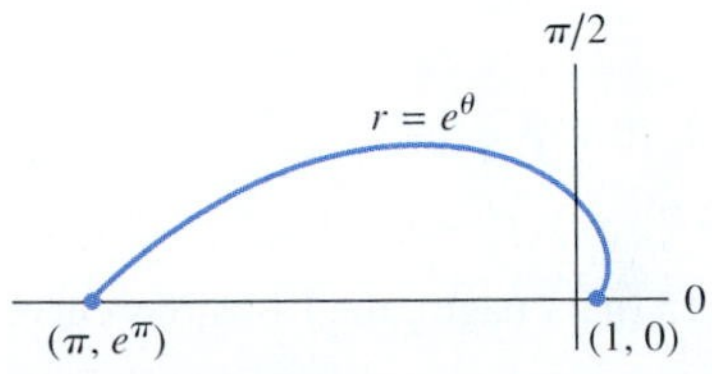

Figure 12.2.10

Solution.

$$L = \int_{\alpha}^{\beta} \sqrt{r^2 + \left(\frac{dr}{d\theta}\right)^2}\, d\theta = \int_0^{\pi} \sqrt{(e^{\theta})^2 + (e^{\theta})^2}\, d\theta$$

$$= \int_0^{\pi} \sqrt{2}\, e^{\theta}\, d\theta = \sqrt{2}\, e^{\theta}\Big]_0^{\pi} = \sqrt{2}(e^{\pi} - 1) \approx 31.3$$

◀

Example 9

Find the total arc length of the cardioid $r = 1 + \cos\theta$.

Solution. The cardioid is traced out once as θ varies from $\theta = 0$ to $\theta = 2\pi$. Thus,

$$\begin{aligned} L = \int_{\alpha}^{\beta} \sqrt{r^2 + \left(\frac{dr}{d\theta}\right)^2}\, d\theta &= \int_0^{2\pi} \sqrt{(1+\cos\theta)^2 + (-\sin\theta)^2}\, d\theta \\ &= \sqrt{2}\int_0^{2\pi} \sqrt{1+\cos\theta}\, d\theta \\ &= 2\int_0^{2\pi} \sqrt{\cos^2 \frac{1}{2}\theta}\, d\theta \\ &= 2\int_0^{2\pi} \left|\cos\frac{1}{2}\theta\right| d\theta \end{aligned}$$

Identity (45) of Appendix E

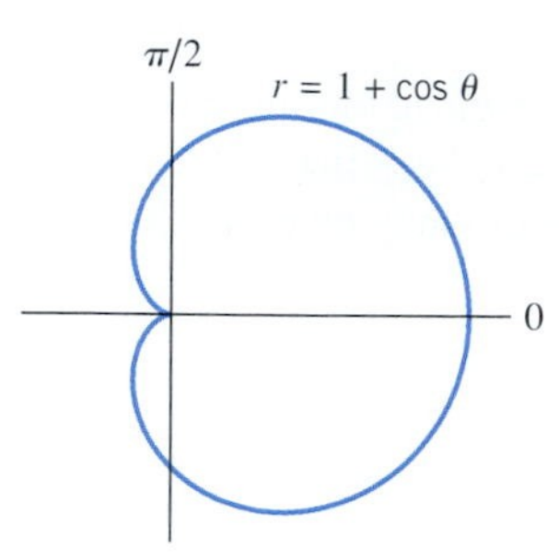

Figure 12.2.11

Since $\cos\frac{1}{2}\theta$ changes sign at π, we must split the last integral into the sum of two integrals: the integral from 0 to π plus the integral from π to 2π. However, the integral from π to 2π is equal to the integral from 0 to π, since the cardioid is symmetric about the polar axis (Figure 12.2.11). Thus,

$$L = 2\int_0^{2\pi} \left|\cos\frac{1}{2}\theta\right| d\theta = 4\int_0^{\pi} \cos\frac{1}{2}\theta\, d\theta = 8\sin\frac{1}{2}\theta\Big]_0^{\pi} = 8$$ ◀

EXERCISE SET 12.2 Graphing Calculator

1. (a) Find the slope of the tangent line to the parametric curve $x = t^2 + 1$, $y = t/2$ at $t = -1$ and at $t = 1$ without eliminating the parameter.
   (b) Check your answers in part (a) by eliminating the parameter and differentiating an appropriate function of x.

2. (a) Find the slope of the tangent line to the parametric curve $x = 3\cos t$, $y = 4\sin t$ at $t = \pi/4$ and at $t = 7\pi/4$ without eliminating the parameter.
   (b) Check your answers in part (a) by eliminating the parameter and differentiating an appropriate function of x.

3. For the parametric curve in Exercise 1, make a conjecture about the sign of d^2y/dx^2 at $t = -1$ and at $t = 1$, and confirm your conjecture without eliminating the parameter.

4. For the parametric curve in Exercise 2, make a conjecture about the sign of d^2y/dx^2 at $t = \pi/4$ and at $t = 7\pi/4$, and confirm your conjecture without eliminating the parameter.

In Exercises 5–10, find dy/dx and d^2y/dx^2 at the given point without eliminating the parameter.

5. $x = \sqrt{t}$, $y = 2t + 4$; $t = 1$
6. $x = \frac{1}{2}t^2$, $y = \frac{1}{3}t^3$; $t = 2$
7. $x = \sec t$, $y = \tan t$; $t = \pi/3$
8. $x = \sinh t$, $y = \cosh t$; $t = 0$
9. $x = 2\theta + \cos\theta$, $y = 1 - \sin\theta$; $\theta = \pi/3$
10. $x = \cos\phi$, $y = 3\sin\phi$; $\phi = 5\pi/6$
11. (a) Find the equation of the tangent line to the curve
    $$x = e^t, \quad y = e^{-t}$$
    at $t = 1$ without eliminating the parameter.
    (b) Check your answer in part (a) by eliminating the parameter.
12. (a) Find the equation of the tangent line to the curve
    $$x = 2t + 4, \quad y = 8t^2 - 2t + 4$$
    at $t = 1$ without eliminating the parameter.
    (b) Check your answer in part (a) by eliminating the parameter.

In Exercises 13 and 14, find all values of t at which the parametric curve has (a) a horizontal tangent line and (b) a vertical tangent line.

13. $x = 2\cos t$, $y = 4\sin t$ $\quad (0 \le t \le 2\pi)$
14. $x = 2t^3 - 15t^2 + 24t + 7$, $y = t^2 + t + 1$
15. As shown in Figure Ex-15 (next page), the Lissajous curve
    $$x = \sin t, \quad y = \sin 2t \qquad (0 \le t \le 2\pi)$$
    crosses itself at the origin. Find equations for the two tangent lines at the origin.

16. As shown in the accompanying figure, the ***prolate cycloid***

$$x = 2 - \pi \cos t, \quad y = 2t - \pi \sin t \qquad (-\pi \le t \le \pi)$$

crosses itself at a point on the x-axis. Find equations for the two tangent lines at that point.

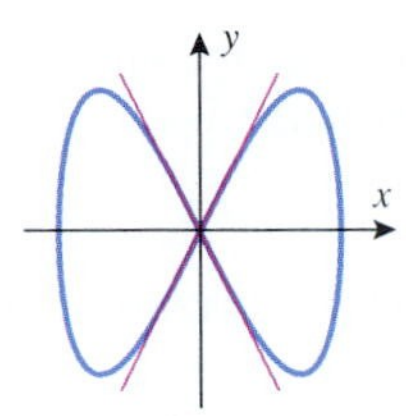

Figure Ex-15

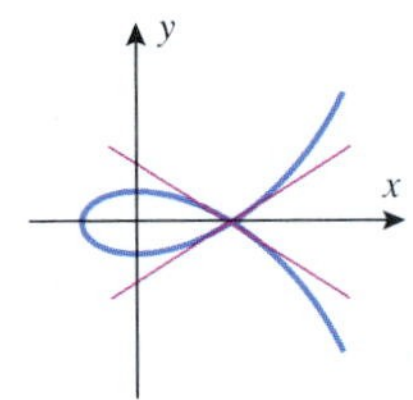

Figure Ex-16

17. Show that the curve $x = t^3 - 4t$, $y = t^2$ intersects itself at the point $(0, 4)$, and find equations for the two tangent lines to the curve at the point of intersection.

18. Show that the curve with parametric equations

$$x = t^2 - 3t + 5, \quad y = t^3 + t^2 - 10t + 9$$

intersects itself at the point $(3, 1)$, and find equations for the two tangent lines to the curve at the point of intersection.

19. (a) Use a graphing utility to generate the graph of the parametric curve

$$x = \cos^3 t, \quad y = \sin^3 t \qquad (0 \le t \le 2\pi)$$

and make a conjecture about the values of t at which singular points occur.

(b) Confirm your conjecture in part (a) by calculating appropriate derivatives.

20. (a) At what values of θ would you expect the cycloid in Figure 1.7.13 to have singular points?

(b) Confirm your answer in part (a) by calculating appropriate derivatives.

In Exercises 21–26, find the slope of the tangent line to the polar curve for the given value of θ.

21. $r = 2\cos\theta$; $\theta = \pi/3$

22. $r = 1 + \sin\theta$; $\theta = \pi/4$

23. $r = 1/\theta$; $\theta = 2$

24. $r = a\sec 2\theta$; $\theta = \pi/6$

25. $r = \cos 3\theta$; $\theta = 3\pi/4$

26. $r = 4 - 3\sin\theta$; $\theta = \pi$

In Exercises 27 and 28, calculate the slopes of the tangent lines indicated in the accompanying figures.

27. $r = 2 + 2\sin\theta$

28. $r = 1 - 2\sin\theta$

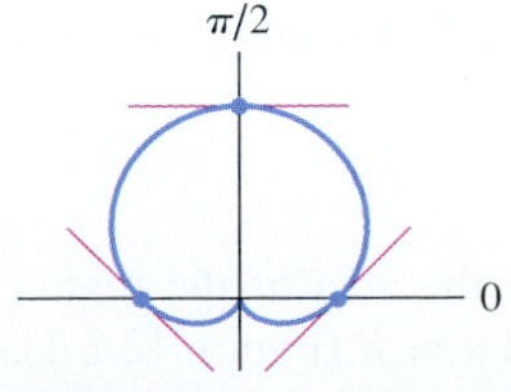

Figure Ex-27

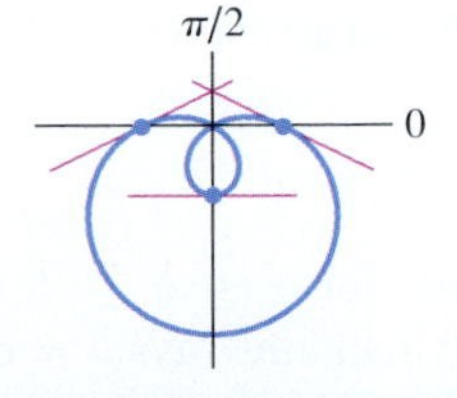

Figure Ex-28

In Exercises 29 and 30, find polar coordinates of all points at which the polar curve has a horizontal or a vertical tangent line.

29. $r = a(1 + \cos\theta)$

30. $r = a\sin\theta$

In Exercises 31 and 32, use a graphing utility to make a conjecture about the number of points on the polar curve at which there is a horizontal tangent line, and confirm your conjecture by finding appropriate derivatives.

31. $r = \sin\theta\cos^2\theta$

32. $r = 1 - 2\sin\theta$

In Exercises 33–38, sketch the polar curve and find polar equations of the tangent lines to the curve at the pole.

33. $r = 2\cos 3\theta$

34. $r = 4\cos\theta$

35. $r = 4\sqrt{\cos 2\theta}$

36. $r = \sin 2\theta$

37. $r = 1 + 2\cos\theta$

38. $r = 2\theta$

In Exercises 39–44, use Formula (8) to calculate the arc length of the polar curve.

39. The entire circle $r = a$

40. The entire circle $r = 2a\cos\theta$

41. The entire cardioid $r = a(1 - \cos\theta)$

42. $r = \sin^2(\theta/2)$ from $\theta = 0$ to $\theta = \pi$

43. $r = e^{3\theta}$ from $\theta = 0$ to $\theta = 2$

44. $r = \sin^3(\theta/3)$ from $\theta = 0$ to $\theta = \pi/2$

45. (a) What is the slope of the tangent line at time t to the trajectory of the paper airplane in Example 2?

(b) What was the airplane's approximate angle of inclination when it crashed into the wall?

46. Suppose that a bee follows the trajectory

$$x = t - 2\sin t, \quad y = 2 - 2\cos t \qquad (t \ge 0)$$

but lands on a wall at time $t = 10$.

(a) At what times was the bee flying horizontally?

(b) At what times was the bee flying vertically?

47. (a) Show that the arc length of one petal of the rose $r = \cos n\theta$ is given by

$$2\int_0^{\pi/(2n)} \sqrt{1 + (n^2 - 1)\sin^2 n\theta}\, d\theta$$

(b) Use the numerical integration capability of a calculating utility to approximate the arc length of one petal of the four-petal rose $r = \cos 2\theta$.

(c) Use the numerical integration capability of a calculating utility to approximate the arc length of one petal of the n-petal rose $r = \cos n\theta$ for $n = 2, 3, 4, \ldots, 20$; then make a conjecture about the limit of these arc lengths as $n \to +\infty$.

48. (a) Sketch the spiral $r = e^{-\theta}$ $(0 \le \theta < +\infty)$.
(b) Find an improper integral for the total arc length of the spiral.
(c) Show that the integral converges and find the total arc length of the spiral.

Exercises 49–54 require the formulas developed in the following discussion: If $f'(t)$ and $g'(t)$ are continuous functions and if no segment of the curve

$$x = f(t), \quad y = g(t) \qquad (a \le t \le b)$$

is traced more than once, then it can be shown that the area of the surface generated by revolving this curve about the x-axis is

$$S = \int_a^b 2\pi y \sqrt{\left(\frac{dx}{dt}\right)^2 + \left(\frac{dy}{dt}\right)^2}\, dt$$

and the area of the surface generated by revolving the curve about the y-axis is

$$S = \int_a^b 2\pi x \sqrt{\left(\frac{dx}{dt}\right)^2 + \left(\frac{dy}{dt}\right)^2}\, dt$$

[The derivations are similar to those used to obtain Formulas (4) and (5) in Section 8.5.]

49. Find the area of the surface generated by revolving $x = t^2$, $y = 2t$ $(0 \le t \le 4)$ about the x-axis.

50. Find the area of the surface generated by revolving the equations $x = e^t \cos t$, $y = e^t \sin t$ $(0 \le t \le \pi/2)$ about the x-axis.

51. Find the area of the surface generated by revolving the equations $x = \cos^2 t$, $y = \sin^2 t$ $(0 \le t \le \pi/2)$ about the y-axis.

52. Find the area of the surface generated by revolving $x = t$, $y = 2t^2$ $(0 \le t \le 1)$ about the y-axis.

53. By revolving the semicircle

$$x = r\cos t, \quad y = r\sin t \qquad (0 \le t \le \pi)$$

about the x-axis, show that the surface area of a sphere of radius r is $4\pi r^2$.

54. The equations

$$x = a\phi - a\sin\phi, \quad y = a - a\cos\phi \qquad (0 \le \phi \le 2\pi)$$

represent one arch of a cycloid. Show that the surface area generated by revolving this curve about the x-axis is given by $S = 64\pi a^2/3$.

55. As illustrated in the accompanying figure, suppose that a rod with one end fixed at the pole of a polar coordinate system rotates counterclockwise at the constant rate of 1 rad/s. At time $t = 0$ a bug on the rod is 10 mm from the pole and is moving outward along the rod at the constant speed of 2 mm/s.
(a) Find an equation of the form $r = f(\theta)$ for the path of motion of the bug, assuming that $\theta = 0$ when $t = 0$.
(b) Find the distance the bug travels along the path in part (a) during the first 5 seconds. Round your answer to the nearest tenth of a millimeter.

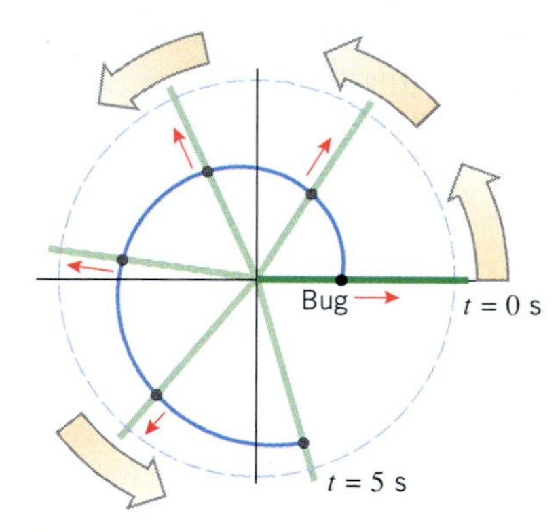

Figure Ex-55

56. Use Formula (6) of Section 8.4 to derive Formula (8).

12.3 AREA IN POLAR COORDINATES

In this section we will show how to find areas of regions that are bounded by polar curves.

AREA IN POLAR COORDINATES

12.3.1 AREA PROBLEM IN POLAR COORDINATES. Suppose that α and β are angles that satisfy the condition

$$\alpha < \beta \le \alpha + 2\pi$$

and suppose that $f(\theta)$ is continuous for $\alpha \le \theta \le \beta$. Find the area of the region R enclosed by the polar curve $r = f(\theta)$ and the rays $\theta = \alpha$ and $\theta = \beta$ (Figure 12.3.1).

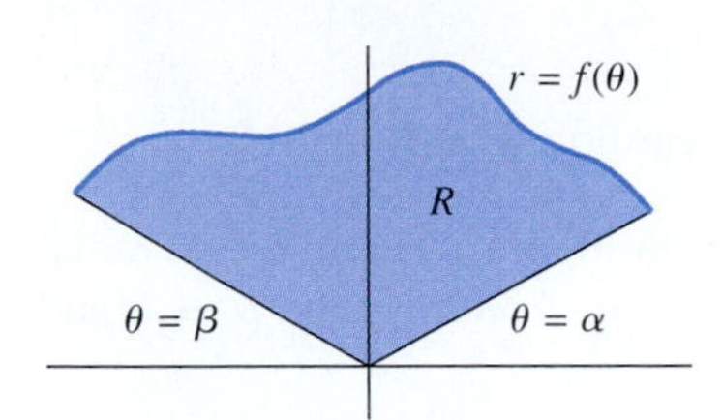

Figure 12.3.1

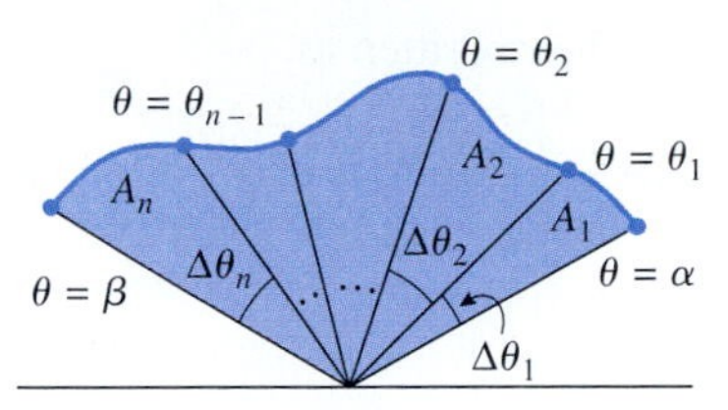

Figure 12.3.2

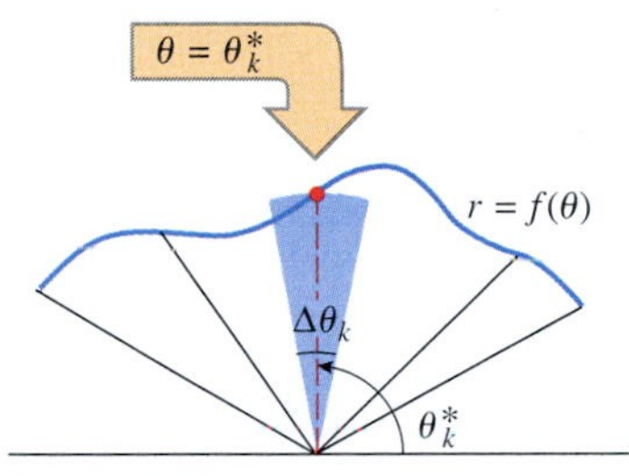

Figure 12.3.3

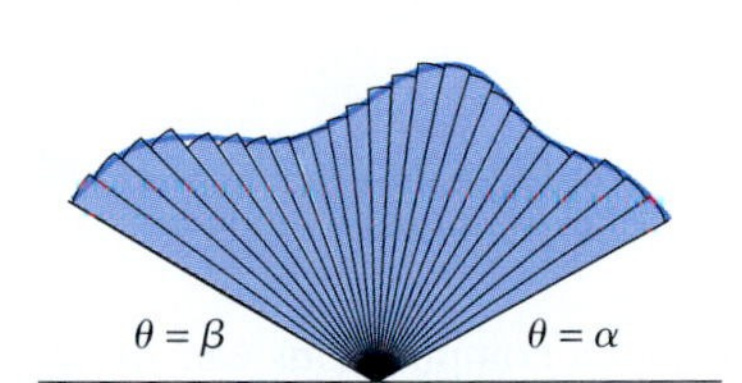

Figure 12.3.4

In rectangular coordinates we solved Area Problem 7.1.1 by dividing the region into an increasing number of vertical strips, approximating the strips by rectangles, and taking a limit. In polar coordinates rectangles are clumsy to work with, and it is better to divide the region into ***wedges*** by using rays

$$\theta = \theta_1,\ \theta = \theta_2,\ \ldots,\ \theta = \theta_{n-1}$$

such that

$$\alpha < \theta_1 < \theta_2 < \cdots < \theta_{n-1} < \beta$$

(Figure 12.3.2). As shown in that figure, the rays divide the region R into n wedges with areas $A_1, A_2, \ldots, A_n$ and central angles $\Delta\theta_1, \Delta\theta_2, \ldots, \Delta\theta_n$. The area of the entire region can be written as

$$A = A_1 + A_2 + \cdots + A_n = \sum_{k=1}^{n} A_k \tag{1}$$

If $\Delta\theta_k$ is small, and if we assume for simplicity that $f(\theta)$ is nonnegative, then we can approximate the area A_k of the kth wedge by the area of a sector with central angle $\Delta\theta_k$ and radius $f(\theta_k^*)$, where $\theta = \theta_k^*$ is any ray that lies in the kth wedge (Figure 12.3.3). Thus, from (1) and Formula (5) of Appendix E for the area of a sector, we obtain

$$A = \sum_{k=1}^{n} A_k \approx \sum_{k=1}^{n} \frac{1}{2}[f(\theta_k^*)]^2 \Delta\theta_k \tag{2}$$

If we now increase n in such a way that $\max \Delta\theta_k \to 0$, then the sectors will become better and better approximations of the wedges and it is reasonable to expect that (2) will approach the exact value of the area A (Figure 12.3.4); that is,

$$A = \lim_{\max \Delta\theta_k \to 0} \sum_{k=1}^{n} \frac{1}{2}[f(\theta_k^*)]^2 \Delta\theta_k = \int_{\alpha}^{\beta} \frac{1}{2}[f(\theta)]^2\, d\theta$$

Thus, we have the following solution of Area Problem 12.3.1.

12.3.2 AREA IN POLAR COORDINATES. If α and β are angles that satisfy the condition

$$\alpha < \beta \le \alpha + 2\pi$$

and if $f(\theta)$ is continuous for $\alpha \le \theta \le \beta$, then the area A of the region R enclosed by the polar curve $r = f(\theta)$ and the rays $\theta = \alpha$ and $\theta = \beta$ is

$$A = \int_{\alpha}^{\beta} \frac{1}{2}[f(\theta)]^2\, d\theta = \int_{\alpha}^{\beta} \frac{1}{2} r^2\, d\theta \tag{3}$$

The hardest part of applying (3) is determining the limits of integration. This can be done as follows:

Step 1. Sketch the region R whose area is to be determined.

Step 2. Draw an arbitrary "radial line" from the pole to the boundary curve $r = f(\theta)$.

Step 3. Ask, "Over what interval of values must θ vary in order for the radial line to sweep out the region R?"

Step 4. Your answer in Step 3 will determine the lower and upper limits of integration.

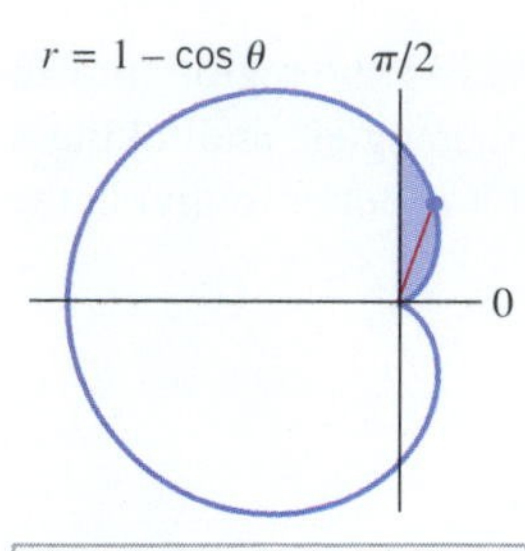

The shaded region is swept out by the radial line as θ varies from 0 to $\pi/2$.

Figure 12.3.5

Example 1

Find the area of the region in the first quadrant within the cardioid $r = 1 - \cos\theta$.

Solution. The region and a typical radial line are shown in Figure 12.3.5. For the radial line to sweep out the region, θ must vary from 0 to $\pi/2$. Thus, from (3) with $\alpha = 0$ and $\beta = \pi/2$, we obtain

$$A = \int_0^{\pi/2} \frac{1}{2}r^2\,d\theta = \int_0^{\pi/2} (1 - \cos\theta)^2\,d\theta = \frac{1}{2}\int_0^{\pi/2} (1 - 2\cos\theta + \cos^2\theta)\,d\theta$$

With the help of the identity $\cos^2\theta = \frac{1}{2}(1 + \cos 2\theta)$, this can be rewritten as

$$A = \frac{1}{2}\int_0^{\pi/2}\left(\frac{3}{2} - 2\cos\theta + \frac{1}{2}\cos 2\theta\right)d\theta = \frac{1}{2}\left[\frac{3}{2}\theta - 2\sin\theta + \frac{1}{4}\sin 2\theta\right]_0^{\pi/2} = \frac{3}{8}\pi - 1 \quad \blacktriangleleft$$

Example 2

Find the entire area within the cardioid of Example 1.

Solution. For the radial line to sweep out the entire cardioid, θ must vary from 0 to 2π. Thus, from (3) with $\alpha = 0$ and $\beta = 2\pi$,

$$A = \int_0^{2\pi} \frac{1}{2}r^2\,d\theta = \frac{1}{2}\int_0^{2\pi} (1 - \cos\theta)^2\,d\theta$$

If we proceed as in Example 1, this reduces to

$$A = \frac{1}{2}\int_0^{2\pi}\left(\frac{3}{2} - 2\cos\theta + \frac{1}{2}\cos 2\theta\right)d\theta = \frac{3\pi}{2}$$

Alternative Solution. Since the cardioid is symmetric about the x-axis, we can calculate the portion of the area above the x-axis and double the result. In the portion of the cardioid above the x-axis, θ ranges from 0 to π, so that

$$A = 2\int_0^{\pi} \frac{1}{2}r^2\,d\theta = \int_0^{\pi} (1 - \cos\theta)^2\,d\theta = \frac{3\pi}{2} \quad \blacktriangleleft$$

USING SYMMETRY

Although Formula (3) is applicable if $r = f(\theta)$ is negative, area computations can sometimes be simplified by using symmetry to restrict the limits of integration to intervals where $r \geq 0$. This is illustrated in the next example.

Example 3

Find the area of the region enclosed by the rose curve $r = \cos 2\theta$.

Solution. Referring to Figure 12.1.10 and using symmetry, the area in the first quadrant that is swept out for $0 \leq \theta \leq \pi/4$ is one-eighth of the total area inside the rose. Thus, from Formula (3)

$$\begin{aligned} A &= 8\int_0^{\pi/4} \frac{1}{2}r^2\,d\theta = 4\int_0^{\pi/4} \cos^2 2\theta\,d\theta \\ &= 4\int_0^{\pi/4} \frac{1}{2}(1 + \cos 4\theta)\,d\theta = 2\int_0^{\pi/4} (1 + \cos 4\theta)\,d\theta \\ &= 2\theta + \frac{1}{2}\sin 4\theta\Big]_0^{\pi/4} = \frac{\pi}{2} \quad \blacktriangleleft \end{aligned}$$

Sometimes the most natural way to satisfy the restriction $\alpha < \beta \leq \alpha + 2\pi$ required by Formula (3) is to use a negative value for α. For example, suppose that we are interested in finding the area of the shaded region in Figure 12.3.6*a*. The first step would be to determine the intersections of the cardioid $r = 4+4\cos\theta$ and the circle $r = 6$, since this information is needed for the limits of integration. To find the points of intersection, we can equate the two expressions for r. This yields

$$4+4\cos\theta = 6 \quad \text{or} \quad \cos\theta = \frac{1}{2}$$

which is satisfied by the positive angles

$$\theta = \frac{\pi}{3} \quad \text{and} \quad \theta = \frac{5\pi}{3}$$

However, there is a problem here because the radial lines to the circle and cardioid do not sweep through the shaded region shown in Figure 12.3.6*b* as θ varies over the interval $\pi/3 \le \theta \le 5\pi/3$. There are two ways to circumvent this problem—one is to take advantage of the symmetry by integrating over the interval $0 \le \theta \le \pi/3$ and doubling the result, and the second is to use a negative lower limit of integration and integrate over the interval $-\pi/3 \le \theta \le \pi/3$ (Figure 12.3.6*c*). The two methods are illustrated in the next example.

Example 4

Find the area of the region that is inside of the cardioid $r = 4+4\cos\theta$ and outside of the circle $r = 6$.

Solution Using a Negative Angle. The area of the region can be obtained by subtracting the areas in Figures 12.3.6*d* and 12.3.6*e*:

$$\begin{aligned} A &= \int_{-\pi/3}^{\pi/3} \frac{1}{2}(4+4\cos\theta)^2\,d\theta - \int_{-\pi/3}^{\pi/3} \frac{1}{2}(6)^2\,d\theta \\ &= \int_{-\pi/3}^{\pi/3} \frac{1}{2}[(4+4\cos\theta)^2 - 36]\,d\theta = \int_{-\pi/3}^{\pi/3} (16\cos\theta + 8\cos^2\theta - 10)\,d\theta \\ &= \Big[16\sin\theta + (4\theta + 2\sin 2\theta) - 10\theta\Big]_{-\pi/3}^{\pi/3} = 18\sqrt{3} - 4\pi \end{aligned}$$

Area inside cardioid minus area inside circle.

Solution Using Symmetry. Using symmetry, we can calculate the area above the polar axis and double it. This yields (verify)

$$A = 2\int_{0}^{\pi/3} \frac{1}{2}[(4+4\cos\theta)^2 - 36]\,d\theta = 2(9\sqrt{3} - 2\pi) = 18\sqrt{3} - 4\pi$$

which agrees with the preceding result. ◀

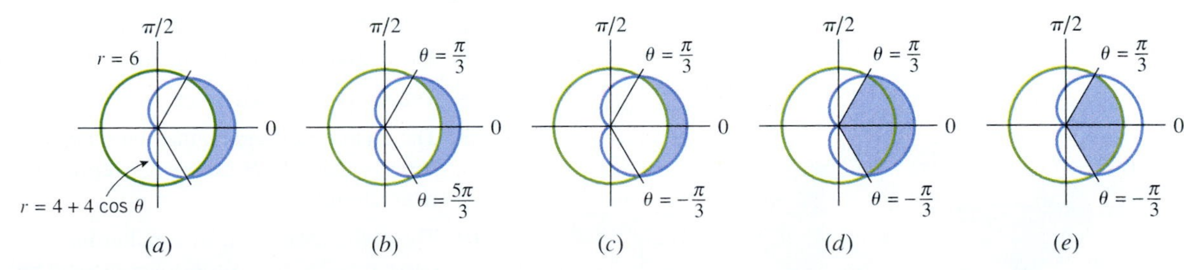

Figure 12.3.6

INTERSECTIONS OF POLAR GRAPHS

In the last example we found the intersections of the cardioid and circle by equating their expressions for r and solving for θ. However, because a point can be represented in different ways in polar coordinates, this procedure will not always produce all of the intersections. For example, the cardioids

$$r = 1 - \cos\theta \quad \text{and} \quad r = 1 + \cos\theta \tag{4}$$

intersect at three points: the pole, the point $(1, \pi/2)$, and the point $(1, 3\pi/2)$ (Figure 12.3.7). Equating the right-hand sides of the equations in (4) yields $1 - \cos\theta = 1 + \cos\theta$

or $\cos\theta = 0$, so

$$\theta = \frac{\pi}{2} + k\pi, \quad k = 0, \pm 1, \pm 2, \ldots$$

Substituting any of these values in (4) yields $r = 1$, so that we have found only two distinct points of intersection, $(1, \pi/2)$ and $(1, 3\pi/2)$; the pole has been missed. This problem occurs because the two cardioids pass through the pole at different values of θ—the cardioid $r = 1 - \cos\theta$ passes through the pole at $\theta = 0$, and the cardioid $r = 1 + \cos\theta$ passes through the pole at $\theta = \pi$.

The situation with the cardioids is analogous to two satellites circling the Earth in intersecting orbits (Figure 12.3.8). The satellites will not collide unless they reach the same point at the same time. In general, when looking for intersections of polar curves, it is a good idea to graph the curves to determine how many intersections there should be.

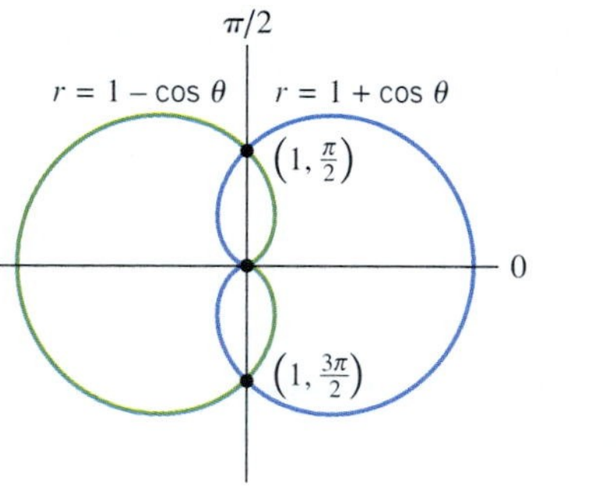

Figure 12.3.7

Figure 12.3.8

EXERCISE SET 12.3 Graphing Calculator C CAS

1. Write down, but do not evaluate, an integral for the area of each shaded region.

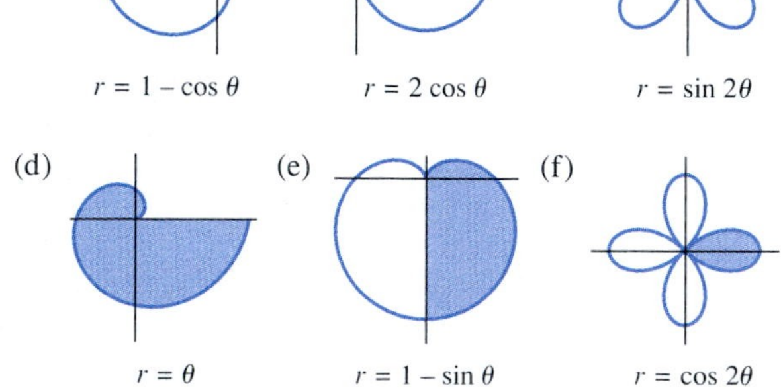

2. Evaluate the integrals you obtained in Exercise 1.

3. In each part, find the area of the circle by integration.
(a) $r = a$ (b) $r = 2a\sin\theta$ (c) $r = 2a\cos\theta$

4. (a) Show that $r = \sin\theta + \cos\theta$ is a circle.
(b) Find the area of the circle using a geometric formula and then by integration.

In Exercises 5–10, find the area of the region described.

5. The region that is enclosed by the cardioid $r = 2 + 2\cos\theta$.

6. The region in the first quadrant within the cardioid $r = 1 + \sin\theta$.

7. The region enclosed by the rose $r = 4\cos 3\theta$.

8. The region enclosed by the rose $r = 2\sin 2\theta$.

9. The region enclosed by the inner loop of the limaçon $r = 1 + 2\cos\theta$. [*Hint:* $r \le 0$ over the interval of integration.]

10. The region swept out by a radial line from the pole to the curve $r = 2/\theta$ as θ varies over the interval $1 \le \theta \le 3$.

In Exercises 11–14, find the area of the shaded region.

11.

12.

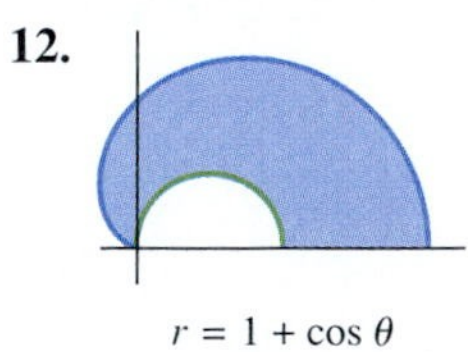

13. 

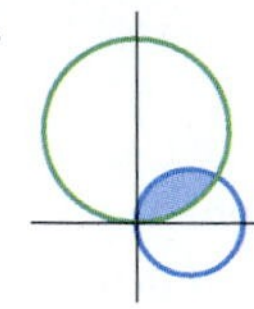

$r = 4\cos t$
$r = 4\sqrt{3}\sin t$

14. 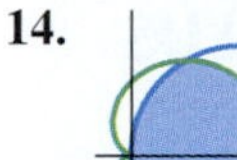

$r = 1 + \cos t$
$r = 3\cos t$

In Exercises 15–22, find the area of the region described.

15. The region inside the circle $r = 5\sin\theta$ and outside the limaçon $r = 2 + \sin\theta$.

16. The region outside the cardioid $r = 2 - 2\cos\theta$ and inside the circle $r = 4$.

17. The region inside the cardioid $r = 2 + 2\cos\theta$ and outside the circle $r = 3$.

18. The region that is common to the circles $r = 4\cos\theta$ and $r = 4\sin\theta$.

19. The region between the loops of the limaçon $r = \frac{1}{2} + \cos\theta$.

20. The region inside the cardioid $r = 2 + 2\cos\theta$ and to the right of the line $r\cos\theta = \frac{3}{2}$.

21. The region inside the circle $r = 10$ and to the right of the line $r = 6\sec\theta$.

22. The region inside the rose $r = 2a\cos 2\theta$ and outside the circle $r = a\sqrt{2}$.

23. (a) Find the error: The area that is inside the lemniscate $r^2 = a^2\cos 2\theta$ is

$$A = \int_0^{2\pi} \frac{1}{2}r^2\,d\theta = \int_0^{2\pi} \frac{1}{2}a^2\cos 2\theta\,d\theta$$
$$= \frac{1}{4}a^2\sin 2\theta\bigg]_0^{2\pi} = 0$$

(b) Find the correct area.

(c) Find the area inside the lemniscate $r^2 = 4\cos 2\theta$ and outside the circle $r = \sqrt{2}$.

24. Find the area inside the curve $r^2 = \sin 2\theta$.

25. A radial line is drawn from the origin to the spiral $r = a\theta$ ($a > 0$ and $\theta \geq 0$). Find the area swept out during the second revolution of the radial line that was not swept out during the first revolution.

26. (a) In the discussion associated with Exercises 49–54 of Section 12.2, formulas were given for the area of the surface of revolution that is generated by revolving a parametric curve about the x-axis or y-axis. Use those formulas to derive the following formulas for the areas of the surfaces of revolution that are generated by revolving the portion of the polar curve $r = f(\theta)$ from $\theta = \alpha$ to $\theta = \beta$ about the polar axis and about the line $\theta = \pi/2$:

$$S = \int_\alpha^\beta 2\pi r\sin\theta\sqrt{r^2 + \left(\frac{dr}{d\theta}\right)^2}\,d\theta \quad \text{About } \theta = 0$$

$$S = \int_\alpha^\beta 2\pi r\cos\theta\sqrt{r^2 + \left(\frac{dr}{d\theta}\right)^2}\,d\theta \quad \text{About } \theta = \pi/2$$

(b) State conditions under which these formulas hold.

In Exercises 27–30, sketch the surface, and use the formulas in Exercise 26 to find the surface area.

27. The surface generated by revolving the circle $r = \cos\theta$ about the line $\theta = \pi/2$.

28. The surface generated by revolving the spiral $r = e^\theta$ ($0 \leq \theta \leq \pi/2$) about the line $\theta = \pi/2$.

29. The "apple" generated by revolving the upper half of the cardioid $r = 1 - \cos\theta$ ($0 \leq \theta \leq \pi$) about the polar axis.

30. The sphere of radius a generated by revolving the semicircle $r = a$ in the upper half-plane about the polar axis.

C **31.** (a) Show that the Folium of Descartes $x^3 - 3xy + y^3 = 0$ can be expressed in polar coordinates as

$$r = \frac{3\sin\theta\cos\theta}{\cos^3\theta + \sin^3\theta}$$

(b) Use a CAS to show that the area inside of the loop is $\frac{3}{2}$ (Figure 4.3.2).

C **32.** (a) What is the area that is enclosed by one petal of the rose $r = a\cos n\theta$ if n is an even integer?

(b) What is the area that is enclosed by one petal of the rose $r = a\cos n\theta$ if n is an odd integer?

(c) Use a CAS to show that the total area enclosed by the rose $r = a\cos n\theta$ is $\pi a^2/2$ if the number of petals is even. [*Hint:* See Exercise 73 of Section 12.1.]

(d) Use a CAS to show that the total area enclosed by the rose $r = a\cos n\theta$ is $\pi a^2/4$ if the number of petals is odd.

33. One of the most famous problems in Greek antiquity was "squaring the circle"; that is, using a straightedge and compass to construct a square whose area is equal to that of a given circle. It was proved in the nineteenth century that no such construction is possible. However, show that the shaded areas in the accompanying figure are equal, thereby "squaring the crescent."

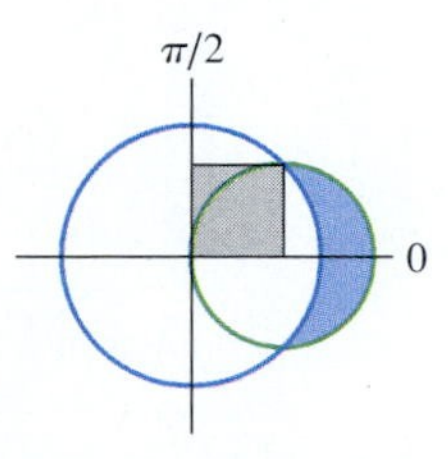

Figure Ex-33

34. Use a graphing utility to generate the polar graph of the equation $r = \cos 3\theta + 2$, and find the area that it encloses.

35. Use a graphing utility to generate the graph of the ***bifolium*** $r = 2\cos\theta\sin^2\theta$, and find the area of the upper loop.

12.4 CONIC SECTIONS IN CALCULUS

In this section we will discuss some of the basic geometric properties of parabolas, ellipses, and hyperbolas. These curves play an important role in calculus and also arise naturally in a broad range of applications in such fields as planetary motion, design of telescopes and antennas, geodetic positioning, and medicine, to name a few.

Some students may already be familiar with the material in this section, in which case it can be treated as a review. Instructors who want to spend some additional time on precalculus review may want to allocate more than one lecture on this material.

CONIC SECTIONS

Circles, ellipses, parabolas, and hyperbolas are called ***conic sections*** or ***conics*** because they can be obtained as intersections of a plane with a double-napped circular cone (Figure 12.4.1). If the plane passes through the vertex of the double-napped cone, then the intersection is a point, a pair of intersecting lines, or a single line. These are called ***degenerate conic sections***.

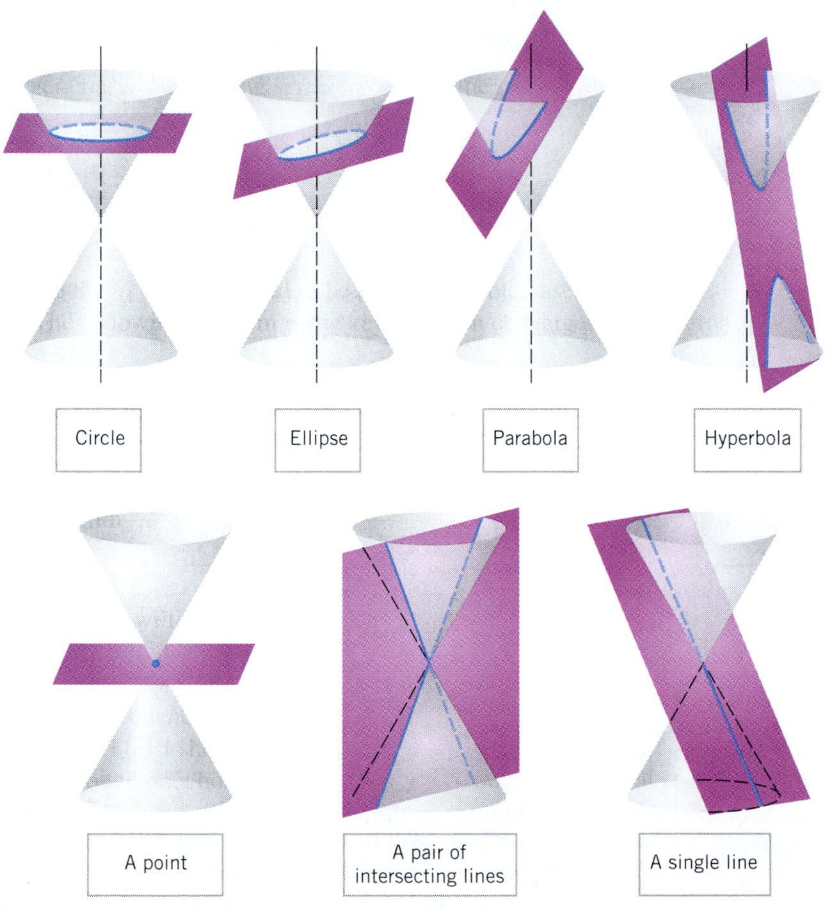

Figure 12.4.1

DEFINITIONS OF THE CONIC SECTIONS

Although we could derive properties of parabolas, ellipses, and hyperbolas by defining them as intersections with a double-napped cone, it will be better suited to calculus if we begin with equivalent definitions that are based on their geometric properties.

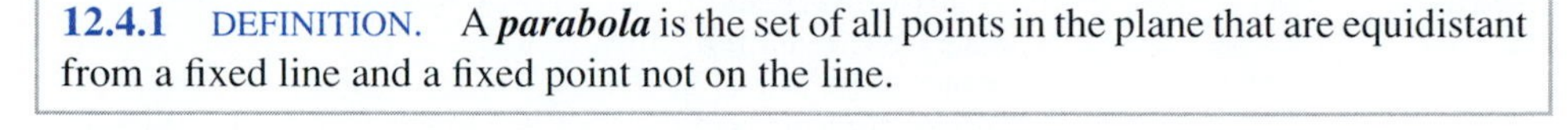

12.4.1 DEFINITION. A ***parabola*** is the set of all points in the plane that are equidistant from a fixed line and a fixed point not on the line.

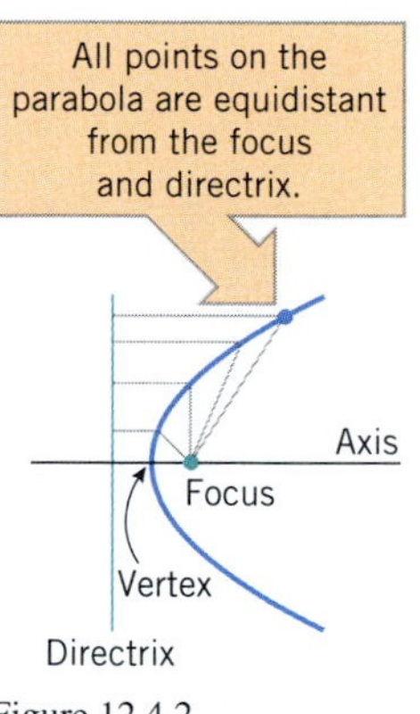

Figure 12.4.2

The line is called the ***directrix*** of the parabola, and the point is called the ***focus*** (Figure 12.4.2). A parabola is symmetric about the line that passes through the focus at right angles to the directrix. This line, called the ***axis*** or the ***axis of symmetry*** of the parabola, intersects the parabola at a point called the ***vertex***.

12.4.2 DEFINITION. An ***ellipse*** is the set of all points in the plane, the sum of whose distances from two fixed points is a given positive constant that is greater than the distance between the fixed points.

The two fixed points are called the ***foci*** (plural of "focus") of the ellipse, and the midpoint of the line segment joining the foci is called the ***center*** (Figure 12.4.3*a*). To help visualize Definition 12.4.2, imagine that two ends of a string are tacked to the foci and a pencil traces a curve as it is held tight against the string (Figure 12.4.3*b*). The resulting curve will be an ellipse since the sum of the distances to the foci is a constant, namely the total length of the string. Note that if the foci coincide, the ellipse reduces to a circle. For ellipses other than circles, the line segment through the foci and across the ellipse is called the ***major axis*** (Figure 12.4.3*c*), and the line segment across the ellipse, through the center, and perpendicular to the major axis is called the ***minor axis***. The endpoints of the major axis are called ***vertices***.

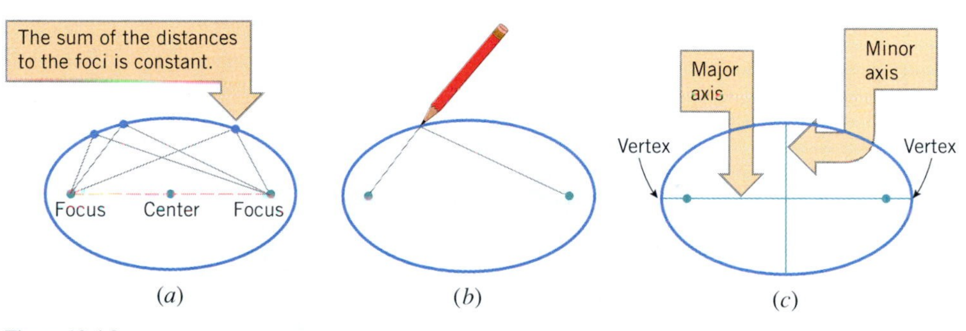

Figure 12.4.3

12.4.3 DEFINITION. A ***hyperbola*** is the set of all points in the plane, the difference of whose distances from two fixed distinct points is a given positive constant that is less than the distance between the fixed points.

The two fixed points are called the ***foci*** of the hyperbola, and the term "difference" that is used in the definition is understood to mean the distance to the farther focus minus the distance to the closer focus. As a result, the points on the hyperbola form two ***branches***, each "wrapping around" the closer focus (Figure 12.4.4*a*). The midpoint of the line segment joining the foci is called the ***center*** of the hyperbola, the line through the foci is called the ***focal axis***, and the line through the center that is perpendicular to the focal axis is called the ***conjugate axis***. The hyperbola intersects the focal axis at two points called the ***vertices***.

Associated with every hyperbola is a pair of lines, called the ***asymptotes*** of the hyperbola. These lines intersect at the center of the hyperbola and have the property that as a point P moves along the hyperbola away from the center, the vertical distance between P and one of the asymptotes approaches zero (Figure 12.4.4*b*).

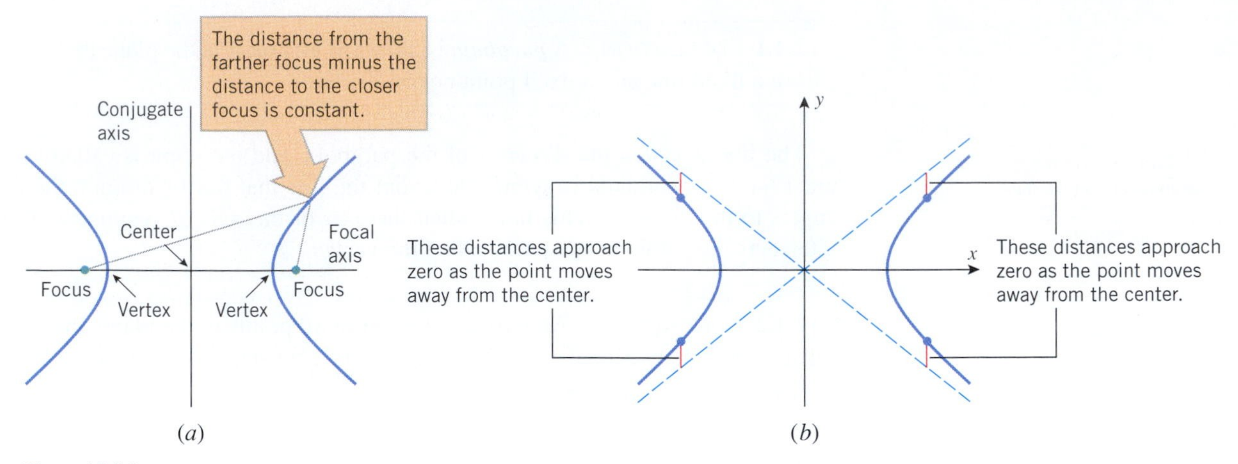

Figure 12.4.4

EQUATIONS OF PARABOLAS IN STANDARD POSITION

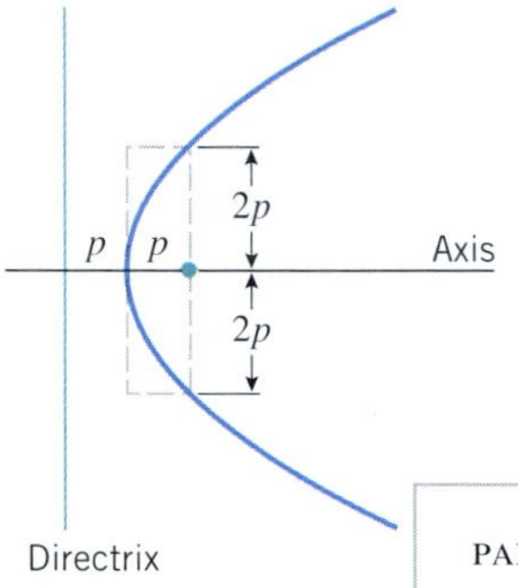

Figure 12.4.5

It is traditional in the study of parabolas to denote the distance between the focus and the vertex by p. The vertex is equidistant from the focus and the directrix, so the distance between the vertex and the directrix is also p; consequently, the distance between the focus and the directrix is $2p$ (Figure 12.4.5). As illustrated in that figure, the parabola passes through two of the corners of a box that extends from the vertex to the focus along the axis of symmetry and extends $2p$ units above and $2p$ units below the axis of symmetry.

The equation of a parabola is simplest if the vertex is the origin and the axis of symmetry is along the x-axis or y-axis. The four possible such orientations are shown in Figure 12.4.6. These are called the ***standard positions*** of a parabola, and the resulting equations are called the ***standard equations*** of a parabola.

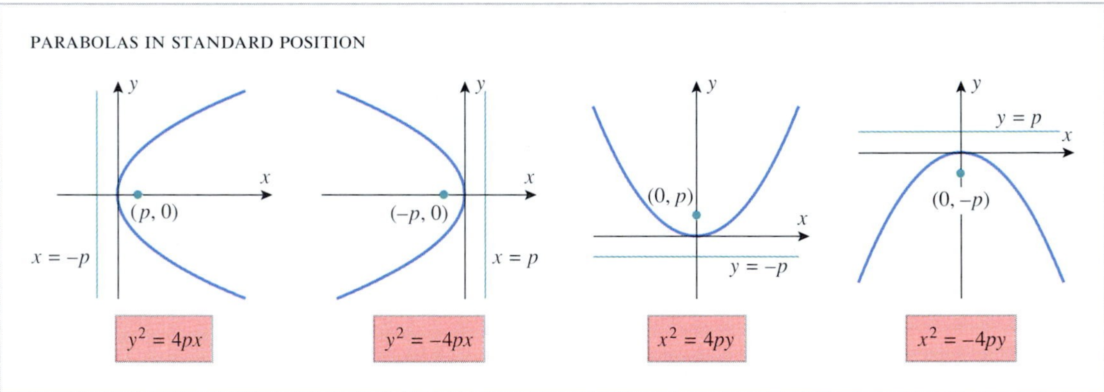

Figure 12.4.6

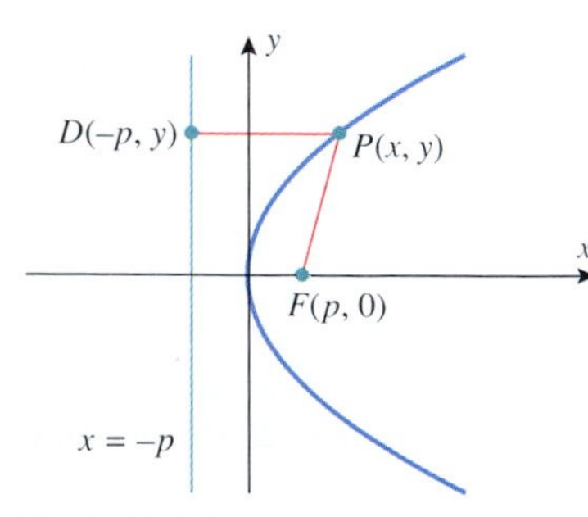

Figure 12.4.7

To illustrate how the equations in Figure 12.4.6 are obtained, we will derive the equation for the parabola with focus $(p, 0)$ and directrix $x = -p$. Let $P(x, y)$ by any point on the parabola. Since P is equidistant from the focus and directrix, the distances PF and PD in Figure 12.4.7 are equal; that is,

$$PF = PD \tag{1}$$

where $D(-p, y)$ is the foot of the perpendicular from P to the directrix. From the distance formula, the distances PF and PD are

$$PF = \sqrt{(x-p)^2 + y^2} \quad \text{and} \quad PD = \sqrt{(x+p)^2} \tag{2}$$

Substituting in (1) and squaring yields

$$(x-p)^2 + y^2 = (x+p)^2 \tag{3}$$

and after simplifying

$$y^2 = 4px \tag{4}$$

The derivations of the other equations in Figure 12.4.6 are similar.

A TECHNIQUE FOR SKETCHING PARABOLAS

Parabolas can be sketched from their *standard equations* using four basic steps:

- Determine whether the axis of symmetry is along the x-axis or the y-axis. Referring to Figure 12.4.6, the axis of symmetry is along the x-axis if the equation has a y^2-term, and it is along the y-axis if it has an x^2-term.
- Determine which way the parabola opens. If the axis of symmetry is along the x-axis, then the parabola opens to the right if the coefficient of x is positive, and it opens to the left if the coefficient is negative. If the axis of symmetry is along the y-axis, then the parabola opens up if the coefficient of y is positive, and it opens down if the coefficient is negative.
- Determine the value of p and draw a box extending p units from the origin along the axis of symmetry in the direction in which the parabola opens and extending $2p$ units on each side of the axis of symmetry.
- Using the box as a guide, sketch the parabola so that its vertex is at the origin and it passes through the corners of the box (Figure 12.4.8).

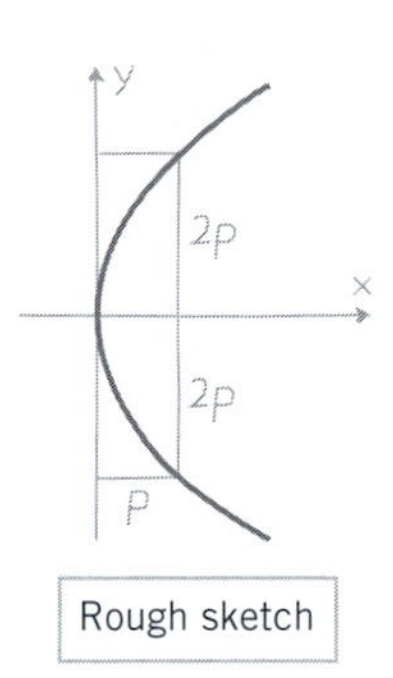

Figure 12.4.8

Example 1

Sketch the graphs of the parabolas

(a) $x^2 = 12y$ (b) $y^2 + 8x = 0$

and show the focus and directrix of each.

Solution (a). This equation involves x^2, so the axis of symmetry is along the y-axis, and the coefficient of y is positive, so the parabola opens upward. From the coefficient of y, we obtain $4p = 12$ or $p = 3$. Drawing a box extending $p = 3$ units up from the origin and $2p = 6$ units to the left and $2p = 6$ units to the right of the y-axis, then using corners of the box as a guide, yields the graph in Figure 12.4.9.

The focus is $p = 3$ units from the vertex along the axis of symmetry in the direction in which the parabola opens, so its coordinates are $(0, 3)$. The directrix is perpendicular to the axis of symmetry at a distance of $p = 3$ units from the vertex on the opposite side from the focus, so its equation is $y = -3$.

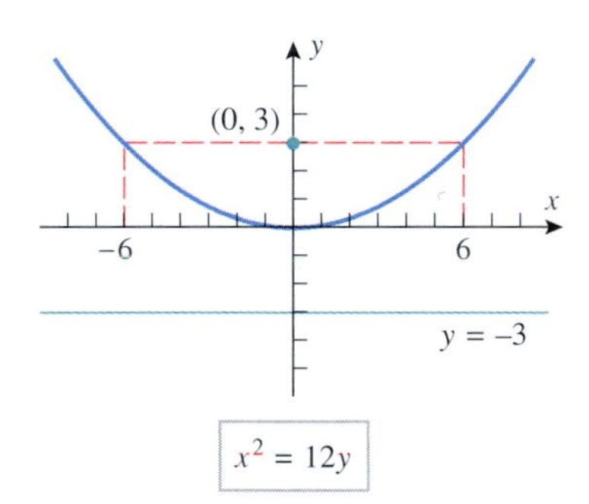

Figure 12.4.9

Solution (b). We first rewrite the equation in the standard form

$$y^2 = -8x$$

This equation involves y^2, so the axis of symmetry is along the x-axis, and the coefficient of x is negative, so the parabola opens to the left. From the coefficient of x we obtain $4p = 8$, so $p = 2$. Drawing a box extending $p = 2$ units left from the origin and $2p = 4$ units above and $2p = 4$ units below the x-axis, then using corners of the box as a guide, yields the graph in Figure 12.4.10. ◀

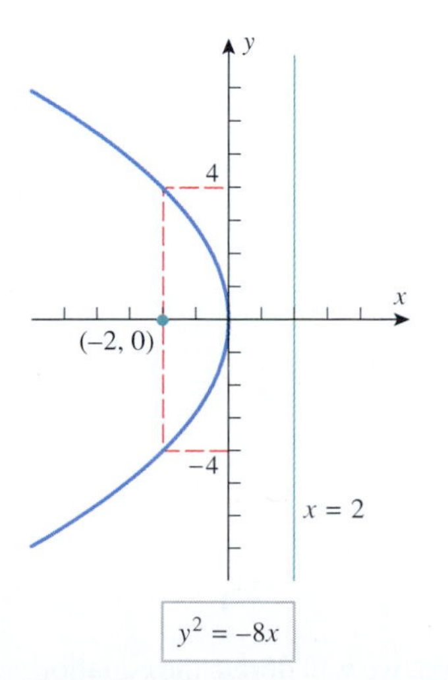

Figure 12.4.10

Example 2

Find an equation of the parabola that is symmetric about the y-axis, has its vertex at the origin, and passes through the point $(5, 2)$.

Solution. Since the parabola is symmetric about the y-axis and has its vertex at the origin, the equation is of the form

$$x^2 = 4py \quad \text{or} \quad x^2 = -4py$$

where the sign depends on whether the parabola opens up or down. But the parabola must open up, since it passes through the point (5, 2), which lies in the first quadrant. Thus, the equation is of the form

$$x^2 = 4py \tag{5}$$

Since the parabola passes through (5, 2), we must have $5^2 = 4p \cdot 2$ or $4p = \frac{25}{2}$. Therefore, (5) becomes

$$x^2 = \tfrac{25}{2}y$$ ◀

EQUATIONS OF ELLIPSES IN STANDARD POSITION

It is traditional in the study of ellipses to denote the length of the major axis by $2a$, the length of the minor axis by $2b$, and the distance between the foci by $2c$ (Figure 12.4.11). The number a is called the ***semimajor axis*** and the number b the ***semiminor axis*** (standard but odd terminology, since a and b are numbers, not geometric axes).

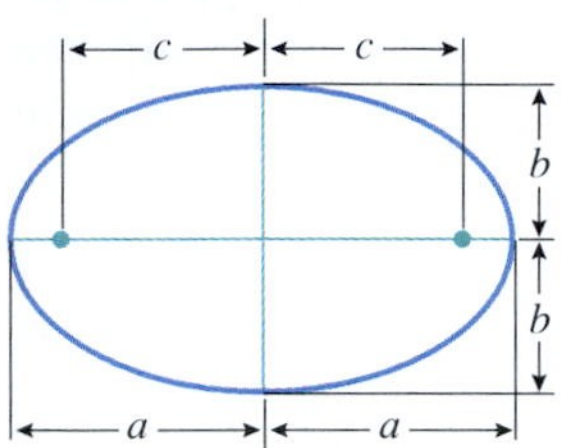
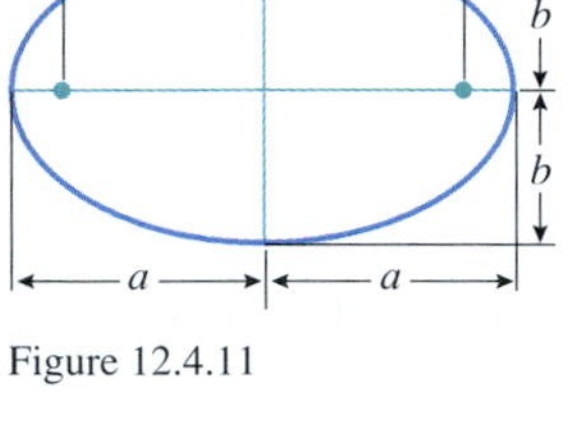

Figure 12.4.11

There is a basic relationship between the numbers a, b, and c that can be obtained by examining the sum of the distances to the foci from a point P at the end of the major axis and from a point Q at the end of the minor axis (Figure 12.4.12). From Definition 12.4.2, these sums must be equal, so we obtain

$$2\sqrt{b^2 + c^2} = (a - c) + (a + c)$$

from which it follows that

$$a = \sqrt{b^2 + c^2} \tag{6}$$

or, equivalently,

$$c = \sqrt{a^2 - b^2} \tag{7}$$

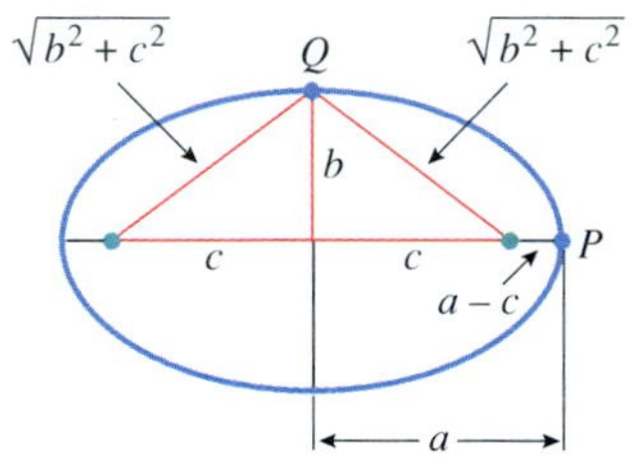

Figure 12.4.12

From (6), the distance from a focus to an end of the minor axis is a (Figure 12.4.13), which implies that for *all* points on the ellipse the sum of the distances to the foci is $2a$.

It also follows from (6) that $a \geq b$ with the equality holding only when $c = 0$. Geometrically, this means that the major axis of an ellipse is at least as large as the minor axis and that the two axes have equal length only when the foci coincide, in which case the ellipse is a circle.

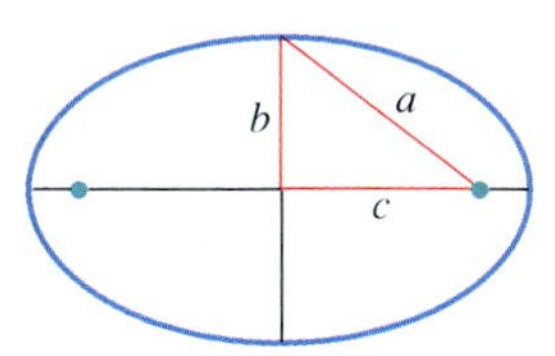

Figure 12.4.13

The equation of an ellipse is simplest if the center of the ellipse is at the origin and the foci are on the x-axis or y-axis. The two possible such orientations are shown in Figure 12.4.14. These are called the ***standard positions*** of an ellipse, and the resulting equations are called the ***standard equations*** of an ellipse.

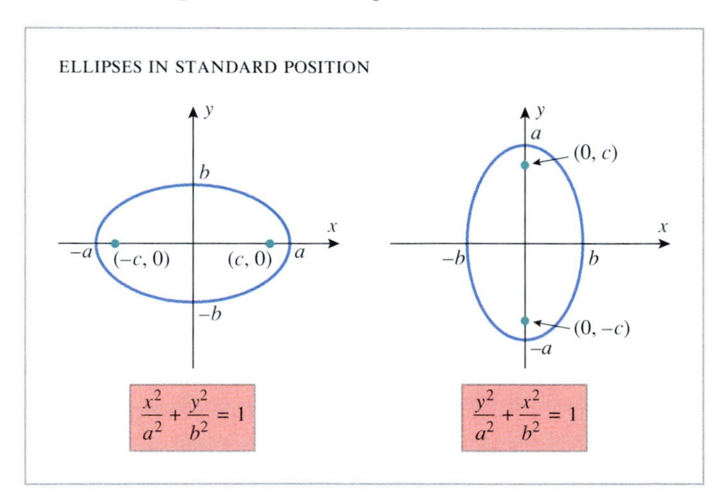

Figure 12.4.14

To illustrate how the equations in Figure 12.4.14 are obtained, we will derive the equation for the ellipse with foci on the x-axis. Let $P(x, y)$ be any point on that ellipse. Since the sum of the distances from P to the foci is $2a$, it follows (Figure 12.4.15) that

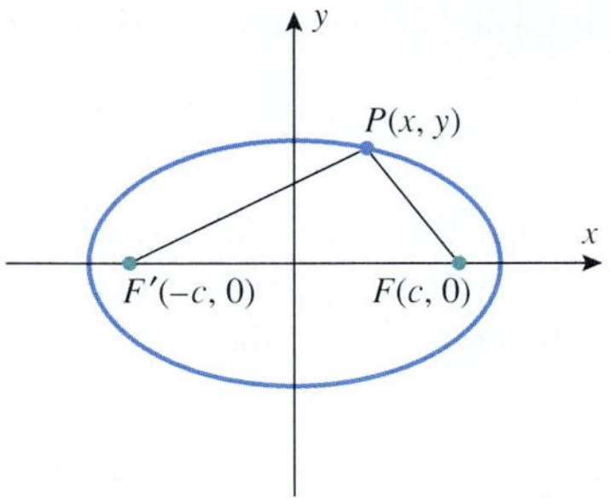

Figure 12.4.15

$$PF' + PF = 2a$$

so

$$\sqrt{(x+c)^2 + y^2} + \sqrt{(x-c)^2 + y^2} = 2a$$

Transposing the second radical to the right side of the equation and squaring yields

$$(x+c)^2 + y^2 = 4a^2 - 4a\sqrt{(x-c)^2 + y^2} + (x-c)^2 + y^2$$

and, on simplifying,

$$\sqrt{(x-c)^2 + y^2} = a - \frac{c}{a}x \tag{8}$$

Squaring again and simplifying yields

$$\frac{x^2}{a^2} + \frac{y^2}{a^2 - c^2} = 1$$

which, by virtue of (6), can be written as

$$\frac{x^2}{a^2} + \frac{y^2}{b^2} = 1 \tag{9}$$

Conversely, it can be shown that any point whose coordinates satisfy (9) has $2a$ as the sum of its distances from the foci, so that such a point is on the ellipse.

A TECHNIQUE FOR SKETCHING ELLIPSES

Ellipses can be sketched from their *standard equations* using three basic steps:

- Determine whether the major axis is on the x-axis or the y-axis. This can be ascertained from the sizes of the denominators in the equation. Referring to Figure 12.4.14, and keeping in mind that $a^2 > b^2$ (since $a > b$), the major axis is along the x-axis if x^2 has the larger denominator, and it is along the y-axis if y^2 has the larger denominator. If the denominators are equal, the ellipse is a circle.
- Determine the values of a and b and draw a box extending a units on each side of the center along the major axis and b units on each side of the center along the minor axis.
- Using the box as a guide, sketch the ellipse so that its center is at the origin and it touches the sides of the box where the sides intersect the coordinate axes (Figure 12.4.16).

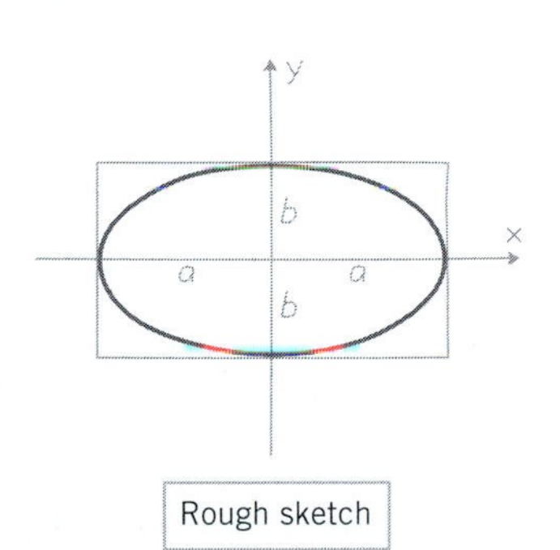

Figure 12.4.16

Example 3

Sketch the graphs of the ellipses

$$\text{(a)}\ \frac{x^2}{9} + \frac{y^2}{16} = 1 \qquad \text{(b)}\ x^2 + 2y^2 = 4$$

showing the foci of each.

Solution (a). Since y^2 has the larger denominator, the major axis is along the y-axis. Moreover, since $a^2 > b^2$, we must have $a^2 = 16$ and $b^2 = 9$, so

$$a = 4 \quad \text{and} \quad b = 3$$

Drawing a box extending 4 units on each side of the origin along the y-axis and 3 units on each side of the origin along the x-axis as a guide yields the graph in Figure 12.4.17.

The foci lie c units on each side of the center along the major axis, where c is given by (7). From the values of a^2 and b^2 above, we obtain

$$c = \sqrt{a^2 - b^2} = \sqrt{16 - 9} = \sqrt{7} \approx 2.6$$

Thus, the coordinates of the foci are $(0, \sqrt{7})$ and $(0, -\sqrt{7})$, since they lie on the y-axis.

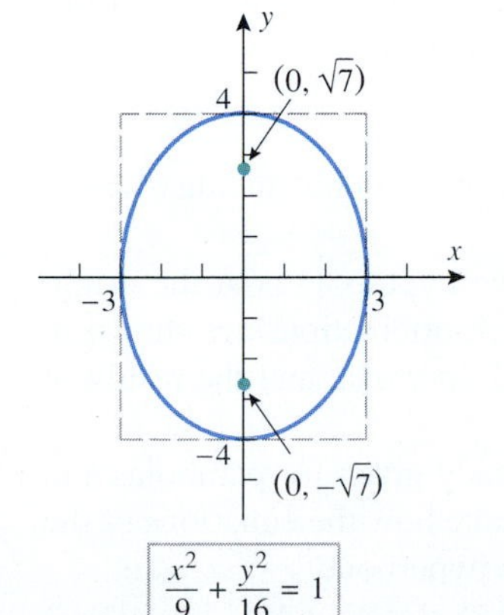

Figure 12.4.17

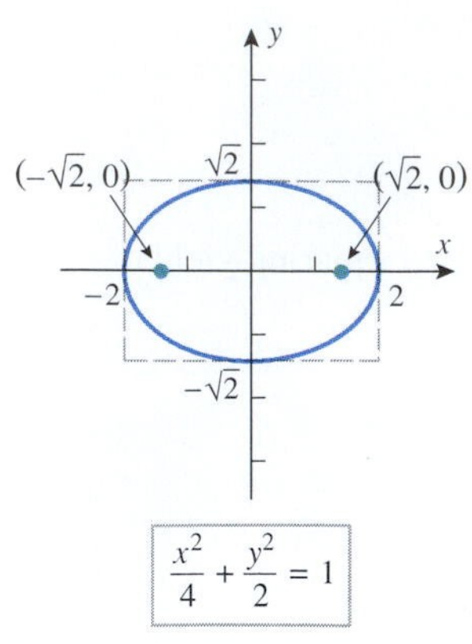

Figure 12.4.18

Solution (b). We first rewrite the equation in the standard form

$$\frac{x^2}{4} + \frac{y^2}{2} = 1$$

Since x^2 has the larger denominator, the major axis lies along the x-axis, and we have $a^2 = 4$ and $b^2 = 2$. Drawing a box extending $a = 2$ on each side of the origin along the x-axis and extending $b = \sqrt{2} \approx 1.4$ units on each side of the origin along the y-axis as a guide yields the graph in Figure 12.4.18.

From (7), we obtain

$$c = \sqrt{a^2 - b^2} = \sqrt{2} \approx 1.4$$

Thus, the coordinates of the foci are $(\sqrt{2}, 0)$ and $(-\sqrt{2}, 0)$, since they lie on the x-axis. ◀

Example 4

Find an equation for the ellipse with foci $(0, \pm 2)$ and major axis with endpoints $(0, \pm 4)$.

Solution. From Figure 12.4.14, the equation has the form

$$\frac{x^2}{b^2} + \frac{y^2}{a^2} = 1$$

and from the given information, $a = 4$ and $c = 2$. It follows from (6) that

$$b^2 = a^2 - c^2 = 16 - 4 = 12$$

so the equation of the ellipse is

$$\frac{x^2}{12} + \frac{y^2}{16} = 1$$ ◀

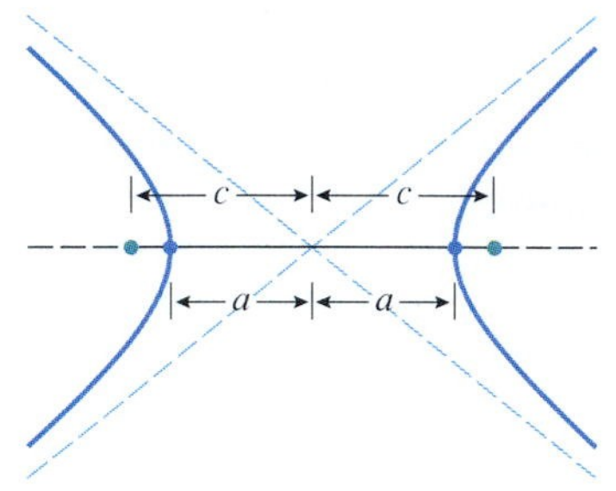

Figure 12.4.19

EQUATIONS OF HYPERBOLAS IN STANDARD POSITION

It is traditional in the study of hyperbolas to denote the distance between the vertices by $2a$, the distance between the foci by $2c$ (Figure 12.4.19), and to define the quantity b as

$$b = \sqrt{c^2 - a^2} \tag{10}$$

This relationship, which can also be expressed as

$$c = \sqrt{a^2 + b^2} \tag{11}$$

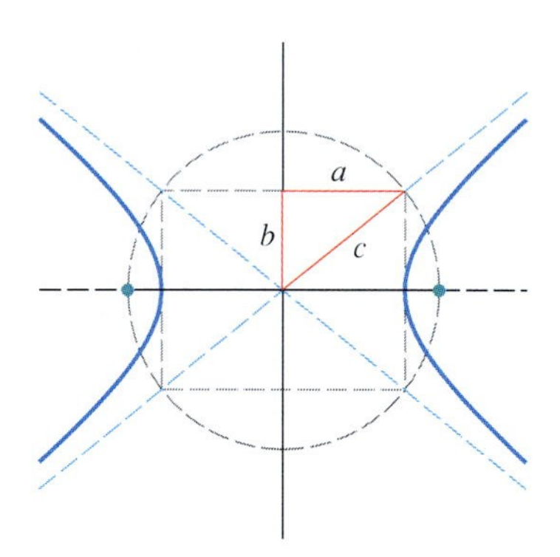

Figure 12.4.20

is pictured geometrically in Figure 12.4.20. As illustrated in that figure, and as we will show later in this section, the asymptotes pass through the corners of a box extending b units on each side of the center along the conjugate axis and a units on each side of the center along the focal axis. The number a is called the ***semifocal axis*** of the hyperbola and the number b the ***semiconjugate axis***. (As with the semimajor and semiminor axes of an ellipse, these are numbers, not geometric axes).

If V is one vertex of a hyperbola, then, as illustrated in Figure 12.4.21, the distance from V to the farther focus minus the distance from V to the closer focus is

$$[(c - a) + 2a] - (c - a) = 2a$$

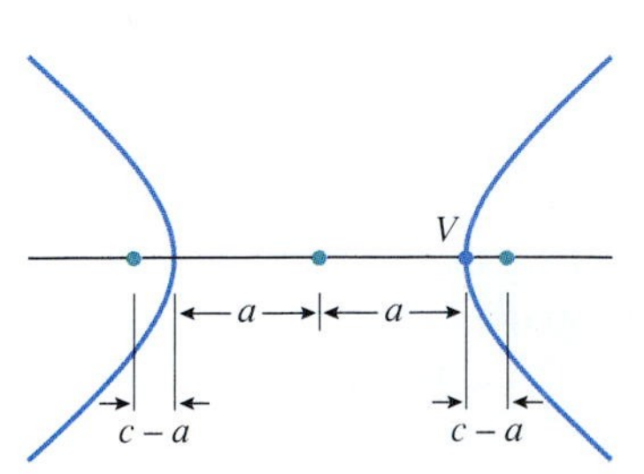

Figure 12.4.21

Thus, for *all* points on a hyperbola, the distance to the farther focus minus the distance to the closer focus is $2a$.

The equation of a hyperbola is simplest if the center of the hyperbola is at the origin and the foci are on the x-axis or y-axis. The two possible such orientations are shown in Figure 12.4.22. These are called the ***standard positions*** of a hyperbola, and the resulting equations are called the ***standard equations*** of a hyperbola.

The derivations of these equations are similar to those already given for parabolas and ellipses, so we will leave them as exercises. However, to illustrate how the equations of the asymptotes are derived, we will derive those equations for the hyperbola

$$\frac{x^2}{a^2} - \frac{y^2}{b^2} = 1$$

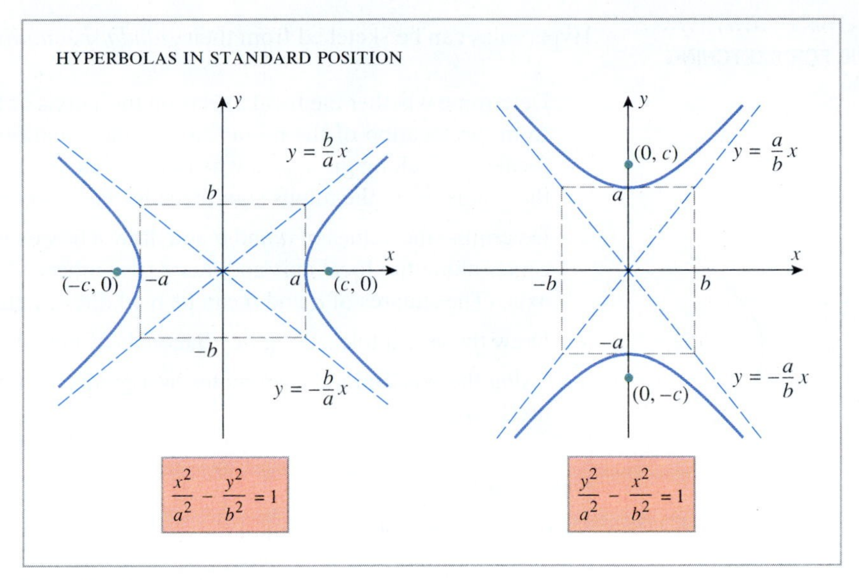

Figure 12.4.22

We can rewrite this equation as

$$y^2 = \frac{b^2}{a^2}(x^2 - a^2)$$

which is equivalent to the pair of equations

$$y = \frac{b}{a}\sqrt{x^2 - a^2} \quad \text{and} \quad y = -\frac{b}{a}\sqrt{x^2 - a^2}$$

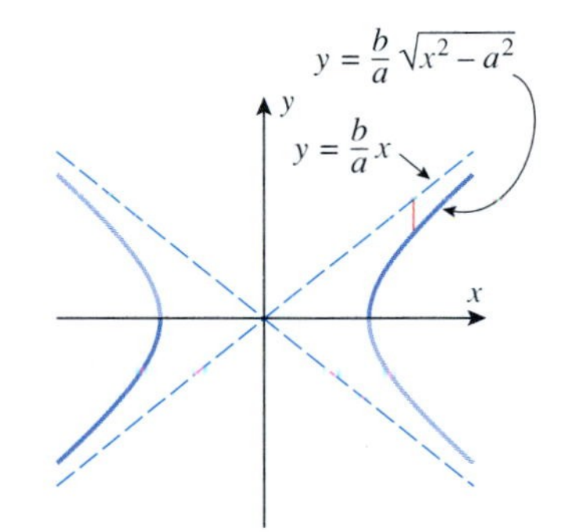

Figure 12.4.23

Thus, in the first quadrant, the vertical distance between the line $y = (b/a)x$ and the hyperbola can be written (Figure 12.4.23) as

$$\frac{b}{a}x - \frac{b}{a}\sqrt{x^2 - a^2}$$

But this distance tends to zero as $x \to +\infty$ since

$$\begin{aligned}
\lim_{x \to +\infty} \left(\frac{b}{a}x - \frac{b}{a}\sqrt{x^2 - a^2}\right) &= \lim_{x \to +\infty} \frac{b}{a}(x - \sqrt{x^2 - a^2}) \\
&= \lim_{x \to +\infty} \frac{b}{a} \frac{(x - \sqrt{x^2 - a^2})(x + \sqrt{x^2 - a^2})}{x + \sqrt{x^2 - a^2}} \\
&= \lim_{x \to +\infty} \frac{ab}{x + \sqrt{x^2 - a^2}} = 0
\end{aligned}$$

The analysis in the remaining quadrants is similar.

A QUICK WAY TO FIND ASYMPTOTES

There is a trick that can be used to avoid memorizing the equations of the asymptotes of a hyperbola. They can be obtained, when needed, by substituting 0 for the 1 on the right side of the hyperbola equation, and then solving for y in terms of x. For example, for the hyperbola

$$\frac{x^2}{a^2} - \frac{y^2}{b^2} = 1$$

we would write

$$\frac{x^2}{a^2} - \frac{y^2}{b^2} = 0 \quad \text{or} \quad y^2 = \frac{b^2}{a^2}x^2 \quad \text{or} \quad y = \pm\frac{b}{a}x$$

which are the equations for the asymptotes.

A TECHNIQUE FOR SKETCHING HYPERBOLAS

Hyperbolas can be sketched from their *standard equations* using four basic steps:

- Determine whether the focal axis is on the x-axis or the y-axis. This can be ascertained from the location of the minus sign in the equation. Referring to Figure 12.4.22, the focal axis is along the x-axis when the minus sign precedes the y^2-term, and it is along the y-axis when the minus sign precedes the x^2-term.
- Determine the values of a and b and draw a box extending a units on either side of the center along the focal axis and b units on either side of the center along the conjugate axis. (The squares of a and b can be read directly from the equation.)
- Draw the asymptotes along the diagonals of the box.
- Using the box and the asymptotes as a guide, sketch the graph of the hyperbola (Figure 12.4.24).

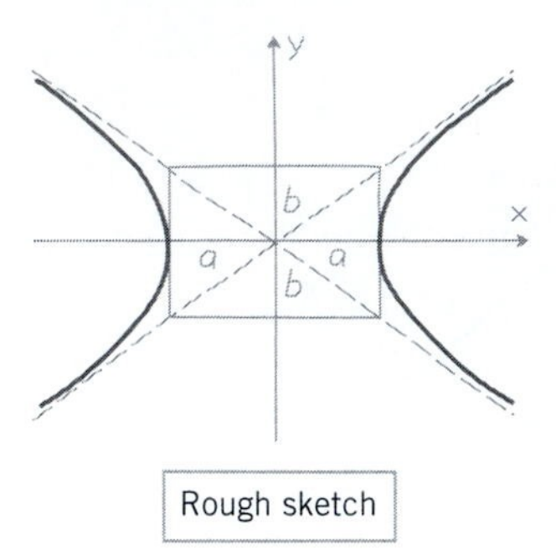

Figure 12.4.24

Example 5

Sketch the graphs of the hyperbolas

$$\text{(a)}\ \frac{x^2}{4} - \frac{y^2}{9} = 1 \qquad \text{(b)}\ y^2 - x^2 = 1$$

showing their vertices, foci, and asymptotes.

Solution (a). The minus sign precedes the y^2-term, so the focal axis is along the x-axis. From the denominators in the equation we obtain

$$a^2 = 4 \quad \text{and} \quad b^2 = 9$$

Since a and b are positive, we must have $a = 2$ and $b = 3$. Recalling that the vertices lie a units on each side of the center on the focal axis, it follows that their coordinates in this case are $(2, 0)$ and $(-2, 0)$. Drawing a box extending $a = 2$ units along the x-axis on each side of the origin and $b = 3$ units on each side of the origin along the y-axis, then drawing the asymptotes along the diagonals of the box as a guide, yields the graph in Figure 12.4.25.

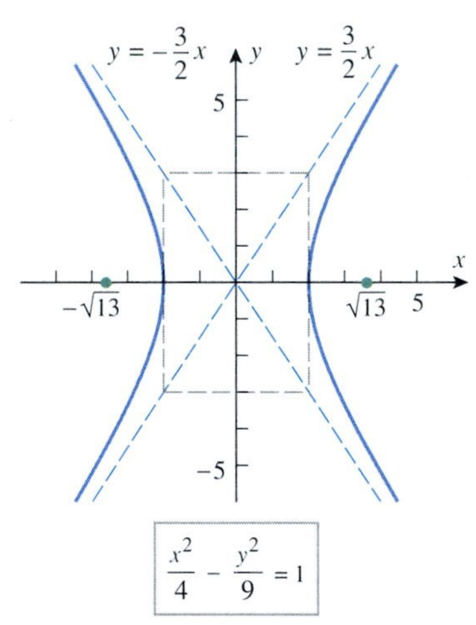

Figure 12.4.25

To obtain equations for the asymptotes, we substitute 0 for 1 in the given equation; this yields

$$\frac{x^2}{4} - \frac{y^2}{9} = 0 \quad \text{or} \quad y = \pm\frac{3}{2}x$$

The foci lie c units on each side of the center along the focal axis, where c is given by (11). From the values of a^2 and b^2 above we obtain

$$c = \sqrt{a^2 + b^2} = \sqrt{4 + 9} = \sqrt{13} \approx 3.6$$

Since the foci lie on the x-axis in this case, their coordinates are $(\sqrt{13}, 0)$ and $(-\sqrt{13}, 0)$.

Solution (b). The minus sign precedes the x^2-term, so the focal axis is along the y-axis. From the denominators in the equation we obtain $a^2 = 1$ and $b^2 = 1$, from which it follows that

$$a = 1 \quad \text{and} \quad b = 1$$

Thus, the vertices are at $(0, -1)$ and $(0, 1)$. Drawing a box extending $a = 1$ unit on either side of the origin along the y-axis and $b = 1$ unit on either side of the origin along the x-axis, then drawing the asymptotes, yields the graph in Figure 12.4.26. Since the box is actually a square, the asymptotes are perpendicular and have equations $y = \pm x$. This can also be seen by substituting 0 for 1 in the given equation, which yields $y^2 - x^2 = 0$ or $y = \pm x$. Also,

$$c = \sqrt{a^2 + b^2} = \sqrt{1 + 1} = \sqrt{2}$$

so the foci, which lie on the y-axis, are $(0, -\sqrt{2})$ and $(0, \sqrt{2})$. ◀

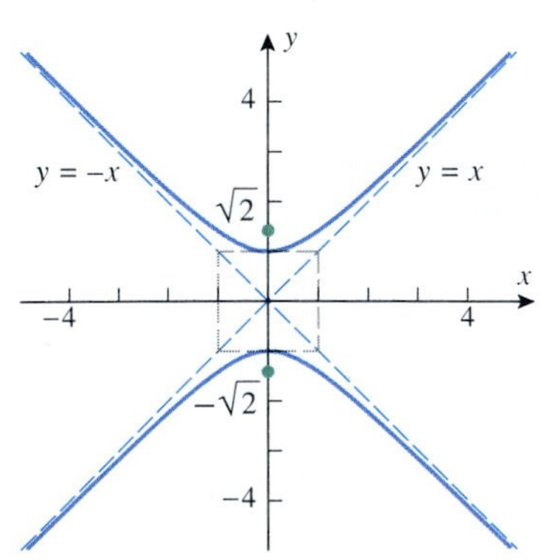

Figure 12.4.26

REMARK. A hyperbola in which $a = b$, as in part (b) of this example, is called an ***equilateral hyperbola***. Such hyperbolas always have perpendicular asymptotes.

Example 6

Find the equation of the hyperbola with vertices $(0, \pm 8)$ and asymptotes $y = \pm\frac{4}{3}x$.

Solution. Since the vertices are on the y-axis, the equation of the hyperbolas has the form $(y^2/a^2) - (x^2/b^2) = 1$ and the asymptotes are

$$y = \pm\frac{a}{b}x$$

From the location of the vertices we have $a = 8$, so the given equations of the asymptotes yield

$$y = \pm\frac{a}{b}x = \pm\frac{8}{b}x = \pm\frac{4}{3}x$$

from which it follows that $b = 6$. Thus, the hyperbola has the equation

$$\frac{y^2}{64} - \frac{x^2}{36} = 1$$ ◀

TRANSLATED CONICS

Equations of conics that are translated from their standard positions can be obtained by replacing x by $x - h$ and y by $y - k$ in their standard equations. For a parabola, this translates the vertex from the origin to the point (h, k); and for ellipses and hyperbolas, this translates the center from the origin to the point (h, k).

Parabolas with vertex (h, k) and axis parallel to x-axis

$$(y - k)^2 = 4p(x - h) \quad \text{[Opens right]} \tag{12}$$

$$(y - k)^2 = -4p(x - h) \quad \text{[Opens left]} \tag{13}$$

Parabolas with vertex (h, k) and axis parallel to y-axis

$$(x - h)^2 = 4p(y - k) \quad \text{[Opens up]} \tag{14}$$

$$(x - h)^2 = -4p(y - k) \quad \text{[Opens down]} \tag{15}$$

Ellipse with center (h, k) and major axis parallel to x-axis

$$\frac{(x - h)^2}{a^2} + \frac{(y - k)^2}{b^2} = 1 \quad [b \le a] \tag{16}$$

Ellipse with center (h, k) and major axis parallel to y-axis

$$\frac{(x - h)^2}{b^2} + \frac{(y - k)^2}{a^2} = 1 \quad [b \le a] \tag{17}$$

Hyperbola with center (h, k) and focal axis parallel to x-axis

$$\frac{(x - h)^2}{a^2} - \frac{(y - k)^2}{b^2} = 1 \tag{18}$$

Hyperbola with center (h, k) and focal axis parallel to y-axis

$$\frac{(y - k)^2}{a^2} - \frac{(x - h)^2}{b^2} = 1 \tag{19}$$

Example 7

Find an equation for the parabola that has its vertex at $(1, 2)$ and its focus at $(4, 2)$.

Solution. Since the focus and vertex are on a horizontal line, and since the focus is to the right of the vertex, the parabola opens to the right and its equation has the form

$$(y - k)^2 = 4p(x - h)$$

Since the vertex and focus are 3 units apart, we have $p = 3$, and since the vertex is at $(h, k) = (1, 2)$, we obtain

$$(y - 2)^2 = 12(x - 1)$$ ◀

Sometimes the equations of translated conics occur in expanded form, in which case we are faced with the problem of identifying the graph of a ***quadratic equation in x and y***:

$$Ax^2 + Cy^2 + Dx + Ey + F = 0 \tag{20}$$

The basic procedure for determining the nature of such a graph is to complete the squares of the quadratic terms and then try to match up the resulting equation with one for the forms of a translated conic.

Example 8

Describe the graph of the equation

$$y^2 - 8x - 6y - 23 = 0$$

Solution. The equation involves quadratic terms in y but none in x, so we first take all of the y-terms to one side:

$$y^2 - 6y = 8x + 23$$

Next, we complete the square on the y-terms by adding 9 to both sides:

$$(y - 3)^2 = 8x + 32$$

Finally, we factor out the coefficient of the x-term to obtain

$$(y - 3)^2 = 8(x + 4)$$

This equation is of form (12) with $h = -4$, $k = 3$, and $p = 2$, so the graph is a parabola with vertex $(-4, 3)$ opening to the right. Since $p = 2$, the focus is 2 units to the right of the vertex, which places it at the point $(-2, 3)$; and the directrix is 2 units to the left of the vertex, which means that its equation is $x = -6$. The parabola is shown in Figure 12.4.27. ◀

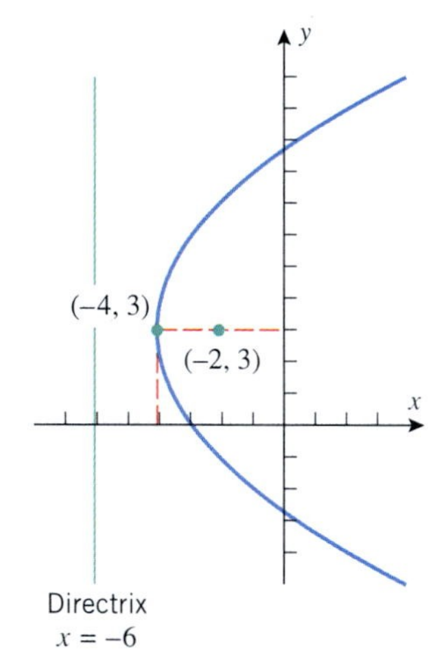

Figure 12.4.27

Example 9

Describe the graph of the equation

$$16x^2 + 9y^2 - 64x - 54y + 1 = 0$$

Solution. This equation involves quadratic terms in both x and y, so we will group the x-terms and the y-terms on one side and put the constant on the other:

$$(16x^2 - 64x) + (9y^2 - 54y) = -1$$

Next, factor out the coefficients of x^2 and y^2 and complete the squares:

$$16(x^2 - 4x + 4) + 9(y^2 - 6y + 9) = -1 + 64 + 81$$

or

$$16(x - 2)^2 + 9(y - 3)^2 = 144$$

Finally, divide through by 144 to introduce a 1 on the right side:

$$\frac{(x - 2)^2}{9} + \frac{(y - 3)^2}{16} = 1$$

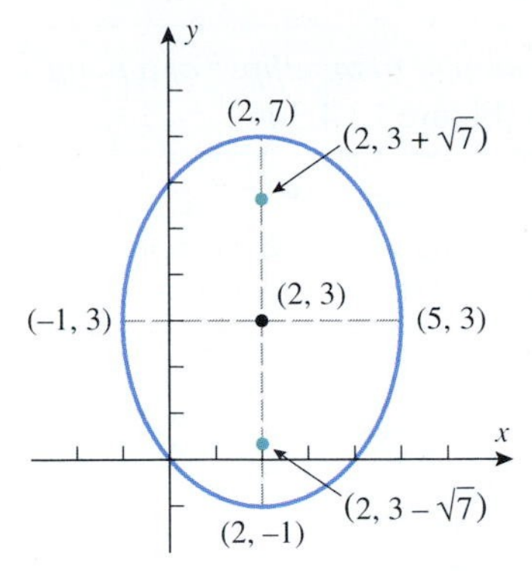

Figure 12.4.28

This is an equation of form (17), with $h = 2$, $k = 3$, $a^2 = 16$, and $b^2 = 9$. Thus, the equation is an ellipse with center $(2, 3)$ and major axis parallel to the y-axis. Since $a = 4$, the major axis extends 4 units above and 4 units below the center, so its endpoints are $(2, 7)$ and $(2, -1)$ (Figure 12.4.28). Since $b = 3$, the minor axis extends 3 units to the left and 3 units to the right of the center, so its endpoints are $(-1, 3)$ and $(5, 3)$. Since

$$c = \sqrt{a^2 - b^2} = \sqrt{16 - 9} = \sqrt{7}$$

the foci lie $\sqrt{7}$ units above and below the center, placing them at the points $(2, 3 + \sqrt{7})$ and $(2, 3 - \sqrt{7})$. ◀

Example 10

Describe the graph of the equation

$$x^2 - y^2 - 4x + 8y - 21 = 0$$

Solution. This equation involves quadratic terms in both x and y, so we will group the x-terms and the y-terms on one side and put the constant on the other:

$$(x^2 - 4x) - (y^2 - 8y) = 21$$

We leave it for you to verify by completing the squares that this equation can be written as

$$\frac{(x-2)^2}{9} - \frac{(y-4)^2}{9} = 1 \tag{21}$$

This is an equation of form (18) with $h = 2$, $k = 4$, $a^2 = 9$, and $b^2 = 9$. Thus, the equation represents a hyperbola with center $(2, 4)$ and focal axis parallel to the x-axis. Since $a = 3$, the vertices are located 3 units to the left and 3 units to the right of the center, or at the points $(-1, 4)$ and $(5, 4)$. From (11), $c = \sqrt{a^2 + b^2} = \sqrt{9 + 9} = 3\sqrt{2}$, so the foci are located $3\sqrt{2}$ units to the left and right of the center, or at the points $(2 - 3\sqrt{2}, 4)$ and $(2 + 3\sqrt{2}, 4)$.

The equations of the asymptotes may be found using the trick of substituting 0 for 1 in (21) to obtain

$$\frac{(x-2)^2}{9} - \frac{(y-4)^2}{9} = 0$$

This can be written as $y - 4 = \pm(x - 2)$, which yields the asymptotes

$$y = x + 2 \quad \text{and} \quad y = -x + 6$$

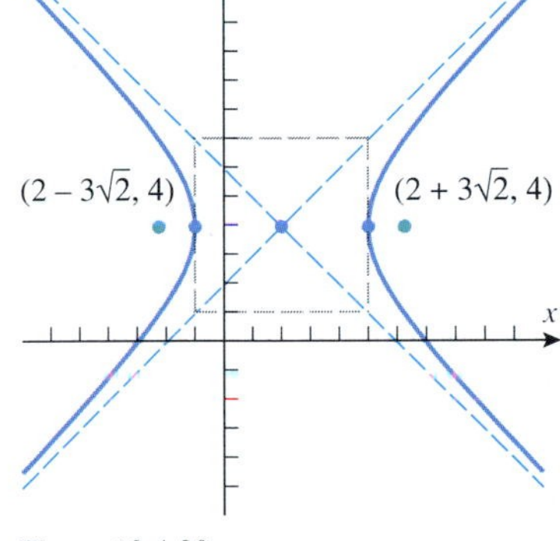

Figure 12.4.29

With the aid of a box extending $a = 3$ units left and right of the center and $b = 3$ units above and below the center, we obtain the sketch in Figure 12.4.29. ◀

ROTATED CONICS

An equation of the form

$$Ax^2 + Bxy + Cy^2 + Dx + Ey + F = 0 \tag{22}$$

is called a ***second-degree equation in x and y.*** The term Bxy in this equation is called the ***cross-product term***. If the cross-product term is absent from the equation ($B = 0$), then the equation reduces to (20), in which case the graph is a conic section (possibly degenerate) that is either in standard position or translated from its standard position. It can be proved that if the cross-product term is present ($B \neq 0$), then the graph is a conic (possibly degenerate) that is *rotated* from its standard orientation. A discussion of rotated conics can be found in the *Student Resources*.

REFLECTION PROPERTIES OF THE CONIC SECTIONS

Parabolas, ellipses, and hyperbolas have certain reflection properties that make them extremely valuable in various applications. In the exercises we will ask you to prove the following results.

> **12.4.4 THEOREM (*Reflection Property of Parabolas*).** *The tangent line at a point P on a parabola makes equal angles with the line through P parallel to the axis of symmetry and the line through P and the focus* (Figure 12.4.30*a*).

12.4.5 THEOREM (*Reflection Property of Ellipses*). *A line tangent to an ellipse at a point P makes equal angles with the lines joining P to the foci* (Figure 12.4.30b).

12.4.6 THEOREM (*Reflection Property of Hyperbolas*). *A line tangent to a hyperbola at a point P makes equal angles with the lines joining P to the foci* (Figure 12.4.30c).

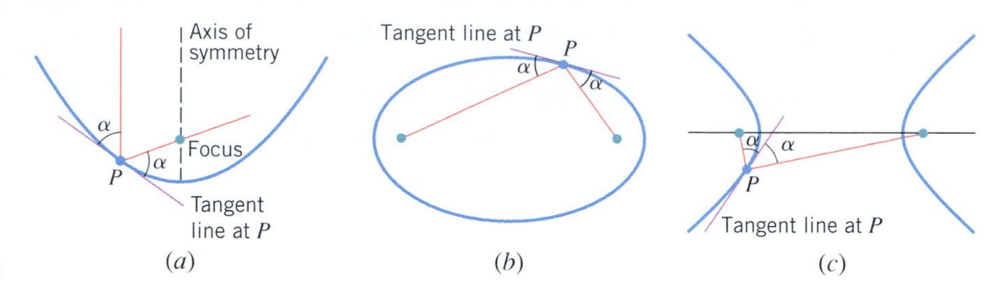

Figure 12.4.30

APPLICATIONS OF THE CONIC SECTIONS

Incoming signals are reflected by the parabolic antenna to the receiver at the focus.

It is a principle of physics that when light is reflected from a point P on a surface the angle between the incoming ray and the tangent line at P is equal to the angle between the outgoing ray and the tangent line at P. Therefore, if a reflecting surface has parabolic cross sections with a common focus and axis, then it follows from Theorem 12.4.4 that all light rays entering parallel to the axis will be reflected to the focus (Figure 12.4.31a); conversely, if a light source is located at the focus, then the reflected rays will all be parallel to the axis (Figure 12.4.31b). This principle is used in certain telescopes to reflect the approximately parallel rays of light from the stars and planets off of a parabolic mirror to an eyepiece at the focus; and the parabolic reflectors in flashlights and automobile headlights utilize this principle to form a parallel beam of light rays from a bulb placed at the focus. The same optical principles apply to radar signals and sound waves, which explains the parabolic shape of many antennas.

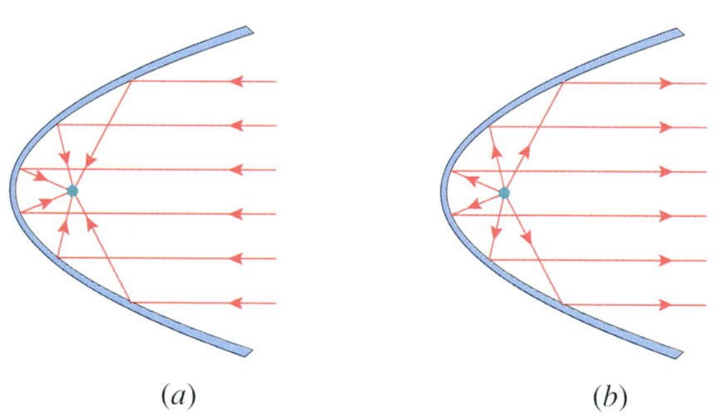

Figure 12.4.31

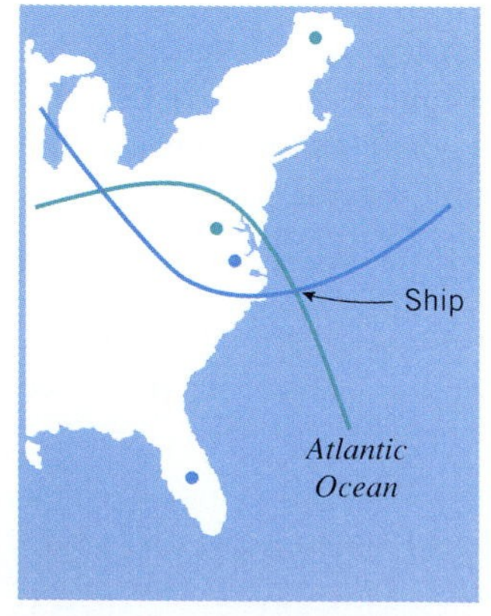

Figure 12.4.32

Visitors to various rooms in the United States Capitol Building and in St. Paul's Cathedral in Rome are often astonished by the "whispering gallery" effect in which two people at opposite ends of the room can hear one another's whispers very clearly. Such rooms have ceilings with elliptical cross sections and common foci. Thus, when the two people stand at the foci, their whispers are reflected directly to one another off of the elliptical ceiling.

Hyperbolic navigation systems, which were developed in World War II as navigational aids to ships, are based on the definition of a hyperbola. With these systems the ship receives synchronized radio signals from two widely spaced transmitters with known positions. The ship's electronic receiver measures the difference in reception times between the signals and then uses that difference to compute the difference $2a$ in its distance between the two transmitters. This information places the ship somewhere on the hyperbola whose foci are at the transmitters and whose points have $2a$ as the difference in their distances from the foci. By repeating the process with a second set of transmitters, the position of the ship can be determined as the intersection of two hyperbolas (Figure 12.4.32).

EXERCISE SET 12.4 Graphing Calculator C CAS

1. In each part, find the equation of the conic.

(a)

(b)

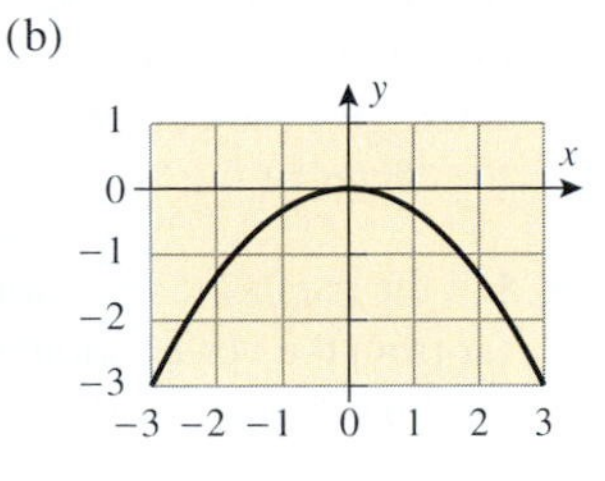

(c)

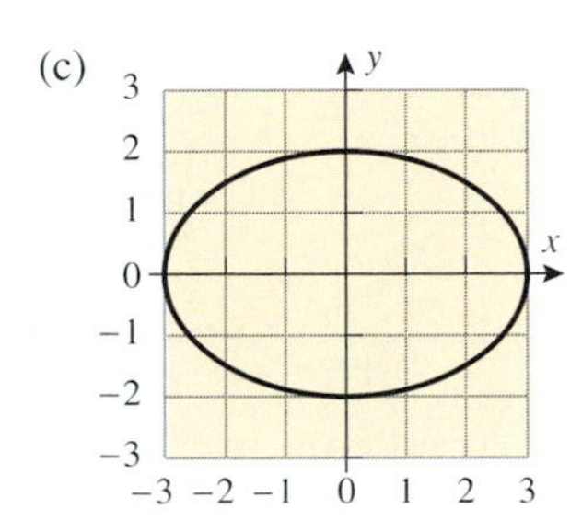

(d)

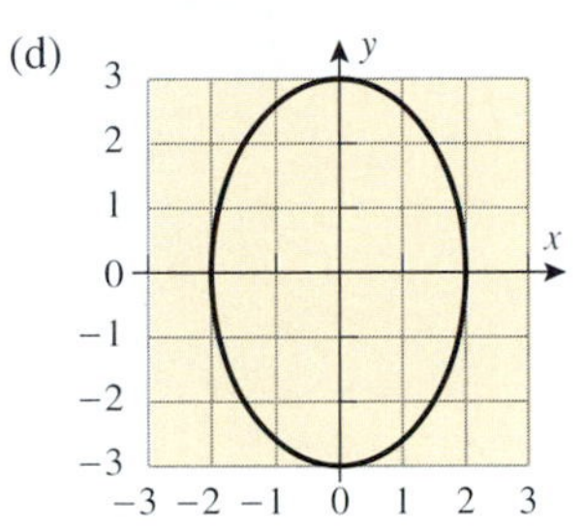

(e)

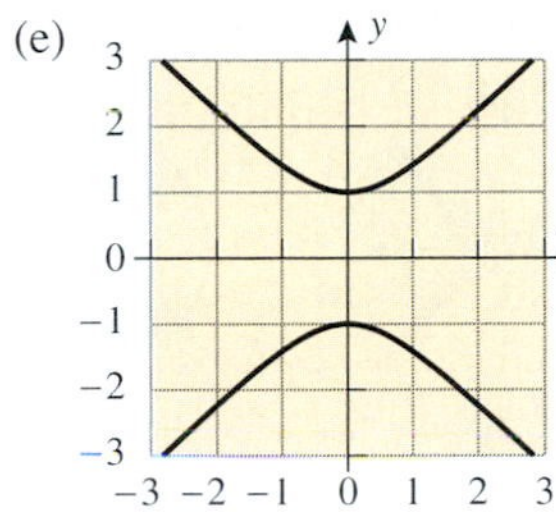

(f)

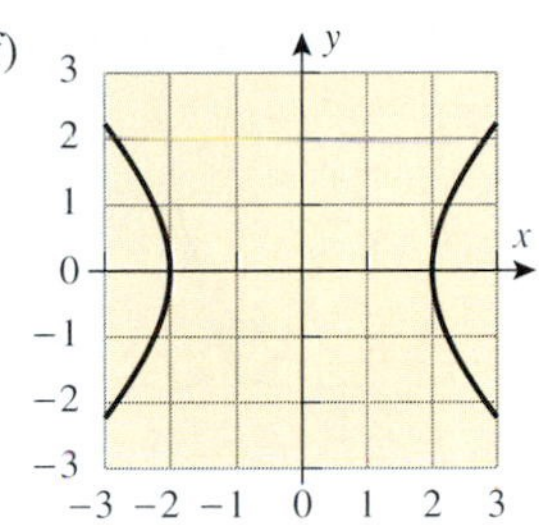

2. (a) Find the focus and directrix of the parabolas that are given in Exercise 1.
   (b) Find the foci of the ellipses in Exercise 1.
   (c) Find the foci and the equations of the asymptotes of the hyperbolas in Exercise 1.

In Exercises 3–8, sketch the parabola, and label the focus, vertex, and directrix.

3. (a) $y^2 = 6x$ (b) $x^2 = -9y$
4. (a) $y^2 = -10x$ (b) $x^2 = 4y$
5. (a) $(y-3)^2 = 6(x-2)$ (b) $(x+2)^2 = -(y+2)$
6. (a) $(y+1)^2 = -7(x-4)$ (b) $\left(x-\frac{1}{2}\right)^2 = 2(y-1)$
7. (a) $x^2 - 4x + 2y = 1$ (b) $x = y^2 - 4y + 2$
8. (a) $y^2 - 6y - 2x + 1 = 0$ (b) $y = 4x^2 + 8x + 5$

In Exercises 9–14, sketch the ellipse, and label the foci, the vertices, and the ends of the minor axis.

9. (a) $\dfrac{x^2}{16} + \dfrac{y^2}{9} = 1$ (b) $9x^2 + y^2 = 9$
10. (a) $\dfrac{x^2}{4} + \dfrac{y^2}{25} = 1$ (b) $4x^2 + 9y^2 = 36$
11. (a) $9(x-1)^2 + 16(y-3)^2 = 144$
    (b) $3(x+2)^2 + 4(y+1)^2 = 12$
12. (a) $(x+3)^2 + 4(y-5)^2 = 16$
    (b) $\frac{1}{4}x^2 + \frac{1}{9}(y+2)^2 - 1 = 0$
13. (a) $x^2 + 9y^2 + 2x - 18y + 1 = 0$
    (b) $4x^2 + y^2 + 8x - 10y = -13$
14. (a) $9x^2 + 4y^2 + 18x - 24y + 9 = 0$
    (b) $5x^2 + 9y^2 - 20x + 54y = -56$

In Exercises 15–20, sketch the hyperbola, and label the vertices, foci, and asymptotes.

15. (a) $\dfrac{x^2}{16} - \dfrac{y^2}{4} = 1$ (b) $9y^2 - 4x^2 = 36$
16. (a) $\dfrac{y^2}{9} - \dfrac{x^2}{25} = 1$ (b) $16x^2 - 25y^2 = 400$
17. (a) $\dfrac{(x-2)^2}{9} - \dfrac{(y-4)^2}{4} = 1$
    (b) $(y+3)^2 - 9(x+2)^2 = 36$
18. (a) $\dfrac{(y+4)^2}{3} - \dfrac{(x-2)^2}{5} = 1$
    (b) $16(x+1)^2 - 8(y-3)^2 = 16$
19. (a) $x^2 - 4y^2 + 2x + 8y - 7 = 0$
    (b) $16x^2 - y^2 - 32x - 6y = 57$
20. (a) $4x^2 - 9y^2 + 16x + 54y - 29 = 0$
    (b) $4y^2 - x^2 + 40y - 4x = -60$

In Exercises 21–26, find an equation for the parabola that satisfies the given conditions.

21. (a) Vertex $(0, 0)$; focus $(3, 0)$.
    (b) Vertex $(0, 0)$; directrix $x = 7$.
22. (a) Vertex $(0, 0)$; focus $(0, -4)$.
    (b) Vertex $(0, 0)$; directrix $y = \frac{1}{2}$.
23. (a) Focus $(0, -3)$; directrix $y = 3$.
    (b) Vertex $(1, 1)$; directrix $y = -2$.
24. (a) Focus $(6, 0)$; directrix $x = -6$.
    (b) Focus $(-1, 4)$; directrix $x = 5$.
25. Axis $y = 0$; passes through $(3, 2)$ and $(2, -3)$.
26. Vertex $(5, -3)$; axis parallel to the y-axis; passes through $(9, 5)$.

In Exercises 27–32, find an equation for the ellipse that satisfies the given conditions.

27. (a) Ends of major axis $(\pm 3, 0)$; ends of minor axis $(0, \pm 2)$.
    (b) Length of major axis 26; foci $(\pm 5, 0)$.

28. (a) Ends of major axis $(0, \pm\sqrt{5})$; ends of minor axis $(\pm 1, 0)$.
(b) Length of minor axis 16; foci $(0, \pm 6)$.

29. (a) Foci $(\pm 1, 0)$; $b = \sqrt{2}$.
(b) $c = 2\sqrt{3}$; $a = 4$; center at the origin; foci on a coordinate axis (two answers).

30. (a) Foci $(\pm 3, 0)$; $a = 4$.
(b) $b = 3$; $c = 4$; center at the origin; foci on a coordinate axis (two answers).

31. (a) Ends of major axis $(\pm 6, 0)$; passes through $(2, 3)$.
(b) Foci $(1, 2)$ and $(1, 4)$; minor axis of length 2.

32. (a) Center at $(0, 0)$; major and minor axes along the coordinate axes; passes through $(3, 2)$ and $(1, 6)$.
(b) Foci $(2, 1)$ and $(2, -3)$; major axis of length 6.

In Exercises 33–38, find an equation for a hyperbola that satisfies the given conditions. (In some cases there may be more than one hyperbola.)

33. (a) Vertices $(\pm 2, 0)$; foci $(\pm 3, 0)$.
(b) Vertices $(\pm 1, 0)$; asymptotes $y = \pm 2x$.

34. (a) Vertices $(0, \pm 3)$; foci $(0, \pm 5)$.
(b) Vertices $(0, \pm 3)$; asymptotes $y = \pm x$.

35. (a) Asymptotes $y = \pm\frac{3}{2}x$; $b = 4$.
(b) Foci $(0, \pm 5)$; asymptotes $y = \pm 2x$.

36. (a) Asymptotes $y = \pm\frac{3}{4}x$; $c = 5$.
(b) Foci $(\pm 3, 0)$; asymptotes $y = \pm 2x$.

37. (a) Vertices $(2, 4)$ and $(10, 4)$; foci 10 units apart.
(b) Asymptotes $y = 2x + 1$ and $y = -2x + 3$; passes through the origin.

38. (a) Foci $(1, 8)$ and $(1, -12)$; vertices 4 units apart.
(b) Vertices $(-3, -1)$ and $(5, -1)$; $b = 4$.

39. (a) As illustrated in the accompanying figure, a parabolic arch spans a road 40 feet wide. How high is the arch if a center section of the road 20 feet wide has a minimum clearance of 12 feet?
(b) How high would the center be if the arch were the upper half of an ellipse?

40. (a) Find an equation for the parabolic arch with base b and height h, shown in the accompanying figure.
(b) Find the area under the arch.

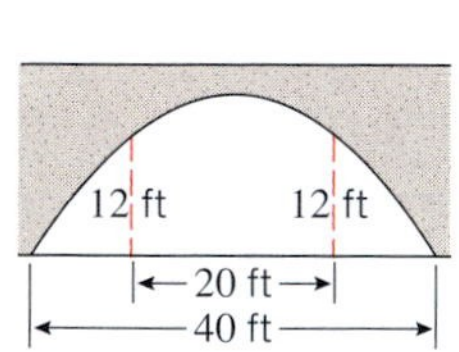

Figure Ex-39

Figure Ex-40

41. Show that the vertex is the closest point on a parabola to the focus. [*Suggestion:* Introduce a convenient coordinate system and use Definition 12.4.1.]

42. As illustrated in the accompanying figure, suppose that a comet moves in a parabolic orbit with the Sun at its focus and that the line from the Sun to the comet makes an angle of 60° with the axis of the parabola when the comet is 40 million miles from the center of the Sun. Use the result in Exercise 41 to determine how close the comet will come to the center of the Sun.

43. For the parabolic reflector in the accompanying figure, how far from the vertex should the light source be placed to produce a beam of parallel rays?

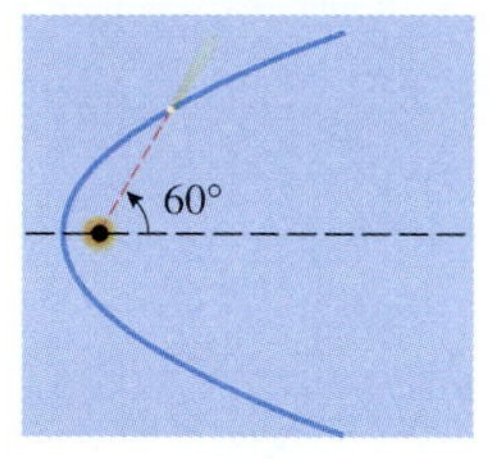

Figure Ex-42

Figure Ex-43

44. In each part, find the shaded area in the figure.

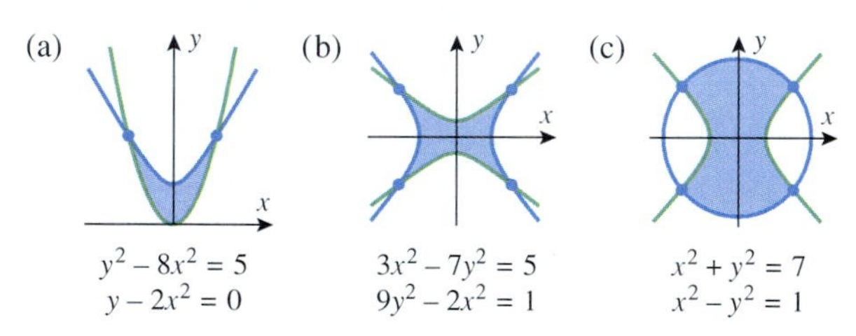

45. (a) The accompanying figure shows an ellipse with semimajor axis a and semiminor axis b. Express the coordinates of the points P, Q, and R in terms of t.
(b) How does the geometric interpretation of the parameter t differ between a circle

$$x = a\cos t, \quad y = a\sin t$$

and an ellipse

$$x = a\cos t, \quad y = b\sin t?$$

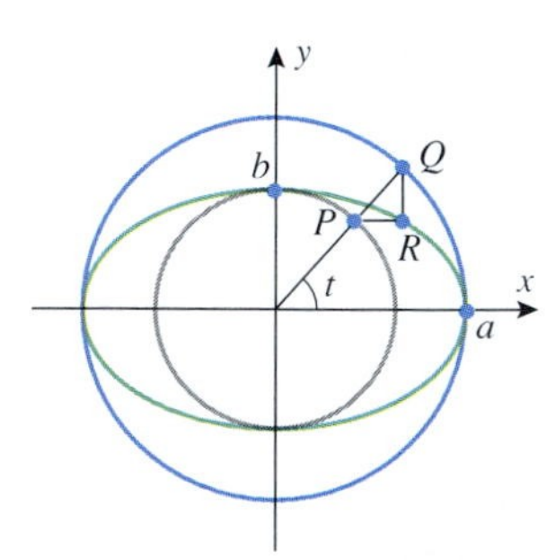

Figure Ex-45

46. (a) Show that the right and left branches of the hyperbola

$$\frac{x^2}{a^2} - \frac{y^2}{b^2} = 1$$

can be represented parametrically as

$$x = a\cosh t,\quad y = b\sinh t \qquad (-\infty < t < +\infty)$$
$$x = -a\cosh t,\quad y = b\sinh t \qquad (-\infty < t < +\infty)$$

(b) Use a graphing utility to generate both branches of the hyperbola $x^2 - y^2 = 1$ on the same screen.

47. (a) Show that the right and left branches of the hyperbola

$$\frac{x^2}{a^2} - \frac{y^2}{b^2} = 1$$

can be represented parametrically as

$$x = a\sec t,\quad y = b\tan t \qquad (-\pi/2 < t < \pi/2)$$
$$x = -a\sec t,\quad y = b\tan t \qquad (-\pi/2 < t < \pi/2)$$

(b) Use a graphing utility to generate both branches of the hyperbola $x^2 - y^2 = 1$ on the same screen.

48. Find an equation of the parabola traced by a point that moves so that its distance from $(-1, 4)$ is the same as its distance to $y = 1$.

49. Find an equation of the ellipse traced by a point that moves so that the sum of its distances to $(4, 1)$ and $(4, 5)$ is 12.

50. Find the equation of the hyperbola traced by a point that moves so that the difference between its distances to $(0, 0)$ and $(1, 1)$ is 1.

51. Suppose that the base of a solid is elliptical with a major axis of length 9 and a minor axis of length 4. Find the volume of the solid if the cross sections perpendicular to the major axis are squares (see the accompanying figure).

52. Suppose that the base of a solid is elliptical with a major axis of length 9 and a minor axis of length 4. Find the volume of the solid if the cross sections perpendicular to the minor axis are equilateral triangles (see the accompanying figure).

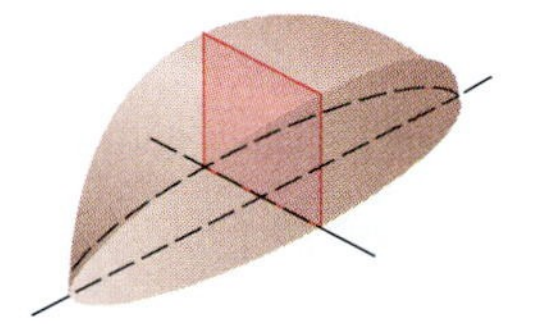

Figure Ex-51

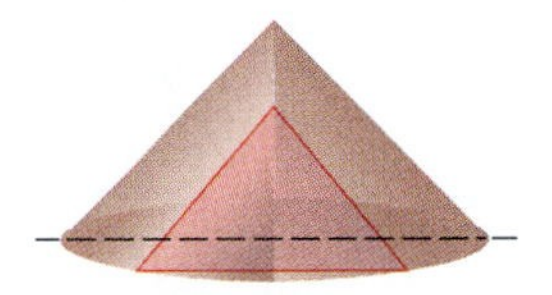

Figure Ex-52

53. Show that an ellipse with semimajor axis a and semiminor axis b has area $A = \pi ab$.

54. (a) Show that the ellipsoid that results when an ellipse with semimajor axis a and semiminor axis b is revolved about the major axis has volume $V = \frac{4}{3}\pi ab^2$.

(b) Show that the ellipsoid that results when an ellipse with semimajor axis a and semiminor axis b is revolved about the minor axis has volume $V = \frac{4}{3}\pi a^2 b$.

55. Show that the ellipsoid that results when an ellipse with semimajor axis a and semiminor axis b is revolved about the major axis has surface area

$$S = 2\pi ab\left(\frac{b}{a} + \frac{a}{c}\sin^{-1}\frac{c}{a}\right)$$

where $c = \sqrt{a^2 - b^2}$.

56. Show that the ellipsoid that results when an ellipse with semimajor axis a and semiminor axis b is revolved about the minor axis has surface area

$$S = 2\pi ab\left(\frac{a}{b} + \frac{b}{c}\ln\frac{a+c}{b}\right)$$

where $c = \sqrt{a^2 - b^2}$.

57. Suppose that you want to draw an ellipse that has given values for the lengths of the major and minor axes by using the method shown in Figure 12.4.3*b*. Assuming that the axes are drawn, explain how a compass can be used to locate the positions for the tacks.

58. The accompanying figure shows Kepler's method for constructing a parabola: a piece of string the length of the left edge of the drafting triangle is tacked to the vertex Q of the triangle and the other end to a fixed point F. A pencil holds the string taut against the base of the triangle as the edge opposite Q slides along a horizontal line L below F. Show that the pencil traces an arc of a parabola with focus F and directrix L.

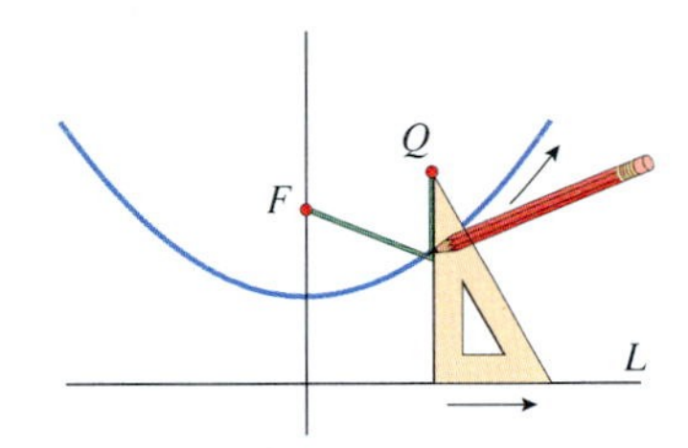

Figure Ex-58

59. The accompanying figure shows a method for constructing a hyperbola: a corner of a ruler is pinned to a fixed point F_1 and the ruler is free to rotate about that point. A piece of string whose length is less than that of the ruler is tacked to a point F_2 and to the free corner Q of the ruler on the same edge as F_1. A pencil holds the string taut against the top edge of the ruler as the ruler rotates about the point F_1. Show that the pencil traces an arc of a hyperbola with foci F_1 and F_2.

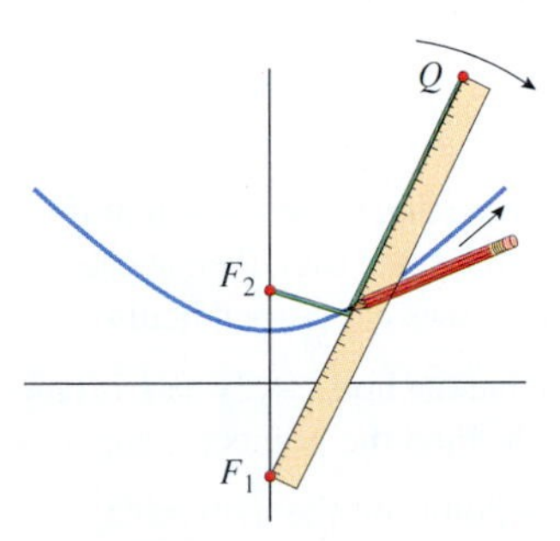

Figure Ex-59

60. Show that if a plane is not parallel to the axis of a right circular cylinder, then the intersection of the plane and cylinder is an ellipse (possibly a circle). [*Hint:* Let θ be the angle shown in Figure Ex-60 (next page), introduce coordinate axes as shown, and express x' and y' in terms of x and y.]

61. As illustrated in the accompanying figure, a carpenter needs to cut an elliptical hole in a sloped roof through which a circular vent pipe of diameter D is to be inserted vertically. The carpenter wants to draw the outline of the hole on the roof using a pencil, two tacks, and a piece of string (as in Figure 12.4.3*b*). The center point of the ellipse is known, and common sense suggests that its major axis must be perpendicular to the drip line of the roof. The carpenter needs to determine the length L of the string and the distance T between a tack and the center point. The architect's plans show that the pitch of the roof is p (pitch = rise over run; see the accompanying figure). Find T and L in terms of D and p. [*Note:* This exercise is based on an article by William H. Enos, which appeared in the *Mathematics Teacher*, Feb. 1991, p. 148.]

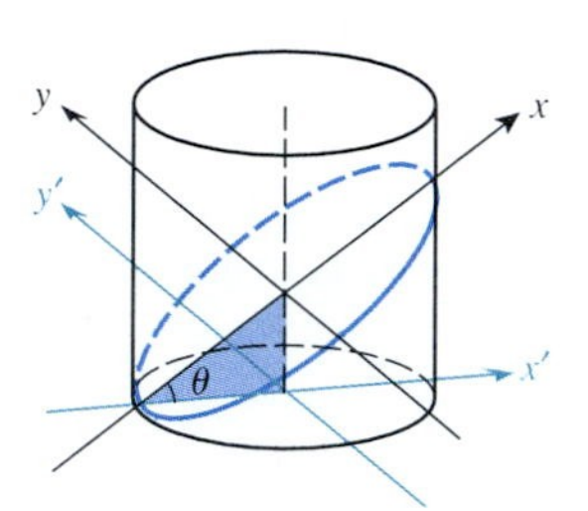

Figure Ex-60

Figure Ex-61

62. Prove: The line tangent to the parabola $x^2 = 4py$ at the point (x_0, y_0) is $x_0x = 2p(y + y_0)$.

63. Prove: The line tangent to the ellipse

$$\frac{x^2}{a^2} + \frac{y^2}{b^2} = 1$$

at the point (x_0, y_0) has the equation

$$\frac{xx_0}{a^2} + \frac{yy_0}{b^2} = 1$$

64. Prove: The line tangent to the hyperbola

$$\frac{x^2}{a^2} - \frac{y^2}{b^2} = 1$$

at the point (x_0, y_0) has the equation

$$\frac{xx_0}{a^2} - \frac{yy_0}{b^2} = 1$$

65. Use the results in Exercises 63 and 64 to show that if an ellipse and a hyperbola have the same foci, then at each point of intersection their tangent lines are perpendicular.

66. Find two values of k such that the line $x + 2y = k$ is tangent to the ellipse $x^2 + 4y^2 = 8$. Find the points of tangency.

67. Find the coordinates of all points on the hyperbola

$$4x^2 - y^2 = 4$$

where the two lines that pass through the point and the foci are perpendicular.

68. A line tangent to the hyperbola $4x^2 - y^2 = 36$ intersects the y-axis at the point $(0, 4)$. Find the point(s) of tangency.

69. As illustrated in the accompanying figure, suppose that two observers are stationed at the points $F_1(c, 0)$ and $F_2(-c, 0)$ in an xy-coordinate system. Suppose also that the sound of an explosion in the xy-plane is heard by the F_1 observer t seconds before it is heard by the F_2 observer. Assuming that the speed of sound is a constant v, show that the explosion occurred somewhere on the hyperbola

$$\frac{x^2}{v^2t^2/4} - \frac{y^2}{c^2 - (v^2t^2/4)} = 1$$

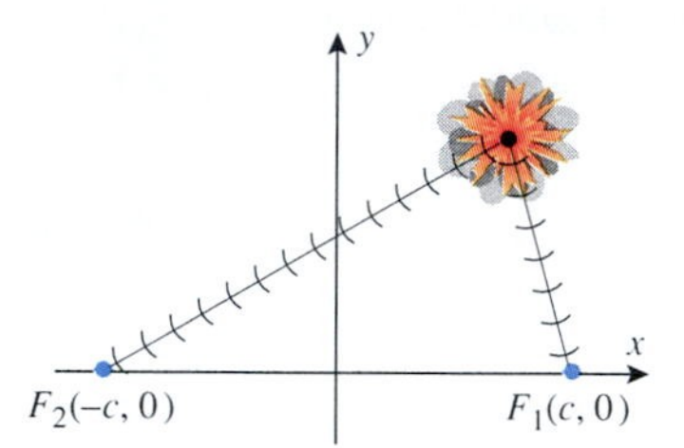

Figure Ex-69

70. As illustrated in the accompanying figure, suppose that two transmitting stations are positioned 100 km apart at points $F_1(50, 0)$ and $F_2(-50, 0)$ on a straight shoreline in an xy-coordinate system. Suppose also that a ship is traveling parallel to the shoreline but 200 km at sea. Find the coordinates of the ship if the stations transmit a pulse simultaneously, but the pulse from station F_1 is received by the ship 0.1 microsecond sooner than the pulse from station F_2. [*Hint:* Use the formula obtained in Exercise 69, assuming that the pulses travel at the speed of light (299,792,458 m/s).]

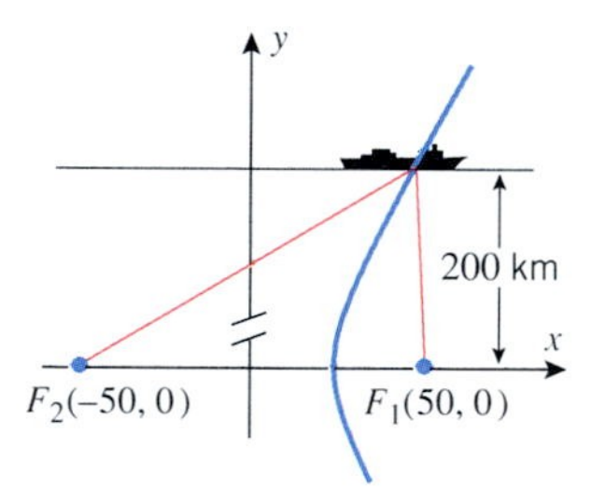

Figure Ex-70

C **71.** As illustrated in Figure Ex-71 (next page), the tank of an oil truck is 18 feet long and has elliptical cross sections that are 6 feet wide and 4 feet high.

(a) Show that the volume V of oil in the tank (in cubic feet) when it is filled to a depth of h feet is

$$V = 27\left[4\sin^{-1}\frac{h-2}{2} + (h-2)\sqrt{4h - h^2} + 2\pi\right]$$

(b) Use the numerical root-finding capability of a CAS to determine how many inches from the bottom of a dipstick the calibration marks should be placed to indicate when the tank is $\frac{1}{4}$, $\frac{1}{2}$, and $\frac{3}{4}$ full.

72. Consider the second-degree equation

$$Ax^2 + Cy^2 + Dx + Ey + F = 0$$

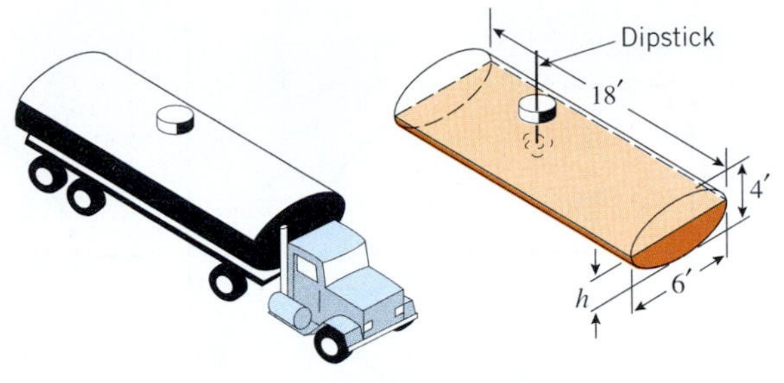

Figure Ex-71

where A and C are not both 0. Show by completing the square:

(a) If $AC > 0$, then the equation represents an ellipse, a circle, a point, or has no graph.

(b) If $AC < 0$, then the equation represents a hyperbola or a pair of intersecting lines.

(c) If $AC = 0$, then the equation represents a parabola, a pair of parallel lines, or has no graph.

73. In each part, use the result in Exercise 72 to make a statement about the graph of the equation, and then check your conclusion by completing the square and identifying the graph.

(a) $x^2 - 5y^2 - 2x - 10y - 9 = 0$

(b) $x^2 - 3y^2 - 6y - 3 = 0$

(c) $4x^2 + 8y^2 + 16x + 16y + 20 = 0$

(d) $3x^2 + y^2 + 12x + 2y + 13 = 0$

(e) $x^2 + 8x + 2y + 14 = 0$

(f) $5x^2 + 40x + 2y + 94 = 0$

74. Derive the equation $x^2 = 4py$ in Figure 12.4.6.

75. Derive the equation $(x^2/b^2) + (y^2/a^2) = 1$ given in Figure 12.4.14.

76. Derive the equation $(x^2/a^2) - (y^2/b^2) = 1$ given in Figure 12.4.22.

12.5 CONIC SECTIONS IN POLAR COORDINATES

It will be shown later in the text that if an object moves in a gravitational field that is directed toward a fixed point (such as the center of the Sun), then the path of that object must be a conic section with the fixed point at a focus. For example, planets in our solar system move along elliptical paths with the Sun at a focus, and the comets move along parabolic, elliptical, or hyperbolic paths with the Sun at a focus, depending on the conditions under which they were born. For applications of this type it is usually desirable to express the equations of the conic sections in polar coordinates with the pole at a focus. In this section we will show how to do this.

THE FOCUS–DIRECTRIX CHARACTERIZATION OF CONICS

To obtain polar equations for the conic sections we will need the following theorem.

12.5.1 THEOREM (***Focus–Directrix Property of Conics***). *Suppose that a point P moves in the plane determined by a fixed point (called the* **focus***) and a fixed line (called the* **directrix***), where the focus does not lie on the directrix. If the point moves in such a way that its distance to the focus divided by its distance to the directrix is some constant e (called the* **eccentricity***), then the curve traced by the point is a conic section. Moreover, the conic is a parabola if $e = 1$, an ellipse if $0 < e < 1$, and a hyperbola if $e > 1$.*

REMARK. It is an unfortunate historical accident that the letter e is used for the base of the natural logarithms and the eccentricity of conic sections. However, the appropriate interpretation will usually be clear from the context in which the letter is used.

We will not give a formal proof of this theorem; rather, we will use the specific cases in Figure 12.5.1 to illustrate the basic ideas. For the parabola, we will take the directrix to be $x = -p$, as usual; and for the ellipse and the hyperbola we will take the directrix to be $x = a^2/c$. We want to show in all three cases that if P is a point on the graph, F is the focus, and D is the directrix, then the ratio PF/PD is some constant e, where $e = 1$ for the parabola, $0 < e < 1$ for the ellipse, and $e > 1$ for the hyperbola. We will give the arguments for the parabola and ellipse and leave the argument for the hyperbola as an exercise.

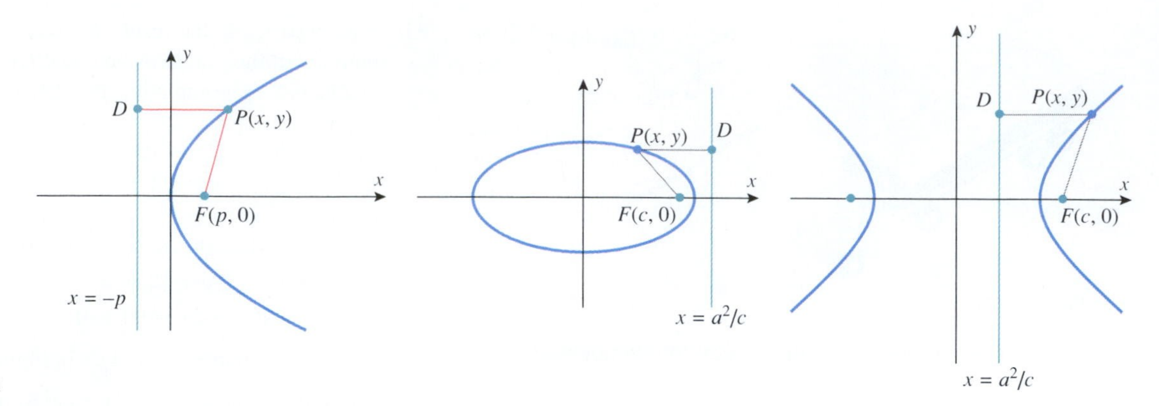

Figure 12.5.1

For the parabola, the distance PF to the focus is equal to the distance PD to the directrix, so that $PF/PD = 1$, which is what we wanted to show. For the ellipse, we rewrite Equation (8) of Section 12.4 as

$$\sqrt{(x-c)^2 + y^2} = a - \frac{c}{a}x = \frac{c}{a}\left(\frac{a^2}{c} - x\right)$$

But the expression on the left side is the distance PF, and the expression in the parentheses on the right side is the distance PD, so we have shown that

$$PF = \frac{c}{a}PD$$

Thus, PF/PD is constant, and the eccentricity is

$$e = \frac{c}{a} \tag{1}$$

If we rule out the degenerate case where $a = 0$ or $c = 0$, then it follows from Formula (7) of Section 12.4 that $0 < c < a$, so $0 < e < 1$, which is what we wanted to show.

We will leave it as an exercise to show that the eccentricity of the hyperbola in Figure 12.5.1 is also given by Formula (1), but in this case it follows from Formula (11) of Section 12.4 that $c > a$, so $e > 1$.

ECCENTRICITY OF AN ELLIPSE AS A MEASURE OF FLATNESS

The eccentricity of an ellipse can be viewed as a measure of its flatness—as e approaches 0 the ellipses become more and more circular, and as e approaches 1 they become more and more flat (Figure 12.5.2). Table 12.5.1 shows the orbital eccentricities of various celestial objects. Note that most of the planets actually have fairly circular orbits.

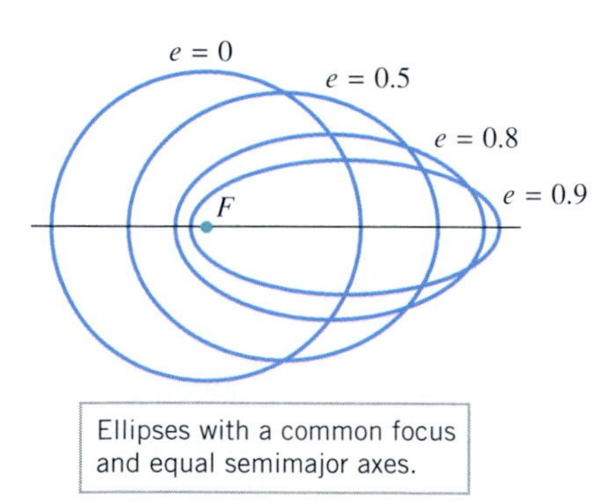

Figure 12.5.2

Table 12.5.1

CELESTIAL BODY	ECCENTRICITY
Mercury	0.206
Venus	0.007
Earth	0.017
Mars	0.093
Jupiter	0.048
Saturn	0.056
Uranus	0.046
Neptune	0.010
Pluto	0.249
Halley's comet	0.970

POLAR EQUATIONS OF CONICS

Our next objective is to derive polar equations for the conic sections from their focus–directrix characterizations. We will assume that the focus is at the pole and the directrix is either parallel or perpendicular to the polar axis. If the directrix is parallel to the polar axis, then it can be above or below the pole; and if the directrix is perpendicular to the polar axis,

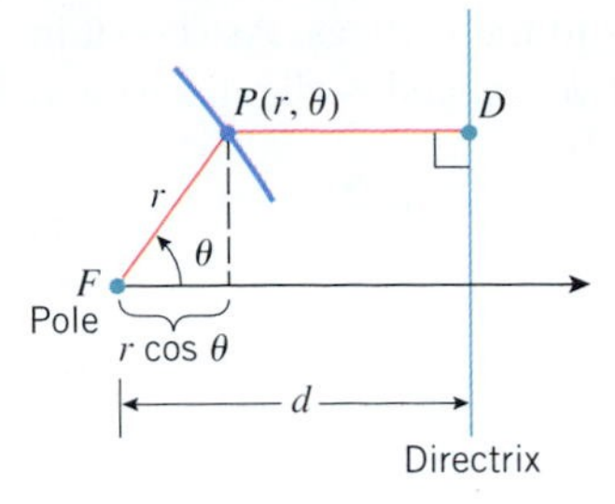

Figure 12.5.3

then it can be to the left or right of the pole. Thus, there are four cases to consider. We will derive the formulas for the case in which the directrix is perpendicular to the polar axis and to the right of the pole.

As illustrated in Figure 12.5.3, let us assume that the directrix is perpendicular to the polar axis and d units to the right of the pole, where the constant d is known. If P is a point on the conic and if the eccentricity of the conic is e, then it follows from Theorem 12.5.1 that $PF/PD = e$ or, equivalently, that

$$PF = ePD \tag{2}$$

However, it is evident from Figure 12.5.3 that $PF = r$ and $PD = d - r\cos\theta$. Thus, (2) can be written as

$$r = e(d - r\cos\theta)$$

which can be solved for r and expressed as

$$r = \frac{ed}{1 + e\cos\theta}$$

(verify). Observe that this single polar equation can represent a parabola, an ellipse, or a hyperbola, depending on the value of e. In contrast, the rectangular equations for these conics all have different forms. The derivations in the other three cases are similar.

12.5.2 THEOREM. *If a conic section with eccentricity e is positioned in a polar coordinate system so that its focus is at the pole and the corresponding directrix is d units from the pole, then the equation of the conic has one of four possible forms, depending on its orientation:*

$$r = \frac{ed}{1 + e\cos\theta} \quad \text{Directrix right of pole} \qquad r = \frac{ed}{1 - e\cos\theta} \quad \text{Directrix left of pole} \tag{3–4}$$

$$r = \frac{ed}{1 + e\sin\theta} \quad \text{Directrix above pole} \qquad r = \frac{ed}{1 - e\sin\theta} \quad \text{Directrix below pole} \tag{5–6}$$

SKETCHING CONICS IN POLAR COORDINATES

Precise graphs of conic sections in polar coordinates can be generated with graphing utilities. However, it is often useful to be able to make quick sketches of these graphs that show their orientation and give some sense of their dimensions. The orientation of a conic relative to the polar axis can be deduced by matching its equation with one of the four forms in Theorem 12.5.2. The key dimensions of a parabola are determined by the constant p (Figure 12.4.5) and those of ellipses and hyperbolas by the constants a, b, and c (Figures 12.4.11 and 12.4.20). Thus, we need to show how these constants can be obtained from the polar equations.

Example 1

Sketch the graph of $r = \dfrac{2}{1 - \cos\theta}$ in polar coordinates.

Solution. The equation is an exact match to (4) with $d = 2$ and $e = 1$. Thus, the graph is a parabola with the focus at the pole and the directrix 2 units to the left of the pole. This tells us that the parabola opens to the right along the polar axis and $p = 1$. Thus, the parabola looks roughly like that sketched in Figure 12.5.4. ◀

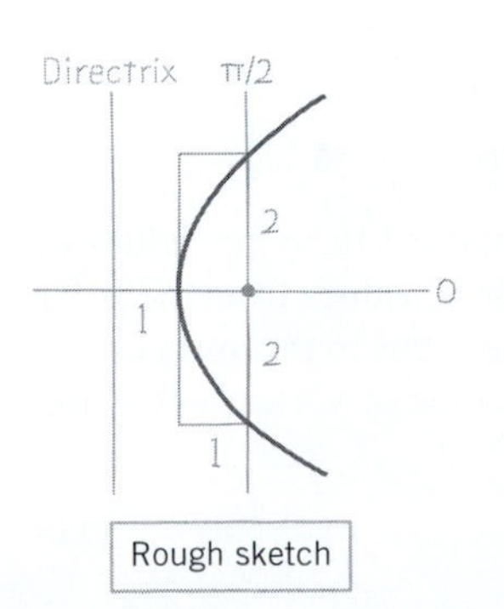

Figure 12.5.4

All of the important geometric information about an ellipse can be obtained from the values of a, b, and c in Figure 12.5.5. One way to find these values from the polar equation

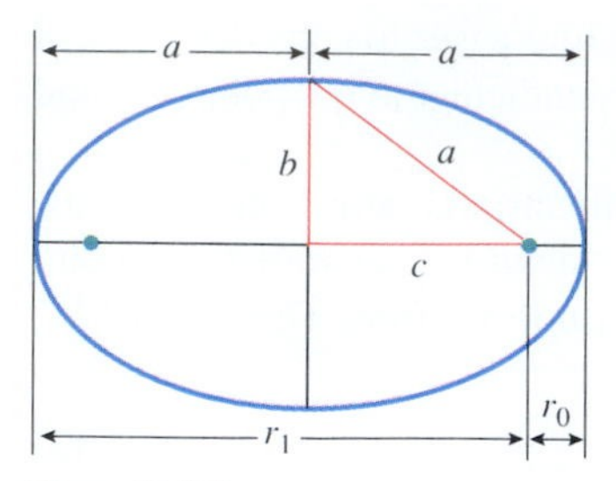

Figure 12.5.5

of an ellipse is based on finding the distances from the focus to the vertices. As shown in the figure, let r_0 be the distance from the focus to the closest vertex and r_1 the distance to the farthest vertex. Thus,

$$r_0 = a - c \quad \text{and} \quad r_1 = a + c \tag{7}$$

from which it follows that

$$a = \tfrac{1}{2}(r_1 + r_0) \tag{8}$$

and

$$c = \tfrac{1}{2}(r_1 - r_0) \tag{9}$$

Moreover, it also follows from (7) that

$$r_0 r_1 = a^2 - c^2 = b^2$$

Thus,

$$b = \sqrt{r_0 r_1} \tag{10}$$

REMARK. In words, Formula (8) states that a is the ***arithmetic average*** (also called the ***arithmetic mean***) of r_0 and r_1, and Formula (10) states that b is the ***geometric mean*** of r_0 and r_1.

Example 2

Sketch the graph of $r = \dfrac{6}{2 + \cos\theta}$ in polar coordinates.

Solution. This equation does not match any of the forms in Theorem 12.5.2 because they all require a constant term of 1 in the denominator. However, we can put the equation into one of these forms by dividing the numerator and denominator by 2 to obtain

$$r = \frac{3}{1 + \frac{1}{2}\cos\theta}$$

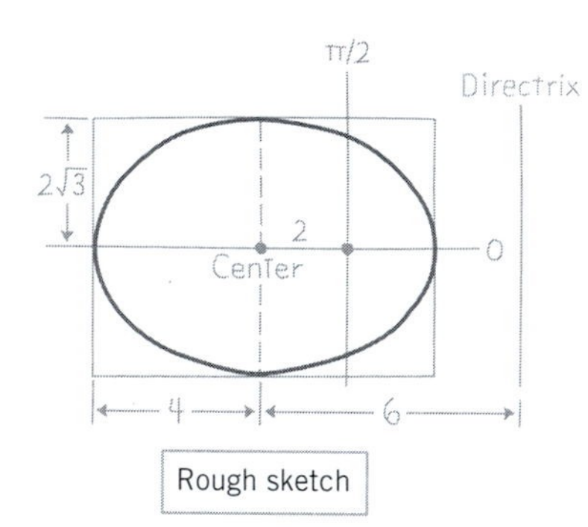

Figure 12.5.6

This is an exact match to (3) with $d = 6$ and $e = \frac{1}{2}$, so the graph is an ellipse with the directrix 6 units to the right of the pole. The distance r_0 from the focus to the closest vertex can be obtained by setting $\theta = 0$ in this equation, and the distance r_1 to the farthest vertex can be obtained by setting $\theta = \pi$. This yields

$$r_0 = \frac{3}{1 + \frac{1}{2}\cos 0} = \frac{3}{\frac{3}{2}} = 2, \quad r_1 = \frac{3}{1 + \frac{1}{2}\cos\pi} = \frac{3}{\frac{1}{2}} = 6$$

Thus, from Formulas (8), (10), and (9), respectively, we obtain

$$a = \tfrac{1}{2}(r_1 + r_0) = 4, \quad b = \sqrt{r_0 r_1} = 2\sqrt{3}, \quad c = \tfrac{1}{2}(r_1 - r_0) = 2$$

Thus, the ellipse looks roughly like that sketched in Figure 12.5.6. ◀

All of the important information about a hyperbola can be obtained from the values of a, b, and c in Figure 12.5.7. As with the ellipse, one way to find these values from the polar equation of a hyperbola is based on finding the distances from the focus to the vertices. As shown in the figure, let r_0 be the distance from the focus to the closest vertex and r_1 the distance to the farthest vertex. Thus,

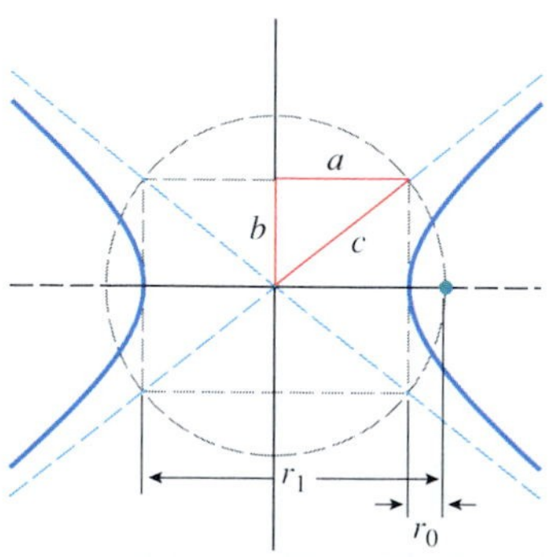

Figure 12.5.7

$$r_0 = c - a \quad \text{and} \quad r_1 = c + a \tag{11}$$

from which it follows that

$$a = \tfrac{1}{2}(r_1 - r_0) \tag{12}$$

and

$$c = \tfrac{1}{2}(r_1 + r_0) \tag{13}$$

Moreover, it also follows from (11) that

$$r_0 r_1 = c^2 - a^2 = b^2$$

from which it follows that

$$b = \sqrt{r_0 r_1} \tag{14}$$

Example 3

Sketch the graph of $r = \dfrac{2}{1 + 2\sin\theta}$ in polar coordinates.

Solution. This equation is an exact match to (5) with $d = 1$ and $e = 2$. Thus, the graph is a hyperbola with its directrix 1 unit above the pole. However, it is not so straightforward to compute the values of r_0 and r_1, since hyperbolas in polar coordinates are generated in a strange way as θ varies from 0 to 2π. This can be seen from Figure 12.5.8*a*, which is the graph of the given equation in rectangular coordinates. It follows from this graph that the corresponding polar graph is generated in pieces (see Figure 12.5.8*b*):

- As θ varies over the interval $0 \le \theta < 7\pi/6$, the value of r is positive and varies from 2 to $+\infty$, which generates part of the lower branch.
- As θ varies over the interval $7\pi/6 < \theta \le 3\pi/2$, the value of r is negative and varies from $-\infty$ to -2, which generates the right part of the upper branch.
- As θ varies over the interval $3\pi/2 \le \theta < 11\pi/6$, the value of r is negative and varies from -2 to $-\infty$, which generates the left part of the upper branch.
- As θ varies over the interval $11\pi/6 < \theta \le 2\pi$, the value of r is positive and varies from $+\infty$ to 2, which fills in the missing piece of the lower right branch.

It is now clear that we can obtain r_0 by setting $\theta = \pi/2$ and r_1 by setting $\theta = 3\pi/2$. Keeping in mind that r_0 and r_1 are positive, this yields

$$r_0 = \frac{2}{1 + 2\sin(\pi/2)} = \frac{2}{3}, \qquad r_1 = \left|\frac{2}{1 + 2\sin(3\pi/2)}\right| = \left|\frac{2}{-1}\right| = 2$$

Thus, from Formulas (12), (14), and (13), respectively, we obtain

$$a = \frac{1}{2}(r_1 - r_0) = \frac{2}{3}, \qquad b = \sqrt{r_0 r_1} = \frac{2\sqrt{3}}{3}, \qquad c = \frac{1}{2}(r_1 + r_0) = \frac{4}{3}$$

Thus, the hyperbola looks roughly like that sketched in Figure 12.5.8*c*. ◀

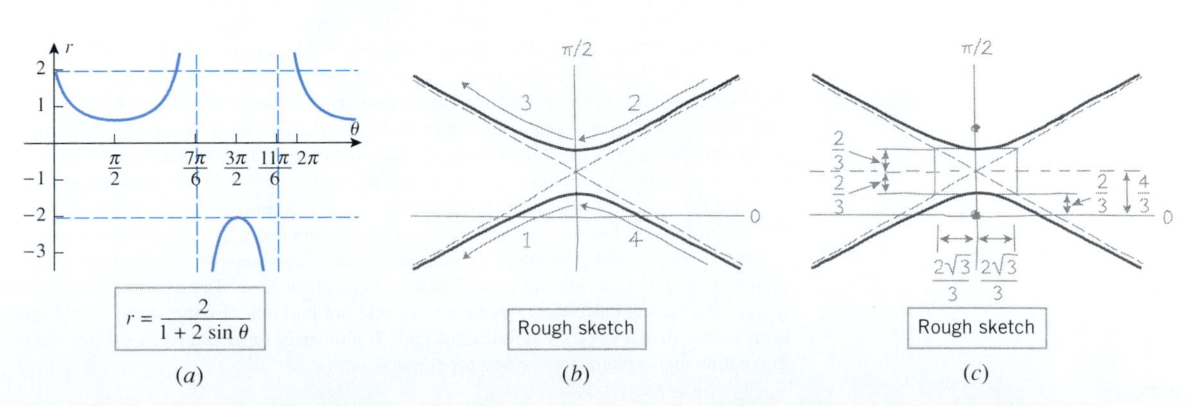

Figure 12.5.8

APPLICATIONS IN ASTRONOMY

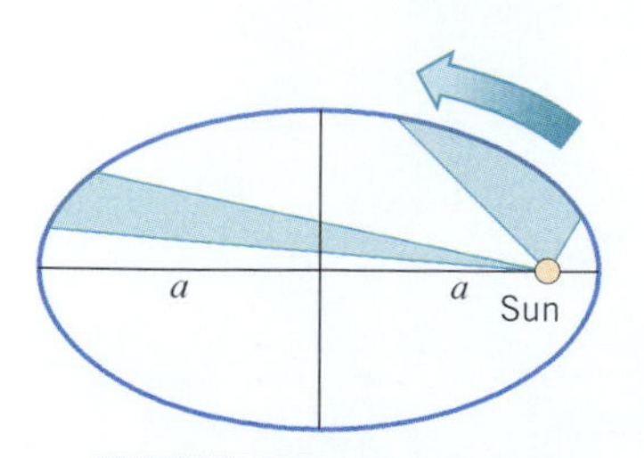

Figure 12.5.9

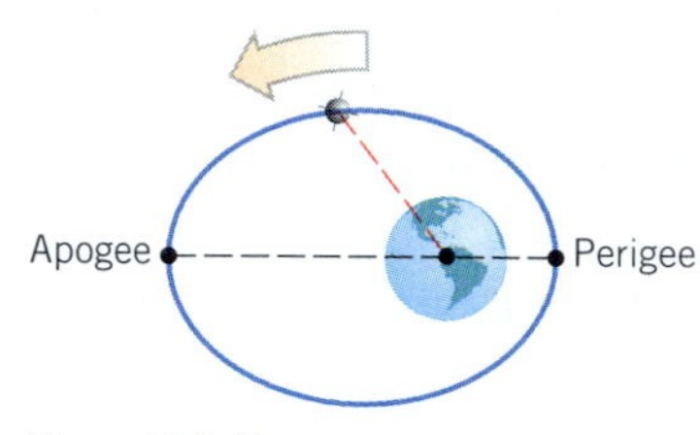

Figure 12.5.10

In 1609 Johannes Kepler* published a book known as *Astronomia Nova* (or sometimes *Commentaries on the Motions of Mars*) in which he succeeded in distilling thousands of years of observational astronomy into three beautiful laws of planetary motion (Figure 12.5.9).

12.5.3 KEPLER'S LAWS.

- First law (***Law of Orbits***). Each planet moves in an elliptical orbit with the Sun at a focus.
- Second law (***Law of Areas***). The radial line from the center of the Sun to the center of a planet sweeps out equal areas in equal times.
- Third law (***Law of Periods***). The square of a planet's period (the time it takes the planet to complete one orbit about the Sun) is proportional to the cube of the semimajor axis of its orbit.

Kepler's laws, although stated for planetary motion around the Sun, apply to all orbiting celestial bodies that are subjected to a *single* central gravitational force—artificial satellites subjected only to the central force of Earth's gravity and moons subjected only to the central gravitational force of a planet, for example. Later in the text we will derive Kepler's laws from basic principles, but for now we will show how they can be used in basic astronomical computations.

In an elliptical orbit, the closest point to the focus is called the ***perigee*** and the farthest point the ***apogee*** (Figure 12.5.10). The distances from the focus to the perigee and apogee are called the ***perigee distance*** and ***apogee distance***, respectively. For orbits around the Sun, it is more common to use the terms ***perihelion*** and ***aphelion***, rather than perigee and apogee, and to measure time in Earth years and distances in astronomical units (AU), where 1 AU is the semimajor axis a of the Earth's orbit (approximately 150×10^6 km or 92.9×10^6 mi). With this choice of units, the constant of proportionality in Kepler's third law is 1, since $a = 1$ AU produces a period of $T = 1$ Earth year. In this case Kepler's third law can be expressed as

$$T = a^{3/2} \tag{15}$$

Shapes of elliptical orbits are often specified by giving the eccentricity e and the semimajor axis a, so it is useful to express the polar equations of an ellipse in terms of these

---

* JOHANNES KEPLER (1571–1630). German astronomer and physicist, Kepler, whose work provided our contemporary view of planetary motion, led a fascinating but ill-starred life. His alcoholic father made him work in a family-owned tavern as a child, later withdrawing him from elementary school and hiring him out as a field laborer, where the boy contracted smallpox, permanently crippling his hands and impairing his eyesight. In later years, Kepler's first wife and several children died, his mother was accused of witchcraft, and being a Protestant he was often subjected to persecution by Catholic authorities. He was often impoverished, eking out a living as an astrologer and prognosticator. Looking back on his unhappy childhood, Kepler described his father as "criminally inclined" and "quarrelsome" and his mother as "garrulous" and "bad-tempered." However, it was his mother who left an indelible mark on the six-year-old Kepler by showing him the comet of 1577; and in later life he personally prepared her defense against the witchcraft charges. Kepler became acquainted with the work of Copernicus as a student at the University of Tübingen, where he received his master's degree in 1591. He continued on as a theological student, but at the urging of the university officials he abandoned his clerical studies and accepted a position as a mathematician and teacher in Graz, Austria. However, he was expelled from the city when it came under Catholic control, and in 1600 he finally moved on to Prague, where he became an assistant at the observatory of the famous Danish astronomer Tycho Brahe. Brahe was a brilliant and meticulous astronomical observer who amassed the most accurate astronomical data known at that time; and when Brahe died in 1601 Kepler inherited the treasure-trove of data. After eight years of intense labor, Kepler deciphered the underlying principles buried in the data and in 1609 published his monumental work, *Astronomia Nova*, in which he stated his first two laws of planetary motion. Commenting on his discovery of elliptical orbits, Kepler wrote, "I was almost driven to madness in considering and calculating this matter. I could not find out why the planet would rather go on an elliptical orbit (rather than a circle). Oh ridiculous me!" It ultimately remained for Isaac Newton to discover the laws of gravitation that explained the reason for elliptical orbits.

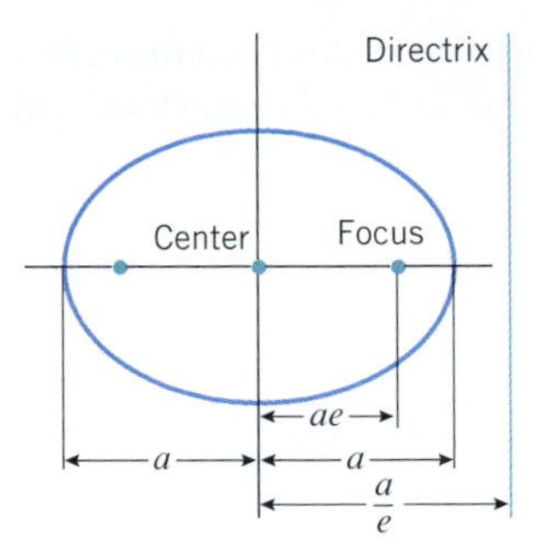

Figure 12.5.11

constants. Figure 12.5.11, which can be obtained from the ellipse in Figure 12.5.1 and the relationship $c = ea$, implies that the distance d between the focus and the directrix is

$$d = \frac{a}{e} - c = \frac{a}{e} - ea = \frac{a(1-e^2)}{e} \tag{16}$$

from which it follows that $ed = a(1-e^2)$. Thus, depending on the orientation of the ellipse, the formulas in Theorem 12.5.2 can be expressed in terms of a and e as

$$r = \frac{a(1-e^2)}{1 \pm e\cos\theta} \qquad r = \frac{a(1-e^2)}{1 \pm e\sin\theta} \tag{17–18}$$

$+$: Directrix right of pole
$-$: Directrix left of pole

$+$: Directrix above pole
$-$: Directrix below pole

Moreover, it is evident from Figure 12.5.11 that the distances from the focus to the closest and farthest vertices can be expressed in terms of a and e as

$$r_0 = a - ea = a(1-e) \quad \text{and} \quad r_1 = a + ea = a(1+e) \tag{19–20}$$

Example 4

Halley's comet (last seen in 1986) has an eccentricity of 0.97 and a semimajor axis of $a = 18.1$ AU.

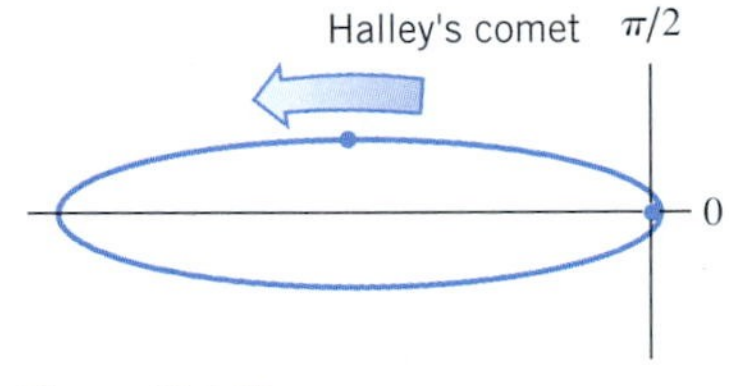

Figure 12.5.12

(a) Find the equation of its orbit in the polar coordinate system shown in Figure 12.5.12.

(b) Find the period of its orbit.

(c) Find its perihelion and aphelion distances.

Solution (a). From (17), the polar equation of the orbit has the form

$$r = \frac{a(1-e^2)}{1 + e\cos\theta}$$

But $a(1-e^2) = 18.1[1 - (0.97)^2] \approx 1.07$. Thus, the equation of the orbit is

$$r = \frac{1.07}{1 + 0.97\cos\theta}$$

Halley's comet photographed April 21, 1910 in Peru

Solution (b). From (15), with $a = 18.1$, the period of the orbit is

$$T = (18.1)^{3/2} \approx 77 \text{ years}$$

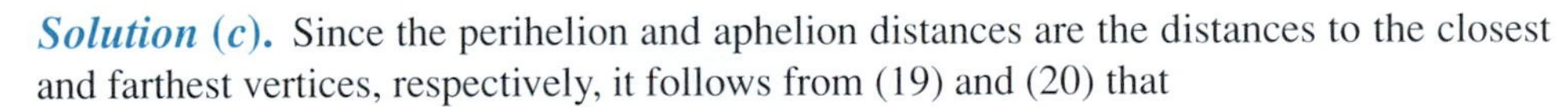

Solution (c). Since the perihelion and aphelion distances are the distances to the closest and farthest vertices, respectively, it follows from (19) and (20) that

$$r_0 = a - ea = a(1-e) = 18.1(1-0.97) \approx 0.543 \text{ AU}$$
$$r_1 = a + ea = a(1+e) = 18.1(1+0.97) \approx 35.7 \text{ AU}$$

or since 1 AU $\approx 150 \times 10^6$ km, the perihelion and aphelion distances in kilometers are

$$r_0 = 18.1(1-0.97)(150 \times 10^6) \approx 81{,}500{,}000 \text{ km}$$
$$r_1 = 18.1(1+0.97)(150 \times 10^6) \approx 5{,}350{,}000{,}000 \text{ km}$$

◀

FOR THE READER. Use the polar equation of the orbit of Halley's comet to check the values of r_0 and r_1.

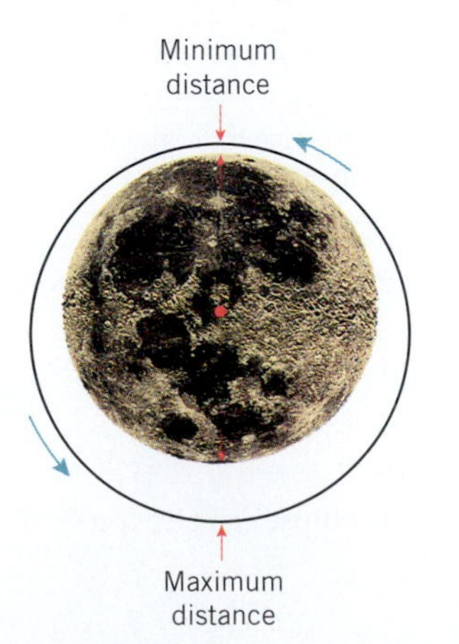

Figure 12.5.13

Example 5

An Apollo lunar lander orbits the Moon in an elliptic orbit with eccentricity $e = 0.12$ and semimajor axis $a = 2015$ km. Assuming the Moon to be a sphere of radius 1740 km, find the minimum and maximum heights of the lander above the lunar surface (Figure 12.5.13).

Solution. If we let r_0 and r_1 denote the minimum and maximum distances from the center of the Moon, then the minimum and maximum distances from the surface of the Moon will be

$$d_{\min} = r_0 - 1740$$
$$d_{\max} = r_1 - 1740$$

or from Formulas (19) and (20)

$$d_{\min} = r_0 - 1740 = a(1-e) - 1740 = 2015(0.88) - 1740 \approx 33.2 \text{ km}$$
$$d_{\max} = r_1 - 1740 = a(1+e) - 1740 = 2015(1.12) \approx 516.8 \text{ km}$$ ◀

EXERCISE SET 12.5 Graphing Calculator C CAS

For the conics in Exercises 1 and 2, find the eccentricity and the distance from the pole to the directrix, and sketch the graph in polar coordinates.

1. (a) $r = \dfrac{3}{2 - 2\cos\theta}$ (b) $r = \dfrac{3}{2 + \sin\theta}$
(c) $r = \dfrac{4}{2 + 3\cos\theta}$ (d) $r = \dfrac{5}{3 + 3\sin\theta}$

2. (a) $r = \dfrac{4}{3 - 2\cos\theta}$ (b) $r = \dfrac{3}{3 - 4\sin\theta}$
(c) $r = \dfrac{1}{3 + 3\sin\theta}$ (d) $r = \dfrac{1}{2 + 6\sin\theta}$

In Exercises 3 and 4, use Formulas (3)–(6) to name and describe the orientation of the conic, and then check your answer by generating the graph with a graphing utility.

3. (a) $r = \dfrac{8}{1 - \sin\theta}$ (b) $r = \dfrac{16}{4 + 3\sin\theta}$
(c) $r = \dfrac{4}{2 - 3\sin\theta}$ (d) $r = \dfrac{12}{4 + \cos\theta}$

4. (a) $r = \dfrac{15}{1 + \cos\theta}$ (b) $r = \dfrac{2}{3 + 3\cos\theta}$
(c) $r = \dfrac{64}{7 - 12\sin\theta}$ (d) $r = \dfrac{12}{3 - 2\cos\theta}$

In Exercises 5–8, find a polar equation for the conic that has its focus at the pole and satisfies the stated conditions. Points are in polar coordinates and directrices in rectangular coordinates for simplicity. (In some cases there may be more than one conic that satisfies the conditions.)

5. (a) Ellipse; $e = \frac{2}{3}$; directrix $x = 1$.
(b) Parabola; directrix $x = -1$.
(c) Hyperbola; $e = \frac{3}{2}$; directrix $y = 1$.

6. (a) Ellipse; $e = \frac{2}{3}$; directrix $y = -1$.
(b) Parabola; directrix $y = 1$.
(c) Hyperbola; $e = \frac{4}{3}$; directrix $x = -1$.

7. (a) Ellipse; vertices $(6, 0)$ and $(4, \pi)$.
(b) Parabola; vertex $(1, 3\pi/2)$.
(c) Hyperbola; vertices $(3, \pi/2)$ and $(7, \pi/2)$.

8. (a) Ellipse; ends of major axis $(1, \pi/2)$ and $(4, 3\pi/2)$.
(b) Parabola; vertex $(3, \pi)$.
(c) Hyperbola; equilateral; vertex $(5, 0)$.

In Exercises 9 and 10, find the distances from the pole to the vertices, and then apply Formulas (8)–(10) to find the equation of the ellipse in rectangular coordinates.

9. (a) $r = \dfrac{6}{2 + \sin\theta}$ (b) $r = \dfrac{1}{2 - \cos\theta}$

10. (a) $r = \dfrac{6}{5 + 2\cos\theta}$ (b) $r = \dfrac{8}{4 - 3\sin\theta}$

In Exercises 11 and 12, find the distances from the pole to the vertices, and then apply Formulas (12)–(14) to find the equation of the hyperbola in rectangular coordinates.

11. (a) $r = \dfrac{2}{1 + 3\sin\theta}$ (b) $r = \dfrac{10}{6 - 9\cos\theta}$

12. (a) $r = \dfrac{4}{1 - 2\sin\theta}$ (b) $r = \dfrac{15}{2 + 8\cos\theta}$

In Exercises 13 and 14, find a polar equation for the ellipse that has its focus at the pole and satisfies the stated conditions.

13. (a) Directrix to the right of the pole; $a = 8$; $e = \frac{1}{2}$.
(b) Directrix below the pole; $a = 4$; $e = \frac{3}{5}$.
(c) Directrix to the left of the pole; $b = 4$; $e = \frac{3}{5}$.
(d) Directrix above the pole; $c = 5$; $e = \frac{1}{5}$.

14. (a) Directrix above the pole; $a = 10$; $e = \frac{1}{2}$.
(b) Directrix to the left of the pole; $a = 6$; $e = \frac{1}{5}$.
(c) Directrix below the pole; $b = 4$; $e = \frac{3}{4}$.
(d) Directrix to the right of the pole; $c = 10$; $e = \frac{4}{5}$.

15. (a) Show that the eccentricity of an ellipse can be expressed in terms of r_0 and r_1 as

$$e = \frac{r_1 - r_0}{r_1 + r_0}$$

(b) Show that

$$\frac{r_1}{r_0} = \frac{1+e}{1-e}$$

16. (a) Show that the eccentricity of a hyperbola can be expressed in terms of r_0 and r_1 as

$$e = \frac{r_1 + r_0}{r_1 - r_0}$$

(b) Show that

$$\frac{r_1}{r_0} = \frac{e+1}{e-1}$$

In Exercises 17–22, use the following values, where needed:

radius of the Earth = 4000 mi = 6440 km
1 year (Earth year) = 365 days (Earth days)
1 AU = 92.9×10^6 mi = 150×10^6 km

17. The planet Pluto has eccentricity $e = 0.249$ and semimajor axis $a = 39.5$ AU.
(a) Find the period T in years.
(b) Find the perihelion and aphelion distances.
(c) Choose a polar coordinate system with the center of the Sun at the pole, and find a polar equation of Pluto's orbit in that coordinate system.
(d) Make a sketch of the orbit with reasonably accurate proportions.

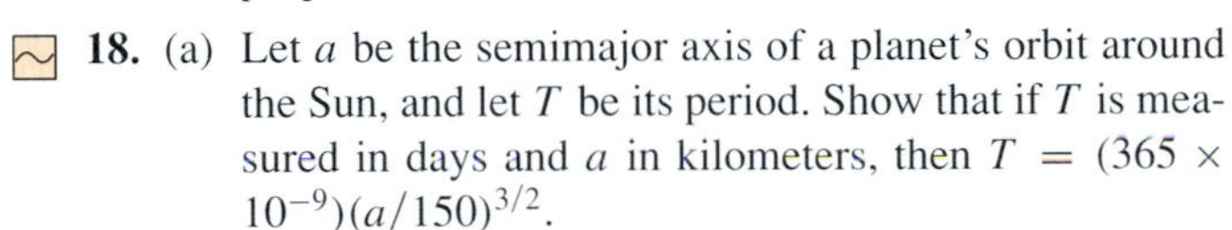

18. (a) Let a be the semimajor axis of a planet's orbit around the Sun, and let T be its period. Show that if T is measured in days and a in kilometers, then $T = (365 \times 10^{-9})(a/150)^{3/2}$.
(b) Use the result in part (a) to find the period of the planet Mercury in days, given that its semimajor axis is $a = 57.95 \times 10^6$ km.
(c) Choose a polar coordinate system with the Sun at the pole, and find an equation for the orbit of Mercury in that coordinate system given that the eccentricity of the orbit is $e = 0.206$.
(d) Use a graphing utility to generate the orbit of Mercury from the equation obtained in part (c).

19. The Hale–Bopp comet, discovered independently on July 23, 1995 by Alan Hale and Thomas Bopp, has an orbital eccentricity of $e = 0.9951$ and a period of 2380 years.
(a) Find its semimajor axis in astronomical units (AU).
(b) Find its perihelion and aphelion distances.
(c) Choose a polar coordinate system with the center of the Sun at the pole, and find an equation for the Hale–Bopp orbit in that coordinate system.
(d) Make a sketch of the Hale–Bopp orbit with reasonably accurate proportions.

20. Mars has a perihelion distance of 204,520,000 km and an aphelion distance of 246,280,000 km.
(a) Use these data to calculate the eccentricity, and compare your answer to the value given in Table 12.5.1.
(b) Find the period of Mars.
(c) Choose a polar coordinate system with the center of the Sun at the pole, and find an equation for the orbit of Mars in that coordinate system.
(d) Use a graphing utility to generate the orbit of Mars from the equation obtained in part (c).

21. *Vanguard 1* was launched in March 1958 into an orbit around the Earth with eccentricity $e = 0.21$ and semimajor axis 8864.5 km. Find the minimum and maximum heights of *Vanguard 1* above the surface of the Earth.

22. The planet Jupiter is believed to have a rocky core of radius 10,000 km surrounded by two layers of hydrogen—a 40,000-km-thick layer of compressed metallic-like hydrogen and a 20,000-km-thick layer of ordinary molecular hydrogen. The visible features, such as the Great Red Spot, are at the outer surface of the molecular hydrogen layer. On November 6, 1997 the spacecraft *Galileo* was placed in a Jovian orbit to study the moon Europa. The orbit had eccentricity 0.814580 and semimajor axis 3,514,918.9 km. Find *Galileo*'s minimum and maximum heights above the molecular hydrogen layer (see the accompanying figure).

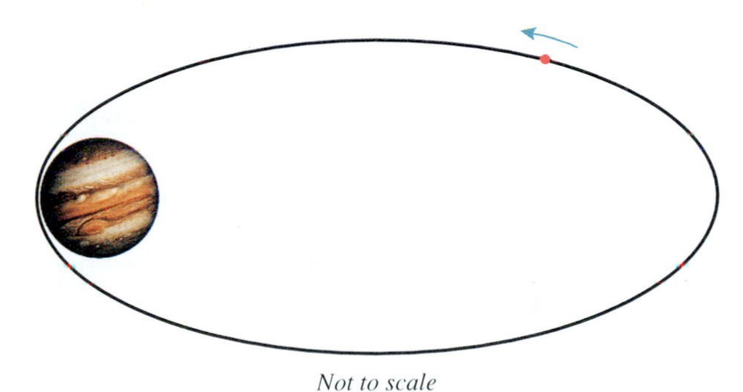

Figure Ex-22

23. What happens to the distance between the directrix and the center of an ellipse if the foci remain fixed and $e \to 0$?

24. (a) Show that the coordinates of the point P on the hyperbola in Figure 12.5.1 satisfy the equation

$$\sqrt{(x-c)^2 + y^2} = \frac{c}{a}x - a$$

(b) Use the result in part (a) to show that $PF/PD = c/a$.

SUPPLEMENTARY EXERCISES

1. Under what conditions does a parametric curve $x = f(t)$, $y = g(t)$ have a horizontal tangent line? A vertical tangent line? A singular point?

2. Express the point whose xy-coordinates are $(-1, 1)$ in polar coordinates with
(a) $r > 0,\ 0 \le \theta < 2\pi$ (b) $r < 0,\ 0 \le \theta < 2\pi$
(c) $r > 0,\ -\pi < \theta \le \pi$ (d) $r < 0,\ -\pi < \theta \le \pi$.

3. In each part, state the name that describes the polar curve most precisely: a rose, a line, a circle, a limaçon, a cardioid, a spiral, a lemniscate, or none of these.
(a) $r = 3\cos\theta$ (b) $r = \cos 3\theta$
(c) $r = \dfrac{3}{\cos\theta}$ (d) $r = 3 - \cos\theta$
(e) $r = 1 - 3\cos\theta$ (f) $r^2 = 3\cos\theta$
(g) $r = (3\cos\theta)^2$ (h) $r = 1 + 3\theta$

4. In each part: (i) Identify the polar graph as a parabola, an ellipse, or a hyperbola; (ii) state whether the directrix is above, below, to the left, or to the right of the pole; and (iii) find the distance from the pole to the directrix.
(a) $r = \dfrac{1}{3 + \cos\theta}$ (b) $r = \dfrac{1}{1 - 3\cos\theta}$
(c) $r = \dfrac{1}{3(1 + \sin\theta)}$ (d) $r = \dfrac{3}{1 - \sin\theta}$

5. The accompanying figure shows the polar graph of the equation $r = f(\theta)$. Sketch the graph of
(a) $r = f(-\theta)$ (b) $r = f\left(\theta - \dfrac{\pi}{2}\right)$
(c) $r = f\left(\theta + \dfrac{\pi}{2}\right)$ (d) $r = -f(\theta)$
(e) $r = f(\theta) + 1$.

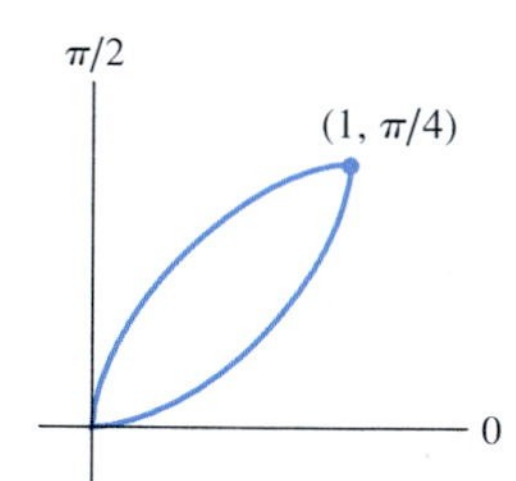

Figure Ex-5

6. Find equations for the two families of circles in the accompanying figure.

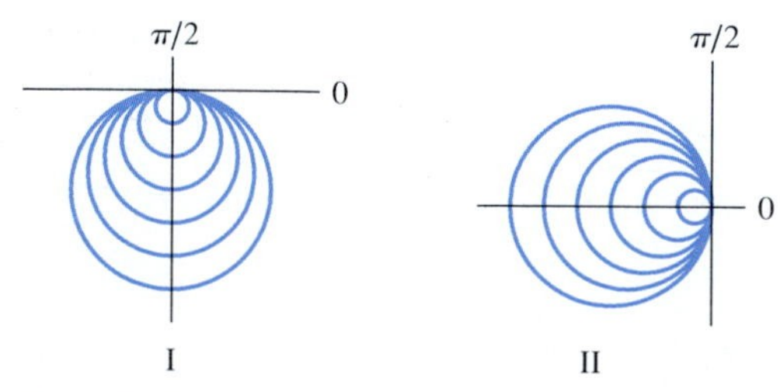

Figure Ex-6

7. In each part, identify the curve by converting the polar equation to rectangular coordinates. Assume that $a > 0$.
(a) $r = a\sec^2\dfrac{\theta}{2}$ (b) $r^2\cos 2\theta = a^2$
(c) $r = 4\csc\left(\theta - \dfrac{\pi}{4}\right)$ (d) $r = 4\cos\theta + 8\sin\theta$

8. Use a graphing utility to investigate how the family of polar curves $r = 1 + a\cos n\theta$ is affected by changing the values of a and n, where a is a positive real number and n is a positive integer. Write a brief paragraph to explain your conclusions.

In Exercises 9 and 10, find an equation in xy-coordinates for the conic section that satisfies the given conditions.

9. (a) Ellipse with eccentricity $e = \frac{2}{7}$ and ends of the minor axis at the points $(0, \pm 3)$.
(b) Parabola with vertex at the origin, focus on the y-axis, and directrix passing through the point $(7, 4)$.
(c) Hyperbola that has the same foci as the ellipse $3x^2 + 16y^2 = 48$ and asymptotes $y - 5 = \pm 8(x + 1)$.

10. (a) Ellipse with center $(-3, 2)$, vertex $(2, 2)$, and eccentricity $e = \frac{4}{5}$.
(b) Parabola with focus $(-2, -2)$ and vertex $(-2, 0)$.
(c) Hyperbola with vertex $(-1, 7)$ and asymptotes $y - 5 = 8(x + 1)$.

11. In each part, sketch the graph of the conic section with reasonably accurate proportions.
(a) $x^2 - 4x + 8y + 36 = 0$
(b) $3x^2 + 4y^2 - 30x - 8y + 67 = 0$
(c) $4x^2 - 5y^2 - 8x - 30y - 21 = 0$

12. If you have a CAS that can graph implicit equations, use it to check your work in Exercise 11.

13. It can be shown that hanging cables form parabolic arcs rather than catenaries if they are subjected to uniformly distributed downward forces along their length. For example, if the weight of the roadway in a suspension bridge is assumed to be uniformly distributed along the supporting cables, then the cables can be modeled by parabolas.
(a) Assuming a parabolic model, find an equation for the cable in the accompanying figure, taking the y-axis to be vertical and the origin at the low point of the cable.
(b) Find the length of the cable between the supports.

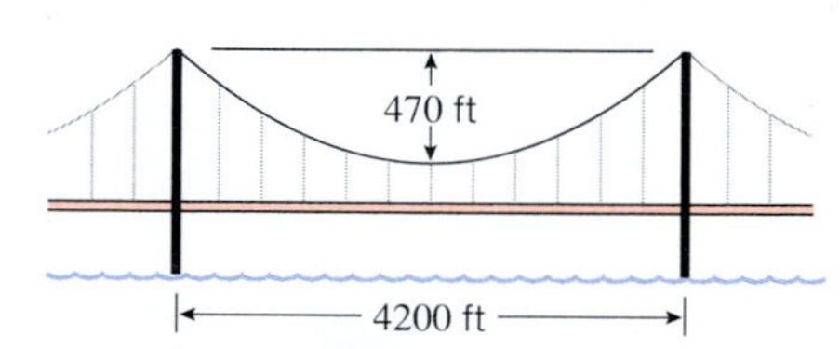

Figure Ex-13

14. A parametric curve of the form

$$x = a\cot t + b\cos t, \quad y = a + b\sin t \qquad (0 < t < 2\pi)$$

is called a ***conchoid of Nicomedes*** (see the accompanying figure for the case $0 < a < b$).

(a) Describe how the conchoid

$$x = \cot t + 4\cos t, \quad y = 1 + 4\sin t$$

is generated as t varies over the interval $0 < t < 2\pi$.

(b) Find the horizontal asymptote of the conchoid given in part (a).

(c) For what values of t does the conchoid in part (a) have a horizontal tangent line? A vertical tangent line?

(d) Find a polar equation $r = f(\theta)$ for the conchoid in part (a), and then find polar equations for the tangent lines to the conchoid at the pole.

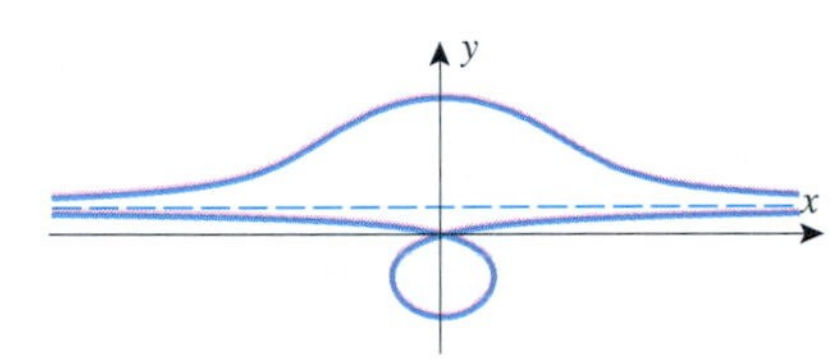

Figure Ex-14

15. Find the area of the region that is common to the circles $r = 1$, $r = 2\cos\theta$, and $r = 2\sin\theta$.

16. Find the area of the region that is inside the cardioid $r = a(1 + \sin\theta)$ and outside the circle $r = a\sin\theta$.

17. (a) Find the arc length of the polar curve $r = 1/\theta$ for $\pi/4 \le \theta \le \pi/2$.

(b) What can you say about the arc length of the portion of the curve that lies inside the circle $r = 1$?

18. (a) If a thread is unwound from a fixed circle while being held taut (i.e., tangent to the circle), then the end of the thread traces a curve called an ***involute of a circle***. Show that if the circle is centered at the origin, has radius a, and the end of the thread is initially at the point $(a, 0)$, then the involute can be expressed parametrically as

$$x = a(\cos\theta + \theta\sin\theta), \quad y = a(\sin\theta - \theta\cos\theta)$$

where θ is the angle shown in part (*a*) of the accompanying figure.

(b) Assuming that the dog in part (*b*) of the accompanying figure unravels its leash while keeping it taut, for what values of θ in the interval $0 \le \theta \le 2\pi$ will the dog be moving North? South? East? West?

(c) Use a graphing utility to generate the curve traced by the dog, and show that it is consistent with your answer in part (b).

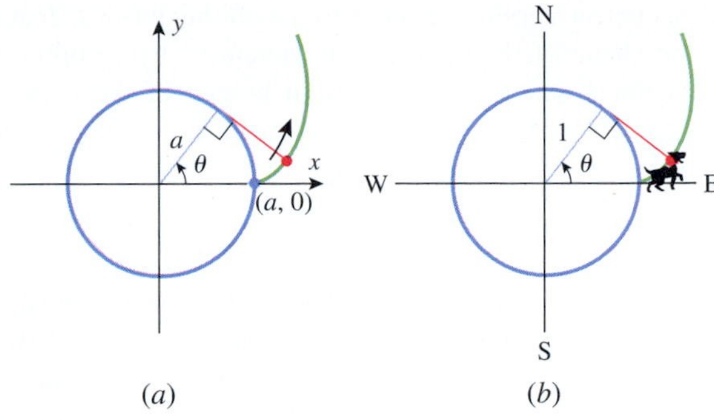

Figure Ex-18

19. Let R be the region that is above the x-axis and enclosed between the curve $b^2x^2 - a^2y^2 = a^2b^2$ and the line $x = \sqrt{a^2 + b^2}$.

(a) Sketch the solid generated by revolving R about the x-axis, and find its volume.

(b) Sketch the solid generated by revolving R about the y-axis, and find its volume.

20. (a) Sketch the curves

$$r = \frac{1}{1 + \cos\theta} \quad \text{and} \quad r = \frac{1}{1 - \cos\theta}$$

(b) Find polar coordinates of the intersections of the curves in part (a).

(c) Show that the curves are *orthogonal*, that is, their tangent lines are perpendicular at the points of intersection.

21. How is the shape of a hyperbola affected as its eccentricity approaches 1? As it approaches $+\infty$? Draw some pictures to illustrate your conclusions.

22. Use the formula obtained in part (a) of Exercise 67 of Section 12.1 to find the distance between successive tips of the three-petal rose $r = \sin 3\theta$, and check your answer using trigonometry.

23. (a) Find the minimum and maximum x-coordinates of points on the cardioid $r = 1 + \cos\theta$.

(b) Find the minimum and maximum y-coordinates of points on the cardioid in part (a).

24. (a) Show that the maximum value of the y-coordinate of points on the curve $r = 1/\sqrt{\theta}$ for θ in the interval $(0, \pi]$ occurs when $\tan\theta = 2\theta$.

(b) Use Newton's Method to solve the equation in part (a) for θ to at least four decimal-place accuracy.

(c) Use the result of part (b) to approximate the maximum value of y for $0 < \theta \le \pi$.

25. Define the width of a petal of a rose curve to be the dimension shown in the accompanying figure. Show that the width w of a petal of the four-petal rose $r = \cos 2\theta$ is $w = 2\sqrt{6}/9$. [*Hint:* Express y in terms of θ, and investigate the maximum value of y.]

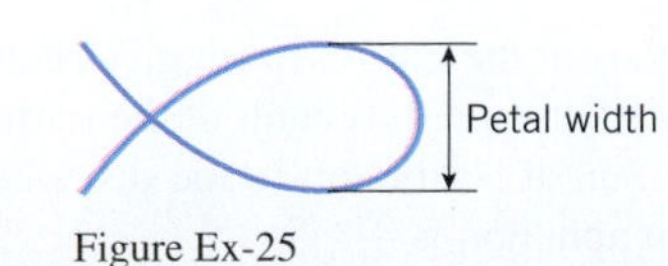

Figure Ex-25

26. A nuclear cooling tower is to have a height of h feet and the shape of the solid that is generated by revolving the region R enclosed by the right branch of the hyperbola $1521x^2 - 225y^2 = 342{,}225$ and the lines $x = 0$, $y = -h/2$, and $y = h/2$ about the y-axis.
(a) Find the volume of the tower.
(b) Find the lateral surface area of the tower.

27. The amusement park rides illustrated in the accompanying figure consist of two connected rotating arms of length 1—an inner arm that rotates counterclockwise at 1 radian per second and an outer arm that can be programmed to rotate either clockwise at 2 radians per second (the Scrambler ride) or counterclockwise at 2 radians per second (the Calypso ride). The center of the rider cage is at the end of the outer arm.
(a) Show that in the Scrambler ride the center of the cage has parametric equations

$$x = \cos t + \cos 2t, \quad y = \sin t - \sin 2t$$

(b) Find parametric equations for the center of the cage in the Calypso ride, and use a graphing utility to confirm that the center traces the curve shown in the accompanying figure.
(c) Do you think that a rider travels the same distance in one revolution of the Scrambler ride as in one revolution of the Calypso ride? Justify your conclusion.

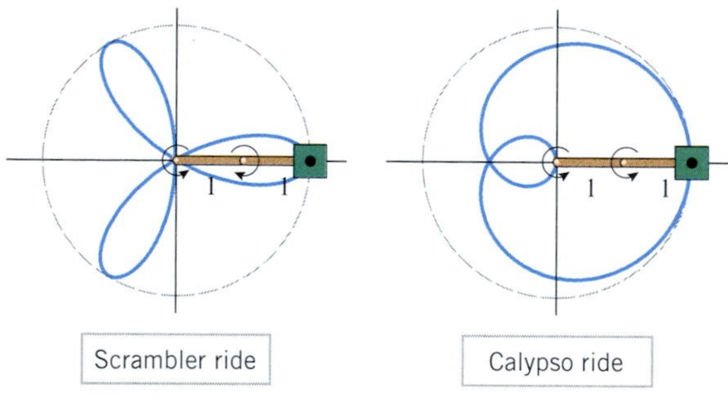

Figure Ex-27

28. Use a graphing utility to explore the effect of changing the rotation rates and the arm lengths in Exercise 27.

29. Use the parametric equations $x = a\cos t$, $y = b\sin t$ to show that the circumference C of an ellipse with semimajor axis a and eccentricity e is

$$C = 4a\int_0^{\pi/2} \sqrt{1 - e^2\sin^2 u}\, du$$

30. Use Simpson's rule or the numerical integration capability of a graphing utility to approximate the circumference of the ellipse $4x^2 + 9y^2 = 36$ from the integral obtained in Exercise 29.

31. (a) Calculate the eccentricity of the Earth's orbit, given that the ratio of the distance between the center of the Earth and the center of the Sun at perihelion to the distance between the centers at aphelion is $\frac{59}{61}$.
(b) Find the distance between the center of the Earth and the center of the Sun at perihelion, given that the average value of the perihelion and aphelion distances between the centers is 93 million miles.
(c) Use the result in Exercise 29 and Simpson's rule or the numerical integration capability of a graphing utility to approximate the distance that the Earth travels in 1 year (one revolution around the Sun).

32. It will be shown later in this text that if a projectile is launched with speed v_0 at an angle α with the horizontal and at a height y_0 above ground level, then the resulting trajectory relative to the coordinate system in the accompanying figure will have parametric equations

$$x = (v_0\cos\alpha)t, \quad y = y_0 + (v_0\sin\alpha)t - \tfrac{1}{2}gt^2$$

(a) Show that the trajectory is a parabola.
(b) Find the coordinates of the vertex.

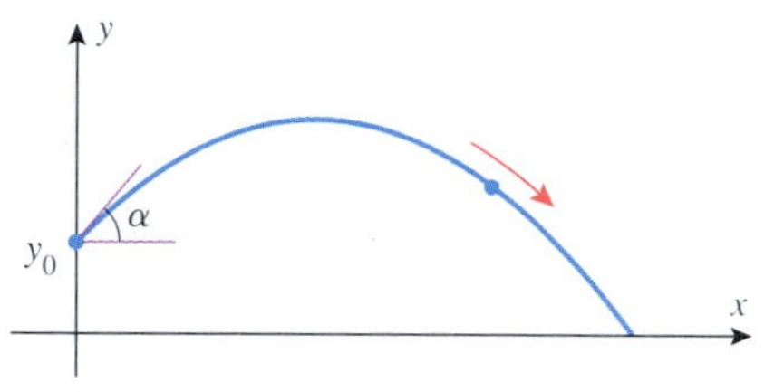

Figure Ex-32

33. Mickey Mantle is recognized as baseball's unofficial king of long home runs. On April 17, 1953 Mantle blasted a pitch by Chuck Stobbs of the hapless Washington Senators out of Griffith Stadium, just clearing the 50-ft wall at the 391-ft marker in left center. Assuming that the ball left the bat at a height of 3 ft above the ground and at an angle of 45°, use the parametric equations in Exercise 32 with $g = 32$ ft/s^2 to find
(a) the speed of the ball as it left the bat
(b) the maximum height of the ball
(c) the distance along the ground from home plate where the ball struck the ground.

C **34.** Recall from Section 7.9 that the Fresnel sine and cosine functions are defined as

$$S(x) = \int_0^x \sin\left(\frac{\pi t^2}{2}\right) dt \quad \text{and} \quad C(x) = \int_0^x \cos\left(\frac{\pi t^2}{2}\right) dt$$

The following parametric curve, which is used to study amplitudes of light waves in optics, is called a ***clothoid*** or ***Cornu spiral*** in honor of the French scientist Marie Alfred Cornu (1841–1902):

$$x = C(t) = \int_0^t \cos\left(\frac{\pi u^2}{2}\right) du \qquad (-\infty < t < +\infty)$$
$$y = S(t) = \int_0^t \sin\left(\frac{\pi u^2}{2}\right) du$$

(a) Use a CAS to graph the cornu spiral.
(b) Describe the behavior of the spiral as $t \to +\infty$ and as $t \to -\infty$.
(c) Find the arc length of the spiral for $-1 \le t \le 1$.

35. As illustrated in the accompanying figure, let $P(r, \theta)$ be a point on the polar curve $r = f(\theta)$, let ψ be the smallest counterclockwise angle from the extended radius OP to the tangent line at P, and let ϕ be the angle of inclination of the tangent line. Derive the formula

$$\tan\psi = \frac{r}{dr/d\theta}$$

by substituting $\tan\phi$ for dy/dx in Formula (7) of Section 12.2 and applying the trigonometric identity

$$\tan(\phi - \theta) = \frac{\tan\phi - \tan\theta}{1 + \tan\phi\tan\theta}$$

In Exercises 36 and 37, use the formula for ψ obtained in Exercise 35.

36. (a) Use the trigonometric identity

$$\tan\frac{\theta}{2} = \frac{1 - \cos\theta}{\sin\theta}$$

to show that if (r, θ) is a point on the cardioid

$$r = 1 - \cos\theta \quad (0 \le \theta < 2\pi)$$

then $\psi = \theta/2$.

(b) Sketch the cardioid and show the angle ψ at the points where the cardioid crosses the y-axis.

(c) Find the angle ψ at the points where the cardioid crosses the y-axis.

37. Show that for a logarithmic spiral $r = ae^{b\theta}$, the angle from the radial line to the tangent line is constant along the spiral (see the accompanying figure). [*Note:* For this reason, logarithmic spirals are sometimes called ***equiangular spirals***.]

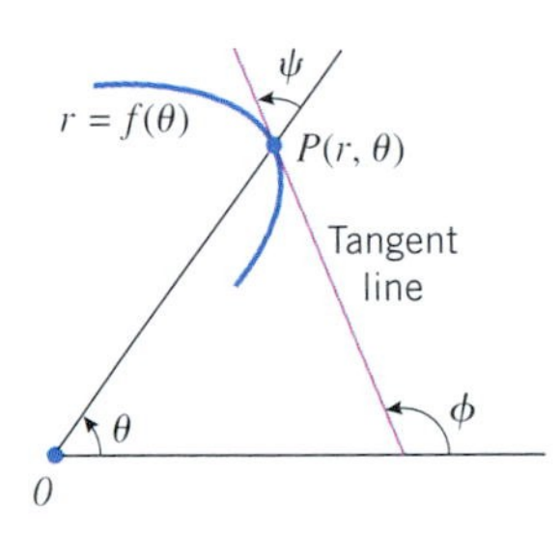

Figure Ex-35

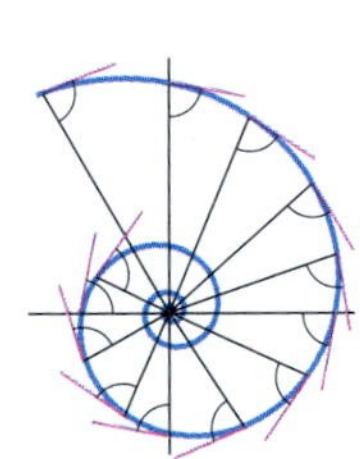
Figure Ex-37

EXPANDING THE CALCULUS HORIZON

Comet Collision

The Earth lives in a cosmic shooting gallery of comets and asteroids. Although the probability that the Earth will be hit by a comet or asteroid in any given year is small, the consequences of such a collision are so catastrophic that the international community is now beginning to track **near Earth objects** *(NEOs). Your job, as part of the international NEO tracking team, is to compute the orbits of incoming comets and asteroids, determine how close they will come to colliding with the Earth, and issue a notification if there is danger of a collision or near miss.*

At the time when the Earth is at its *aphelion* (its farthest point from the Sun), your NEO tracking team receives a notification from the NASA/Caltech Jet Propulsion Laboratory that a previously unknown comet (designation Rogue 2000) is hurtling in the direction of the Earth. You immediately transmit a request to NASA for the orbital parameters and the current positions of the Earth and Rogue 2000 and receive the following report:

ORBITAL PARAMETERS

EARTH	ROGUE 2000
Eccentricity: $e_1 = 0.017$	Eccentricity: $e_2 = 0.98$
Semimajor axis: $a_1 = 1$ AU $= 1.496 \times 10^8$ km	Semimajor axis: $a_2 = 5$ AU $= 7.48 \times 10^8$ km
Period: $T_1 = 1$ year	Period: $T_2 = 5\sqrt{5}$ years

INITIAL POSITION INFORMATION

The major axes of Earth and Rogue 2000 coincide.
The aphelions of Earth and Rogue 2000 are on the same side of the Sun.
Initial polar angle of Earth: $\theta = 0$ radians.
Initial polar angle of Rogue 2000: $\theta = 0.45$ radian.

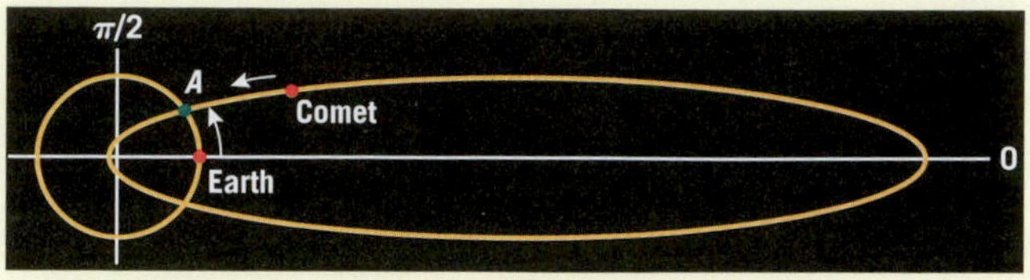

Initial configuration of Earth and Rogue 2000

Figure 1

The Calculation Strategy

Since the immediate concern is a possible collision at intersection A in Figure 1, your team works out the following plan:

Step 1. Find the polar equations for Earth and Rogue 2000.
Step 2. Find the polar coordinates of intersection A.
Step 3. Determine how long it will take the Earth to reach intersection A.
Step 4. Determine where Rogue 2000 will be when the Earth reaches intersection A.
Step 5. Determine how far Rogue 2000 will be from the Earth when the Earth is at intersection A.

Polar Equations of the Orbits

Exercise 1 Write polar equations of the form

$$r = \frac{a(1-e^2)}{1-e\cos\theta}$$

for the orbits of Earth and Rogue 2000 using AU units for r.

Exercise 2 Use a graphing utility to generate the two orbits on the same screen.

Intersection of the Orbits

The second step in your team's calculation plan is to find the polar coordinates of intersection A in Figure 1.

Exercise 3 For simplicity, let $k_1 = a_1(1-e_1^2)$ and $k_2 = a_2(1-e_2^2)$, and use the polar equations obtained in Exercise 1 to show that the angle θ at intersection A satisfies the equation

$$\cos\theta = \frac{k_1 - k_2}{k_1 e_2 - k_2 e_1}$$

Exercise 4 Use the result in Exercise 3 and the inverse cosine capability of a calculating utility to show that the angle θ at intersection A in Figure 1 is $\theta = 0.607$ radian.

Exercise 5 Use the result in Exercise 4 and either polar equation obtained in Exercise 1 to show that if r is in AU units, then the polar coordinates of intersection A are $(r, \theta) = (1.014, 0.607)$.

Time Required for Earth to Reach Intersection A

According to Kepler's second law (see 12.5.3), the radial line from the center of the Sun to the center of an object orbiting around it sweeps out equal areas in equal times. Thus, if t is the time that it takes for the radial line to sweep out an "elliptic sector" from some initial angle θ_I to some final angle θ_F (Figure 2), and if T is the period of the object (the time for one complete revolution), then

$$\frac{t}{T} = \frac{\text{area of the "elliptic sector"}}{\text{area of the entire ellipse}} \tag{1}$$

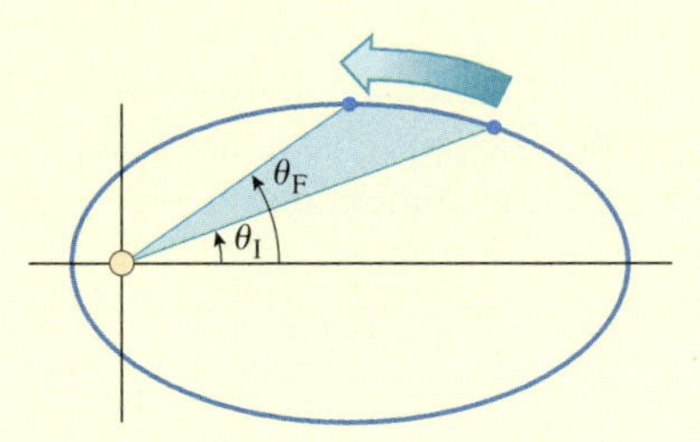

Figure 2

Exercise 6 Use Formula (1) to show that

$$t = \frac{T\displaystyle\int_{\theta_I}^{\theta_F} r^2\,d\theta}{2\pi a^2\sqrt{1-e^2}} \tag{2}$$

Exercise 7 Use a calculating utility with a numerical integration capability, Formula (2), and the polar equation for the orbit of the Earth obtained in Exercise 1 to find the time t (in years) required for the Earth to move from its initial position to intersection A.

Position of Rogue 2000 When the Earth Is at Intersection A

The fourth step in your team's calculation strategy is to determine the position of Rogue 2000 when the Earth reaches intersection A.

Exercise 8 During the time that it takes for the Earth to move from its initial position to intersection A, the polar angle of Rogue 2000 will change from its initial value $\theta_I = 0.45$ radian to some final value θ_F that remains to be determined. Apply Formula (2) using the orbital data for Rogue 2000 and the time t obtained in Exercise 7 to show that θ_F satisfies the equation

$$\int_{0.45}^{\theta_F}\left[\frac{a_2(1-e_2^2)}{1-e_2\cos\theta}\right]^2 d\theta = \frac{2t\pi a_2^2\sqrt{1-e_2^2}}{5\sqrt{5}} \tag{3}$$

Your team is now faced with the problem of solving Equation (3) for the unknown upper limit θ_F. Some members of the team plan to use a CAS to perform the integration, some plan to use integration tables, and others plan to use hand calculation by making the substitution $u = \tan(\theta/2)$ and applying the formulas in (5) of Section 9.6.

Exercise 9

(a) Evaluate the integral in (3) using a CAS or by hand calculation.

(b) Use the root-finding capability of a calculating utility to find the polar angle of Rogue 2000 when the Earth is at intersection A.

Calculating the Critical Distance

It is the policy of your NEO tracking team to issue a notification to various governmental agencies for any asteroid or comet that will be within 4 million kilometers of the Earth at an orbital intersection. (This distance is roughly 10 times that between the Earth and the Moon.) Accordingly, the final step in your team's plan is to calculate the distance between the Earth and Rogue 2000 when the Earth is at intersection A, and then determine whether a notification should be issued.

Exercise 10 Use the polar equation of Rogue 2000 obtained in Exercise 1 and the result in Exercise 9(b) to find polar coordinates of Rogue 2000 with r in AU units when the Earth is at intersection A.

Exercise 11 Use the distance formula in Exercise 67(a) of Section 12.1 to calculate the distance between the Earth and Rogue 2000 in AU units when the Earth is at intersection A, and then use the conversion factor $1\text{ AU} = 1.496 \times 10^8$ km to determine whether a government notification should be issued.

Note: One of the closest near misses in recent history occurred on October 30, 1937 when the asteroid Hermes passed within 900,000 km of the Earth. More recently, on June 14, 1968 the asteroid Icarus passed within 23,000,000 km of the Earth.

Module by Mary Ann Connors, USMA, West Point, and Howard Anton, Drexel University

Additional material for this module can be found on the World Wide Web at http://www.wiley.com/college/anton

APPENDIX A

Real Numbers, Intervals, and Inequalities

REAL NUMBERS

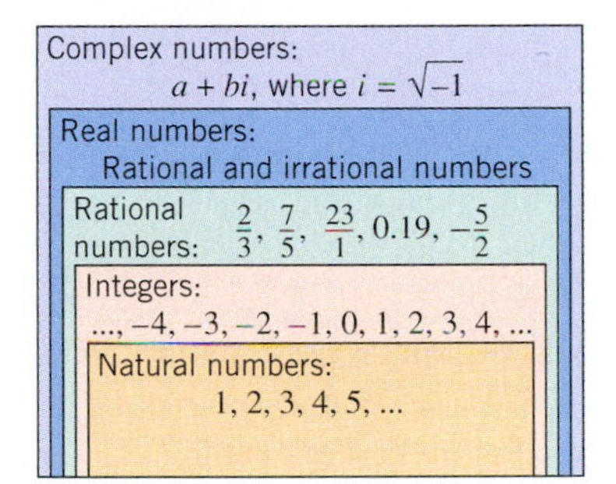

Figure A.1

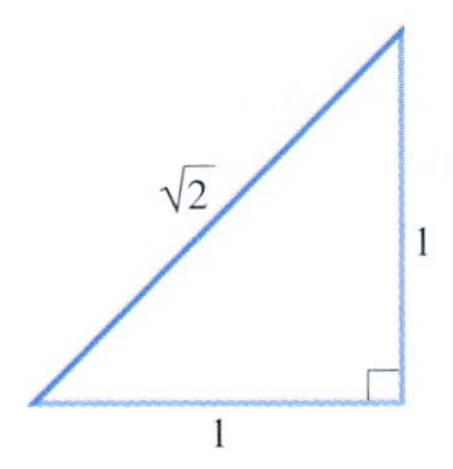

Figure A.2

Figure A.1 describes the various categories of numbers that we will encounter in this text. The simplest numbers are the ***natural numbers***

$$1,\ 2,\ 3,\ 4,\ 5, \ldots$$

These are a subset of the ***integers***

$$\ldots,\ -4,\ -3,\ -2,\ -1,\ 0,\ 1,\ 2,\ 3,\ 4, \ldots$$

and these in turn are a subset of the ***rational numbers***, which are the numbers formed by taking ratios of integers (avoiding division by 0). Some examples are

$$\frac{2}{3},\quad \frac{7}{5},\quad 23 = \frac{23}{1},\quad 0.19 = \frac{19}{100},\quad -\frac{5}{2} = \frac{-5}{2} = \frac{5}{-2}$$

The early Greeks believed that every measurable quantity had to be a rational number. However, this idea was overturned in the fifth century B.C. by Hippasus of Metapontum* who demonstrated the existence of ***irrational numbers***, that is, numbers that cannot be expressed as the ratio of two integers. Using geometric methods, he showed that the length of the hypotenuse of the triangle in Figure A.2 could not be expressed as a ratio of integers, thereby proving that $\sqrt{2}$ is an irrational number. Some other examples of irrational numbers are

$$\sqrt{3},\quad \sqrt{5},\quad 1+\sqrt{2},\quad \sqrt[3]{7},\quad \pi,\quad \cos 19^\circ$$

The rational and irrational numbers together comprise what is called the ***real number system***, and both the rational and irrational numbers are called ***real numbers***.

COMPLEX NUMBERS

Because the square of a real number cannot be negative, the equation

$$x^2 = -1$$

has no solutions in the real number system. In the eighteenth century mathematicians remedied this problem by inventing a new number, which they denoted by

$$i = \sqrt{-1}$$

and which they defined to have the property $i^2 = -1$. This, in turn, led to the development

* HIPPASUS OF METAPONTUM (circa 500 B.C.). A Greek Pythagorean philosopher. According to legend, Hippasus made his discovery at sea and was thrown overboard by fanatic Pythagoreans because his result contradicted their doctrine. The discovery of Hippasus is one of the most fundamental in the entire history of science.

of the ***complex numbers***, which are numbers of the form

$$a + bi$$

where a and b are real numbers. Some examples are

$2 + 3i$	$3 - 4i$	$6i$	$\frac{2}{3}$
$[a = 2, b = 3]$	$[a = 3, b = -4]$	$[a = 0, b = 6]$	$[a = \frac{2}{3}, b = 0]$

Observe that every real number a is also a complex number because it can be written as

$$a = a + 0i$$

Thus, the real numbers are a subset of the complex numbers. Those complex numbers that are not real numbers are called ***imaginary numbers***. Although we will be concerned primarily with real numbers in this text, imaginary numbers will arise in the course of solving equations. For example, the solutions of the quadratic equation

$$ax^2 + bx + c = 0$$

which are given by the ***quadratic formula***

$$x = \frac{-b \pm \sqrt{b^2 - 4ac}}{2a}$$

are imaginary if the quantity $b^2 - 4ac$ is negative.

DIVISION BY ZERO

Division by zero is not allowed in numerical computations because it leads to mathematical inconsistencies. For example, if $1/0$ were assigned some numerical value, say $1/0 = p$, then it would follow that $0 \cdot p = 1$, which is incorrect.

DECIMAL REPRESENTATION OF REAL NUMBERS

Rational and irrational numbers can be distinguished by their decimal representations. Rational numbers have decimals that are ***repeating***, by which we mean that at some point in the decimal some fixed block of numbers begins to repeat indefinitely. For example,

$\frac{4}{3} = 1.333\ldots,$	$\frac{3}{11} = .272727\ldots,$	$\frac{1}{2} = .50000\ldots,$	$\frac{5}{7} = .714285714285714285\ldots$
3 repeats	27 repeats	0 repeats	714285 repeats

Decimals in which zero repeats from some point on are called ***terminating decimals***. For brevity, it is usual to omit the repetitive zeros in terminating decimals and for other repeating decimals to write the repeating digits only once but with a bar over them to indicate the repetition. For example,

$$\frac{1}{2} = .5, \quad \frac{12}{4} = 3, \quad \frac{8}{25} = .32, \quad \frac{4}{3} = 1.\overline{3}, \quad \frac{3}{11} = .\overline{27}, \quad \frac{5}{7} = .\overline{714285}$$

Irrational numbers have nonrepeating decimals, so we can be certain that the decimals

$$\sqrt{2} = 1.414213562373095\ldots \quad \text{and} \quad \pi = 3.141592653589793\ldots$$

do not repeat from some point on. Moreover, if we stop the decimal expansion of an irrational number at some point, we get only an approximation to the number, never an exact value. For example, even if we compute π to 1000 decimal places, as in Figure A.3, we still have only an approximation.

```
3.141592653589793238462643383279502884197169
39937510582097494459230781640628620899862803
48253421170679821480865132823066470938446095
50582231725359408128481117450284102701938521
10555964462294895493038196442881097566593344
61284756482337867831652712019091456485669234
60348610454326648213393607260249141273724587
00660631558817488152092096282925409171536436
78925903600113305305488204665213841469519415
11609433057270365759591953092186117381932611
79310511854807446237996274956735188575272489
12279381830119491298336733624406566430860213
94946395224737190702179860943702770539217176
29317675238467481846766940513200056812714526
35608277857713427577896091736371787214684409
01224953430146549585371050792279689258923542
01995611212902196086403441815981362977477130
99605187072113499999983729780499510597317328
16096318595024459455346908302642522308253344
68503526193118817101000313783875288658753320
83814206171776691473035982534904287554687311
59562863882353787593751957781857780532171226
8066130019278766111959092164201989
```

Figure A.3

REMARK. Beginning mathematics students are sometimes taught to approximate π by $\frac{22}{7}$. Keep in mind, however, that this is only an approximation, since

$$\frac{22}{7} = 3.\overline{142857}$$

is a rational number whose decimal representation begins to differ from π in the third decimal place.

COORDINATE LINES

In 1637 René Descartes* published a philosophical work called *Discourse on the Method of Rightly Conducting the Reason*. In the back of that book was an appendix that the British philosopher John Stuart Mill described as "the greatest single step ever made in the progress of the exact sciences." In that appendix René Descartes linked together algebra and geometry, thereby creating a new subject called ***analytic geometry***; it gave a way of describing algebraic formulas by geometric curves and, conversely, geometric curves by algebraic formulas.

The key step in analytic geometry is to establish a correspondence between real numbers and points on a line. To do this, choose any point on the line as a reference point, and call it the ***origin***; and then arbitrarily choose one of the two directions along the line to be the ***positive direction***, and let the other be the ***negative direction***. It is usual to mark the positive direction with an arrowhead, as in Figure A.4, and to take the positive direction to the right when the line is horizontal. Next, choose a convenient unit of measure, and represent each positive number r by the point that is r units from the origin in the positive direction, each negative number $-r$ by the point that is r units from the origin in the negative direction from the origin, and 0 by the origin itself (Figure A.5). The number associated with a point P is called the ***coordinate*** of P, and the line is called a ***coordinate line***, a ***real number line***, or a ***real line***.

– Origin +

Figure A.4

Figure A.5

INEQUALITY NOTATION

The real numbers can be ordered by size as follows: If $b - a$ is positive, then we write either $a < b$ (read "a is less than b") or $b > a$ (read "b is greater than a"). We write $a \leq b$ to mean $a < b$ or $a = b$, and we write $a < b < c$ to mean that $a < b$ and $b < c$. As one traverses a coordinate line in the positive direction, the real numbers increase in size, so on a horizontal coordinate line the inequality $a < b$ implies that a is to left of b, and the inequalities $a < b < c$ imply that a is to the left of c, and b lies between a and c. The meaning of such symbols as

$$a \leq b < c, \quad a \leq b \leq c, \quad \text{and} \quad a < b < c < d$$

should be clear. For example, you should be able to confirm that all of the following are true statements:

$$3 < 8, \quad -7 < 1.5, \quad -12 \leq -\pi, \quad 5 \leq 5, \quad 0 \leq 2 \leq 4,$$
$$8 \geq 3, \quad 1.5 > -7, \quad -\pi > -12, \quad 5 \geq 5, \quad 3 > 0 > -1 > -3$$

REVIEW OF SETS

In the following discussion we will be concerned with certain sets of real numbers, so it will be helpful to review the basic ideas about sets. Recall that a ***set*** is a collection of objects, called ***elements*** or ***members*** of the set. In this text we will be concerned primarily with sets whose members are numbers or points that lie on a line, a plane, or in three-dimensional

* RENÉ DESCARTES (1596–1650). Descartes, a French aristocrat, was the son of a government official. He graduated from the University of Poitiers with a law degree at age 20. After a brief probe into the pleasures of Paris he became a military engineer, first for the Dutch Prince of Nassau and then for the German Duke of Bavaria. It was during his service as a soldier that Descartes began to pursue mathematics seriously and develop his analytic geometry. After the wars, he returned to Paris where he stalked the city as an eccentric, wearing a sword in his belt and a plumed hat. He lived in leisure, seldom arose before 11 A.M., and dabbled in the study of human physiology, philosophy, glaciers, meteors, and rainbows. He eventually moved to Holland, where he published his *Discourse on the Method*, and finally to Sweden where he died while serving as tutor to Queen Christina. Descartes is regarded as a genius of the first magnitude. In addition to major contributions in mathematics and philosophy, he is considered, along with William Harvey, to be a founder of modern physiology.

space. We will denote sets by capital letters and elements by lowercase letters. To indicate that a is a member of the set A we will write $a \in A$ (read "a belongs to A"), and to indicate that a is not a member of the set A we will write $a \notin A$ (read "a does not belong to A"). For example, if A is the set of positive integers, then $5 \in A$, but $-5 \notin A$. Sometimes sets arise that have no members (e.g., the set of odd integers that are divisible by 2). A set with no members is called an ***empty set*** or a ***null set*** and is denoted by the symbol $\varnothing$.

Some sets can be described by listing their members between braces. The order in which the members are listed does not matter, so, for example, the set A of positive integers that are less than 6 can be expressed as

$$A = \{1, 2, 3, 4, 5\} \quad \text{or} \quad A = \{2, 3, 1, 5, 4\}$$

We can also write A in *set-builder notation* as

$$A = \{x : x \text{ is an integer and } 0 < x < 6\}$$

which is read "A is the set of all x such that x is an integer and $0 < x < 6$." In general, to express a set S in set-builder notation we write $S = \{x : \_\_\_\_\_\_\}$ in which the line is replaced by a property that uniquely defines the set S.

INTERVALS

In calculus we will be concerned with sets of real numbers, called ***intervals***, that correspond to line segments on a coordinate line. For example, if $a < b$, then the ***open interval*** from a to b, denoted by (a, b), is the line segment extending from a to b, *excluding* the endpoints; and the ***closed interval*** from a to b, denoted by $[a, b]$, is the line segment extending from a to b, *including* the endpoints (Figure A.6). These sets can be expressed in set-builder notation as

$$(a, b) = \{x : a < x < b\}$$ The open interval from a to b

$$[a, b] = \{x : a \le x \le b\}$$ The closed interval from a to b

The open interval (a, b)

a b

a b

The closed interval $[a, b]$

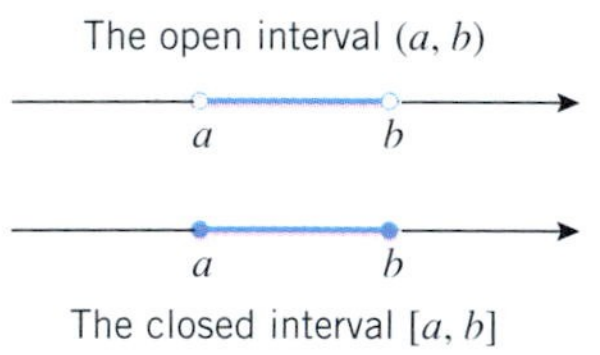

Figure A.6

REMARK. Observe that in this notation and in the corresponding Figure A.6, parentheses and open dots mark endpoints that are excluded from the interval, whereas brackets and closed dots mark endpoints that are included in the interval. Observe also, that in set-builder notation for the intervals, it is understood that x is a real number, even though it is not stated explicitly.

As shown in Table 1, an interval can include one endpoint and not the other; such intervals are called ***half-open*** (or sometimes ***half-closed***). Moreover, the table also shows that it is possible for an interval to extend indefinitely in one or both directions. To indicate that an interval extends indefinitely in the positive direction we write $+\infty$ (read "positive infinity") in place of a right endpoint, and to indicate that an interval extends indefinitely in the negative direction we write $-\infty$ (read "negative infinity") in place of a left endpoint. Intervals that extend between two real numbers are called ***finite intervals***, whereas intervals that extend indefinitely in one or both directions are called ***infinite intervals***.

REMARK. By convention, infinite intervals of the form $[a, +\infty)$ or $(-\infty, b]$ are considered to be closed because they contain their endpoint, and intervals of the form $(a, +\infty)$ and $(-\infty, b)$ are considered to be open because they do not include their endpoint. The interval $(-\infty, +\infty)$, which is the set of all real numbers, has no endpoints and can be regarded as either open or closed, as convenient. This set is often denoted by the special symbol $\mathbb{R}$. To distinguish verbally between the open interval $(0, +\infty) = \{x : x > 0\}$ and the closed interval $[0, +\infty) = \{x : x \ge 0\}$, we will call x ***positive*** if $x > 0$ and ***nonnegative*** if $x \ge 0$. Thus, a positive number must be nonnegative, but a nonnegative number need not be positive, since it might possibly be 0.

Table 1

INTERVAL NOTATION	SET NOTATION	GEOMETRIC PICTURE	CLASSIFICATION
(a, b)	$\{x : a < x < b\}$		Finite; open
$[a, b]$	$\{x : a \leq x \leq b\}$		Finite; closed
$[a, b)$	$\{x : a \leq x < b\}$		Finite; half-open
$(a, b]$	$\{x : a < x \leq b\}$		Finite; half-open
$(-\infty, b]$	$\{x : x \leq b\}$		Infinite; closed
$(-\infty, b)$	$\{x : x < b\}$		Infinite; open
$[a, +\infty)$	$\{x : x \geq a\}$		Infinite; closed
$(a, +\infty)$	$\{x : x > a\}$		Infinite; open
$(-\infty, +\infty)$	$\mathbb{R}$		Infinite; open and closed

UNIONS AND INTERSECTIONS OF INTERVALS

If A and B are sets, then the ***union*** of A and B (denoted by $A \cup B$) is the set whose members belong to A or B (or both), and the ***intersection*** of A and B (denoted by $A \cap B$) is the set whose members belong to both A and B. For example,

$$\{x : 0 < x < 5\} \cup \{x : 1 < x < 7\} = \{x : 0 < x < 7\}$$

$$\{x : x < 1\} \cap \{x : x \geq 0\} = \{x : 0 \leq x < 1\}$$

$$\{x : x < 0\} \cap \{x : x > 0\} = \varnothing$$

or in interval notation,

$$(0, 5) \cup (1, 7) = (0, 7)$$

$$(-\infty, 1) \cap [0, +\infty) = [0, 1)$$

$$(-\infty, 0) \cap (0, +\infty) = \varnothing$$

ALGEBRAIC PROPERTIES OF INEQUALITIES

The following algebraic properties of inequalities will be used frequently in this text. We omit the proofs.

> **A.1** THEOREM (***Properties of Inequalities***). *Let a, b, c, and d be real numbers.*
>
> (*a*) *If $a < b$ and $b < c$, then $a < c$.*
>
> (*b*) *If $a < b$, then $a + c < b + c$ and $a - c < b - c$.*
>
> (*c*) *If $a < b$, then $ac < bc$ when c is positive and $ac > bc$ when c is negative.*
>
> (*d*) *If $a < b$ and $c < d$, then $a + c < b + d$.*
>
> (*e*) *If a and b are both positive or both negative and $a < b$, then $1/a > 1/b$.*

If we call the direction of an inequality its *sense*, then these properties can be paraphrased as follows:

(*b*) *The sense of an inequality is unchanged if the same number is added to or subtracted from both sides.*

(*c*) *The sense of an inequality is unchanged if both sides are multiplied by the same positive number, but the sense is reversed if both sides are multiplied by the same negative number.*

(*d*) *Inequalities with the same sense can be added.*

(*e*) *If both sides of an inequality have the same sign, then the sense of the inequality is reversed by taking the reciprocal of each side.*

REMARK. These properties remain true if the symbols $<$ and $>$ are replaced by $\leq$ and $\geq$ in Theorem A.1.

Example 1

STARTING INEQUALITY	OPERATION	RESULTING INEQUALITY
$-2 < 6$	Add 7 to both sides.	$5 < 13$
$-2 < 6$	Subtract 8 from both sides.	$-10 < -2$
$-2 < 6$	Multiply both sides by 3.	$-6 < 18$
$-2 < 6$	Multiply both sides by -3.	$6 > -18$
$3 < 7$	Multiply both sides by 4.	$12 < 28$
$3 < 7$	Multiply both sides by -4.	$-12 > -28$
$3 < 7$	Take reciprocals of both sides.	$\frac{1}{3} > \frac{1}{7}$
$-8 < -6$	Take reciprocals of both sides.	$-\frac{1}{8} > -\frac{1}{6}$
$4 < 5, -7 < 8$	Add corresponding sides.	$-3 < 13$

◀

SOLVING INEQUALITIES

A ***solution*** of an inequality in an unknown x is a value for x that makes the inequality a true statement. For example, $x = 1$ is a solution of the inequality $x < 5$, but $x = 7$ is not. The set of all solutions of an inequality is called its ***solution set***. It can be shown that if one does not multiply both sides of an inequality by zero or an expression involving an unknown, then the operations in Theorem A.1 will not change the solution set of the inequality. The process of finding the solution set of an inequality is called ***solving*** the inequality.

Example 2

Solve $3 + 7x \leq 2x - 9$.

Solution. We will use the operations of Theorem A.1 to isolate x on one side of the inequality.

$$3 + 7x \leq 2x - 9 \qquad \text{Given.}$$

$$7x \leq 2x - 12 \qquad \text{We subtracted 3 from both sides.}$$

$$5x \leq -12 \qquad \text{We subtracted } 2x \text{ from both sides.}$$

$$x \leq -\tfrac{12}{5} \qquad \text{We multiplied both sides by } \tfrac{1}{5}.$$

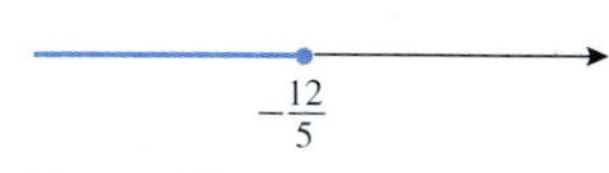

Figure A.7

Because we have not multiplied by any expressions involving the unknown x, the last inequality has the same solution set as the first. Thus, the solution set is the interval $\left(-\infty, -\frac{12}{5}\right]$ shown in Figure A.7. ◀

Example 3

Solve $7 \leq 2 - 5x < 9$.

Solution. The given inequality is actually a combination of the two inequalities

$$7 \leq 2 - 5x \quad \text{and} \quad 2 - 5x < 9$$

We could solve the two inequalities separately, then determine the values of x that satisfy both by taking the intersection of the two solution sets. However, it is possible to work with the combined inequalities in this problem:

$7 \le 2 - 5x < 9$	Given.
$5 \le -5x < 7$	We subtracted 2 from each member.
$-1 \ge x > -\dfrac{7}{5}$	We multiplied by $-\frac{1}{5}$ and reversed the sense of the inequalities.
$-\dfrac{7}{5} < x \le -1$	For clarity, we rewrote the inequalities with the smaller number on the left.

$-\frac{7}{5}$ -1

Figure A.8

Thus, the solution set is the interval $\left(-\frac{7}{5}, -1\right]$ shown in Figure A.8. ◀

Example 4

Solve $x^2 - 3x > 10$.

Solution. By subtracting 10 from both sides, the inequality can be rewritten as

$$x^2 - 3x - 10 > 0$$

Factoring the left side yields

$$(x+2)(x-5) > 0$$

The values of x for which $x + 2 = 0$ or $x - 5 = 0$ are $x = -2$ and $x = 5$. These points divide the coordinate line into three open intervals,

$$(-\infty, -2), \quad (-2, 5), \quad (5, +\infty)$$

on each of which the product $(x+2)(x-5)$ has constant sign. To determine those signs we will choose an *arbitrary* point in each interval at which we will determine the sign; these are called ***test points***. As shown in Figure A.9, we will use -3, 0, and 6 as our test points. The results can be organized as follows:

INTERVAL	TEST POINT	SIGN OF $(x+2)(x-5)$ AT THE TEST POINT
$(-\infty, -2)$	-3	$(-)(-) = +$
$(-2, 5)$	0	$(+)(-) = -$
$(5, +\infty)$	6	$(+)(+) = +$

The pattern of signs in the intervals is shown on the number line in the middle of Figure A.9. We deduce that the solution set is $(-\infty, -2) \cup (5, +\infty)$, which is shown at the bottom of Figure A.9. ◀

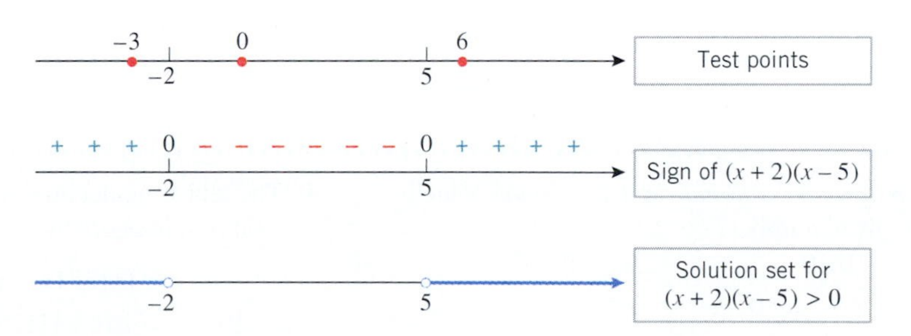

Figure A.9

Example 5

Solve $\dfrac{2x-5}{x-2} < 1$.

Solution. We could start by multiplying both sides by $x - 2$ to eliminate the fraction. However, this would require us to consider the cases $x - 2 > 0$ and $x - 2 < 0$ separately because the sense of the inequality would be reversed in the second case, but not the first. The following approach is simpler:

$$\frac{2x-5}{x-2} < 1 \quad \boxed{\text{Given.}}$$

$$\frac{2x-5}{x-2} - 1 < 0 \quad \boxed{\text{We subtracted 1 from both sides to obtain a 0 on the right.}}$$

$$\frac{(2x-5)-(x-2)}{x-2} < 0 \quad \boxed{\text{We combined terms.}}$$

$$\frac{x-3}{x-2} < 0 \quad \boxed{\text{We simplified.}}$$

The quantity $x - 3$ is zero if $x = 3$, and the quantity $x - 2$ is zero if $x = 2$. These points divide the coordinate line into three open intervals,

$$(-\infty, 2), \quad (2, 3), \quad (3, +\infty)$$

on each of which the quotient $(x - 3)/(x - 2)$ has constant sign. Using 0, 2.5, and 4 as test points (Figure A.10), we obtain the following results:

INTERVAL	TEST POINT	SIGN OF $(x-3)(x-2)$ AT THE TEST POINT
$(-\infty, 2)$	0	$(-)/(-) = +$
$(2, 3)$	2.5	$(-)/(+) = -$
$(3, +\infty)$	4	$(+)/(+) = +$

The signs of the quotient are shown in the middle of Figure A.10. From the figure we see that the solution set consists of all real values of x such that $2 < x < 3$. This is the interval $(2, 3)$ shown at the bottom of Figure A.10. ◀

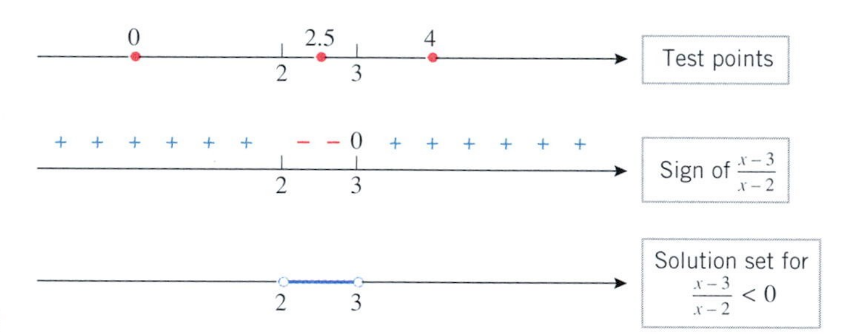

Figure A.10

Exercise Set A

1. Among the terms *integer*, *rational*, and *irrational*, which ones apply to the given number?
(a) $-\frac{3}{4}$ (b) 0 (c) $\frac{24}{8}$
(d) 0.25 (e) $-\sqrt{16}$ (f) $2^{1/2}$
(g) 0.020202 . . . (h) 7.000 . . .

2. Which of the terms *integer*, *rational*, and *irrational* apply to the given number?
(a) 0.31311311131111 . . . (b) 0.729999 . . .
(c) 0.376237623762 . . . (d) $17\frac{4}{5}$

3. The repeating decimal 0.137137137 . . . can be expressed as a ratio of integers by writing

$$x = 0.137137137\ldots$$
$$1000x = 137.137137137\ldots$$

and subtracting to obtain $999x = 137$ or $x = \frac{137}{999}$. Use this idea, where needed, to express the following decimals as ratios of integers.
(a) 0.123123123 . . . (b) 12.7777 . . .
(c) 38.07818181 . . . (d) 0.4296000 . . .

4. Show that the repeating decimal 0.99999... represents the number 1. Since 1.000... is also a decimal representation of 1, this problem shows that a real number can have two different decimal representations. [*Hint:* Use the technique of Exercise 3.]

5. The Rhind Papyrus, which is a fragment of Egyptian mathematical writing from about 1650 B.C., is one of the oldest known examples of written mathematics. It is stated in the papyrus that the area A of a circle is related to its diameter D by

$$A = \left(\tfrac{8}{9}D\right)^2$$

(a) What approximation to π were the Egyptians using?
(b) Use a calculating utility to determine if this approximation is better or worse than the approximation of $\frac{22}{7}$.

6. The following are all famous approximations to π:

$\frac{333}{106}$ — Adrian Athoniszoon, c. 1583

$\frac{355}{113}$ — Tsu Chung-Chi and others

$\frac{63}{25}\left(\frac{17+15\sqrt{5}}{7+15\sqrt{5}}\right)$ — Ramanujan

$\frac{22}{7}$ — Archimedes

$\frac{223}{71}$ — Archimedes

(a) Use a calculating utility to order these approximations according to size.
(b) Which of these approximations is closest to but larger than π?
(c) Which of these approximations is closest to but smaller than π?
(d) Which of these approximations is most accurate?

7. In each line of the table in the accompanying figure, check the blocks, if any, that describe a valid relationship between the real numbers a and b. The first line is already completed as an illustration.

a	b	$a < b$	$a \le b$	$a > b$	$a \ge b$	$a = b$
1	6	✓	✓			
6	1					
−3	5					
5	−3					
−4	−4					
0.25	$\frac{1}{3}$					
$-\frac{1}{4}$	$-\frac{3}{4}$					

Figure Ex-7

8. In each line of the table in the accompanying figure, check the blocks, if any, that describe a valid relationship between the real numbers a, b, and c.

a	b	c	$a < b < c$	$a \le b \le c$	$a < b \le c$	$a \le b < c$
−1	0	2				
2	4	−3				
$\frac{1}{2}$	$\frac{1}{2}$	$\frac{3}{4}$				
−5	−5	−5				
0.75	1.25	1.25				

Figure Ex-8

9. Which of the following are always correct if $a \le b$?
(a) $a - 3 \le b - 3$ (b) $-a \le -b$
(c) $3 - a \le 3 - b$ (d) $6a \le 6b$
(e) $a^2 \le ab$ (f) $a^3 \le a^2 b$

10. Which of the following are always correct if $a \le b$ and $c \le d$?
(a) $a + 2c \le b + 2d$ (b) $a - 2c \le b - 2d$
(c) $a - 2c \ge b - 2d$

11. For what values of a are the following inequalities valid?
(a) $a \le a$ (b) $a < a$

12. If $a \le b$ and $b \le a$, what can you say about a and b?

13. (a) If $a < b$ is true, does it follow that $a \le b$ must also be true?
(b) If $a \le b$ is true, does it follow that $a < b$ must also be true?

14. In each part, list the elements in the set.
(a) $\{x : x^2 - 5x = 0\}$
(b) $\{x : x \text{ is an integer satisfying } -2 < x < 3\}$

15. In each part, express the set in the notation $\{x : \_\_\_\_\_\_\_\_\}$.
(a) $\{1, 3, 5, 7, 9, \ldots\}$
(b) the set of even integers
(c) the set of irrational numbers
(d) $\{7, 8, 9, 10\}$

16. Let $A = \{1, 2, 3\}$. Which of the following sets are equal to A?
(a) $\{0, 1, 2, 3\}$ (b) $\{3, 2, 1\}$
(c) $\{x : (x - 3)(x^2 - 3x + 2) = 0\}$

17. In the accompanying figure, let

S = the set of points inside the square
T = the set of points inside the triangle
C = the set of points inside the circle

and let a, b, and c be the points shown. Answer the following as true or false.

(a) $T \subset C$ (b) $T \subset S$
(c) $a \notin T$ (d) $a \notin S$
(e) $b \in T$ and $b \in C$ (f) $a \in C$ or $a \in T$
(g) $c \in T$ and $c \notin C$

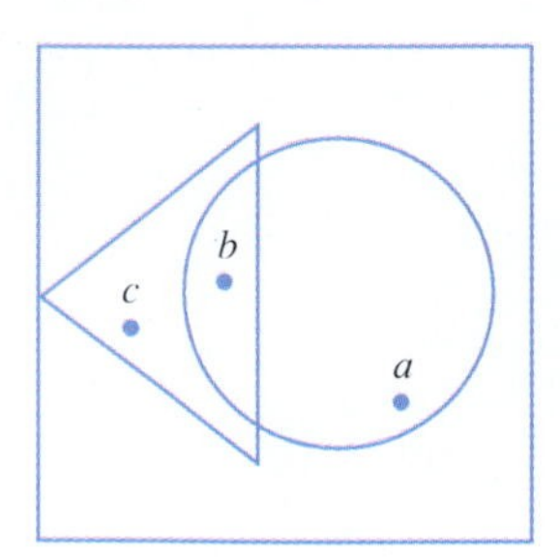

Figure Ex-17

18. List all subsets of
(a) $\{a_1, a_2, a_3\}$ (b) $\varnothing$.

19. In each part, sketch on a coordinate line all values of x that satisfy the stated condition.
(a) $x \le 4$ (b) $x \ge -3$ (c) $-1 \le x \le 7$
(d) $x^2 = 9$ (e) $x^2 \le 9$ (f) $x^2 \ge 9$

20. In parts (a)–(d), sketch on a coordinate line all values of x, if any, that satisfy the stated conditions.
(a) $x > 4$ and $x \le 8$
(b) $x \le 2$ or $x \ge 5$
(c) $x > -2$ and $x \ge 3$
(d) $x \le 5$ and $x > 7$

21. Express in interval notation.
(a) $\{x : x^2 \le 4\}$ (b) $\{x : x^2 > 4\}$

22. In each part, sketch the set on a coordinate line.
(a) $[-3, 2] \cup [1, 4]$ (b) $[4, 6] \cup [8, 11]$
(c) $(-4, 0) \cup (-5, 1)$ (d) $[2, 4) \cup (4, 7)$
(e) $(-2, 4) \cap (0, 5]$ (f) $[1, 2.3) \cup (1.4, \sqrt{2})$
(g) $(-\infty, -1) \cup (-3, +\infty)$ (h) $(-\infty, 5) \cap [0, +\infty)$

In Exercises 23–44, solve the inequality and sketch the solution on a coordinate line.

23. $3x - 2 < 8$
24. $\frac{1}{5}x + 6 \ge 14$
25. $4 + 5x \le 3x - 7$
26. $2x - 1 > 11x + 9$
27. $3 \le 4 - 2x < 7$
28. $-2 \ge 3 - 8x \ge -11$
29. $\dfrac{x}{x-3} < 4$
30. $\dfrac{x}{8-x} \ge -2$
31. $\dfrac{3x+1}{x-2} < 1$
32. $\dfrac{\frac{1}{2}x - 3}{4+x} > 1$
33. $\dfrac{4}{2-x} \le 1$
34. $\dfrac{3}{x-5} \le 2$
35. $x^2 > 9$
36. $x^2 \le 5$
37. $(x-4)(x+2) > 0$
38. $(x-3)(x+4) < 0$
39. $x^2 - 9x + 20 \le 0$
40. $2 - 3x + x^2 \ge 0$
41. $\dfrac{2}{x} < \dfrac{3}{x-4}$
42. $\dfrac{1}{x+1} \ge \dfrac{3}{x-2}$
43. $x^3 - x^2 - x - 2 > 0$
44. $x^3 - 3x + 2 \le 0$

In Exercises 45 and 46, find all values of x for which the given expression yields a real number.

45. $\sqrt{x^2 + x - 6}$
46. $\sqrt{\dfrac{x+2}{x-1}}$

47. Fahrenheit and Celsius temperatures are related by the formula $C = \frac{5}{9}(F - 32)$. If the temperature in degrees Celsius ranges over the interval $25 \le C \le 40$ on a certain day, what is the temperature range in degrees Fahrenheit that day?

48. Every integer is either even or odd. The even integers are those that are divisible by 2, so n is even if and only if $n = 2k$ for some integer k. Each odd integer is one unit larger than an even integer, so n is odd if and only if $n = 2k + 1$ for some integer k. Show:
(a) If n is even, then so is n^2
(b) If n is odd, then so is n^2.

49. Prove the following results about sums of rational and irrational numbers:
(a) rational + rational = rational
(b) rational + irrational = irrational.

50. Prove the following results about products of rational and irrational numbers:
(a) rational · rational = rational
(b) rational · irrational = irrational (provided the rational factor is nonzero).

51. Show that the sum or product of two irrational numbers can be rational or irrational.

52. Classify the following as rational or irrational and justify your conclusion.
(a) $3 + \pi$ (b) $\frac{3}{4}\sqrt{2}$
(c) $\sqrt{8}\sqrt{2}$ (d) $\sqrt{\pi}$
(See Exercises 49 and 50.)

53. Prove: The average of two rational numbers is a rational number, but the average of two irrational numbers can be rational or irrational.

54. Can a rational number satisfy $10^x = 3$?

55. Solve: $8x^3 - 4x^2 - 2x + 1 < 0$.

56. Solve: $12x^3 - 20x^2 \ge -11x + 2$.

57. Prove: If a, b, c, and d are positive numbers such that $a < b$ and $c < d$, then $ac < bd$. (This result gives conditions under which inequalities can be "multiplied together.")

58. Is the number represented by the decimal

$$0.101001000100001000001\ldots$$

rational or irrational? Explain your reasoning.

APPENDIX B

Absolute Value

ABSOLUTE VALUE

> **B.1 DEFINITION.** The ***absolute value*** or ***magnitude*** of a real number a is denoted by $|a|$ and is defined by
>
> $$|a| = \begin{cases} a & \text{if} \quad a \geq 0 \\ -a & \text{if} \quad a < 0 \end{cases}$$

Example 1

$$|5| = 5 \qquad \left|-\tfrac{4}{7}\right| = -\left(-\tfrac{4}{7}\right) = \tfrac{4}{7} \qquad |0| = 0$$

Since $5 > 0$ Since $-\frac{4}{7} < 0$ Since $0 \geq 0$ ◀

Note that the effect of taking the absolute value of a number is to strip away the minus sign if the number is negative and to leave the number unchanged if it is nonnegative.

Example 2

Solve $|x - 3| = 4$.

Solution. Depending on whether $x - 3$ is positive or negative, the equation $|x - 3| = 4$ can be written as

$$x - 3 = 4 \quad \text{or} \quad x - 3 = -4$$

Solving these two equations gives $x = 7$ and $x = -1$. ◀

Example 3

Solve $|3x - 2| = |5x + 4|$.

Solution. Because two numbers with the same absolute value are either equal or differ in sign, the given equation will be satisfied if either

$$3x - 2 = 5x + 4 \quad \text{or} \quad 3x - 2 = -(5x + 4)$$

Solving the first equation yields $x = -3$ and solving the second yields $x = -\frac{1}{4}$; thus, the given equation has the solutions $x = -3$ and $x = -\frac{1}{4}$. ◀

RELATIONSHIP BETWEEN SQUARE ROOTS AND ABSOLUTE VALUES

Recall from algebra that a number is called a ***square root*** of a if its square is a. Recall also that every positive real number has two square roots, one positive and one negative; the positive square root is denoted by $\sqrt{a}$ and the negative square root by $-\sqrt{a}$. For example, the positive square root of 9 is $\sqrt{9} = 3$, and the negative square root of 9 is $-\sqrt{9} = -3$.

REMARK. Readers who may have been taught to write $\sqrt{9} = \pm 3$ should stop doing so, since it is incorrect.

It is a common error to write $\sqrt{a^2} = a$. Although this equality is correct when a is nonnegative, it is false for negative a. For example, if $a = -4$, then

$$\sqrt{a^2} = \sqrt{(-4)^2} = \sqrt{16} = 4 \neq a$$

A result that is correct for all a is given in the following theorem.

B.2 THEOREM. *For any real number a,*

$$\sqrt{a^2} = |a|$$

Proof. Since $a^2 = (+a)^2 = (-a)^2$, the numbers $+a$ and $-a$ are square roots of a^2. If $a \geq 0$, then $+a$ is the nonnegative square root of a^2, and if $a < 0$, then $-a$ is the nonnegative square root of a^2. Since $\sqrt{a^2}$ denotes the nonnegative square root of a^2, it follows that

$$\sqrt{a^2} = +a \quad \text{if} \quad a \geq 0$$
$$\sqrt{a^2} = -a \quad \text{if} \quad a < 0$$

That is, $\sqrt{a^2} = |a|$. ∎

PROPERTIES OF ABSOLUTE VALUE

B.3 THEOREM. *If a and b are real numbers, then*

(*a*) $|-a| = |a|$ — A number and its negative have the same absolute value.

(*b*) $|ab| = |a||b|$ — The absolute value of a product is the product of the absolute values.

(*c*) $|a/b| = |a|/|b|$ — The absolute value of a ratio is the ratio of the absolute values.

We will prove parts (*a*) and (*b*) only.

Proof (a). From Theorem B.2,

$$|-a| = \sqrt{(-a)^2} = \sqrt{a^2} = |a|$$

Proof (b). From Theorem B.2 and a basic property of square roots,

$$|ab| = \sqrt{(ab)^2} = \sqrt{a^2 b^2} = \sqrt{a^2}\sqrt{b^2} = |a||b|$$ ∎

REMARK. In part (*c*) of Theorem B.3 we did not explicitly state that $b \neq 0$, but this must be so since division by zero is not allowed. Whenever divisions occur in this text, it will be assumed that the denominator is not zero, even if we do not mention it explicitly.

The result in part (*b*) of Theorem B.3 can be extended to three or more factors. More precisely, for any n real numbers, $a_1, a_2, \ldots, a_n$, it follows that

$$|a_1 a_2 \cdots a_n| = |a_1||a_2| \cdots |a_n| \tag{1}$$

In the special case where $a_1, a_2, \ldots, a_n$ have the same value, a, it follows from (1) that

$$|a^n| = |a|^n \tag{2}$$

GEOMETRIC INTERPRETATION OF ABSOLUTE VALUE

The notion of absolute value arises naturally in distance problems. For example, suppose that A and B are points on a coordinate line that have coordinates a and b, respectively. Depending on the relative positions of the points, the distance d between them will be $b - a$ or $a - b$ (Figure B.1). In either case, the distance can be written as $d = |b - a|$, so we have the following result.

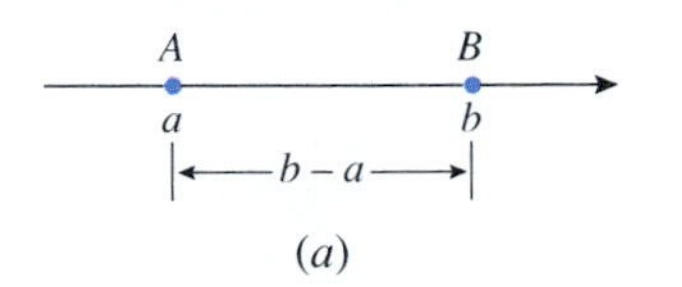

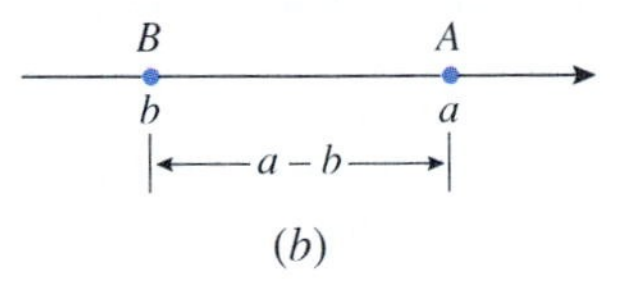

Figure B.1

> **B.4 THEOREM (*Distance Formula*).** *If A and B are points on a coordinate line with coordinates a and b, respectively, then the distance d between A and B is $d = |b - a|$.*

This theorem provides useful geometric interpretations of some common mathematical expressions:

EXPRESSION	GEOMETRIC INTERPRETATION ON A COORDINATE LINE
$\|x - a\|$	The distance between x and a
$\|x + a\|$	The distance between x and $-a$ (since $\|x + a\| = \|x - (-a)\|$)
$\|x\|$	The distance between x and the origin (since $\|x\| = \|x - 0\|$)

INEQUALITIES WITH ABSOLUTE VALUES

Inequalities of the form $|x - a| < k$ and $|x - a| > k$ arise so often that we have summarized the key facts about them in Table 1.

Table 1

INEQUALITY ($k > 0$)	GEOMETRIC INTERPRETATION	FIGURE	ALTERNATIVE FORMS OF THE INEQUALITY
$\|x - a\| < k$	x is within k units of a.	k units, k units; $a - k$, a, x, $a + k$	$-k < x - a < k$ $a - k < x < a + k$
$\|x - a\| > k$	x is more than k units away from a.	k units, k units; $a - k$, a, $a + k$, x	$x - a < -k$ or $x - a > k$ $x < a - k$ or $x > a + k$

REMARK. The statements in this table remain true if $<$ is replaced by $\leq$ and $>$ by $\geq$, and if the open dots are replaced by closed dots in the illustrations.

Example 4

Solve

(a) $|x - 3| < 4$ (b) $|x + 4| \geq 2$ (c) $\dfrac{1}{|2x - 3|} > 5$

Solution (a). The inequality $|x - 3| < 4$ can be rewritten as

$$-4 < x - 3 < 4$$

Adding 3 throughout yields

$$-1 < x < 7$$

which can be written in interval notation as $(-1, 7)$. Observe that this solution set consists of all x that are within 4 units of 3 on a number line (Figure B.2), which is consistent with Table 1.

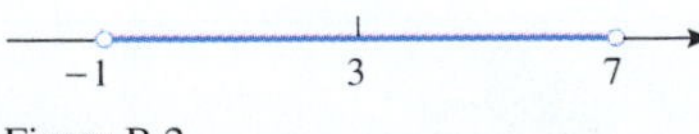

Figure B.2

Solution (b). The inequality $|x+4| \geq 2$ will be satisfied if

$$x+4 \leq -2 \quad \text{or} \quad x+4 \geq 2$$

Solving for x in the two cases yields

$$x \leq -6 \quad \text{or} \quad x \geq -2$$

which can be expressed in interval notation as

$$(-\infty, -6] \cup [-2, +\infty)$$

Figure B.3

Observe that the solution set consists of all x that are at least 2 units away from -4 on a number line (Figure B.3), which is consistent with Table 1 and the remark that follows it.

Solution (c). Observe first that $x = \frac{3}{2}$ results in a division by zero, so this value of x cannot be in the solution set. Putting this aside for the moment, we will begin by taking reciprocals on both sides and reversing the sense of the inequality in accordance with Theorem A.1(d) of Appendix A; then we will use Theorem B.3 to rewrite the inequality $1/|2x-3| > 5$ in a more familiar form:

$$|2x-3| < \tfrac{1}{5}$$

$$|2||x-\tfrac{3}{2}| < \tfrac{1}{5} \qquad \text{Theorem B.3}(b)$$

$$|x-\tfrac{3}{2}| < \tfrac{1}{10} \qquad \text{We multiplied both sides by } 1/|2| = 1/2.$$

$$-\tfrac{1}{10} < x-\tfrac{3}{2} < \tfrac{1}{10} \qquad \text{Table 1}$$

$$\tfrac{7}{5} < x < \tfrac{8}{5} \qquad \text{We added } 3/2 \text{ throughout.}$$

As noted earlier, we must eliminate $x = \frac{3}{2}$ to avoid a division by zero, so the solution set is

$$\tfrac{7}{5} < x < \tfrac{3}{2} \quad \text{or} \quad \tfrac{3}{2} < x < \tfrac{8}{5}$$

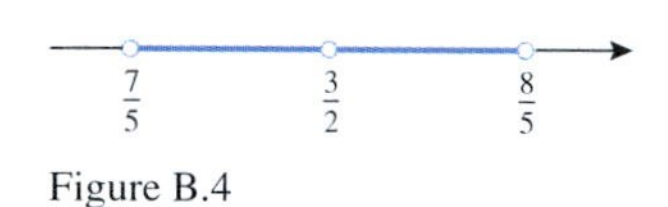

Figure B.4

which can be expressed in interval notation as $\left(\frac{7}{5}, \frac{3}{2}\right) \cup \left(\frac{3}{2}, \frac{8}{5}\right)$. (See Figure B.4.) ◀

AN INEQUALITY FROM CALCULUS

One of the most important inequalities in calculus is

$$0 < |x-a| < \delta \tag{3}$$

where δ (Greek "delta") is a positive real number. This is equivalent to the two inequalities

$$0 < |x-a| \quad \text{and} \quad |x-a| < \delta$$

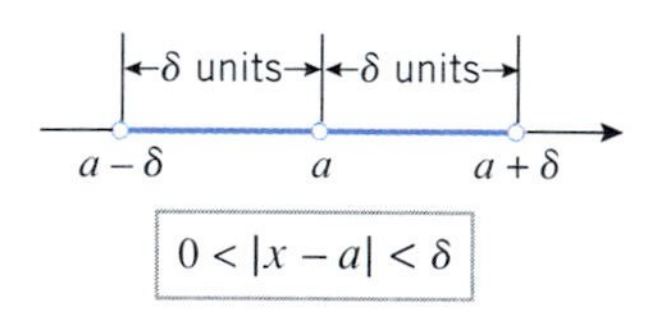

Figure B.5

the first of which is satisfied by all x except $x = a$, and the second of which is satisfied by all x that are within δ units of a on a coordinate line. Combining these two restrictions, we conclude that the solution set of (3) consists of all x in the interval $(a-\delta, a+\delta)$ except $x = a$ (Figure B.5). Stated another way, the solution set of (3) is

$$(a-\delta, a) \cup (a, a+\delta) \tag{4}$$

THE TRIANGLE INEQUALITY

It is *not* generally true that $|a+b| = |a|+|b|$. For example, if $a = 1$ and $b = -1$, then $|a+b| = 0$, whereas $|a|+|b| = 2$. It is true, however, that *the absolute value of a sum is always less than or equal to the sum of the absolute values*. This is the content of the following useful theorem, called the ***triangle inequality***.

B.5 THEOREM (*Triangle Inequality*). *If a and b are any real numbers, then*

$$|a+b| \leq |a|+|b| \tag{5}$$

Proof. Observe first that a satisfies the inequality

$$-|a| \leq a \leq |a|$$

because either $a = |a|$ or $a = -|a|$, depending on the sign of a. The corresponding inequality for b is

$$-|b| \le b \le |b|$$

Adding the two inequalities we obtain

$$-(|a| + |b|) \le a + b \le (|a| + |b|) \tag{6}$$

Let us now consider the cases $a + b \ge 0$ and $a + b < 0$ separately. In the first case, $a + b = |a + b|$, so the right-hand inequality in (6) yields the triangle inequality (5). In the second case, $a + b = -|a + b|$, so the left-hand inequality in (6) can be written as

$$-(|a| + |b|) \le -|a + b|$$

which yields the triangle inequality (5) on multiplying by -1. ■

REMARK. The name "triangle inequality" arises from a geometric interpretation of the inequality that can be made when a and b are complex numbers. A more detailed explanation is outside the scope of this text.

EXERCISE SET B

1. Compute $|x|$ if
(a) $x = 7$ (b) $x = -\sqrt{2}$
(c) $x = k^2$ (d) $x = -k^2$.

2. Rewrite $\sqrt{(x-6)^2}$ without using a square root or absolute value sign.

In Exercises 3–10, find all values of x for which the given statement is true.

3. $|x - 3| = 3 - x$
4. $|x + 2| = x + 2$
5. $|x^2 + 9| = x^2 + 9$
6. $|x^2 + 5x| = x^2 + 5x$
7. $|3x^2 + 2x| = x|3x + 2|$
8. $|6 - 2x| = 2|x - 3|$
9. $\sqrt{(x+5)^2} = x + 5$
10. $\sqrt{(3x-2)^2} = 2 - 3x$

11. Verify $\sqrt{a^2} = |a|$ for $a = 7$ and $a = -7$.

12. Verify the inequalities $-|a| \le a \le |a|$ for $a = 2$ and for $a = -5$.

13. Let A and B be points with coordinates a and b. In each part find the distance between A and B.
(a) $a = 9,\ b = 7$ (b) $a = 2,\ b = 3$
(c) $a = -8,\ b = 6$ (d) $a = \sqrt{2},\ b = -3$
(e) $a = -11,\ b = -4$ (f) $a = 0,\ b = -5$

14. Is the equality $\sqrt{a^4} = a^2$ valid for all values of a? Explain.

15. Let A and B be points with coordinates a and b. In each part, use the given information to find b.
(a) $a = -3$, B is to the left of A, and $|b - a| = 6$.
(b) $a = -2$, B is to the right of A, and $|b - a| = 9$.
(c) $a = 5$, $|b - a| = 7$, and $b > 0$.

16. Let E and F be points with coordinates e and f. In each part, determine whether E is to the left or to the right of F on a coordinate line.
(a) $f - e = 4$ (b) $e - f = 4$
(c) $f - e = -6$ (d) $e - f = -7$

In Exercises 17–24, solve for x.

17. $|6x - 2| = 7$
18. $|3 + 2x| = 11$
19. $|6x - 7| = |3 + 2x|$
20. $|4x + 5| = |8x - 3|$
21. $|9x| - 11 = x$
22. $2x - 7 = |x + 1|$
23. $\left|\dfrac{x+5}{2-x}\right| = 6$
24. $\left|\dfrac{x-3}{x+4}\right| = 5$

In Exercises 25–36, solve for x and express the solution in terms of intervals.

25. $|x + 6| < 3$
26. $|7 - x| \le 5$
27. $|2x - 3| \le 6$
28. $|3x + 1| < 4$
29. $|x + 2| > 1$
30. $|\tfrac{1}{2}x - 1| \ge 2$
31. $|5 - 2x| \ge 4$
32. $|7x + 1| > 3$
33. $\dfrac{1}{|x-1|} < 2$
34. $\dfrac{1}{|3x+1|} \ge 5$
35. $\dfrac{3}{|2x-1|} \ge 4$
36. $\dfrac{2}{|x+3|} < 1$

37. For which values of x is $\sqrt{(x^2 - 5x + 6)^2} = x^2 - 5x + 6$?

38. Solve $3 \le |x - 2| \le 7$ for x.

39. Solve $|x - 3|^2 - 4|x - 3| = 12$ for x. [*Hint:* Begin by letting $u = |x - 3|$.]

40. Verify the triangle inequality $|a + b| \le |a| + |b|$ (Theorem B.5) for
(a) $a = 3,\ b = 4$ (b) $a = -2,\ b = 6$
(c) $a = -7,\ b = -8$ (d) $a = -4,\ b = 4$.

41. Prove: $|a - b| \le |a| + |b|$.

42. Prove: $|a| - |b| \le |a - b|$.

43. Prove: $\big|\,|a| - |b|\,\big| \le |a - b|$. [*Hint:* Use Exercise 42.]

APPENDIX C

Coordinate Planes and Lines

RECTANGULAR COORDINATE SYSTEMS

Just as points on a coordinate line can be associated with real numbers, so points in a plane can be associated with pairs of real numbers by introducing a ***rectangular coordinate system*** (also called a ***Cartesian coordinate system***). A rectangular coordinate system consists of two perpendicular coordinate lines, called ***coordinate axes***, that intersect at their origins. Usually, but not always, one axis is horizontal with its positive direction to the right, and the other is vertical with its positive direction up. The intersection of the axes is called the ***origin*** of the coordinate system.

It is common to call the horizontal axis the ***x-axis*** and the vertical axis the ***y-axis***, in which case the plane and the axes together are referred to as the ***xy-plane*** (Figure C.1). Although labeling the axes with the letters x and y is common, other letters may be more appropriate in specific applications. Figure C.2 shows a uv-plane and a ts-plane—the first letter in the name of the plane always refers to the horizontal axis and the second to the vertical axis.

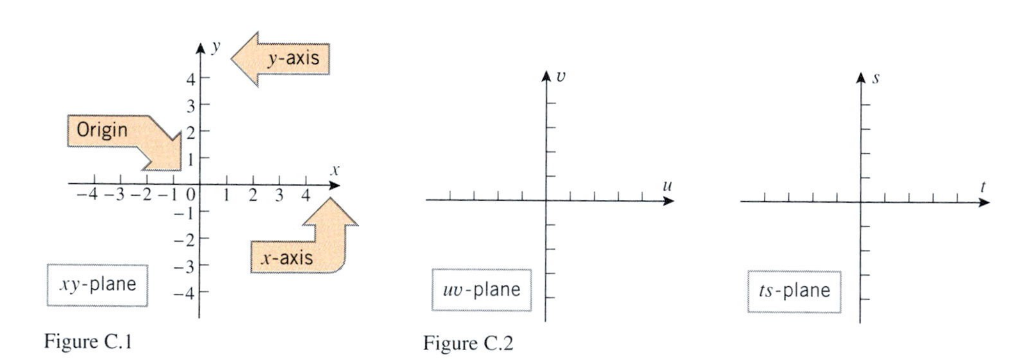

Figure C.1

Figure C.2

COORDINATES

Every point P in a coordinate plane can be associated with a unique ordered pair of real numbers by drawing two lines through P, one perpendicular to the x-axis and the other perpendicular to the y-axis (Figure C.3). If the first line intersects the x-axis at the point with coordinate a and the second line intersects the y-axis at the point with coordinate b, then we associate the ordered pair of real numbers (a, b) with the point P. The number a is called the ***x-coordinate*** or ***abscissa*** of P and the number b is called the ***y-coordinate*** or ***ordinate*** of P. We will say that P has ***coordinates*** (a, b) and write $P(a, b)$ when we want to emphasize that the coordinates of P are (a, b). We can also reverse the above procedure and find the point P associated with the coordinates (a, b) by locating the intersection of the dashed

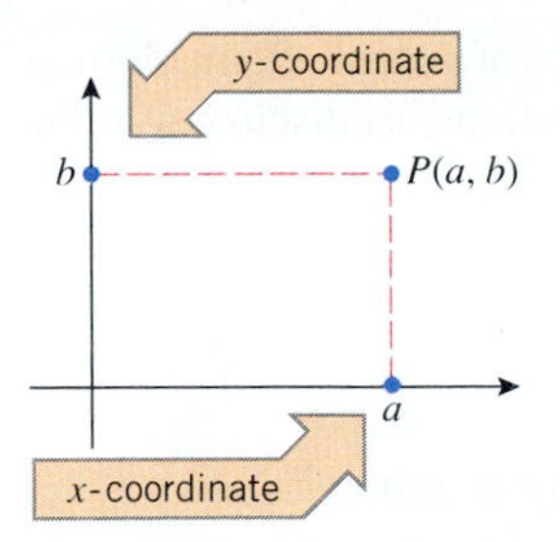

Figure C.3

lines in Figure C.3. Because of this one-to-one correspondence between coordinates and points, we will sometimes blur the distinction between points and ordered pairs of numbers by talking about the *point* (a, b).

REMARK. Recall that the symbol (a, b) also denotes the open interval between a and b; the appropriate interpretation will usually be clear from the context.

In a rectangular coordinate system the coordinate axes divide the plane into four regions called ***quadrants***. These are numbered counterclockwise with roman numerals as shown in Figure C.4. As indicated in that figure, it is easy to determine the quadrant in which a given point lies from the signs of its coordinates: a point with two positive coordinates $(+, +)$ lies in Quadrant I, a point with a negative x-coordinate and a positive y-coordinate $(-, +)$ lies in Quadrant II, and so forth. Points with a zero x-coordinate lie on the y-axis and points with a zero y-coordinate lie on the x-axis.

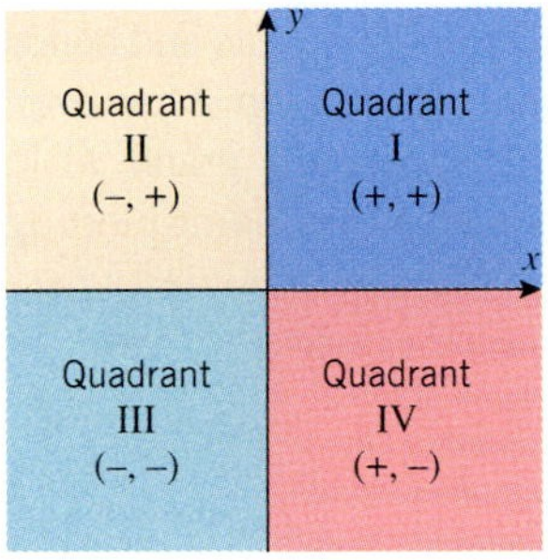

Quadrant II (−, +)
Quadrant I (+, +)
Quadrant III (−, −)
Quadrant IV (+, −)

Figure C.4

To ***plot*** a point $P(a, b)$ means to locate the point with coordinates (a, b) in a coordinate plane. For example, in Figure C.5 we have plotted the points

$$P(2, 5), \quad Q(-4, 3), \quad R(-5, -2), \quad \text{and} \quad S(4, -3)$$

Observe how the signs of the coordinates identify the quadrants in which the points lie.

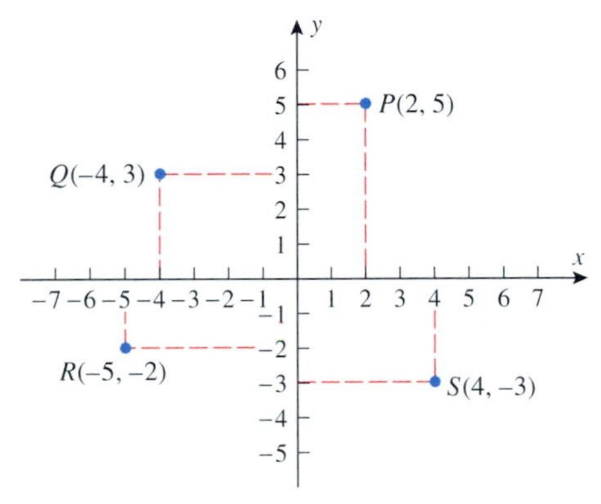

Figure C.5

GRAPHS

The correspondence between points in a plane and ordered pairs of real numbers makes it possible to visualize algebraic equations as geometric curves, and, conversely, to represent geometric curves by algebraic equations. To understand how this is done, suppose that we have an xy-coordinate system and an equation involving two variables x and y, say

$$6x - 4y = 10, \quad y = \sqrt{x}, \quad x = y^3 + 1, \quad \text{or} \quad x^2 + y^2 = 1$$

We define a ***solution*** of such an equation to be any ordered pair of real numbers (a, b) whose coordinates satisfy the equation when we substitute $x = a$ and $y = b$. For example, the ordered pair $(3, 2)$ is a solution of the equation $6x - 4y = 10$, since the equation is satisfied by $x = 3$ and $y = 2$ (verify). However, the ordered pair $(2, 0)$ is not a solution of this equation, since the equation is not satisfied by $x = 2$ and $y = 0$ (verify).

The following definition makes the association between equations in x and y and curves in the xy-plane.

C.1 DEFINITION. The set of all solutions of an equation in x and y is called the ***solution set*** of the equation, and the set of all points in the xy-plane whose coordinates are members of the solution set is called the ***graph*** of the equation.

One of the main themes in calculus is to identify the exact shape of a graph. Point plotting is one approach to obtaining a graph, but this method has limitations, as discussed in the following example.

Example 1

Sketch the graph of $y = x^2$.

Solution. The solution set of the equation has infinitely many members, since we can substitute an arbitrary value for x into the right side of $y = x^2$ and compute the associated y to obtain a point (x, y) in the solution set. The fact that the solution set has infinitely many members means that we cannot obtain the *entire* graph of $y = x^2$ by point plotting. However, we can obtain an *approximation* to the graph by plotting some sample members of the solution set and connecting them with a smooth curve, as in Figure C.6. The problem with this method is that we cannot be sure how the graph behaves *between* the plotted points. For example, the curves in Figure C.7 also pass through the plotted points and hence are legitimate candidates for the graph in the absence of additional information. Moreover, even if we use a graphing calculator or a computer program to generate the graph, as in Figure C.8, we have the same problem because graphing technology uses point-plotting algorithms to generate graphs. Indeed, in Section 1.3 of the text we see examples where graphing technology can be fooled into producing grossly inaccurate graphs. ◀

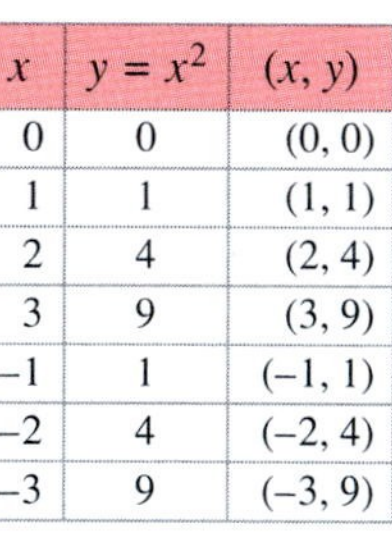

x	$y = x^2$	(x, y)
0	0	(0, 0)
1	1	(1, 1)
2	4	(2, 4)
3	9	(3, 9)
−1	1	(−1, 1)
−2	4	(−2, 4)
−3	9	(−3, 9)

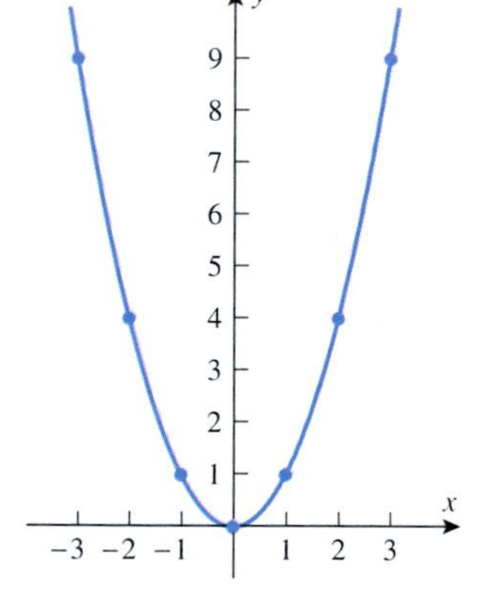

Figure C.6

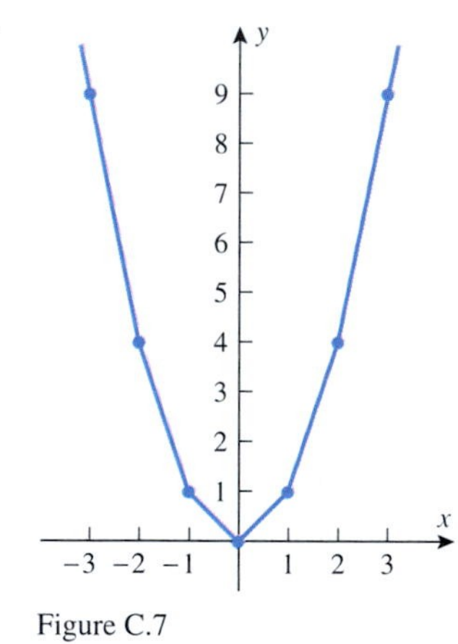

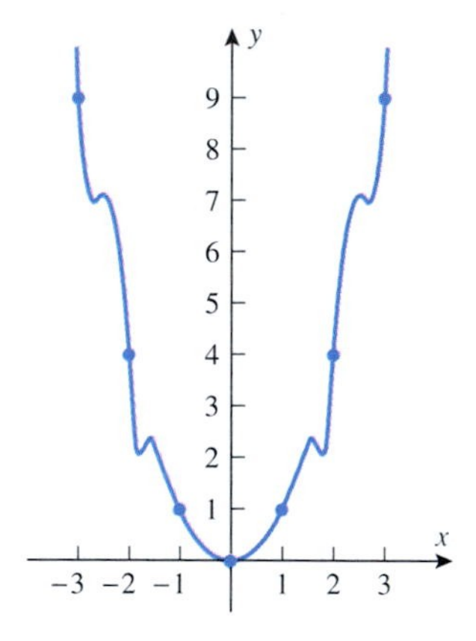

Figure C.7

[−4, 4] × [0, 10]
xScl = 1, yScl = 2

Figure C.8

In spite of its limitations, point plotting by hand or with the help of graphing technology can be useful, so here are two more examples.

Example 2

Sketch the graph of $y = \sqrt{x}$.

Solution. If $x < 0$, then $\sqrt{x}$ is an imaginary number. Thus, we can only plot points for which $x \geq 0$, since points in the xy-plane have real coordinates. Figure C.9 shows the graph obtained by point plotting and a graph obtained with a graphing calculator. ◀

Example 3

Sketch the graph of $y^2 - 2y - x = 0$.

Solution. To calculate coordinates of points on the graph of an equation in x and y, it is desirable to have y expressed in terms of x or of x in terms of y. In this case it is easier to

x	$y = \sqrt{x}$	(x, y)
0	0	(0, 0)
1	1	(1, 1)
2	$\sqrt{2}$	$(2, \sqrt{2}) \approx (2, 1.4)$
3	$\sqrt{3}$	$(3, \sqrt{3}) \approx (3, 1.7)$
4	2	(4, 2)

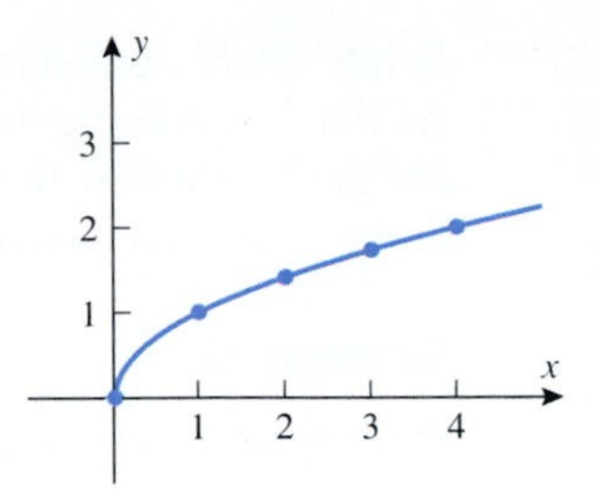

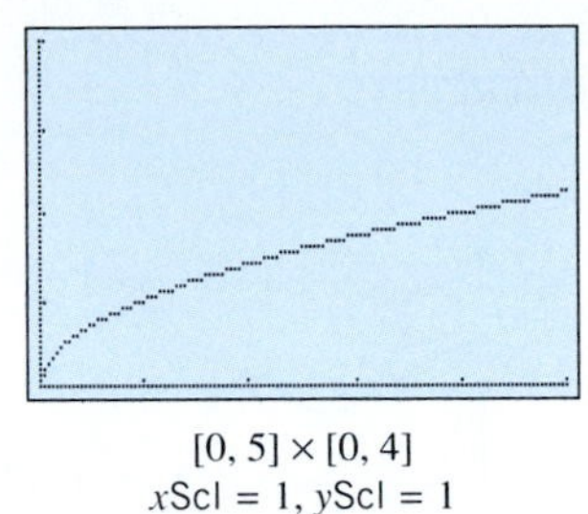

[0, 5] × [0, 4]
xScl = 1, yScl = 1

Figure C.9

express x in terms of y, so we rewrite the equation as

$$x = y^2 - 2y$$

Members of the solution set can be obtained from this equation by substituting arbitrary values for y in the right side and computing the associated values of x (Figure C.10). ◀

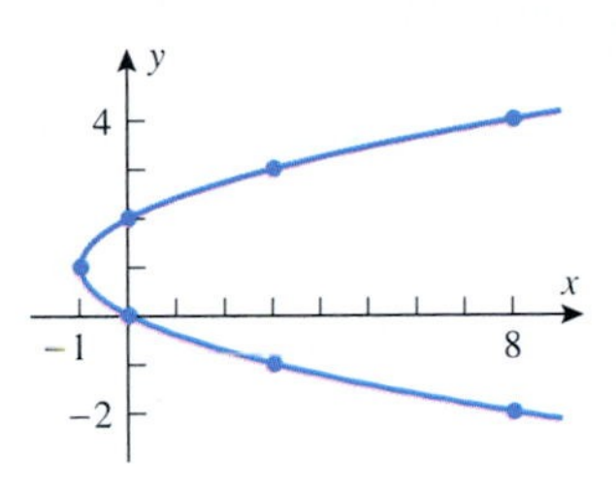

y	$x = y^2 - 2y$	(x, y)
−2	8	(8, −2)
−1	3	(3, −1)
0	0	(0, 0)
1	−1	(−1, 1)
2	0	(0, 2)
3	3	(3, 3)
4	8	(8, 4)

Figure C.10

REMARK. Most graphing calculators and computer graphing programs require that y be expressed in terms of x to generate a graph in the xy-plane. In Section 1.7 we discuss a method for circumventing this restriction.

Example 4

Sketch the graph of $y = 1/x$.

Solution. Because $1/x$ is undefined at $x = 0$, we can only plot points for which $x \neq 0$. This forces a break, called a *discontinuity*, in the graph at $x = 0$ (Figure C.11). ◀

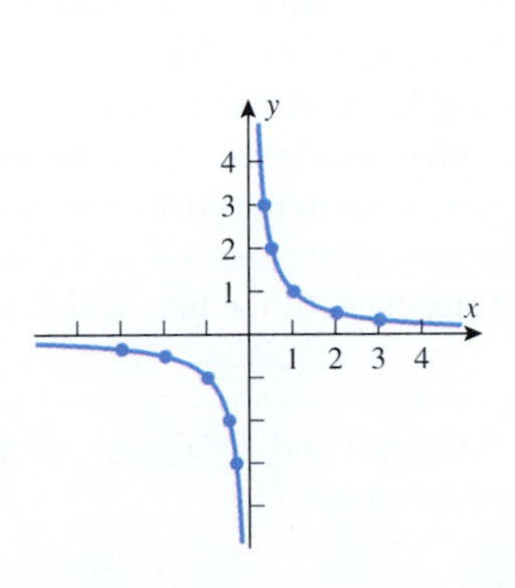

x	$y = 1/x$	(x, y)
$\frac{1}{3}$	3	$\left(\frac{1}{3}, 3\right)$
$\frac{1}{2}$	2	$\left(\frac{1}{2}, 2\right)$
1	1	(1, 1)
2	$\frac{1}{2}$	$\left(2, \frac{1}{2}\right)$
3	$\frac{1}{3}$	$\left(3, \frac{1}{3}\right)$
$-\frac{1}{3}$	−3	$\left(-\frac{1}{3}, -3\right)$
$-\frac{1}{2}$	−2	$\left(-\frac{1}{2}, -2\right)$
−1	−1	(−1, −1)
−2	$-\frac{1}{2}$	$\left(-2, -\frac{1}{2}\right)$
−3	$-\frac{1}{3}$	$\left(-3, -\frac{1}{3}\right)$

Figure C.11

INTERCEPTS

Points where a graph intersects the coordinate axes are of special interest in many problems. As illustrated in Figure C.12, intersections of a graph with the x-axis have the form $(a, 0)$ and intersections with the y-axis have the form $(0, b)$. The number a is called an ***x*-intercept** of the graph and the number b a ***y*-intercept**.

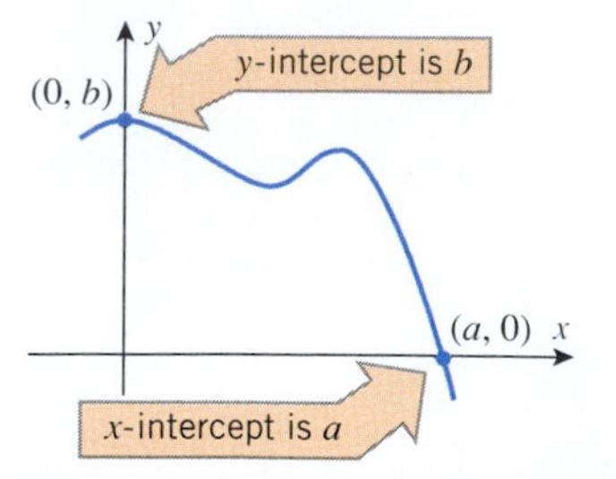

Figure C.12

Example 5

Find all intercepts of

(a) $3x + 2y = 6$ (b) $x = y^2 - 2y$ (c) $y = 1/x$

Solution (a). To find the x-intercepts we set $y = 0$ and solve for x:

$$3x = 6 \quad \text{or} \quad x = 2$$

To find the y-intercepts we set $x = 0$ and solve for y:

$$2y = 6 \quad \text{or} \quad y = 3$$

As we will see later, the graph of $3x + 2y = 6$ is the line shown in Figure C.13.

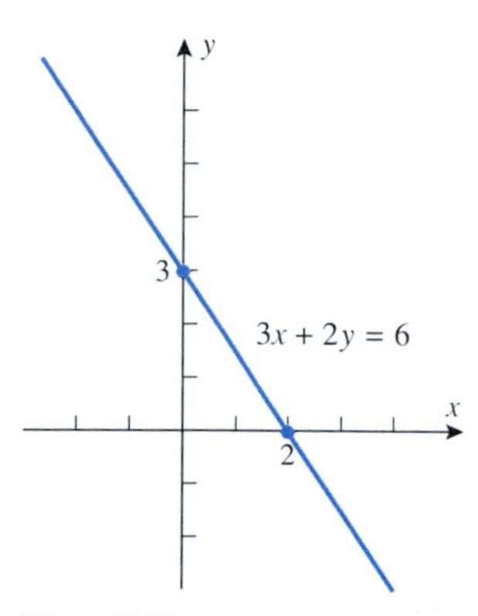

Figure C.13

Solution (b). To find the x-intercepts, set $y = 0$ and solve for x:

$$x = 0$$

Thus, $x = 0$ is the only x-intercept. To find the y-intercepts, set $x = 0$ and solve for y:

$$y^2 - 2y = 0$$

$$y(y - 2) = 0$$

So the y-intercepts are $y = 0$ and $y = 2$. The graph is shown in Figure C.10.

Solution (c). To find the x-intercepts, set $y = 0$:

$$\frac{1}{x} = 0$$

This equation has no solutions (why?), so there are no x-intercepts. To find y-intercepts we would set $x = 0$ and solve for y. But, substituting $x = 0$ leads to a division by zero, which is not allowed, so there are no y-intercepts either. The graph of the equation is shown in Figure C.11. ◀

SLOPE

To obtain equations of lines we will first need to discuss the concept of *slope*, which is a numerical measure of the "steepness" of a line.

Consider a particle moving left to right along a *nonvertical* line from a point $P_1(x_1, y_1)$ to a point $P_2(x_2, y_2)$. As shown in Figure C.14, the particle moves $y_2 - y_1$ units in the y-direction as it travels $x_2 - x_1$ units in the positive x-direction. The vertical change $y_2 - y_1$ is called the ***rise***, and the horizontal change $x_2 - x_1$ the ***run***. The ratio of the rise over the run can be used to measure the steepness of the line, which leads us to the following definition.

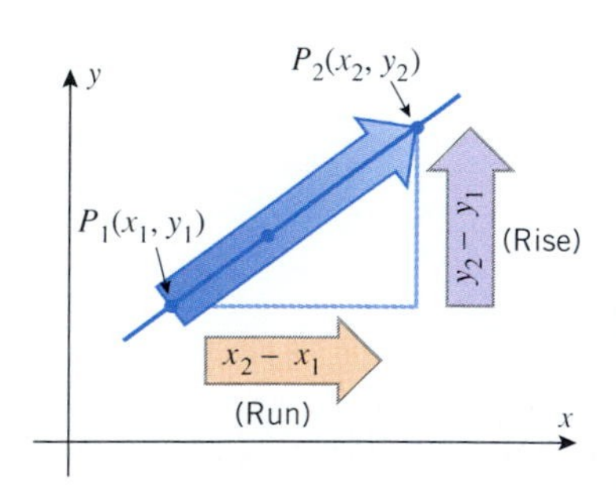

Figure C.14

C.2 DEFINITION. If $P_1(x_1, y_1)$ and $P_2(x_2, y_2)$ are points on a nonvertical line, then the ***slope*** m of the line is defined by

$$m = \frac{\text{rise}}{\text{run}} = \frac{y_2 - y_1}{x_2 - x_1} \tag{1}$$

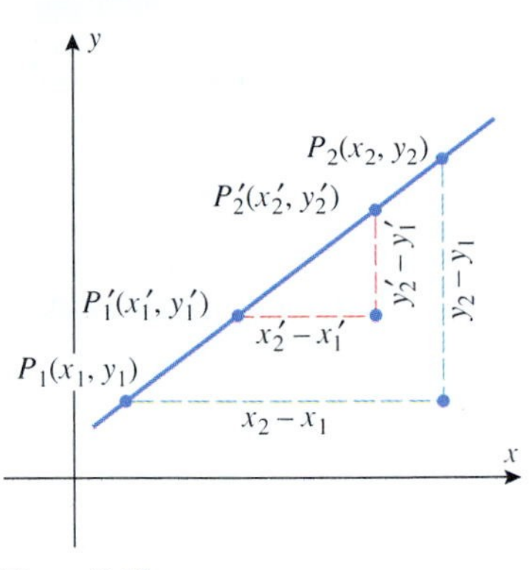

Figure C.15

REMARK. Observe that this definition does not apply to vertical lines. For such lines we have $x_2 = x_1$ (a zero run), which means that the formula for m involves a division by zero. For this reason, the slope of a vertical line is ***undefined***, which is sometimes described informally by stating that a vertical line has ***infinite slope***.

When calculating the slope of a nonvertical line from Formula (1), it does not matter which two points on the line you use for the calculation, as long as they are distinct. This can be proved using Figure C.15 and similar triangles to show that

$$m = \frac{y_2 - y_1}{x_2 - x_1} = \frac{y_2' - y_1'}{x_2' - x_1'}$$

Moreover, once you choose two points to use for the calculation, it does not matter which one you call P_1 and which one you call P_2 because reversing the points reverses the sign of both the numerator and denominator of (1) and hence has no effect on the ratio.

Example 6

In each part find the slope of the line through

(a) the points (6, 2) and (9, 8)

(b) the points (2, 9) and (4, 3)

(c) the points (−2, 7) and (5, 7).

Solution.

(a) $m = \dfrac{8-2}{9-6} = \dfrac{6}{3} = 2$ (b) $m = \dfrac{3-9}{4-2} = \dfrac{-6}{2} = -3$ (c) $m = \dfrac{7-7}{5-(-2)} = 0$ ◀

Example 7

Figure C.16 shows the three lines determined by the points in Example 6 and explains the significance of their slopes. ◀

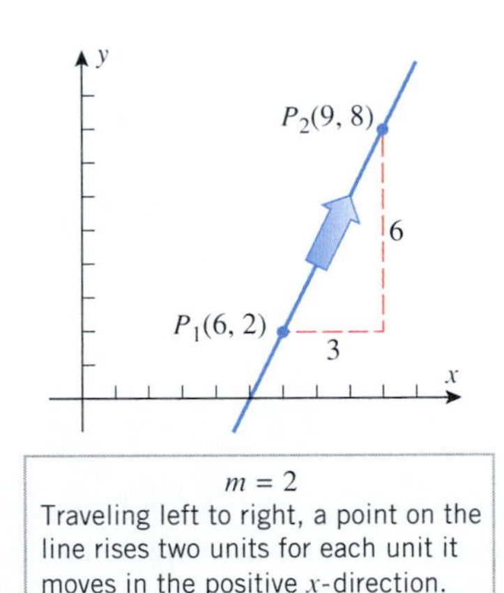

$m = 2$
Traveling left to right, a point on the line rises two units for each unit it moves in the positive x-direction.

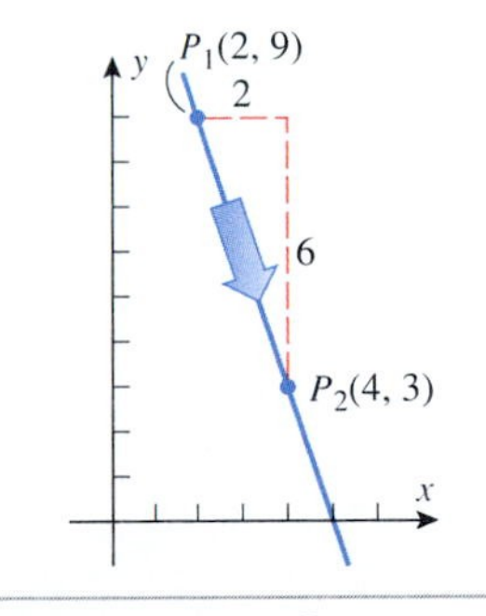

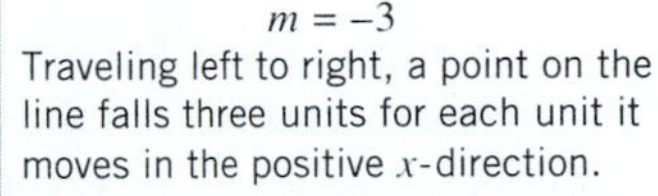

$m = -3$
Traveling left to right, a point on the line falls three units for each unit it moves in the positive x-direction.

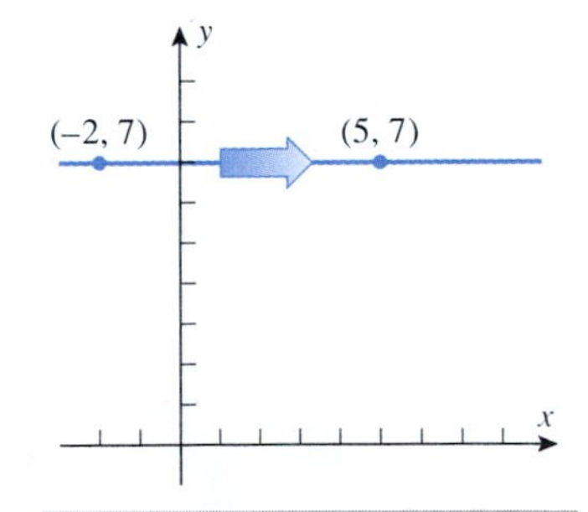

$m = 0$
Traveling left to right, a point on the line neither rises nor falls.

Figure C.16

As illustrated in this example, the slope of a line can be positive, negative, or zero. A positive slope means that the line is inclined upward to the right, a negative slope means that the line is inclined downward to the right, and a zero slope means that the line is horizontal.

An undefined slope means that the line is vertical. Figure C.17 shows various lines through the origin with their slopes.

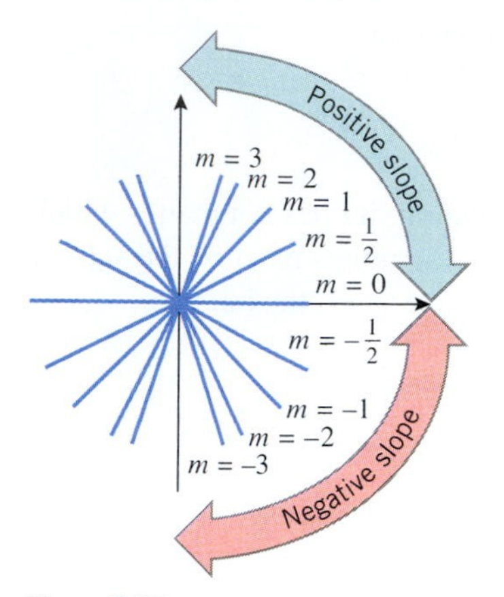

Figure C.17

PARALLEL AND PERPENDICULAR LINES

The following theorem shows how slopes can be used to tell whether two lines are parallel or perpendicular.

C.3 THEOREM.

(*a*) *Two nonvertical lines with slopes m_1 and m_2 are parallel if and only if they have the same slope, that is,*

$$m_1 = m_2$$

(*b*) *Two nonvertical lines with slopes m_1 and m_2 are perpendicular if and only if the product of their slopes is -1, that is,*

$$m_1 m_2 = -1$$

This relationship can also be expressed as $m_1 = -1/m_2$ or $m_2 = -1/m_1$, which states that nonvertical lines are perpendicular if and only if their slopes are negative reciprocals of one another.

A complete proof of this theorem is a little tedious, but it is not hard to motivate the results informally. Let us start with part (*a*).

Suppose that L_1 and L_2 are nonvertical parallel lines with slopes m_1 and m_2, respectively. If the lines are parallel to the x-axis, then $m_1 = m_2 = 0$, and we are done. If they are not parallel to the x-axis, then both lines intersect the x-axis; and for simplicity assume that they are oriented as in Figure C.18*a*. On each line choose the point whose run relative to the point of intersection with the x-axis is 1. On line L_1 the corresponding rise will be m_1 and on L_2 it will be m_2. However, because the lines are parallel, the shaded triangles in the figure must be congruent (verify), so $m_1 = m_2$. Conversely, the condition $m_1 = m_2$ can be used to show that the shaded triangles are congruent, from which it follows that the lines make the same angle with the x-axis and hence are parallel (verify).

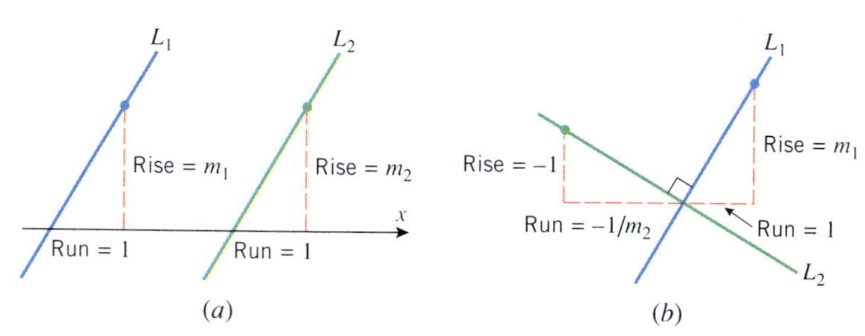

Figure C.18

Now suppose that L_1 and L_2 are nonvertical perpendicular lines with slopes m_1 and m_2, respectively; and for simplicity assume that they are oriented as in Figure C.18*b*. On line L_1 choose the point whose run relative to the point of intersection of the lines is 1, in which case the corresponding rise will be m_1; and on line L_2 choose the point whose rise relative to the point of intersection is -1, in which case the corresponding run will be $-1/m_2$. Because the lines are perpendicular, the shaded triangles in the figure must be congruent (verify), and hence the ratios of corresponding sides of the triangles must be equal. Taking into account that for line L_2 the vertical side of the triangle has length 1 and the horizontal side has length $-1/m_2$ (since m_2 is negative), the congruence of the triangles implies that

$m_1/1 = (-1/m_2)/1$ or $m_1 m_2 = -1$. Conversely, the condition $m_1 = -1/m_2$ can be used to show that the shaded triangles are congruent, from which it can be deduced that the lines are perpendicular (verify).

Example 8

Use slopes to show that the points $A(1, 3)$, $B(3, 7)$, and $C(7, 5)$ are vertices of a right triangle.

Solution. We will show that the line through A and B is perpendicular to the line through B and C. The slopes of these lines are

$$m_1 = \frac{7-3}{3-1} = 2 \qquad \text{and} \qquad m_2 = \frac{5-7}{7-3} = -\frac{1}{2}$$

Slope of the line through A and B — Slope of the line through B and C

Since $m_1 m_2 = -1$, the line through A and B is perpendicular to the line through B and C; thus, ABC is a right triangle (Figure C.19). ◀

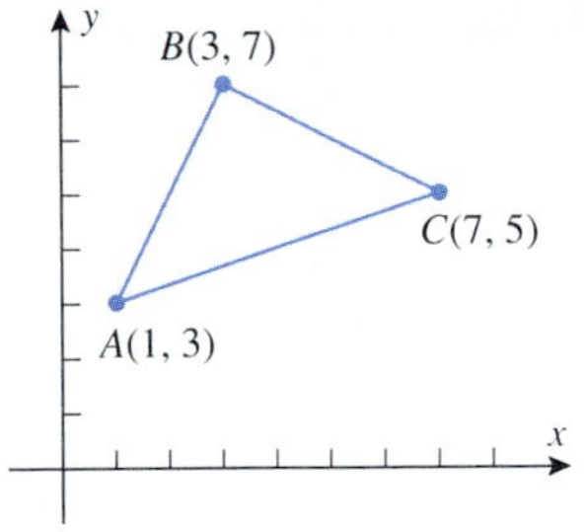

Figure C.19

LINES PARALLEL TO THE COORDINATE AXES

We now turn to the problem of finding equations of lines that satisfy specified conditions. The simplest cases are lines parallel to the coordinate axes. A line parallel to the y-axis intersects the x-axis at some point $(a, 0)$. This line consists precisely of those points whose x-coordinate is equal to a (Figure C.20). Similarly, a line parallel to the x-axis intersects the y-axis at some point $(0, b)$. This line consists precisely of those points whose y-coordinate is equal to b (Figure C.20). Thus, we have the following theorem.

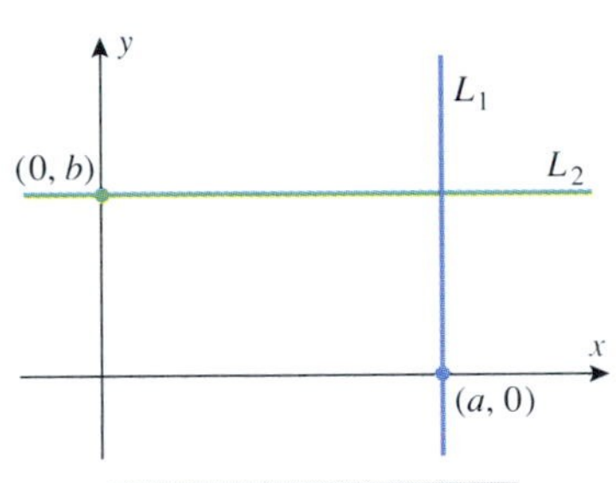

Every point on L_1 has an x-coordinate of a and every point on L_2 has a y-coordinate of b.

Figure C.20

C.4 THEOREM. *The vertical line through* $(a, 0)$ *and the horizontal line through* $(0, b)$ *are represented, respectively, by the equations*

$$x = a \qquad \text{and} \qquad y = b$$

Example 9

The graph of $x = -5$ is the vertical line through $(-5, 0)$, and the graph of $y = 7$ is the horizontal line through $(0, 7)$ (Figure C.21). ◀

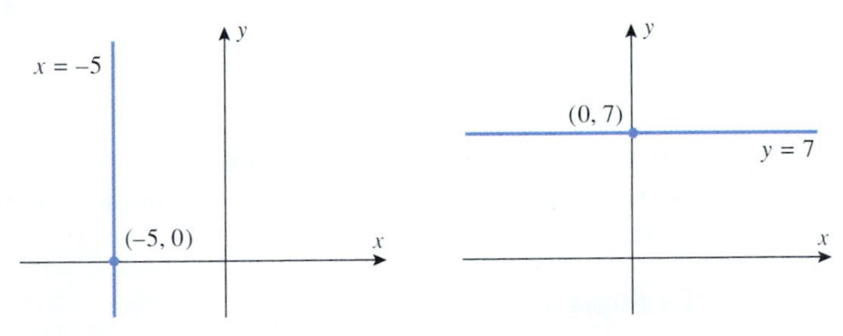

Figure C.21

LINES DETERMINED BY POINT AND SLOPE

There are infinitely many lines that pass through any given point in the plane. However, if we specify the slope of the line in addition to a point on it, then the point and the slope together determine a unique line (Figure C.22).

Let us now consider how to find an equation of a nonvertical line L that passes through a point $P_1(x_1, y_1)$ and has slope m. If $P(x, y)$ is any point on L, different from P_1, then the

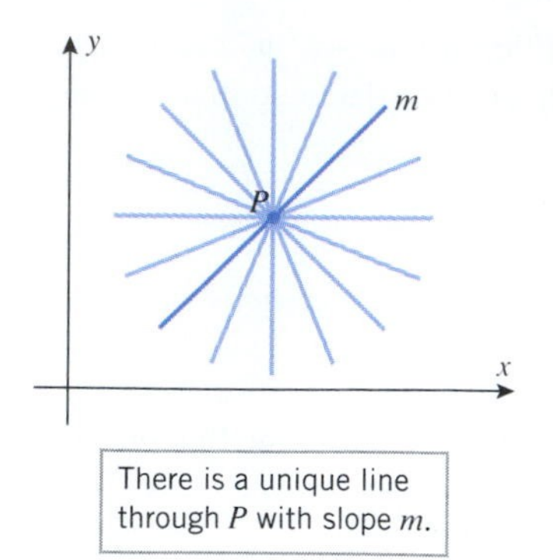

Figure C.22

slope m can be obtained from the points $P(x, y)$ and $P_1(x_1, y_1)$; this gives

$$m = \frac{y - y_1}{x - x_1}$$

which can be rewritten as

$$y - y_1 = m(x - x_1) \tag{2}$$

With the possible exception of (x_1, y_1), we have shown that every point on L satisfies (2). But $x = x_1$, $y = y_1$ satisfies (2), so that all points on L satisfy (2). We leave it as an exercise to show that every point satisfying (2) lies on L.

In summary, we have the following theorem.

C.5 THEOREM. *The line passing through $P_1(x_1, y_1)$ and having slope m is given by the equation*

$$y - y_1 = m(x - x_1) \tag{3}$$

*This is called the **point-slope form** of the line.*

Example 10

Find the point-slope form of the line through $(4, -3)$ with slope 5.

Solution. Substituting the values $x_1 = 4$, $y_1 = -3$, and $m = 5$ in (3) yields the point-slope form $y + 3 = 5(x - 4)$. ◀

LINES DETERMINED BY SLOPE AND y-INTERCEPT

A nonvertical line crosses the y-axis at some point $(0, b)$. If we use this point in the point-slope form of its equation, we obtain

$$y - b = m(x - 0)$$

which we can rewrite as $y = mx + b$. To summarize:

C.6 THEOREM. *The line with y-intercept b and slope m is given by the equation*

$$y = mx + b \tag{4}$$

*This is called the **slope-intercept form** of the line.*

REMARK. Note that y is alone on one side of Equation (4). When the equation of a line is written in this way the slope of the line and its y-intercept can be determined by inspection of the equation—the slope is the coefficient of x and the y-intercept is the constant term (Figure C.23).

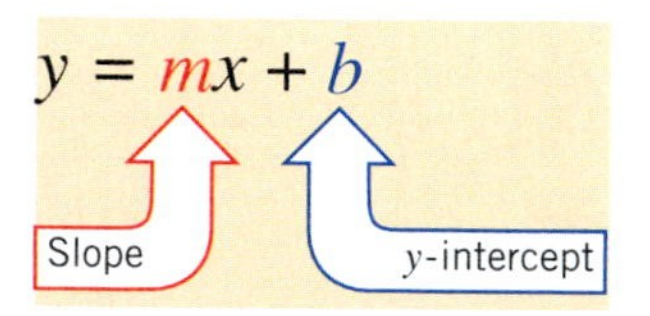

Figure C.23

Example 11

EQUATION	SLOPE	y-INTERCEPT
$y = 3x + 7$	$m = 3$	$b = 7$
$y = -x + \frac{1}{2}$	$m = -1$	$b = \frac{1}{2}$
$y = x$	$m = 1$	$b = 0$
$y = \sqrt{2}x - 8$	$m = \sqrt{2}$	$b = -8$
$y = 2$	$m = 0$	$b = 2$

◀

Example 12

Find the slope-intercept form of the equation of the line that satisfies the stated conditions:

(a) slope is -9; crosses the y-axis at $(0, -4)$

(b) slope is 1; passes through the origin

(c) passes through $(5, -1)$; perpendicular to $y = 3x + 4$

(d) passes through $(3, 4)$ and $(2, -5)$.

Solution (a). From the given conditions we have $m = -9$ and $b = -4$, so (4) yields $y = -9x - 4$.

Solution (b). From the given conditions $m = 1$ and the line passes through $(0, 0)$, so $b = 0$. Thus, it follows from (4) that $y = x + 0$ or $y = x$.

Solution (c). The given line has slope 3, so the line to be determined will have slope $m = -\frac{1}{3}$. Substituting this slope and the given point in the point-slope form (3) and then simplifying yields

$$y - (-1) = -\tfrac{1}{3}(x - 5)$$
$$y = -\tfrac{1}{3}x + \tfrac{2}{3}$$

Solution (d). We will first find the point-slope form, then solve for y in terms of x to obtain the slope-intercept form. From the given points the slope of the line is

$$m = \frac{-5 - 4}{2 - 3} = 9$$

We can use either of the given points for (x_1, y_1) in (3). We will use $(3, 4)$. This yields the point-slope form

$$y - 4 = 9(x - 3)$$

Solving for y in terms of x yields the slope-intercept form

$$y = 9x - 23$$

We leave it for the reader to show that the same equation results if $(2, -5)$ rather than $(3, 4)$ is used for (x_1, y_1) in (3). ◀

THE GENERAL EQUATION OF A LINE

An equation that is expressible in the form

$$Ax + By + C = 0 \qquad (5)$$

where A, B, and C are constants and A and B are not both zero, is called a ***first-degree equation*** in x and y. For example,

$$4x + 6y - 5 = 0$$

is a first-degree equation in x and y since it has form (5) with

$$A = 4, \quad B = 6, \quad C = -5$$

In fact, all the equations of lines studied in this section are first-degree equations in x and y.

The following theorem states that the first-degree equations in x and y are precisely the equations whose graphs in the xy-plane are straight lines.

> **C.7 THEOREM.** *Every first-degree equation in x and y has a straight line as its graph and, conversely, every straight line can be represented by a first-degree equation in x and y.*

Because of this theorem, (5) is sometimes called the ***general equation*** of a line or a ***linear equation*** in x and y.

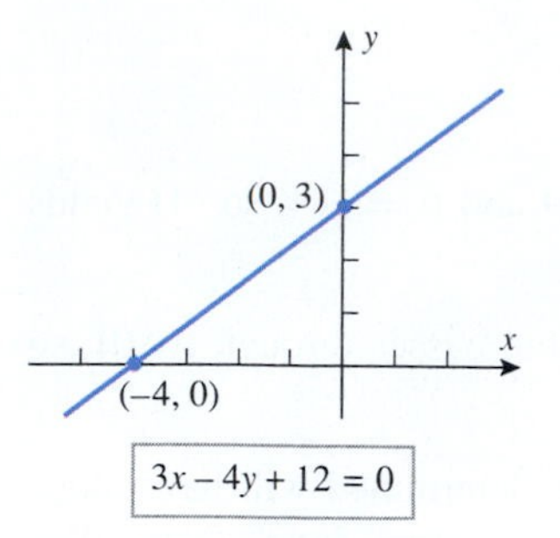

Figure C.24

Example 13

Graph the equation $3x - 4y + 12 = 0$.

Solution. Since this is a linear equation in x and y, its graph is a straight line. Thus, to sketch the graph we need only plot any two points on the graph and draw the line through them. It is particularly convenient to plot the points where the line crosses the coordinate axes. These points are $(0, 3)$ and $(-4, 0)$ (verify), so the graph is the line in Figure C.24. ◀

Example 14

Find the slope of the line in Example 13.

Solution. Solving the equation for y yields

$$y = \tfrac{3}{4}x + 3$$

which is the slope-intercept form of the line. Thus, the slope is $m = \frac{3}{4}$. ◀

Exercise Set C

1. Draw the rectangle, three of whose vertices are $(6, 1)$, $(-4, 1)$, and $(6, 7)$, and find the coordinates of the fourth vertex.

2. Draw the triangle whose vertices are $(-3, 2)$, $(5, 2)$, and $(4, 3)$, and find its area.

In Exercises 3 and 4, draw a rectangular coordinate system and sketch the set of points whose coordinates (x, y) satisfy the given conditions.

3. (a) $x = 2$ (b) $y = -3$ (c) $x \geq 0$
(d) $y = x$ (e) $y \geq x$ (f) $|x| \geq 1$

4. (a) $x = 0$ (b) $y = 0$
(c) $y < 0$ (d) $x \geq 1$ and $y \leq 2$
(e) $x = 3$ (f) $|x| = 5$

In Exercises 5–12, sketch the graph of the equation. (A calculating utility will be helpful in some of these problems.)

5. $y = 4 - x^2$
6. $y = 1 + x^2$
7. $y = \sqrt{x - 4}$
8. $y = -\sqrt{x + 1}$
9. $x^2 - x + y = 0$
10. $x = y^3 - y^2$
11. $x^2 y = 2$
12. $xy = -1$

13. Find the slope of the line through
(a) $(-1, 2)$ and $(3, 4)$ (b) $(5, 3)$ and $(7, 1)$
(c) $(4, \sqrt{2})$ and $(-3, \sqrt{2})$ (d) $(-2, -6)$ and $(-2, 12)$.

14. Find the slopes of the sides of the triangle with vertices $(-1, 2)$, $(6, 5)$, and $(2, 7)$.

15. Use slopes to determine whether the given points lie on the same line.
(a) $(1, 1)$, $(-2, -5)$, and $(0, -1)$
(b) $(-2, 4)$, $(0, 2)$, and $(1, 5)$

16. Draw the line through $(4, 2)$ with slope
(a) $m = 3$ (b) $m = -2$ (c) $m = -\frac{3}{4}$.

17. Draw the line through $(-1, -2)$ with slope
(a) $m = \frac{3}{5}$ (b) $m = -1$ (c) $m = \sqrt{2}$.

18. An equilateral triangle has one vertex at the origin, another on the x-axis, and the third in the first quadrant. Find the slopes of its sides.

19. List the lines in the accompanying figure in the order of increasing slope.

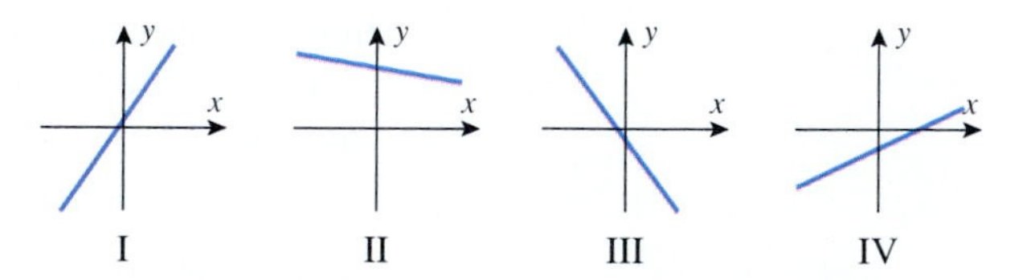

20. List the lines in the accompanying figure in the order of increasing slope.

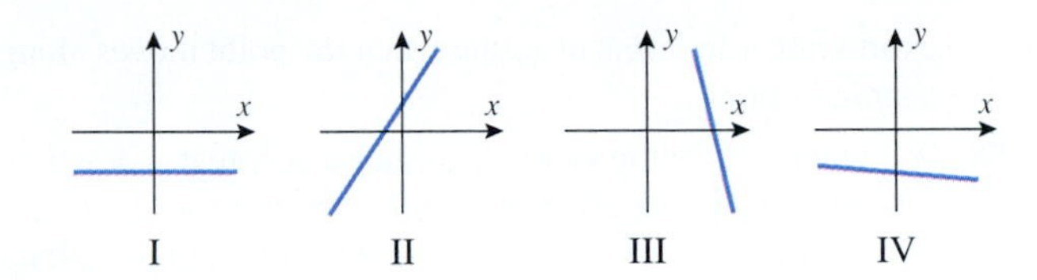

21. A particle, initially at (1, 2), moves along a line of slope $m = 3$ to a new position (x, y).
(a) Find y if $x = 5$. (b) Find x if $y = -2$.

22. A particle, initially at (7, 5), moves along a line of slope $m = -2$ to a new position (x, y).
(a) Find y if $x = 9$. (b) Find x if $y = 12$.

23. Let the point $(3, k)$ lie on the line of slope $m = 5$ through $(-2, 4)$; find k.

24. Given that the point $(k, 4)$ is on the line through (1, 5) and $(2, -3)$, find k.

25. Find x if the slope of the line through (1, 2) and $(x, 0)$ is the negative of the slope of the line through (4, 5) and $(x, 0)$.

26. Find x and y if the line through (0, 0) and (x, y) has slope $\frac{1}{2}$, and the line through (x, y) and (7, 5) has slope 2.

27. Use slopes to show that $(3, -1)$, (6, 4), $(-3, 2)$, and $(-6, -3)$ are vertices of a parallelogram.

28. Use slopes to show that (3, 1), (6, 3), and (2, 9) are vertices of a right triangle.

29. Graph the equations
(a) $2x + 5y = 15$ (b) $x = 3$
(c) $y = -2$ (d) $y = 2x - 7$.

30. Graph the equations
(a) $\dfrac{x}{3} - \dfrac{y}{4} = 1$ (b) $x = -8$
(c) $y = 0$ (d) $x = 3y + 2$.

31. Graph the equations
(a) $y = 2x - 1$ (b) $y = 3$
(c) $y = -2x$.

32. Graph the equations
(a) $y = 2 - 3x$ (b) $y = \frac{1}{4}x$
(c) $y = -\sqrt{3}$.

33. Find the slope and y-intercept of
(a) $y = 3x + 2$ (b) $y = 3 - \frac{1}{4}x$
(c) $3x + 5y = 8$ (d) $y = 1$
(e) $\dfrac{x}{a} + \dfrac{y}{b} = 1$.

34. Find the slope and y-intercept of
(a) $y = -4x + 2$ (b) $x = 3y + 2$
(c) $\dfrac{x}{2} + \dfrac{y}{3} = 1$ (d) $y - 3 = 0$
(e) $a_0x + a_1y = 0 \quad (a_1 \neq 0)$.

In Exercises 35 and 36, use the graph to find the equation of the line in slope-intercept form.

35.

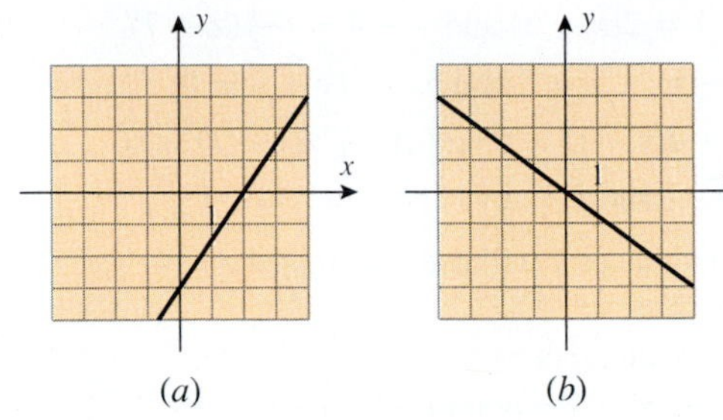

Figure Ex-35

36.

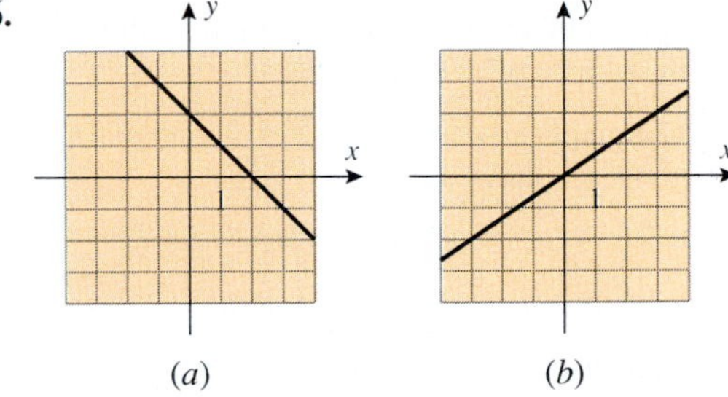

Figure Ex-36

In Exercises 37–48, find the slope-intercept form of the line satisfying the given conditions.

37. Slope $= -2$, y-intercept $= 4$.

38. $m = 5$, $b = -3$.

39. The line is parallel to $y = 4x - 2$ and its y-intercept is 7.

40. The line is parallel to $3x + 2y = 5$ and passes through $(-1, 2)$.

41. The line is perpendicular to $y = 5x + 9$ and its y-intercept is 6.

42. The line is perpendicular to $x - 4y = 7$ and passes through $(3, -4)$.

43. The line passes through (2, 4) and $(1, -7)$.

44. The line passes through $(-3, 6)$ and $(-2, 1)$.

45. The y-intercept is 2 and the x-intercept is -4.

46. The y-intercept is b and the x-intercept is a.

47. The line is perpendicular to the y-axis and passes through $(-4, 1)$.

48. The line is parallel to $y = -5$ and passes through $(-1, -8)$.

49. In each part, classify the lines as parallel, perpendicular, or neither.
(a) $y = 4x - 7$ and $y = 4x + 9$
(b) $y = 2x - 3$ and $y = 7 - \frac{1}{2}x$
(c) $5x - 3y + 6 = 0$ and $10x - 6y + 7 = 0$
(d) $Ax + By + C = 0$ and $Bx - Ay + D = 0$
(e) $y - 2 = 4(x - 3)$ and $y - 7 = \frac{1}{4}(x - 3)$

50. In each part, classify the lines as parallel, perpendicular, or neither.
(a) $y = -5x + 1$ and $y = 3 - 5x$

(b) $y - 1 = 2(x - 3)$ and $y - 4 = -\frac{1}{2}(x + 7)$
(c) $4x + 5y + 7 = 0$ and $5x - 4y + 9 = 0$
(d) $Ax + By + C = 0$ and $Ax + By + D = 0$
(e) $y = \frac{1}{2}x$ and $x = \frac{1}{2}y$

51. For what value of k will the line $3x + ky = 4$
(a) have slope 2
(b) have y-intercept 5
(c) pass through the point $(-2, 4)$
(d) be parallel to the line $2x - 5y = 1$
(e) be perpendicular to the line $4x + 3y = 2$?

52. Sketch the graph of $y^2 = 3x$ and explain how this graph is related to the graphs of $y = \sqrt{3x}$ and $y = -\sqrt{3x}$.

53. Sketch the graph of $(x - y)(x + y) = 0$ and explain how it is related to the graphs of $x - y = 0$ and $x + y = 0$.

54. Graph $F = \frac{9}{5}C + 32$ in a CF-coordinate system.

55. Graph $u = 3v^2$ in a uv-coordinate system.

56. Graph $Y = 4X + 5$ in a YX-coordinate system.

57. A point moves in the xy-plane in such a way that at any time t its coordinates are given by $x = 5t + 2$ and $y = t - 3$. By expressing y in terms of x, show that the point moves along a straight line.

58. A point moves in the xy-plane in such a way that at any time t its coordinates are given by $x = 1 + 3t^2$ and $y = 2 - t^2$. By expressing y in terms of x, show that the point moves along a straight-line path and specify the values of x for which the equation is valid.

59. Find the area of the triangle formed by the coordinate axes and the line through $(1, 4)$ and $(2, 1)$.

60. Draw the graph of $4x^2 - 9y^2 = 0$.

61. In each part, name an appropriate coordinate system for graphing the equation [e.g., an $\alpha\beta$-coordinate system in part (a)], and state whether the graph of the equation is a line in that coordinate system.
(a) $3\alpha - 2\beta = 5$
(b) $A = 2000(1 + 0.06t)$
(c) $A = \pi r^2$
(d) $E = mc^2$ (c constant)
(e) $V = C(1 - rt)$ (r and C constant)
(f) $V = \frac{1}{3}\pi r^2 h$ (r constant)
(g) $V = \frac{1}{3}\pi r^2 h$ (h constant)

APPENDIX D

Distance, Circles, and Quadratic Equations

DISTANCE BETWEEN TWO POINTS IN THE PLANE

Suppose that we are interested in finding the distance d between two points $P_1(x_1, y_1)$ and $P_2(x_2, y_2)$ in the xy-plane. If, as in Figure D.1, we form a right triangle with P_1 and P_2 as vertices, then it follows from Theorem B.4 in Appendix B that the sides of that triangle have lengths $|x_2 - x_1|$ and $|y_2 - y_1|$. Thus, it follows from the Theorem of Pythagoras that

$$d = \sqrt{|x_2 - x_1|^2 + |y_2 - y_1|^2} = \sqrt{(x_2 - x_1)^2 + (y_2 - y_1)^2}$$

and hence we have the following result.

D.1 THEOREM. *The distance d between two points $P_1(x_1, y_1)$ and $P_2(x_2, y_2)$ in a coordinate plane is given by*

$$d = \sqrt{(x_2 - x_1)^2 + (y_2 - y_1)^2} \tag{1}$$

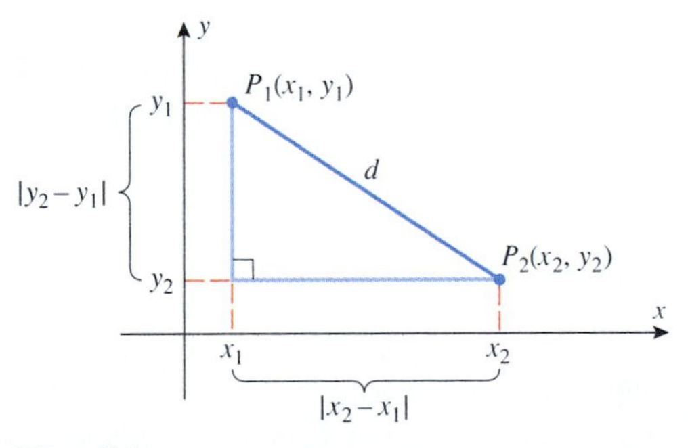

Figure D.1

REMARK. To apply Formula (1) the scales on the coordinate axes must be the same; otherwise, we would not have been able to use the Theorem of Pythagoras in the derivation. Moreover, when using Formula (1) it does not matter which point is labeled P_1 and which one is labeled P_2, since reversing the points changes the signs of $x_2 - x_1$ and $y_2 - y_1$; this has no effect on the value of d because these quantities are squared in the formula. When it is important to emphasize the points, the distance between P_1 and P_2 is denoted by $d(P_1, P_2)$ or $d(P_2, P_1)$.

Example 1

Find the distance between the points $(-2, 3)$ and $(1, 7)$.

Solution. If we let (x_1, y_1) be $(-2, 3)$ and let (x_2, y_2) be $(1, 7)$, then (1) yields

$$d = \sqrt{[1-(-2)]^2 + [7-3]^2} = \sqrt{3^2+4^2} = \sqrt{25} = 5$$

◀

Example 2

It can be shown that the converse of the Theorem of Pythagoras is true; that is, if the sides of a triangle satisfy the relationship $a^2 + b^2 = c^2$, then the triangle must be a right triangle. Use this result to show that the points $A(4, 6)$, $B(1, -3)$, and $C(7, 5)$ are vertices of a right triangle.

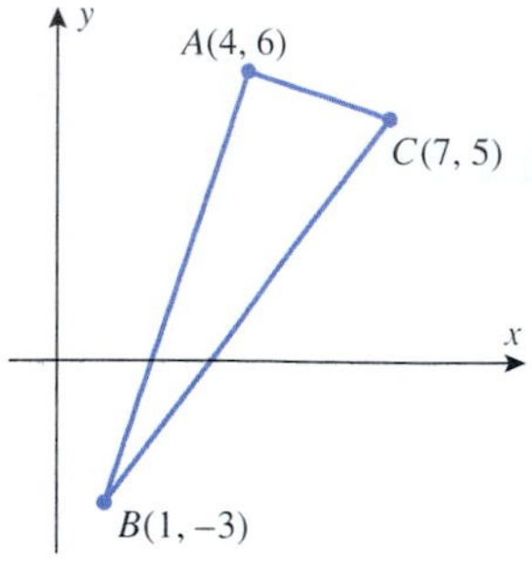

Figure D.2

Solution. The points and the triangle are shown in Figure D.2. From (1), the lengths of the sides of the triangles are

$$d(A, B) = \sqrt{(1-4)^2 + (-3-6)^2} = \sqrt{9+81} = \sqrt{90}$$

$$d(A, C) = \sqrt{(7-4)^2 + (5-6)^2} = \sqrt{9+1} = \sqrt{10}$$

$$d(B, C) = \sqrt{(7-1)^2 + [5-(-3)]^2} = \sqrt{36+64} = \sqrt{100} = 10$$

Since

$$[d(A, B)]^2 + [d(A, C)]^2 = [d(B, C)]^2$$

it follows that ΔABC is a right triangle with hypotenuse BC. ◀

THE MIDPOINT FORMULA

It is often necessary to find the coordinates of the midpoint of a line segment joining two points in the plane. To derive the midpoint formula, we will start with two points on a coordinate line. If we assume that the points have coordinates a and b and that $a \le b$, then, as shown in Figure D.3, the distance between a and b is $b - a$, and the coordinate of the midpoint between a and b is

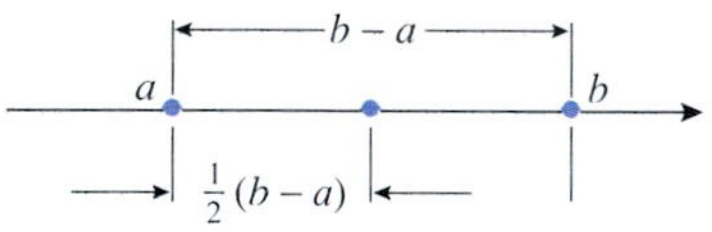

Figure D.3

$$a + \tfrac{1}{2}(b-a) = \tfrac{1}{2}a + \tfrac{1}{2}b = \tfrac{1}{2}(a+b)$$

which is the arithmetic average of a and b. Had the points been labeled with $b \le a$, the same formula would have resulted (verify). Therefore, *the midpoint of two points on a coordinate line is the arithmetic average of their coordinates, regardless of their relative positions.*

If we now let $P_1(x_1, y_1)$ and $P_2(x_2, y_2)$ be any two points in the plane and $M(x, y)$ the midpoint of the line segment joining them (Figure D.4), then it can be shown using similar triangles that x is the midpoint of x_1 and x_2 on the x-axis and y is the midpoint of y_1 and y_2 on the y-axis, so

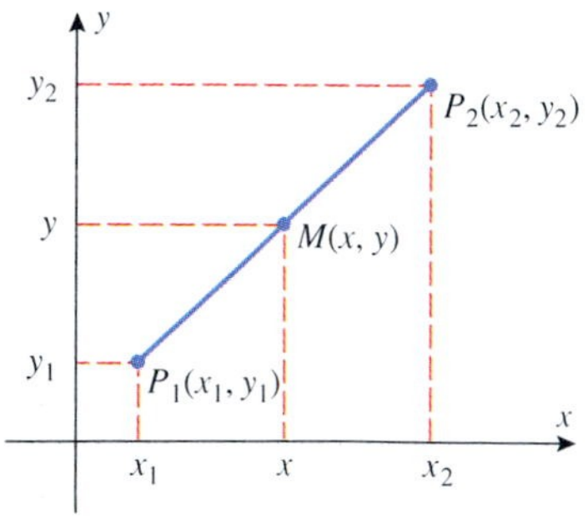

$$x = \tfrac{1}{2}(x_1 + x_2) \quad \text{and} \quad y = \tfrac{1}{2}(y_1 + y_2)$$

Figure D.4

Thus, we have the following result.

> **D.2 THEOREM (*The Midpoint Formula*).** *The midpoint of the line segment joining two points (x_1, y_1) and (x_2, y_2) in a coordinate plane is*
>
> $$\left(\tfrac{1}{2}(x_1 + x_2), \tfrac{1}{2}(y_1 + y_2)\right) \tag{2}$$

Example 3

Find the midpoint of the line segment joining $(3, -4)$ and $(7, 2)$.

Solution. From (2) the midpoint is

$$\left(\tfrac{1}{2}(3+7), \tfrac{1}{2}(-4+2)\right) = (5, -1)$$

◀

CIRCLES

If (x_0, y_0) is a fixed point in the plane, then the circle of radius r centered at (x_0, y_0) is the set of all points in the plane whose distance from (x_0, y_0) is r (Figure D.5). Thus, a point (x, y) will lie on this circle if and only if

$$\sqrt{(x - x_0)^2 + (y - y_0)^2} = r$$

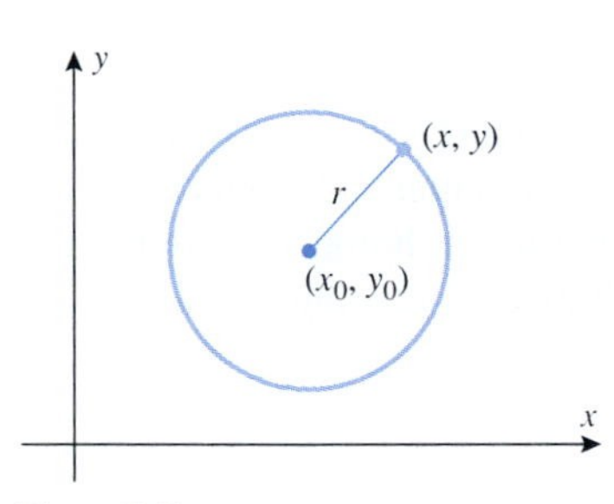

Figure D.5

or equivalently,

$$(x - x_0)^2 + (y - y_0)^2 = r^2 \tag{3}$$

This is called the ***standard form of the equation of a circle***.

Example 4

Find an equation for the circle of radius 4 centered at $(-5, 3)$.

Solution. From (3) with $x_0 = -5$, $y_0 = 3$, and $r = 4$ we obtain

$$(x + 5)^2 + (y - 3)^2 = 16$$

If desired, this equation can be written in an expanded form by squaring the terms and then simplifying:

$$(x^2 + 10x + 25) + (y^2 - 6y + 9) - 16 = 0$$
$$x^2 + y^2 + 10x - 6y + 18 = 0$$

◀

Example 5

Find an equation for the circle with center $(1, -2)$ that passes through $(4, 2)$.

Solution. The radius r of the circle is the distance between $(4, 2)$ and $(1, -2)$, so

$$r = \sqrt{(1 - 4)^2 + (-2 - 2)^2} = 5$$

We now know the center and radius, so we can use (3) to obtain the equation

$$(x - 1)^2 + (y + 2)^2 = 25 \quad \text{or} \quad x^2 + y^2 - 2x + 4y - 20 = 0$$

◀

FINDING THE CENTER AND RADIUS OF A CIRCLE

When you encounter an equation of form (3), you will know immediately that its graph is a circle; its center and radius can then be found from the constants that appear in the equation:

$$(x - x_0)^2 \quad + \quad (y - y_0)^2 \quad = \quad r^2$$

$(x - x_0)^2$	$(y - y_0)^2$	r^2
x-coordinate of the center is x_0	y-coordinate of the center is y_0	radius squared

Example 6

EQUATION OF A CIRCLE	CENTER (x_0, y_0)	RADIUS r
$(x - 2)^2 + (y - 5)^2 = 9$	$(2, 5)$	3
$(x + 7)^2 + (y + 1)^2 = 16$	$(-7, -1)$	4
$x^2 + y^2 = 25$	$(0, 0)$	5
$(x - 4)^2 + y^2 = 5$	$(4, 0)$	$\sqrt{5}$

◀

$x^2 + y^2 = 1$

The unit circle

Figure D.6

The circle $x^2 + y^2 = 1$, which is centered at the origin and has radius 1, is of special importance; it is called the ***unit circle*** (Figure D.6).

OTHER FORMS FOR THE EQUATION OF A CIRCLE

An alternative version of Equation (3) can be obtained by squaring the terms and simplifying. This yields an equation of the form

$$x^2 + y^2 + dx + ey + f = 0 \tag{4}$$

where d, e, and f are constants. (See the final equations in Examples 4 and 5.)

Still another version of the equation of a circle can be obtained by multiplying both sides of (4) by a nonzero constant A. This yields an equation of the form

$$Ax^2 + Ay^2 + Dx + Ey + F = 0 \tag{5}$$

where A, D, E, and F are constants and $A \neq 0$.

If the equation of a circle is given by (4) or (5), then the center and radius can be found by first rewriting the equation in standard form, then reading off the center and radius from that equation. The following example shows how to do this using the technique of ***completing the square***. However, in preparation for the example, recall that completing the square is a method for rewriting an expression of the form

$$x^2 + bx$$

as a difference of two squares. The procedure is to take half the coefficient of x, square it, and then add and subtract that result from the original expression to obtain

$$x^2 + bx = x^2 + bx + (b/2)^2 - (b/2)^2 = [x + (b/2)]^2 - (b/2)^2$$

Example 7

Find the center and radius of the circle with equation

(a) $x^2 + y^2 - 8x + 2y + 8 = 0$ (b) $2x^2 + 2y^2 + 24x - 81 = 0$

Solution (a). First, group the x-terms, group the y-terms, and take the constant to the right side:

$$(x^2 - 8x) + (y^2 + 2y) = -8$$

Next we want to add the appropriate constant within each set of parentheses to complete the square, and subtract the same constant outside the parentheses to maintain equality. The appropriate constant is obtained by taking half the coefficient of the first-degree term and squaring it. This yields

$$(x^2 - 8x + 16) - 16 + (y^2 + 2y + 1) - 1 = -8$$

from which we obtain

$$(x - 4)^2 + (y + 1)^2 = -8 + 16 + 1 \qquad \text{or} \qquad (x - 4)^2 + (y + 1)^2 = 9$$

Thus from (3), the circle has center $(4, -1)$ and radius 3.

Solution (b). The given equation is of form (5). We will first divide through by 2 (the coefficient of the squared terms) to reduce the equation to form (4). Then we will proceed as in part (a) of this example. The computations are as follows:

$$x^2 + y^2 + 12x - \tfrac{81}{2} = 0 \qquad \text{We divided through by 2.}$$

$$(x^2 + 12x) + y^2 = \tfrac{81}{2}$$

$$(x^2 + 12x + 36) + y^2 = \tfrac{81}{2} + 36 \qquad \text{We completed the square.}$$

$$(x + 6)^2 + y^2 = \tfrac{153}{2}$$

From (3), the circle has center $(-6, 0)$ and radius $\sqrt{\frac{153}{2}}$. ◀

DEGENERATE CASES OF A CIRCLE

There is no guarantee that an equation of form (5) represents a circle. For example, suppose that we divide both sides of (5) by A, then complete the squares to obtain

$$(x - x_0)^2 + (y - y_0)^2 = k$$

Depending on the value of k, the following situations occur:

- $(k > 0)$ The graph is a circle with center (x_0, y_0) and radius $\sqrt{k}$.
- $(k = 0)$ The only solution of the equation is $x = x_0$, $y = y_0$, so the graph is the single point (x_0, y_0).
- $(k < 0)$ The equation has no real solutions and consequently no graph.

Example 8

Describe the graphs of

(a) $(x-1)^2 + (y+4)^2 = -9$ (b) $(x-1)^2 + (y+4)^2 = 0$

Solution (a). There are no real values of x and y that will make the left side of the equation negative. Thus, the solution set of the equation is empty, and the equation has no graph.

Solution (b). The only values of x and y that will make the left side of the equation 0 are $x = 1$, $y = -4$. Thus, the graph of the equation is the single point $(1, -4)$. ◀

The following theorem summarizes our observations.

D.3 THEOREM. *An equation of the form*

$$Ax^2 + Ay^2 + Dx + Ey + F = 0 \tag{6}$$

where $A \neq 0$, represents a circle, or a point, or else has no graph.

REMARK. The last two cases in Theorem D.3 are called ***degenerate cases***. In spite of the fact that these degenerate cases can occur, (6) is often called the ***general equation of a circle***.

THE GRAPH of $y = ax^2 + bx + c$

An equation of the form

$$y = ax^2 + bx + c \quad (a \neq 0) \tag{7}$$

is called a ***quadratic equation in x***. Depending on whether a is positive or negative, the graph, which is called a ***parabola***, has one of the two forms shown in Figure D.7. In both cases the parabola is symmetric about a vertical line parallel to the y-axis. This line of symmetry cuts the parabola at a point called the ***vertex***. The vertex is the low point on the curve if $a > 0$ and the high point if $a < 0$.

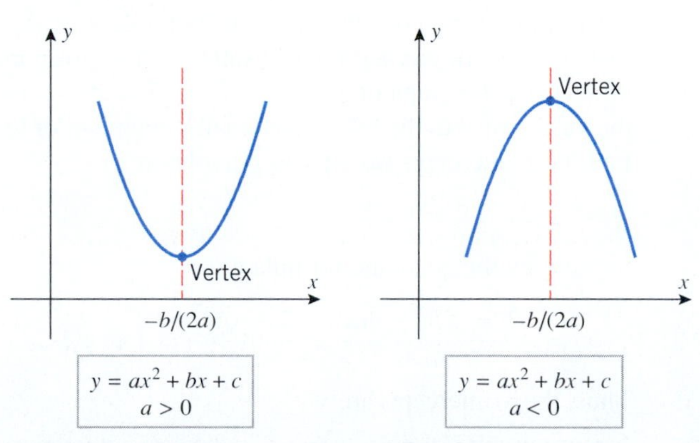

Figure D.7

In the exercises (Exercise 78) we will help the reader show that the x-coordinate of the vertex is given by the formula

$$x = -\frac{b}{2a} \tag{8}$$

With the aid of this formula, a reasonably accurate graph of a quadratic equation in x can be obtained by plotting the vertex and two points on each side of it.

Example 9

Sketch the graph of

(a) $y = x^2 - 2x - 2$ (b) $y = -x^2 + 4x - 5$

Solution (a). The equation is of form (7) with $a = 1$, $b = -2$, and $c = -2$, so by (8) the x-coordinate of the vertex is

$$x = -\frac{b}{2a} = 1$$

Using this value and two additional values on each side, we obtain Figure D.8.

x	$y = x^2 - 2x - 2$
−1	1
0	−2
1	−3
2	−2
3	1

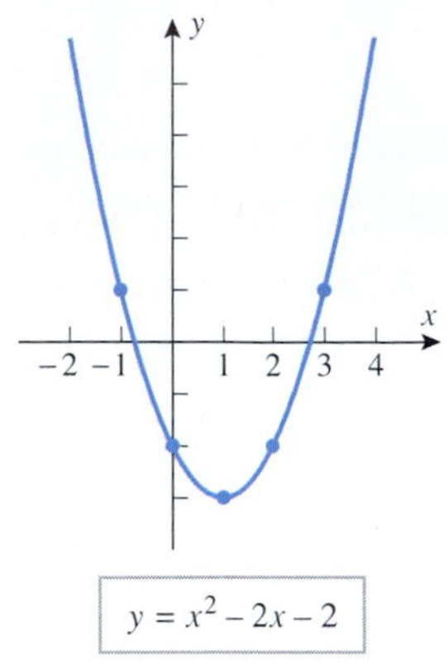

Figure D.8

Solution (b). The equation is of form (7) with $a = -1$, $b = 4$, and $c = -5$, so by (8) the x-coordinate of the vertex is

$$x = -\frac{b}{2a} = 2$$

Using this value and two additional values on each side, we obtain the table and graph in Figure D.9. ◀

x	$y = -x^2 + 4x - 5$
0	−5
1	−2
2	−1
3	−2
4	−5

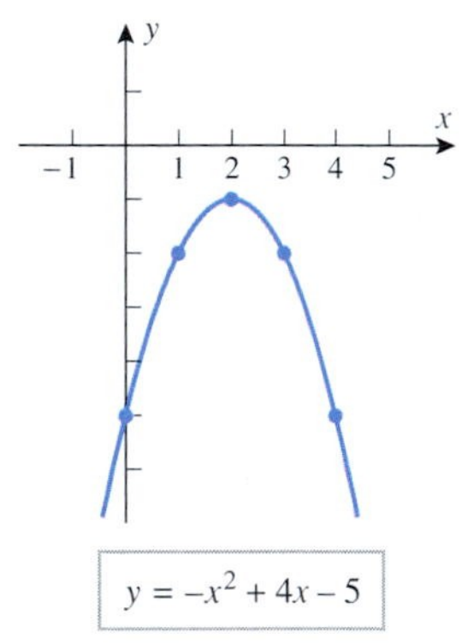

Figure D.9

Quite often the intercepts of a parabola $y = ax^2 + bx + c$ are important to know. The y-intercept, $y = c$, results immediately by setting $x = 0$. However, in order to obtain the x-intercepts, if any, we must set $y = 0$ and then solve the resulting quadratic equation $ax^2 + bx + c = 0$.

Example 10

Solve the inequality

$$x^2 - 2x - 2 > 0$$

Solution. Because the left side of the inequality does not have readily discernible factors, the test-point method illustrated in Example 4 of Appendix A is not convenient to use. Instead, we will give a graphical solution. The given inequality is satisfied for those values of x where the graph of $y = x^2 - 2x - 2$ is above the x-axis. From Figure D.8 those are the values of x to the left of the smaller intercept or to the right of the larger intercept. To find these intercepts we set $y = 0$ to obtain

$$x^2 - 2x - 2 = 0$$

Solving by the quadratic formula gives

$$x = \frac{-b \pm \sqrt{b^2 - 4ac}}{2a} = \frac{2 \pm \sqrt{12}}{2} = 1 \pm \sqrt{3}$$

Thus, the x-intercepts are

$$x = 1 + \sqrt{3} \approx 2.7 \quad \text{and} \quad x = 1 - \sqrt{3} \approx -0.7$$

and the solution set of the inequality is

$$(-\infty, 1 - \sqrt{3}) \cup (1 + \sqrt{3}, +\infty)$$ ◀

REMARK. Note that the decimal approximations of the intercepts calculated in the preceding example agree with the graph in Figure D.8. Observe, however, that we used the exact values of the intercepts to express the solution. The choice of exact versus approximate values is often a matter of judgment that depends on the purpose for which the values are to be used. Numerical approximations often provide a sense of size that exact values do not, but they can introduce severe errors if not used with care.

Example 11

From Figure D.9 we see that the parabola $y = -x^2 + 4x - 5$ has no x-intercepts. This can also be seen algebraically by solving for the x-intercepts. Setting $y = 0$ and solving the resulting equation

$$-x^2 + 4x - 5 = 0$$

by the quadratic formula yields

$$y = \frac{-4 \pm \sqrt{16 - 20}}{-2} = 2 \pm i$$

Because the solutions are complex numbers, there are no (real) x-intercepts. ◀

Example 12

A ball is thrown straight up from the surface of the Earth at time $t = 0$ s with an initial velocity of 24.5 m/s. If air resistance is ignored, it can be shown that the distance s (in meters) of the ball above the ground after t seconds is given by

$$s = 24.5t - 4.9t^2 \qquad (9)$$

(a) Graph s versus t, making the t-axis horizontal and the s-axis vertical.

(b) How high does the ball rise above the ground?

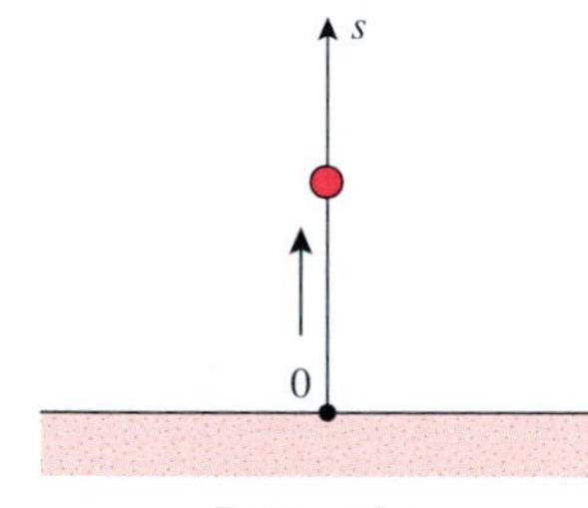

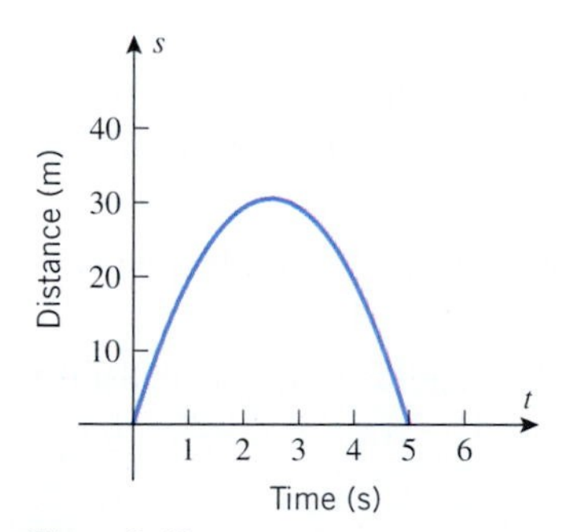

Figure D.10

Solution (a). Equation (9) is of form (7) with $a = -4.9$, $b = 24.5$, and $c = 0$, so by (8) the t-coordinate of the vertex is

$$t = -\frac{b}{2a} = -\frac{24.5}{2(-4.9)} = 2.5 \text{ s}$$

and consequently the s-coordinate of the vertex is

$$s = 24.5(2.5) - 4.9(2.5)^2 = 30.625 \text{ m}$$

The factored form of (9) is

$$s = 4.9t(5 - t)$$

so the graph has t-intercepts $t = 0$ and $t = 5$. From the vertex and the intercepts we obtain the graph shown in Figure D.10.

Solution (b). From the s-coordinate of the vertex we deduce that the ball rises 30.625 m above the ground. ◀

THE GRAPH of $x = ay^2 + by + c$

If x and y are interchanged in (7), the resulting equation,

$$x = ay^2 + by + c$$

is called a ***quadratic equation in y***. The graph of such an equation is a parabola with its line of symmetry parallel to the x-axis and its vertex at the point with y-coordinate $y = -b/(2a)$ (Figure D.11). Some problems relating to such equations appear in the exercises.

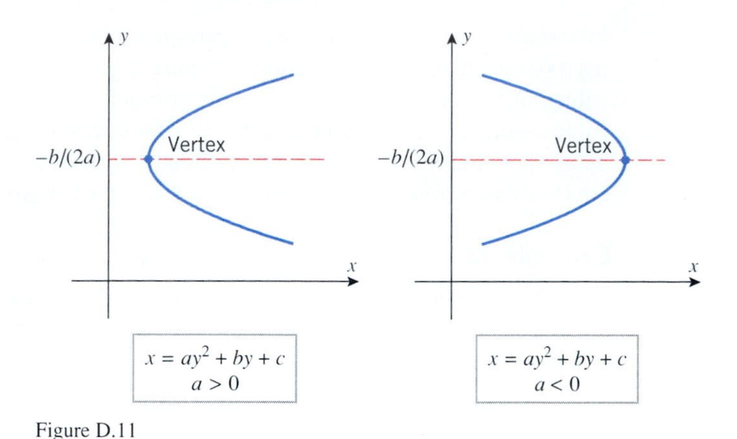

Figure D.11

Exercise Set D

1. Where in this section did we use the fact that the same scale was used on both coordinate axes?

In Exercises 2–5, find
(a) the distance between A and B
(b) the midpoint of the line segment joining A and B.

2. $A(2, 5)$, $B(-1, 1)$

3. $A(7, 1)$, $B(1, 9)$

4. $A(2, 0)$, $B(-3, 6)$

5. $A(-2, -6)$, $B(-7, -4)$

In Exercises 6–10, use the distance formula to solve the given problem.

6. Prove that $(1, 1)$, $(-2, -8)$, and $(4, 10)$ lie on a straight line.

7. Prove that the triangle with vertices $(5, -2)$, $(6, 5)$, $(2, 2)$ is isosceles.

8. Prove that $(1, 3)$, $(4, 2)$, and $(-2, -6)$ are vertices of a right triangle and then specify the vertex at which the right angle occurs.

9. Prove that $(0, -2)$, $(-4, 8)$, and $(3, 1)$ lie on a circle with center $(-2, 3)$.

10. Prove that for all values of t the point $(t, 2t - 6)$ is equidistant from $(0, 4)$ and $(8, 0)$.

11. Find k, given that $(2, k)$ is equidistant from $(3, 7)$ and $(9, 1)$.

12. Find x and y if $(4, -5)$ is the midpoint of the line segment joining $(-3, 2)$ and (x, y).

In Exercises 13 and 14, find an equation of the given line.

13. The line is the perpendicular bisector of the line segment joining $(2, 8)$ and $(-4, 6)$.

14. The line is the perpendicular bisector of the line segment joining $(5, -1)$ and $(4, 8)$.

15. Find the point on the line $4x - 2y + 3 = 0$ that is equidistant from $(3, 3)$ and $(7, -3)$. [*Hint:* First find an equation of the line that is the perpendicular bisector of the line segment joining $(3, 3)$ and $(7, -3)$.]

16. Find the distance from the point $(3, -2)$ to the line
(a) $y = 4$ (b) $x = -1$.

17. Find the distance from $(2, 1)$ to the line $4x - 3y + 10 = 0$. [*Hint:* Find the foot of the perpendicular dropped from the point to the line.]

18. Find the distance from $(8, 4)$ to the line $5x + 12y - 36 = 0$. [*Hint:* See the hint in Exercise 17.]

19. Use the method described in Exercise 17 to prove that the distance d from (x_0, y_0) to the line $Ax + By + C = 0$ is

$$d = \frac{|Ax_0 + By_0 + C|}{\sqrt{A^2 + B^2}}$$

20. Use the formula in Exercise 19 to solve Exercise 17.

21. Use the formula in Exercise 19 to solve Exercise 18.

22. Prove: For any triangle, the perpendicular bisectors of the sides meet at a point. [*Hint:* Position the triangle with one vertex on the y-axis and the opposite side on the x-axis, so that the vertices are $(0, a)$, $(b, 0)$, and $(c, 0)$.]

In Exercises 23 and 24, find the center and radius of each circle.

23. (a) $x^2 + y^2 = 25$
(b) $(x-1)^2 + (y-4)^2 = 16$
(c) $(x+1)^2 + (y+3)^2 = 5$
(d) $x^2 + (y+2)^2 = 1$

24. (a) $x^2 + y^2 = 9$
(b) $(x-3)^2 + (y-5)^2 = 36$
(c) $(x+4)^2 + (y+1)^2 = 8$
(d) $(x+1)^2 + y^2 = 1$

In Exercises 25–32, find the standard equation of the circle satisfying the given conditions.

25. Center $(3, -2)$; radius = 4.

26. Center $(1, 0)$; diameter = $\sqrt{8}$.

27. Center $(-4, 8)$; circle is tangent to the x-axis.

28. Center $(5, 8)$; circle is tangent to the y-axis.

29. Center $(-3, -4)$; circle passes through the origin.

30. Center $(4, -5)$; circle passes through $(1, 3)$.

31. A diameter has endpoints $(2, 0)$ and $(0, 2)$.

32. A diameter has endpoints $(6, 1)$ and $(-2, 3)$.

In Exercises 33–44, determine whether the equation represents a circle, a point, or no graph. If the equation represents a circle, find the center and radius.

33. $x^2 + y^2 - 2x - 4y - 11 = 0$

34. $x^2 + y^2 + 8x + 8 = 0$

35. $2x^2 + 2y^2 + 4x - 4y = 0$

36. $6x^2 + 6y^2 - 6x + 6y = 3$

37. $x^2 + y^2 + 2x + 2y + 2 = 0$

38. $x^2 + y^2 - 4x - 6y + 13 = 0$

39. $9x^2 + 9y^2 = 1$

40. $(x^2/4) + (y^2/4) = 1$

41. $x^2 + y^2 + 10y + 26 = 0$

42. $x^2 + y^2 - 10x - 2y + 29 = 0$

43. $16x^2 + 16y^2 + 40x + 16y - 7 = 0$

44. $4x^2 + 4y^2 - 16x - 24y = 9$

45. Find an equation of
(a) the bottom half of the circle $x^2 + y^2 = 16$
(b) the top half of the circle $x^2 + y^2 + 2x - 4y + 1 = 0$.

46. Find an equation of
(a) the right half of the circle $x^2 + y^2 = 9$
(b) the left half of the circle $x^2 + y^2 - 4x + 3 = 0$.

47. Graph
(a) $y = \sqrt{25 - x^2}$ (b) $y = \sqrt{5 + 4x - x^2}$.

48. Graph
(a) $x = -\sqrt{4 - y^2}$ (b) $x = 3 + \sqrt{4 - y^2}$.

49. Find an equation of the line that is tangent to the circle

$$x^2 + y^2 = 25$$

at the point $(3, 4)$ on the circle.

50. Find an equation of the line that is tangent to the circle at the point P on the circle
(a) $x^2 + y^2 + 2x = 9$; $P(2, -1)$
(b) $x^2 + y^2 - 6x + 4y = 13$; $P(4, 3)$.

51. For the circle $x^2 + y^2 = 20$ and the point $P(-1, 2)$:
(a) Is P inside, outside, or on the circle?
(b) Find the largest and smallest distances between P and points on the circle.

52. Follow the directions of Exercise 51 for the circle

$$x^2 + y^2 - 2y - 4 = 0$$

and the point $P\left(3, \frac{5}{2}\right)$.

53. Referring to the accompanying figure, find the coordinates of the points T and T', where the lines L and L' are tangent to the circle of radius 1 with center at the origin.

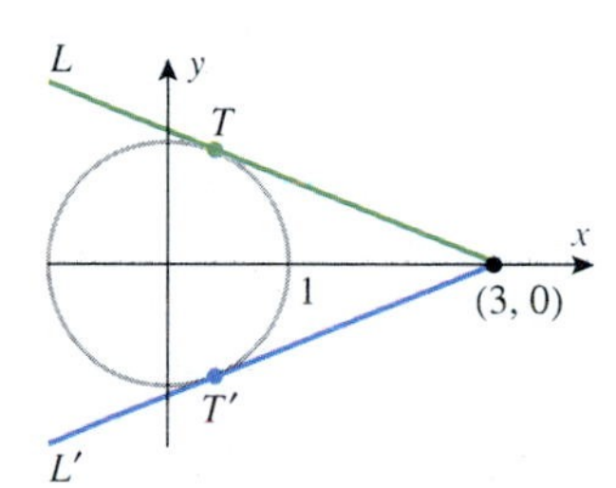

Figure Ex-53

54. A point (x, y) moves so that its distance to $(2, 0)$ is $\sqrt{2}$ times its distance to $(0, 1)$.
(a) Show that the point moves along a circle.
(b) Find the center and radius.

55. A point (x, y) moves so that the sum of the squares of its distances from $(4, 1)$ and $(2, -5)$ is 45.
(a) Show that the point moves along a circle.
(b) Find the center and radius.

56. Find all values of c for which the system of equations

$$\begin{cases} x^2 - y^2 = 0 \\ (x - c)^2 + y^2 = 1 \end{cases}$$

has 0, 1, 2, 3, or 4 solutions. [*Hint:* Sketch a graph.]

In Exercises 57–70, graph the parabola and label the coordinates of the vertex and the intersections with the coordinate axes.

57. $y = x^2 + 2$

58. $y = x^2 - 3$

59. $y = x^2 + 2x - 3$

60. $y = x^2 - 3x - 4$

61. $y = -x^2 + 4x + 5$

62. $y = -x^2 + x$

63. $y = (x - 2)^2$

64. $y = (3 + x)^2$

65. $x^2 - 2x + y = 0$

66. $x^2 + 8x + 8y = 0$

67. $y = 3x^2 - 2x + 1$

68. $y = x^2 + x + 2$

69. $x = -y^2 + 2y + 2$

70. $x = y^2 - 4y + 5$

71. Find an equation of
(a) the right half of the parabola $y = 3 - x^2$
(b) the left half of the parabola $y = x^2 - 2x$.

72. Find an equation of
(a) the upper half of the parabola $x = y^2 - 5$
(b) the lower half of the parabola $x = y^2 - y - 2$.

73. Graph
(a) $y = \sqrt{x+5}$ (b) $x = -\sqrt{4-y}$.

74. Graph
(a) $y = 1 + \sqrt{4-x}$ (b) $x = 3 + \sqrt{y}$.

75. If a ball is thrown straight up with an initial velocity of 32 ft/s, then after t seconds the distance s above its starting height, in feet, is given by $s = 32t - 16t^2$.
(a) Graph this equation in a ts-coordinate system (t-axis horizontal).
(b) At what time t will the ball be at its highest point, and how high will it rise?

76. A rectangular field is to be enclosed with 500 ft of fencing along three sides and by a straight stream on the fourth side. Let x be the length of each side perpendicular to the stream, and let y be the length of the side parallel to the stream.
(a) Express y in terms of x.
(b) Express the area A of the field in terms of x.
(c) What is the largest area that can be enclosed?

77. A rectangular plot of land is to be enclosed using two kinds of fencing. Two opposite sides will have heavy-duty fencing costing \$3/ft, while the other two sides will have standard fencing costing \$2/ft. A total of \$600 is available for the fencing. Let x be the length of each side with the heavy-duty fencing, and let y be the length of each side with the standard fencing.
(a) Express y in terms of x.
(b) Find a formula for the area A of the rectangular plot in terms of x.
(c) What is the largest area that can be enclosed?

78. (a) By completing the square, show that the quadratic equation $y = ax^2 + bx + c$ can be rewritten as

$$y = a\left(x + \frac{b}{2a}\right)^2 + \left(c - \frac{b^2}{4a}\right)$$

if $a \neq 0$.
(b) Use the result in part (a) to show that the graph of the quadratic equation $y = ax^2 + bx + c$ has its high point at $x = -b/(2a)$ if $a < 0$ and its low point there if $a > 0$.

In Exercises 79 and 80, solve the given inequality.

79. (a) $2x^2 + 5x - 1 < 0$ (b) $x^2 - 2x + 3 > 0$

80. (a) $x^2 + x - 1 > 0$ (b) $x^2 - 4x + 6 < 0$

81. At time $t = 0$ a ball is thrown straight up from a height of 5 ft above the ground. After t seconds its distance s, in feet, above the ground is given by $s = 5 + 40t - 16t^2$.
(a) Find the maximum height of the ball above the ground.
(b) Find, to the nearest tenth of a second, the time when the ball strikes the ground.
(c) Find, to the nearest tenth of a second, how long the ball will be more than 12 ft above the ground.

82. Find all values of x at which points on the parabola $y = x^2$ lie below the line $y = x + 3$.

APPENDIX E

Trigonometry Review

TRIGONOMETRIC FUNCTIONS AND IDENTITIES

ANGLES

Angles in the plane can be generated by rotating a ray about its endpoint. The starting position of the ray is called the ***initial side*** of the angle, the final position is called the ***terminal side*** of the angle, and the point at which the initial and terminal sides meet is called the ***vertex*** of the angle. We allow for the possibility that the ray may make more than one complete revolution. Angles are considered to be ***positive*** if generated counterclockwise and ***negative*** if generated clockwise (Figure E.1).

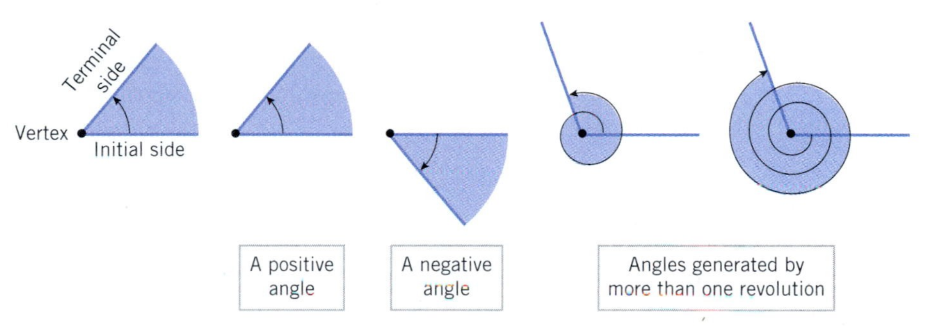

Figure E.1

There are two standard measurement systems for describing the size of an angle: ***degree measure*** and ***radian measure***. In degree measure, one degree (written 1°) is the measure of an angle generated by $1/360$ of one revolution. Thus, there are 360° in an angle of one revolution, 180° in an angle of one-half revolution, 90° in an angle of one-quarter revolution (a *right angle*), and so forth. Degrees are divided into sixty equal parts, called ***minutes***, and minutes are divided into sixty equal parts, called ***seconds***. Thus, one minute (written $1'$) is $1/60$ of a degree, and one second (written $1''$) is $1/60$ of a minute. Smaller subdivisions of a degree are expressed as fractions of a second.

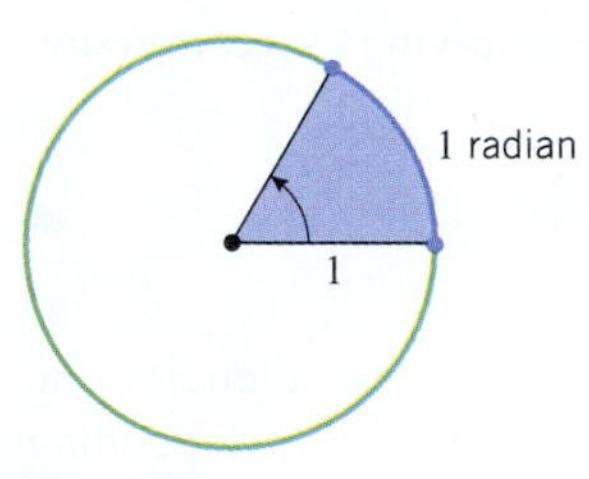

Figure E.2

In radian measure, angles are measured by the length of the arc that the angle subtends on a circle of radius 1 when the vertex is at the center. One unit of arc on a circle of radius 1 is called one ***radian*** (written 1 radian or 1 rad) (Figure E.2), and hence the entire circumference of a circle of radius 1 is 2π radians. It follows that an angle of 360° subtends an arc of 2π radians, an angle of 180° subtends an arc of π radians, an angle of 90° subtends an arc of $\pi/2$ radians, and so forth. Figure E.3 and Table 1 show the relationship between degree measure and radian measure for some important positive angles.

REMARK. Observe that in Table 1, angles in degrees are designated by the degree symbol, but angles in radians have no units specified. This is standard practice—when no units are specified for an angle, it is understood that the units are radians.

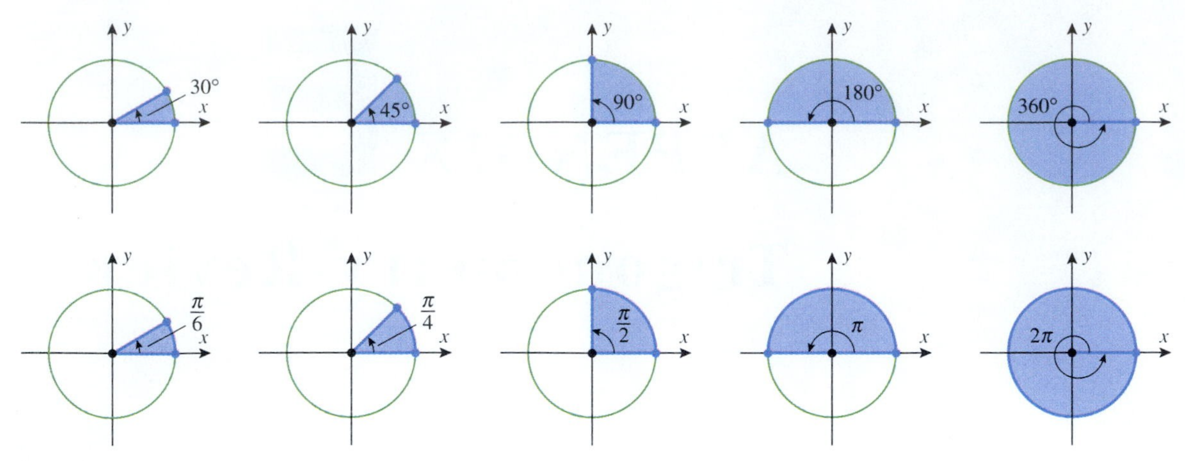

Figure E.3

Table 1

DEGREES	30°	45°	60°	90°	120°	135°	150°	180°	270°	360°
RADIANS	$\frac{\pi}{6}$	$\frac{\pi}{4}$	$\frac{\pi}{3}$	$\frac{\pi}{2}$	$\frac{2\pi}{3}$	$\frac{3\pi}{4}$	$\frac{5\pi}{6}$	π	$\frac{3\pi}{2}$	2π

From the fact that π radians corresponds to 180°, we obtain the following formulas, which are useful for converting from degrees to radians and conversely.

$$1^\circ = \frac{\pi}{180}\text{rad} \approx 0.01745 \text{ rad} \tag{1}$$

$$1 \text{ rad} = \left(\frac{180}{\pi}\right)^\circ \approx 57^\circ 17' 44.8'' \tag{2}$$

Example 1

(a) Express 146° in radians. (b) Express 3 radians in degrees.

Solution (a). From (1), degrees can be converted to radians by multiplying by a conversion factor of $\pi/180$. Thus,

$$146^\circ = \left(\frac{\pi}{180} \cdot 146\right) \text{rad} = \frac{73\pi}{90} \text{ rad} \approx 2.5482 \text{ rad}$$

Solution (b). From (2), radians can be converted to degrees by multiplying by a conversion factor of $180/\pi$. Thus,

$$3 \text{ rad} = \left(3 \cdot \frac{180}{\pi}\right)^\circ = \left(\frac{540}{\pi}\right)^\circ \approx 171.9^\circ$$ ◀

RELATIONSHIPS BETWEEN ARC LENGTH, ANGLE, RADIUS, AND AREA

There is a theorem from plane geometry which states that for two concentric circles, the ratio of the arc lengths subtended by a central angle is equal to the ratio of the corresponding radii (Figure E.4). In particular, if s is the arc length subtended on a circle of radius r by a central angle of θ radians, then by comparison with the arc length subtended by that angle on a circle of radius 1 we obtain

$$\frac{s}{\theta} = \frac{r}{1}$$

from which we obtain the following relationships between the central angle θ, the radius r, and the subtended arc length s when θ is in radians (Figure E.5):

$$\theta = s/r \quad \text{and} \quad s = r\theta \tag{3–4}$$

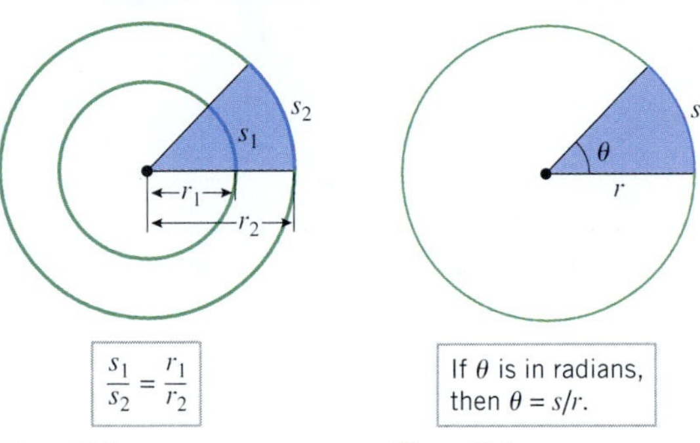

Figure E.4

Figure E.5

The shaded region in Figure E.5 is called a ***sector***. It is a theorem from plane geometry that the ratio of the area A of this sector to the area of the entire circle is the same as the ratio of the central angle of the sector to the central angle of the entire circle; thus, if the angles are in radians, we have

$$\frac{A}{\pi r^2} = \frac{\theta}{2\pi}$$

Solving for A yields the following formula for the area of a sector in terms of the radius r and the angle θ in radians:

$$A = \tfrac{1}{2}r^2\theta \tag{5}$$

TRIGONOMETRIC FUNCTIONS FOR RIGHT TRIANGLES

The ***sine***, ***cosine***, ***tangent***, ***cosecant***, ***secant***, and ***cotangent*** of a positive acute angle θ can be defined as ratios of the sides of a right triangle. Using the notation from Figure E.6, these definitions take the following form:

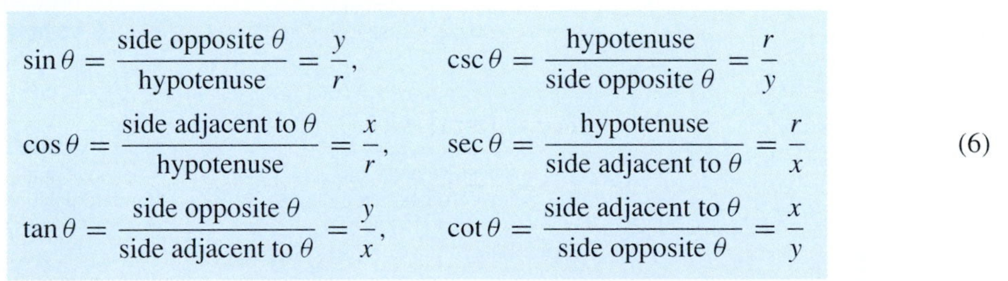

$$\sin\theta = \frac{\text{side opposite } \theta}{\text{hypotenuse}} = \frac{y}{r}, \qquad \csc\theta = \frac{\text{hypotenuse}}{\text{side opposite } \theta} = \frac{r}{y}$$
$$\cos\theta = \frac{\text{side adjacent to } \theta}{\text{hypotenuse}} = \frac{x}{r}, \qquad \sec\theta = \frac{\text{hypotenuse}}{\text{side adjacent to } \theta} = \frac{r}{x} \tag{6}$$
$$\tan\theta = \frac{\text{side opposite } \theta}{\text{side adjacent to } \theta} = \frac{y}{x}, \qquad \cot\theta = \frac{\text{side adjacent to } \theta}{\text{side opposite } \theta} = \frac{x}{y}$$

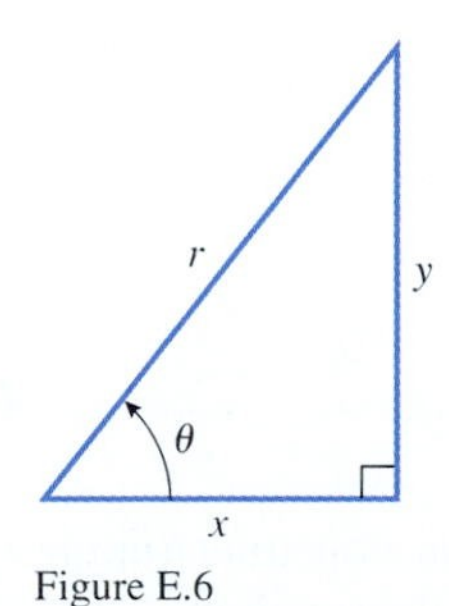

Figure E.6

We will call sin, cos, tan, csc, sec, and cot the ***trigonometric functions***. Because similar triangles have proportional sides, the values of the trigonometric functions depend only on the size of θ and not on the particular right triangle used to compute the ratios. Moreover, in these definitions it does not matter whether θ is measured in degrees or radians.

Example 2

Recall from geometry that the two legs of a $45°–45°–90°$ triangle are of equal size and that the hypotenuse of a $30°–60°–90°$ triangle is twice the shorter leg, where the shorter leg is opposite the $30°$ angle. These facts and the Theorem of Pythagoras yield Figure E.7. From that figure we obtain the results in Table 2.

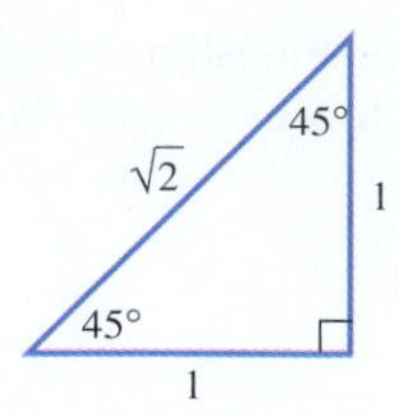

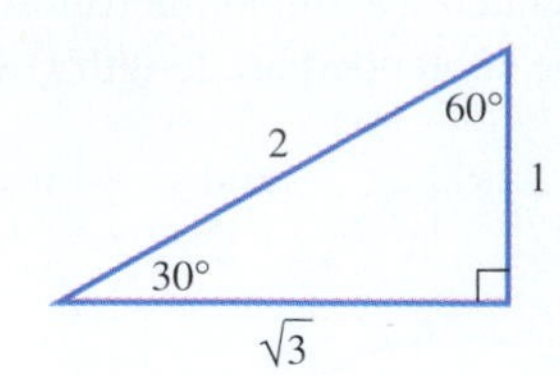

Figure E.7

Table 2

$\sin 45° = 1/\sqrt{2}$,	$\cos 45° = 1/\sqrt{2}$,	$\tan 45° = 1$
$\csc 45° = \sqrt{2}$,	$\sec 45° = \sqrt{2}$,	$\cot 45° = 1$
$\sin 30° = 1/2$,	$\cos 30° = \sqrt{3}/2$,	$\tan 30° = 1/\sqrt{3}$
$\csc 30° = 2$,	$\sec 30° = 2/\sqrt{3}$,	$\cot 30° = \sqrt{3}$
$\sin 60° = \sqrt{3}/2$,	$\cos 60° = 1/2$,	$\tan 60° = \sqrt{3}$
$\csc 60° = 2/\sqrt{3}$,	$\sec 60° = 2$,	$\cot 60° = 1/\sqrt{3}$

◀

ANGLES IN RECTANGULAR COORDINATE SYSTEMS

Because the angles of a right triangle are between 0° and 90°, the formulas in (6) are not directly applicable to negative angles or to angles greater than 90°. To extend the trigonometric functions to include these cases, it will be convenient to consider angles in rectangular coordinate systems. An angle is said to be in ***standard position*** in an xy-coordinate system if its vertex is at the origin and its initial side is on the positive x-axis (Figure E.8).

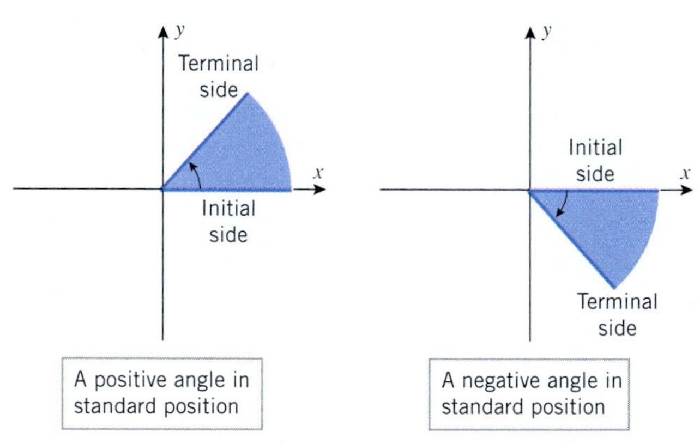

Figure E.8

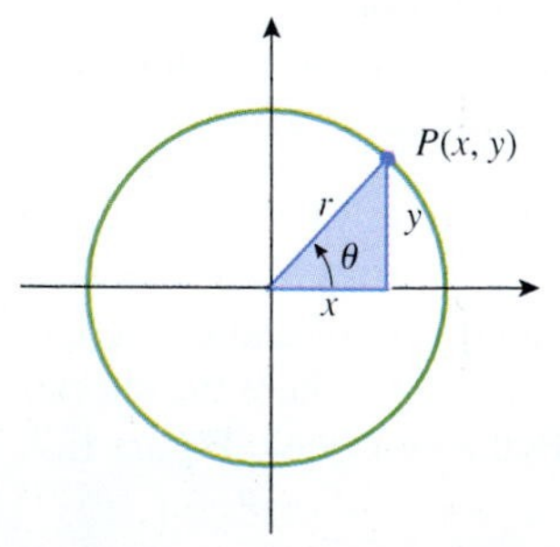

Figure E.9

To define the trigonometric functions of an angle θ in standard position, construct a circle of radius r, centered at the origin, and let $P(x, y)$ be the intersection of the terminal side of θ with this circle (Figure E.9). We make the following definition.

E.1 DEFINITION.

$$\sin\theta = \frac{y}{r}, \quad \cos\theta = \frac{x}{r}, \quad \tan\theta = \frac{y}{x}$$

$$\csc\theta = \frac{r}{y}, \quad \sec\theta = \frac{r}{x}, \quad \cot\theta = \frac{x}{y}$$

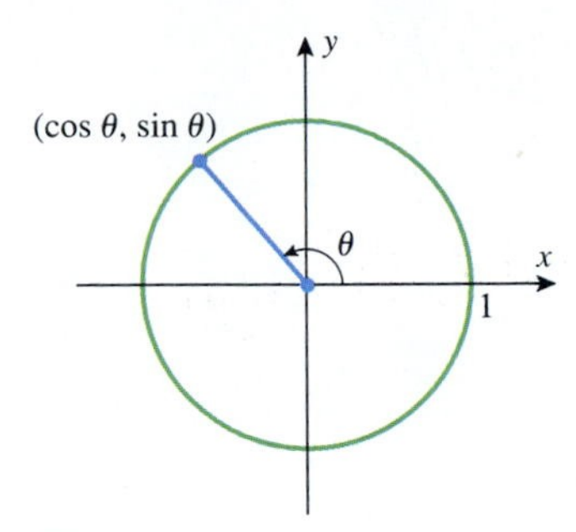

Figure E.10

Note that the formulas in this definition agree with those in (6), so there is no conflict with the earlier definition of the trigonometric functions for triangles. However, this definition applies to all angles (except for cases where a zero denominator occurs).

In the special case where $r = 1$, we have $\sin\theta = y$ and $\cos\theta = x$, so the terminal side of the angle θ intersects the unit circle at the point $(\cos\theta, \sin\theta)$ (Figure E.10). It follows from Definition E.1 that the remaining trigonometric functions of θ are expressible as (verify)

$$\tan\theta = \frac{\sin\theta}{\cos\theta}, \quad \cot\theta = \frac{\cos\theta}{\sin\theta} = \frac{1}{\tan\theta}, \quad \sec\theta = \frac{1}{\cos\theta}, \quad \csc\theta = \frac{1}{\sin\theta} \tag{7–10}$$

These observations suggest the following procedure for evaluating the trigonometric functions of common angles:

- Construct the angle θ in standard position in an xy-coordinate system.
- Find the coordinates of the intersection of the terminal side of the angle and the unit circle; the x- and y-coordinates of this intersection are the values of $\cos\theta$ and $\sin\theta$, respectively.
- Use Formulas (7) through (10) to find the values of the remaining trigonometric functions from the values of $\cos\theta$ and $\sin\theta$.

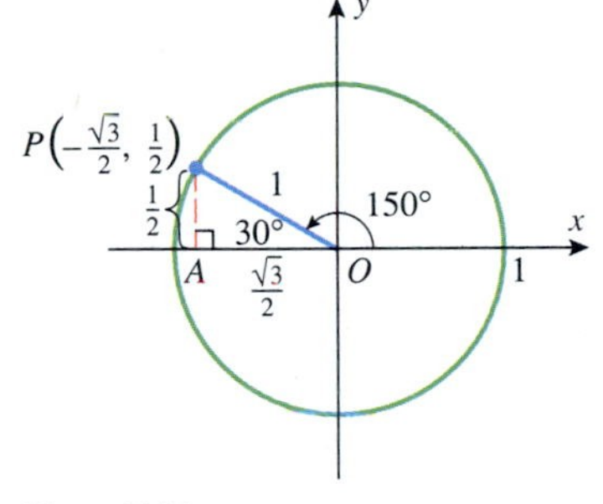

Figure E.11

Example 3

Evaluate the trigonometric functions of $\theta = 150°$.

Solution. Construct a unit circle and place the angle $\theta = 150°$ in standard position (Figure E.11). Since $\angle AOP$ is 30° and $\triangle OAP$ is a 30°–60°–90° triangle, the leg AP has length $\frac{1}{2}$ (half the hypotenuse) and the leg OA has length $\sqrt{3}/2$ by the Theorem of Pythagoras. Thus, the coordinates of P are $(-\sqrt{3}/2, 1/2)$, from which we obtain

$$\sin 150° = \frac{1}{2}, \quad \cos 150° = -\frac{\sqrt{3}}{2}, \quad \tan 150° = \frac{\sin 150°}{\cos 150°} = \frac{1/2}{-\sqrt{3}/2} = -\frac{1}{\sqrt{3}}$$

$$\csc 150° = \frac{1}{\sin 150°} = 2, \quad \sec 150° = \frac{1}{\cos 150°} = -\frac{2}{\sqrt{3}}$$

$$\cot 150° = \frac{1}{\tan 150°} = -\sqrt{3}$$

◀

Example 4

Evaluate the trigonometric functions of $\theta = 5\pi/6$.

Solution. Since $5\pi/6 = 150°$, this problem is equivalent to that of Example 3. From that example we obtain

$$\sin\frac{5\pi}{6} = \frac{1}{2}, \quad \cos\frac{5\pi}{6} = -\frac{\sqrt{3}}{2}, \quad \tan\frac{5\pi}{6} = -\frac{1}{\sqrt{3}}$$

$$\csc\frac{5\pi}{6} = 2, \quad \sec\frac{5\pi}{6} = -\frac{2}{\sqrt{3}}, \quad \cot\frac{5\pi}{6} = -\sqrt{3}$$

◀

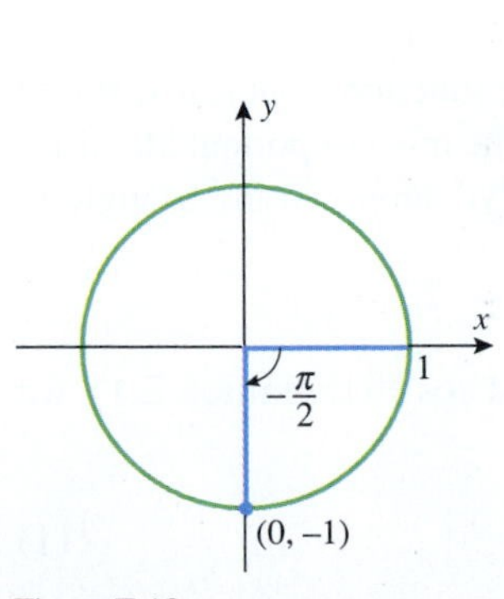

Figure E.12

Example 5

Evaluate the trigonometric functions of $\theta = -\pi/2$.

Solution. As shown in Figure E.12, the terminal side of $\theta = -\pi/2$ intersects the unit circle at the point $(0, -1)$, so

$$\sin(-\pi/2) = -1, \quad \cos(-\pi/2) = 0$$

and from Formulas (7) through (10),

$$\tan(-\pi/2) = \frac{\sin(-\pi/2)}{\cos(-\pi/2)} = \frac{-1}{0} \quad \text{(undefined)}$$

$$\cot(-\pi/2) = \frac{\cos(-\pi/2)}{\sin(-\pi/2)} = \frac{0}{-1} = 0$$

$$\sec(-\pi/2) = \frac{1}{\cos(-\pi/2)} = \frac{1}{0} \quad \text{(undefined)}$$

$$\csc(-\pi/2) = \frac{1}{\sin(-\pi/2)} = \frac{1}{-1} = -1$$

◀

The reader should be able to obtain all of the results in Table 3 by the methods illustrated in the last three examples. The dashes indicate quantities that are undefined.

Table 3

	$\theta = 0$ (0°)	$\pi/6$ (30°)	$\pi/4$ (45°)	$\pi/3$ (60°)	$\pi/2$ (90°)	$2\pi/3$ (120°)	$3\pi/4$ (135°)	$5\pi/6$ (150°)	π (180°)	$3\pi/2$ (270°)	2π (360°)
$\sin\theta$	0	1/2	$1/\sqrt{2}$	$\sqrt{3}/2$	1	$\sqrt{3}/2$	$1/\sqrt{2}$	1/2	0	−1	0
$\cos\theta$	1	$\sqrt{3}/2$	$1/\sqrt{2}$	1/2	0	−1/2	$-1/\sqrt{2}$	$-\sqrt{3}/2$	−1	0	1
$\tan\theta$	0	$1/\sqrt{3}$	1	$\sqrt{3}$	—	$-\sqrt{3}$	−1	$-1/\sqrt{3}$	0	—	0
$\csc\theta$	—	2	$\sqrt{2}$	$2/\sqrt{3}$	1	$2/\sqrt{3}$	$\sqrt{2}$	2	—	−1	—
$\sec\theta$	1	$2/\sqrt{3}$	$\sqrt{2}$	2	—	−2	$-\sqrt{2}$	$-2/\sqrt{3}$	−1	—	1
$\cot\theta$	—	$\sqrt{3}$	1	$1/\sqrt{3}$	0	$-1/\sqrt{3}$	−1	$-\sqrt{3}$	—	0	—

REMARK. It is only in special cases that exact values for trigonometric functions can be obtained; usually, a calculating utility or a computer program will be required.

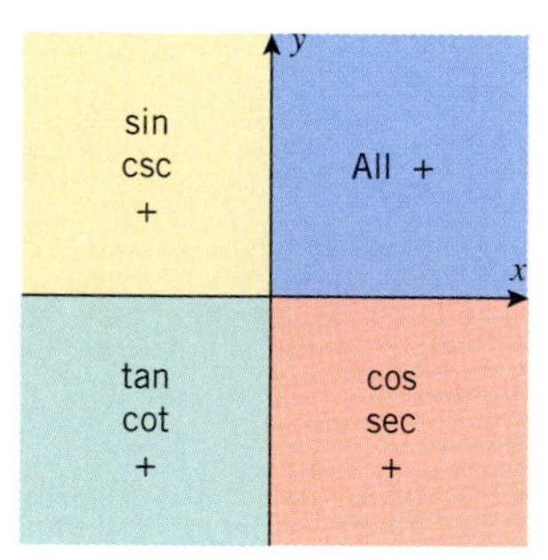

Figure E.13

The signs of the trigonometric functions of an angle are determined by the quadrant in which the terminal side of the angle falls. For example, if the terminal side falls in the first quadrant, then x and y are positive in Definition E.1, so all of the trigonometric functions have positive values. If the terminal side falls in the second quadrant, then x is negative and y is positive, so sin and csc are positive, but all other trigonometric functions are negative. The diagram in Figure E.13 shows which trigonometric functions are positive in the various quadrants. The reader will find it instructive to check that the results in Table 3 are consistent with Figure E.13.

TRIGONOMETRIC IDENTITIES

A ***trigonometric identity*** is an equation involving trigonometric functions that is true for all angles for which both sides of the equation are defined. One of the most important identities in trigonometry can be derived by applying the Theorem of Pythagoras to the triangle in Figure E.9 to obtain

$$x^2 + y^2 = r^2$$

Dividing both sides by r^2 and using the definitions of $\sin\theta$ and $\cos\theta$ (Definition E.1), we obtain the following fundamental result:

$$\sin^2\theta + \cos^2\theta = 1 \tag{11}$$

The following identities can be obtained from (11) by dividing through by $\cos^2\theta$ and $\sin^2\theta$, respectively, then applying Formulas (7) through (10):

$$\tan^2\theta + 1 = \sec^2\theta \tag{12}$$

$$1 + \cot^2\theta = \csc^2\theta \tag{13}$$

If (x, y) is a point on the unit circle, then the points $(-x, y)$, $(-x, -y)$, and $(x, -y)$ also lie on the unit circle (why?), and the four points form corners of a rectangle with sides parallel to the coordinate axes (Figure E.14*a*). The x- and y-coordinates of each corner represent the cosine and sine of an angle in standard position whose terminal side passes through the corner; hence we obtain the identities in parts (*b*), (*c*), and (*d*) of Figure E.14 for sine and cosine. Dividing those identities leads to identities for the tangent. In summary:

$$\sin(\pi-\theta) = \sin\theta, \quad \sin(\pi+\theta) = -\sin\theta, \quad \sin(-\theta) = -\sin\theta \tag{14–16}$$

$$\cos(\pi-\theta) = -\cos\theta, \quad \cos(\pi+\theta) = -\cos\theta, \quad \cos(-\theta) = \cos\theta \tag{17–19}$$

$$\tan(\pi-\theta) = -\tan\theta, \quad \tan(\pi+\theta) = \tan\theta, \quad \tan(-\theta) = -\tan\theta \tag{20–22}$$

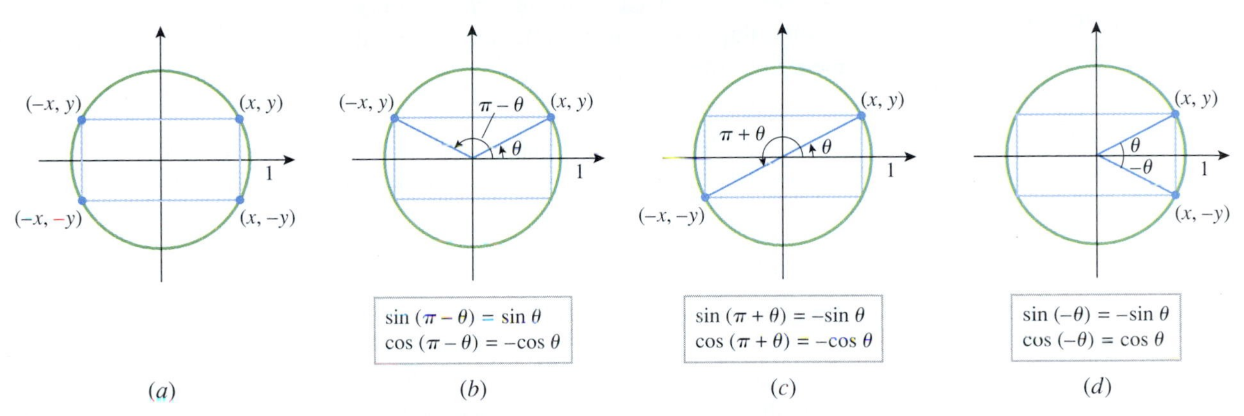

Figure E.14

Two angles in standard position that have the same terminal side must have the same values for their trigonometric functions since their terminal sides intersect the unit circle at the same point. In particular, two angles whose radian measures differ by a multiple of 2π have the same terminal side and hence have the same values for their trigonometric functions. This yields the identities

$$\sin\theta = \sin(\theta + 2\pi) = \sin(\theta - 2\pi) \tag{23}$$

$$\cos\theta = \cos(\theta + 2\pi) = \cos(\theta - 2\pi) \tag{24}$$

and more generally,

$$\sin\theta = \sin(\theta \pm 2n\pi), \quad n = 0, 1, 2, \ldots \tag{25}$$

$$\cos\theta = \cos(\theta \pm 2n\pi), \quad n = 0, 1, 2, \ldots \tag{26}$$

Identities (20) through (22) imply that

$$\tan\theta = \tan(\theta + \pi) \quad \text{and} \quad \tan\theta = \tan(\theta - \pi) \tag{27–28}$$

Identity (27) is just (21) with the terms in the sum reversed, and identity (28) follows from (20) and (22) (verify). These two identities state that adding or subtracting π from an angle does not affect the value of the tangent of the angle. It follows that the same is true for any multiple of π; thus,

$$\tan\theta = \tan(\theta \pm n\pi), \quad n = 0, 1, 2, \ldots \tag{29}$$

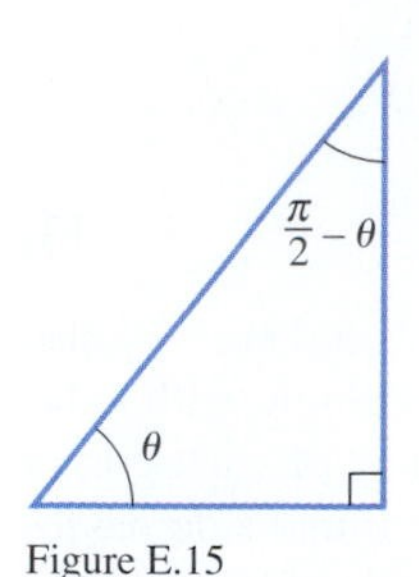

Figure E.15

Figure E.15 shows complementary angles θ and $(\pi/2) - \theta$ of a right triangle. It follows from (6) that

$$\sin\theta = \frac{\text{side opposite } \theta}{\text{hypotenuse}} = \frac{\text{side adjacent to } (\pi/2) - \theta}{\text{hypotenuse}} = \cos\left(\frac{\pi}{2} - \theta\right)$$

$$\cos\theta = \frac{\text{side adjacent to } \theta}{\text{hypotenuse}} = \frac{\text{side opposite } (\pi/2) - \theta}{\text{hypotenuse}} = \sin\left(\frac{\pi}{2} - \theta\right)$$

which yields the identities

$$\sin\left(\frac{\pi}{2} - \theta\right) = \cos\theta, \quad \cos\left(\frac{\pi}{2} - \theta\right) = \sin\theta, \quad \tan\left(\frac{\pi}{2} - \theta\right) = \cot\theta \tag{30–32}$$

where the third identity results from dividing the first two. These identities are also valid for angles that are not acute and for negative angles as well.

THE LAW OF COSINES

The next theorem, called the ***law of cosines***, generalizes the Theorem of Pythagoras. This result is important in its own right and is also the starting point for some important trigonometric identities.

> **E.2 THEOREM (*Law of Cosines*).** *If the sides of a triangle have lengths a, b, and c, and if θ is the angle between the sides with lengths a and b, then*
>
> $$c^2 = a^2 + b^2 - 2ab\cos\theta$$

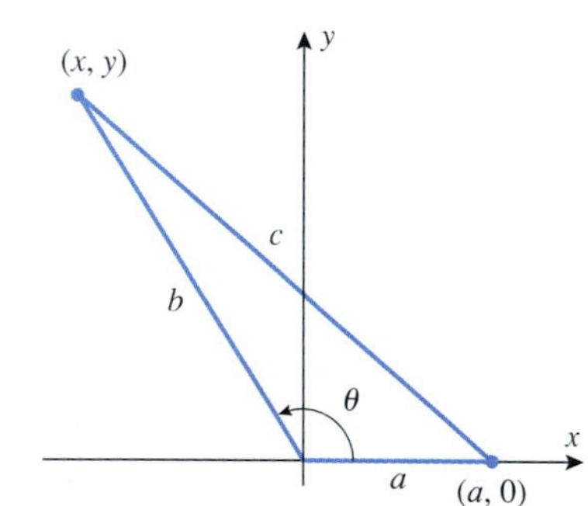

Figure E.16

Proof. Introduce a coordinate system so that θ is in standard position and the side of length a falls along the positive x-axis. As shown in Figure E.16, the side of length a extends from the origin to $(a, 0)$ and the side of length b extends from the origin to some point (x, y). From the definition of $\sin\theta$ and $\cos\theta$ we have $\sin\theta = y/b$ and $\cos\theta = x/b$, so

$$y = b\sin\theta, \quad x = b\cos\theta \tag{33}$$

From the distance formula in Theorem D.1 of Appendix D, we obtain

$$c^2 = (x - a)^2 + (y - 0)^2$$

so that, from (33),

$$\begin{aligned} c^2 &= (b\cos\theta - a)^2 + b^2\sin^2\theta \\ &= a^2 + b^2(\cos^2\theta + \sin^2\theta) - 2ab\cos\theta \\ &= a^2 + b^2 - 2ab\cos\theta \end{aligned}$$

which completes the proof. ∎

We will now show how the law of cosines can be used to obtain the following identities, called the ***addition formulas*** for sine and cosine:

$$\sin(\alpha + \beta) = \sin\alpha\cos\beta + \cos\alpha\sin\beta \tag{34}$$

$$\cos(\alpha + \beta) = \cos\alpha\cos\beta - \sin\alpha\sin\beta \tag{35}$$

$$\sin(\alpha - \beta) = \sin\alpha\cos\beta - \cos\alpha\sin\beta \tag{36}$$

$$\cos(\alpha - \beta) = \cos\alpha\cos\beta + \sin\alpha\sin\beta \tag{37}$$

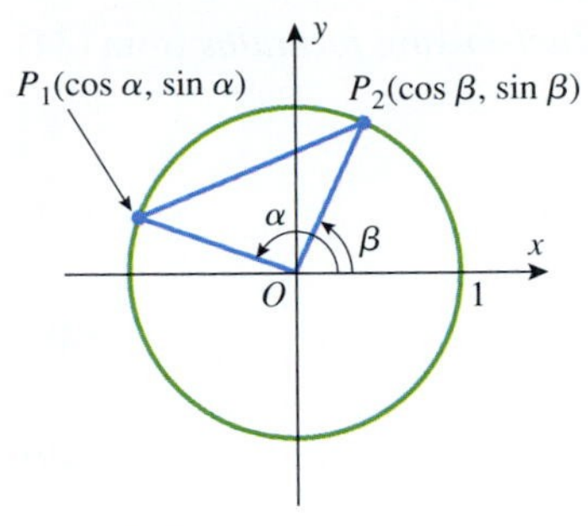

Figure E.17

We will derive (37) first. In our derivation we will assume that $0 \leq \beta < \alpha < 2\pi$ (Figure E.17). As shown in the figure, the terminal sides of α and β intersect the unit circle at the points $P_1(\cos\alpha, \sin\alpha)$ and $P_2(\cos\beta, \sin\beta)$. If we denote the lengths of the sides of triangle OP_1P_2 by OP_1, P_1P_2, and OP_2, then $OP_1 = OP_2 = 1$ and, from the distance formula in Theorem D.1 of Appendix D,

$$\begin{aligned}(P_1P_2)^2 &= (\cos\beta - \cos\alpha)^2 + (\sin\beta - \sin\alpha)^2 \\ &= (\sin^2\alpha + \cos^2\alpha) + (\sin^2\beta + \cos^2\beta) - 2(\cos\alpha\cos\beta + \sin\alpha\sin\beta) \\ &= 2 - 2(\cos\alpha\cos\beta + \sin\alpha\sin\beta)\end{aligned}$$

But angle $P_2OP_1 = \alpha - \beta$, so that the law of cosines yields

$$\begin{aligned}(P_1P_2)^2 &= (OP_1)^2 + (OP_2)^2 - 2(OP_1)(OP_2)\cos(\alpha - \beta) \\ &= 2 - 2\cos(\alpha - \beta)\end{aligned}$$

Equating the two expressions for $(P_1P_2)^2$ and simplifying, we obtain

$$\cos(\alpha - \beta) = \cos\alpha\cos\beta + \sin\alpha\sin\beta$$

which completes the derivation of (37).

We can use (31) and (37) to derive (36) as follows:

$$\begin{aligned}\sin(\alpha - \beta) &= \cos\left[\frac{\pi}{2} - (\alpha - \beta)\right] = \cos\left[\left(\frac{\pi}{2} - \alpha\right) - (-\beta)\right] \\ &= \cos\left(\frac{\pi}{2} - \alpha\right)\cos(-\beta) + \sin\left(\frac{\pi}{2} - \alpha\right)\sin(-\beta) \\ &= \cos\left(\frac{\pi}{2} - \alpha\right)\cos\beta - \sin\left(\frac{\pi}{2} - \alpha\right)\sin\beta \\ &= \sin\alpha\cos\beta - \cos\alpha\sin\beta\end{aligned}$$

Identities (34) and (35) can be obtained from (36) and (37) by substituting $-\beta$ for β and using the identities

$$\sin(-\beta) = -\sin\beta, \quad \cos(-\beta) = \cos\beta$$

We leave it for the reader to derive the identities

$$\tan(\alpha + \beta) = \frac{\tan\alpha + \tan\beta}{1 - \tan\alpha\tan\beta} \qquad \tan(\alpha - \beta) = \frac{\tan\alpha - \tan\beta}{1 + \tan\alpha\tan\beta} \tag{38–39}$$

Identity (38) can be obtained by dividing (34) by (35) and then simplifying. Identity (39) can be obtained from (38) by substituting $-\beta$ for β and simplifying.

In the special case where $\alpha = \beta$, identities (34), (35), and (38) yield the ***double-angle formulas***

$$\sin 2\alpha = 2\sin\alpha\cos\alpha \tag{40}$$

$$\cos 2\alpha = \cos^2\alpha - \sin^2\alpha \tag{41}$$

$$\tan 2\alpha = \frac{2\tan\alpha}{1 - \tan^2\alpha} \tag{42}$$

By using the identity $\sin^2\alpha + \cos^2\alpha = 1$, (41) can be rewritten in the alternative forms

$$\cos 2\alpha = 2\cos^2\alpha - 1 \quad \text{and} \quad \cos 2\alpha = 1 - 2\sin^2\alpha \tag{43–44}$$

If we replace α by $\alpha/2$ in (43) and (44) and use some algebra, we obtain the ***half-angle formulas***

$$\cos^2\frac{\alpha}{2} = \frac{1 + \cos\alpha}{2} \quad \text{and} \quad \sin^2\frac{\alpha}{2} = \frac{1 - \cos\alpha}{2} \tag{45–46}$$

We leave it for the exercises to derive the following ***product-to-sum formulas*** from (34) through (37):

$$\sin\alpha\cos\beta = \frac{1}{2}[\sin(\alpha-\beta)+\sin(\alpha+\beta)] \tag{47}$$

$$\sin\alpha\sin\beta = \frac{1}{2}[\cos(\alpha-\beta)-\cos(\alpha+\beta)] \tag{48}$$

$$\cos\alpha\cos\beta = \frac{1}{2}[\cos(\alpha-\beta)+\cos(\alpha+\beta)] \tag{49}$$

We also leave it for the exercises to derive the following ***sum-to-product formulas***:

$$\sin\alpha+\sin\beta = 2\sin\frac{\alpha+\beta}{2}\cos\frac{\alpha-\beta}{2} \tag{50}$$

$$\sin\alpha-\sin\beta = 2\cos\frac{\alpha+\beta}{2}\sin\frac{\alpha-\beta}{2} \tag{51}$$

$$\cos\alpha+\cos\beta = 2\cos\frac{\alpha+\beta}{2}\cos\frac{\alpha-\beta}{2} \tag{52}$$

$$\cos\alpha-\cos\beta = -2\sin\frac{\alpha+\beta}{2}\sin\frac{\alpha-\beta}{2} \tag{53}$$

FINDING AN ANGLE FROM THE VALUE OF ITS TRIGONOMETRIC FUNCTIONS

There are numerous situations in which it is necessary to find an unknown angle from a known value of one of its trigonometric functions. The following example illustrates a method for doing this.

Example 6

Find θ if $\sin\theta = \frac{1}{2}$.

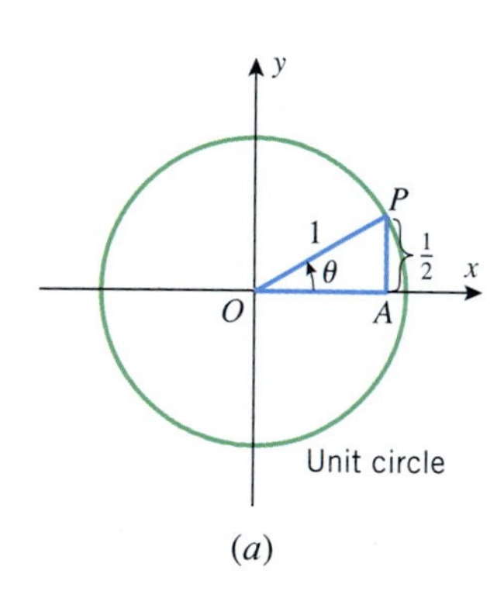

(*a*)

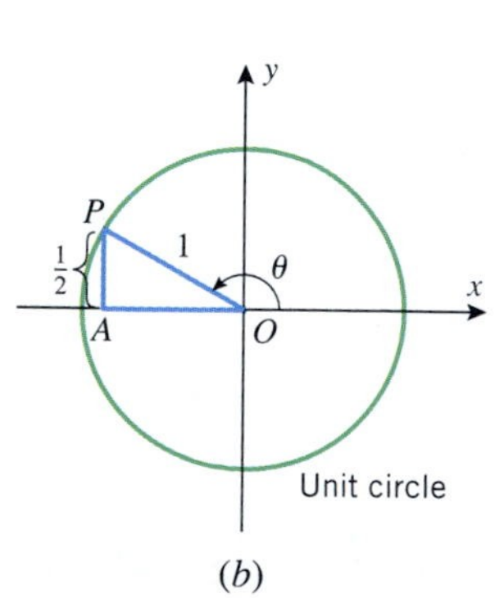

(*b*)

Figure E.18

Solution. We begin by looking for positive angles that satisfy the equation. Because $\sin\theta$ is positive, the angle θ must terminate in the first or second quadrant. If it terminates in the first quadrant, then the hypotenuse of $\triangle OAP$ in Figure E.18*a* is double the leg AP, so

$$\theta = 30^\circ = \frac{\pi}{6} \text{ radians}$$

If θ terminates in the second quadrant (Figure E.18*b*), then the hypotenuse of $\triangle OAP$ is double the leg AP, so $\angle AOP = 30^\circ$, which implies that

$$\theta = 180^\circ - 30^\circ = 150^\circ = \frac{5\pi}{6} \text{ radians}$$

Now that we have found these two solutions, all other solutions are obtained by adding or subtracting multiples of 360° (2π radians) to them. Thus, the entire set of solutions is given by the formulas

$$\theta = 30^\circ \pm n\cdot 360^\circ, \quad n = 0, 1, 2, \ldots$$

and

$$\theta = 150^\circ \pm n\cdot 360^\circ, \quad n = 0, 1, 2, \ldots$$

or in radian measure,

$$\theta = \frac{\pi}{6} \pm n\cdot 2\pi, \quad n = 0, 1, 2, \ldots$$

and

$$\theta = \frac{5\pi}{6} \pm n\cdot 2\pi, \quad n = 0, 1, 2, \ldots$$

◀

EXERCISE SET E

In Exercises 1 and 2, express the angles in radians.

1. (a) $75°$ (b) $390°$ (c) $20°$ (d) $138°$

2. (a) $420°$ (b) $15°$ (c) $225°$ (d) $165°$

In Exercises 3 and 4, express the angles in degrees.

3. (a) $\pi/15$ (b) 1.5 (c) $8\pi/5$ (d) 3π

4. (a) $\pi/10$ (b) 2 (c) $2\pi/5$ (d) $7\pi/6$

In Exercises 5 and 6, find the exact values of all six trigonometric functions of θ.

5. (a)

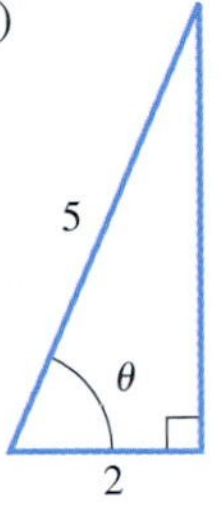

(b)

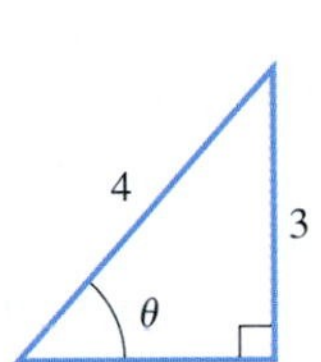

(c) 3, θ, 1

6. (a)

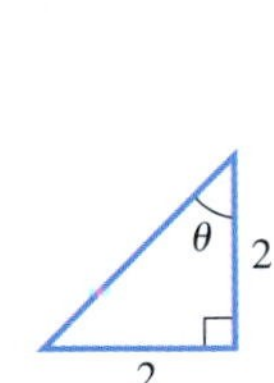

(b)

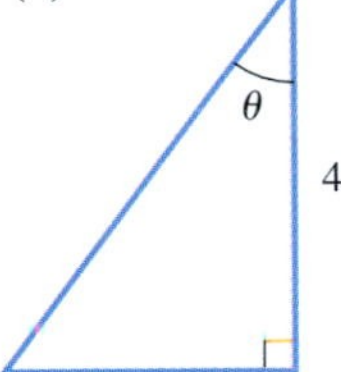

(c)

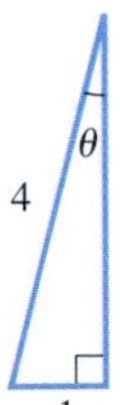

In Exercises 7–12, the angle θ is an acute angle of a right triangle. Solve the problems by drawing an appropriate right triangle. Do *not* use a calculator.

7. Find $\sin\theta$ and $\cos\theta$ given that $\tan\theta = 3$.

8. Find $\sin\theta$ and $\tan\theta$ given that $\cos\theta = \frac{2}{3}$.

9. Find $\tan\theta$ and $\csc\theta$ given that $\sec\theta = \frac{5}{2}$.

10. Find $\cot\theta$ and $\sec\theta$ given that $\csc\theta = 4$.

11. Find the length of the side adjacent to θ given that the hypotenuse has length 6 and $\cos\theta = 0.3$.

12. Find the length of the hypotenuse given that the side opposite θ has length 2.4 and $\sin\theta = 0.8$.

In Exercises 13 and 14, the value of an angle θ is given. Find the values of all six trigonometric functions of θ without using a calculator.

13. (a) $225°$ (b) $-210°$ (c) $5\pi/3$ (d) $-3\pi/2$

14. (a) $330°$ (b) $-120°$ (c) $9\pi/4$ (d) -3π

In Exercises 15 and 16, use the information to find the exact values of the remaining five trigonometric functions of θ.

15. (a) $\cos\theta = \frac{3}{5},\ 0 < \theta < \pi/2$
(b) $\cos\theta = \frac{3}{5},\ -\pi/2 < \theta < 0$
(c) $\tan\theta = -1/\sqrt{3},\ \pi/2 < \theta < \pi$
(d) $\tan\theta = -1/\sqrt{3},\ -\pi/2 < \theta < 0$
(e) $\csc\theta = \sqrt{2},\ 0 < \theta < \pi/2$
(f) $\csc\theta = \sqrt{2},\ \pi/2 < \theta < \pi$

16. (a) $\sin\theta = \frac{1}{4},\ 0 < \theta < \pi/2$
(b) $\sin\theta = \frac{1}{4},\ \pi/2 < \theta < \pi$
(c) $\cot\theta = \frac{1}{3},\ 0 < \theta < \pi/2$
(d) $\cot\theta = \frac{1}{3},\ \pi < \theta < 3\pi/2$
(e) $\sec\theta = -\frac{5}{2},\ \pi/2 < \theta < \pi$
(f) $\sec\theta = -\frac{5}{2},\ \pi < \theta < 3\pi/2$

In Exercises 17 and 18, use a calculating utility to find x to four decimal places.

17. (a)

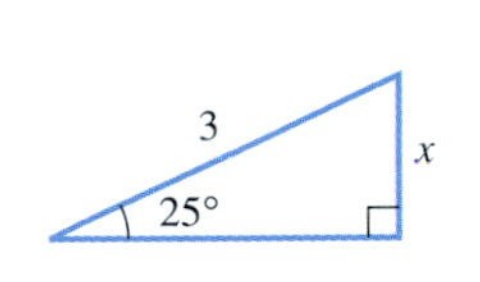

(b)

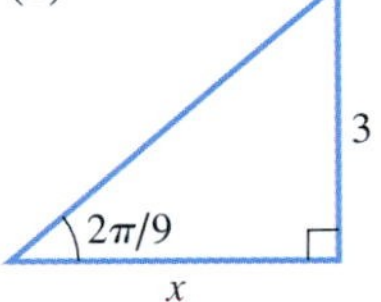

18. (a)

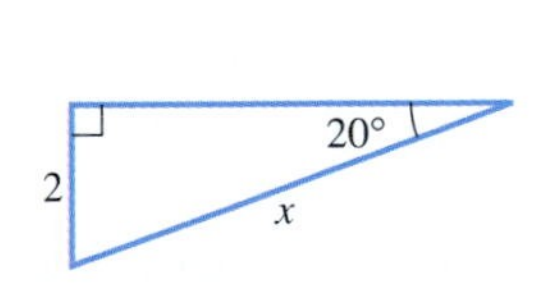

(b)

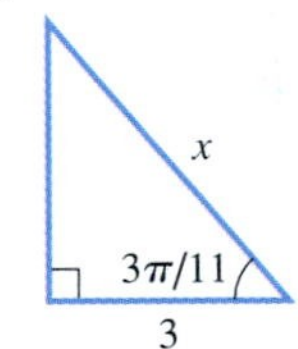

19. In each part, let θ be an acute angle of a right triangle. Express the remaining five trigonometric functions in terms of a.
(a) $\sin\theta = a/3$ (b) $\tan\theta = a/5$ (c) $\sec\theta = a$

In Exercises 20–27, find all values of θ (in radians) that satisfy the given equation. Do not use a calculator.

20. (a) $\cos\theta = -1/\sqrt{2}$ (b) $\sin\theta = -1/\sqrt{2}$

21. (a) $\tan\theta = -1$ (b) $\cos\theta = \frac{1}{2}$

22. (a) $\sin\theta = -\frac{1}{2}$ (b) $\tan\theta = \sqrt{3}$

23. (a) $\tan\theta = 1/\sqrt{3}$ (b) $\sin\theta = -\sqrt{3}/2$

24. (a) $\sin\theta = -1$ (b) $\cos\theta = -1$

25. (a) $\cot\theta = -1$ (b) $\cot\theta = \sqrt{3}$

26. (a) $\sec\theta = -2$ (b) $\csc\theta = -2$

27. (a) $\csc\theta = 2/\sqrt{3}$ (b) $\sec\theta = 2/\sqrt{3}$

In Exercises 28 and 29, find the values of all six trigonometric functions of θ.

28.

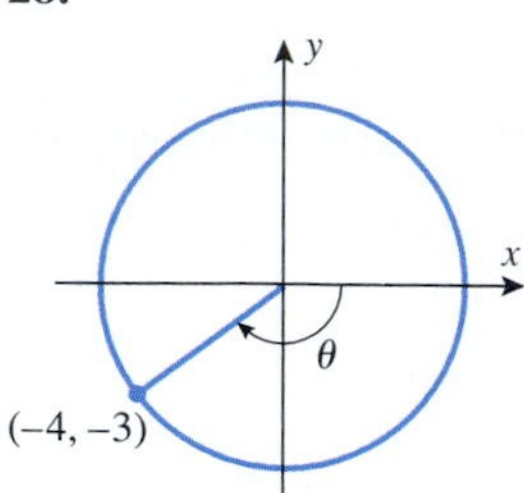

29.

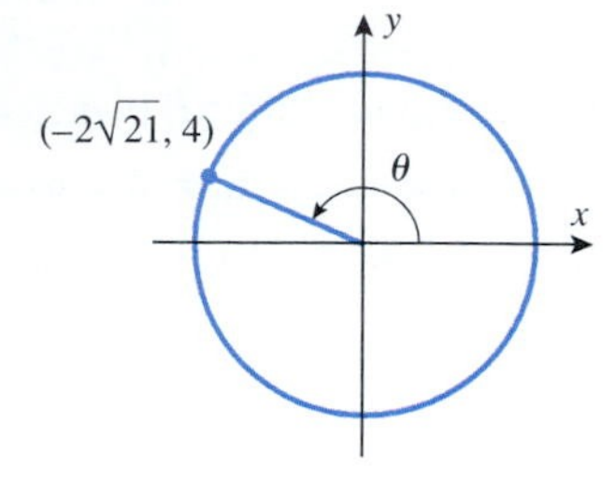

30. Find all values of θ (in radians) such that
(a) $\sin\theta = 1$ (b) $\cos\theta = 1$ (c) $\tan\theta = 1$
(d) $\csc\theta = 1$ (e) $\sec\theta = 1$ (f) $\cot\theta = 1$.

31. Find all values of θ (in radians) such that
(a) $\sin\theta = 0$ (b) $\cos\theta = 0$ (c) $\tan\theta = 0$
(d) $\csc\theta$ is undefined (e) $\sec\theta$ is undefined
(f) $\cot\theta$ is undefined.

32. How could you use a ruler and protractor to approximate $\sin 17^\circ$ and $\cos 17^\circ$?

33. Find the length of the circular arc on a circle of radius 4 cm subtended by an angle of
(a) $\pi/6$ (b) 150°.

34. Find the radius of a circular sector that has an angle of $\pi/3$ and a circular arc length of 7 units.

35. A point P moving counterclockwise on a circle of radius 5 cm traverses an arc length of 2 cm. What is the angle swept out by a radius from the center to P?

36. Find a formula for the area A of a circular sector in terms of its radius r and arc length s.

37. As shown in the accompanying figure, a right circular cone is made from a circular piece of paper of radius R by cutting out a sector of angle θ radians and gluing the cut edges of the remaining piece together. Find
(a) the radius r of the base of the cone in terms of R and θ
(b) the height h of the cone in terms of R and θ.

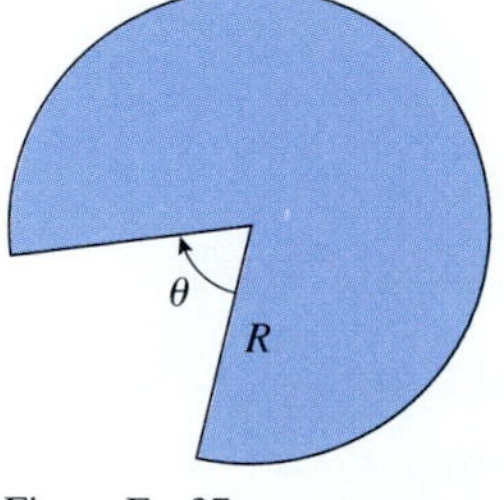

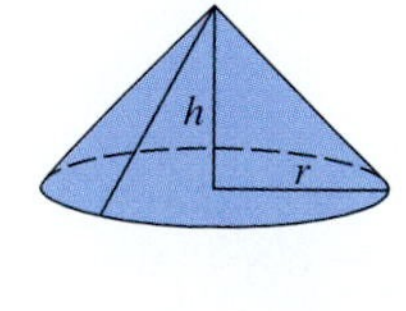

Figure Ex-37

38. As shown in the accompanying figure, let r and L be the radius of the base and the slant height of a right circular cone. Show that the lateral surface area, S, of the cone is $S = \pi r L$. [*Hint:* As shown in the figure in Exercise 37, the lateral surface of the cone becomes a circular sector when cut along a line from the vertex to the base and flattened.]

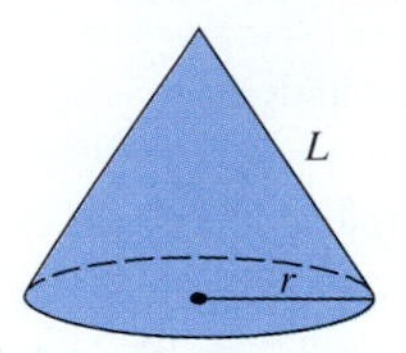

Figure Ex-38

39. Two sides of a triangle have lengths of 3 cm and 7 cm and meet at an angle of 60°. Find the area of the triangle.

40. Let ABC be a triangle whose angles at A and B are 30° and 45°. If the side opposite the angle B has length 9, find the lengths of the remaining sides and the size of the angle C.

41. A 10-foot ladder leans against a house and makes an angle of 67° with level ground. How far is the top of the ladder above the ground? Express your answer to the nearest tenth of a foot.

42. From a point 120 feet on level ground from a building, the angle of elevation to the top of the building is 76°. Find the height of the building. Express your answer to the nearest foot.

43. An observer on level ground is at a distance d from a building. The angles of elevation to the bottom of the windows on the second and third floors are α and β, respectively. Find the distance h between the bottoms of the windows in terms of α, β, and d.

44. From a point on level ground, the angle of elevation to the top of a tower is α. From a point that is d units closer to the tower, the angle of elevation is β. Find the height h of the tower in terms of α, β, and d.

In Exercises 45 and 46, do *not* use a calculator.

45. If $\cos\theta = \frac{2}{3}$ and $0 < \theta < \pi/2$, find
(a) $\sin 2\theta$ (b) $\cos 2\theta$.

46. If $\tan\alpha = \frac{3}{4}$ and $\tan\beta = 2$, where $0 < \alpha < \pi/2$ and $0 < \beta < \pi/2$, find
(a) $\sin(\alpha - \beta)$ (b) $\cos(\alpha + \beta)$.

47. Express $\sin 3\theta$ and $\cos 3\theta$ in terms of $\sin\theta$ and $\cos\theta$.

In Exercises 48–58, derive the given identities.

48. $\dfrac{\cos\theta\sec\theta}{1 + \tan^2\theta} = \cos^2\theta$

49. $\dfrac{\cos\theta\tan\theta + \sin\theta}{\tan\theta} = 2\cos\theta$

50. $2\csc 2\theta = \sec\theta\csc\theta$

51. $\tan\theta + \cot\theta = 2\csc 2\theta$

52. $\dfrac{\sin 2\theta}{\sin\theta} - \dfrac{\cos 2\theta}{\cos\theta} = \sec\theta$

53. $\dfrac{\sin\theta + \cos 2\theta - 1}{\cos\theta - \sin 2\theta} = \tan\theta$

54. $\sin 3\theta + \sin\theta = 2\sin 2\theta\cos\theta$

55. $\sin 3\theta - \sin\theta = 2\cos 2\theta\sin\theta$

56. $\tan\dfrac{\theta}{2} = \dfrac{1-\cos\theta}{\sin\theta}$

57. $\tan\dfrac{\theta}{2} = \dfrac{\sin\theta}{1+\cos\theta}$

58. $\cos\left(\dfrac{\pi}{3}+\theta\right) + \cos\left(\dfrac{\pi}{3}-\theta\right) = \cos\theta$

Exercises 59 and 60 refer to an arbitrary triangle ABC in which the side of length a is opposite angle A, the side of length b is opposite angle B, and the side of length c is opposite angle C.

59. Prove: The area of a triangle ABC can be written as

$$\text{area} = \tfrac{1}{2}bc\sin A$$

Find two other similar formulas for the area.

60. Prove the ***law of sines***: In any triangle, the ratios of the sides to the sines of the opposite angles are equal; that is,

$$\frac{a}{\sin A} = \frac{b}{\sin B} = \frac{c}{\sin C}$$

61. Use identities (34) through (37) to express each of the following in terms of $\sin\theta$ or $\cos\theta$.

(a) $\sin\left(\dfrac{\pi}{2}+\theta\right)$ (b) $\cos\left(\dfrac{\pi}{2}+\theta\right)$

(c) $\sin\left(\dfrac{3\pi}{2}-\theta\right)$ (d) $\cos\left(\dfrac{3\pi}{2}+\theta\right)$

62. Derive identities (38) and (39).

63. Derive identity
(a) (47) (b) (48) (c) (49).

64. If $A = \alpha + \beta$ and $B = \alpha - \beta$, then $\alpha = \frac{1}{2}(A+B)$ and $\beta = \frac{1}{2}(A-B)$ (verify). Use this result and identities (47) through (49) to derive identity
(a) (50) (b) (52) (c) (53).

65. Substitute $-\beta$ for β in identity (50) to derive identity (51).

66. (a) Express $3\sin\alpha + 5\cos\alpha$ in the form

$$C\sin(\alpha+\phi)$$

(b) Show that a sum of the form

$$A\sin\alpha + B\cos\alpha$$

can be rewritten in the form $C\sin(\alpha+\phi)$.

67. Show that the length of the diagonal of the parallelogram in the accompanying figure is

$$d = \sqrt{a^2+b^2+2ab\cos\theta}$$

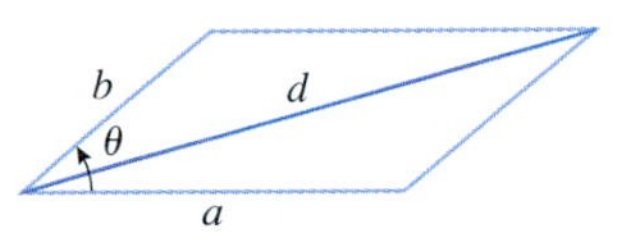

Figure Ex-67

APPENDIX F

Solving Polynomial Equations

In the subsection of Section 5.3 entitled A Brief Review of Polynomials, we reviewed some of the basic ideas and terminology concerning polynomials. We will assume in this appendix that you have read that material, and we will also assume that you know how to divide polynomials using long division and synthetic division. If you need to review those techniques, refer to an algebra book.

THE REMAINDER THEOREM

When two positive integers are divided, the numerator can be expressed as the quotient plus the remainder over the divisor, where the remainder is less than the divisor. For example,

$$\frac{17}{5} = 3 + \frac{2}{5}$$

If we multiply this equation through by 5, we obtain

$$17 = 5 \cdot 3 + 2$$

which states that the *numerator is the divisor times the quotient plus the remainder.*

The following theorem, which we state without proof, is an analogous result for division of polynomials.

F.1 THEOREM. *If $p(x)$ and $s(x)$ are polynomials, and if $s(x)$ is not the zero polynomial, then $p(x)$ can be expressed as*

$$p(x) = s(x)q(x) + r(x)$$

where $q(x)$ and $r(x)$ are the quotient and remainder that result when $p(x)$ is divided by $s(x)$, and the degree of $r(x)$ is less than the degree of $s(x)$.

In the special case where $p(x)$ is divided by a first-degree polynomial of the form $x - c$, the degree of the remainder must be 0, since it is less than the degree of $x - c$, which is 1. This implies that the remainder is a constant, say r. Thus, Theorem F.1 implies that

$$p(x) = (x - c)q(x) + r$$

and this in turn implies that $p(c) = r$. In summary, we have the following theorem.

F.2 THEOREM (*Remainder Theorem*). *If a polynomial $p(x)$ is divided by $x - c$, then the remainder is $p(c)$.*

Example 1

According to the Remainder Theorem, the remainder on dividing

$$p(x) = 2x^3 + 3x^2 - 4x - 3$$

by $x + 4$ should be

$$p(-4) = 2(-4)^3 + 3(-4)^2 - 4(-4) - 3 = -67$$

Show that this is so.

Solution. By long division

$$\begin{array}{r|l} & 2x^2 - 5x + 16 \\ \hline x + 4 & 2x^3 + 3x^2 - 4x - 3 \\ & \underline{2x^3 + 8x^2} \\ & \quad -5x^2 - 4x \\ & \quad \underline{-5x^2 - 20x} \\ & \qquad\qquad 16x - 3 \\ & \qquad\qquad \underline{16x + 64} \\ & \qquad\qquad\quad -67 \end{array}$$

which shows that the remainder is -67.

Alternative Solution. Because we are dividing by an expression of the form $x - c$ (where $c = -4$), we can use synthetic division rather than long division. The computations are

$$\begin{array}{r|rrrr} -4 & 2 & 3 & -4 & -3 \\ & & -8 & 20 & -64 \\ \hline & 2 & -5 & 16 & -67 \end{array}$$

which again shows that the remainder is -67. ◀

THE FACTOR THEOREM

To ***factor*** a polynomial $p(x)$ is to write it as a product of lower-degree polynomials, called ***factors*** of $p(x)$. For $s(x)$ to be a factor of $p(x)$ there must be no remainder when $p(x)$ is divided by $s(x)$. For example, if $p(x)$ can be factored as

$$p(x) = s(x)q(x) \tag{1}$$

then

$$\frac{p(x)}{s(x)} = q(x) \tag{2}$$

so dividing $p(x)$ by $s(x)$ produces a quotient $q(x)$ with no remainder. Conversely, (2) implies (1), so $s(x)$ is a factor of $p(x)$ if there is no remainder when $p(x)$ is divided by $s(x)$.

In the special case where $x - c$ is a factor of $p(x)$, the polynomial $p(x)$ can be expressed as

$$p(x) = (x - c)q(x)$$

which implies that $p(c) = 0$. Conversely, if $p(c) = 0$, then the Remainder Theorem implies that $x - c$ is a factor of $p(x)$, since the remainder is 0 when $p(x)$ is divided by $x - c$. These results are summarized in the following theorem.

> **F.3 THEOREM (*Factor Theorem*).** *A polynomial $p(x)$ has a factor $x - c$ if and only if $p(c) = 0$.*

It follows from this theorem that the statements below say the same thing in different ways:

- $x - c$ is a factor of $p(x)$.
- $p(c) = 0$.
- c is a zero of $p(x)$.
- c is a root of the equation $p(x) = 0$.
- c is a solution of the equation $p(x) = 0$.
- c is an x-intercept of $y = p(x)$.

Example 2

Confirm that $x - 1$ is a factor of

$$p(x) = x^3 - 3x^2 - 13x + 15$$

by dividing $x - 1$ into $p(x)$ and checking that the remainder is zero.

Solution. By long division

$$\begin{array}{r|l}
 & x^2 - 2x - 15 \\ \hline
x - 1 & x^3 - 3x^2 - 13x + 15 \\
 & \underline{x^3 - x^2} \\
 & \quad -2x^2 - 13x \\
 & \quad \underline{-2x^2 + 2x} \\
 & \qquad\quad -15x + 15 \\
 & \qquad\quad \underline{-15x + 15} \\
 & \qquad\qquad\qquad 0
\end{array}$$

which shows that the remainder is zero.

Alternative Solution. Because we are dividing by an expression of the form $x - c$, we can use synthetic division rather than long division. The computations are

$$\begin{array}{r|rrrr}
1 & 1 & -3 & -13 & 15 \\
 & & 1 & -2 & -15 \\ \hline
 & 1 & -2 & -15 & 0
\end{array}$$

which again confirms that the remainder is zero. ◀

USING ONE FACTOR TO FIND OTHER FACTORS

If $x - c$ is a factor of $p(x)$, and if $q(x) = p(x)/(x - c)$, then

$$p(x) = (x - c)q(x) \tag{3}$$

so that additional linear factors of $p(x)$ can be obtained by factoring the quotient $q(x)$.

Example 3

Factor

$$p(x) = x^3 - 3x^2 - 13x + 15 \tag{4}$$

completely into linear factors.

Solution. We showed in Example 2 that $x - 1$ is a factor of $p(x)$ and we also showed that $p(x)/(x - 1) = x^2 - 2x - 15$. Thus,

$$x^3 - 3x^2 - 13x + 15 = (x - 1)(x^2 - 2x - 15)$$

Factoring $x^2 - 2x - 15$ by inspection yields

$$x^3 - 3x^2 - 13x + 15 = (x - 1)(x - 5)(x + 3)$$

which is the complete linear factorization of $p(x)$. ◀

METHODS FOR FINDING ROOTS

A general quadratic equation $ax^2 + bx + c = 0$ can be solved by using the quadratic formula to express the solutions of the equation in terms of the coefficients. Versions of this formula were known since Babylonian times, and by the seventeenth century formulas had been obtained for solving general cubic and quartic equations. However, attempts to find formulas for the solutions of general fifth-degree equations and higher proved fruitless. The reason for this became clear in 1829 when the French mathematician Evariste Galois (1811–1832) proved that it is impossible to express the solutions of a general fifth-degree equation or higher in terms of its coefficients using algebraic operations.

Today, we have powerful computer programs for finding the zeros of specific polynomials. For example, it takes only seconds for a computer algebra system, such as *Mathematica*, *Maple*, or *Derive*, to show that the zeros of the polynomial

$$p(x) = 10x^4 - 23x^3 - 10x^2 + 29x + 6 \tag{5}$$

are

$$x = -1, \quad x = -\tfrac{1}{5}, \quad x = \tfrac{3}{2}, \quad \text{and} \quad x = 2 \tag{6}$$

The algorithms that these programs use to find the integer and rational zeros of a polynomial, if any, are based on the following theorem, which is proved in advanced algebra courses.

F.4 THEOREM. *Suppose that*

$$p(x) = c_n x^n + c_{n-1}x^{n-1} + \cdots + c_1 x + c_0$$

is a polynomial with integer coefficients.

(*a*) *If r is an integer zero of $p(x)$, then r must be a divisor of the constant term c_0.*

(*b*) *If $r = a/b$ is a rational zero of $p(x)$ in which all common factors of a and b have been canceled, then a must be a divisor of the constant term c_0, and b must be a divisor of the leading coefficient c_n.*

For example, in (5) the constant term is 6 (which has divisors $\pm 1, \pm 2, \pm 3$, and ± 6) and the leading coefficient is 10 (which has divisors $\pm 1, \pm 2, \pm 5$, and ± 10). Thus, the only possible integer zeros of $p(x)$ are

$$\pm 1, \ \pm 2, \ \pm 3, \ \pm 6$$

and the only possible noninteger rational zeros are

$$\pm\tfrac{1}{2}, \ \pm\tfrac{1}{5}, \ \pm\tfrac{1}{10}, \ \pm\tfrac{2}{5}, \ \pm\tfrac{3}{2}, \ \pm\tfrac{3}{5}, \ \pm\tfrac{3}{10}, \ \pm\tfrac{6}{5}$$

Using a computer, it is a simple matter to evaluate $p(x)$ at each of the numbers in these lists to show that its only zeros are the numbers in (6).

Example 4

Solve the equation $x^3 + 3x^2 - 7x - 21 = 0$.

Solution. The solutions of the equation are the zeros of the polynomial

$$p(x) = x^3 + 3x^2 - 7x - 21$$

We will look for integer zeros first. All such zeros must divide the constant term, so the only possibilities are $\pm 1, \pm 3, \pm 7$, and ± 21. Substituting these values into $p(x)$ (or using the method of Exercise 6) shows that $x = -3$ is an integer zero. This tells us that $x + 3$ is a factor of $p(x)$ and that $p(x)$ can be written as

$$x^3 + 3x^2 - 7x - 21 = (x + 3)q(x)$$

where $q(x)$ is the quotient that results when $x^3 + 3x^2 - 7x - 21$ is divided by $x + 3$. We leave it for you to perform the division and show that $q(x) = x^2 - 7$; hence,

$$x^3 + 3x^2 - 7x - 21 = (x + 3)(x^2 - 7) = (x + 1)(x + \sqrt{7})(x - \sqrt{7})$$

which tells us that the solutions of the given equation are $x = 3$, $x = \sqrt{7} \approx 2.65$, and $x = -\sqrt{7} \approx -2.65$. ◀

EXERCISE SET F C CAS

In Exercises 1 and 2, find the quotient $q(x)$ and the remainder $r(x)$ that result when $p(x)$ is divided by $s(x)$.

1. (a) $p(x) = x^4 + 3x^3 - 5x + 10$; $s(x) = x^2 - x + 2$
(b) $p(x) = 6x^4 + 10x^2 + 5$; $s(x) = 3x^2 - 1$
(c) $p(x) = x^5 + x^3 + 1$; $s(x) = x^2 + x$

2. (a) $p(x) = 2x^4 - 3x^3 + 5x^2 + 2x + 7$; $s(x) = x^2 - x + 1$
(b) $p(x) = 2x^5 + 5x^4 - 4x^3 + 8x^2 + 1$; $s(x) = 2x^2 - x + 1$
(c) $p(x) = 5x^6 + 4x^2 + 5$; $s(x) = x^3 + 1$

In Exercises 3 and 4, use synthetic division to find the quotient $q(x)$ and the remainder r that result when $p(x)$ is divided by $s(x)$.

3. (a) $p(x) = 3x^3 - 4x - 1$; $s(x) = x - 2$
(b) $p(x) = x^4 - 5x^2 + 4$; $s(x) = x + 5$
(c) $p(x) = x^5 - 1$; $s(x) = x - 1$

4. (a) $p(x) = 2x^3 - x^2 - 2x + 1$; $s(x) = x - 1$
(b) $p(x) = 2x^4 + 3x^3 - 17x^2 - 27x - 9$; $s(x) = x + 4$
(c) $p(x) = x^7 + 1$; $s(x) = x - 1$

5. Let $p(x) = 2x^4 + x^3 - 3x^2 + x - 4$. Use synthetic division and the Remainder Theorem to find $p(0)$, $p(1)$, $p(-3)$, and $p(7)$.

6. Let $p(x)$ be the polynomial in Example 4. Use synthetic division and the Remainder Theorem to evaluate $p(x)$ at $x = \pm 1, \pm 3, \pm 7$, and ± 21.

7. Let $p(x) = x^3 + 4x^2 + x - 6$. Find a polynomial $q(x)$ and a constant r such that
(a) $p(x) = (x - 2)q(x) + r$
(b) $p(x) = (x + 1)q(x) + r$.

8. Let $p(x) = x^5 - 1$. Find a polynomial $q(x)$ and a constant r such that
(a) $p(x) = (x + 1)q(x) + r$
(b) $p(x) = (x - 1)q(x) + r$.

9. In each part, make a list of all possible candidates for the rational zeros of $p(x)$.
(a) $p(x) = x^7 + 3x^3 - x + 24$
(b) $p(x) = 3x^4 - 2x^2 + 7x - 10$
(c) $p(x) = x^{35} - 17$

10. Find all integer zeros of

$$p(x) = x^6 + 5x^5 - 16x^4 - 15x^3 - 12x^2 - 38x - 21$$

In Exercises 11–15, factor the polynomials completely.

11. $p(x) = x^3 - 2x^2 - x + 2$

12. $p(x) = 3x^3 + x^2 - 12x - 4$

13. $p(x) = x^4 + 10x^3 + 36x^2 + 54x + 27$

14. $p(x) = 2x^4 + x^3 - 19x^2 + 9$

15. $p(x) = x^5 + 4x^4 - 4x^3 - 34x^2 - 45x - 18$

C **16.** For each of the factorizations that you obtained in Exercises 11–15, check your answer using a CAS.

In Exercises 17–21, find all real solutions of the equations.

17. $x^3 + 3x^2 + 4x + 12 = 0$

18. $2x^3 - 5x^2 - 10x + 3 = 0$

19. $3x^4 + 14x^3 + 14x^2 - 8x - 8 = 0$

20. $2x^4 - x^3 - 14x^2 - 5x + 6 = 0$

21. $x^5 - 2x^4 - 6x^3 + 5x^2 + 8x + 12 = 0$

C **22.** For each of the equations you solved in Exercises 17–21, check your answer using a CAS.

23. Find all values of k for which $x - 1$ is a factor of the polynomial $p(x) = k^2x^3 - 7kx + 10$.

24. Is $x + 3$ a factor of $x^7 + 2187$? Justify your answer.

C **25.** A 3-cm-thick slice is cut from a cube, leaving a volume of 196 cm^3. Use a CAS to find the length of a side of the original cube.

26. (a) Show that there is no positive rational number that exceeds its cube by 1.
(b) Does there exist a real number that exceeds its cube by 1? Justify your answer.

27. Use the Factor Theorem to show each of the following.
(a) $x - y$ is a factor of $x^n - y^n$ for all positive integer values of n.
(b) $x + y$ is a factor of $x^n - y^n$ for all positive even integer values of n.
(c) $x + y$ is a factor of $x^n + y^n$ for all positive odd integer values of n.

APPENDIX G

Selected Proofs

PROOFS OF BASIC LIMIT THEOREMS

An extensive excursion into proofs of limit theorems would be too time consuming to undertake, so we have selected a few proofs of results from Section 2.2 that illustrate some of the basic ideas.

G.1 THEOREM. *Let k be a constant, and suppose that $\lim_{x \to a} f(x) = L_1$ and that $\lim_{x \to a} g(x) = L_2$. Then*

(a) $\lim_{x \to a} k = k$

(b) $\lim_{x \to a} [f(x) + g(x)] = \lim_{x \to a} f(x) + \lim_{x \to a} g(x) = L_1 + L_2$

(c) $\lim_{x \to a} [f(x)g(x)] = \lim_{x \to a} f(x) \lim_{x \to a} g(x) = L_1 L_2$

Proof (a). We will apply Definition 2.3.3 with $f(x) = k$ and $L = k$. Thus, given $\epsilon > 0$, we must find a number $\delta > 0$ such that

$$|k - k| < \epsilon \quad \text{if} \quad 0 < |x - a| < \delta$$

or equivalently,

$$0 < \epsilon \quad \text{if} \quad 0 < |x - a| < \delta$$

But the condition on the left side of this statement is *always* true, no matter how δ is chosen. Thus, any positive value for δ will suffice.

Proof (b). We must show that given $\epsilon > 0$ we can find a number $\delta > 0$ such that

$$|(f(x) + g(x)) - (L_1 + L_2)| < \epsilon \quad \text{if} \quad 0 < |x - a| < \delta \tag{1}$$

However, from the limits of f and g in the hypothesis of the theorem we can find numbers δ_1 and δ_2 such that

$$|f(x) - L_1| < \epsilon/2 \quad \text{if} \quad 0 < |x - a| < \delta_1$$

$$|g(x) - L_2| < \epsilon/2 \quad \text{if} \quad 0 < |x - a| < \delta_2$$

Moreover, the inequalities on the left sides of these statements *both* hold if we replace δ_1 and δ_2 by any positive number δ that is less than both δ_1 and δ_2. Thus, for any such δ it follows that

$$|f(x) - L_1| + |g(x) - L_2| < \epsilon \quad \text{if} \quad 0 < |x - a| < \delta \tag{2}$$

However, it follows from the triangle inequality [Theorem 1.2.2(*d*)] that

$$\begin{aligned} |(f(x) + g(x)) - (L_1 + L_2)| &= |(f(x) - L_1) + (g(x) - L_2)| \\ &\leq |f(x) - L_1| + |g(x) - L_2| \end{aligned}$$

so that (1) follows from (2).

Proof (c). We must show that given $\epsilon > 0$ we can find a number $\delta > 0$ such that

$$|f(x)g(x) - L_1L_2| < \epsilon \quad \text{if} \quad 0 < |x - a| < \delta \tag{3}$$

To find δ it will be helpful to express (3) in a different form. If we rewrite $f(x)$ and $g(x)$ as

$$f(x) = L_1 + (f(x) - L_1) \quad \text{and} \quad g(x) = L_2 + (g(x) - L_2)$$

then the inequality on the left side of (3) can be expressed as (verify)

$$|L_1(g(x) - L_2) + L_2(f(x) - L_1) + (f(x) - L_1)(g(x) - L_2)| < \epsilon \tag{4}$$

Since

$$\lim_{x\to a} f(x) = L_1 \quad \text{and} \quad \lim_{x\to a} g(x) = L_2$$

we can find positive numbers δ_1, δ_2, δ_3, and δ_4 such that

$$\begin{aligned}
|f(x) - L_1| &< \sqrt{\epsilon/3} && \text{if} \quad 0 < |x - a| < \delta_1 \\
|f(x) - L_1| &< \frac{\epsilon}{3(1 + |L_2|)} && \text{if} \quad 0 < |x - a| < \delta_2 \\
|g(x) - L_2| &< \sqrt{\epsilon/3} && \text{if} \quad 0 < |x - a| < \delta_3 \\
|g(x) - L_2| &< \frac{\epsilon}{3(1 + |L_1|)} && \text{if} \quad 0 < |x - a| < \delta_4
\end{aligned} \tag{5}$$

Moreover, the inequalities on the left sides of these four statements *all* hold if we replace δ_1, δ_2, δ_3, and δ_4 by any number δ that is smaller than δ_1, δ_2, δ_3, and δ_4. Thus, for any such δ it follows with the help of the triangle inequality that

$$\begin{aligned}
&|L_1(g(x) - L_2) + L_2(f(x) - L_1) + (f(x) - L_1)(g(x) - L_2)| \\
&\quad \le |L_1(g(x) - L_2)| + |L_2(f(x) - L_1)| + |(f(x) - L_1)(g(x) - L_2)| \\
&\quad = |L_1||g(x) - L_2| + |L_2||f(x) - L_1| + |f(x) - L_1||g(x) - L_2| \\
&\quad < |L_1|\frac{\epsilon}{3(1 + |L_1|)} + |L_2|\frac{\epsilon}{3(1 + |L_2|)} + \sqrt{\epsilon/3}\sqrt{\epsilon/3} && \text{From (5)} \\
&\quad = \frac{\epsilon}{3}\frac{|L_1|}{1 + |L_1|} + \frac{\epsilon}{3}\frac{|L_2|}{1 + |L_2|} + \frac{\epsilon}{3} \\
&\quad < \frac{\epsilon}{3} + \frac{\epsilon}{3} + \frac{\epsilon}{3} = \epsilon && \text{Since } \frac{|L_1|}{1 + |L_1|} < 1 \text{ and } \frac{|L_2|}{1 + |L_2|} < 1
\end{aligned}$$

which shows that (4) holds for the δ selected. ∎

REMARK. Do not be alarmed if the proof of part (*c*) seems difficult; it takes some experience with proofs of this type to develop a feel for choosing the right δ. Your initial goal should be to understand the ideas and the computations.

PROOF OF A BASIC CONTINUITY PROPERTY

Next, we will prove Theorem 2.4.5 for two-sided limits.

> **G.2** THEOREM (*Theorem 2.4.5*). *If* $\lim_{x\to c} g(x) = L$ *and if the function* f *is continuous at* L, *then* $\lim_{x\to c} f(g(x)) = f(L)$; *that is,* $\lim_{x\to c} f(g(x)) = f(\lim_{x\to c} g(x))$.

Proof. We must show that given $\epsilon > 0$, we can find a number $\delta > 0$ such that

$$|f(g(x)) - f(L)| < \epsilon \quad \text{if} \quad 0 < |x - c| < \delta \tag{6}$$

Since f is continuous at L, we have

$$\lim_{u\to L} f(u) = f(L)$$

and hence we can find a number $\delta_1 > 0$ such that

$$|f(u) - f(L)| < \epsilon \quad \text{if} \quad |u - L| < \delta_1$$

In particular, if $u = g(x)$, then

$$|f(g(x)) - f(L)| < \epsilon \quad \text{if} \quad |g(x) - L| < \delta_1 \tag{7}$$

But $\lim_{x\to c} g(x) = L$, and hence there is a number $\delta > 0$ such that

$$|g(x) - L| < \delta_1 \quad \text{if} \quad 0 < |x - c| < \delta \tag{8}$$

Thus, if x satisfies the condition on the right side of statement (8), then it follows that $g(x)$ satisfies the condition on the right side of statement (7), and this implies that the condition on the left side of statement (6) is satisfied, completing the proof. ■

PROOF OF THE CHAIN RULE

Next, we will prove the chain rule (Theorem 3.5.2), but first we need a preliminary result.

> **G.3 THEOREM.** *If f is differentiable at x and if $y = f(x)$, then*
>
> $$\Delta y = f'(x)\Delta x + \epsilon\Delta x$$
>
> *where $\epsilon \to 0$ as $\Delta x \to 0$ and $\epsilon = 0$ if $\Delta x = 0$.*

Proof. Define

$$\epsilon = \begin{cases} \dfrac{f(x+\Delta x) - f(x)}{\Delta x} - f'(x) & \text{if } \Delta x \neq 0 \\ 0 & \text{if } \Delta x = 0 \end{cases} \tag{9}$$

If $\Delta x \neq 0$, it follows from (9) that

$$\epsilon\Delta x = [f(x+\Delta x) - f(x)] - f'(x)\Delta x \tag{10}$$

But

$$\Delta y = f(x+\Delta x) - f(x) \tag{11}$$

so (10) can be written as

$$\epsilon\Delta x = \Delta y - f'(x)\Delta x$$

or

$$\Delta y = f'(x)\Delta x + \epsilon\Delta x \tag{12}$$

If $\Delta x = 0$, then (12) still holds (why?), so (12) is valid for all values of Δx. It remains to show that $\epsilon \to 0$ as $\Delta x \to 0$. But this follows from the assumption that f is differentiable at x, since

$$\lim_{\Delta x\to 0} \epsilon = \lim_{\Delta x\to 0}\left[\frac{f(x+\Delta x) - f(x)}{\Delta x} - f'(x)\right] = f'(x) - f'(x) = 0$$ ■

We are now ready to prove the chain rule.

> **G.4 THEOREM (*Theorem 3.5.2*).** *If g is differentiable at the point x and f is differentiable at the point $g(x)$, then the composition $f \circ g$ is differentiable at the point x. Moreover, if $y = f(g(x))$ and $u = g(x)$, then*
>
> $$\frac{dy}{dx} = \frac{dy}{du} \cdot \frac{du}{dx}$$

Proof. Since g is differentiable at x and $u = g(x)$, it follows from Theorem G.3 that

$$\Delta u = g'(x)\Delta x + \epsilon_1 \Delta x \tag{13}$$

where $\epsilon_1 \to 0$ as $\Delta x \to 0$. And since $y = f(u)$ is differentiable at $u = g(x)$, it follows from Theorem G.3 that

$$\Delta y = f'(u)\Delta u + \epsilon_2 \Delta u \tag{14}$$

where $\epsilon_2 \to 0$ as $\Delta u \to 0$.

Factoring out the Δu in (14) and then substituting (13) yields

$$\Delta y = [f'(u) + \epsilon_2][g'(x)\Delta x + \epsilon_1 \Delta x]$$

or

$$\Delta y = [f'(u) + \epsilon_2][g'(x) + \epsilon_1]\Delta x$$

or if $\Delta x \neq 0$,

$$\frac{\Delta y}{\Delta x} = [f'(u) + \epsilon_2][g'(x) + \epsilon_1] \tag{15}$$

But (13) implies that $\Delta u \to 0$ as $\Delta x \to 0$, and hence $\epsilon_1 \to 0$ and $\epsilon_2 \to 0$ as $\Delta x \to 0$. Thus, from (15)

$$\lim_{\Delta x \to 0} \frac{\Delta y}{\Delta x} = f'(u)g'(x)$$

or

$$\frac{dy}{dx} = f'(u)g'(x) = \frac{dy}{du} \cdot \frac{du}{dx}$$ ■

PROOF THAT RELATIVE EXTREMA OCCUR AT CRITICAL POINTS

In this subsection we will prove Theorem 5.2.2, which states that the relative extrema of a function occur at critical points.

G.5 THEOREM (*Theorem 5.2.2*). *If a function f has any relative extrema, then they occur either at points where $f'(x) = 0$ or at points where f is not differentiable.*

Proof. There are two possibilities—either f is differentiable at a point x_0 or it is not. If it is not, then x_0 is a critical point for f and we are done. If f is differentiable at x_0, then we must show that $f'(x_0) = 0$. We will do this by showing that $f'(x_0) \geq 0$ and $f'(x_0) \leq 0$, from which it follows that $f'(x_0) = 0$. From the definition of a derivative we have

$$f'(x_0) = \lim_{h \to 0} \frac{f(x_0 + h) - f(x_0)}{h}$$

so that

$$f'(x_0) = \lim_{h \to 0^+} \frac{f(x_0 + h) - f(x_0)}{h} \tag{16}$$

and

$$f'(x_0) = \lim_{h \to 0^-} \frac{f(x_0 + h) - f(x_0)}{h} \tag{17}$$

Because f has a relative maximum at x_0, there is an open interval (a, b) containing x_0 in which $f(x) \leq f(x_0)$ for all x in (a, b).

Assume that h is sufficiently small so that $x_0 + h$ lies in the interval (a, b). Thus,

$$f(x_0 + h) \leq f(x_0) \quad \text{or equivalently,} \quad f(x_0 + h) - f(x_0) \leq 0$$

Thus, if h is negative,

$$\frac{f(x_0+h)-f(x_0)}{h} \geq 0 \tag{18}$$

and if h is positive,

$$\frac{f(x_0+h)-f(x_0)}{h} \leq 0 \tag{19}$$

But an expression that never assumes negative values cannot approach a negative limit and an expression that never assumes positive values cannot approach a positive limit, so that

$$f'(x_0) = \lim_{h \to 0^-} \frac{f(x_0+h)-f(x_0)}{h} \geq 0 \qquad \text{From (17) and (18)}$$

and

$$f'(x_0) = \lim_{h \to 0^+} \frac{f(x_0+h)-f(x_0)}{h} \leq 0 \qquad \text{From (16) and (19)}$$

Since $f'(x_0) \geq 0$ and $f'(x_0) \leq 0$, it must be that $f'(x_0) = 0$. ∎

PROOF OF THE LIMIT COMPARISON TEST

G.6 THEOREM (***Theorem 11.6.4***). *Let $\sum a_k$ and $\sum b_k$ be a series with positive terms and suppose that*

$$\rho = \lim_{k \to +\infty} \frac{a_k}{b_k}$$

If ρ is finite and $\rho > 0$, then the series both converge or both diverge.

Proof. We need only show that $\sum b_k$ converges when $\sum a_k$ converges and that $\sum b_k$ diverges when $\sum a_k$ diverges, since the remaining cases are logical implications of these (why?). The idea of the proof is to apply the comparison test to $\sum a_k$ and suitable multiples of $\sum b_k$. For this purpose let ϵ be any positive number. Since

$$\rho = \lim_{k \to +\infty} \frac{a_k}{b_k}$$

it follows that eventually the terms in the sequence $\{a_k/b_k\}$ must be within ϵ units of ρ; that is, there is a positive integer K such that for $k \geq K$ we have

$$\rho - \epsilon < \frac{a_k}{b_k} < \rho + \epsilon$$

In particular, if we take $\epsilon = \rho/2$, then for $k \geq K$ we have

$$\frac{1}{2}\rho < \frac{a_k}{b_k} < \frac{3}{2}\rho \quad \text{or} \quad \frac{1}{2}\rho b_k < a_k < \frac{3}{2}\rho b_k$$

Thus, by the comparison test we can conclude that

$$\sum_{k=K}^{\infty} \frac{1}{2}\rho b_k \quad \text{converges if} \quad \sum_{k=K}^{\infty} a_k \quad \text{converges} \tag{20}$$

$$\sum_{k=K}^{\infty} \frac{3}{2}\rho b_k \quad \text{diverges if} \quad \sum_{k=K}^{\infty} a_k \quad \text{diverges} \tag{21}$$

But the convergence or divergence of a series is not affected by deleting finitely many terms or by multiplying the general term by a nonzero constant, so (20) and (21) imply that

$$\sum_{k=1}^{\infty} b_k \quad \text{converges if} \quad \sum_{k=1}^{\infty} a_k \quad \text{converges}$$

$$\sum_{k=1}^{\infty} b_k \quad \text{diverges if} \quad \sum_{k=1}^{\infty} a_k \quad \text{diverges}$$

∎

PROOF OF THE RATIO TEST

G.7 THEOREM (*Theorem 11.6.5*). *Let $\sum u_k$ be a series with positive terms and suppose that*

$$\rho = \lim_{k\to+\infty} \frac{u_{k+1}}{u_k}$$

(*a*) *If $\rho < 1$, the series converges.*
(*b*) *If $\rho > 1$ or $\rho = +\infty$, the series diverges.*
(*c*) *If $\rho = 1$, the series may converge or diverge, so that another test must be tried.*

Proof (a). The number ρ must be nonnegative since it is the limit of u_{k+1}/u_k, which is positive for all k. In this part of the proof we assume that $\rho < 1$, so that $0 \le \rho < 1$.

We will prove convergence by showing that the terms of the given series are eventually less than the terms of a convergent geometric series. For this purpose, choose any real number r such that $0 < \rho < r < 1$. Since the limit of u_{k+1}/u_k is ρ, and $\rho < r$, the terms of the sequence $\{u_{k+1}/u_k\}$ must eventually be less than r. Thus, there is a positive integer K such that for $k \ge K$ we have

$$\frac{u_{k+1}}{u_k} < r \quad \text{or} \quad u_{k+1} < ru_k$$

This yields the inequalities

$$\begin{aligned} &u_{K+1} < ru_K \\ &u_{K+2} < ru_{K+1} < r^2u_K \\ &u_{K+3} < ru_{K+2} < r^3u_K \\ &u_{K+4} < ru_{K+3} < r^4u_K \\ &\qquad\vdots \end{aligned} \tag{22}$$

But $0 < r < 1$, so

$$ru_K + r^2u_K + r^3u_K + \cdots$$

is a convergent geometric series. From the inequalities in (22) and the comparison test it follows that

$$u_{K+1} + u_{K+2} + u_{K+3} + \cdots$$

must also be a convergent series. Thus, $u_1 + u_2 + u_3 + \cdots + u_k + \cdots$ converges by Theorem 11.4.3(*c*).

Proof (b). In this part we will prove divergence by showing that the limit of the general term is not zero. Since the limit of u_{k+1}/u_k is ρ and $\rho > 1$, the terms in the sequence $\{u_{k+1}/u_k\}$ must eventually be greater than 1. Thus, there is a positive integer K such that for $k \ge K$ we have

$$\frac{u_{k+1}}{u_k} > 1 \quad \text{or} \quad u_{k+1} > u_k$$

This yields the inequalities

$$\begin{aligned} &u_{K+1} > u_K \\ &u_{K+2} > u_{K+1} > u_K \\ &u_{K+3} > u_{K+2} > u_K \\ &u_{K+4} > u_{K+3} > u_K \\ &\qquad\vdots \end{aligned} \tag{23}$$

Since $u_K > 0$, it follows from the inequalities in (23) that $\lim_{k\to+\infty} u_k \ne 0$, and thus the series

$u_1 + u_2 + \cdots + u_k + \cdots$ diverges by part (*a*) of Theorem 11.4.1. The proof in the case where $\rho = +\infty$ is omitted.

***Proof*(*c*).** The divergent harmonic series and the convergent p-series with $p = 2$ both have $\rho = 1$ (verify), so the ratio test does not distinguish between convergence and divergence when $\rho = 1$. ∎

PROOF OF THE REMAINDER ESTIMATION THEOREM

G.8 THEOREM (*Theorem 11.9.3*). *If the function f can be differentiated $n + 1$ times on an interval I containing the point x_0, and if $|f^{(n+1)}(x)| \le M$ for all x in I, then*

$$|R_n(x)| \le \frac{M}{(n+1)!}|x - x_0|^{n+1}$$

for all x in I.

Proof. We are assuming that f can be differentiated $n+1$ times on an interval I containing the point x_0 and that

$$|f^{(n+1)}(x)| \le M \tag{24}$$

for all x in I. We want to show that

$$|R_n(x)| \le \frac{M}{(n+1)!}|x - x_0|^{n+1} \tag{25}$$

for all x in I, where

$$R_n(x) = f(x) - \sum_{k=0}^{n} \frac{f^{(k)}(x_0)}{k!}(x - x_0)^k \tag{26}$$

In our proof we will need the following two properties of $R_n(x)$:

$$R_n(x_0) = R_n'(x_0) = \cdots = R_n^{(n)}(x_0) = 0 \tag{27}$$

$$R_n^{(n+1)}(x) = f^{(n+1)}(x) \quad \text{for all } x \text{ in } I \tag{28}$$

These properties can be obtained by analyzing what happens if the expression for $R_n(x)$ in Formula (26) is differentiated j times and x_0 is then substituted in that derivative. If $j < n$, then the jth derivative of the summation in Formula (26) consists of a constant term $f^{(j)}(x_0)$ plus terms involving powers of $x - x_0$ (verify). Thus, $R_n^{(j)}(x_0) = 0$ for $j < n$, which proves all but the last equation in (27). For the last equation, observe that the nth derivative of the summation in (26) is the constant $f^{(n)}(x_0)$, so $R_n^{(n)}(x_0) = 0$. Formula (28) follows from the observation that the $(n + 1)$-st derivative of the summation in (26) is zero (why?).

Now to the main part of the proof. For simplicity we will give the proof for the case where $x \ge x_0$ and leave the case where $x < x_0$ for the reader. It follows from (24) and (28) that $|R_n^{(n+1)}(x)| \le M$, and hence

$$-M \le R_n^{(n+1)}(x) \le M$$

Thus,

$$\int_{x_0}^{x} -M\,dt \le \int_{x_0}^{x} R_n^{(n+1)}(t)\,dt \le \int_{x_0}^{x} M\,dt \tag{29}$$

However, it follows from (27) that $R_n^{(n)}(x_0) = 0$, so

$$\int_{x_0}^{x} R_n^{(n+1)}(t)\,dt = R_n^{(n)}(t)\Big]_{x_0}^{x} = R_n^{(n)}(x)$$

Thus, performing the integrations in (29) we obtain the inequalities

$$-M(x - x_0) \le R_n^{(n)}(x) \le M(x - x_0)$$

Now we will integrate again. Replacing x by t in these inequalities, integrating from x_0 to x, and using $R_n^{(n-1)}(x_0) = 0$ yields

$$-\frac{M}{2}(x - x_0)^2 \leq R_n^{(n-1)}(x) \leq \frac{M}{2}(x - x_0)^2$$

If we keep repeating this process, then after n integrations we will obtain

$$-\frac{M}{(n+1)!}(x - x_0)^{n+1} \leq R_n(x) \leq \frac{M}{(n+1)!}(x - x_0)^{n+1}$$

which we can rewrite as

$$|R_n(x)| \leq \frac{M}{(n+1)!}(x - x_0)^{n+1}$$

This completes the proof of (25), since the absolute value signs can be omitted in that formula when $x \geq x_0$ (which is the case we are considering). ∎

ANSWERS TO ODD-NUMBERED EXERCISES

▶ Exercise Set 7.1 (Page 382)

1. $A = 1/2$

n	1	2	3	4	5
A_n	1.0000	0.7500	0.6666	0.6250	0.6000

n	6	7	8	9	10
A_n	0.5833	0.5714	0.5625	0.5556	0.5500

3. $A = 16$

n	1	2	3	4	5
A_n	28.0000	22.0000	20.0000	19.0000	18.4000

n	6	7	8	9	10
A_n	18.0000	17.7143	17.5000	17.3333	17.2000

5. $\frac{1}{2}$ 7. 16 9. $e-1$

▶ Exercise Set 7.2 (Page 389)

1. **(a)** $\int \frac{x}{\sqrt{1+x^2}}\,dx = \sqrt{1+x^2}+C$ **(b)** $\int (x+1)e^x\,dx = xe^x + C$

3. $\frac{d}{dx}\left[\sqrt{x^3+5}\right] = \frac{3x^2}{2\sqrt{x^3+5}}$, so $\int \frac{3x^2}{2\sqrt{x^3+5}}\,dx = \sqrt{x^3+5}+C.$

5. $\frac{d}{dx}\left[\sin(2\sqrt{x})\right] = \frac{\cos(2\sqrt{x})}{\sqrt{x}}$, so $\int \frac{\cos(2\sqrt{x})}{\sqrt{x}}\,dx = \sin(2\sqrt{x})+C.$

7. **(a)** $(x^9/9)+C$ **(b)** $\frac{7}{12}x^{12/7}+C$ **(c)** $\frac{2}{9}x^{9/2}+C$ 9. **(a)** $-\frac{1}{4}x^{-2}+C$ **(b)** $(u^4/4)-u^2+7u+C$

11. $-\frac{1}{2}x^{-2}+\frac{2}{3}x^{3/2}-\frac{12}{5}x^{5/4}+\frac{1}{3}x^3+C$ 13. $(x^2/2)+(x^5/5)+C$ 15. $3x^{4/3}-\frac{12}{7}x^{7/3}+\frac{3}{10}x^{10/3}+C$ 17. $\frac{x^2}{2}-\frac{2}{x}+\frac{1}{3x^3}+C$

19. $2\ln x+3e^x+C$ 21. $-4\cos x+2\sin x+C$ 23. $\tan x+\sec x+C$ 25. $\ln\theta-2e^\theta+\cot\theta+C$ 27. $\sec x+C$

29. $\theta-\cos\theta+C$ 31. $\tan x-\sec x+C$ 33. **(a)** **(b)** $f(x)=(x^2/2)+5$ 35.

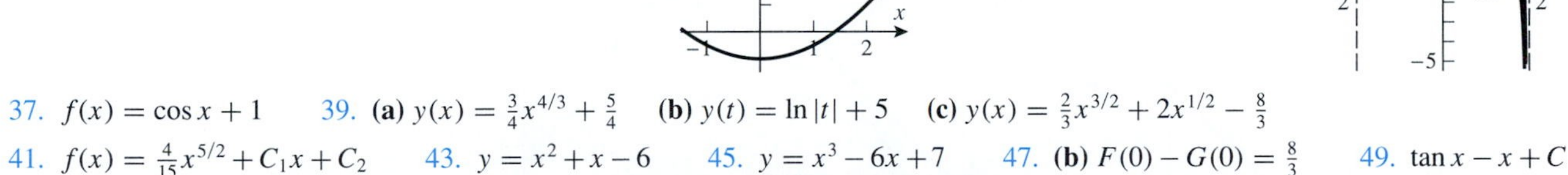

37. $f(x)=\cos x+1$ 39. **(a)** $y(x)=\frac{3}{4}x^{4/3}+\frac{5}{4}$ **(b)** $y(t)=\ln|t|+5$ **(c)** $y(x)=\frac{2}{3}x^{3/2}+2x^{1/2}-\frac{8}{3}$

41. $f(x)=\frac{4}{15}x^{5/2}+C_1x+C_2$ 43. $y=x^2+x-6$ 45. $y=x^3-6x+7$ 47. **(b)** $F(0)-G(0)=\frac{8}{3}$ 49. $\tan x-x+C$

51. **(a)** $\frac{1}{2}(x-\sin x)+C$ **(b)** $\frac{1}{2}(x+\sin x)+C$ 53. $v=\frac{1087}{\sqrt{273}}T^{1/2}$ ft/s

▶ Exercise Set 7.3 (Page 395)

1. **(a)** $\frac{(x^2+1)^{24}}{24}+C$ **(b)** $-\frac{\cos^4 x}{4}+C$ **(c)** $-2\cos\sqrt{x}+C$ **(d)** $\frac{3}{4}\sqrt{4x^2+5}+C$ **(e)** $\frac{1}{3}\ln(x^3-4)+C$

3. **(a)** $-\frac{1}{2}\cot^2 x+C$ **(b)** $\frac{1}{10}(1+\sin t)^{10}+C$ **(c)** $\ln|\ln x|+C$ **(d)** $-\frac{1}{5}e^{-5x}+C$ **(e)** $-\frac{1}{3}\ln|(1+\cos 3\theta)|+C$

5. $\frac{1}{2}e^{2x}+C$ 7. $-\frac{(2-x^2)^4}{8}+C$ 9. $\frac{1}{8}\sin 8x+C$ 11. $\frac{1}{4}\sec 4x+C$ 13. $\frac{1}{21}(7t^2+12)^{3/2}+C$ 15. $\frac{2}{3}\sqrt{x^3+1}+C$

17. $-\frac{1}{16}(4x^2+1)^{-2}+C$ 19. $e^{\sin x}+C$ 21. $-\frac{1}{6}e^{-2x^3}+C$ 23. $\frac{1}{5}\cos(5/x)+C$ 25. $\frac{1}{3}\tan(x^3)+C$ 27. $-e^{-x}+C$

29. $\frac{1}{18}\sin^6 3t + C$ 31. $-\frac{1}{6}(2-\sin 4\theta)^{3/2} + C$ 33. $\frac{1}{6}\sec^3 2x + C$ 35. $2e^{\sqrt{y}} + C$ 39. $\dfrac{1}{b(n+1)}\sin^{n+1}(a+bx) + C$

41. $\frac{2}{5}(x-3)^{5/2} + 2(x-3)^{3/2} + C$ 43. $\frac{1}{3}(\tan 3\theta - 3\theta) + C$ 45. $t + \ln|t| + C$ 47. $\int [\ln(e^x) + \ln(e^{-x})]\, dx = C$

49. **(a)** with $u = \sin x$, $\frac{1}{2}\sin^2 x + C_1$; with $u = \cos x$, $-\frac{1}{2}\cos^2 x + C_2$ **(b)** because they differ by a constant

51. $y(x) = \frac{2}{9}(3x+1)^{3/2} + \frac{29}{9}$ 53. $f(x) = \frac{2}{9}(3x+1)^{3/2} + \frac{7}{9}$ 55. 100,416

▶ Exercise Set 7.4 (Page 402)

1. **(a)** 36 **(b)** 55 **(c)** 40 **(d)** 6 **(e)** 11 **(f)** 0 3. $\sum_{k=1}^{10} k$ 5. $\sum_{k=1}^{49} k(k+1)$ 7. $\sum_{k=1}^{10} 2k$ 9. $\sum_{k=1}^{6} (-1)^{k+1}(2k-1)$

11. $\sum_{k=1}^{5} (-1)^k \frac{1}{k}$ 13. **(a)** $\sum_{1}^{50} 2k$ **(b)** $\sum_{1}^{50} (2k-1)$ 15. 5050 17. 2870 19. 1728 21. 214,365 23. $2n^2 - n$

25. $\frac{3}{2}(n+1)$ 27. $\frac{1}{4}(n-1)^2$ 31. **(a)** $\sum_{k=0}^{19} 3^{k+1} = \frac{3}{2}(3^{20}-1)$ **(b)** $\sum_{k=0}^{25} 2^{k+5} = 2^{31} - 2^5$ **(c)** $\sum_{k=0}^{100} (-1)\left(\frac{-1}{2}\right)^k = -\frac{2}{3}\left(1 + \frac{1}{2^{101}}\right)$

33. $\dfrac{n+1}{2n}$; $\frac{1}{2}$ 35. $\dfrac{5(n+1)}{2n}$; $\frac{5}{2}$ 37. **(a)** $\sum_{j=0}^{5} 2^j$ **(b)** $\sum_{j=1}^{6} 2^{j-1}$ **(c)** $\sum_{j=2}^{7} 2^{j-2}$ 39. **(a)** $\sum_{k=1}^{18} \sin\left(\frac{\pi}{k}\right)$ **(b)** $\sum_{k=0}^{6} e^k = \dfrac{e^7 - 1}{e - 1}$

43. $3^{17} - 3^4$ 45. $-\frac{399}{400}$ 47. **(b)** $\frac{1}{2}$ 49. Both identities are valid. 55. 18,755

▶ Exercise Set 7.5 (Page 414)

1. **(a)** 8 **(b)** greater than A **(c)** less than A **(d)** equal to A

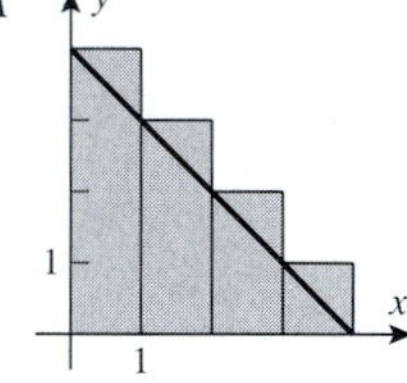

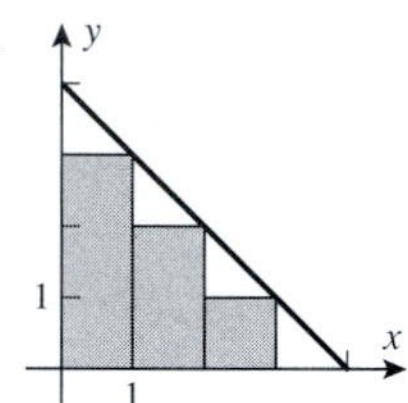

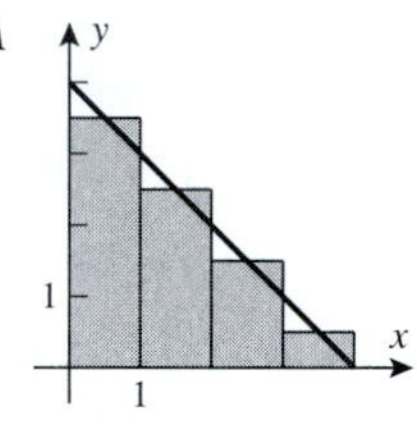

3. 35, 60, 46.25 5. $\dfrac{(1+\sqrt{2})\pi}{4} \approx 1.896$, $\dfrac{(1+\sqrt{2})\pi}{4} \approx 1.896$, $\dfrac{\pi\sqrt{2}\cos(\pi/8)}{2} \approx 2.052$

7. left endpoints: $A \approx (2+3+2+1)(1) = 8$; right endpoints: $A \approx (3+2+1+2)(1) = 8$

9. 0.718771403, 0.668771403, 0.692835360 11. 0.919403170, 1.07648280, 1.001028825

13. 0.351220577, 0.420535296, 0.386502483

15. **(a)** 0.693097198, 0.666154270, 1.000164512, 5.336963538, 0.386327689, 1.718167282
(b) 0.693134682, 0.666538346, 1.000041125, 5.334644416, 0.386302694, 1.718253191
(c) 0.693144056, 0.666634573, 1.000010281, 5.333803776, 0.3862964444, 1.718274669

17. **(a)** $A = \frac{9}{2}$ **(b)** $-A = -\frac{3}{2}$ **(c)** $-A_1 + A_2 = \frac{15}{2}$ **(d)** $-A_1 + A_2 = 0$

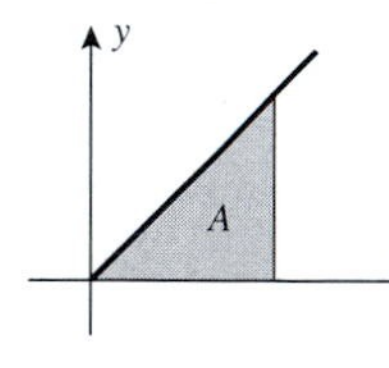

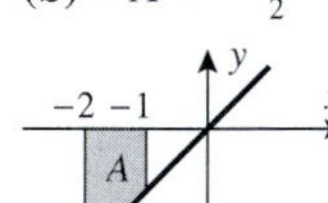

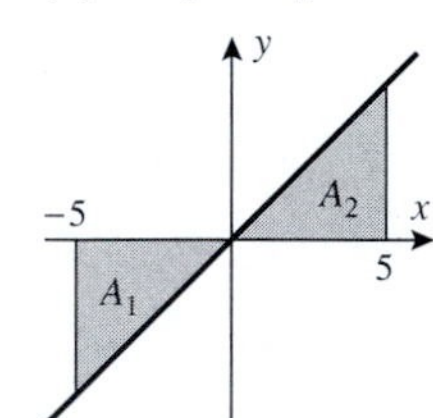

19. **(a)** $A = 10$ **(b)** $A_1 - A_2 = 0$ by symmetry **(c)** $A_1 + A_2 = \frac{13}{2}$ **(d)** $\pi/2$

21. **(a)** 0.8 **(b)** −2.6 **(c)** −1.8 **(d)** −0.3

23. −1 25. 3 27. **(a)** $(1+\pi)/2$ **(b)** −4 29. **(a)** negative **(b)** positive 31. $25\pi/2$

33. **(a)** $\int_{-3}^{3} 4x(1-3x)\,dx$ **(b)** $\int_{0}^{1} e^x\,dx$ 35. $\frac{5}{2}$

37. **(a)** $\lim_{\max \Delta x_k \to 0} \sum_{k=1}^{n} 2x_k^* \Delta x_k;\ a=1, b=2$ **(b)** $\lim_{\max \Delta x_k \to 0} \sum_{k=1}^{n} \frac{x_k^*}{x_k^*+1} \Delta x_k;\ a=0, b=1$

39. **(d)** $\frac{3}{2}$ 41. $\frac{1}{3}$ 43. 320 45. **(a)** yes **(b)** yes **(c)** no **(d)** yes

---

▶ Exercise Set 7.6 (Page 425)

1. **(a)** $\int_{0}^{2} (2-x)\,dx = 2$ **(b)** $\int_{-1}^{1} 2\,dx = 4$ **(c)** $\int_{1}^{3} (x+1)\,dx = 6$ 3. $\frac{65}{4}$ 5. $\frac{52}{3}$ 7. $e^3 - e$ 9. 48 11. $\frac{2}{3}$

13. $\frac{844}{5}$ 15. 0 17. $\sqrt{2}$ 19. $5e^3 - 10$ 21. $-\frac{55}{3}$ 23. $\frac{\pi^2}{9} + 2\sqrt{3}$ 27. **(a)** $\frac{5}{2}$ **(b)** $2 - \frac{\sqrt{2}}{2}$ 29. $-\frac{11}{6}$

31. 0.665867079; $\frac{2}{3}$ 33. 1.098242635; $\ln 3 \approx 1.098612289$ 35. 12 37. $\frac{9}{2}$

39. $A_1 = \frac{23}{6}, A_2 = \frac{343}{6}, A_3 = \frac{243}{6}, A = \frac{203}{2}$

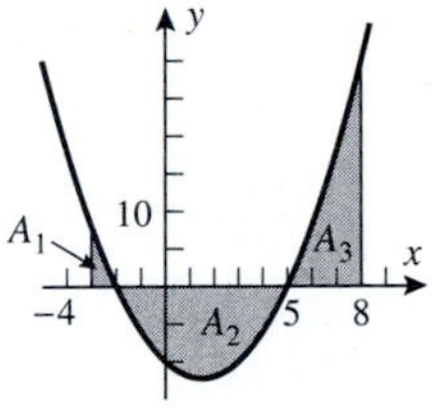

41. **(a)** The integral is zero.
(c) $\int_{-a}^{a} f(x)\,dx = 2\int_{0}^{a} f(x)\,dx$

43. **(a)** $x^3 + 1$

45. **(a)** $\sin\sqrt{x}$ **(b)** e^{x^2} 47. $-\frac{x}{\cos x}$ 49. **(a)** 0 **(b)** $\sqrt{13}$ **(c)** $6/\sqrt{13}$

51. **(a)** $x = 3$ **(b)** increasing on $[3, +\infty)$, decreasing on $(-\infty, 3]$ **(c)** concave up on $(-1, 7)$, concave down on $(-\infty, -1)$ and $(7, +\infty)$

53. **(a)** $(0, +\infty)$ **(b)** $x = 1$ 55. **(a)** $x^* = 4$ **(b)** $x^* = e - 1$ 57. $3\sqrt{2} \le \int_{0}^{3} \sqrt{x^3+2}\,dx \le 3\sqrt{29}$

▶ Exercise Set 7.7 (Page 437)

1. **(a)** the increase in height in inches, during the first 10 years
(b) the change in the radius in cm, during the time interval $t = 1$ to $t = 2$ seconds
(c) the change in the speed of sound in ft/s, during an increase in temperature from $t = 32°$F to $t = 100°$F
(d) the displacement of the particle in cm, during the time interval $t = t_1$ to $t = t_2$ seconds

3. **(a)** displacement $= -\frac{1}{2}$; distance $= \frac{1}{2}$ **(b)** displacement $= 5$; distance $= \frac{5}{2}$

5. **(a)** 31.3 m/s **(b)** 55.15 m/s 7. **(a)** $\frac{1}{4}t^4 - \frac{2}{3}t^3 + t + 1$ **(b)** $-\cos 2t - t - 2$ 9. **(a)** $t^2 - 3t + 7$ **(b)** $-\cos t + t - (\pi/2)$

11. **(a)** displacement $= 1$; distance $= 1$ **(b)** displacement $= -1$; distance $= 3$

13. **(a)** displacement $= \frac{9}{4}$; distance $= \frac{11}{4}$ **(b)** displacement $= e^3 - 7$; distance $= e^3 - 9 + 4\ln 2$

15. displacement $= -6$; distance $= \frac{13}{2}$ 17. displacement $= \frac{204}{25}$; distance $= \frac{204}{25}$

19. **(a)** $s = 2/\pi, v = 1, |v| = 1, a = 0$ **(b)** $s = \frac{1}{2}, v = -\frac{3}{2}, |v| = \frac{3}{2}, a = -3$ 21. $\frac{22}{3}$ 23. $(1/e) + e - 2$

25. **(a)**

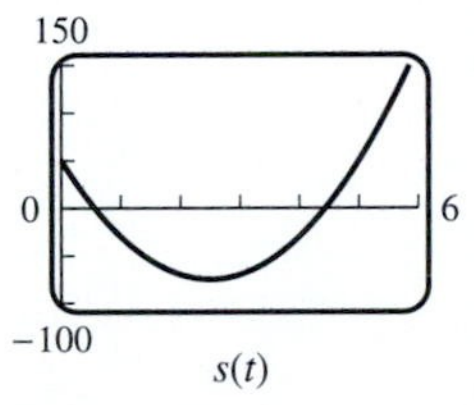

$s(t)$

(b)

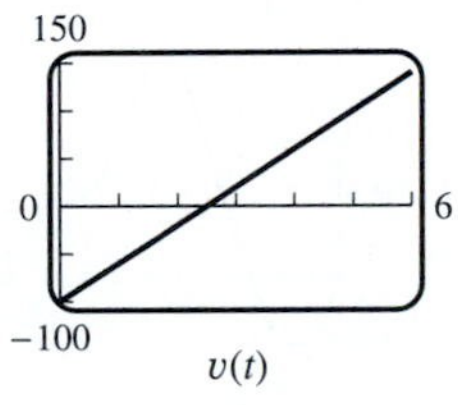

$v(t)$

(c)

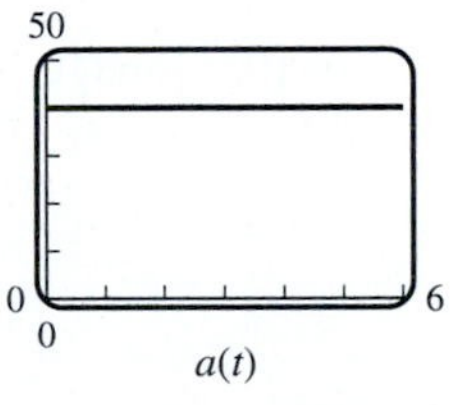

$a(t)$

27. **(a)** The displacement is always positive.

29. **(a)** $a(t) = \begin{cases} 0, & t < 4 \\ -10, & t > 4 \end{cases}$ **(b)** $v(t) = \begin{cases} 25, & t < 4 \\ 65 - 10t, & t > 4 \end{cases}$ **(c)** $x(t) = \begin{cases} 25t, & t < 4 \\ 65t - 5t^2 - 80, & t > 4 \end{cases}$ **(d)** $x(6.5) = 131.25$

so $x(8) = 120$, $x(12) = -20$.

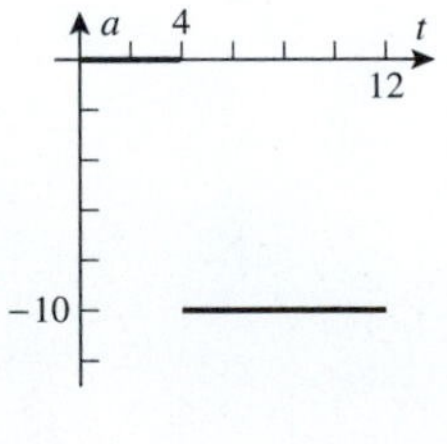

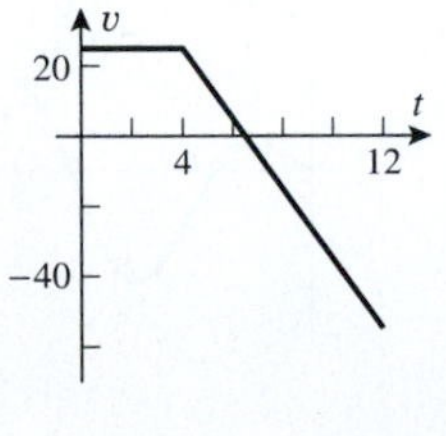

31. **(a)** $-\frac{22}{15}$ ft/s^2 **(b)** $\frac{1}{7200}$ km/s^2 33. **(a)** $-\frac{121}{5}$ ft/s^2 **(b)** $\frac{70}{33}$ s **(c)** $\frac{60}{11}$ s
35. 280 m 37. 100 s; 10,000 ft 39. **(a)** -48 ft/s **(b)** 196 ft **(c)** 112 ft/s
41. **(a)** 1 s **(b)** $\frac{1}{2}$ s 43. **(a)** $(5+5\sqrt{33})/8$ s **(b)** $20\sqrt{33}$ ft/s
45. **(a)** 5 s **(b)** 272.5 m **(c)** 10 s **(d)** -49 m/s **(e)** 12.46 s **(f)** 73.1 m/s 47. 4.04 m/s 49. 6 51. $2/\pi$ 53. $\dfrac{1}{e-1}$
55. **(a)** $\frac{4}{3}$ **(b)** $2/\sqrt{3}$ **(c)**

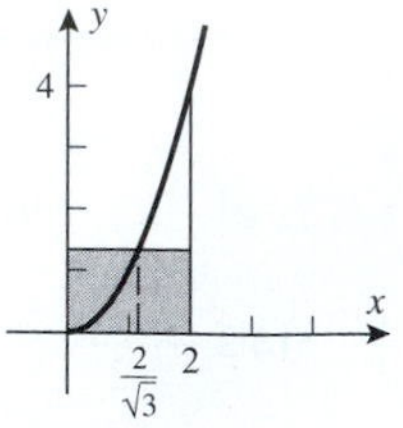

57. **(a)** $\frac{263}{4}$ **(b)** 31 59. 1404π lb 61. **(a)** 120 gal **(b)** 420 gal **(c)** 2076.36 gal 63. **(b)** no

▶ Exercise Set 7.8 (Page 444)

1. **(a)** $\displaystyle\int_1^3 u^7\,du$ **(b)** $\displaystyle -\frac{1}{2}\int_7^4 u^{1/2}du$ **(c)** $\displaystyle\frac{1}{\pi}\int_{-\pi}^{\pi}\sin u\,du$ **(d)** $\displaystyle\int_{-3}^{0}(u+5)u^{20}du$ 3. $\frac{121}{5}$ 5. 10 7. $\frac{1192}{15}$
9. $8-(4\sqrt{2})$ 11. $\ln\frac{21}{13}$ 13. $\frac{25}{12}\pi$ 15. $\pi/8$ 17. $2/\pi$ 19. $\frac{1}{24}$ 21. $\dfrac{1-e^{-8}}{8}$ 23. $\frac{2}{3}$ 25. $\frac{2}{3}(\sqrt{10}-2\sqrt{2})$
27. $2(\sqrt{7}-\sqrt{3})$ 29. 0 31. 0 33. $(\sqrt{3}-1)/3$ 35. $\frac{106}{405}$ 37. $\ln 2$ 41. **(a)** $\frac{5}{3}$ **(b)** $\frac{5}{3}$ **(c)** $-\frac{1}{2}$
45. 48,233,525,650 47. **(a)** 328.69 ft **(b)** yes 49. **(b)** 169.7 V 51. **(b)** $\frac{3}{2}$ **(c)** $\pi/4$ 53. $2/\pi$

▶ Exercise Set 7.9 (Page 451)

1. **(a)** **(b)** 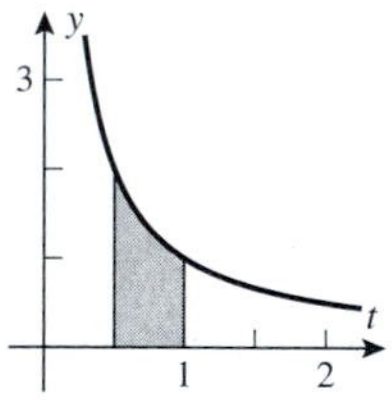 **(c)**

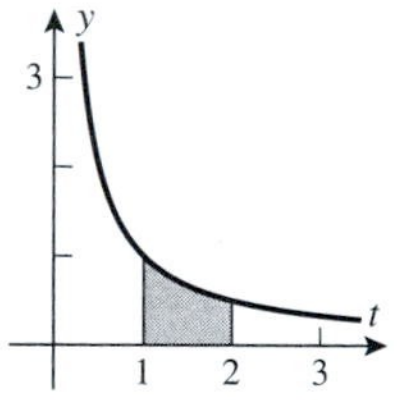

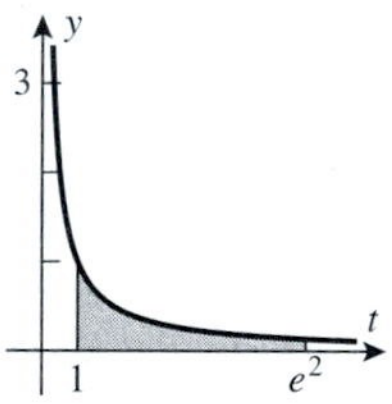

3. **(a)** 7 **(b)** -5 **(c)** -3 **(d)** 6
5. 1.603210678; magnitude of error is < 0.0063 7. **(a)** $x^{-1}, x>0$ **(b)** $x^2, x\neq 0$ **(c)** $-x^2, -\infty<x<+\infty$ **(d)** $-x, -\infty<x<+\infty$ **(e)** $x^3, x>0$ **(f)** $\ln x + x, x>0$ **(g)** $x-\sqrt[3]{x}, -\infty<x<+\infty$ **(h)** $\dfrac{e^x}{x}, x>0$
9. **(a)** $e^{\pi\ln 3}$ **(b)** $e^{\sqrt{2}\ln 2}$ 11. **(a)** e^2 **(b)** e^2 13. x^2-x 15. **(a)** $3/x$ **(b)** 1 17. **(a)** 0 **(b)** $\frac{1}{3}$ **(c)** 0
19. **(a)** $2x^3\sqrt{1+x^2}$ **(b)** $-\dfrac{2}{3}(x^2+1)^{3/2}+\dfrac{2}{5}(x^2+1)^{5/2}-\dfrac{4\sqrt{2}}{15}$ 21. **(a)** $-\sin x^2$ **(b)** $-\tan^2 x$
23. $-3\dfrac{3x-1}{9x^2+1}+2x\dfrac{x^2-1}{x^4+1}$ 25. **(a)** $3x^2\sin^2(x^3)-2x\sin^2(x^2)$ **(b)** $\dfrac{2}{1-x^2}$
27. **(a)** $F(0)=0, F(3)=0, F(5)=6, F(7)=6, F(10)=3$
(b) increasing on $[\frac{3}{2}, 6]$ and $[\frac{37}{4}, 10]$, decreasing on $[0, \frac{3}{2}]$ and $[6, \frac{37}{4}]$
(c) maximum $\frac{15}{2}$ at $x=6$, minimum $-\frac{9}{4}$ at $x=\frac{3}{2}$
(d)

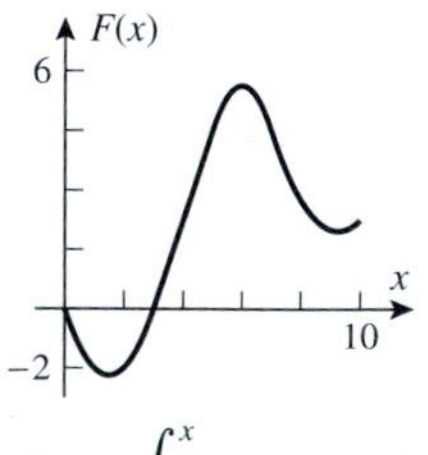

29. $F(x)=\begin{cases}(1-x^2)/2, & x<0\\(1+x^2)/2, & x\geq 0\end{cases}$
31. $y(x)=\frac{5}{4}+\frac{3}{4}x^{4/3}$ 33. $y(x)=\tan x+\cos x-(\sqrt{2}/2)$ 35. $P(x)=P_0+\displaystyle\int_0^x r(t)\,dt$ individuals
37. I is the derivative of II. 39. **(a)** $t=3$ **(b)** $t=1$ **(c)** $t=5$ **(d)** $t=3$ **(e)** F is concave up on $(0, \frac{1}{2})$ and $(2, 4)$, concave down on $(\frac{1}{2}, 2)$ and $(4, 5)$. **(f)**

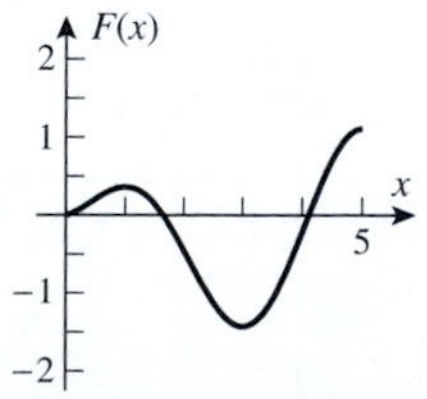

41. **(a)** relative maxima at $x = \pm\sqrt{4k+1}$, $k = 0, 1, \ldots$; relative minima at $x = \pm\sqrt{4k-1}$, $k = 1, 2, \ldots$
(b) $x = \pm\sqrt{2k}$, $k = 1, 2, \ldots$, and at $x = 0$

43. $f(x) = 3e^{3x}$, $a = \frac{1}{3}\ln 2$

45. 0.06

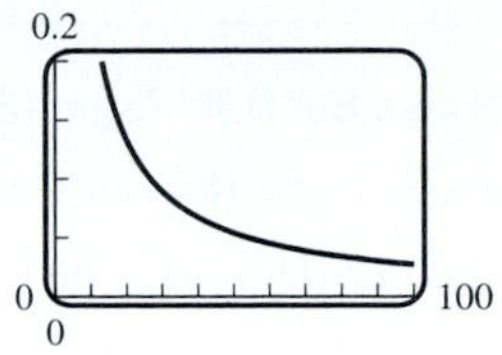

▶ Chapter 7 Supplementary Exercises (Page 454)

5. $s(t) = \frac{1}{2}at^2 + v_0 t + s_0$, $v(t) = a(t) + v_0$

7. **(a)** $\frac{3}{4}$ **(b)** $-\frac{3}{2}$ **(c)** $-\frac{35}{4}$ **(d)** -2 **(e)** not enough information **(f)** not enough information

9. **(a)** $2 + (\pi/2)$ **(b)** $\frac{1}{3}(10^{3/2} - 1) - \frac{9\pi}{4}$ **(c)** $\pi/8$

11. $35\pi/128$

15. **(d)** $n \geq 1000$

21. $(x^{2/3} + 1)^{3/2} + C$

23. **(a)** $\int_1^x \frac{1}{1+t^2}\,dt$ **(b)** $\int_{\tan[(\pi/4)-2]}^x \frac{1}{1+t^2}\,dt$

27. **(a)** $F(x)$ is 0 if $x = 1$, positive if $x > 1$, and negative if $x < 1$.
(b) $F(x)$ is 0 if $x = -1$, positive if $-1 < x \leq 2$, and negative if $-2 \leq x < -1$.

29. **(a)** 37,773.06 kW **(b)** 2200.32 kW/h **(c)**

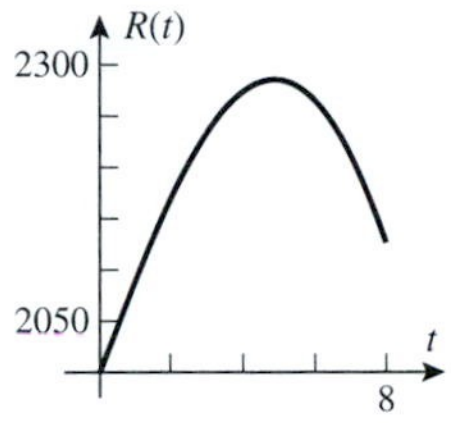

(d) 2285.32 kW/h

31. **(a)** no **(b)** $25 < t < 40$ **(c)** 3.54 ft/s **(d)** 141.5 ft **(e)** no **(f)** no

33. $\frac{1}{3}\sqrt{5 + 2\sin 3x} + C$

35. $-\frac{1}{3a^2x^3 + 3ab} + C$

37. C

39. $\ln 2$

41. $\frac{3}{8} + \frac{1}{2}\left(\sin 1 - \sin\frac{1}{4}\right)$

43. 1.007514

45. **(a)** $k = 2.073948$ **(b)** $k = 1.837992$

47. **(a)**

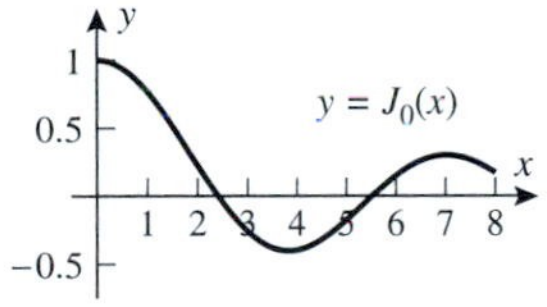

(b) 0.7651976866 **(c)** $x = 2.404826$

49. The integral is better.

▶ Exercise Set 8.1 (Page 467)

1. 9/2 3. 1 5. **(a)** 32/3 **(b)** 32/3 7. 49/192 9. 1/2 11. $\sqrt{2}$ 13. 1/2 15. 24
17. 37/12 19. $4\sqrt{2}$ 21. 1/2 23. 9152/105 25. $9/\sqrt[3]{4}$ 27. **(a)** 4/3 **(b)** $m = 2 - \sqrt[3]{4}$ 31. 1.180898334
33. **(a)** 1800 ft **(b)** $\frac{3}{2}T^2 - \frac{1}{60}T^3$ ft 35. $a^2/6$

▶ Exercise Set 8.2 (Page 473)

1. 8π 3. $13\pi/6$ 5. $32\pi/5$ 7. $(1 - \sqrt{2}/2)\pi$ 9. $256\pi/3$ 11. 4π 13. $2048\pi/15$ 15. $3\pi/5$ 17. 8π
19. 2π 21. $72\pi/5$ 23. $4\pi ab^2/3$ 25. π 27. $648\pi/5$ 29. $\pi/2$ 31. $40{,}000\pi$ ft^3 33. 1/30
35. **(a)** $2\pi/3$ **(b)** 16/3 **(c)** $4\sqrt{3}/3$ 37. 0.710172176 41. **(b)** left ≈ 11.157; right ≈ 11.771; $V \approx$ average $= 11.464$ cm^3

43. $V = \begin{cases} 3\pi h^2, & 0 \leq h < 2 \\ \frac{1}{3}\pi(12h^2 - h^3 - 4), & 2 \leq h \leq 4 \end{cases}$ 45. $r^3 \tan\theta$ 47. $16r^3/3$

▶ Exercise Set 8.3 (Page 479)

1. $15\pi/2$ 3. $\pi/3$ 5. $2\pi/5$ 7. 4π 9. $20\pi/3$ 11. $\pi(e^3 - e)$ 13. $\pi/2$ 15. $\pi/5$ 17. $2\pi^2$
19. **(a)** $7\pi/30$ **(b)** easier 21. $9\pi/14$ 23. $\pi r^2 h/3$ 25. $V = \frac{4\pi}{3}[r^3 - (r^2 - a^2)^{3/2}]$ 27. $b = 1$

▶ Exercise Set 8.4 (Page 483)

1. $L = \sqrt{5}$ 3. $(85\sqrt{85} - 8)/243$ 5. $\frac{1}{27}(80\sqrt{10} - 13\sqrt{13})$ 7. $(e^3 - e^{-3})/2$ 9. $(2\sqrt{2} - 1)/3$ 11. π
13. $\sqrt{2}(e^{\pi/2} - 1)$ 15. $\ln(1 + \sqrt{2})$ 19. **(a)**

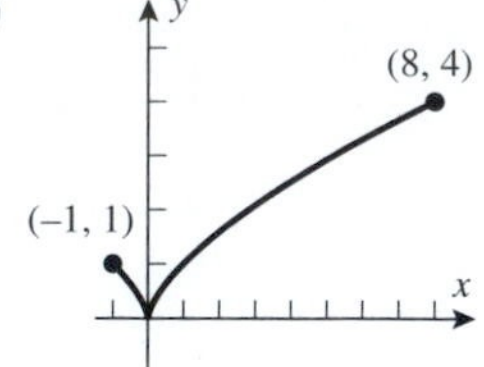

(b) dy/dx does not exist at $x = 0$.
(c) $L = (13\sqrt{13} + 80\sqrt{10} - 16)/27$

21. 4.645975301 23. 3.820197788 27. **(b)** 9.69 **(c)** 5.16 cm 29. $k = 1.83$

▶ Exercise Set 8.5 (Page 488)

1. $35\pi\sqrt{2}$ 3. 8π 5. $40\pi\sqrt{82}$ 7. 24π 9. $16\pi/9$ 11. $16{,}911\pi/1024$ 13. 22.94 15. 7.05
21. $\dfrac{2\sqrt{2}\pi}{5}(2e^{\pi} + 1)$ 23. $\dfrac{\pi}{24}(17\sqrt{17} - 1)$ 29. **(b)** for $f(x)$ constant on $[a, b]$

▶ Exercise Set 8.6 (Page 494)

1. **(a)** 210 ft·lb **(b)** 5/6 ft·lb 3. 100 ft·lb 5. 160 J 7. 20 lb/ft 9. $900\pi\rho$ ft·lb 11. 261,600 J
13. **(a)** 926,640 ft·lb **(b)** hp of motor = 0.468 15. 75,000 ft·lb
17. **(a)** $2{,}400{,}000{,}000/x^2$ lb **(b)** $(9.6 \times 10^{10})/(x + 4000)^2$ lb **(c)** 2.5344×10^{10} ft·lb 19. $v_f = 100$ m/s
21. **(a)** decrease of 4.5×10^{14} J **(b)** ≈ 0.107 **(c)** ≈ 8.24 bombs

▶ Exercise Set 8.7 (Page 499)

1. **(a)** $F = 31{,}200$ lb; $P = 312$ lb/ft^2 **(b)** $F = 2{,}452{,}500$ N; $P = 98.1$ kPa 3. 499.2 lb 5. 8.175×10^5 N
7. 1,098,720 N 9. yes 11. $\rho a^3/\sqrt{2}$ lb 13. $14{,}976\sqrt{17}$ lb 15. **(b)** $80\rho_0$ lb/min

▶ Exercise Set 8.8 (Page 508)

1. **(a)** 10.0179 **(b)** 3.7622 **(c)** $15/17 \approx 0.8824$ **(d)** -1.4436 **(e)** 1.7627 **(f)** 0.9730
3. **(a)** $\frac{4}{3}$ **(b)** $\frac{5}{4}$ **(c)** $\frac{312}{313}$ **(d)** $-\frac{63}{16}$
5.

	$\sinh x_0$	$\cosh x_0$	$\tanh x_0$	$\coth x_0$	$\operatorname{sech} x_0$	$\operatorname{csch} x_0$
(a)	2	$\sqrt{5}$	$2/\sqrt{5}$	$\sqrt{5}/2$	$1/\sqrt{5}$	1/2
(b)	3/4	5/4	3/5	5/3	4/5	4/3
(c)	4/3	5/3	4/5	5/4	3/5	3/4

9. $4\cosh(4x - 8)$ 11. $-\dfrac{1}{x}\operatorname{csch}^2(\ln x)$ 13. $\dfrac{1}{x^2}\operatorname{csch}\left(\dfrac{1}{x}\right)\coth\left(\dfrac{1}{x}\right)$ 15. $\dfrac{2 + 5\cosh(5x)\sinh(5x)}{\sqrt{4x + \cosh^2(5x)}}$
17. $x^{5/2}\tanh(\sqrt{x})\operatorname{sech}^2(\sqrt{x}) + 3x^2\tanh^2(\sqrt{x})$ 19. $\dfrac{1}{\sqrt{9 + x^2}}$ 21. $\dfrac{1}{(\cosh^{-1} x)\sqrt{x^2 - 1}}$ 23. $\dfrac{-(\tanh^{-1} x)^{-2}}{1 - x^2}$
25. $\dfrac{\sinh x}{|\sinh x|} = \begin{cases} 1, & x > 0 \\ -1, & x < 0 \end{cases}$ 27. $-\dfrac{e^x}{2x\sqrt{1 - x}} + e^x \operatorname{sech}^{-1} x$ 31. $\frac{1}{7}\sinh^7 x + C$ 33. $\frac{2}{3}(\tanh x)^{3/2} + C$ 35. $\ln(\cosh x) + C$
37. 37/375 39. $\frac{1}{3}\sinh^{-1} 3x + C$ 41. $-\operatorname{sech}^{-1}(e^x) + C$ 43. $-\operatorname{csch}^{-1}|2x| + C$ 45. $\frac{1}{2}\ln 3$ 49. 16/9 51. 5π
53. $\frac{3}{4}$ 61. $|u| < 1$: $\tanh^{-1} u + C$; $|u| > 1$: $\tanh^{-1}(1/u) + C$ 63. **(a)** $+\infty$ **(b)** $-\infty$ **(c)** 1 **(d)** -1 **(e)** $+\infty$ **(f)** $+\infty$
71. 405.9 ft

▶ Chapter 8 Supplementary Exercises (Page 510)

7. **(a)** $\displaystyle\int_a^b (f(x) - g(x))\,dx + \int_b^c (g(x) - f(x))\,dx + \int_c^d (f(x) - g(x))\,dx$ **(b)** $\frac{11}{4}$ 9. $9a/8$

13. Set $a = 68.7672$, $b = 0.0100333$, $c = 693.8597$, $d = 299.2239$.
(a)

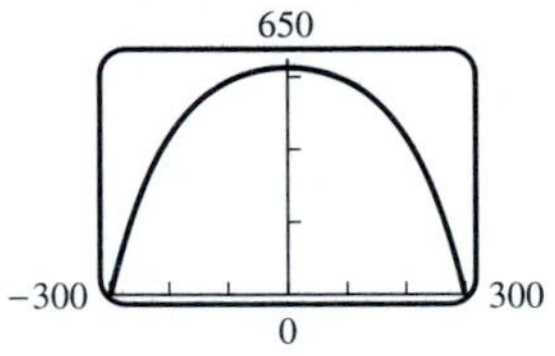

(b) 1480.2798 ft
(c) 283.6249 ft
(d) 82°

15. **(a)** $\sinh^{-1}(x/2) + C$
(b) $\cosh^{-1}(x/3) + C$
(c) $\begin{cases} \dfrac{1}{\sqrt{2}}\tanh^{-1}\left(\dfrac{x}{\sqrt{2}}\right) + C, & |x| < \sqrt{2} \\ \dfrac{1}{\sqrt{2}}\coth^{-1}\left(\dfrac{x}{\sqrt{2}}\right) + C, & |x| > \sqrt{2} \end{cases}$ or $\dfrac{1}{2\sqrt{2}}\ln\left|\dfrac{\sqrt{2}+x}{\sqrt{2}-x}\right| + C$
(d) $\dfrac{1}{\sqrt{5}}\sinh^{-1}\left(\dfrac{\sqrt{5}x}{4}\right) + C$

17. **(a)** $W = \frac{1}{16}$ J **(b)** 5 m

19. **(a)** $F = \displaystyle\int_0^1 \rho x 3\,dx$ N
(b) $F = \displaystyle\int_1^4 \rho(1+x)2x\,dx$ lb/ft²
(c) $\displaystyle\int_{-10}^{0} 9810|y|2\sqrt{\tfrac{125}{8}(y+10)}\,dy$ N

21. **(a)**

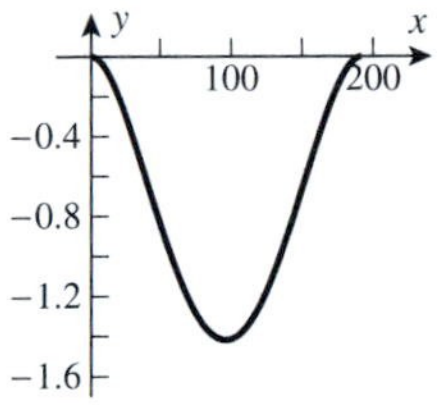

(b) 1.42 in
(c) The length of the centerline is 192.026 in.

23. $k \approx 0.724611$ 25. **(a)** $\displaystyle\int_a^{a+2} \frac{x}{\sqrt{1+x^3}}\,dx$ **(b)** $a \approx 0.683772$; maximum work $= 1.347655$ J

▶ Exercise Set 9.1 (Page 515)

1. $-\frac{1}{8}(3-2x)^4 + C$ 3. $\frac{1}{2}\tan(x^2) + C$ 5. $-\frac{1}{3}\ln(2+\cos 3x) + C$ 7. $\cosh e^x + C$ 9. $-e^{\cot x} + C$
11. $-\frac{1}{42}\cos^6 7x + C$ 13. $\ln(e^x + \sqrt{e^{2x}+4}) + C$ 15. $2e^{\sqrt{x-2}} + C$ 17. $2\sinh\sqrt{x} + C$ 19. $-\dfrac{2}{\ln 3}3^{-\sqrt{x}} + C$
21. $\dfrac{1}{2}\coth\dfrac{2}{x} + C$ 23. $-\dfrac{1}{4}\ln\left|\dfrac{2+e^{-x}}{2-e^{-x}}\right| + C$ 25. $\sin^{-1} e^x + C$ 27. $\frac{1}{2}\sin(x^2) + C$ 29. $-\dfrac{1}{\ln 16}4^{-x^2} + C$

▶ Exercise Set 9.2 (Page 521)

1. $-xe^{-x} - e^{-x} + C$ 3. $x^2e^x - 2xe^x + 2e^x + C$ 5. $-\frac{1}{2}x\cos 2x + \frac{1}{4}\sin 2x + C$ 7. $x^2\sin x + 2x\cos x - 2\sin x + C$
9. $\frac{2}{3}x^{3/2}\ln x - \frac{4}{9}x^{3/2} + C$ 11. $x(\ln x)^2 - 2x\ln x + 2x + C$ 13. $x\ln(2x+3) - x + \frac{3}{2}\ln(2x+3) + C$
15. $x\sin^{-1} x + \sqrt{1-x^2} + C$ 17. $x\tan^{-1}(2x) - \frac{1}{4}\ln(1+4x^2) + C$ 19. $\frac{1}{2}e^x(\sin x - \cos x) + C$
21. $\dfrac{e^{ax}}{a^2+b^2}(a\sin bx - b\cos bx) + C$ 23. $(x/2)[\sin(\ln x) - \cos(\ln x)] + C$ 25. $x\tan x + \ln|\cos x| + C$
27. $\frac{1}{2}x^2e^{x^2} - \frac{1}{2}e^{x^2} + C$ 29. $(1-6e^{-5})/25$ 31. $(2e^3+1)/9$ 33. $5\ln 5 - 4$ 35. $\dfrac{5\pi}{6} - \sqrt{3} + 1$ 37. $-\pi/8$
39. $\dfrac{1}{3}\left(2\sqrt{3}\pi - \dfrac{\pi}{2} - 2 + \ln 2\right)$ 41. **(a)** $2(\sqrt{x}-1)e^{\sqrt{x}} + C$ **(b)** $2\sqrt{x}\sin\sqrt{x} + 2\cos\sqrt{x} + C$ 43. **(a)** $A = 1$ **(b)** $V = \pi(e-2)$
45. $V = 2\pi^2$ 47. distance $= -37e^{-5} + 2$ 49. **(a)** $-\frac{1}{3}\sin^2 x\cos x - \frac{2}{3}\cos x + C$ **(b)** $\dfrac{3\pi}{32} - \dfrac{1}{4}$
53. **(a)** $\frac{1}{3}\tan^3 x - \tan x + x + C$ **(b)** $\frac{1}{3}\sec^2 x\tan x + \frac{2}{3}\tan x + C$ **(c)** $x^3e^x - 3x^2e^x + 6xe^x - 6e^x + C$

▶ Exercise Set 9.3 (Page 529)

1. $-\frac{1}{6}\cos^6 x + C$ 3. $\dfrac{1}{2a}\sin^2 ax + C,\ \ a \neq 0$ 5. $\frac{1}{2}\theta - \frac{1}{20}\sin 10\theta + C$ 7. $\sin\theta - \frac{2}{3}\sin^3\theta + \frac{1}{5}\sin^5\theta + C$
9. $\frac{1}{6}\sin^3 2t - \frac{1}{10}\sin^5 2t + C$ 11. $\frac{1}{8}x - \frac{1}{32}\sin 4x + C$ 13. $-\frac{1}{6}\cos 3x + \frac{1}{2}\cos x + C$ 15. $-\frac{1}{3}\cos(3x/2) - \cos(x/2) + C$
17. $(5\sqrt{2})/12$ 19. 0 21. $\frac{1}{24}$ 23. $\frac{1}{3}\tan(3x+1) + C$ 25. $\frac{1}{2}\ln|\cos(e^{-2x})| + C$ 27. $\frac{1}{2}\ln|\sec 2x + \tan 2x| + C$
29. $\frac{1}{3}\tan^3 x + C$ 31. $\frac{1}{16}\tan^4 4x + \frac{1}{24}\tan^6 4x + C$ 33. $\frac{1}{7}\sec^7 x - \frac{1}{5}\sec^5 x + C$
35. $\frac{1}{4}\sec^3 x\tan x - \frac{5}{8}\sec x\tan x + \frac{3}{8}\ln|\sec x + \tan x| + C$ 37. $\frac{1}{6}\sec^3 2t + C$ 39. $\tan x + \frac{1}{3}\tan^3 x + C$
41. $\frac{1}{3}\tan^3 x - \tan x + x + C$ 43. $\frac{2}{3}\tan^{3/2} x + \frac{2}{7}\tan^{7/2} x + C$ 45. $\dfrac{\sqrt{3}}{2} - \dfrac{\pi}{6}$ 47. $-\frac{1}{2} + \ln 2$
49. $-\frac{1}{5}\csc^5 x + \frac{1}{3}\csc^3 x + C$ 51. $-\frac{1}{2}\csc^2 x - \ln|\sin x| + C$ 55. $L = \ln(\sqrt{2}+1)$ 57. $V = \pi/2$
63. $-\dfrac{1}{\sqrt{a^2+b^2}}\ln\left|\dfrac{\sqrt{a^2+b^2} + a\cos x - b\sin x}{a\sin x + b\cos x}\right| + C$ 65. **(a)** $\frac{2}{3}$ **(b)** $3\pi/16$ **(c)** $\frac{8}{15}$ **(d)** $5\pi/32$

▶ Exercise Set 9.4 (Page 535)

1. $2\sin^{-1}(x/2) + \frac{1}{2}x\sqrt{4-x^2} + C$ 3. $\frac{9}{2}\sin^{-1}(x/3) - \frac{1}{2}x\sqrt{9-x^2} + C$ 5. $\frac{1}{16}\tan^{-1}(x/2) + \dfrac{x}{8(4+x^2)} + C$

7. $\sqrt{x^2-9} - 3\sec^{-1}(x/3) + C$ 9. $-2\sqrt{2-x^2} + \frac{1}{3}(2-x^2)^{3/2} + C$ 11. $\dfrac{\sqrt{4x^2-9}}{9x} + C$ 13. $\dfrac{x}{\sqrt{1-x^2}} + C$

15. $\ln|x + \sqrt{x^2-1}| + C$ 17. $-(x/\sqrt{9x^2-1}) + C$ 19. $\frac{1}{2}\sin^{-1}(e^x) + \frac{1}{2}e^x\sqrt{1-e^{2x}} + C$ 21. $\frac{2048}{15}$ 23. $(\sqrt{3}-\sqrt{2})/2$

25. $\dfrac{10\sqrt{3}+18}{243}$ 27. $\frac{1}{2}\ln(x^2+4) + C$ 29. $L = \sqrt{5} - \sqrt{2} + \ln\dfrac{2+2\sqrt{2}}{1+\sqrt{5}}$ 31. $S = \dfrac{\pi}{32}[18\sqrt{5} - \ln(2+\sqrt{5})]$

33. **(a)** $\sinh^{-1}(x/3) + C$ **(b)** $\ln\left(\dfrac{\sqrt{x^2+9}}{3} + \dfrac{x}{3}\right) + C$ **(c)** $\frac{1}{2}x\sqrt{x^2-1} - \frac{1}{2}\cosh^{-1}x + C$ 35. $\dfrac{1}{3}\tan^{-1}\left(\dfrac{x-2}{3}\right) + C$

37. $\sin^{-1}\left(\dfrac{x-1}{3}\right) + C$ 39. $\ln(x - 3 + \sqrt{(x-3)^2+1}) + C$ 41. $2\sin^{-1}\left(\dfrac{x+1}{2}\right) + \dfrac{1}{2}(x+1)\sqrt{3-2x-x^2} + C$

43. $\dfrac{1}{\sqrt{10}}\tan^{-1}\sqrt{\frac{2}{5}}\,(x+1) + C$ 45. $\pi/6$ 49. $u = \sin^2 x,\ \dfrac{1}{2}\displaystyle\int\sqrt{1-u^2}\,du = \dfrac{1}{4}[\sin^2 x\sqrt{1-\sin^4 x} + \sin^{-1}(\sin^2 x)] + C$

▶ Exercise Set 9.5 (Page 542)

1. $\dfrac{A}{(x-2)} + \dfrac{B}{(x+5)}$ 3. $\dfrac{A}{x} + \dfrac{B}{x^2} + \dfrac{C}{x-1}$ 5. $\dfrac{A}{x} + \dfrac{B}{x^2} + \dfrac{C}{x^3} + \dfrac{Dx+E}{x^2+1}$ 7. $\dfrac{Ax+B}{x^2+5} + \dfrac{Cx+D}{(x^2+5)^2}$

9. $\dfrac{1}{5}\ln\left|\dfrac{x-1}{x+4}\right| + C$ 11. $\frac{5}{2}\ln|2x-1| + 3\ln|x+4| + C$ 13. $\ln\left|\dfrac{x(x+3)^2}{x-3}\right| + C$ 15. $\frac{1}{2}x^2 - 2x + 6\ln|x+2| + C$

17. $3x + 12\ln|x-2| - \dfrac{2}{x-2} + C$ 19. $\dfrac{1}{3}x^3 + x + \ln\left|\dfrac{(x+1)(x-1)^2}{x}\right| + C$ 21. $3\ln|x| - \ln|x-1| - \dfrac{5}{x-1} + C$

23. $\ln\dfrac{(x-3)^2}{|x+1|} + \dfrac{1}{x-3} + C$ 25. $\ln|x+2| + \dfrac{4}{x+2} - \dfrac{2}{(x+2)^2} + C$ 27. $-\frac{7}{34}\ln|4x-1| + \frac{6}{17}\ln(x^2+1) + \frac{3}{17}\tan^{-1}x + C$

29. $3\tan^{-1}x + \frac{1}{2}\ln(x^2+3) + C$ 31. $\frac{1}{2}x^2 - 3x + \frac{1}{2}\ln(x^2+1) + C$ 33. $\dfrac{1}{6}\ln\left(\dfrac{1-\sin\theta}{5+\sin\theta}\right) + C$ 35. $V = \pi\left(\frac{19}{5} - \frac{9}{4}\ln 5\right)$

37. $\dfrac{1}{\sqrt{2}}\tan^{-1}\left(\dfrac{x+1}{\sqrt{2}}\right) + \dfrac{1}{x^2+2x+3} + C$ 39. $\frac{1}{8}\ln|x-1| - \frac{1}{5}\ln|x-2| + \frac{1}{12}\ln|x-3| - \frac{1}{120}\ln|x+3| + C$

▶ Exercise Set 9.6 (Page 551)

1. Formula (60): $\frac{3}{16}[4x + \ln|-1+4x|] + C$ 3. Formula (65): $\dfrac{1}{5}\ln\left|\dfrac{x}{5+2x}\right| + C$ 5. Formula (102): $\frac{1}{5}(x+1)(-3+2x)^{3/2} + C$

7. Formula (108): $\dfrac{1}{2}\ln\left|\dfrac{\sqrt{4-3x}-2}{\sqrt{4-3x}+2}\right| + C$ 9. Formula (69): $\dfrac{1}{2\sqrt{5}}\ln\left|\dfrac{x+\sqrt{5}}{x-\sqrt{5}}\right| + C$

11. Formula (73): $\dfrac{x}{2}\sqrt{x^2-3} - \dfrac{3}{2}\ln|x + \sqrt{x^2-3}| + C$ 13. Formula (95): $\dfrac{x}{2}\sqrt{x^2+4} - 2\ln(x + \sqrt{x^2+4}) + C$

15. Formula (74): $\dfrac{x}{2}\sqrt{9-x^2} + \dfrac{9}{2}\sin^{-1}\dfrac{x}{3} + C$ 17. Formula (79): $\sqrt{3-x^2} - \sqrt{3}\ln\left|\dfrac{\sqrt{3}+\sqrt{9-x^2}}{x}\right| + C$

19. Formula (38): $-\frac{1}{10}\sin(5x) + \frac{1}{2}\sin x + C$ 21. Formula (50): $\dfrac{x^4}{16}[4\ln x - 1] + C$

23. Formula (42): $\dfrac{e^{-2x}}{13}[-2\sin(3x) - 3\cos(3x)] + C$ 25. Formula (62): $\dfrac{1}{2}\displaystyle\int\dfrac{u\,du}{(4-3u)^2} = \dfrac{1}{18}\left[\dfrac{4}{4-3e^{2x}} + \ln\left|4-3e^{2x}\right|\right] + C$

27. Formula (68): $\dfrac{2}{3}\displaystyle\int\dfrac{du}{u^2+4} = \dfrac{1}{3}\tan^{-1}\dfrac{3\sqrt{x}}{2} + C$ 29. Formula (76): $\dfrac{1}{3}\displaystyle\int\dfrac{du}{\sqrt{u^2-4}} = \dfrac{1}{3}\ln|3x + \sqrt{9x^2-4}| + C$

31. Formula (81): $\dfrac{1}{54}\displaystyle\int\dfrac{u^2\,du}{\sqrt{5-u^2}} = -\dfrac{x^2}{36}\sqrt{5-9x^4} + \dfrac{5}{108}\sin^{-1}\dfrac{3x^2}{\sqrt{5}} + C$

33. Formula (26): $\displaystyle\int\sin^2 u\,du = \dfrac{1}{2}\ln x + \dfrac{1}{4}\sin(2\ln x) + C$ 35. Formula (51): $\dfrac{1}{4}\displaystyle\int ue^u\,du = \dfrac{1}{4}(-2x-1)e^{-2x} + C$

37. $u = \cos 3x$, Formula (67): $-\displaystyle\int\dfrac{du}{u(u+1)^2} = -\dfrac{1}{3}\left[\dfrac{1}{1+\cos 3x} + \ln\left|\dfrac{\cos 3x}{1+\cos 3x}\right|\right] + C$

39. $u = 4x^2$, Formula (70): $\dfrac{1}{8}\displaystyle\int\dfrac{du}{u^2-1} = \dfrac{1}{16}\ln\left|\dfrac{4x^2-1}{4x^2+1}\right| + C$

41. $u = 2e^x$, Formula (74): $\frac{1}{2}\int \sqrt{3-u^2}\,du = \frac{1}{2}e^x\sqrt{3-4e^{2x}} + \frac{3}{4}\sin^{-1}\left(\frac{2e^x}{\sqrt{3}}\right) + C$

43. $u = 3x$, Formula (112): $\frac{1}{3}\int \sqrt{\frac{5}{3}u - u^2}\,du = \frac{18x-5}{36}\sqrt{5x-9x^2} + \frac{25}{216}\sin^{-1}\left(\frac{18x-5}{5}\right) + C$

45. $u = 3x$, Formula (44): $\frac{1}{9}\int u\sin u\,du = \frac{1}{9}(\sin 3x - 3x\cos 3x) + C$

47. $u = -\sqrt{x}$, Formula (51): $2\int ue^u\,du = -2(\sqrt{x}+1)e^{-\sqrt{x}} + C$

49. $x^2 + 4x - 5 = (x+2)^2 - 9$; $u = x + 2$, Formula (70): $\int \frac{du}{u^2-9} = \frac{1}{6}\ln\left|\frac{x-1}{x+5}\right| + C$

51. $x^2 - 4x - 5 = (x-2)^2 - 9$, $u = x - 2$, Formula (77): $\int \frac{u+2}{\sqrt{9-u^2}}\,du = -\sqrt{5+4x-x^2} + 2\sin^{-1}\left(\frac{x-2}{3}\right) + C$

53. $u = \sqrt{x-2}$, $\frac{2}{5}(x-2)^{5/2} + \frac{4}{3}(x-2)^{3/2} + C$ 55. $u = \sqrt{x^3+1}$, $\frac{2}{3}\int u^2(u^2-1)\,du = \frac{2}{15}(x^3+1)^{5/2} - \frac{2}{9}(x^3+1)^{3/2} + C$

57. $u = x^{1/6}$, $\int \frac{6u^5}{u^3+u^2}\,du = 2x^{1/2} - 3x^{1/3} + 6x^{1/6} - 6\ln(x^{1/6}+1) + C$ 59. $u = x^{1/4}$, $4\int \frac{1}{u(1-u)}\,du = 4\ln\frac{x^{1/4}}{|1-x^{1/4}|} + C$

61. $u = x^{1/6}$, $6\int \frac{u^3}{u-1}\,du = 2x^{1/2} + 3x^{1/3} + 6x^{1/6} + 6\ln|x^{1/6}-1| + C$

63. $u = \sqrt{1+x^2}$, $\int (u^2-1)\,du = \frac{1}{3}(1+x^2)^{3/2} - (1+x^2)^{1/2} + C$ 65. $u = \sqrt{x}$, $2\int u\sin u\,du = 2\sin\sqrt{x} - 2\sqrt{x}\cos\sqrt{x} + C$

67. $\int \frac{1}{1 + \frac{2u}{1+u^2} + \frac{1-u^2}{1+u^2}}\frac{2}{1+u^2}\,du = \int \frac{1}{u+1}\,du = \ln|\tan(x/2)+1| + C$

69. $\int \frac{d\theta}{1-\cos\theta} = \int \frac{1}{u^2}\,du = -\frac{1}{u} + C = -\cot(\theta/2) + C$ 71. $2\int \frac{1-u^2}{(3u^2+1)(u^2+1)}\,du$, $\frac{4}{\sqrt{3}}\tan^{-1}[\sqrt{3}\tan(x/2)] - x + C$

73. $x \approx 3.523188312$ 75. $A \approx 17.59119023$ 77. $A \approx 0.054930614$ 79. $V \approx 3.586419094$ 81. $V \approx 5.031899801$

83. $L \approx 8.409316783$ 85. $S \approx 14.42359945$ 87. **(a)** $s(t) = 2 + \int_0^t 20\cos^6 u\sin^3 u\,du = -\frac{20}{9}\sin^2 t\cos^7 t - \frac{40}{63}\cos^7 t + \frac{166}{63}$

(b)

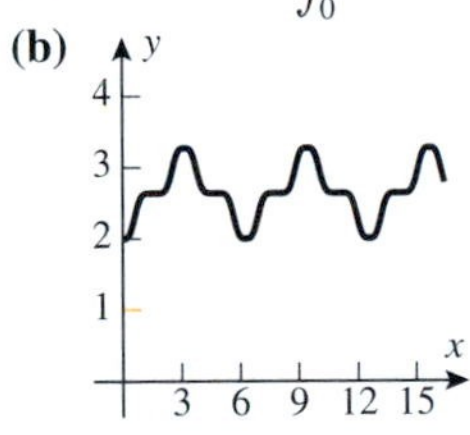

91. $\frac{2}{\sqrt{3}}\tan^{-1}\left(\frac{2\tanh(x/2)+1}{\sqrt{3}} + C\right)$

▶ Exercise Set 9.7 (Page 562)

1. exact value $= 14/3 \approx 4.666666667$
   (a) 4.667600663, $|E_M| \approx 0.000933996$
   (b) 4.664795679, $|E_T| \approx 0.001870988$
   (c) 4.666651630, $|E_S| \approx 0.000015037$

3. exact value $= 2$
   (a) 2.008248408, $|E_M| \approx 0.008248408$
   (b) 1.983523538, $|E_T| \approx 0.016476462$
   (c) 2.000109517, $|E_S| \approx 0.000109517$

5. exact value $= e^{-1} - e^{-3} \approx 0.318092373$
   (a) 0.317562837, $|E_M| \approx 0.000529536$
   (b) 0.319151975, $|E_T| \approx 0.001059602$
   (c) 0.318095187, $|E_S| \approx 0.000002814$

7. **(a)** $|E_M| \le \frac{27}{2400}(1/4) = 0.002812500$ **(b)** $|E_T| \le \frac{27}{1200}(1/4) = 0.005625000$ **(c)** $|E_S| \le \frac{243}{180\times 10^4}(15/16) \approx 0.000126563$

9. **(a)** $|E_M| \le \frac{\pi^3}{2400}(1) \approx 0.012919282$ **(b)** $|E_T| \le \frac{\pi^3}{1200}(1) \approx 0.025838564$ **(c)** $|E_S| \le \frac{\pi^5}{180\times 10^4}(1) \approx 0.000170011$

11. **(a)** $|E_M| \le \frac{8}{2400}(e^{-1}) \approx 0.001226265$ **(b)** $|E_T| \le \frac{8}{1200}(e^{-1}) \approx 0.002452530$ **(c)** $|E_S| \le \frac{32}{180\times 10^4}(e^{-1}) \approx 0.000006540$

13. **(a)** $n = 24$ **(b)** $n = 34$ **(c)** $n = 8$ 15. **(a)** $n = 36$ **(b)** $n = 51$ **(c)** $n = 8$ 17. **(a)** $n = 351$ **(b)** $n = 496$ **(c)** $n = 16$

19. 0.746824948, 0.746824133 21. 2.129861595, 2.129861293 23. 0.805376152, 0.804776489

25. **(a)** 3.142425985, $|E_M| \approx 0.000833331$ **(b)** 3.139925989, $|E_T| \approx 0.001666665$ **(c)** 3.141592614, $|E_S| \approx 0.000000040$

27. $S_{14} = 0.693147984$, $|E_S| \approx 0.000000803 = 8.03\times 10^{-7}$ 29. $n = 116$ 33. $L \approx 3.820187623$ 35. 1604 ft

37. 37.9 mi 39. 9.3 L 43. **(a)** $\max|f''(x)| \approx 3.844880$ **(b)** $n = 18$ **(c)** 0.904741

45. **(a)** $\max|f^{(4)}(x)| \approx 42.551816$ **(b)** $n = 8$ **(c)** 0.904524

▶ Exercise Set 9.8 (Page 571)

1. **(a)** improper; infinite discontinuity at $x = 3$ **(b)** not improper **(c)** improper; infinite discontinuity at $x = 0$ **(d)** improper; infinite interval of integration **(e)** improper; infinite interval of integration and infinite discontinuity at $x = 1$ **(f)** not improper
3. 1 5. $\ln \frac{5}{3}$ 7. $\frac{1}{2}$ 9. $-\frac{1}{4}$ 11. $\frac{1}{3}$ 13. divergent 15. 0 17. divergent 19. divergent 21. $\pi/2$
23. 1 25. divergent 27. $\frac{9}{2}$ 29. divergent 31. 2 33. 2 37. $\frac{1}{2}$
39. **(a)** 2.726585 **(b)** 2.804364 **(c)** 0.219384 **(d)** 0.504067 41. -1 43. $\frac{1}{9}$
45. **(a)** $V = \pi/2$ **(b)** $S = \pi[\sqrt{2} + \ln(1 + \sqrt{2})]$ 47. **(b)** $1/e$ **(c)** It is convergent. 51. $\dfrac{8\sqrt{2}}{5}$
53. $\dfrac{2\pi NI}{kr}\left(1 - \dfrac{a}{\sqrt{r^2 + a^2}}\right)$ 55. **(b)** 2.4×10^7 mi·lb 57. **(a)** $\dfrac{1}{s^2}$ **(b)** $\dfrac{2}{s^3}$ **(c)** $\dfrac{e^{-3s}}{s}$ 61. **(a)** 1.047 65. 1.809

▶ Chapter 9 Supplementary Exercises (Page 574)

1. **(a)** parts **(b)** substitution **(c)** reduction formula **(d)** substitution **(e)** substitution **(f)** substitution **(g)** parts **(h)** substitution **(i)** $u = 4 - x^2$ 5. **(a)** 40 **(b)** 57 **(c)** 113 **(d)** 108 **(e)** 52 **(f)** 71
7. **(a)** $-\frac{1}{8}\sin^3 2x \cos 2x - \frac{3}{16}\sin 2x \cos 2x + \frac{3}{8}x + C$
(b) $\frac{1}{8}\cos^3(x^2)\sin(x^2) + \frac{3}{16}\cos(x^2)\sin(x^2) + \frac{3}{16}x^2 + C$
9. 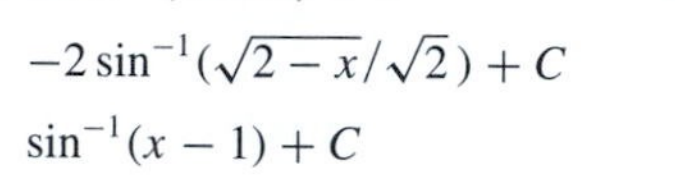
$2\sin^{-1}(\sqrt{x}/2) + C$
$-2\sin^{-1}(\sqrt{2-x}/\sqrt{2}) + C$
$\sin^{-1}(x-1) + C$
11. 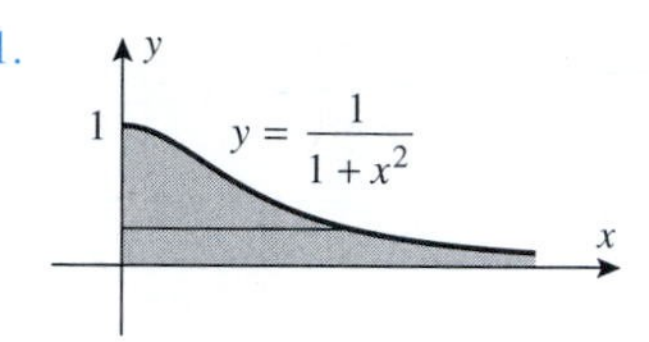

13. $V = 2\pi$ 15. $-\frac{2}{3}\cos^{3/2}\theta + C$ 17. $\frac{1}{6}\tan^3(x^2) + C$ 19. $\dfrac{x}{3\sqrt{3 + x^2}} + C$
21. $\sqrt{x^2 + 2x + 2} + 2\ln(\sqrt{x^2 + 2x + 2} + x + 1) + C$ 23. $-\frac{1}{6}\ln|x - 1| + \frac{1}{15}\ln|x + 2| + \frac{1}{10}\ln|x - 3| + C$ 25. $4 - \pi$
27. $\ln \dfrac{\sqrt{e^x + 1} - 1}{\sqrt{e^x + 1} + 1} + C$ 29. $\dfrac{1}{2(a^2 + 1)}$ 31. $\frac{1}{4}\sin^{-1}(x^4) + C$ 33. $\dfrac{\sqrt{2}}{3}[(x + 2)^{3/2} - (x - 2)^{3/2}] + C$
35. **(a)** $(x + 4)(x - 5)(x^2 + 1)^2$; $\dfrac{A}{x + 4} + \dfrac{B}{x - 5} + \dfrac{Cx + D}{x^2 + 1} + \dfrac{Ex + F}{(x^2 + 1)^2}$
(b) $-\dfrac{3}{x + 4} + \dfrac{2}{x - 5} - \dfrac{x - 2}{x^2 + 1} - \dfrac{3}{(x^2 + 1)^2}$ **(c)** $-3\ln|x + 4| + 2\ln|x - 5| + 2\tan^{-1}x - \dfrac{1}{2}\ln(x^2 + 1) - \dfrac{3}{2}\left(\dfrac{x}{x^2 + 1} + \tan^{-1}x\right)$

▶ Exercise Set 10.1 (Page 589)

3. **(a)** first order **(b)** second order 7. **(a)** $y = Ce^{-3x}$ **(b)** $y = Ce^{2t}$ 9. $y = Cx$ 11. $y = Ce^{-\sqrt{1+x^2}} - 1$
13. $\ln|y| + y^2/2 = e^x + C$ and $y = 0$ 15. $y = \ln(\sec x + C)$ 17. $y = \dfrac{1}{1 - C(\csc x - \cot x)}$ and $y = 0$ 19. $y = e^{-2x} + Ce^{-3x}$
21. $y = e^{-x}\sin(e^x) + Ce^{-x}$ 23. $y = \dfrac{C}{\sqrt{x^2 + 1}}$ 25. **(a)** $y = \dfrac{x}{2} + \dfrac{3}{2x}$ **(b)** $y = \dfrac{x}{2} - \dfrac{5}{2x}$
27. $y = -1 + 4e^{x^2/2}$ 29. $3y^2 + 6\sin y = 8x^3 + 3\pi^2 - 8$ 31. $y^2 - 2y = t^2 + t + 3$
33. **(a)** **(b)** $x = 2y^2$ 35. $y = \dfrac{C}{\sqrt{x^2 + 4}}$ 37. $x^3 + y^3 - 3y = C$

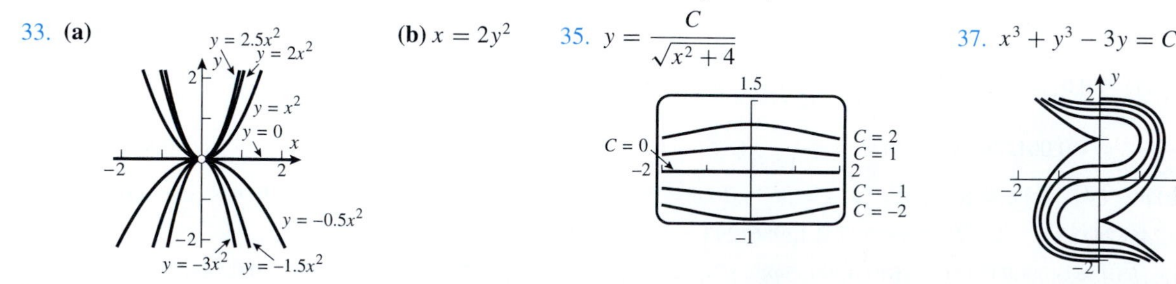

41. $x^2 + 2e^{-y} = 6$ 43. **(a)** $200 - 175e^{-t/25}$ oz **(b)** 136 oz 45. 25 lb 49. **(a)** $I(t) = \frac{6}{5}(1 - e^{-5t/2})$ A **(b)** It tends to $\frac{6}{5}$ A.
51. **(a)** $v = c\ln\dfrac{m_0}{m_0 - kt} - gt$ **(b)** 3044 m/s 53. **(a)** $h \approx (2 - 0.003979t)^2$ **(b)** 8.4 min
55. $v = \dfrac{50}{2t + 1}$ cm/s, $x = 25\ln(2t + 1)$ cm 57. $\dfrac{dy}{dx} = -\sin x + e^{-x^2}$, $y(0) = 1$

▶ Exercise Set 10.2 (Page 597)

1. 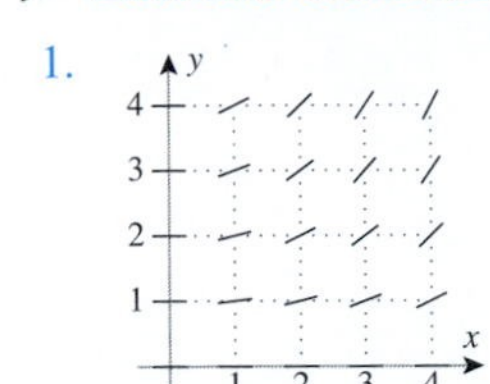

3. 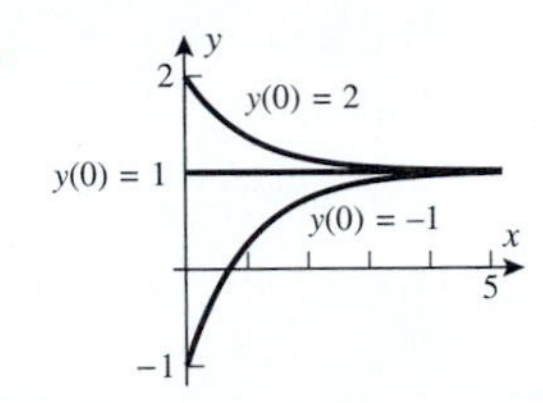

5. 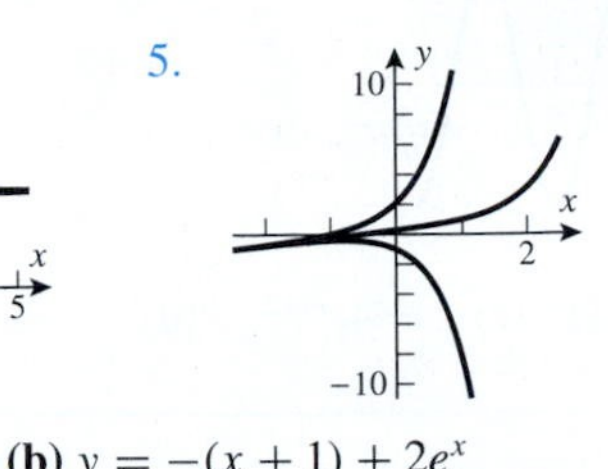

9. **(a)** IV **(b)** VI **(c)** V **(d)** II **(e)** I **(f)** III

11. **(a)**

n	0	1	2	3	4	5
x_n	0	0.2	0.4	0.6	0.8	1.0
y_n	1	1.20	1.48	1.86	2.35	2.98

(b) $y = -(x+1) + 2e^x$

x_n	0	0.2	0.4	0.6	0.8	1.0
$y(x_n)$	1	1.24	1.58	2.04	2.65	3.44
absolute error	0	0.04	0.10	0.19	0.30	0.46
percentage error	0	3	7	9	11	13

(c) 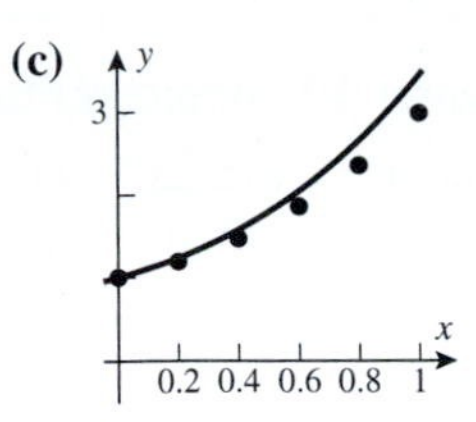

13.

n	0	1	2	3	4	5	6	7	8
x_n	0	0.5	1	1.5	2	2.5	3	3.5	4
y_n	1	1.50	2.11	2.84	3.68	4.64	5.72	6.91	8.23

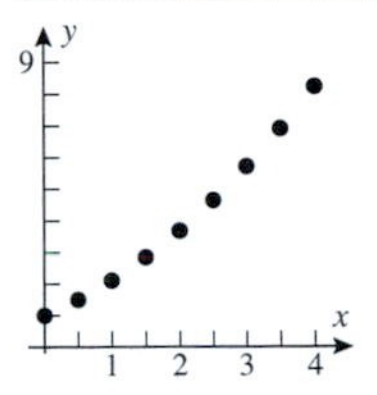

15. 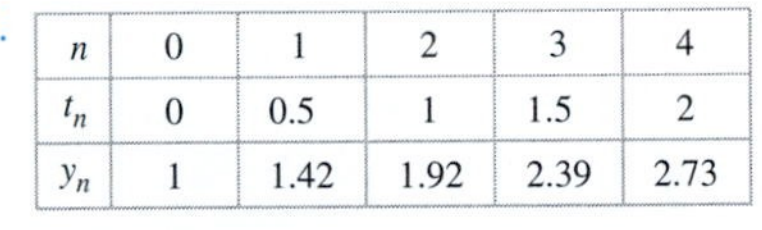

n	0	1	2	3	4
t_n	0	0.5	1	1.5	2
y_n	1	1.42	1.92	2.39	2.73

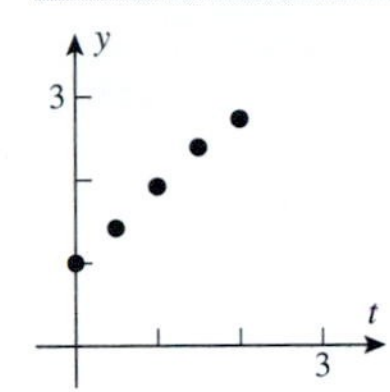

17. $y_5 \approx 1$

n	0	1	2	3	4	5
t_n	0	0.2	0.4	0.6	0.8	1.0
y_n	1.00	1.06	0.90	0.74	0.80	1.00

19. **(b)** $y(1/2) = \sqrt{3}/2$

▶ Exercise Set 10.3 (Page 609)

1. **(a)** $\dfrac{dy}{dt} = ky^2,\ y(0) = y_0 (k > 0)$ **(b)** $\dfrac{dy}{dt} = -ky^2,\ y(0) = y_0 (k > 0)$ 3. **(a)** $\dfrac{ds}{dt} = \dfrac{1}{2}s$ **(b)** $\dfrac{d^2s}{dt^2} = 2\dfrac{ds}{dt}$

5. **(a)** $\dfrac{dy}{dt} = 0.01y,\ y_0 = 10{,}000$ **(b)** $y = 10{,}000e^{t/100}$ **(c)** 69.31 h **(d)** 150.41 h

7. **(a)** $\dfrac{dy}{dt} = -ky,\ k \approx 0.1810$ **(b)** $y = 5.0 \times 10^7 e^{-0.181t}$ **(c)** 219,297 atoms **(d)** 12.72 days 9. 196 days 11. 3.30 days

13. **(a)** $y \approx 2e^{0.1386t}$ **(b)** $y = 5e^{0.015t}$ **(c)** $y \approx 0.5995e^{0.5117t}$ **(d)** $y \approx 0.8706e^{0.1386t}$ 17. **(b)** 70 years **(c)** 20 years **(d)** 7%

21. $y_0 \approx 2,\ L \approx 8,\ k \approx 0.5493$ 23. **(a)** $y_0 = 5$ **(b)** $L = 12$ **(c)** $k = 1$ **(d)** $t = 0.3365$ **(e)** $\dfrac{dy}{dt} = \dfrac{1}{12}y(12 - y),\ y(0) = 5$

25. **(a)** $L = 10$ **(b)** $k = 10$ **(c)** $y = 5$

27. Assume that $y(t)$ students have had the flu t days after semester break. Then $y(0) = 20,\ y(5) = 35$.
(a) $\dfrac{dy}{dt} = ky(1000 - y),\ y_0 = 20$
(b) $y = \dfrac{1000}{1 + 49e^{-1000kt}};\ k = 0.000115$
(c)

t	0	1	2	3	4	5	6	7	8	9	10	11	12	13	14
$y(t)$	20	22	25	28	31	35	39	44	49	54	61	67	75	83	93

(d) 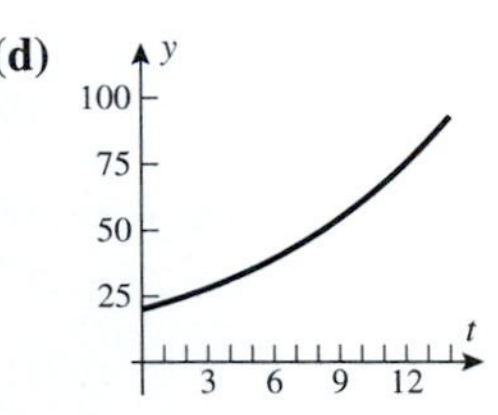

29. **(a)** $\dfrac{dT}{dt} = -k(T - 21),\ T(0) = 95$
$T = 21 + 74e^{-kt}$
(b) 6.22 min

33. **(a)** $y = 0.3\cos(t/2)$
(b) $T = 4\pi$ s, $f = 1/(4\pi)$ Hz
(c) 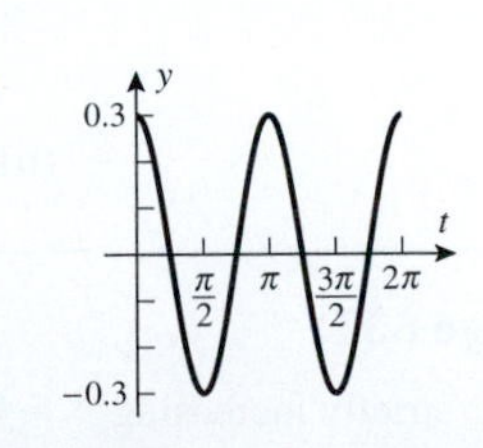
(d) $t = \pi$ s
(e) $t = 2\pi$ s

35. **(a)** $y = -0.12\cos 14t$ **(b)** $T = \pi/7$ s, $f = 7/\pi$ Hz **(c)**

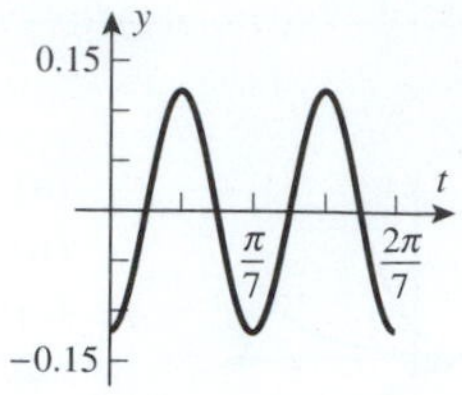

(d) $t = \pi/28$ s **(e)** $t = \pi/14$ s

37. **(a)** Maximum speed occurs when $y = 0$. **(b)** Minimum speed occurs when $y = \pm y_0$.

39. $mx''(t) + kx(t) = 0, x(0) = x_0, x'(0) = 0$ 41. **(c)** $y = 4e^{t\ln 2}$ **(d)** $y = 4e^{-t\ln 2}$

▶ Chapter 10 Supplementary Exercises (Page 612)

5. **(a)** linear **(b)** both **(c)** separable **(d)** neither 7. $y = L/2$ 9. $r = 4 - t$ m

11. **(a)** $P = 4(1 - e^{-t/12{,}000})$ percent **(b)** 35.95 min 13. $y^{-4} + 4\ln(x/y) = 1$ 15. $y = \dfrac{1}{3 - 2\tan 2x}$

17. **(a)** $y = \left(-\dfrac{3}{10}x - \dfrac{3}{50}\right)\cos 3x + \left(-\dfrac{1}{10}x + \dfrac{2}{25}\right)\sin 3x + \dfrac{53}{50}e^x$
(c)

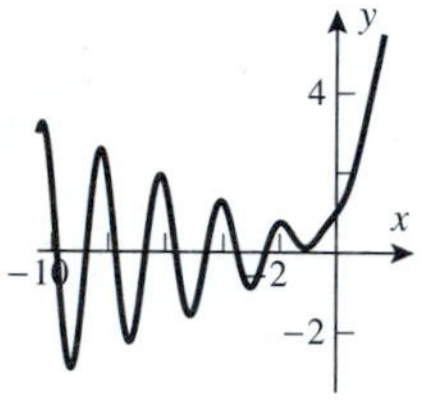

19. **(a)** no **(b)** $100\,\left|e^{ktr/(100+r)} - 1\right|$ percent

21. **(b)** $y = C_1e^x + C_2e^{-x}$ **(c)** $y = e^x$

23. **(a)** 7.77 years **(b)** $\dfrac{dy}{dt} = k\left(1 - \dfrac{y}{95}\right)y,\ y(0) = 19$ 27. **(a)** \$1491.82 **(b)** \$4493.29 **(c)** 8.7 years

▶ Exercise Set 11.1 (Page 624)

1. **(a)** $\dfrac{1}{3^{n-1}}$ **(b)** $\dfrac{(-1)^{n-1}}{3^{n-1}}$ **(c)** $\dfrac{2n-1}{2n}$ **(d)** $\dfrac{n^2}{\pi^{1/(n+1)}}$ 3. **(a)** 2, 0, 2, 0 **(b)** 1, −1, 1, −1 **(c)** $2(1 + (-1)^n)$; $2 + 2\cos n\pi$

5. $\frac{1}{3}, \frac{2}{4}, \frac{3}{5}, \frac{4}{6}, \frac{5}{7}$; converges, $\lim\limits_{n\to+\infty} \dfrac{n}{n+2} = 1$ 7. 2, 2, 2, 2, 2; converges, $\lim\limits_{n\to+\infty} 2 = 2$

9. $\dfrac{\ln 1}{1}, \dfrac{\ln 2}{2}, \dfrac{\ln 3}{3}, \dfrac{\ln 4}{4}, \dfrac{\ln 5}{5}$; converges, $\lim\limits_{n\to+\infty} \dfrac{\ln n}{n} = 0$ 11. 0, 2, 0, 2, 0; diverges

13. $-1, \frac{16}{9}, -\frac{54}{28}, \frac{128}{65}, -\frac{250}{126}$; diverges 15. $\dfrac{6}{2}, \dfrac{12}{8}, \dfrac{20}{18}, \dfrac{30}{32}, \dfrac{42}{50}$; converges, $\lim\limits_{n\to+\infty} \dfrac{1}{2}\left(1 + \dfrac{1}{n}\right)\left(1 + \dfrac{2}{n}\right) = \dfrac{1}{2}$

17. $\cos 3, \cos\frac{3}{2}, \cos 1, \cos\frac{3}{4}, \cos\frac{3}{5}$; converges, $\lim\limits_{n\to+\infty} \cos(3/n) = 1$

19. $e^{-1}, 4e^{-2}, 9e^{-3}, 16e^{-4}, 25e^{-5}$; converges, $\lim\limits_{n\to+\infty} n^2e^{-n} = 0$ 21. $2, \left(\dfrac{5}{3}\right)^2, \left(\dfrac{6}{4}\right)^3, \left(\dfrac{7}{5}\right)^4, \left(\dfrac{8}{6}\right)^5$; converges, $\lim\limits_{n\to+\infty}\left[\dfrac{n+3}{n+1}\right]^n = e^2$

23. $\left\{\dfrac{2n-1}{2n}\right\}_{n=1}^{+\infty}$; converges, $\lim\limits_{n\to+\infty} \dfrac{2n-1}{2n} = 1$ 25. $\left\{\dfrac{1}{3^n}\right\}_{n=1}^{+\infty}$; converges, $\lim\limits_{n\to+\infty} \dfrac{1}{3^n} = 0$

27. $\left\{\dfrac{1}{n} - \dfrac{1}{n+1}\right\}_{n=1}^{+\infty}$; converges, $\lim\limits_{n\to+\infty}\left(\dfrac{1}{n} - \dfrac{1}{n+1}\right) = 0$ 29. $\{\sqrt{n+1} - \sqrt{n+2}\}_{n=1}^{+\infty}$; converges, $\lim\limits_{n\to+\infty} (\sqrt{n+1} - \sqrt{n+2}) = 0$

33. **(a)** 1, 2, 1, 4, 1, 6 **(b)** $a_n = \begin{cases} n, & n \text{ odd} \\ 1/2^n, & n \text{ even} \end{cases}$ **(c)** $a_n = \begin{cases} 1/n, & n \text{ odd} \\ 1/(n+1), & n \text{ even} \end{cases}$ **(d)** (a) diverges; (b) diverges; (c) $\lim\limits_{n\to+\infty} a_n = 0$

37. **(a)** $1, \frac{3}{4}, \frac{2}{3}, \frac{5}{8}$ 41. **(a)** $(0.5)^{2^n}$ **(d)** $-1 \le a_0 \le 1$

43. **(a)**

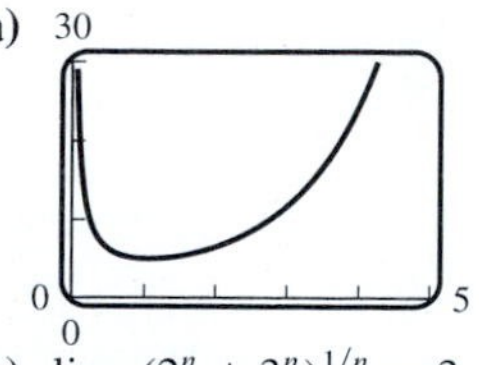

(b) $\lim\limits_{n\to+\infty} (2^n + 3^n)^{1/n} = 3$

45. converges to 0 47. **(a)** $N = 3$ **(b)** $N = 11$ **(c)** $N = 1001$

▶ Exercise Set 11.2 (Page 631)

1. strictly decreasing 3. strictly increasing 5. strictly decreasing 7. strictly increasing 9. strictly decreasing

11. strictly increasing 13. strictly increasing 15. strictly decreasing 17. strictly decreasing

19. eventually strictly increasing 21. eventually strictly decreasing 23. eventually strictly increasing
25. **(a)** Yes; the limit lies in the interval $[1, 2]$. **(b)** No, but if so, then the limit is ≤ 2. 27. $\sqrt{2}, \sqrt{2+\sqrt{2}}, \sqrt{2+\sqrt{2+\sqrt{2}}}$

▶ Exercise Set 11.3 (Page 638)

1. **(a)** $2, \frac{12}{5}, \frac{62}{25}, \frac{312}{125}, \frac{5}{2}\left(1-\left(\frac{1}{5}\right)^n\right), \lim_{n\to+\infty} s_n = \frac{5}{2}$, converges **(b)** $\frac{1}{4}, \frac{3}{4}, \frac{7}{4}, \frac{15}{4}, -\frac{1}{4}(1-2^n), \lim_{n\to+\infty} s_n = +\infty$, diverges
(c) $\frac{1}{6}, \frac{1}{4}, \frac{3}{10}, \frac{1}{3}, \frac{1}{2} - \frac{1}{n+2}, \lim_{n\to+\infty} s_n = \frac{1}{2}$, converges 3. $\frac{4}{7}$ 5. 6 7. $\lim_{n\to+\infty} s_n = \frac{1}{3}$
9. $\lim_{n\to+\infty} s_n = \frac{1}{6}$ 11. diverges 13. $\frac{448}{3}$ 15. $\frac{4}{9}$ 17. $\frac{532}{99}$ 19. $\frac{869}{1111}$ 23. 70 m
25. **(a)** $s_n = -\ln(n+1), \lim_{n\to+\infty} s_n = -\infty$, diverges **(b)** $s_n = \sum_{k=2}^{n+1}\left[\ln\frac{k-1}{k} - \ln\frac{k}{k+1}\right], \lim_{n\to+\infty} s_n = -\ln 2$
27. **(a)** converges for $|x| < 1$; $S = \frac{x}{1+x^2}$ **(b)** converges for $|x| > 2$; $S = \frac{1}{x^2 - 2x}$ **(c)** converges for $x > 0$; $S = \frac{1}{e^x - 1}$
33. $a_n = \frac{1}{2^{n-1}}a_1 + \frac{1}{2^{n-1}} + \frac{1}{2^{n-2}} + \cdots + \frac{1}{2}, \lim_{n\to+\infty} a_n = 1$ 35. The series converges only if $-1 < x < 1$.
39. **(b)** $A = 1, B = -2$ **(c)** $s_n = 2 - \frac{2^{n+1}}{3^{n+1} - 2^{n+1}}, \lim_{n\to+\infty} s_n = \lim_{n\to+\infty}\left[2 - \frac{(2/3)^{n+1}}{1-(2/3)^{n+1}}\right] = 2$

▶ Exercise Set 11.4 (Page 645)

1. **(a)** $\frac{4}{3}$ **(b)** $-\frac{3}{4}$ 3. **(a)** $p = 3$, converges **(b)** $p = \frac{1}{2}$, diverges **(c)** $p = 1$, diverges **(d)** $p = \frac{2}{3}$, diverges
5. **(a)** diverges **(b)** diverges **(c)** diverges **(d)** no information 7. **(a)** diverges **(b)** converges 9. diverges
11. diverges 13. diverges 15. diverges 17. diverges 19. converges 21. diverges 23. converges
25. converges for $p > 1$ 27. **(a)** $\left(\frac{\pi^2}{2}\right) - \left(\frac{\pi^4}{90}\right)$ **(b)** $\left(\frac{\pi^2}{6}\right) - \left(\frac{5}{4}\right)$ **(c)** $\pi^4/90$ 29. **(a)** diverge **(b)** diverges **(c)** converges
31. **(c)** $\frac{1}{11} < \frac{1}{6}\pi^2 - s_{10} < \frac{1}{10}$ 33. **(b)** $n = 5$ **(c)** $S \approx 1.203$ 35. **(b)** $13 < s_{1,000,000} < 15$ **(d)** $n > 2.69 \times 10^{43}$ 37. converges

▶ Exercise Set 11.5 (Page 655)

1. **(a)** $1 - x + \frac{1}{2}x^2, 1 - x$ **(b)** $1 - \frac{1}{2}x^2, 1$ **(c)** $1 - \frac{1}{2}(x - \pi/2)^2, 1$ **(d)** $1 + \frac{1}{2}(x-1) - \frac{1}{8}(x-1)^2, 1 + \frac{1}{2}(x-1)$
3. **(a)** $1 + \frac{1}{2}(x-1) - \frac{1}{8}(x-1)^2$ **(b)** 1.04875 5. 1.80397443
7. $p_0(x) = 1,\ p_1(x) = 1 - x,\ p_2(x) = 1 - x + \frac{1}{2}x^2,$
$p_3(x) = 1 - x + \frac{1}{2}x^2 - \frac{1}{3!}x^3,\ p_4(x) = 1 - x + \frac{1}{2}x^2 - \frac{1}{3!}x^3 + \frac{1}{4!}x^4;\ \sum_{k=0}^{\infty}\frac{(-1)^k}{k!}x^k$
9. $p_0(x) = 1,\ p_1(x) = 1,\ p_2(x) = 1 - \frac{\pi^2}{2!}x^2;\ p_3(x) = 1 - \frac{\pi^2}{2!}x^2,\ p_4(x) = 1 - \frac{\pi^2}{2!}x^2 + \frac{\pi^4}{4!}x^4;\ \sum_{k=0}^{\infty}\frac{(-1)^k\pi^{2k}}{(2k)!}x^{2k}$
11. $p_0(x) = 0,\ p_1(x) = x,\ p_2(x) = x - \frac{1}{2}x^2,\ p_3(x) = x - \frac{1}{2}x^2 + \frac{1}{3}x^3,\ p_4(x) = x - \frac{1}{2}x^2 + \frac{1}{3}x^3 - \frac{1}{4}x^4;\ \sum_{k=1}^{\infty}\frac{(-1)^{k+1}}{k}x^k$
13. $p_0(x) = 1,\ p_1(x) = 1,\ p_2(x) = 1 + \frac{x^2}{2},\ p_3(x) = 1 + \frac{x^2}{2},\ p_4(x) = 1 + \frac{x^2}{2} + \frac{x^4}{4!};\ \sum_{k=0}^{\infty}\frac{1}{(2k)!}x^{2k}$
15. $p_0(x) = 0,\ p_1(x) = 0,\ p_2(x) = x^2,\ p_3(x) = x^2,\ p_4(x) = x^2 - \frac{1}{6}x^4;\ \sum_{k=0}^{\infty}\frac{(-1)^k}{(2k+1)!}x^{2k+2}$ 17. **(a)** $1 + 2x - x^2 + x^3$ **(b)** $c_0 + c_1x + c_2x^2 + \cdots + c_nx^n$
19. $p_0(x) = e,\ p_1(x) = e + e(x-1),\ p_2(x) = e + e(x-1) + \frac{e}{2}(x-1)^2,\ p_3(x) = e + e(x-1) + \frac{e}{2}(x-1)^2 + \frac{e}{3!}(x-1)^3,$
$p_4(x) = e + e(x-1) + \frac{e}{2}(x-1)^2 + \frac{e}{3!}(x-1)^3 + \frac{e}{4!}(x-1)^4;\ \sum_{k=0}^{\infty}\frac{e}{k!}(x-1)^k$
21. $p_0(x) = -1;\ p_1(x) = -1 - (x+1);\ p_2(x) = -1 - (x+1) - (x+1)^2;\ p_3(x) = -1 - (x+1) - (x+1)^2 - (x+1)^3;$
$p_4(x) = -1 - (x+1) - (x+1)^2 - (x+1)^3 - (x+1)^4;\ \sum_{k=0}^{\infty}(-1)(x+1)^k$
23. $p_0(x) = p_1(x) = 1,\ p_2(x) = p_3(x) = 1 - \frac{\pi^2}{2}\left(x - \frac{1}{2}\right)^2,$
$p_4(x) = 1 - \frac{\pi^2}{2}\left(x - \frac{1}{2}\right)^2 + \frac{\pi^4}{4!}\left(x - \frac{1}{2}\right)^4;\ \sum_{k=0}^{\infty}\frac{(-1)^k\pi^{2k}}{(2k)!}\left(x - \frac{1}{2}\right)^{2k}$

25. $p_0(x) = 0,\ p_1(x) = (x-1);\ p_2(x) = (x-1) - \frac{1}{2}(x-1)^2;\ p_3(x) = (x-1) - \frac{1}{2}(x-1)^2 + \frac{1}{3}(x-1)^3,$
$p_4(x) = (x-1) - \frac{1}{2}(x-1)^2 + \frac{1}{3}(x-1)^3 - \frac{1}{4}(x-1)^4;\ \sum_{k=1}^{\infty} \frac{(-1)^{k-1}}{k}(x-1)^k$

27. **(a)** $1 + 2(x-1) - (x-1)^2 + (x-1)^3$ **(b)** $c_0 + c_1(x-x_0) + c_2(x-x_0)^2 + \cdots + c_n(x-x_0)^n$

29. $p_0(x) = 1,\ p_1(x) = 1 - 2x,$
$p_2(x) = 1 - 2x + 2x^2,\ p_3(x) = 1 - 2x + 2x^2 - \frac{4}{3}x^3$

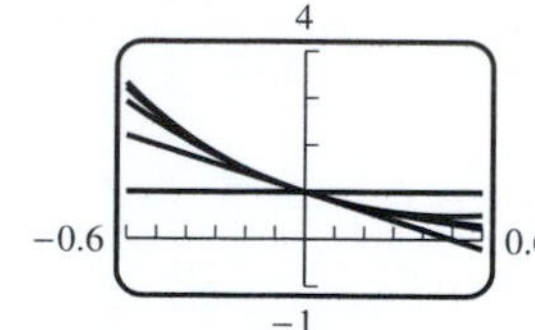

31. $p_0(x) = -1,\ p_2(x) = -1 + \frac{1}{2}(x-\pi)^2,$
$p_4(x) = -1 + \frac{1}{2}(x-\pi)^2 - \frac{1}{24}(x-\pi)^4,$
$p_6(x) = -1 + \frac{1}{2}(x-\pi)^2 - \frac{1}{24}(x-\pi)^4 + \frac{1}{720}(x-\pi)^6$

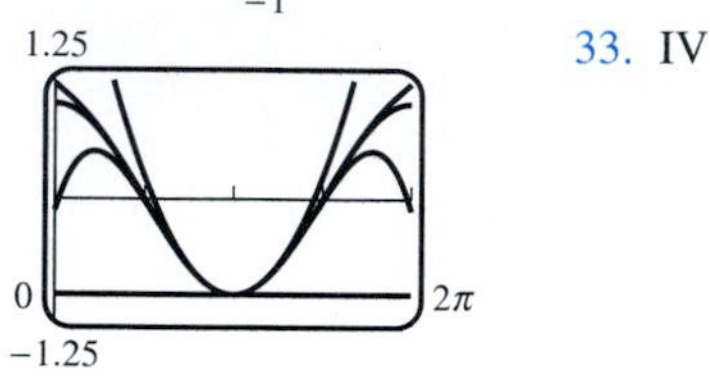

33. IV

37. **(a)**

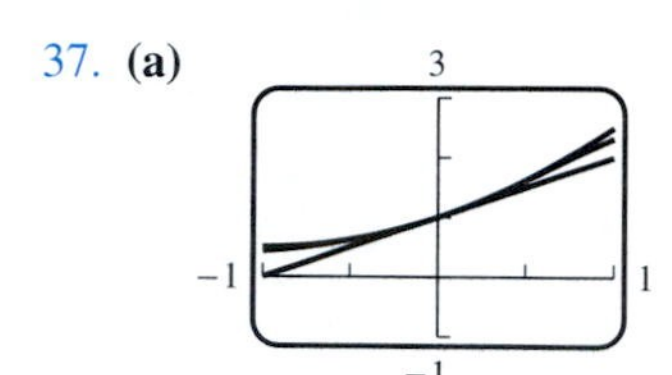

(b)

x	−1.000	−0.750	−0.500	−0.250	0.000	0.250	0.500	0.750	1.000
$f(x)$	0.431	0.506	0.619	0.781	1.000	1.281	1.615	1.977	2.320
$p_1(x)$	0.000	0.250	0.500	0.750	1.000	1.250	1.500	1.750	2.000
$p_2(x)$	0.500	0.531	0.625	0.781	1.000	1.281	1.625	2.031	2.500

(c) $|e^{\sin x} - (1+x)| < 0.01$ for $-0.14 < x < 0.14$

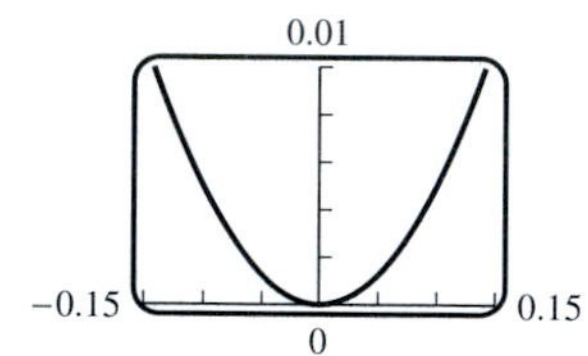

(d) $|e^{\sin x} - \left(1 + x + \frac{x^2}{2}\right)| < 0.01$ for $-0.50 < x < 0.50$

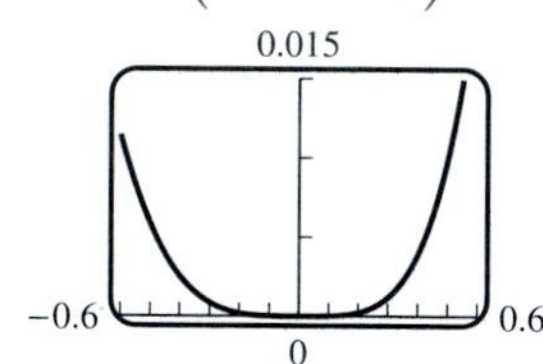

▶ Exercise Set 11.6 (Page 661)

1. **(a)** converges **(b)** diverges 3. **(a)** converges **(b)** converges 5. converges 7. converges 9. diverges 11. converges 13. inconclusive 15. diverges 17. diverges 19. converges 21. converges 23. converges 25. converges 27. converges 29. diverges 31. converges 33. diverges 35. converges 37. converges 39. diverges 41. converges 43. converges 45. $u_k = \frac{k!}{1 \cdot 3 \cdot 5 \cdots (2k-1)},\ \rho = \lim_{k\to+\infty} \frac{k+1}{2k+1} = \frac{1}{2}$; converges 47. converges 51. **(a)** converges **(b)** diverges

▶ Exercise Set 11.7 (Page 669)

3. diverges 5. converges 7. converges absolutely 9. diverges 11. converges absolutely 13. conditionally convergent 15. divergent 17. conditionally convergent 19. conditionally convergent 21. divergent 23. conditionally convergent 25. absolutely convergent 27. conditionally convergent 29. absolutely convergent 31. $|\text{error}| < 0.125$ 33. $|\text{error}| < 0.1$ 35. $n = 9999$ 37. $n = 39{,}999$ 39. $|\text{error}| < 0.00074$; $s_{10} \approx 0.4995$; $S = 0.5$ 41. 0.84 43. 0.41 45. **(c)** $n = 50$ 51. **(a)** $124.58 < d < 124.77$ **(b)** $1243 < s < 1424$

▶ Exercise Set 11.8 (Page 675)

1. $-1 < x < 1,\ \frac{1}{1+x}$ 3. $1 < x < 3,\ \frac{1}{3-x}$ 5. **(a)** $-2 < x < 2$ **(b)** $f(0) = 1;\ f(1) = \frac{2}{3}$ 7. $R = 1, [-1, 1)$ 9. $R = +\infty, (-\infty, +\infty)$ 11. $R = \frac{1}{5}, [-\frac{1}{5}, \frac{1}{5}]$ 13. $R = 1, [-1, 1]$ 15. $R = 1, (-1, 1]$ 17. $R = +\infty, (-\infty, +\infty)$ 19. $R = +\infty, (-\infty, +\infty)$ 21. $R = 1, [-1, 1]$ 23. $R = 1, (-2, 0]$ 25. $R = \frac{4}{3}, (-\frac{19}{3}, -\frac{11}{3})$ 27. $R = 1, [-2, 0]$

29. $R = +\infty$, $(-\infty, +\infty)$ 31. $(-\infty, +\infty)$ 33.

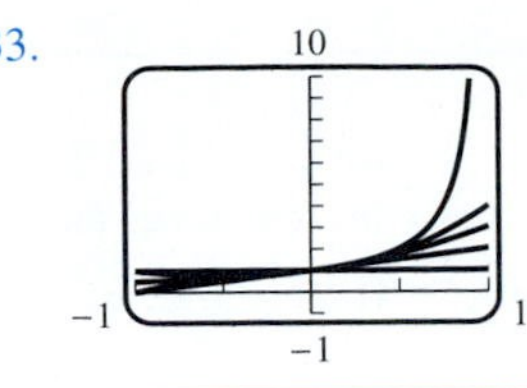

37. **(a)** radius $= R$
(b) radius $= R$
(c) radius $\geq \min(R_1, R_2)$

▶ Exercise Set 11.9 (Page 684)

1. 0.069756 3. 1.64872 5. 0.995004 7. 0.99619 9. 0.5208 13. **(a)** $\sum_{k=1}^{\infty} 2\frac{(1/9)^{2k-1}}{2k-1}$ **(b)** 0.2231

15. **(a)** 0.4635, 0.3218
(b) 3.1412
(c) no

17. **(a)** $(-0.569, 0.569)$

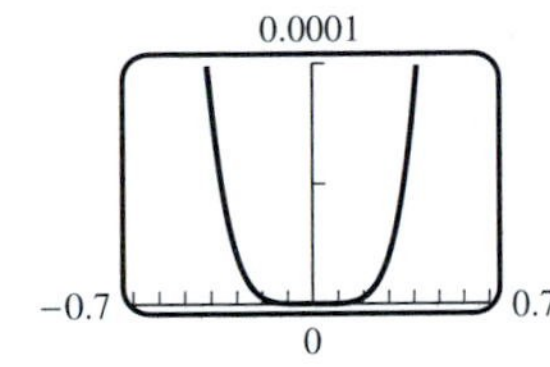

19. **(a)** $|R_5(x)| < 9 \times 10^{-8}$

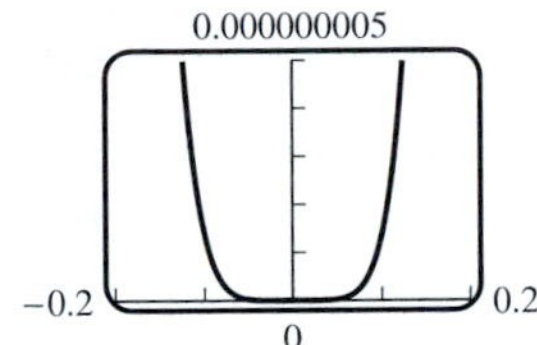

21. **(a)** $\sum_{k=0}^{\infty}(-1)^k x^k$
(b) $1 + \frac{x}{3} + \sum_{k=2}^{\infty}(-1)^{k-1}\frac{2 \cdot 5 \cdots (3k-4)}{3^k k!}x^k$
(c) $\sum_{k=0}^{\infty}(-1)^k \frac{(k+2)(k+1)}{2}x^k$

27. **(a)** 0.78539816339744830961566609
(b)

n	s_n
0	0.3183098 78 . . .
1	0.3183098 861837906 067 . . .
2	0.3183098 861837906 7153776 695 . . .
3	0.3183098 861837906 7153776 752674502 34 . . .
$1/\pi$	0.3183098 861837906 7153776 752674502 87 . . .

▶ Exercise Set 11.10 (Page 693)

1. **(a)** $1 - x + x^2 - \cdots + (-1)^k x^k + \cdots;\ R = 1$
(b) $1 + x^2 + x^4 + \cdots + x^{2k} + \cdots;\ R = 1$
(c) $1 + 2x + 4x^2 + \cdots + 2^k x^k + \cdots;\ R = \frac{1}{2}$
(d) $\frac{1}{2} + \frac{1}{2^2}x + \frac{1}{2^3}x^2 + \cdots + \frac{1}{2^{k+1}}x^k + \cdots;\ R = 2$

3. **(a)** $(2+x)^{-1/2} = \frac{1}{2^{1/2}} - \frac{1}{2^{5/2}}x + \frac{1 \cdot 3}{2^{9/2} \cdot 2!}x^2 - \frac{1 \cdot 3 \cdot 5}{2^{13/2} \cdot 3!}x^3 + \cdots$
(b) $(1 - x^2)^{-2} = 1 + 2x^2 + 3x^4 + 4x^6 + \cdots$

5. **(a)** $2x - \frac{2^3}{3!}x^3 + \frac{2^5}{5!}x^5 - \frac{2^7}{7!}x^7 + \cdots;\ R = +\infty$
(b) $1 - 2x + 2x^2 - \frac{4}{3}x^3 + \cdots;\ R = +\infty$
(c) $1 + x^2 + \frac{1}{2!}x^4 + \frac{1}{3!}x^6 + \cdots;\ R = +\infty$
(d) $x^2 - \frac{\pi^2}{2}x^4 + \frac{\pi^4}{4!}x^6 - \frac{\pi^6}{6!}x^8 + \cdots;\ R = +\infty$

7. **(a)** $x^2 - 3x^3 + 9x^4 - 27x^5 + \cdots;\ R = \frac{1}{3}$
(b) $2x^2 + \frac{2^3}{3!}x^4 + \frac{2^5}{5!}x^6 + \frac{2^7}{7!}x^8 + \cdots;\ R = +\infty$
(c) $x - \frac{3}{2}x^3 + \frac{3}{8}x^5 + \frac{1}{16}x^7 + \cdots;\ R = 1$

9. **(a)** $x^2 - \frac{2^3}{4!}x^4 + \frac{2^5}{6!}x^6 - \frac{2^7}{8!}x^8 + \cdots$ **(b)** $12x^3 - 6x^6 + 4x^9 - 3x^{12} + \cdots$

11. **(a)** $1 - (x-1) + (x-1)^2 - \cdots + (-1)^k(x-1)^k + \cdots$ **(b)** $(0, 2)$

13. **(a)** $x + x^2 + \frac{x^3}{3} - \frac{x^5}{30} + \cdots$ **(b)** $x - \frac{x^3}{24} + \frac{x^4}{24} - \frac{71}{1920}x^5 + \cdots$

15. **(a)** $1 + \frac{1}{2}x^2 + \frac{5}{24}x^4 + \frac{61}{720}x^6 + \cdots$ **(b)** $x - x^2 + \frac{1}{3}x^3 - \frac{1}{30}x^5 + \cdots$ 19. $2 - 4x + 2x^2 - 4x^3 + 2x^4 + \cdots$

25. **(a)** $\sum_{k=0}^{\infty} x^{2k+1}$ **(b)** $f^{(5)}(0) = 5,\ f^{(6)}(0) = 0$ **(c)** $f^{(n)}(0) = n!c_n = \begin{cases} n! & \text{if } n \text{ odd} \\ 0 & \text{if } n \text{ even} \end{cases}$ 27. **(a)** 1 **(b)** $-\frac{1}{3}$ 29. 0.3103

31. 0.200 35. **(a)** $x - \frac{1}{6}x^3 + \frac{3}{40}x^5 - \frac{5}{112}x^7 + \cdots$ **(b)** $x + \sum_{k=1}^{\infty}(-1)^k \frac{1 \cdot 3 \cdot 5 \cdots (2k-1)}{2^k k!(2k+1)}x^{2k+1}$ **(c)** $R = 1$

37. **(a)** $y(t) = y_0 \sum_{k=0}^{\infty} \frac{(-1)^k (0.000121)^k t^k}{k!}$ **(c)** $0.9998790073 y_0$ 39. **(a)** $T \approx 2.00709$ **(b)** $T \approx 2.008044621$ **(c)** 2.008045644

41. **(a)** $F = mg\left(1 - \frac{2h}{R} + \frac{3h^2}{R^2} - \frac{4h^3}{R^3} + \cdots\right)$ **(d)** about 0.27% less

▶ Chapter 11 Supplementary Exercises (Page 696)

9. **(a)** true **(b)** sometimes false **(c)** sometimes false **(d)** true **(e)** sometimes false **(f)** sometimes false **(g)** false **(h)** sometimes false **(i)** true **(j)** true **(k)** sometimes false **(l)** sometimes false 11. **(a)** converges **(b)** converges **(c)** diverges

13. **(a)** converges **(b)** diverges **(c)** converges 15. $\dfrac{1}{4 \cdot 5^{99}}$

17. **(a)** $p_0(x) = 1$, $p_1(x) = 1 - 7x$, $p_2(x) = 1 - 7x + 5x^2$, $p_3(x) = 1 - 7x + 5x^2 - 4x^3$, $p_4(x) = 1 - 7x + 5x^2 - 4x^3$

21. **(a)** converges **(b)** diverges 23. **(a)** $u_{100} = \dfrac{1}{9900}$ **(b)** 0 **(c)** 2 25. **(a)** $e^2 - 1$ **(b)** 0 **(c)** $\cos e$ **(d)** $\frac{1}{3}$

27. 22.07% 31. **(a)** $x + \frac{1}{2}x^2 + \frac{3}{14}x^3 + \frac{3}{35}x^4$; $R = 3$ **(b)** $-x^3 + \frac{2}{3}x^5 - \frac{2}{5}x^7 + \frac{8}{35}x^9$; $R = \sqrt{2}$

▶ Exercise Set 12.1 (Page 710)

1.

3. **(a)** $(3\sqrt{3}, 3)$ **(b)** $(-7/2, 7\sqrt{3}/2)$ **(c)** $(3\sqrt{3}, 3)$ **(d)** $(0, 0)$ **(e)** $(-7\sqrt{3}/2, 7/2)$ **(f)** $(-5, 0)$

5. **(a)** both $(5, \pi)$ **(b)** $(4, 11\pi/6)$, $(4, -\pi/6)$ **(c)** $(2, 3\pi/2)$, $(2, -\pi/2)$ **(d)** $(8\sqrt{2}, 5\pi/4)$, $(8\sqrt{2}, -3\pi/4)$ **(d)** both $(6, 2\pi/3)$ **(d)** both $(\sqrt{2}, \pi/4)$

7. **(a)** $(5, 0.6435)$ **(b)** $(\sqrt{29}, 5.0929)$ **(c)** $(1.2716, 0.6658)$

9. **(a)** circle **(b)** line **(c)** circle **(d)** line

11. **(a)** $r\cos\theta = 7$ **(b)** $r = 3$ **(c)** $r = 6\sin\theta$ **(d)** $r^2 \sin 2\theta = 9/2$

13.

15.

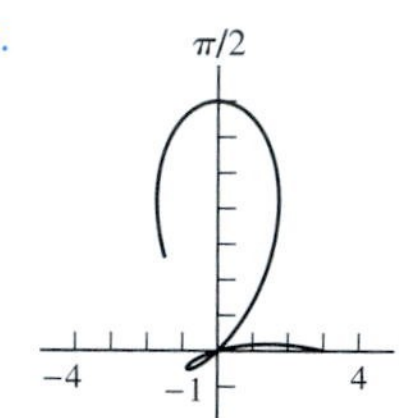

17. **(a)** $r = 5$ **(b)** $r = 6\cos\theta$ **(c)** $r = 1 - \cos\theta$

19. **(a)** $r = 3\sin 2\theta$ **(b)** $r = 3 + 2\sin\theta$ **(c)** $r^2 = 9\cos 2\theta$

21.

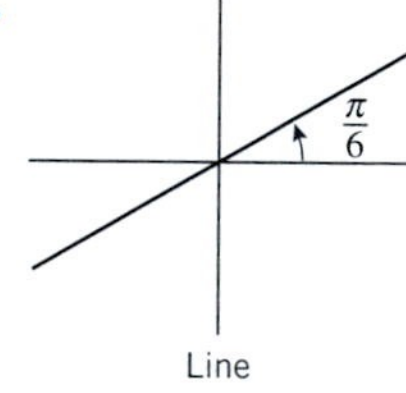

Line

23.

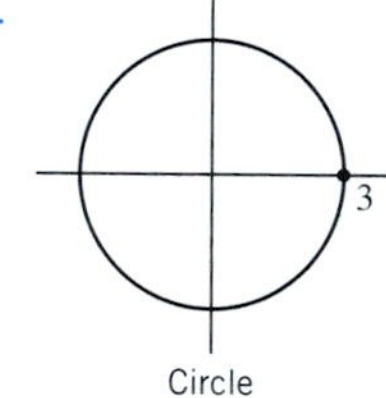

Circle

25.

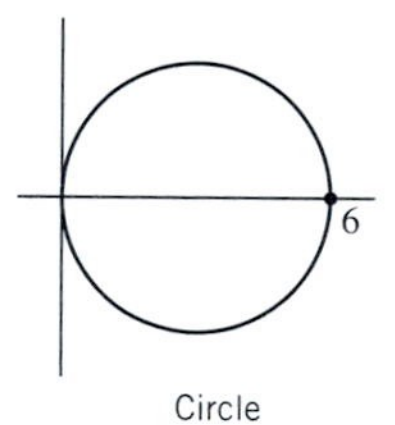

Circle

27.

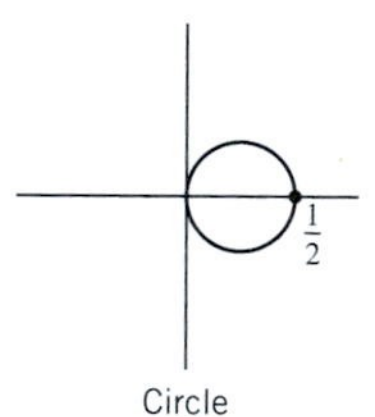

Circle

29.

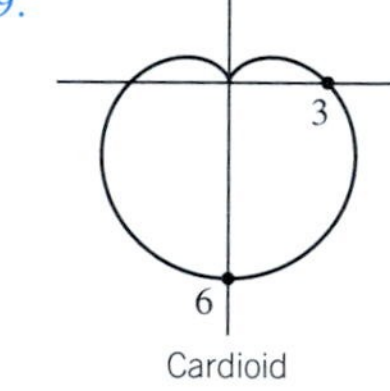

Cardioid

31.

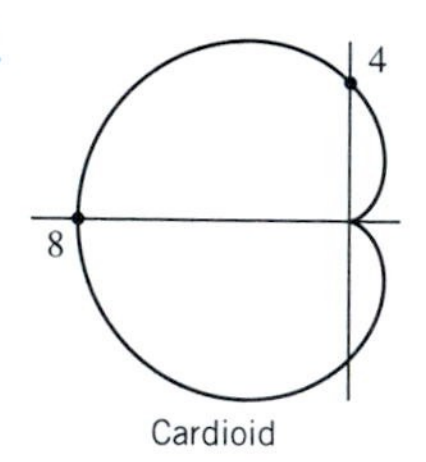

Cardioid

33.

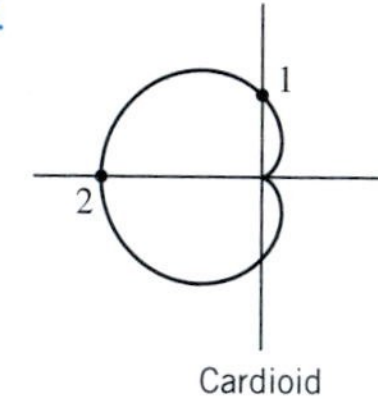

Cardioid

35.

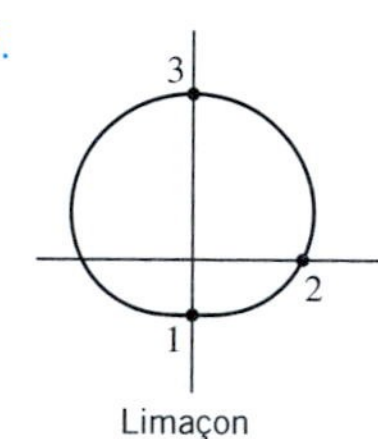

Limaçon

37.

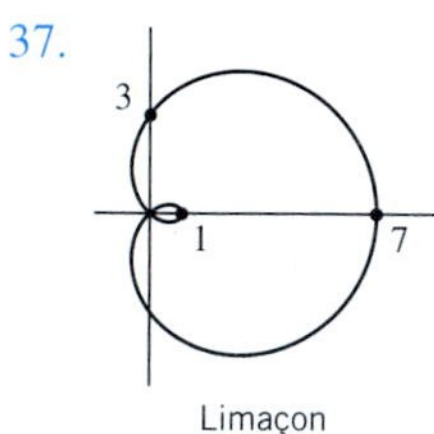

Limaçon

39.

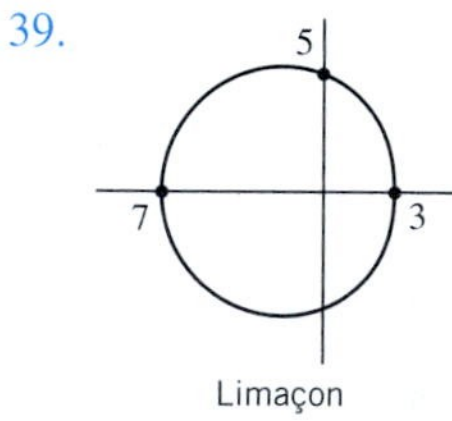

Limaçon

41.

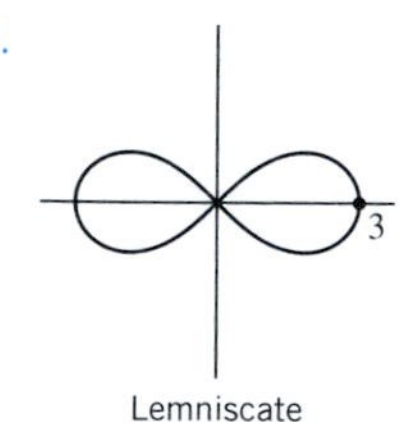

Lemniscate

43.

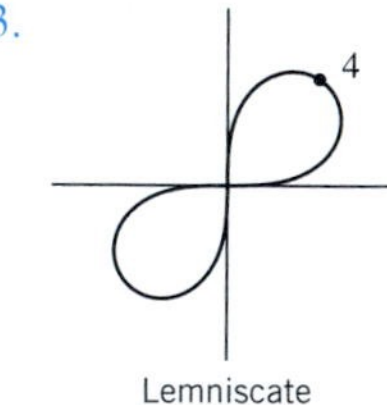

Lemniscate

45.

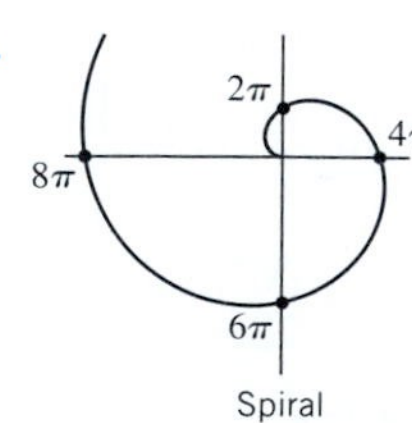

Spiral

47.

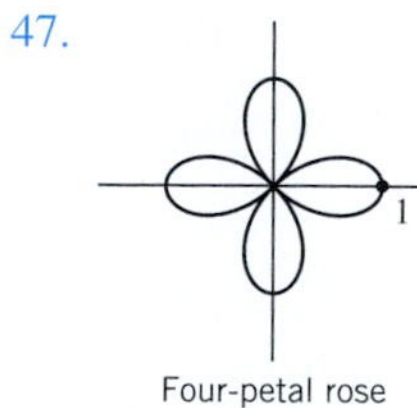

Four-petal rose

49.

Eight-petal rose

53.

55.

57. $-4\pi < \theta < 4\pi$

61. **(a)** $r = 1 + \frac{\sqrt{2}}{2}(\cos\theta + \sin\theta)$
(b) $r = 1 + \sin\theta$
(c) $r = 1 - \cos\theta$
(d) $r = 1 - \frac{\sqrt{2}}{2}(\cos\theta + \sin\theta)$

63.

65. $(3/2, \pi/3)$

67. **(c)** $\sqrt{13 - 6\sqrt{3}} \approx 1.615$ **(d)** $A = 1$

71.

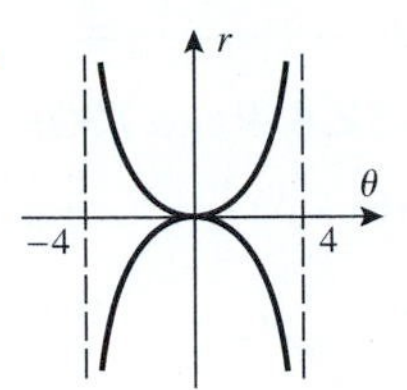

▶ Exercise Set 12.2 (Page 718)

1. **(a)** $-1/4, 1/4$ 3. positive when $t = -1$, negative when $t = 1$ 5. $4, 4$ 7. $2/\sqrt{3}, -1/(3\sqrt{3})$ 9. $\dfrac{-1}{4-\sqrt{3}}, \dfrac{8}{(4-\sqrt{3})^3}$

11. **(a)** $y = -e^{-2}x + 2e^{-1}$ 13. **(a)** $t = \pi/2 + n\pi$ for $n = 0, \pm 1, \cdots$ **(b)** $t = n\pi$ for $n = 0, \pm 1, \cdots$ 15. $y = -2x, y = 2x$ 19.

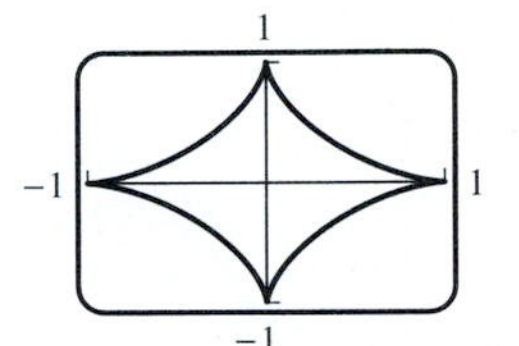

21. $1/\sqrt{3}$ 23. $\dfrac{\tan 2 - 2}{2\tan 2 + 1}$ 25. -2 27. $1, 0, -1$ 29. horizontal: $(3a/2, \pi/3), (0, \pi), (3a/2, 5\pi/3)$; vertical: $(2a, 0), (a/2, 2\pi/3), (a/2, 4\pi/3)$

31. $(0, 0), (\sqrt{2}/4, \pi/4), (\sqrt{2}/4, 3\pi/4)$ 33.

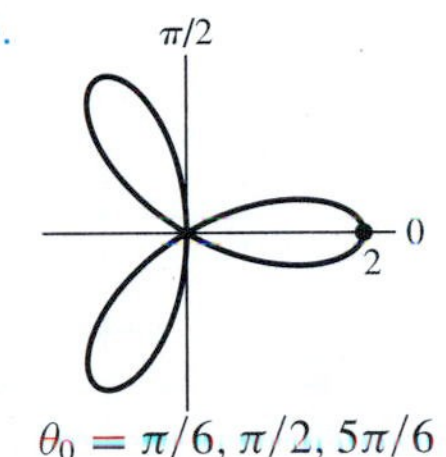

$\theta_0 = \pi/6, \pi/2, 5\pi/6$

35.

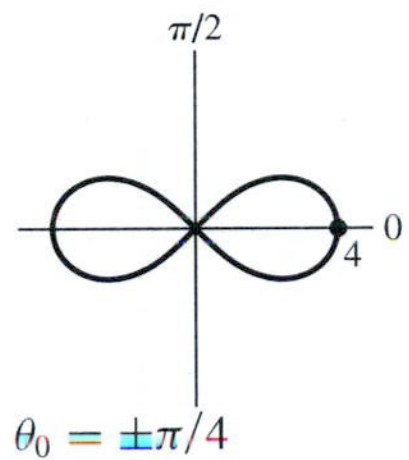

$\theta_0 = \pm\pi/4$

37.

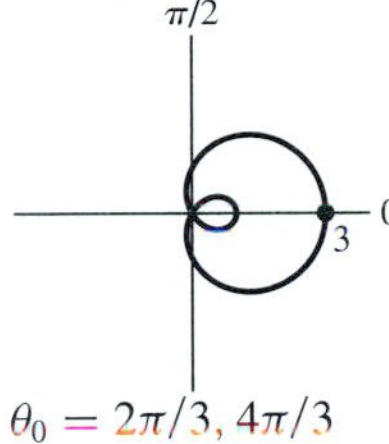

$\theta_0 = 2\pi/3, 4\pi/3$

39. $L = 2\pi a$ 41. $L = 8a$ 43. $L = \sqrt{10}(e^6 - 1)/3$ 45. **(a)** $\dfrac{dy}{dx} = \dfrac{3\sin t}{1 - 3\cos t}$ **(b)** $\theta = -0.4344$

47. **(b)** ≈ 2.42 **(c)**

n	2	3	4	5	6	7	8	9	10	11
L	2.42211	2.22748	2.14461	2.10100	2.07501	2.05816	2.04656	2.03821	2.03199	2.02721

n	12	13	14	15	16	17	18	19	20
L	2.02346	2.02046	2.01802	2.01600	2.01431	2.01288	2.01167	2.01062	2.00971

49. $S = \frac{8\pi}{3}(17\sqrt{17} - 1)$

51. $S = \sqrt{2}\pi$ 55. **(a)** $r = 2\theta + 10$ **(b)** 75.7 mm

▶ Exercise Set 12.3. (Page 724)

1. **(a)** $\int_{\pi/2}^{\pi} \frac{1}{2}(1 - \cos\theta)^2\, d\theta$ **(b)** $\int_0^{\pi/2} \frac{1}{2} 4\cos^2\theta\, d\theta$ **(c)** $\int_0^{\pi/2} \frac{1}{2}\sin^2 2\theta\, d\theta$ **(d)** $\int_0^{2\pi} \frac{1}{2}\theta^2\, d\theta$ **(e)** $\int_{-\pi/2}^{\pi/2} \frac{1}{2}(1 - \sin\theta)^2\, d\theta$ **(f)** $\int_{\pi}^{\pi/3} \frac{1}{2}(1 + \cos 3\theta)^2\, d\theta$

3. **(a)** πa^2 **(b)** πa^2 **(c)** πa^2 5. 6π 7. 4π 9. $\pi - 3\sqrt{3}/2$ 11. $\pi/2 - \frac{1}{4}$

13. $10\pi/3 - 4\sqrt{3}$ 15. $8\pi/3 + \sqrt{3}$ 17. $9\sqrt{3}/2 - \pi$ 19. $(\pi + 3\sqrt{3})/4$ 21. $100\cos^{-1}(3/5) - 48$ 23. **(b)** a^2 **(c)** $2\sqrt{3} - \frac{2\pi}{3}$

25. $8\pi^3 a^2$ 27. π^2

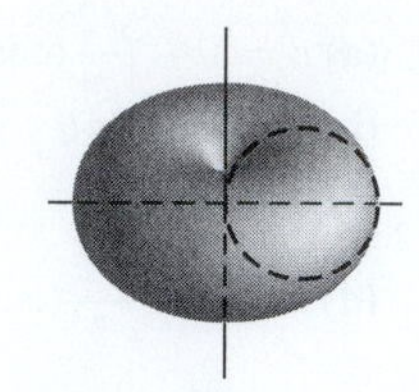

29. $32\pi/5$

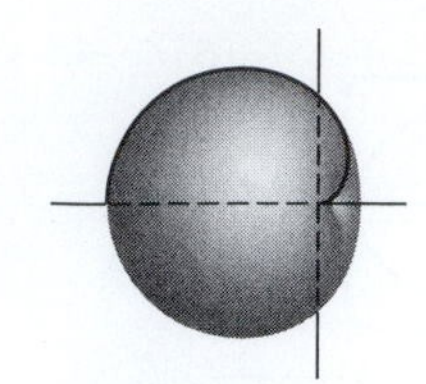

35. $\pi/16$

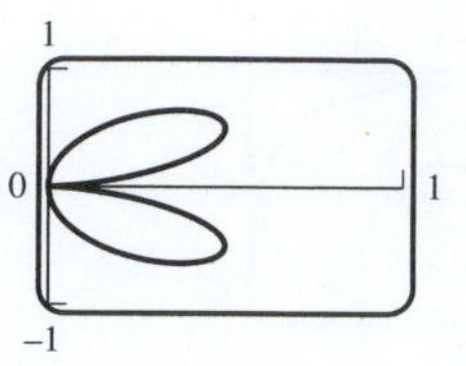

▶ Exercise Set 12.4 (Page 739)

1. **(a)** $x = y^2$ **(b)** $-3y = x^2$ **(c)** $\dfrac{x^2}{9} + \dfrac{y^2}{4} = 1$ **(d)** $\dfrac{x^2}{4} + \dfrac{y^2}{9} = 1$ **(e)** $y^2 - x^2 = 1$ **(f)** $\dfrac{x^2}{4} - \dfrac{y^2}{4} = 1$

3. **(a)**

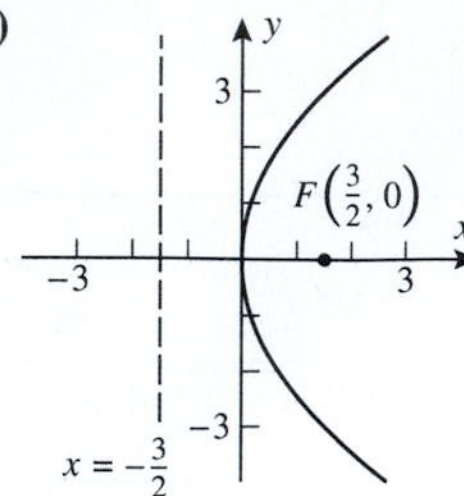

(b)

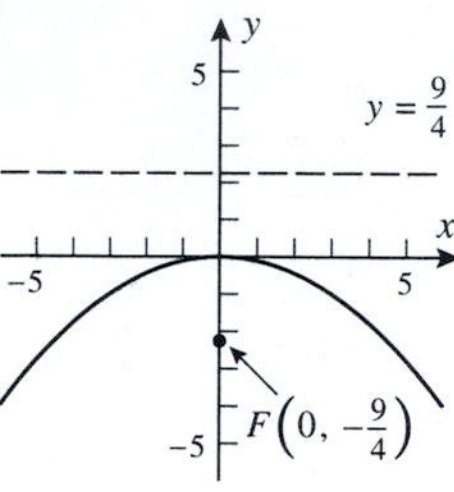

5. **(a)**

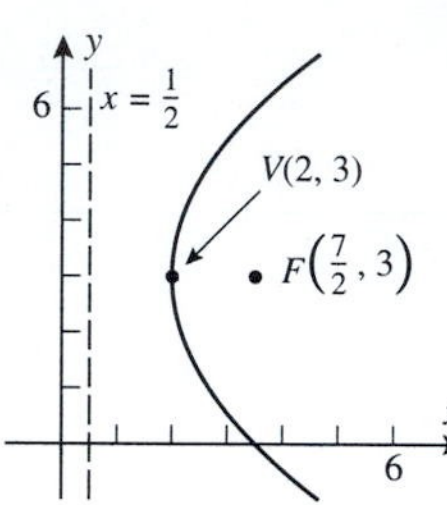

(b)

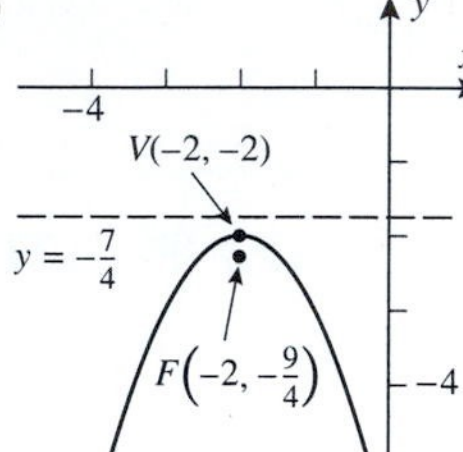

7. **(a)**

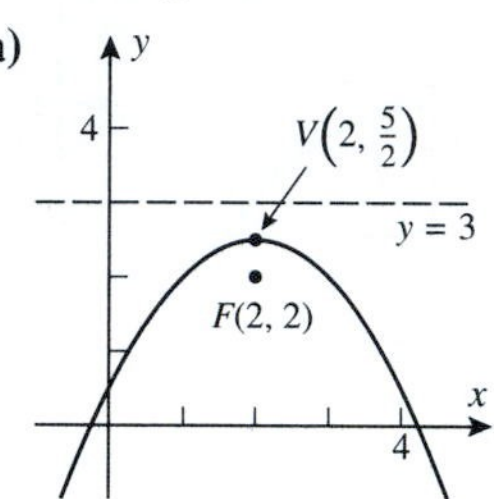

(b)

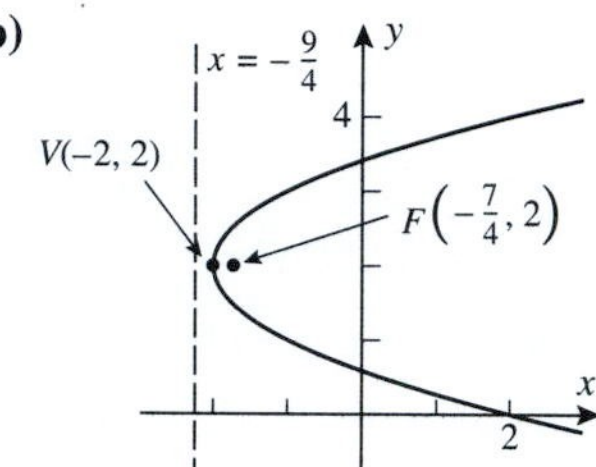

9. **(a)**

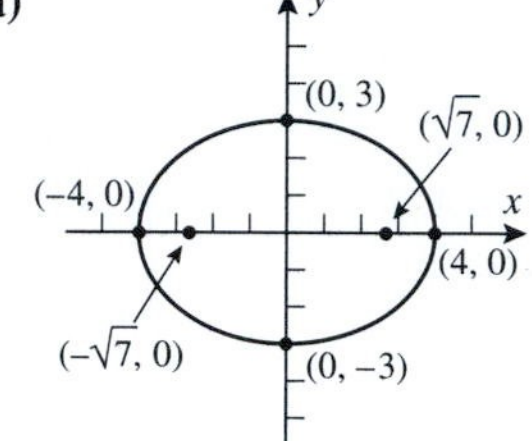

(b)

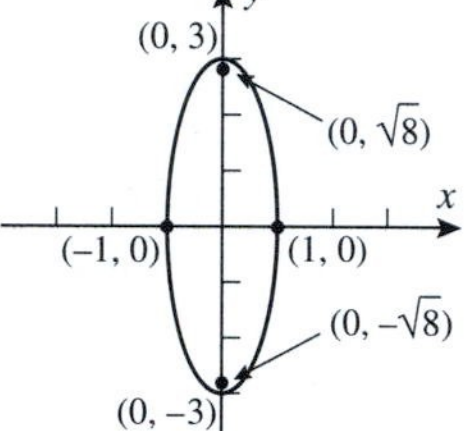

11. **(a)**

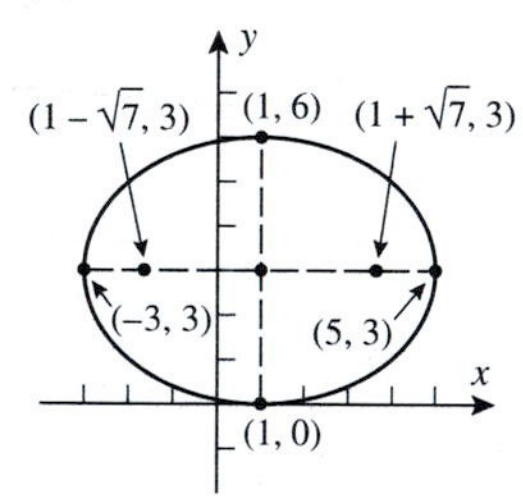

(b)

(–2, –1 + √3)
(–3, –1)
(–1, –1)
(0, –1)
(–4, –1)
(–2, –1 – √3)

13. **(a)**

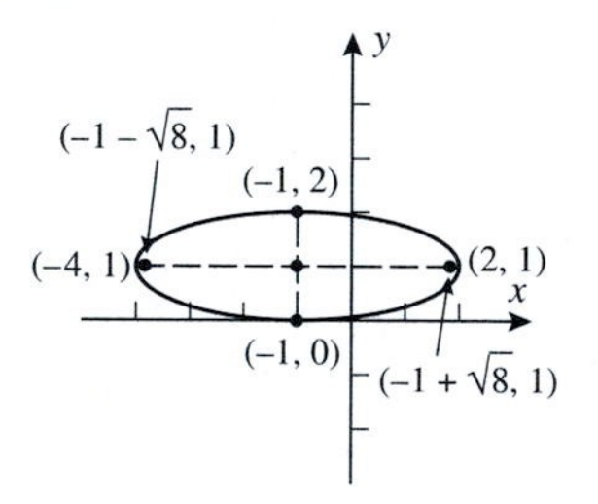

(b)

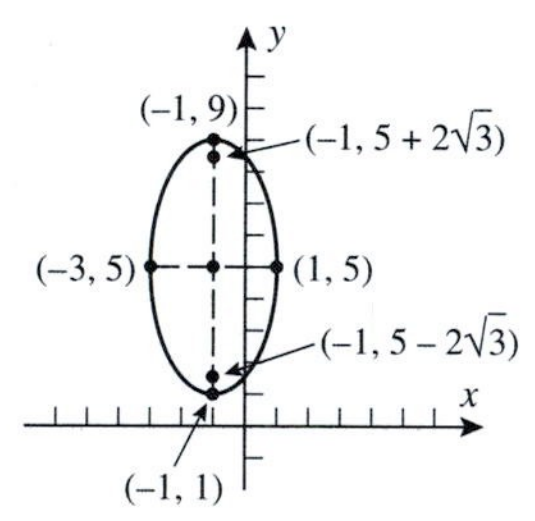

15. **(a)**

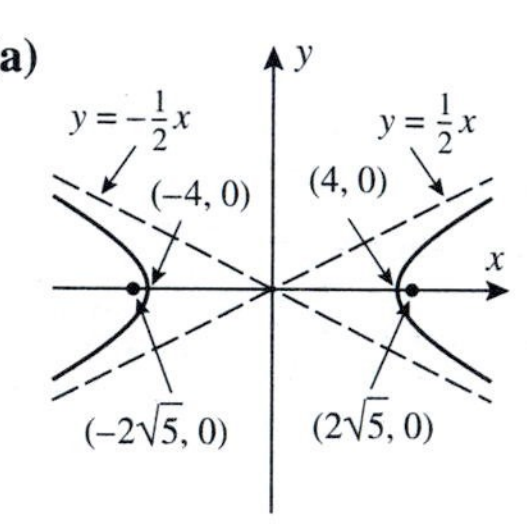

(b)

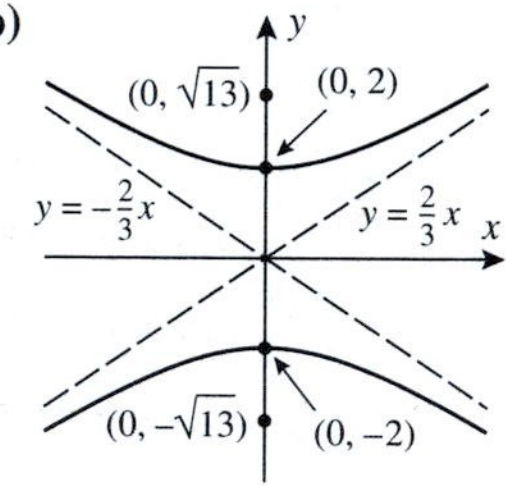

17. **(a)**

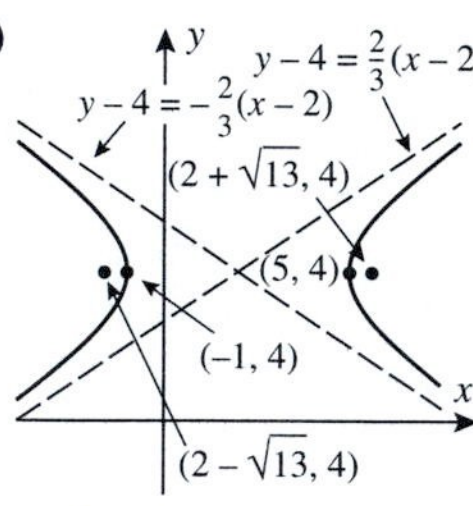

(b)

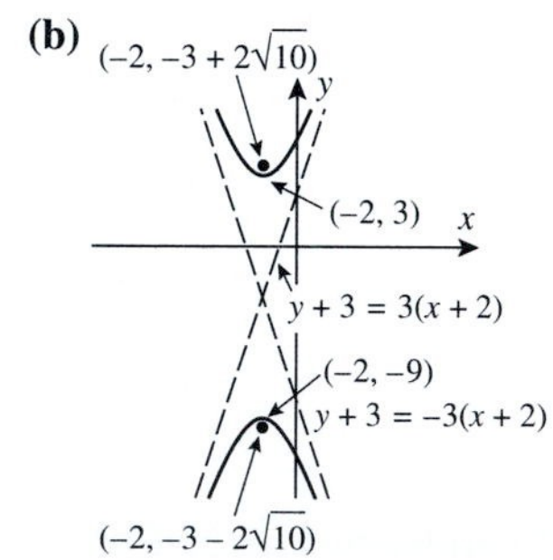

19. **(a)**

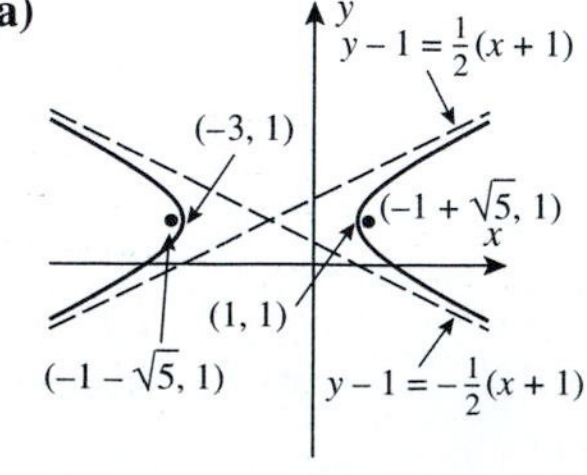

(b)

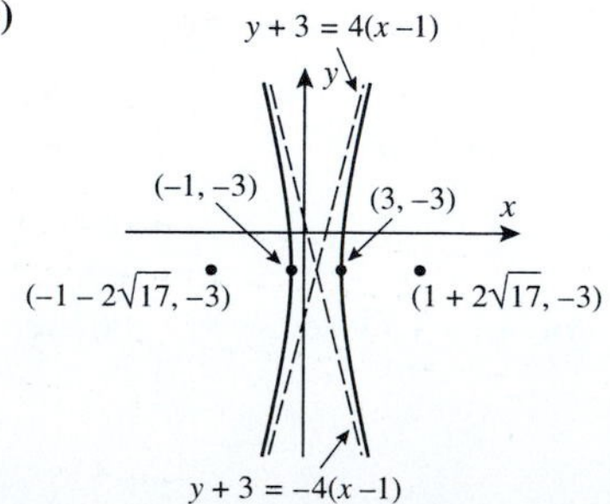

21. **(a)** $y^2 = 12x$
(b) $y^2 = -28x$

23. **(a)** $x^2 = -12y$
(b) $(x - 1)^2 = 12(y - 1)$

25. $y^2 = -5\left(x - \frac{19}{5}\right)$ 27. **(a)** $\frac{1}{9}x^2 + \frac{1}{4}y^2 = 1$ **(b)** $\frac{1}{169}x^2 + \frac{1}{144}y^2 = 1$ 29. **(a)** $\frac{1}{3}x^2 + \frac{1}{2}y^2 = 1$ **(b)** $\frac{1}{4}x^2 + \frac{1}{16}y^2 = 1$

31. **(a)** $\frac{1}{36}x^2 + \frac{8}{81}y^2 = 1$ **(b)** $(x-1)^2 + \frac{1}{2}(y-3)^2 = 1$ 33. **(a)** $\frac{1}{4}x^2 - \frac{1}{5}y^2 = 1$ **(b)** $x^2 - \frac{1}{4}y^2 = 1$

35. **(a)** $\frac{9}{64}x^2 - \frac{1}{16}y^2 = 1$, $\frac{1}{36}y^2 - \frac{1}{16}x^2 = 1$ **(b)** $\frac{1}{20}y^2 - \frac{1}{5}x^2 = 1$

37. **(a)** $\frac{1}{16}(x-6)^2 - \frac{1}{9}(y-4)^2 = 1$ **(b)** $\frac{1}{3}(y-2)^2 - \frac{4}{3}\left(x - \frac{1}{2}\right)^2 = 1$

39. **(a)** 16 ft **(b)** $8\sqrt{3}$ ft

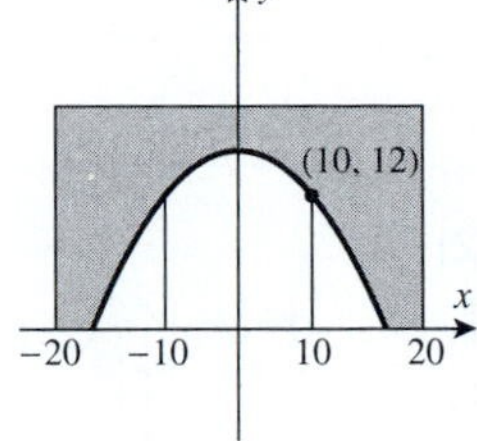

43. $\frac{1}{16}$ ft 45. **(a)** $P: (b\cos t, b\sin t)$; $Q: (a\cos t, a\sin t)$; $R: (a\cos t, b\sin t)$

49. $\frac{1}{32}(x-4)^2 + \frac{1}{36}(y-3)^2 = 1$ 51. 96 61. $L = D\sqrt{1+p^2}$, $T = \frac{1}{2}pD$

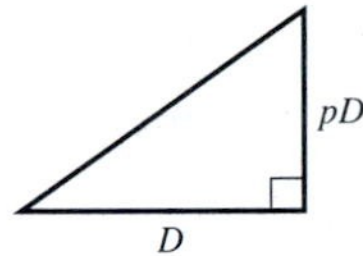

67. $\left(\pm\frac{3}{\sqrt{5}}, \frac{4}{\sqrt{5}}\right), \left(\pm\frac{3}{\sqrt{5}}, -\frac{4}{\sqrt{5}}\right)$

71. **(b)** 14.30465, 24, 33.69535 in

73. **(a)** $(x-1)^2 - 5(y+1)^2 = 5$, hyperbola **(b)** $x^2 - 3(y+1)^2 = 0$, $x = \pm\sqrt{3}(y+1)$, two lines **(c)** $4(x+2)^2 + 8(y+1)^2 = 4$, ellipse **(d)** $3(x+2)^2 + (y+1)^2 = 0$, the point $(-2, -1)$ (degenerate case) **(e)** $(x+4)^2 + 2y = 2$, parabola **(f)** $5(x+4)^2 + 2y = -14$, parabola

▶ Exercise Set 12.5 (Page 750)

1. **(a)** $e = 1, d = \frac{3}{2}$ **(b)** $e = \frac{1}{2}, d = 3$ **(c)** $e = \frac{3}{2}, d = \frac{4}{3}$ **(d)** $e = 1, d = \frac{5}{3}$

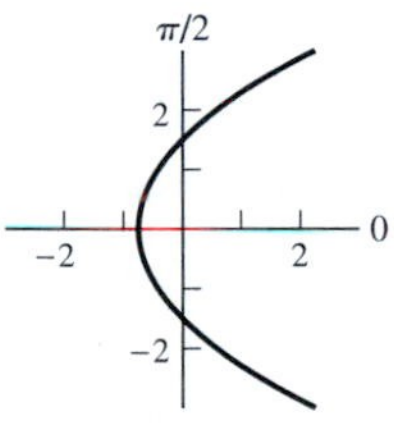

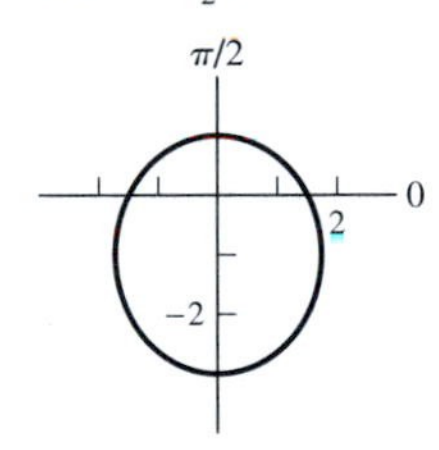

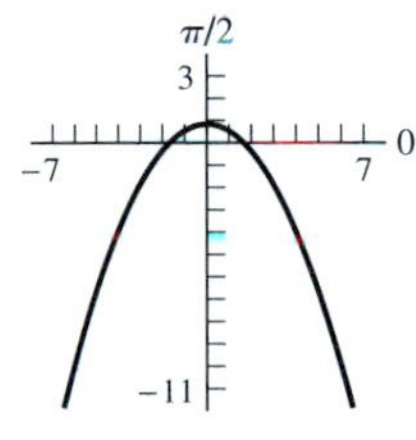

3. **(a)** parabola, opens up **(b)** ellipse, directrix above the pole **(c)** hyperbola, directrix below the pole **(d)** ellipse, directrix to the right of the pole

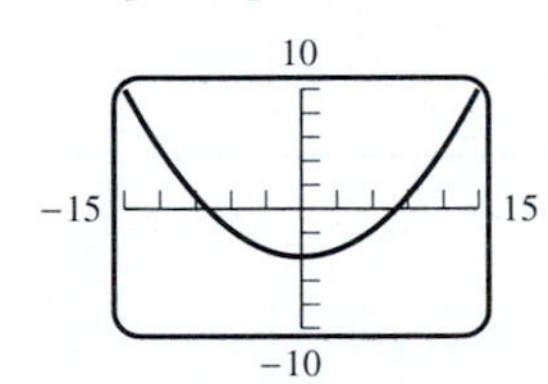

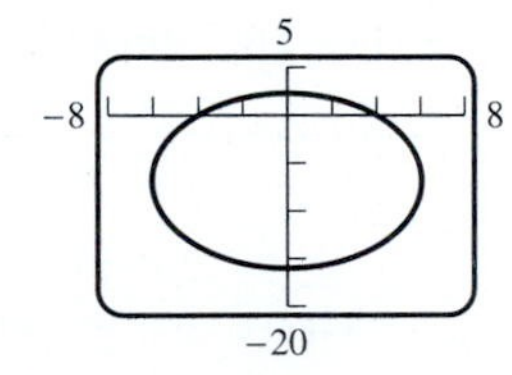

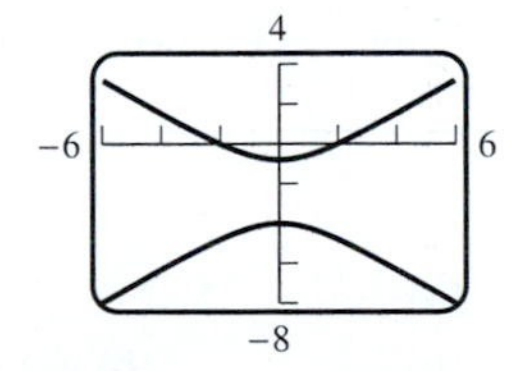

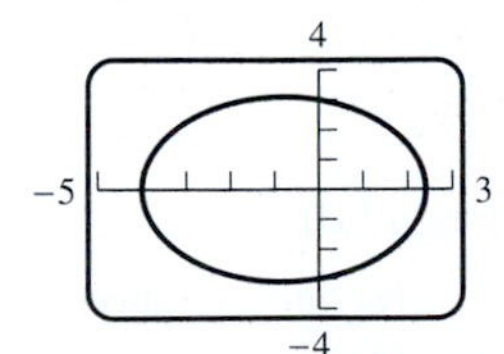

5. **(a)** $r = \dfrac{2}{3 + 2\cos\theta}$ **(b)** $r = \dfrac{1}{1 - \cos\theta}$ **(c)** $r = \dfrac{3}{2 + 3\sin\theta}$

7. **(a)** $r = \dfrac{24}{5 - 5\cos\theta}$ **(b)** $r = \dfrac{2}{1 - \sin\theta}$ **(c)** $r = \dfrac{21}{2 + 5\sin\theta}$

9. **(a)** $d = 6$, $\frac{1}{12}x^2 + \frac{1}{16}(y+2)^2 = 1$ **(b)** $d = 1$, $\frac{9}{4}\left(x - \frac{1}{3}\right)^2 + 3y^2 = 1$

11. **(a)** $d = \frac{2}{3}$, $-2x^2 + 16\left(y - \frac{3}{4}\right)^2 = 1$ **(b)** $d = \frac{10}{9}$, $\frac{9}{16}(x+2)^2 - \frac{9}{20}y^2 = 1$

13. **(a)** $r = \dfrac{12}{2 + \cos\theta}$ **(b)** $r = \dfrac{64}{25 - 15\sin\theta}$ **(c)** $r = \dfrac{16}{5 - 3\cos\theta}$ **(d)** $r = \dfrac{120}{5 + \sin\theta}$

17. **(a)** $T \approx 248$ yr
(b) $r_0 \approx 4{,}449{,}675{,}000$ km
$r_1 \approx 7{,}400{,}325{,}000$ km
(c) $r \approx \dfrac{37.05}{1 + 0.249\cos\theta}$ AU
(d)

19. **(a)** $a \approx 178.26$ AU
(b) $r_0 \approx 0.8735$ AU, $r_1 \approx 355.64$ AU
(c) $r \approx \dfrac{1.74}{1 + 0.9951\cos\theta}$ AU
(d)

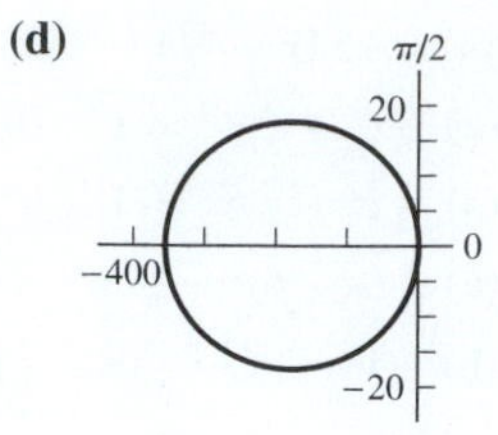

21. 563 km, 4286 km

▶ Chapter 12 Supplementary Exercises (Page 752)

3. **(a)** circle **(b)** rose **(c)** line **(d)** limaçon **(e)** limaçon **(f)** none **(g)** none **(h)** spiral

5. **(a)** **(b)** **(c)** **(d)**

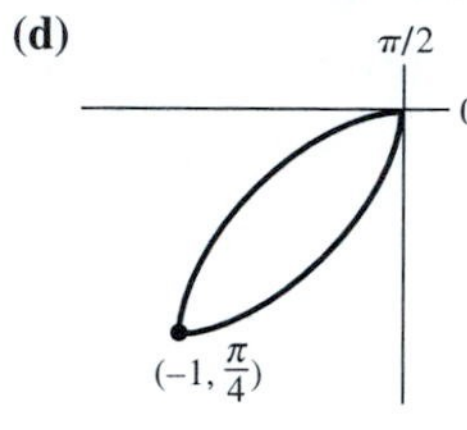

(e)

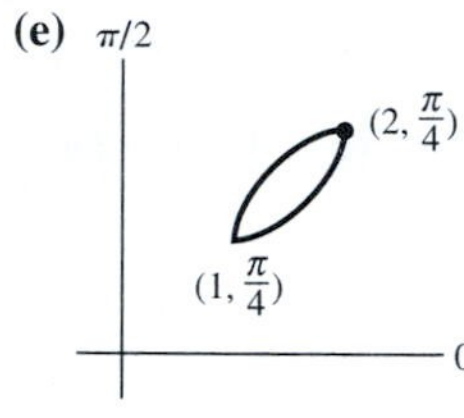

7. **(a)** parabola
(b) hyperbola
(c) line
(d) circle

9. **(a)** $\frac{5}{49}x^2 + \frac{1}{9}y^2 = 1$
(b) $x^2 = -16y$
(c) $\dfrac{x^2}{9} - \dfrac{y^2}{4} = 1$

11. **(a)** **(b)** **(c)**

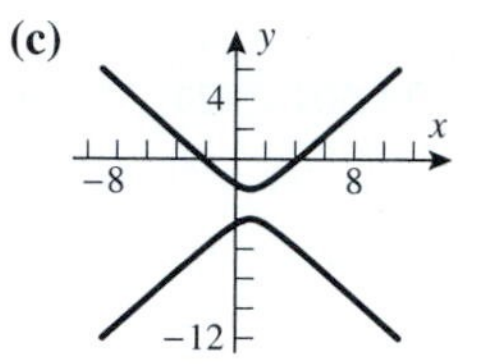

13. **(a)** $y = \dfrac{470}{2100^2}x^2$ **(b)** $L \approx 4336.3$ ft

15. $A = \dfrac{5\pi}{12} - \dfrac{\sqrt{3}}{2}$

17. **(a)** $L = \displaystyle\int_{\pi/4}^{\pi/2} \frac{1}{\theta^2}\sqrt{1+\theta^2}\,d\theta \approx 0.9457$ **(b)** The arc length is infinite.

19. **(a)** $V = \dfrac{\pi b^2}{3a^2}(b^2 - 2a^2)\sqrt{a^2+b^2} + \dfrac{2}{3}ab^2\pi$ **(b)** $V = \dfrac{2b^4}{3a}\pi$

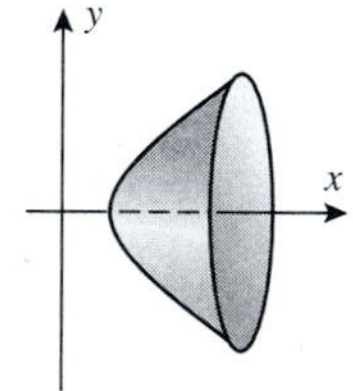

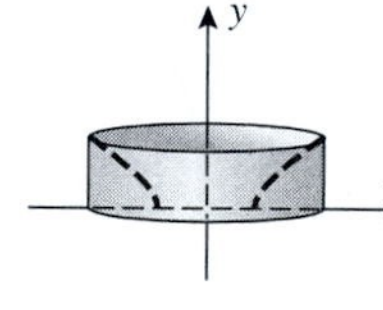

23. **(a)** $-1/4, 2$
(b) $-3\sqrt{3}/4, 3\sqrt{3}/4$

27. **(b)** $x = \cos t + \cos 2t$,
$y = \sin t + \sin 2t$
(c) yes

31. **(a)** $1/60$ **(b)** 91,450,000 mi **(c)** 584,295,652.5 mi

33. **(a)** 119.3 ft/s **(b)** 114.2 ft **(c)** 447.9 ft

▶ Exercise Set A (Page A8)

1. **(a)** rational
(b) integer, rational
(c) integer, rational
(d) rational
(e) integer, rational
(f) irrational
(g) rational
(h) integer, rational

3. **(a)** $\frac{41}{333}$
(b) $\frac{115}{9}$
(c) $\frac{20943}{550}$
(d) $\frac{537}{1250}$

5. **(a)** $\frac{256}{81}$
(b) worse

7.

Line	2	3	4	5	6	7
Blocks	3, 4	1, 2	3, 4	2, 4, 5	1, 2	3, 4

9. (a), (d), (f)

11. **(a)** all values **(b)** none

13. **(a)** yes **(b)** no

15. **(a)** $\{x : x$ is a positive odd integer$\}$ **(b)** $\{x : x$ is an even integer$\}$ **(c)** $\{x : x$ is irrational$\}$ **(d)** $\{x : x$ is an integer and $7 \le x \le 10\}$

17. **(a)** false **(b)** true **(c)** true **(d)** false **(e)** true **(f)** true **(g)** true

19. **(a)** **(b)** **(c)**
(d) **(e)** **(f)**

21. **(a)** $[-2, 2]$ **(b)** $(-\infty, -2) \cup (2, +\infty)$ 23. $\left(-\infty, \frac{10}{3}\right)$ 25. $\left(-\infty, -\frac{11}{2}\right]$ 27. $\left(-\frac{3}{2}, \frac{1}{2}\right]$ 29. $(-\infty, 3) \cup (4, +\infty)$ 31. $\left(-\frac{3}{2}, 2\right)$ 33. $(-\infty, -2] \cup (2, +\infty)$ 35. $(-\infty, -3) \cup (3, +\infty)$ 37. $(-\infty, -2) \cup (4, +\infty)$ 39. $[4, 5]$ 41. $(-8, 0) \cup (4, +\infty)$ 43. $(2, +\infty)$ 45. $(-\infty, -3) \cup [2, +\infty)$ 47. $77 \le F \le 104$ 55. $\left(-\infty, -\frac{1}{2}\right)$

▶ Exercise Set B (Page A15)

1. **(a)** 7 **(b)** $\sqrt{2}$ **(c)** k^2 **(d)** k^2 3. $x \le 3$ 5. all real x 7. $x \ge 0$ or $x = -\frac{2}{3}$ 9. $x \ge -5$ 13. **(a)** 2 **(b)** 1 **(c)** 14 **(d)** $3 + \sqrt{2}$ **(e)** 7 **(f)** 5 15. **(a)** -9 **(b)** 7 **(c)** 12 17. $-\frac{5}{6}, \frac{3}{2}$ 19. $\frac{1}{2}, \frac{5}{2}$ 21. $-\frac{11}{10}, \frac{11}{8}$ 23. $1, \frac{17}{5}$ 25. $(-9, -3)$ 27. $\left[-\frac{3}{2}, \frac{9}{2}\right]$ 29. $(-\infty, -3) \cup (-1, +\infty)$ 31. $\left(-\infty, \frac{1}{2}\right] \cup \left[\frac{9}{2}, +\infty\right)$ 33. $\left(-\infty, \frac{1}{2}\right) \cup \left(\frac{3}{2}, +\infty\right)$ 35. $\left[\frac{1}{8}, \frac{1}{2}\right) \cup \left(\frac{1}{2}, \frac{7}{8}\right]$ 37. $x \in (-\infty, 2] \cup [3, +\infty)$ 39. $-3, 9$

▶ Exercise Set 26C (Page A26)

1. $(-4, 7)$ 3. **(a)** **(b)** **(c)**

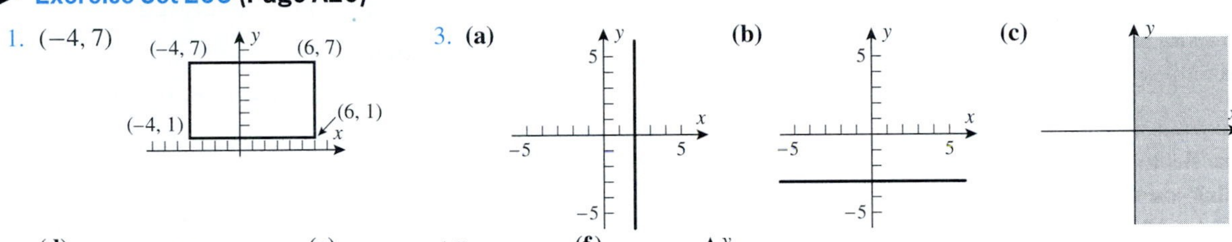

(d) **(e)** **(f)**

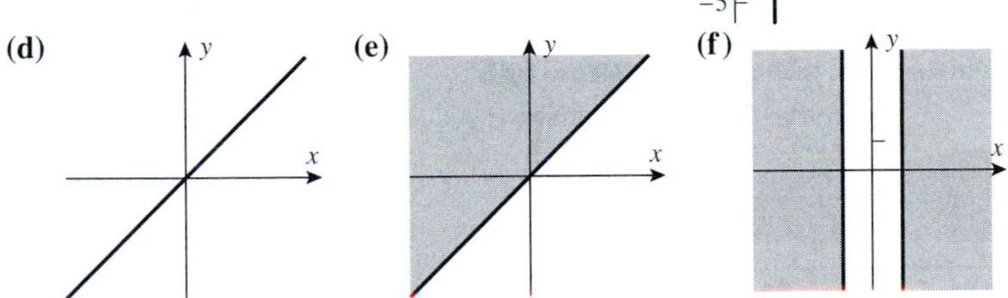

5. 7. 9. 11. 13. **(a)** $\frac{1}{2}$ **(b)** -1 **(c)** 0 **(d)** not defined

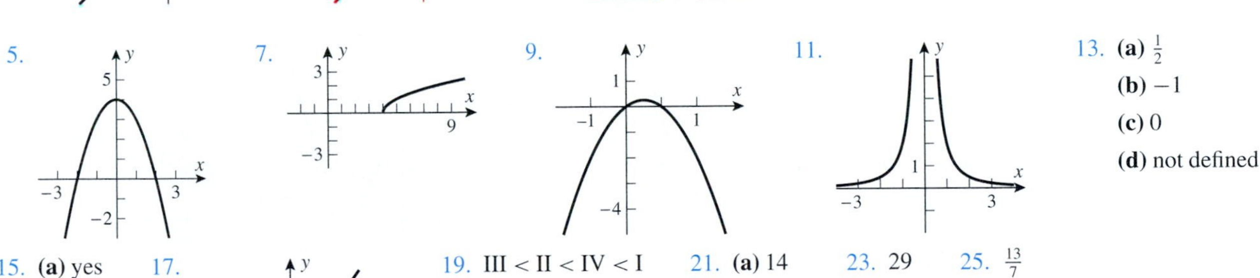

15. **(a)** yes **(b)** no 17.

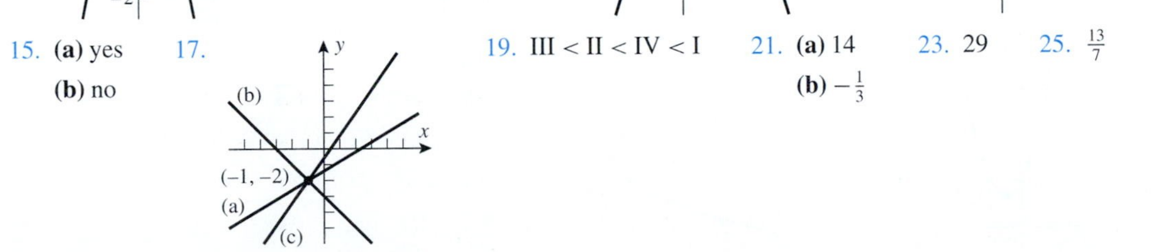

19. III < II < IV < I 21. **(a)** 14 **(b)** $-\frac{1}{3}$ 23. 29 25. $\frac{13}{7}$

29. **(a)** **(b)** **(c)** **(d)**

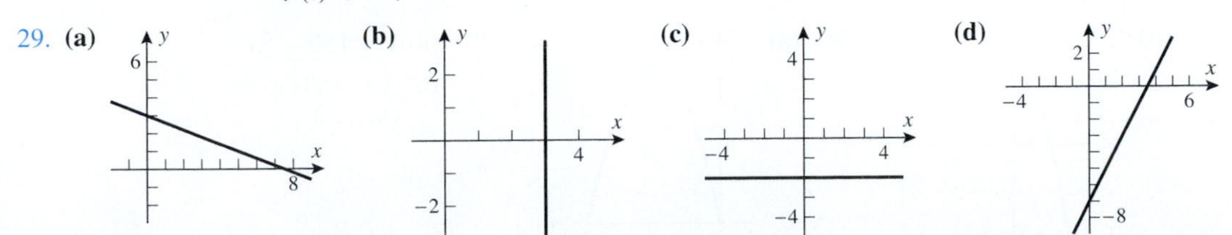

31. **(a)**

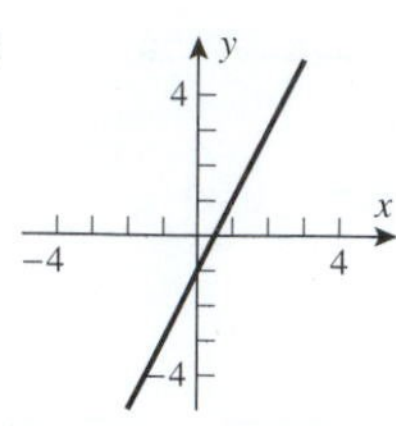

(b)

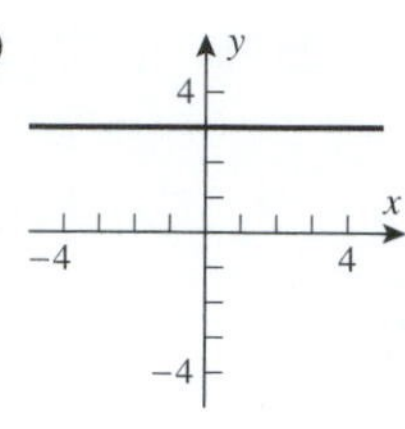

(c)

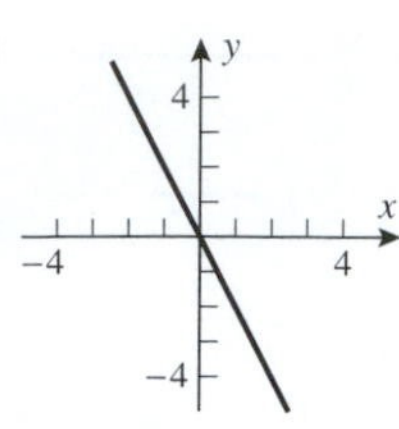

33.

	(a)	(b)	(c)	(d)	(e)
Slope	3	–1/4	–3/5	0	$-b/a$
y-intercept	2	3	8/5	1	b

35. **(a)** $y = \frac{3}{2}x - 3$ **(b)** $y = -\frac{3}{4}x$ 37. $y = -2x + 4$ 39. $y = 4x + 7$ 41. $y = -\frac{1}{5}x + 6$ 43. $y = 11x - 18$
45. $y = \frac{1}{2}x + 2$ 47. $y = 1$ 49. **(a)** parallel **(b)** perpendicular **(c)** parallel **(d)** perpendicular **(e)** neither

51. **(a)** $-\frac{3}{2}$
(b) $\frac{4}{5}$
(c) $\frac{5}{2}$
(d) $-\frac{15}{2}$
(e) -4

53. the union of the graphs of $x - y = 0$ and $x + y = 0$

55.

59. $\frac{49}{6}$

61. **(a)** yes
(b) yes
(c) no
(d) yes
(e) yes
(f) yes
(g) no

▶ Exercise Set D (Page A36)

1. in the proof of Theorem D.1 3. **(a)** 10 **(b)** (4, 5) 5. **(a)** $\sqrt{29}$ **(b)** $\left(-\frac{9}{2}, -5\right)$ 11. 0 13. $y = -3x + 4$
15. $\left(-\frac{29}{8}, -\frac{23}{4}\right)$ 17. 3 21. 4 23. **(a)** (0, 0); 5 **(b)** (1, 4); 4 **(c)** (–1, –3); $\sqrt{5}$ **(d)** (0, –2); 1
25. $(x-3)^2 + (y+2)^2 = 16$ 27. $(x+4)^2 + (y-8)^2 = 64$ 29. $(x+3)^2 + (y+4)^2 = 25$ 31. $(x-1)^2 + (y-1)^2 = 2$
33. circle; center (1, 2), radius 4 35. circle; center (–1, 1), radius $\sqrt{2}$ 37. the point (–1, –1) 39. circle; center (0, 0), radius $\frac{1}{3}$
41. no graph 43. circle; center $\left(-\frac{5}{4}, -\frac{1}{2}\right)$, radius $\frac{3}{2}$ 45. **(a)** $y = -\sqrt{16 - x^2}$ **(b)** $y = 2 + \sqrt{3 - 2x - x^2}$

47. **(a)** **(b)**

49. $y = -\frac{3}{4}x + \frac{25}{4}$ 51. **(a)** inside **(b)** largest $3\sqrt{5}$, smallest $\sqrt{5}$ 53. $(1/3, \pm\sqrt{8}/3)$

55. **(a)** equation: $2x^2 + 2y^2 - 12x + 8y + 1 = 0$ **(b)** center (3, –2), radius $5/\sqrt{2}$

57.

59.

61.

63.

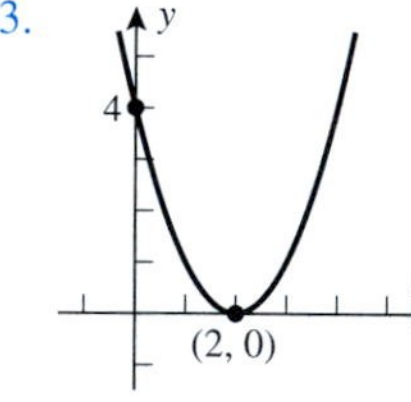

65.

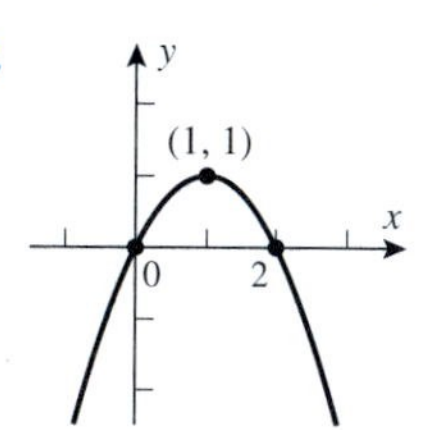

67.

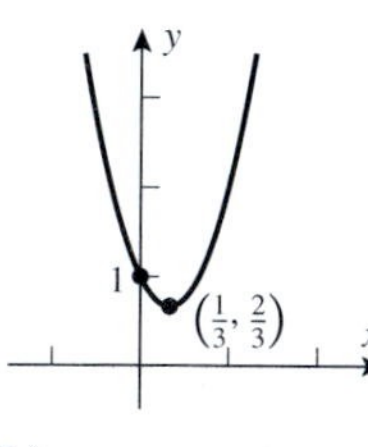

69.

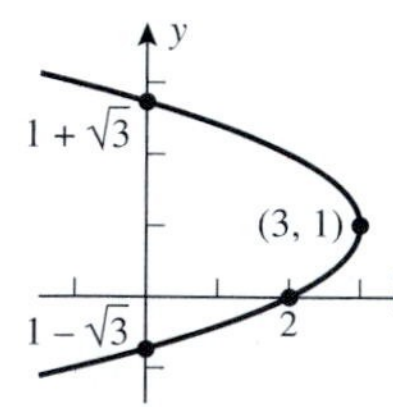

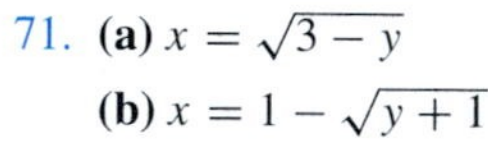

71. **(a)** $x = \sqrt{3 - y}$
(b) $x = 1 - \sqrt{y + 1}$

73. **(a)**

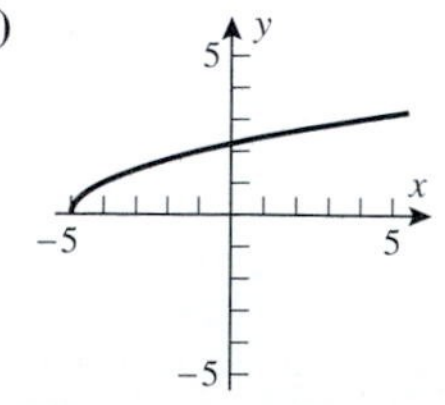

(b)

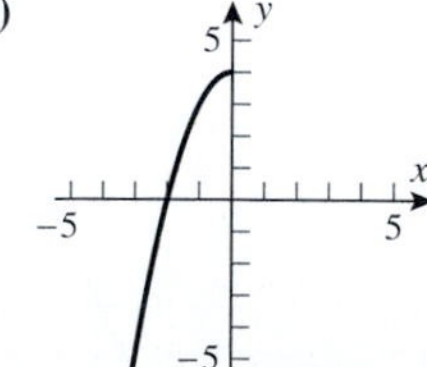

75. **(a)**

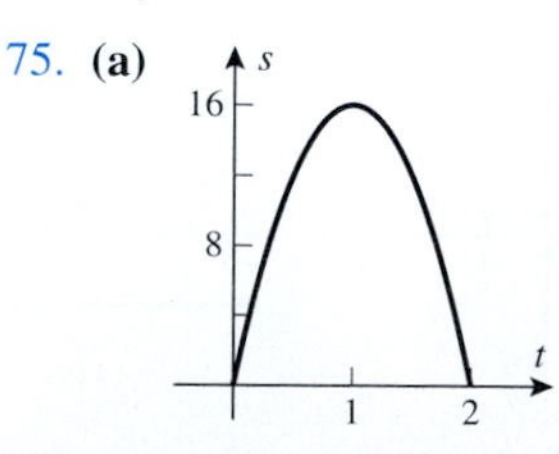

77. **(a)** $y = 150 - \frac{3}{2}x$
(b) $A = 150x - \frac{3}{2}x^2$
(c) 3750 ft^2

79. **(a)** $(-5 - \sqrt{33})/4 < x < (-5 + \sqrt{33})/4$ **(b)** $-\infty < x < +\infty$ 81. **(a)** 30 ft **(b)** 2.6 s **(c)** 2.1 s

▶ Exercise Set E (Page A49)

1. **(a)** $\frac{5}{12}\pi$ **(b)** $\frac{13}{6}\pi$ **(c)** $\frac{1}{9}\pi$ **(d)** $\frac{23}{30}\pi$

3. **(a)** $12°$ **(b)** $(270/\pi)°$ **(c)** $288°$ **(d)** $540°$

5.

	$\sin\theta$	$\cos\theta$	$\tan\theta$	$\csc\theta$	$\sec\theta$	$\cot\theta$
(a)	$\sqrt{21}/5$	$2/5$	$\sqrt{21}/2$	$5/\sqrt{21}$	$5/2$	$2/\sqrt{21}$
(b)	$3/4$	$\sqrt{7}/4$	$3/\sqrt{7}$	$4/3$	$4/\sqrt{7}$	$\sqrt{7}/3$
(c)	$3/\sqrt{10}$	$1/\sqrt{10}$	3	$\sqrt{10}/3$	$\sqrt{10}$	$1/3$

7. $\sin\theta = 3/\sqrt{10}, \cos\theta = 1/\sqrt{10}$

9. $\tan\theta = \sqrt{21}/2, \csc\theta = 5/\sqrt{21}$

11. 1.8

13.

	θ	$\sin\theta$	$\cos\theta$	$\tan\theta$	$\csc\theta$	$\sec\theta$	$\cot\theta$
(a)	$225°$	$-1/\sqrt{2}$	$-1/\sqrt{2}$	1	$-\sqrt{2}$	$-\sqrt{2}$	1
(b)	$-210°$	$1/2$	$-\sqrt{3}/2$	$-1/\sqrt{3}$	2	$-2/\sqrt{3}$	$-\sqrt{3}$
(c)	$5\pi/3$	$-\sqrt{3}/2$	$1/2$	$-\sqrt{3}$	$-2/\sqrt{3}$	2	$-1/\sqrt{3}$
(d)	$-3\pi/2$	1	0	—	1	—	0

15.

	$\sin\theta$	$\cos\theta$	$\tan\theta$	$\csc\theta$	$\sec\theta$	$\cot\theta$
(a)	$4/5$	$3/5$	$4/3$	$5/4$	$5/3$	$3/4$
(b)	$-4/5$	$3/5$	$-4/3$	$-5/4$	$5/3$	$-3/4$
(c)	$1/2$	$-\sqrt{3}/2$	$-1/\sqrt{3}$	2	$-2/\sqrt{3}$	$-\sqrt{3}$
(d)	$-1/2$	$\sqrt{3}/2$	$-1/\sqrt{3}$	-2	$2/\sqrt{3}$	$-\sqrt{3}$
(e)	$1/\sqrt{2}$	$1/\sqrt{2}$	1	$\sqrt{2}$	$\sqrt{2}$	1
(f)	$1/\sqrt{2}$	$-1/\sqrt{2}$	-1	$\sqrt{2}$	$-\sqrt{2}$	-1

17. **(a)** 1.2679 **(b)** 3.5753

19.

	$\sin\theta$	$\cos\theta$	$\tan\theta$	$\csc\theta$	$\sec\theta$	$\cot\theta$
(a)	$a/3$	$\sqrt{9-a^2}/3$	$a/\sqrt{9-a^2}$	$3/a$	$3/\sqrt{9-a^2}$	$\sqrt{9-a^2}/a$
(b)	$a/\sqrt{a^2+25}$	$5/\sqrt{a^2+25}$	$a/5$	$\sqrt{a^2+25}/a$	$\sqrt{a^2+25}/5$	$5/a$
(c)	$\sqrt{a^2-1}/a$	$1/a$	$\sqrt{a^2-1}$	$a/\sqrt{a^2-1}$	a	$1/\sqrt{a^2-1}$

21. **(a)** $3\pi/4 \pm n\pi, n = 0, 1, 2, \ldots$ **(b)** $\pi/3 \pm 2n\pi$ and $5\pi/3 \pm 2n\pi, n = 0, 1, 2, \ldots$

23. **(a)** $\pi/6 \pm n\pi, n = 0, 1, 2, \ldots$ **(b)** $4\pi/3 \pm 2n\pi$ and $5\pi/3 \pm 2n\pi, n = 0, 1, 2, \ldots$

25. **(a)** $3\pi/4 \pm n\pi, n = 0, 1, 2, \ldots$ **(b)** $\pi/6 \pm n\pi, n = 0, 1, 2, \ldots$

27. **(a)** $\pi/3 \pm 2n\pi$ and $2\pi/3 \pm 2n\pi, n = 0, 1, 2, \ldots$ **(b)** $\pi/6 \pm 2n\pi$ and $11\pi/6 \pm 2n\pi, n = 0, 1, 2, \ldots$

29. $\sin\theta = 2/5, \cos\theta = -\sqrt{21}/5, \tan\theta = -2/\sqrt{21}, \csc\theta = 5/2, \sec\theta = -5/\sqrt{21}, \cot\theta = -\sqrt{21}/2$

31. **(a)** $\theta = \pm n\pi, n = 0, 1, 2, \ldots$ **(b)** $\theta = \pi/2 \pm n\pi, n = 0, 1, 2, \ldots$ **(c)** $\theta = \pm n\pi, n = 0, 1, 2, \ldots$ **(d)** $\theta = \pm n\pi, n = 0, 1, 2, \ldots$ **(e)** $\theta = \pi/2 \pm n\pi, n = 0, 1, 2, \ldots$ **(f)** $\theta = \pm n\pi, n = 0, 1, 2, \ldots$

33. **(a)** $2\pi/3$ cm **(b)** $10\pi/3$ cm

35. $\frac{2}{5}$

37. **(a)** $\dfrac{2\pi-\theta}{2\pi}R$ **(b)** $\dfrac{\sqrt{4\pi\theta-\theta^2}}{2\pi}R$

39. $\frac{21}{4}\sqrt{3}$

41. 9.2 ft

43. $h = d(\tan\beta - \tan\alpha)$

45. **(a)** $4\sqrt{5}/9$ **(b)** $-\frac{1}{9}$

47. $\sin 3\theta = 3\sin\theta\cos^2\theta - \sin^3\theta, \cos 3\theta = \cos^3\theta - 3\sin^2\theta\cos\theta$

61. **(a)** $\cos\theta$ **(b)** $-\sin\theta$ **(c)** $-\cos\theta$ **(d)** $\sin\theta$

▶ Exercise Set F (Page A56)

1. **(a)** $x^2 + 4x + 2, -11x + 6$ **(b)** $2x^2 + 4, 9$ **(c)** $x^3 - x^2 + 2x - 2, 2x + 1$

3. **(a)** $3x^2 + 6x + 8, 15$ **(b)** $x^3 - 5x^2 + 20x - 100, 504$ **(c)** $x^4 + x^3 + x^2 + x + 1, 0$

5.

x	0	1	−3	7
$p(x)$	−4	−3	101	5001

7. **(a)** $x^2 + 6x + 13, 20$ **(b)** $x^2 + 3x - 2, -4$

9. **(a)** $\pm1, \pm2, \pm3, \pm4, \pm6, \pm8, \pm12, \pm24$ **(b)** $\pm1, \pm2, \pm5, \pm10, \pm\frac{1}{3}, \pm\frac{2}{3}, \pm\frac{5}{3}, \pm\frac{10}{3}$ **(c)** $\pm1, \pm17$

11. $(x+1)(x-1)(x-2)$

13. $(x+3)^3(x+1)$

15. $(x+3)(x+2)(x+1)^2(x-3)$

17. -3

19. $-2, -\frac{2}{3}$

21. $-2, 2, 3$

23. $2, 5$

25. 7 cm

INDEX

PHOTO CREDITS

Chapter 7
Page 377 (signature): Courtesy New York Public Library. Page 377 (portrait): Granger Collection. Page 377 (bottom): Cyber image/Tony Stone Images/New York, Inc. Page 379: Courtesy New York Public Library. Page 383: Reproduced from C. I. Gerhardt's *Briefwechsel von G.W. Leibniz mit Mathematikern* (1899). Page 457: J. Yulsman/The Image Bank.

Chapter 8
Page 461 (signature): Courtesy Smithsonian Institution. Page 461 (portrait): Corbis-Bettmann. Page 461 (bottom): Jane Sterrett/The Image Bank. Page 490: Stephen Sutton/Duomo Photography, Inc. Page 502: Glen Allison/Tony Stone Images/New York, Inc.

Chapter 9
Page 513 (signature): Courtesy Smithsonian Institution. Page 513 (portrait): Corbis-Bettmann. Page 513 (bottom): Reproduced from C. I. Gerhardt's *Briefwechsel von G.W. Leibniz mit Mathematikern* (1899). Page 528: Courtesy Catherine Watson & David Young/NASA Langley Research Center, Hampton, VA.

Chapter 10
Page 579 (signature): Courtesy Smithsonian Institution. Page 579 (portrait): Corbis-Bettmann. Page 579 (bottom): Alan Hicks/Tony Stone Images/New York, Inc. Page 604: Patrick Mesner/Gamma Liaison.

Chapter 11
Page 615 (signature): Courtesy The British Library. Page 615 (portrait): Corbis-Bettmann. Page 615 (bottom): Nick Dolding/Tony Stone Images/New York, Inc. Page 638: Courtesy Lilly Library, Indiana University.

Chapter 12
Page 699 (signature): Granger Collection. Page 699 (portrait): Archive Photos. Page 699 (bottom): Paul Morrell/Tony Stone Images/New York, Inc. Page 709 (left): Thomas Taylor/Photo Researchers. Page 709 (right): Rex Ziak/AllStock/Picture Network International, Ltd. Page 738: John Mead/Science Photo Library/Photo Researchers. Page 749 (top): Science Photo Library/Photo Researchers. Page 749 (bottom): ©Lick Observatory/Astronomical Society of the Pacific. Page 751: Courtesy NASA.

RATIONAL FUNCTIONS CONTAINING POWERS OF $a + bu$ IN THE DENOMINATOR

60. $\displaystyle\int \frac{u\,du}{a+bu} = \frac{1}{b^2}[bu - a\ln|a+bu|] + C$

61. $\displaystyle\int \frac{u^2\,du}{a+bu} = \frac{1}{b^3}\left[\frac{1}{2}(a+bu)^2 - 2a(a+bu) + a^2\ln|a+bu|\right] + C$

62. $\displaystyle\int \frac{u\,du}{(a+bu)^2} = \frac{1}{b^2}\left[\frac{a}{a+bu} + \ln|a+bu|\right] + C$

63. $\displaystyle\int \frac{u^2\,du}{(a+bu)^2} = \frac{1}{b^3}\left[bu - \frac{a^2}{a+bu} - 2a\ln|a+bu|\right] + C$

64. $\displaystyle\int \frac{u\,du}{(a+bu)^3} = \frac{1}{b^2}\left[\frac{a}{2(a+bu)^2} - \frac{1}{a+bu}\right] + C$

65. $\displaystyle\int \frac{du}{u(a+bu)} = \frac{1}{a}\ln\left|\frac{u}{a+bu}\right| + C$

66. $\displaystyle\int \frac{du}{u^2(a+bu)} = -\frac{1}{au} + \frac{b}{a^2}\ln\left|\frac{a+bu}{u}\right| + C$

67. $\displaystyle\int \frac{du}{u(a+bu)^2} = \frac{1}{a(a+bu)} + \frac{1}{a^2}\ln\left|\frac{u}{a+bu}\right| + C$

RATIONAL FUNCTIONS CONTAINING $a^2 \pm u^2$ IN THE DENOMINATOR ($a > 0$)

68. $\displaystyle\int \frac{du}{a^2+u^2} = \frac{1}{a}\tan^{-1}\frac{u}{a} + C$

69. $\displaystyle\int \frac{du}{a^2-u^2} = \frac{1}{2a}\ln\left|\frac{u+a}{u-a}\right| + C$

70. $\displaystyle\int \frac{du}{u^2-a^2} = \frac{1}{2a}\ln\left|\frac{u-a}{u+a}\right| + C$

71. $\displaystyle\int \frac{bu+c}{a^2+u^2}\,du = \frac{b}{2}\ln(a^2+u^2) + \frac{c}{a}\tan^{-1}\frac{u}{a} + C$

INTEGRALS OF $\sqrt{a^2+u^2}$, $\sqrt{a^2-u^2}$, $\sqrt{u^2-a^2}$ AND THEIR RECIPROCALS ($a > 0$)

72. $\displaystyle\int \sqrt{u^2+a^2}\,du = \frac{u}{2}\sqrt{u^2+a^2} + \frac{a^2}{2}\ln(u+\sqrt{u^2+a^2}) + C$

73. $\displaystyle\int \sqrt{u^2-a^2}\,du = \frac{u}{2}\sqrt{u^2-a^2} - \frac{a^2}{2}\ln|u+\sqrt{u^2-a^2}| + C$

74. $\displaystyle\int \sqrt{a^2-u^2}\,du = \frac{u}{2}\sqrt{a^2-u^2} + \frac{a^2}{2}\sin^{-1}\frac{u}{a} + C$

75. $\displaystyle\int \frac{du}{\sqrt{u^2+a^2}} = \ln(u+\sqrt{u^2+a^2}) + C$

76. $\displaystyle\int \frac{du}{\sqrt{u^2-a^2}} = \ln|u+\sqrt{u^2-a^2}| + C$

77. $\displaystyle\int \frac{du}{\sqrt{a^2-u^2}} = \sin^{-1}\frac{u}{a} + C$

POWERS OF u MULTIPLYING OR DIVIDING $\sqrt{a^2-u^2}$ OR ITS RECIPROCAL

78. $\displaystyle\int u^2\sqrt{a^2-u^2}\,du = \frac{u}{8}(2u^2-a^2)\sqrt{a^2-u^2} + \frac{a^4}{8}\sin^{-1}\frac{u}{a} + C$

79. $\displaystyle\int \frac{\sqrt{a^2-u^2}\,du}{u} = \sqrt{a^2-u^2} - a\ln\left|\frac{a+\sqrt{a^2-u^2}}{u}\right| + C$

80. $\displaystyle\int \frac{\sqrt{a^2-u^2}\,du}{u^2} = -\frac{\sqrt{a^2-u^2}}{u} - \sin^{-1}\frac{u}{a} + C$

81. $\displaystyle\int \frac{u^2\,du}{\sqrt{a^2-u^2}} = -\frac{u}{2}\sqrt{a^2-u^2} + \frac{a^2}{2}\sin^{-1}\frac{u}{a} + C$

82. $\displaystyle\int \frac{du}{u\sqrt{a^2-u^2}} = -\frac{1}{a}\ln\left|\frac{a+\sqrt{a^2-u^2}}{u}\right| + C$

83. $\displaystyle\int \frac{du}{u^2\sqrt{a^2-u^2}} = -\frac{\sqrt{a^2-u^2}}{a^2u} + C$

POWERS OF u MULTIPLYING OR DIVIDING $\sqrt{u^2 \pm a^2}$ OR THEIR RECIPROCALS

84. $\displaystyle\int u\sqrt{u^2+a^2}\,du = \frac{1}{3}(u^2+a^2)^{3/2} + C$

85. $\displaystyle\int u\sqrt{u^2-a^2}\,du = \frac{1}{3}(u^2-a^2)^{3/2} + C$

86. $\displaystyle\int \frac{du}{u\sqrt{u^2+a^2}} = -\frac{1}{a}\ln\left|\frac{a+\sqrt{u^2+a^2}}{u}\right| + C$

87. $\displaystyle\int \frac{du}{u\sqrt{u^2-a^2}} = \frac{1}{a}\sec^{-1}\left|\frac{u}{a}\right| + C$

88. $\displaystyle\int \frac{\sqrt{u^2-a^2}\,du}{u} = \sqrt{u^2-a^2} - a\sec^{-1}\left|\frac{u}{a}\right| + C$

89. $\displaystyle\int \frac{\sqrt{u^2+a^2}\,du}{u} = \sqrt{u^2+a^2} - a\ln\left|\frac{a+\sqrt{u^2+a^2}}{u}\right| + C$

90. $\displaystyle\int \frac{du}{u^2\sqrt{u^2\pm a^2}} = \mp\frac{\sqrt{u^2\pm a^2}}{a^2u} + C$

91. $\displaystyle\int u^2\sqrt{u^2+a^2}\,du = \frac{u}{8}(2u^2+a^2)\sqrt{u^2+a^2} - \frac{a^4}{8}\ln(u+\sqrt{u^2+a^2}) + C$

92. $\displaystyle\int u^2\sqrt{u^2-a^2}\,du = \frac{u}{8}(2u^2-a^2)\sqrt{u^2-a^2} - \frac{a^4}{8}\ln|u+\sqrt{u^2-a^2}| + C$

93. $\displaystyle\int \frac{\sqrt{u^2+a^2}}{u^2}\,du = -\frac{\sqrt{u^2+a^2}}{u} + \ln(u+\sqrt{u^2+a^2}) + C$

94. $\displaystyle\int \frac{\sqrt{u^2-a^2}}{u^2}\,du = -\frac{\sqrt{u^2-a^2}}{u} + \ln|u+\sqrt{u^2-a^2}| + C$

95. $\displaystyle\int \frac{u^2}{\sqrt{u^2+a^2}}\,du = \frac{u}{2}\sqrt{u^2+a^2} - \frac{a^2}{2}\ln(u+\sqrt{u^2+a^2}) + C$

96. $\displaystyle\int \frac{u^2}{\sqrt{u^2-a^2}}\,du = \frac{u}{2}\sqrt{u^2-a^2} + \frac{a^2}{2}\ln|u+\sqrt{u^2-a^2}| + C$

INTEGRALS CONTAINING $(a^2+u^2)^{3/2}$, $(a^2-u^2)^{3/2}$, $(u^2-a^2)^{3/2}$ ($a > 0$)

97. $\displaystyle\int \frac{du}{(a^2-u^2)^{3/2}} = \frac{u}{a^2\sqrt{a^2-u^2}} + C$

98. $\displaystyle\int \frac{du}{(u^2\pm a^2)^{3/2}} = \pm\frac{u}{a^2\sqrt{u^2\pm a^2}} + C$

99. $\displaystyle\int (a^2-u^2)^{3/2}\,du = -\frac{u}{8}(2u^2-5a^2)\sqrt{a^2-u^2} + \frac{3a^4}{8}\sin^{-1}\frac{u}{a} + C$

100. $\displaystyle\int (u^2+a^2)^{3/2}\,du = \frac{u}{8}(2u^2+5a^2)\sqrt{u^2+a^2} + \frac{3a^4}{8}\ln(u+\sqrt{u^2+a^2}) + C$

101. $\displaystyle\int (u^2-a^2)^{3/2}\,du = \frac{u}{8}(2u^2-5a^2)\sqrt{u^2-a^2} + \frac{3a^4}{8}\ln|u+\sqrt{u^2-a^2}| + C$

POWERS OF u MULTIPLYING OR DIVIDING $\sqrt{a+bu}$ OR ITS RECIPROCAL

102. $\int u\sqrt{a+bu}\,du = \frac{2}{15b^2}(3bu-2a)(a+bu)^{3/2}+C$

103. $\int u^2\sqrt{a+bu}\,du = \frac{2}{105b^3}(15b^2u^2-12abu+8a^2)(a+bu)^{3/2}+C$

104. $\int u^n\sqrt{a+bu}\,du = \frac{2u^n(a+bu)^{3/2}}{b(2n+3)} - \frac{2an}{b(2n+3)}\int u^{n-1}\sqrt{a+bu}\,du$

105. $\int \frac{u\,du}{\sqrt{a+bu}} = \frac{2}{3b^2}(bu-2a)\sqrt{a+bu}+C$

106. $\int \frac{u^2\,du}{\sqrt{a+bu}} = \frac{2}{15b^3}(3b^2u^2-4abu+8a^2)\sqrt{a+bu}+C$

107. $\int \frac{u^n\,du}{\sqrt{a+bu}} = \frac{2u^n\sqrt{a+bu}}{b(2n+1)} - \frac{2an}{b(2n+1)}\int \frac{u^{n-1}\,du}{\sqrt{a+bu}}$

108. $\int \frac{du}{u\sqrt{a+bu}} = \begin{cases} \frac{1}{\sqrt{a}}\ln\left|\frac{\sqrt{a+bu}-\sqrt{a}}{\sqrt{a+bu}+\sqrt{a}}\right|+C & (a>0) \\ \frac{2}{\sqrt{-a}}\tan^{-1}\sqrt{\frac{a+bu}{-a}}+C & (a<0) \end{cases}$

109. $\int \frac{du}{u^n\sqrt{a+bu}} = -\frac{\sqrt{a+bu}}{a(n-1)u^{n-1}} - \frac{b(2n-3)}{2a(n-1)}\int \frac{du}{u^{n-1}\sqrt{a+bu}}$

110. $\int \frac{\sqrt{a+bu}\,du}{u} = 2\sqrt{a+bu}+a\int \frac{du}{u\sqrt{a+bu}}$

111. $\int \frac{\sqrt{a+bu}\,du}{u^n} = -\frac{(a+bu)^{3/2}}{a(n-1)u^{n-1}} - \frac{b(2n-5)}{2a(n-1)}\int \frac{\sqrt{a+bu}\,du}{u^{n-1}}$

POWERS OF u MULTIPLYING OR DIVIDING $\sqrt{2au-u^2}$ OR ITS RECIPROCAL

112. $\int \sqrt{2au-u^2}\,du = \frac{u-a}{2}\sqrt{2au-u^2}+\frac{a^2}{2}\sin^{-1}\left(\frac{u-a}{a}\right)+C$

113. $\int u\sqrt{2au-u^2}\,du = \frac{2u^2-au-3a^2}{6}\sqrt{2au-u^2}+\frac{a^3}{2}\sin^{-1}\left(\frac{u-a}{a}\right)+C$

114. $\int \frac{\sqrt{2au-u^2}\,du}{u} = \sqrt{2au-u^2}+a\sin^{-1}\left(\frac{u-a}{a}\right)+C$

115. $\int \frac{\sqrt{2au-u^2}\,du}{u^2} = -\frac{2\sqrt{2au-u^2}}{u}-\sin^{-1}\left(\frac{u-a}{a}\right)+C$

116. $\int \frac{du}{\sqrt{2au-u^2}} = \sin^{-1}\left(\frac{u-a}{a}\right)+C$

117. $\int \frac{du}{u\sqrt{2au-u^2}} = -\frac{\sqrt{2au-u^2}}{au}+C$

118. $\int \frac{u\,du}{\sqrt{2au-u^2}} = -\sqrt{2au-u^2}+a\sin^{-1}\left(\frac{u-a}{a}\right)+C$

119. $\int \frac{u^2\,du}{\sqrt{2au-u^2}} = -\frac{(u+3a)}{2}\sqrt{2au-u^2}+\frac{3a^2}{2}\sin^{-1}\left(\frac{u-a}{a}\right)+C$

INTEGRALS CONTAINING $(2au-u^2)^{3/2}$

120. $\int \frac{du}{(2au-u^2)^{3/2}} = \frac{u-a}{a^2\sqrt{2au-u^2}}+C$

121. $\int \frac{u\,du}{(2au-u^2)^{3/2}} = \frac{u}{a\sqrt{2au-u^2}}+C$

THE WALLIS FORMULA

122. $\int_0^{\pi/2}\sin^n u\,du = \int_0^{\pi/2}\cos^n u\,du = \begin{cases} \frac{1\cdot3\cdot5\cdot\cdots\cdot(n-1)}{2\cdot4\cdot6\cdot\cdots\cdot n}\cdot\frac{\pi}{2} & \left(\begin{array}{l} n \text{ an even} \\ \text{integer and} \\ n\geq 2 \end{array}\right) \\ \frac{2\cdot4\cdot6\cdot\cdots\cdot(n-1)}{3\cdot5\cdot7\cdot\cdots\cdot n} & \left(\begin{array}{l} n \text{ an odd} \\ \text{integer and} \\ n\geq 3 \end{array}\right) \end{cases}$

TRIGONOMETRIC IDENTITIES

PYTHAGOREAN IDENTITIES

$\sin^2\theta+\cos^2\theta=1 \qquad \tan^2\theta+1=\sec^2\theta \qquad 1+\cot^2\theta=\csc^2\theta$

SIGN IDENTITIES

$\sin(-\theta)=-\sin\theta \qquad \cos(-\theta)=\cos\theta \qquad \tan(-\theta)=-\tan\theta$

$\csc(-\theta)=-\csc\theta \qquad \sec(-\theta)=\sec\theta \qquad \cot(-\theta)=-\cot\theta$

COMPLEMENT IDENTITIES

$\sin\left(\frac{\pi}{2}-\theta\right)=\cos\theta \qquad \cos\left(\frac{\pi}{2}-\theta\right)=\sin\theta \qquad \tan\left(\frac{\pi}{2}-\theta\right)=\cot\theta$

$\csc\left(\frac{\pi}{2}-\theta\right)=\sec\theta \qquad \sec\left(\frac{\pi}{2}-\theta\right)=\csc\theta \qquad \cot\left(\frac{\pi}{2}-\theta\right)=\tan\theta$

SUPPLEMENT IDENTITIES

$\sin(\pi-\theta)=\sin\theta \qquad \cos(\pi-\theta)=-\cos\theta \qquad \tan(\pi-\theta)=-\tan\theta$

$\csc(\pi-\theta)=\csc\theta \qquad \sec(\pi-\theta)=-\sec\theta \qquad \cot(\pi-\theta)=-\cot\theta$

$\sin(\pi+\theta)=-\sin\theta \qquad \cos(\pi+\theta)=-\cos\theta \qquad \tan(\pi+\theta)=\tan\theta$

$\csc(\pi+\theta)=-\csc\theta \qquad \sec(\pi+\theta)=-\sec\theta \qquad \cot(\pi+\theta)=\cot\theta$

ADDITION FORMULAS

$\sin(\alpha+\beta)=\sin\alpha\cos\beta+\cos\alpha\sin\beta$

$\sin(\alpha-\beta)=\sin\alpha\cos\beta-\cos\alpha\sin\beta$

$\cos(\alpha+\beta)=\cos\alpha\cos\beta-\sin\alpha\sin\beta$

$\cos(\alpha-\beta)=\cos\alpha\cos\beta+\sin\alpha\sin\beta$

$\tan(\alpha+\beta)=\frac{\tan\alpha+\tan\beta}{1-\tan\alpha\tan\beta}$

$\tan(\alpha-\beta)=\frac{\tan\alpha-\tan\beta}{1+\tan\alpha\tan\beta}$

DOUBLE-ANGLE FORMULAS

$\sin 2\alpha=2\sin\alpha\cos\alpha \qquad \cos 2\alpha=2\cos^2\alpha-1$

$\cos 2\alpha=\cos^2\alpha-\sin^2\alpha \qquad \cos 2\alpha=1-2\sin^2\alpha$

HALF-ANGLE FORMULAS

$\sin^2\frac{\alpha}{2}=\frac{1-\cos\alpha}{2} \qquad \cos^2\frac{\alpha}{2}=\frac{1+\cos\alpha}{2}$